Continuous-Time Signals and Systems

Drawing on author's 30+ years of teaching experience, "Continuous-Time Signals and Systems: A MATLAB Integrated Approach" represents a novel and comprehensive approach to understanding signals and systems theory. Many textbooks use MATLAB as a computational tool, but Alkin's text employs MATLAB both computationally and pedagogically to provide interactive, visual reinforcement of fundamental concepts important in the study of continuous-time signals and systems.

In addition to 210 traditional end-of-chapter problems and 168 solved examples, the book includes hands-on MATLAB modules consisting of:

- 77 MATLAB-based homework problems and projects (coordinated with the traditional end-of-chapter problems)
- 106 live scripts and GUI-based interactive apps that animate key figures and bring core concepts to life
- Downloadable MATLAB code for most of the solved examples
- 64 fully detailed MATLAB exercises that involve step by step development of code to simulate the relevant signal and/or system being discussed, including some case studies on topics such as synthesizers, simulating instrument sounds, pulse-width modulation, etc.
- The ebook+ version includes clickable links that allow running MATLAB code associated with solved examples and exercises in a browser, using the online version of MATLAB. It also includes audio files for some of the examples.

Each module or application is linked to a specific segment of the text to ensure seamless integration between learning and doing. The aim is to not simply give the student just another toolbox of MATLAB functions, but to use the development of MATLAB code as part of the learning process, or as a litmus test of students' understanding of the key concepts. All relevant MATLAB code is freely available from the publisher. In addition, a solutions manual, figures, presentation slides and other ancillary materials are available for instructors with qualifying course adoption.

Continuous-Time Signals and Systems

A MATLAB-Integrated Approach

Second Edition

Oktay Alkin

CRC Press
Taylor & Francis Group
Boca Raton London New York

CRC Press is an imprint of the
Taylor & Francis Group, an **informa** business

Second edition published 2025
by CRC Press
2385 NW Executive Center Drive, Suite 320, Boca Raton FL 33431

and by CRC Press
4 Park Square, Milton Park, Abingdon, Oxon, OX14 4RN

CRC Press is an imprint of Taylor & Francis Group, LLC

ISBN: 978-1-032-67523-7 (hbk)
ISBN: 978-1-032-67524-4 (pbk)
ISBN: 978-1-032-67525-1 (ebk)
ISBN: 978-1-032-67526-8 (eBook+)

DOI: 10.1201/9781032675251

Typeset in Utopia Regular font
by KnowledgeWorks Global Ltd.

Publisher's note: This book has been prepared from camera-ready copy provided by the authors.

Access the Instructor and Student Resources: Will be provided later

To Esin

Contents

PREFACE

The subject of signals and systems is an integral component of any undergraduate program in electrical engineering. It also has applications in computer engineering, mechanical engineering, aerospace engineering, and bioengineering. It encompasses analysis, design, and implementation of systems as well as problems involving signal-system interaction. A solid background in continuous-time signals and systems is essential for the student to be able to venture into fields such as signal and image processing, communications, control systems, VLSI (very large-scale integration) circuit design, power generation and distribution, and so on.

This textbook was written with the goal of providing a modern treatment of continuous-time signals and systems at the undergraduate level. It can be used as the main textbook for a one-semester course. It can also be used for self-study. Writing style is student-friendly, starting with the basics of each topic and advancing gradually. There is not a lot of hand-waving; not many paragraphs start with "It can be shown that ..." without actually showing how or why something is the way it is. On the other hand, proofs and derivations that may safely be skipped at a first reading are clearly indicated through the use of color-coded frames. Also, important concepts and conclusions are highlighted to stand out. No prior signals and systems knowledge is assumed. The level of presentation is appropriate for a second or third year engineering student with differential calculus background. Some knowledge of introductory circuit theory is also helpful.

There are many excellent textbooks available for use in undergraduate level courses in this area of study. Some have matured over long periods of time, decades rather than just years, and have gained great popularity and traction among instructors as well as students. Consequently, the writing of a new textbook in this area is a difficult task as one must consider the questions of what new ideas the book would employ, and what value it would add to the teaching of the subject. This textbook resulted from the author's efforts over the past three decades in trying to find ways to incorporate software into the teaching of the material not only as a computational tool but also as a pedagogical one. It utilizes MATLAB software due to its popularity in the engineering community and its availability for a variety of operating systems. Basic familiarity with MATLAB software is assumed. For readers who have had no prior exposure to MATLAB, it would be helpful to spend a couple of hours going through some of the beginner-level tutorials freely available on the website of MathWorks, the publisher of MATLAB.

Each chapter contains a number of fully solved pencil-and-paper type examples as well as a set of traditional end-of-chapter problems. In addition, software use is integrated into the material at several levels:

1. Interactive MATLAB apps and live scripts: Interactive apps are graphical user interface-based MATLAB programs that allow key concepts of signals and systems to be visualized. We are all visual learners; we tend to remember an interesting scene in a movie much better than something we hear on the radio or something we read in a textbook. Taking this one step further, if we were also able to control the scene in a movie through our actions, we would have an even stronger grasp of the cause-effect relationships between our actions and the results they create. This is perhaps the main reason why children can become very proficient in video games. A large number of interactive apps are available for download with this textbook. Some allow a static figure in the text to come alive on a computer where the student can change key parameters and observe the results of such changes to understand cause-effect relationships. Some take a solved example in the text and expand it in an open-ended manner to allow the student to set up "what if" scenarios. Some of the interactive apps provide animations for concepts that are difficult to teach. Examples of this group include convolution, Fourier series representation of signals, three-dimensional visualization of Laplace transform, and steady-state response of systems. Live scripts serve a similar purpose. Each one is like a MATLAB-based notebook that develops a key concept section by section, displaying results of calculations numerically and graphically. Modifying one or more parameters results in all calculations and graphics being updated as appropriate.
2. MATLAB code for solved examples: Most of the solved examples in each chapter have associated MATLAB listings available for download. These listings include detailed comments, and are useful in a variety of ways: They help students check their work against a computer solution. They reinforce good coding practices. They also allow experimentation by changing parameter values and running the code again.
3. MATLAB exercises: In addition to the code listings associated with solved examples, there is a section at the end of each chapter which contains stand-alone MATLAB exercises that take the student through exploration and/or simulation of a concept by developing the necessary code step by step. Intermediate steps are explained in detail, and good coding practices are enforced throughout. Exercises are designed to help students become more proficient with MATLAB while working on problems in the context of signals and systems. Furthermore, MATLAB exercises are synchronized with the coverage of the material. At specific points in the text the reader is referred to MATLAB exercises relevant to the topic being discussed. The goal is to not just provide cookbook style quick solutions to problems, but rather to develop additional insight and deeper understanding of the material through the use of software. Some of the MATLAB exercises designated as *case studies* are somewhat longer and more complex than others, and highlight practical applications of the concepts covered in that particular chapter.

   Overall, MATLAB exercises were designed to reinforce students' understanding of the theory of signals and systems, and to develop software skills that would be useful in more advanced courses and research projects. It is the author's hope that they can also be motivating and fun.
4. MATLAB-based problems: In addition to traditional end-of-chapter problems, each chapter contains a section with problems that require MATLAB solutions.
5. MATLAB projects: Most chapters contain a section with project ideas that are MATLAB-based. These can be used by instructors as the basis of computer assignments.

6. MATLAB files: MATLAB files downloaded from the support website of the textbook integrate with the help browser of MATLAB software upon installation. This allows the student to have the textbook and a computer running MATLAB side by side while studying. Additionally, it gives the instructor the freedom to display MATLAB exercises on a projector while lecturing, without the need to type and run any code, if preferred.
7. For the ebook version: At the end of each solved example or Matlab exercise, there are clickable links for the script files used. Clicking on a link runs the online version of MATLAB in the default browser for the system, and loads the selected script ready to run. It should be noted that a user account needs to be created with MathWorks in order to use the online version of MATLAB. At the time of writing this text, mobile phones and tablets were unable to run MATLAB in a browser. A desktop or a laptop computer is needed for that. Also, some scripts that produce real-time audio do not work in a browser due to limitations of browsers to interact with the audio hardware of the system.

While software is an integral part of the textbook, and one that is meant to distinguish it from other works in the same field, it "never gets in the way". If desired, one can ignore all MATLAB-related content and use this textbook as a traditional one with which to teach or learn the theory of signals and systems. Obviously, this would not be the recommended use. The main point is that the coverage of theory is not cluttered with code segments. Instead, all MATLAB exercises, problems, and projects are presented in their own sections at the end of each chapter, with appropriate references provided within the narrative. Apart from the software use, the textbook includes plenty of solved pencil-and-paper type examples and traditional end-of-chapter problems.

Organization of the Material

In Chapter 1 we deal with mathematical modeling of continuous-time signals. Basic building blocks for signals are presented as well as mathematical operations applied to signals. Classification methods for signals are discussed and definitions of signal energy and signal power are given. The idea of impulse decomposition of a continuous-time signal is presented in preparation for the discussion of convolution operation in later chapters. Even and odd symmetry properties for real-valued signals are introduced, and the idea of decomposing a signal into its even and odd components is discussed. Notions of even and odd symmetry are generalized to complex-valued signals in the form of conjugate symmetry and conjugate antisymmetry. Finally, phasor representation of sinusoidal signals is explored.

In Chapter 2 the concept of a continuous-time system is introduced. Simplifying assumptions of linearity and time-invariance are discussed. The chapter proceeds with the use of constant-coefficient linear differential equations for describing the behavior of linear systems. Rather than assuming prior background in solving differential equations, or simply referring the student to math textbooks, differential equation solution methods are explained at a level sufficient for working with linear and time-invariant systems. Even if students have previously been exposed to a course on differential equations, it may still be helpful to use this chapter for a quick review, so that the connection between differential equations and other linear system representations such as system functions or state-space models is not missed. The concept of steady-state response is introduced. Representation of a differential equation by a block diagram is discussed. The concepts of impulse response and convolution are developed and their link to the differential equation of the system is shown. Definitions of stability and causality as well as the time-domain conditions for a system to achieve them are given.

Chapter 3 provides a treatment of Fourier analysis of continuous-time signals and systems. It begins with the analysis of periodic signals in terms of their frequency content. First, the idea of finding the best approximation to a periodic signal through the use of a few trigonometric functions is explored in order to build intuition. Afterward trigonometric, exponential, and compact variants of the Fourier series are discussed. The Fourier transform for non-periodic signals is then introduced by generalizing the exponential Fourier series representation of a periodic signal. A discussion of Parseval's theorem is provided leading to the concepts of energy and power spectral density for deterministic signals. System function concept is introduced for continuous-time systems. Response of linear and time invariant systems to both periodic and non-periodic input signals is studied.

In Chapter 4 the Laplace transform is discussed as an analysis tool. Its relationship to the Fourier transform is explored. The bulk of the chapter focuses on the more general, bilateral, variant of the Laplace transform. Convergence characteristics of the transform as well as the region of convergence concept are introduced. Fundamental properties of the Laplace transform are detailed, and solved examples of utilizing those properties are given. The method of finding the inverse Laplace transform through the use of partial fraction expansion is explained. The chapter proceeds with application of the Laplace transform to the analysis of linear time-invariant systems. Special emphasis is placed on making the connections between the s-domain system function, the differential equation, and the impulse response. Graphical interpretation of pole-zero diagrams for the purpose of determining the frequency response of continuous-time linear and time-invariant (CTLTI) systems is illustrated. The chapter also provides a discussion of causality and stability concepts as they relate to the s-domain system function. Characteristics of all-pass systems, minimum-phase systems, and inverse systems are outlined, and an introduction to Bode plots is given. Direct-form, cascade, and parallel simulation structures for CTLTI systems are derived from the s-domain system function. Finally, the unilateral version of the Laplace transform is introduced as a practical tool for use in solving constant-coefficient linear differential equations with specified initial conditions.

In Chapter 5 we introduce state-space analysis for continuous-time systems. The concept of state variables is developed by starting with a differential equation and expressing it in terms of simpler first-order differential equations. Methods of obtaining state space models from a physical description of the system, from a system function, or from a block diagram are also detailed. Various solution methods for the state equations are developed. State transition matrix and its properties are discussed. Methods for obtaining alternative state-space models through the use of similarity transformations are detailed.

Chapter 6 focuses on the application of the material presented so far to the analysis and design of continuous-time filters. The concept of distortionless transmission is explained, and rules for ideal filter behavior are established. Filter design through experimental placement of poles and zeros of the system function is discussed in order to create intuition about the relationship between pole and zero locations and filter behavior. Methods for analog lowpass filter design using Butterworth, Chebyshev, inverse Chebyshev, and elliptic approximation functions are given. Finally, transformation functions for converting lowpass filters to other filter types are derived.

Chapter 7 provides an introductory treatment of another application of the material in the area of communication systems, namely amplitude modulation. It begins by introducing the idea of modulation and the need for it. Generation of amplitude modulated signals and their analysis in the frequency domain using the techniques developed in earlier chapters are discussed. Techniques for demodulation of amplitude-modulated signals are summarized. Two variants of amplitude modulation, namely double sideband suppressed carrier (DSB-SC) modulation and single sideband (SSB) modulation, are introduced. Techniques for generating DSB-SC and SSB signals are discussed. Demodulation of DSB-SC and SSB signals are briefly covered.

A number of appendices are provided for complex numbers and Euler's formula, various mathematical relations, proofs of orthogonality for the basis functions used in Fourier analysis, partial fraction expansion, a brief review of matrix algebra, and solutions to selected end-of-chapter problems.

Supplementary Materials

The following supplementary materials are available to the instructors who adopt the textbook for classroom use:

- A solutions manual in pdf format that contains solutions to the problems at the end of each chapter, including the MATLAB problems,
- Presentation slides in pdf format created using LaTeX and Beamer,
- Image files in pdf format for the figures in the book.

The following supplementary materials are available to all users of the textbook:

- A downloadable archive containing MATLAB code files for GUI-based interactive demo programs, exercises and examples in the textbook as well as the files needed for integrating them into MATLAB help browser.

CHAPTER 1

CONTINUOUS-TIME SIGNAL REPRESENTATION AND MODELING

Chapter Objectives

- Understand the concept of a *signal* and learn how to work with mathematical models of signals.
- Discuss fundamental signal types and signal operations used in the study of signals and systems. Experiment with methods of simulating continuous-time signals with MATLAB.
- Learn various ways of classifying signals and working with symmetry properties.
- Learn basic building blocks for continuous-time signals and ways to express other signals in terms of these building blocks.
- Explore characteristics of sinusoidal signals. Learn *phasor* representation of sinusoidal signals, and how phasors help with analysis.
- Understand the decomposition of continuous-time signals using unit impulse functions.
- Learn energy and power definitions.

1.1 Introduction

Signals are part of our daily lives. We work with a wide variety of signals on a day-to-day basis whether we realize it or not. When a commentator in a radio broadcast studio speaks into a microphone, the sound waves create changes in the air pressure around the microphone. If we happen to be in the same studio, membranes in our ears detect the changes in the air pressure, and we perceive them as sound. Pressure changes also cause vibrations on the membrane within

DOI: 10.1201/9781032675251-1

the microphone. In response, the microphone produces a time-varying electrical voltage in such a way that the variations of the electrical voltage mimic the variations of air pressure in the studio. Thus, the microphone acts as a *transducer* which converts an *acoustic signal* to an *electrical signal.* This is just one step in the process of radio broadcast. The electrical signal produced by a microphone is modified and enhanced in a variety of ways, and its strength is boosted. It is then applied to a transmitting antenna which converts it to an *electromagnetic signal* suitable for transmission over the air for long distances. These electromagnetic waves fill the air space that we live in. An electromagnetic signal picked up by the antenna of a radio receiver is converted back to the form of an electrical voltage signal which is further processed within the radio receiver and sent to a loudspeaker. The loudspeaker, which acts as another transducer, converts this electrical voltage signal to vibrations which recreate the air pressure variations similar to those that started the whole process at the broadcast studio.

The broadcast example discussed above includes acoustic, electrical, and electromagnetic signals. Examples of different physical signals can also be found around us. Time variations of quantities such as force, torque, speed, and acceleration can be taken as examples of mechanical signals. For example, in an automobile, a tachometer measures the velocity of the vehicle as a signal and produces an electrical voltage the strength of which is proportional to it at each time instant. Subsequently, the electrical signal produced by the tachometer can be used in a speed control system to regulate the speed of the vehicle.

Variations of the temperature in a room can be taken as another example of a signal. A thermostat measures the temperature at each instant of time, and uses these measurements to produce control actions such as turning a furnace or an air conditioner on or off as needed.

1.2 Mathematical Modeling of Signals

In the previous section we have discussed several examples of signals that we encounter in our daily lives. These signals are in one or another physical form; some are electrical signals, some are in the form of air pressure variations over time, some correspond to time variations of a force or a torque in a mechanical system, and some represent temperature variations over time. There are other physical forms as well. In working with signals, we have one or both of the following goals in mind:

1. *Understand the characteristics of the signal in terms of its behavior in time and in terms of the frequencies it contains (signal analysis).* For example, consider the electrical signal produced by a microphone. Two people uttering the same word into the microphone do not necessarily produce the same exact signal. Would it be possible to recognize the speaker from his or her signal? Would it be possible to identify the spoken word?
2. *Develop methods of creating signals with desired characteristics (signal synthesis).* Consider an electronic keyboard that synthesizes sounds of different musical instruments such as the piano and the violin. Naturally, to create the sound of either instrument we need to understand the characteristics of the sounds these instruments make. Other examples are the synthesis of human speech from printed text by text-to-speech software and the synthesis of test signals in a laboratory for testing and calibrating specialized equipment.

In addition, signals are often used in conjunction with systems. A signal is applied to a system as *input,* and the system responds to the signal by producing another signal called the *output.* For example, the signal from an electric guitar can be applied (connected via cable) to a system that boosts its strength or enhances the sound acoustically. In problems that involve the interaction of a signal with a system we may have additional goals:

3. *Understand how a system responds to a signal and why (system analysis).* What characteristics of the input signal does the system retain from its input to its output? What characteristics of the input signal are modified by the system, and in what ways?
4. *Develop methods of constructing a system that responds to a signal in some prescribed way (system synthesis).* Understanding different characteristics of signals will also help us in the design of systems that affect signals in desired ways. An example is the design of a system to enhance the sound of a guitar and give it the "feel" of being played in a concert hall with certain acoustic properties. Another example is the automatic speed control system of an automobile that is required to accelerate or decelerate automobile to the desired speed within prescribed constraints.

In analyzing signals of different physical forms we need to develop a uniform framework. It certainly would not be practical to devise different analysis methods for different types of physical signals. Therefore, we work with *mathematical models* of physical signals.

Most of the examples utilized in our discussion so far involved signals in the form of a physical quantity varying over time. There are exceptions; some signals could be in the form of variations of a physical quantity over a spatial variable. Examples of this are the distribution of force along the length of a steel beam, or the distribution of light intensity at different points in an image. To simplify the discussion, we will focus our attention on signals that are functions of time. The techniques that will be developed are equally applicable to signals that use other independent variables as well.

Mathematical models:

The mathematical model for a signal is in the form of a formula, function, algorithm, or a graph that approximately describes the time variations of the physical signal.

1.3 Continuous-Time Signals

Consider $x(t)$, a mathematical function of time chosen to approximate the strength of the physical quantity at the time instant t. In this relationship t is the independent variable, and x is the dependent variable, often referred to as the *amplitude*. A *continuous-time* signal, also referred to as an *analog* signal, is one in which the amplitude is defined for all values of the time variable t. An example of a continuous-time signal is shown in Fig. 1.1.

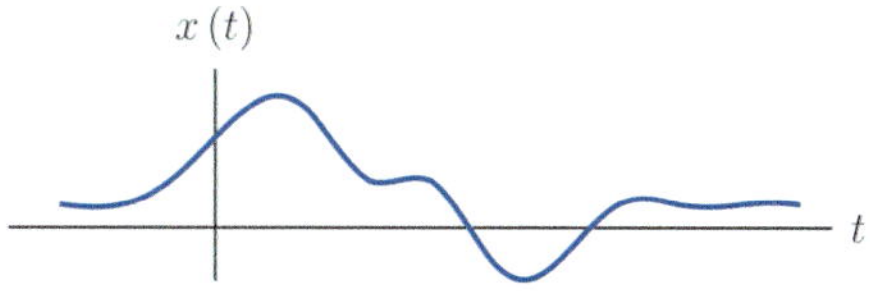

Figure 1.1 – A continuous-time signal.

Fig. 1.2 illustrates the mathematical model for a 25 ms segment of the voice signal that corresponds to the vowel "o" in the word "hello". Similarly, a 25 ms segment of the sound from a violin playing the note A3 is shown in Fig. 1.3.

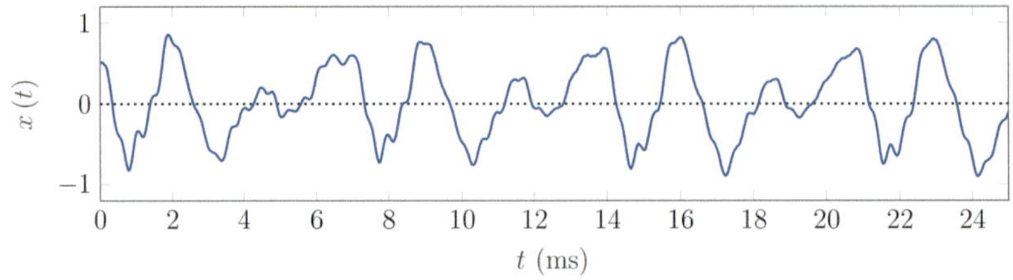

Figure 1.2 – A segment from the vowel "o" of the word "hello".

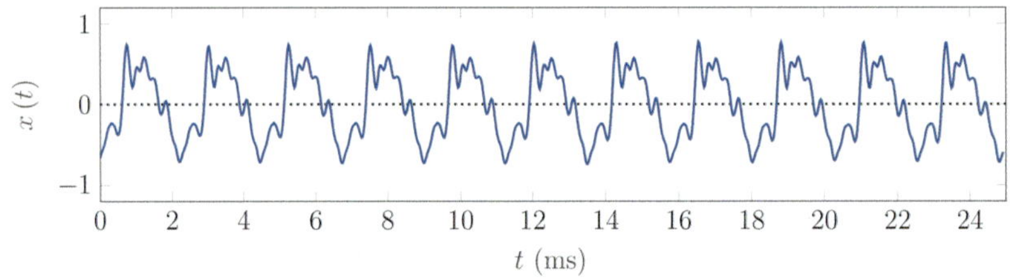

Figure 1.3 – A segment from the sound of a violin.

Some signals can be described analytically. For example, the function

$$x(t) = 5\sin(12t)$$

describes a sinusoidal signal. Sometimes it is convenient to describe a signal model by patching together multiple signal segments, each modeled with a different function. An example of this is given by

$$x(t) = \begin{cases} e^{-3t} - e^{-6t}\,, & t \ge 0 \\ 0\,, & t < 0 \end{cases}$$

Software resource: See MATLAB Exercise 1.1.

In contrast with a continuous-time signal, in some signals the amplitudes may be defined only at some time instants, and not at others. A common practice is to define amplitude values only at time instants that are integer multiples of some step size T_s, that is, for $t = nT_s$. Such a signal is referred to as a *discrete-time* signal and is usually expressed as a function of the index n rather than the time variable t.

In the above discussion we have focused on continuous or discrete nature of the time variable, and assumed that amplitudes are allowed to take on any value. Amplitude values may also be classified as continuous or discrete. If only certain amplitude values are allowed, the resulting signal is said to be a *discrete-amplitude* signal. A signal that is both discrete-time and discrete-amplitude is said to be a *digital signal.*

Discrete-time signals and digital signals are beyond the scope of this text and are covered in the companion textbook *Discrete-Time Signals and Systems: A MATLAB Integrated Approach.*

1.4 Signal Operations

In this section we will discuss time-domain operations that are commonly used in the course of working with continuous-time signals. Some are simple arithmetic operators. Examples of these are addition and multiplication of signals in the time domain, addition of a constant offset to a signal, and multiplication of a signal with a constant gain factor. Other signal operations that are useful in defining signal relationships are time shifting, time scaling, and time reversal. More advanced signal operations such as convolution and sampling will be covered in their own right in later parts of this textbook.

1.4.1 Arithmetic operations

Constant offset

Addition of a constant offset A to the signal $x(t)$ is expressed as

$$g(t) = x(t) + A$$

At each time instant the amplitude of the result $g(t)$ is equal to the amplitude of the signal $x(t)$ plus the value of the constant offset. A practical example of adding a constant offset to a signal is the removal of the dc component at the output of a class-A amplifier circuit through capacitive coupling. Fig. 1.4 illustrates the addition of positive and negative offset values to a signal.

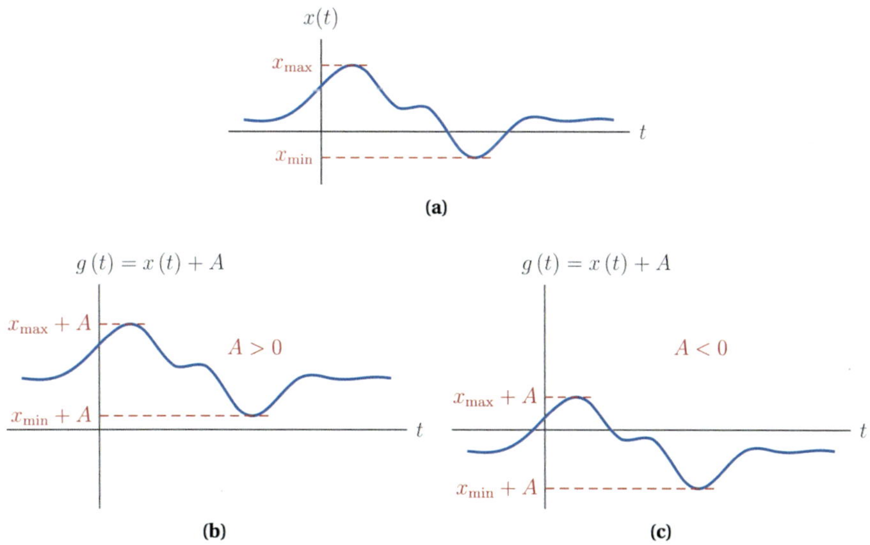

Figure 1.4 – Adding an offset A to signal $x(t)$: **(a)** Original signal $x(t)$, **(b)** $g(t) = x(t) + A$ with $A > 0$, and **(c)** $g(t) = x(t) + A$ with $A < 0$.

Constant gain factor

A signal can also be multiplied with a constant gain factor:

$$g(t) = B\,x(t)$$

The result of this operation is a signal $g(t)$, the amplitude of which is equal to the product of the amplitude of the signal $x(t)$ and the constant gain factor B at each time instant. This is illustrated in Fig. 1.5. A practical example for this relationship is a constant-gain amplifier used for amplifying the voltage signal produced by a microphone.

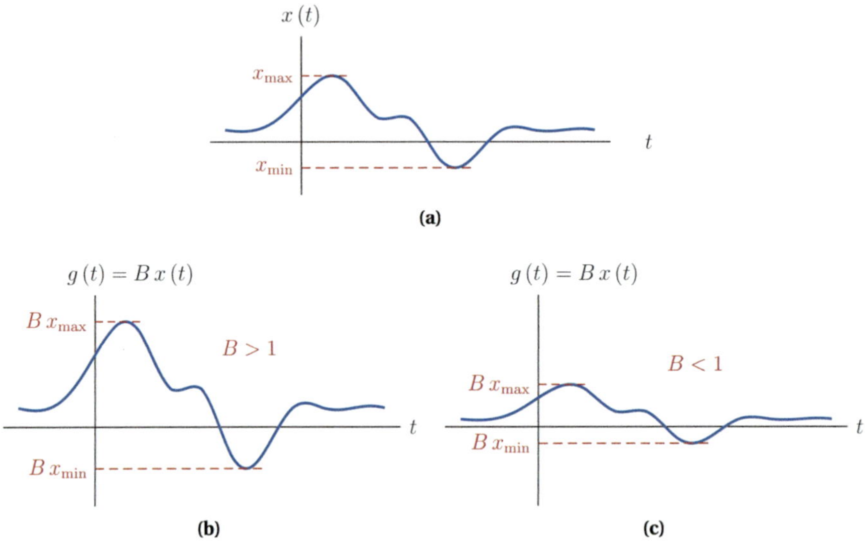

Figure 1.5 – Multiplying signal $x(t)$ with a constant gain factor: **(a)** Original signal $x(t)$, **(b)** $g(t) = B\,x(t)$ with $B > 1$, and **(c)** $g(t) = B\,x(t)$ with $B < 1$.

Adding two signals

Addition of two signals is accomplished by adding the amplitudes of the two signals at each time instant. For two signals $x_1(t)$ and $x_2(t)$, the sum

$$g(t) = x_1(t) + x_2(t)$$

is computed at each time instant $t = t_1$ by adding the values $x_1(t_1)$ and $x_2(t_1)$. This is illustrated in Fig. 1.6. Signal adders are useful in a wide variety of applications. As an example, in an audio recording studio, signals coming from several microphones can be added together to create a complex piece of music.

Multiplying two signals

Multiplication of two signals is carried out in a similar manner. Given two signals $x_1(t)$ and $x_2(t)$, the product

$$g(t) = x_1(t)\,x_2(t)$$

is a signal, the amplitude of which at any time instant $t = t_1$ is equal to the product of the amplitudes of the signals $x_1(t)$ and $x_2(t)$ at the same time instant, that is, $g(t_1) = x_1(t_1)\,x_2(t_1)$. This is illustrated in Fig. 1.7. One example of a practical application of multiplying two signals is in modulating the amplitude of a sinusoidal carrier signal with a message signal. This will be explored when we discuss amplitude modulation in Chapter 7.

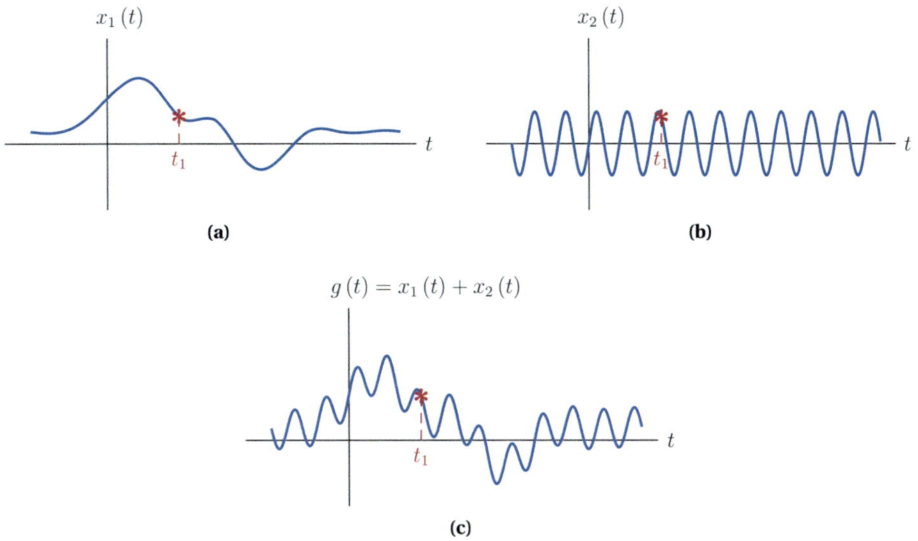

Figure 1.6 – Adding continuous-time signals: **(a)** The signal $x_1(t)$, **(b)** the signal $x_2(t)$, and **(c)** the sum signal $g(t) = x_1(t) + x_2(t)$.

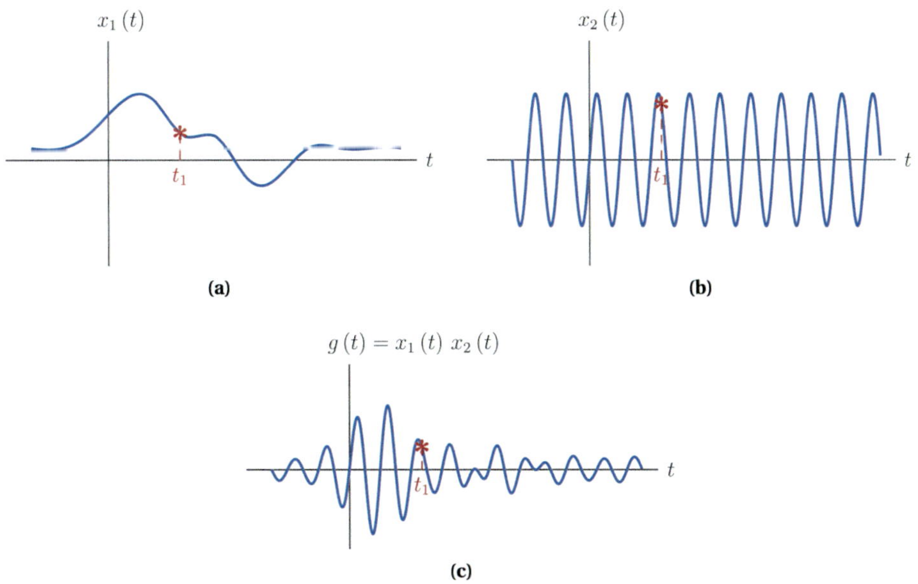

Figure 1.7 – Multiplying continuous-time signals: **(a)** The signal $x_1(t)$, **(b)** the signal $x_2(t)$, and **(c)** the product signal $g(t) = x_1(t)\, x_2(t)$.

Example 1.1: Constant offset and gain

Consider the signal shown in Fig. 1.8. Sketch the signals

a. $g_1(t) = 1.5\,x(t) - 1$

b. $g_2(t) = -1.3\,x(t) + 1$

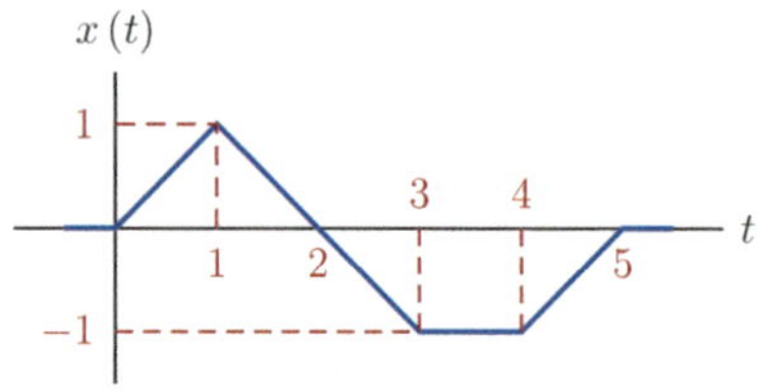

Figure 1.8 – The signal $x(t)$ for Example 1.1.

Solution:

a. The intermediate signal $g_{1a}(t) = 1.5\,x(t)$ is shown in Fig. 1.9(a). In the next step, the signal $g_1(t) = g_{1a}(t) - 1$ is obtained by lowering the graph of the signal $g_{1a}(t)$ by 1 at all time instants. The result is shown in Fig. 1.9(b).

b. The intermediate signal $g_{2a}(t) = -1.3\,x(t)$ is shown in Fig. 1.10(a). In the next step, the signal $g_2(t) = g_{2a}(t) + 1$ is obtained by raising the graph of the signal $g_{2a}(t)$ by 1 at all time instants. The result is shown in Fig. 1.10(b).

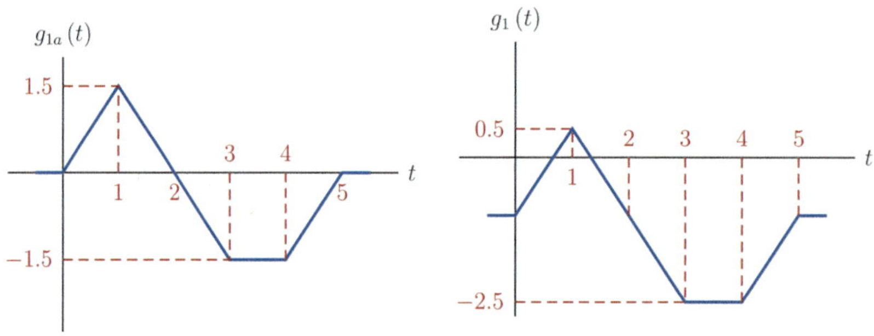

(a) **(b)**

Figure 1.9 – Signals for Example 1.1: **(a)** $g_{1a}(t) = 1.5\,x(t)$ and **(b)** $g_1(t) = 1.5\,x(t) - 1$.

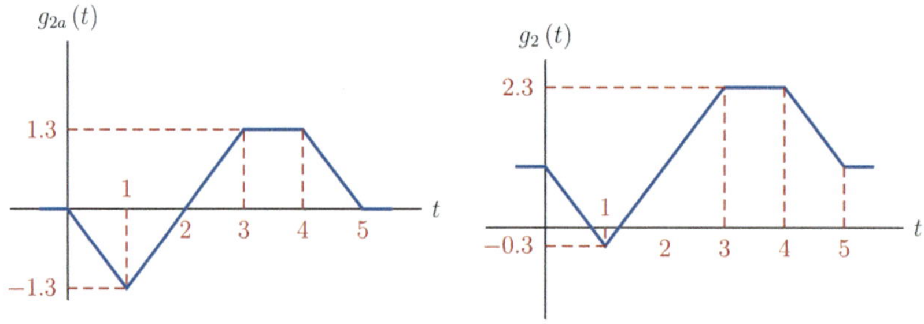

(a) **(b)**

Figure 1.10 – Signals for Example 1.1: **(a)** $g_{2a}(t) = -1.3\,x(t)$ and **(b)** $g_2(t) = -1,3\,x(t) + 1$.

Software resource: See MATLAB Exercise 1.2.

Interactive App: Elementary signal operations – I

The interactive app in `appSigOp1.m` allows experimentation with elementary signal operations involving constant offset and constant gain factor. Specifically, the signal $x(t)$ shown in Figs. 1.4 and 1.5 is used as the basis of the app, and the signal

$$g(t) = B\,x(t) + A$$

is computed and graphed. Values of A and B can be adjusted through the use of slider controls.

a. Let $B = 1$. Using the slider on the left, vary the value of constant offset A, and observe its effect on the signal $g(t)$. Try this with $A = -2, -1, 1, 2$.

b. Let $A = 0$. Using the slider on the right, vary the value of constant gain factor B, and observe the changes in the signal $g(t)$. Try $B = 0.5$ and $B = 2$.

c. Try the previous part with negative gain factor values $B = -0.5$ and $B = -2$.

d. Now vary both parameters as needed to obtain the signals $g(t) = 0.7\,x(t) - 1$ and $g(t) = -1.2\,x(t) + 0.8$. Comment on the results.

Software resource: `appSigOp1.m`

Example 1.2: Arithmetic operations with continuous-time signals

Two signals $x_1(t)$ and $x_2(t)$ are shown in Fig. 1.11. Sketch the signals

a. $g_1(t) = x_1(t) + x_2(t)$

b. $g_2(t) = x_1(t)\,x_2(t)$

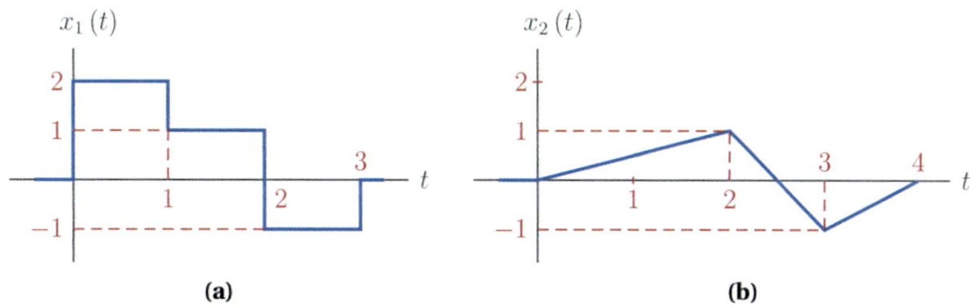

Figure 1.11 – Signals $x_1(t)$ and $x_2(t)$ for Example 1.2.

Solution: The following analytical descriptions of the signals $x_1(t)$ and $x_2(t)$ can be inferred from the graphical descriptions given in Fig. 1.11:

$$x_1(t)=\begin{cases} 2\,, & 0<t<1\\ 1\,, & 1<t<2\\ -1\,, & 2<t<3\\ 0\,, & \text{otherwise}\end{cases} \quad \text{and} \quad x_2(t)=\begin{cases} \frac{1}{2}t\,, & 0<t<2\\ -2t+5\,, & 2<t<3\\ t-4\,, & 3<t<4\\ 0\,, & \text{otherwise.}\end{cases}$$

a. The addition of the two signals can be accomplished by adding the values in each segment separately. The analytical form of the sum is

$$g_1(t)=\begin{cases} \frac{1}{2}t+2\,, & 0<t<1\\ \frac{1}{2}t+1\,, & 1<t<2\\ -2t+4\,, & 2<t<3\\ t-4\,, & 3<t<4\\ 0\,, & \text{otherwise}\end{cases}$$

and is shown graphically in Fig. 1.12(a).

b. The product of the two signals can also computed for each segment, and is obtained as

$$g_2(t)=\begin{cases} t\,, & 0<t<1\\ \frac{1}{2}t\,, & 1<t<2\\ 2t-5\,, & 2<t<3\\ 0\,, & \text{otherwise.}\end{cases}$$

This result is shown in Fig. 1.12(b).

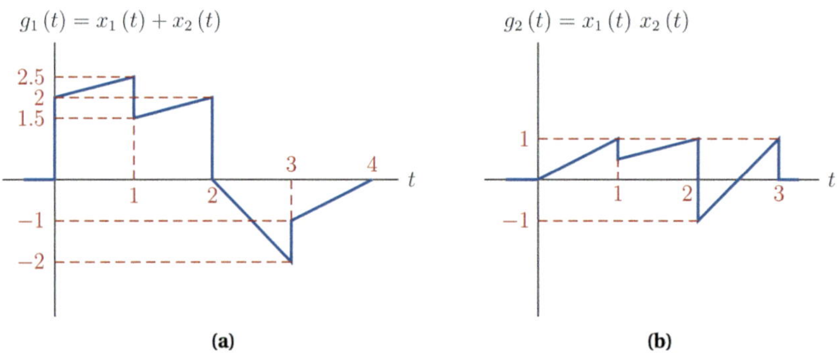

Figure 1.12 – Signals $g_1(t)$ and $g_2(t)$ for Example 1.2.

Software resource: exct_1_2.m

1.4.2 Time shifting

A *time-shifted* version of the signal $x(t)$ can be obtained through

$$g(t) = x(t - t_d)$$

where t_d is any positive or negative constant. Fig. 1.13 illustrates this relationship. In part (a) of Fig. 1.13 the amplitude of $x(t)$ at the time instant $t = t_1$ is marked with a star-shaped marker. Let that marker represent a special event that takes place within the signal $x(t)$. Using $t = t_1 + t_d$ in the signal $g(t)$ we obtain

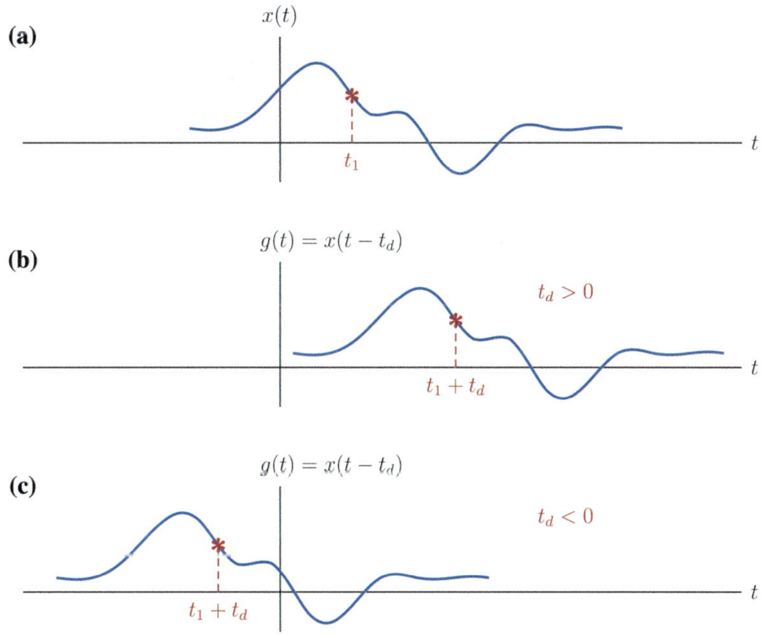

Figure 1.13 – Time shifting a signal: **(a)** Original signal $x(t)$, **(b)** time-shifted signal $g(t)$ for $t_d > 0$, and **(c)** time-shifted signal $g(t)$ for $t_d < 0$.

$$g(t_1 + t_d) = x(t_1)$$

Thus, in the signal $g(t)$, the same event takes place at the time instant $t = t_1 + t_d$. If t_d is positive, $g(t)$ is a delayed version of $x(t)$ and t_d is the amount of time delay. A negative t_d, on the other hand, corresponds to advancing the signal in time by an amount equal to $-t_d$.

1.4.3 Time scaling

A *time-scaled* version of the signal $x(t)$ is obtained through the relationship

$$g(t) = x(at)$$

where a is a positive constant. Depending on the value of a, the signal $g(t)$ is either a compressed or an expanded version of $x(t)$. Fig. 1.14 illustrates this relationship. We will use a logic similar to the one employed in explaining the time-shifting behavior: In part (a) of Fig. 1.14, let the star-shaped marker at the time instant $t = t_1$ represent a special event in the signal $x(t)$. Using $t = t_1/a$ in the signal $g(t)$ we obtain

$$g\left(\frac{t_1}{a}\right) = x(t_1)\,.$$

The corresponding event in $g(t)$ takes place at the time instant $t = t_1/a$. Thus, a scaling parameter value of $a > 1$ results in the signal $g(t)$ being a compressed version of $x(t)$. Conversely, $a < 1$ leads to a signal $g(t)$ that is an expanded version of $x(t)$.

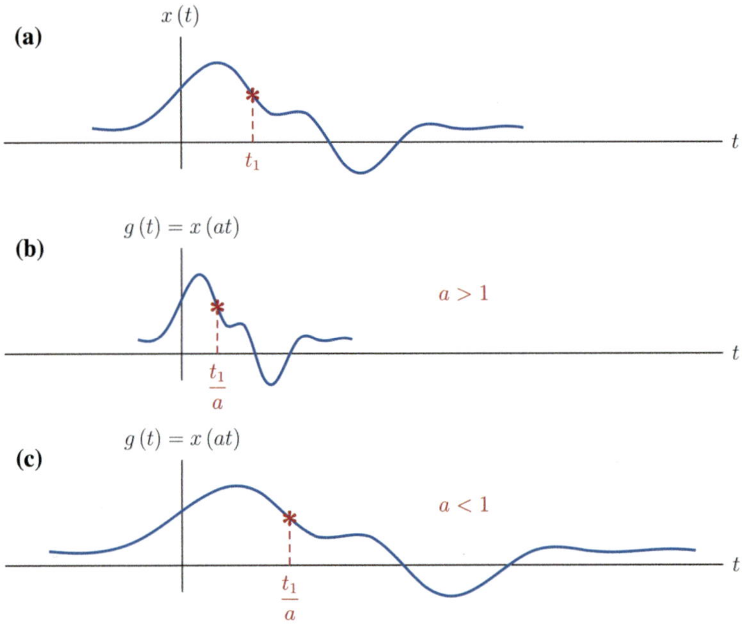

Figure 1.14 – Time scaling a signal: **(a)** Original signal $x(t)$, **(b)** time-scaled signal $g(t)$ for $a > 0$, and **(c)** time-scaled signal $g(t)$ for $a < 0$.

1.4.4 Time reversal

A *time-reversed* version of the signal $x(t)$ is obtained through

$$g(t) = x(-t)$$

which is illustrated in Fig. 1.15. An event that takes place at the time instant $t = t_1$ in the signal $x(t)$ is duplicated at the time instant $t = -t_1$ in the signal $g(t)$. Graphically this corresponds to folding or flipping the signal around the vertical axis. It is possible to combine time shifting, time scaling, and time reversal operations to create more intricate relationships between signals. Two such combinations will be explored in Example 1.3.

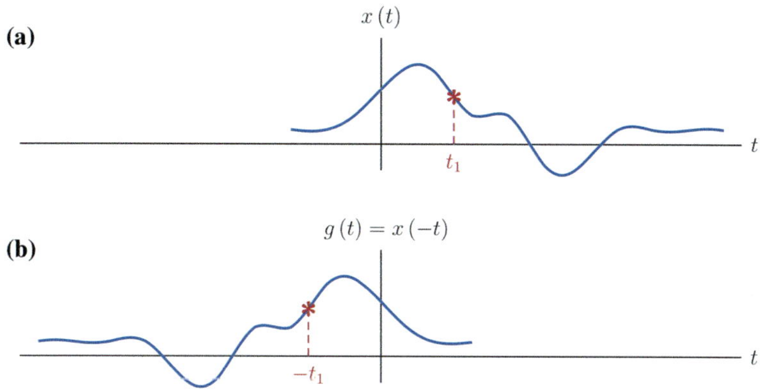

Figure 1.15 – Time reversing a signal: **(a)** Original signal $x(t)$ and **(b)** time reversed signal $g(t)$.

Interactive App: Elementary signal operations – II

The interactive app in `appSigOp2.m` is based on the discussion above and allows experimentation with elementary signal operations, namely shifting, scaling, and reversal of a signal as well as addition and multiplication of two signals. The desired signal operation is selected from the drop down list control. When applicable, slider control for the time delay parameter t_d or the time scaling parameter a becomes available to facilitate adjustment of the corresponding value. Also, the star shaped marker used in Figs. 1.13 through 1.15 is also shown on the graphs for $x(t)$ and $g(t)$ when it is relevant to the signal operation chosen.

- **a.** From the drop-down list, select "Add two signals". Observe the sum signal $g(t) = x(t) + y(t)$.
- **b.** From the drop-down list, select "Multiply two signals". Observe the product signal $g(t) = x(t)\, y(t)$.
- **c.** From the drop-down list, select "Time shifting". Use the time delay slider to obtain the signals $g(t) = x(t-4)$ and $g(t) = x(t+3.5)$. Comment on the results.
- **d.** From the drop-down list, select "Time scaling". Use the scale parameter slider to obtain the signals $g(t) = x(4t)$ and $g(t) = x(t/3)$. Comment on the effects of time scaling with $a > 1$ and $a < 1$.
- **e.** From the drop-down list, select "Time reversal". Observe the resulting signal $g(t) = x(-t)$.

Software resource: `appSigOp2.m`

Example 1.3: Basic operations for continuous-time signals

Consider the signal $x(t)$ shown in Fig. 1.16. Sketch the following signals:

a. $g(t) = x(2t-5)$

b. $h(t) = x(-4t+2)$

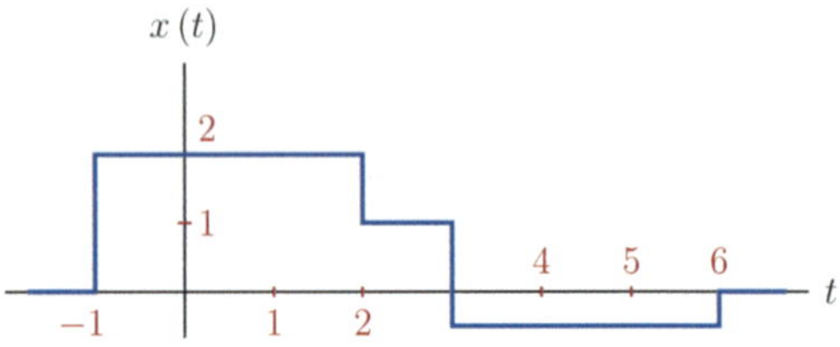

Figure 1.16 – The signal $x(t)$ for Example 1.3.

Solution:

a. We will obtain $g(t)$ in two steps: Let an intermediate signal be defined as $g_1(t) = x(2t)$, a time-scaled version of $x(t)$, shown in Fig. 1.17(b). The signal $g(t)$ can be expressed as

$$g(t) = g_1(t-2.5) = x(2[t-2.5]) = x(2t-5)$$

and is shown in Fig. 1.17(c).

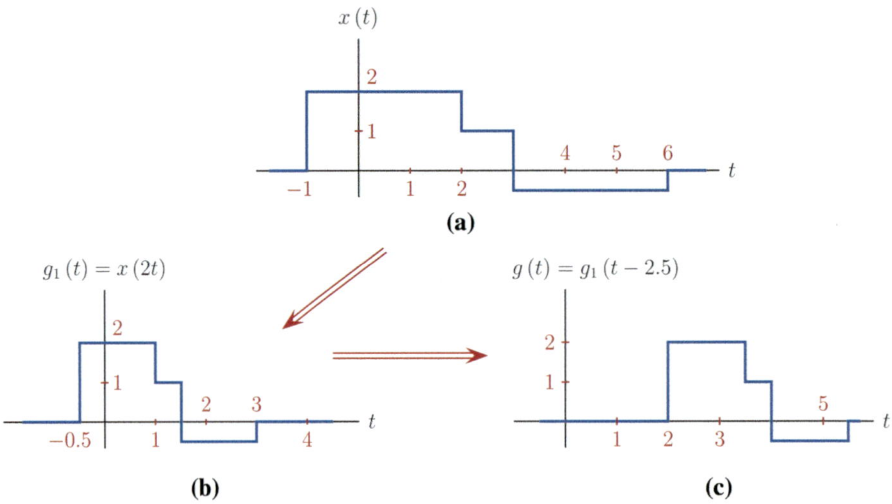

Figure 1.17 – **(a)** The original signal $x(t)$, **(b)** the intermediate signal $g_1(t)$, and **(c)** the signal $g(t)$ for Example 1.3.

b. In this case we will use two intermediate signals: Let $h_1(t) = x(4t)$. A second intermediate signal $h_2(t)$ can be obtained by time shifting $h_1(t)$ so that

$$h_2(t) = h_1(t+0.5) = x(4[t+0.5]) = x(4t+2)$$

Finally, $h(t)$ can be obtained through time reversal of $h_2(t)$:

$$h(t) = h_2(-t) = x(-4t+2)$$

The steps involved in sketching $h(t)$ are shown in Fig. 1.18.

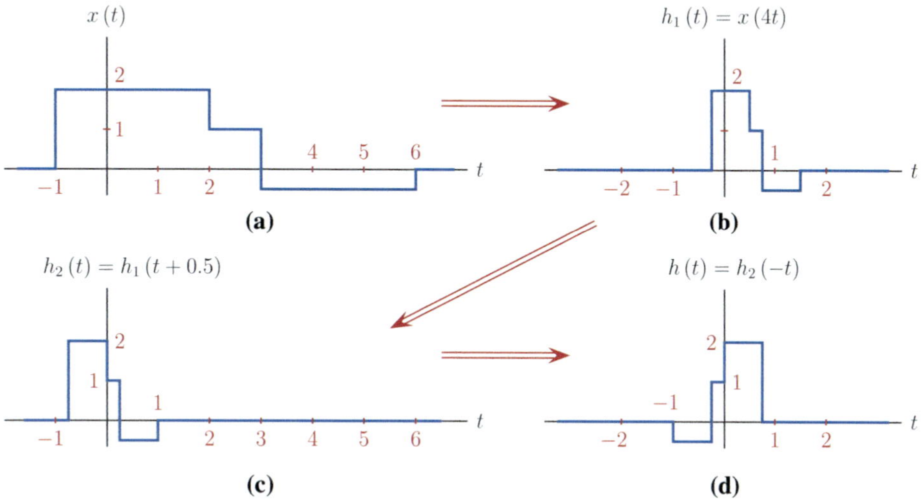

Figure 1.18 – **(a)** The original signal $x(t)$, **(b)** intermediate signal $h_1(t)$, **(c)** intermediate signal $h_2(t)$, and **(d)** the signal $h(t)$ for Example 1.3.

Software resource: See MATLAB Exercise 1.4.

1.4.5 Integration and differentiation

Integration and differentiation operations are used extensively in the study of linear systems. Given a continuous-time signal $x(t)$, a new signal $g(t)$ may be defined as its time derivative in the form

$$g(t) = \frac{dx(t)}{dt}$$

A practical example of this is the relationship between the current $i_C(t)$ and the voltage $v_C(t)$ of an ideal capacitor with capacitance C as given by

$$i_C(t) = C\,\frac{dv_C(t)}{dt}$$

For example, if the voltage in the shape of a periodic triangular waveform shown in Fig. 1.19(a) is applied to an ideal capacitor, the current that flows through the capacitor would be proportional to its derivative, resulting in a square-wave signal as shown in Fig. 1.19(b). Similarly, a signal can be defined as the integral of another signal in the form

$$g(t) = \int_{-\infty}^{t} x(\lambda)\, d\lambda$$

The relationship between the current $i_L(t)$ and the voltage $v_L(t)$ of an ideal inductor can serve as an example of this. Specifically we have

$$i_L(t) = \frac{1}{L}\int_{-\infty}^{t} v_L(\lambda)\, d\lambda$$

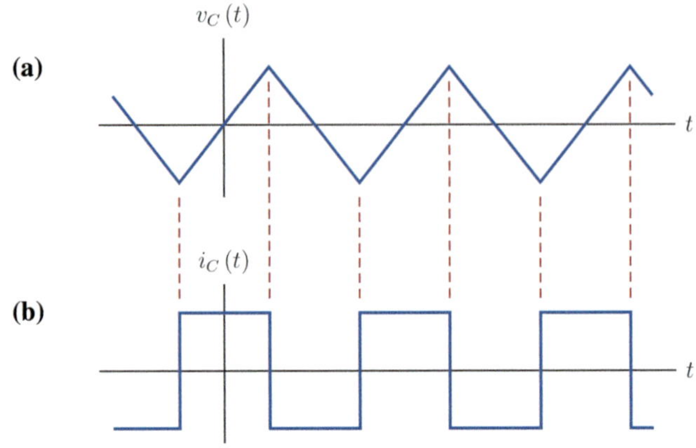

Figure 1.19 – **(a)** A periodic triangular waveform $v_C(t)$ used as the voltage of a capacitor and **(b)** the periodic square-wave current signal $i_C(t)$ that results.

If the voltage in the shape of a periodic square-wave shown in Fig. 1.20(a) is applied to an ideal inductor, the current that flows through it would be proportional to its integral, resulting in a periodic triangular waveform as shown in Fig. 1.20(b).

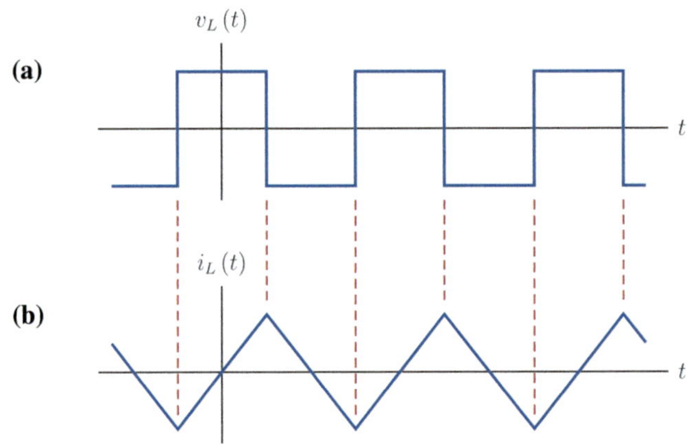

Figure 1.20 – **(a)** A periodic square-wave signal $v_L(t)$ used as the voltage of an inductor and **(b)** the periodic triangular current signal $i_L(t)$ that results.

1.5 Basic Building Blocks for Continuous-Time Signals

There are certain basic signal forms that can be used as building blocks for describing signals with higher complexity. In this section we will study some of these signals. Mathematical models for more advanced signals can be developed by combining these basic building blocks through the use of the signal operations described in Section 1.4.

1.5.1 Unit impulse function

We will see in the rest of this chapter as well as the chapters to follow that the unit impulse function plays an important role in mathematical modeling and analysis of signals and linear systems. It is defined by the following equations:

Unit impulse function:

$$\delta(t) = \begin{cases} 0\,, & \text{if } t \neq 0 \\ \text{undefined}\,, & \text{if } t = 0 \end{cases} \tag{1.1}$$

$$\int_{-\infty}^{\infty} \delta(t)\, dt = 1 \tag{1.2}$$

Note that Eqn. (1.1) by itself represents an incomplete definition of the function $\delta(t)$ since the amplitude of it is defined only when $t \neq 0$, and is undefined at the time instant $t = 0$. Eqn. (1.2) fills this void by defining the area under the function $\delta(t)$ to be unity. The impulse function cannot be graphed using the common practice of graphing amplitude as a function of time, since the only amplitude of interest occurs at a single time instant $t = 0$, and it has an undefined value. Instead, we use an arrow to indicate the location of that undefined amplitude, as shown in Fig. 1.21.

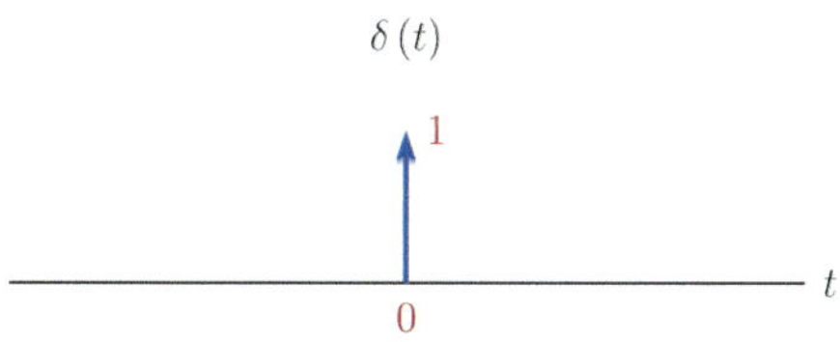

Figure 1.21 – Unit impulse function.

In a number of problems in this text we will utilize shifted and scaled versions of impulse function. The function $a\delta(t - t_1)$ represents a unit impulse function that is time-shifted by t_1 and scaled by constant scale factor a. It is described through the equations

$$a\delta(t - t_1) = \begin{cases} 0\,, & \text{if } t \neq t_1 \\ \text{undefined}\,, & \text{if } t = t_1 \end{cases} \tag{1.3}$$

and

$$\int_{-\infty}^{\infty} a\delta(t - t_1)\, dt = a \tag{1.4}$$

and is shown in Fig. 1.22.

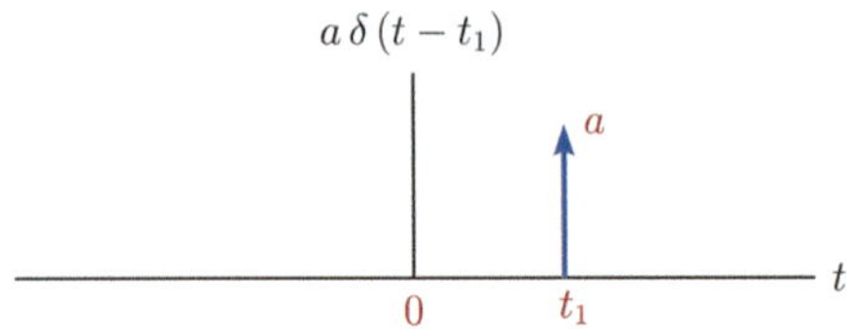

Figure 1.22 – Scaled and time-shifted impulse function.

In Figs. 1.21 and 1.22, the value displayed next to the up arrow is not an amplitude value. Rather, it represents the area of the impulse function. Furthermore, in Eqns. (1.2) and (1.4), the integration limits do not need to be infinite; any set of integration limits that includes the impulse would produce the same result. With any time increment $\Delta t > 0$, Eqn. (1.2) can be written in the alternative form

$$\int_{-\Delta t}^{\Delta t} \delta(t)\, dt = 1$$

and Eqn. (1.4) can be written in the alternative form

$$\int_{t_1-\Delta t}^{t_1+\Delta t} a\,\delta(t-t_1)\, dt = a$$

How can a function that has zero amplitude everywhere except at one point have a non-zero area under it? One way of visualizing how this might work is to start with a rectangular pulse centered at the origin, with a duration of a and an amplitude of $1/a$ as shown in Fig. 1.23(a). Mathematically, the pulse $q(t)$ is defined as

$$q(t) = \begin{cases} \dfrac{1}{a}, & |t| < \dfrac{a}{2} \\ 0, & |t| > \dfrac{a}{2} \end{cases} \tag{1.5}$$

The area under the pulse is clearly equal to unity independent of the value of a. Now imagine the parameter a getting smaller, resulting in the pulse becoming narrower. As the pulse becomes narrower, it also becomes taller as shown in Fig. 1.23(b). The area under the pulse remains unity. In the limit, as the parameter a approaches zero, the pulse approaches an impulse, that is

$$\delta(t) = \lim_{a\to 0}\left[q(t)\right] \tag{1.6}$$

This is illustrated in Fig. 1.23(c).

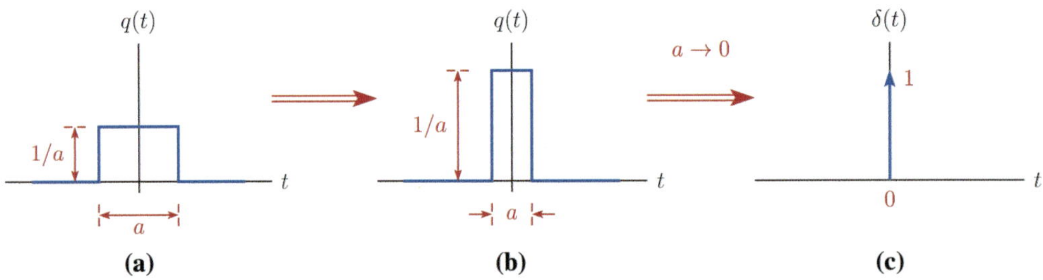

Figure 1.23 – Obtaining a unit impulse as the limit of a pulse with unit area. **(a)** The pulse $q(t)$, **(b)** a narrower and taller version obtained by reducing a, and **(c)** unit impulse function obtained as the limit.

Alternative derivations of the unit impulse function

It is also possible to obtain the unit impulse function as the limit of other functions besides a rectangular pulse. One example is the *Gaussian* function. Consider the limit

$$\delta(t) = \lim_{a \to 0} \left[\frac{1}{a\sqrt{\pi}} e^{-t^2/a^2} \right] \tag{1.7}$$

The Gaussian function has unit area, and becomes narrower as the parameter a is made smaller. This is illustrated in Fig. 1.24. Another example is the *squared sinc pulse.* Consider the limit

$$\delta(t) = \lim_{a \to 0} \left[\frac{a}{(\pi t)^2} \sin^2\left(\frac{\pi t}{a}\right) \right] \tag{1.8}$$

The convergence of this limit to the unit impulse signal is illustrated in Fig. 1.25.

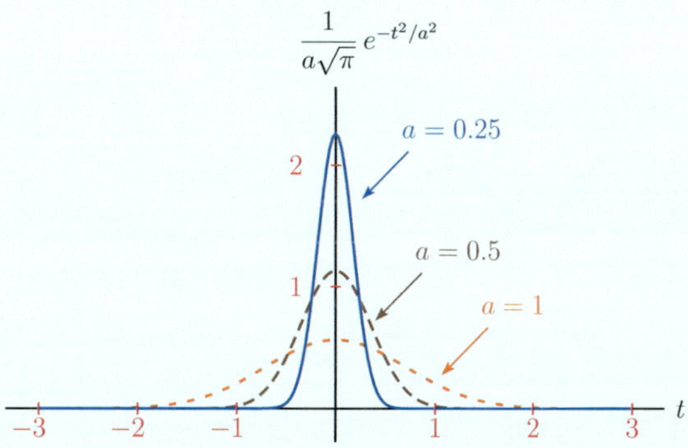

Figure 1.24 – Obtaining unit impulse function as the limit of a Gaussian pulse.

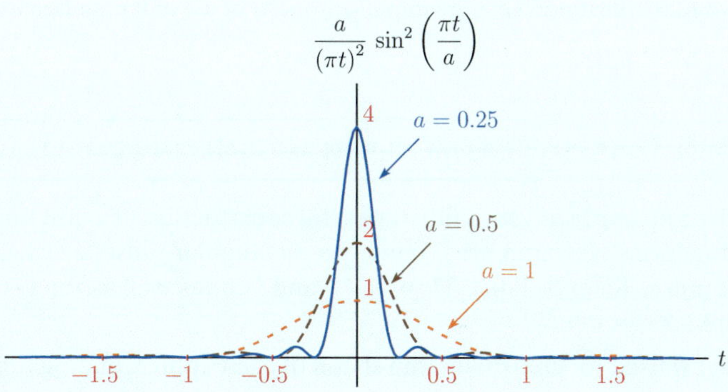

Figure 1.25 – Obtaining a unit impulse function as the limit of a squared sinc pulse.

The impulse function has two fundamental properties that are useful. The first one, referred to as the *sampling property*[1] of the impulse function, is stated as

$$f(t)\,\delta(t-t_1)=f(t_1)\,\delta(t-t_1) \tag{1.9}$$

where $f(t)$ is any function of time that is continuous at $t=t_1$. Intuitively, since $\delta(t-t_1)$ is equal to zero for all $t \neq t_1$, the only value of the function $f(t)$ that should have any significance in terms of the product $f(t)\,\delta(t-t_1)$ is its value at $t=t_1$. The sampling property of the unit impulse function is illustrated in Fig. 1.26.

A consequence of Eqn. (1.9) is the *sifting property* of the impulse function which is expressed as

$$\int_{-\infty}^{\infty} f(t)\,\delta(t-t_1)\,dt=f(t_1) \tag{1.10}$$

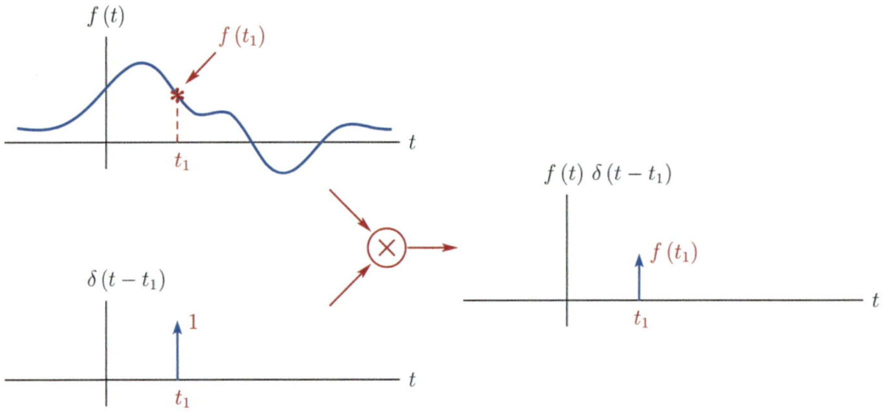

Figure 1.26 – Illustration of the sampling property of the unit impulse function.

Interactive App: Constructing a unit impulse as a limit case of various functions

The interactive app `appImpulse.m` illustrates the construction of a unit impulse function as the limit of functions with unit area, namely a rectangular pulse, a Gaussian pulse, and a squared sinc pulse. Refer to Eqns. (1.6), (1.7), and (1.8) as well as the corresponding Figs. 1.23, 1.24, and 1.25 for details.

The parameter a used for controlling the shape of each approximating function. Using the slider control, gradually lower the value of a, starting with $a=1$ and moving toward $a=0$. Observe how each function exhibits increased similarity to the unit impulse as a is made smaller.

Software resource: `appImpulse.m`

[1] A thorough mathematical study of singularity functions such as the unit impulse function is quite advanced, and is well beyond the scope of this text. From a theoretical standpoint, one could set up conditions under which the sampling property given by Eqn. (1.9) may not hold. However, it will be valid and useful for our purposes in formulating signal models for engineering applications.

The integral of the product of a function $f(t)$ and a time-shifted unit impulse function is equal to the value of $f(t)$ evaluated at the location of the unit impulse. The sifting property expressed by Eqn. (1.10) is easy to justify through the use of the sampling property. Substituting Eqn. (1.9) into Eqn. (1.10), we obtain

$$\begin{aligned}\int_{-\infty}^{\infty} f(t)\,\delta(t-t_1)\,dt &= \int_{-\infty}^{\infty} f(t_1)\,\delta(t-t_1)\,dt \\ &= f(t_1)\int_{-\infty}^{\infty} \delta(t-t_1)\,dt \\ &= f(t_1) \end{aligned} \tag{1.11}$$

As discussed above, the limits of the integral in Eqn. (1.10) do not need to be infinitely large for the result to be valid; any integration interval that includes the impulse will work. Given a time increment $\Delta t > 0$, the following is also valid:

$$\int_{t_1-\Delta t}^{t_1+\Delta t} f(t)\,\delta(t-t_1)\,dt = f(t_1) \tag{1.12}$$

It should be clear from the foregoing discussion that the unit impulse function is a mathematical idealization. No physical signal is capable of occurring without taking any time and still providing a non-zero result when integrated. Therefore, no physical signal that utilizes the unit impulse function as a mathematical model can be obtained in practice. This realization naturally prompts the question "why do we need the unit impulse signal?" or the question "how is the unit impulse signal useful?" The answer to these questions lies in the fact that the unit impulse function is a fundamental building block that is useful in approximating events that occur quickly. The use of the unit impulse function allows us to look at a signal as a series of events that occur momentarily.

Consider the sifting property of the impulse function which was described by Eqn. (1.10). Through the use of the sifting property, we can focus on the behavior of a signal at one specific instant in time. By using the sifting property repeatedly at every imaginable time instant, we can analyze a signal in terms of its behavior as a function of time. This idea will become clearer when we discuss *impulse decomposition of a signal* in the next section which will later become the basis of the development leading to the convolution operation in Chapter 2. Even though the impulse signal is not obtainable in a practical sense, it can be approximated using a number of functions; the three signals we have considered in Figs. 1.23, 1.24, and 1.25 are possible candidates when the parameter a is reasonably small.

1.5.2 Unit step function

The unit step function is useful in situations where we need to model a signal that is turned on or off at a specific time instant. It is defined as follows:

Unit step function:

$$u(t) = \begin{cases} 1, & t > 0 \\ 0, & t < 0 \end{cases} \tag{1.13}$$

The function $u(t)$ is illustrated in Fig. 1.27a. A time-shifted version of the unit step function can be written as

$$u(t-t_1)=\begin{cases}1\,, & t>t_1\\ 0\,, & t<t_1\end{cases}$$

and is shown in Fig. 1.27b.

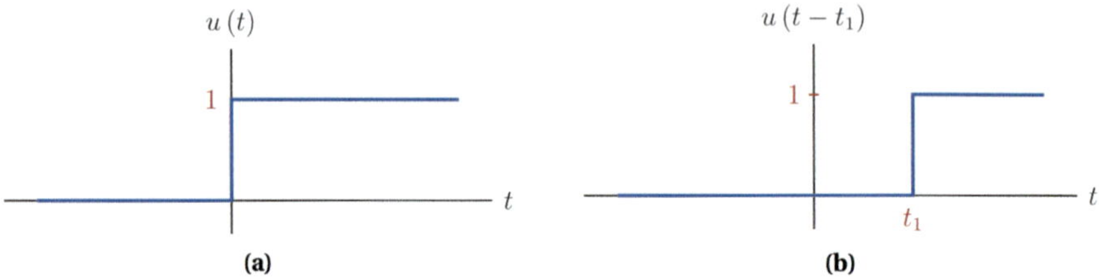

Figure 1.27 – **(a)** Unit step function and **(b)** unit step function with a time shift.

Consider a sinusoidal signal given by

$$x(t)=\sin\left(2\pi f_0 t\right)$$

which oscillates for all values of t. If we need to represent a sinusoidal signal that is switched on at time $t=0$, we can use the unit step function, and write

$$x(t)=\sin\left(2\pi f_0 t\right)\,u(t)=\begin{cases}\sin\left(2\pi f_0 t\right)\,, & t>0\\ 0\,, & t<0\end{cases}$$

Alternatively, the signal can be switched on at an arbitrary time instant $t=t_1$ through the use of a time-shifted unit step function:

$$x(t)=\sin\left(2\pi f_0 t\right)\,u(t-t_1)=\begin{cases}\sin\left(2\pi f_0 t\right)\,, & t>t_1\\ 0\,, & t<t_1\end{cases}$$

Fig. 1.28 illustrates this use of the unit step function for switching a signal on at a specified time instant.

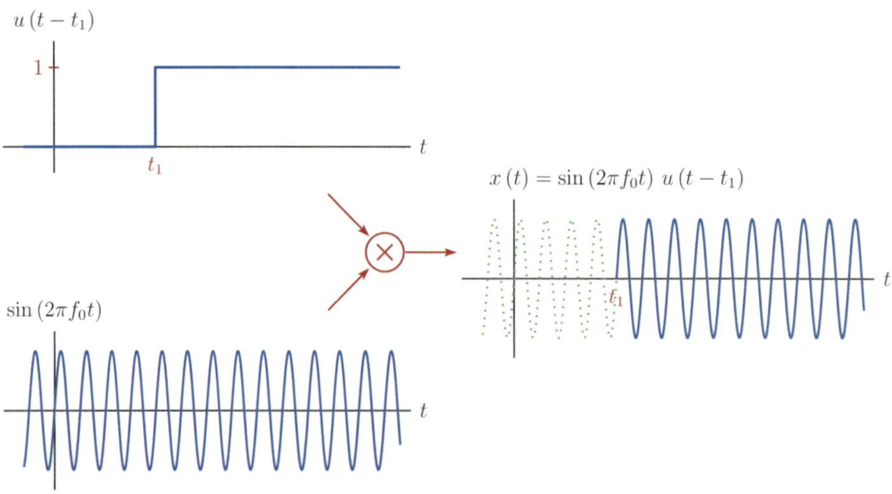

Figure 1.28 – The use of the unit step function for switching a signal on at a specified time.

Interactive App: Unit step function for signal switching – I

The interactive app in `appStep1.m` illustrates the use of the unit step function for switching a signal on at a specified time. It is based on Fig. 1.28. The value of the delay parameter t_1 for the unit step function can be varied by using the slider control. In addition, the frequency and the phase of the sinusoidal signal can be varied using slider controls.

a. Set the frequency and the phase of the sinusoidal signal to $f_0 = 3$ Hz and $\theta_0 = 0$ degrees, respectively. Set the time delay to $t_1 = 0$. Observe the shapes of signals $\sin(2\pi f_0 t)$ and $\sin(2\pi f_0 t)\, u(t)$.

b. Increase the frequency from $f_0 = 3$ Hz to $f_0 = 5$ Hz. Observe the changes in the signals $\sin(2\pi f_0 t)$ and $\sin(2\pi f_0 t)\, u(t)$.

c. Increase the phase from $\theta_0 = 0$ degrees to $\theta_0 = 90$ degrees. Observe the changes in the signals $\sin(2\pi f_0 t + \theta_0)$ and $\sin(2\pi f_0 t + \theta_0)\, u(t)$. Repeat for $\theta_0 = 180$ degrees.

d. With the frequency and the phase set back to $f_0 = 3$ Hz and $\theta_0 = 0$ degrees, respectively, gradually increase the time delay t_1 from 0 to 3 seconds. Observe the changes in the signal $\sin(2\pi f_0 t)\, u(t - t_1)$

Software resource: `appStep1.m`

Conversely, suppose the sinusoidal signal has been on for a long time, and we would like to turn it off at time instant $t = t_1$. This could be accomplished by multiplication of the sinusoidal signal with a time reversed and shifted unit step function in the form

$$x(t) = \sin(2\pi f_0 t)\, u(-t + t_1) = \begin{cases} \sin(2\pi f_0 t)\,, & t < t_1 \\ 0\,, & t > t_1 \end{cases}$$

and is illustrated in Fig. 1.29.

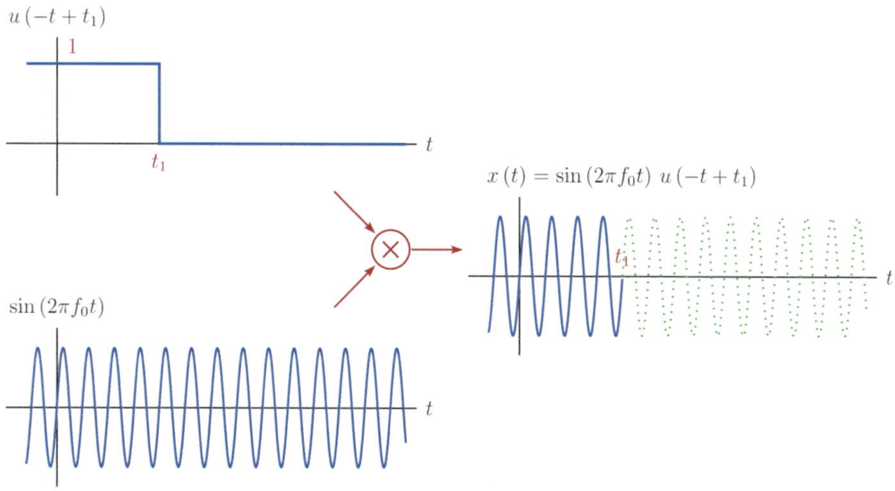

Figure 1.29 – The use of the unit step function for switching a signal off at a specified time.

Interactive App: Unit step function for signal switching – II

The interactive app in `appStep2.m` illustrates the use of the unit step function for switching a signal off at a specified time. It is based on Fig. 1.29. Similar to `appStep1.m`, the value of the delay parameter t_1 for the unit step function can be varied by using the slider control. In addition, the frequency and the phase of the sinusoidal signal can be varied using slider controls.

a. Set the time delay to $t_1 = 0$. Vary the frequency and the phase of the sinusoidal signal, and observe the changes in signals $\sin\left(2\pi f_0 t + \theta_0\right)$ and $\sin\left(2\pi f_0 t + \theta_0\right)\, u(-t)$.

b. With the frequency and the phase set to $f_0 = 3$ Hz and $\theta_0 = 0$ degrees, respectively, gradually increase the time delay t_1 from 0 to 3 seconds. Observe the changes in the signal $\sin\left(2\pi f_0 t\right)\, u(-t + t_1)$

Software resource: `appStep2.m`

The relationship between the unit step function and the unit impulse function is important. The unit step function can be expressed as a *running integral* of the unit impulse function:

$$u(t) = \int_{-\infty}^{t} \delta(\lambda)\, d\lambda \tag{1.14}$$

In Eqn. (1.14) we have changed the independent variable to λ to avoid any confusion with the upper limit t of integration. Recall that the integral of the unit impulse function equals unity provided that the impulse is within the integration limits. This is indeed the case in Eqn. (1.14) if the upper limit of the integral is positive, that is, if $t = t_0 > 0$. Conversely, if $t = t_0 < 0$, then the integration interval stops short of the location of the impulse, and the result of the integral is zero. This is illustrated in Fig. 1.30.

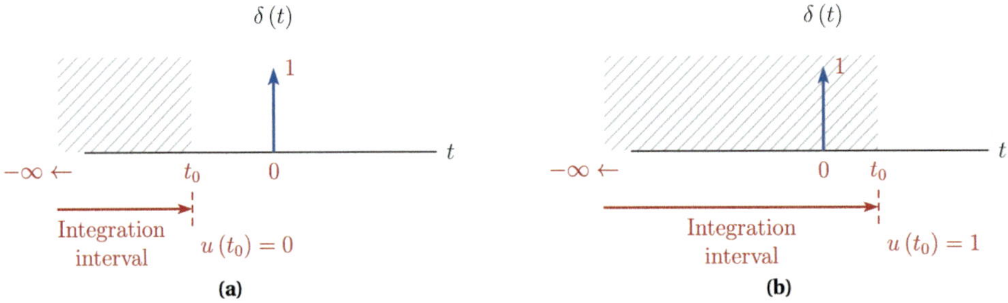

Figure 1.30 – Obtaining a unit step through the running integral of the unit impulse: **(a)** $t = t_0 < 0$ and **(b)** $t = t_0 > 0$.

In Fig. 1.23 and Eqns. (1.5) and (1.6) we have represented the unit impulse function as the limit of a rectangular pulse with unit area. Fig. 1.31 illustrates the running integral of such a pulse. In the limit, the result of the running integral approaches the unit step function as $a \rightarrow 0$.

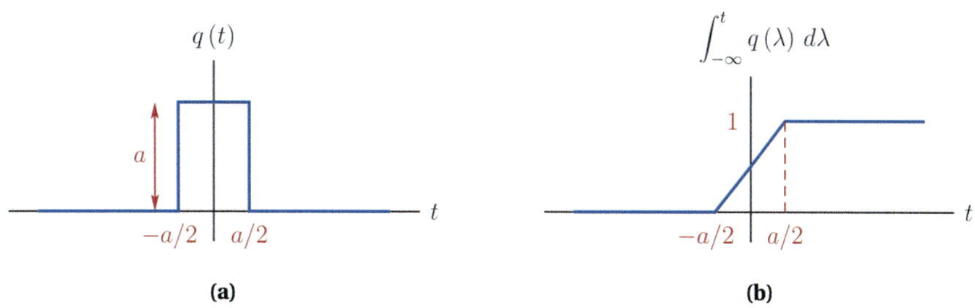

Figure 1.31 – **(a)** Rectangular pulse approximation to a unit impulse and **(b)** running integral of the rectangular pulse as an approximation to a unit step.

The unit impulse function can be written as the derivative of the unit step function, that is,

$$\delta(t) = \frac{du}{dt} \tag{1.15}$$

consistent with Eqn. (1.14).

1.5.3 Unit pulse function

We will define the unit pulse function as a rectangular pulse with unit width and unit amplitude, centered around the origin. Mathematically, it can be expressed as follows:

Unit pulse function:

$$\Pi(t) = \begin{cases} 1, & |t| < \frac{1}{2} \\ 0, & |t| > \frac{1}{2} \end{cases} \tag{1.16}$$

and is shown in Fig. 1.32.

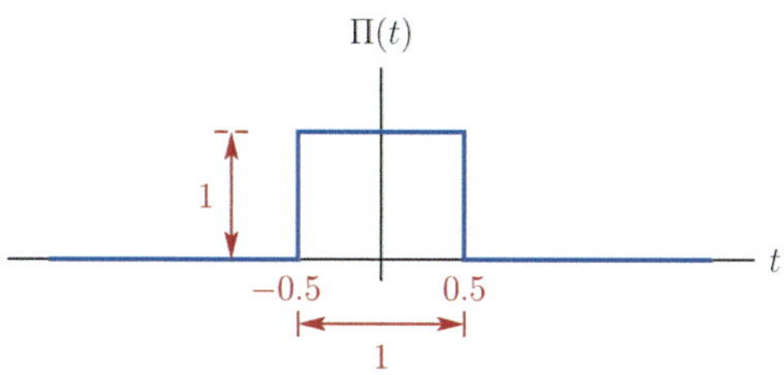

Figure 1.32 – Unit pulse function.

The unit pulse function can be expressed in terms of time-shifted unit step functions as

$$\Pi(t) = u\left(t+\tfrac{1}{2}\right) - u\left(t-\tfrac{1}{2}\right) \tag{1.17}$$

The steps in constructing a unit pulse signal from unit step functions are illustrated in Fig. 1.33.

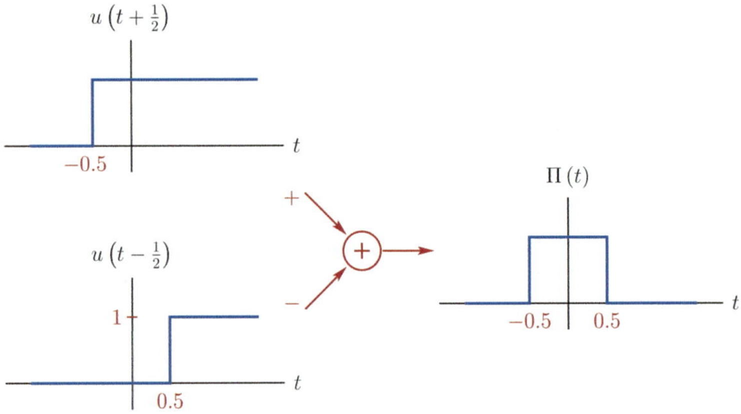

Figure 1.33 – Constructing a unit pulse signal using unit step functions.

It is also possible to construct a unit pulse signal using unit impulse functions. Let us substitute Eqn. (1.14) into Eqn. (1.17) to obtain

$$\Pi(t) = \int_{-\infty}^{t+1/2} \delta(\lambda)\, d\lambda - \int_{-\infty}^{t-1/2} \delta(\lambda)\, d\lambda = \int_{t-1/2}^{t+1/2} \delta(\lambda)\, d\lambda \tag{1.18}$$

The result in Eqn. (1.18) can be interpreted as follows: For the result of the integral to be equal to unity, the impulse needs to appear between the integration limits. This requires that the lower limit be negative and the upper limit be positive. Thus we have

$$\int_{t-1/2}^{t+1/2} \delta(\lambda)\, d\lambda = \begin{cases} 1, & \text{if } \; t-\frac{1}{2}<0 \text{ and } t+\frac{1}{2}>0 \\ 0, & \text{otherwise} \end{cases}$$

$$= \begin{cases} 1, & -\frac{1}{2}<t<\frac{1}{2} \\ 0, & \text{otherwise} \end{cases} \tag{1.19}$$

which matches the definition of the unit pulse signal given in Eqn. (1.16). A graphical illustration of Eqn. (1.18) is given in Fig. 1.34.

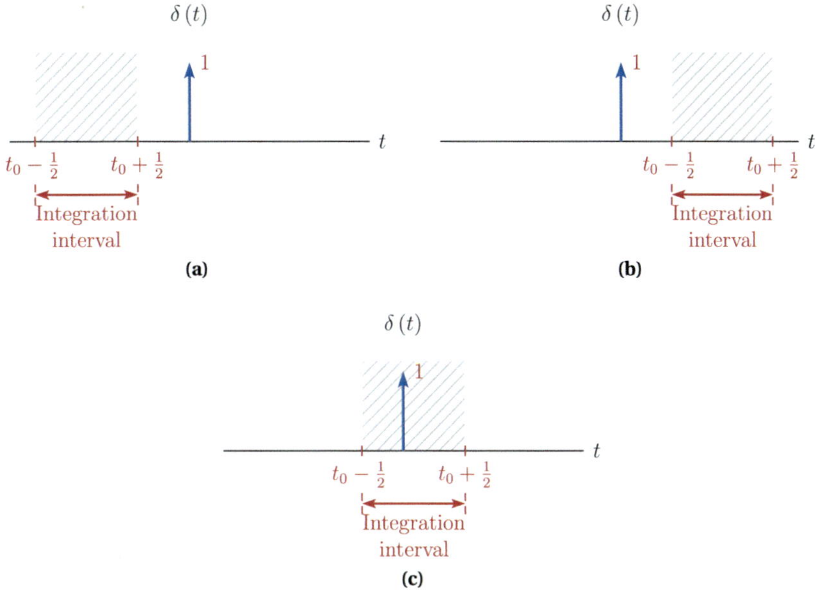

Figure 1.34 – Constructing a unit pulse by integrating a unit impulse: **(a)** $t = t_0 < -1/2$, **(b)** $t = t_0 > 1/2$, and **(c)** $-1/2 < t < 1/2$.

1.5.4 Unit ramp function

The unit ramp function has zero amplitude for $t < 0$, and unit slope for $t \geq 0$. Its mathematical definition is as follows:

Unit ramp function:

$$r(t) = \begin{cases} t, & t \geq 0 \\ 0, & t < 0 \end{cases} \tag{1.20}$$

This behavior is illustrated in Fig. 1.35.

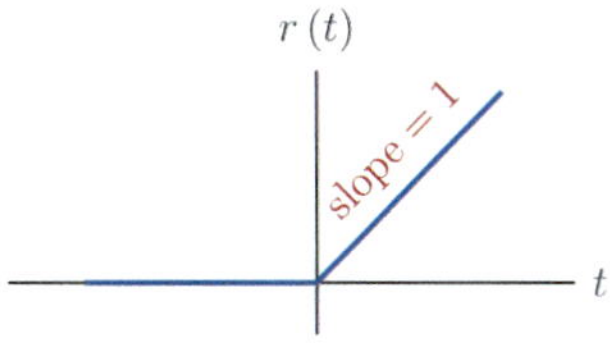

Figure 1.35 – Unit ramp function.

An equivalent definition of the unit ramp function can be written using the product of a linear signal $g(t) = t$ and the unit step function as

$$r(t) = t\, u(t) \tag{1.21}$$

as illustrated in Fig. 1.36.

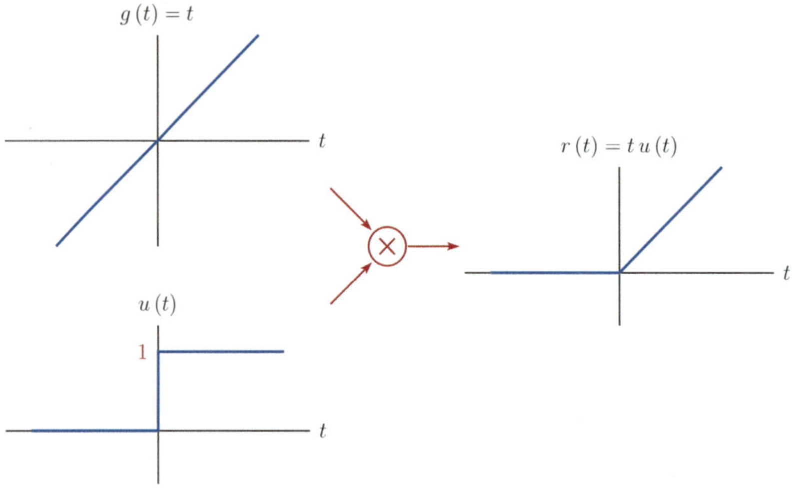

Figure 1.36 – Constructing a unit ramp function.

Alternatively, the unit ramp function may be recognized as the integral of the unit step function as

$$r(t) = \int_{-\infty}^{t} u(\lambda)\, d\lambda \tag{1.22}$$

The integral relationship between unit step and unit ramp functions is illustrated in Fig. 1.37.

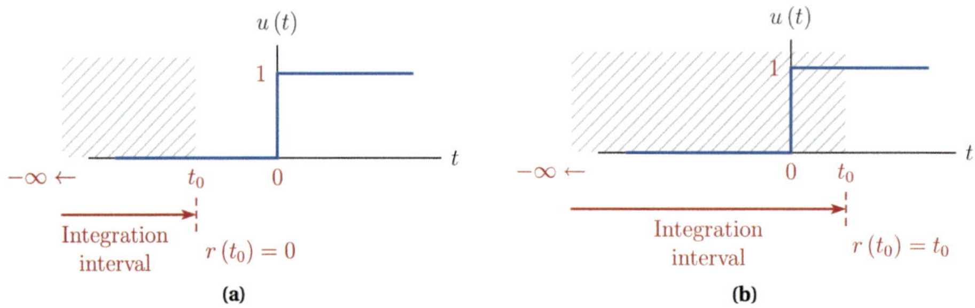

Figure 1.37 – Obtaining a unit ramp function as the running integral of a unit step.

We will find time-shifted and time-scaled versions of the unit ramp function useful in constructing signals that have piecewise linear segments.

1.5.5 Unit triangle function

The unit triangle function is defined as follows:

Unit triangle function:

$$\Lambda(t) = \begin{cases} t+1, & -1 \le t < 0 \\ -t+1, & 0 \le t < 1 \\ 0, & \text{otherwise} \end{cases} \tag{1.23}$$

The signal $\Lambda(t)$ is shown in Fig. 1.38. It can also be expressed in terms of shifted and scaled unit ramp functions as

$$\Lambda(t) = r(t+1) - 2\,r(t) + r(t-1) \tag{1.24}$$

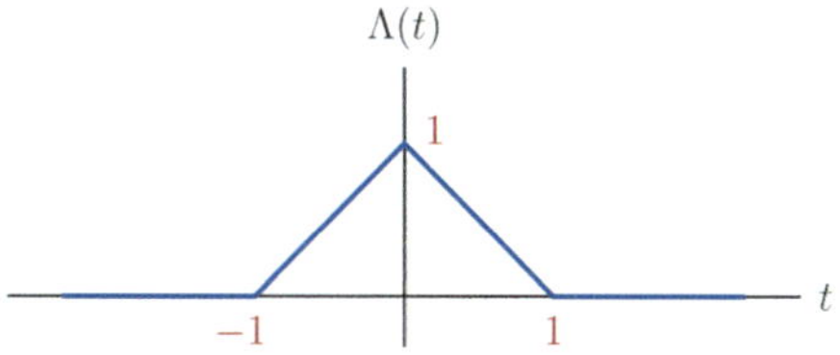

Figure 1.38 – Unit triangle function.

In order to understand Eqn. (1.24) we will observe the slope of the unit triangle waveform in each segment. The first term $r(t+1)$ has a slope of 0 for $t < -1$, and a slope of 1 for $t > -1$. The addition of the term $-2r(t)$ changes the slope to -1 for $t > 0$. The last term, $r(t-1)$, makes the slope 0 for $t > 1$. The process of constructing a unit triangle from shifted and scaled unit ramp functions is illustrated in Fig. 1.39.

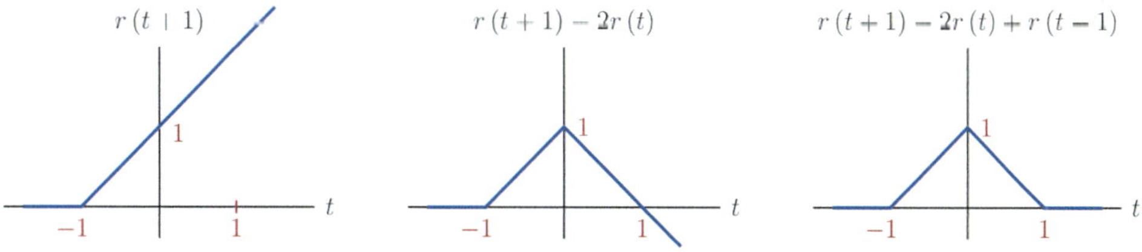

Figure 1.39 – Construction of a unit triangle using unit ramp functions.

Interactive App: Waveform explorer

The interactive app in `appWaveExplorerCT.m` allows experimentation with the basic signal building blocks discussed in Section 1.5. The signal $x(t)$ is constructed using up to 5 terms in the form

$$x(t) = x_1(t) + x_2(t) + x_3(t) + x_4(t) + x_5(t)$$

and is graphed in the lower part of the user interface. Each term $x_i(t)$, $i = 1, \ldots, 5$ in the expression for $x(t)$ can be a scaled and time-shifted step, ramp, pulse or triangle function. Optionally, step and ramp functions can be time reversed; pulse and triangle functions can be time scaled. If you do not need all five terms for constructing $x(t)$, unneeded terms can be kept off.

a. Construct the signal $x(t)$ in Fig. 1.8 using only ramp functions with appropriate scaling and time shifting. Ideally, work this out with pencil and paper first, and then use the app to verify your solution.

b. Repeat the same exercise, but using only unit triangle functions with appropriate scaling and time shifting.

c. Use the app to verify Eqn. 1.17 for constructing a unit pulse signal using time-shifted unit step signals.

d. Use the app to verify Eqn. 1.24 for constructing a unit triangle signal using scaled and time-shifted unit ramp signals.

Software resource: `appWaveExplorerCT.m`

1.5.6 Sinusoidal signals

Sinusoidal signals play a very important role in the study of signals and systems. The general form of a sinusoidal signal is given below.

Sinusoidal signal:

$$x(t) = A\cos(\omega_0 t + \theta) \tag{1.25}$$

There are three parameters of interest: A is the *amplitude* of the signal and ω_0 is the *radian frequency* which has the unit of *radians per second*, abbreviated as rad/s. The parameter θ is the *initial phase angle* in radians. The radian frequency ω_0 can be expressed as

$$\omega_0 = 2\pi f_0 \tag{1.26}$$

where f_0 is the *frequency* in Hz. The resulting signal is shown in Fig. 1.40.

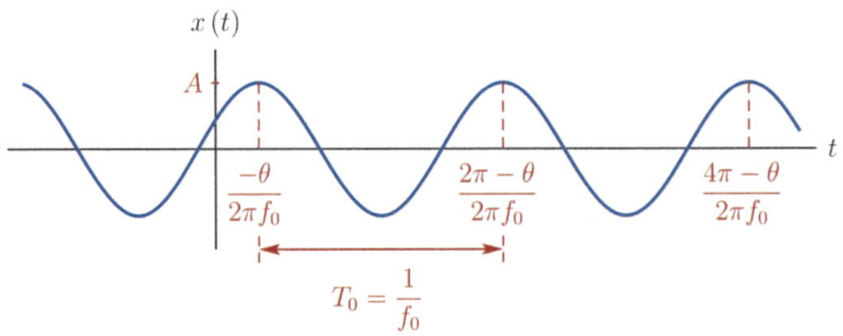

Figure 1.40 – The sinusoidal signal $x(t) = A\cos(2\pi f_0 t + \theta)$.

An examination of Fig. 1.40 reveals that the amplitude parameter A controls positive and negative peak values of the signal. The phase θ affects the locations of the peaks, and is related to time delay.

Timing of the peaks of $x(t) = \cos(2\pi f_0 t + \theta)$:

- If $\theta = 0$, the cosine waveform has a peak at the origin $t_0 = 0$. Other positive peaks of the waveform are at time instants that are solutions the equation

$$2\pi f_0 t_k = 2\pi k, \quad k\text{: integer} \tag{1.27}$$

which leads to $t_k = k/f_0$.

- With a different value of θ, the first peak is shifted to $t_0 = -\theta/(2\pi f_0)$, the time instant that satisfies the equation $2\pi f_0 t_0 + \theta = 0$. Other positive peaks of the waveform are shifted to time instants $t_k = (2\pi k - \theta)/(2\pi f_0)$ which are solutions of the equation

$$2\pi f_0 t_k + \theta = 2\pi k, \quad k\text{: integer} \tag{1.28}$$

Finally, the frequency f_0 controls the repetition of the waveform. Let's refer to the part of the signal from one positive peak to the next as one *cycle* as shown in Fig. 1.40. When expressed in Hz, f_0 is the number cycles (or periods) of the signal per second. Consequently, each cycle or period spans $T_0 = 1/f_0$ seconds.

Interactive App: Continuous-time sinusoidal signal parameters

The interactive app in `appSin1.m` illustrates the key properties of a sinusoidal signal based on the preceding discussion. Refer to Eqn. (1.25) and Fig. 1.40. Slider controls allow the amplitude A, the frequency f_0, and the phase θ to be specified. The resulting sinusoidal signal is shown in the graph area. Values of the period T_0 and the time delay t_d are also displayed on the graph.

a. Observe the value of the peak amplitude of the waveform changing in response to changes in parameter A.

b. Pay attention to the reciprocal relationship between the frequency f_0 and the period T_0.

c. Set $f_0 = 250$ Hz and $\theta = 0$ degrees. Count the number of complete cycles over 40 ms in the time interval -20 ms $\le t \le 20$ ms. How does the number of full cycles correspond to the value of f_0?

d. When $\theta = 0$, the cosine waveform has a peak at $t = 0$. Change the phase θ slightly in the negative direction by moving the slider control to the left, and observe the middle peak of the waveform move to the right. Observe the value of this *time delay* caused by the change in θ, and relate it to Eqn. (1.28).

Software resource: `appSin1.m`

1.6 Impulse Decomposition for Continuous-Time Signals

Sifting property of the unit impulse function is given by Eqn. (1.10). Using this property, a signal $x(t)$ can be expressed in the form

$$x(t) = \int_{-\infty}^{\infty} x(\lambda)\, \delta(t-\lambda)\, d\lambda \tag{1.29}$$

This is referred to as *impulse decomposition* of the signal $x(t)$.

Gaining insight into Eqn. (1.29):

We will explore this relationship in a bit more detail to gain further insight. Suppose we wanted to approximate the signal $x(t)$ using a series of rectangles that are each Δ wide. Let each rectangle be centered around the time instant $t_n = n\Delta$ where n is the integer index, and also let the height of each rectangle be adjusted to the amplitude of the signal $x(t)$ at the mid-point of the rectangle. Our approximation of $x(t)$ would be in the form

$$\hat{x}(t) = \sum_{n=-\infty}^{\infty} x(n\Delta)\, \Pi\left(\frac{t-n\Delta}{\Delta}\right) \tag{1.30}$$

which is graphically depicted in Fig. 1.41. Let's multiply and divide the term in the summation on the right side of Eqn. (1.30) by Δ to obtain

$$\hat{x}(t) = \sum_{n=-\infty}^{\infty} x(n\Delta)\, \frac{1}{\Delta}\, \Pi\left(\frac{t-n\Delta}{\Delta}\right) \Delta \tag{1.31}$$

Furthermore, we will define $p(t)$ as

$$p(t) = \frac{1}{\Delta}\, \Pi\left(\frac{t}{\Delta}\right) \tag{1.32}$$

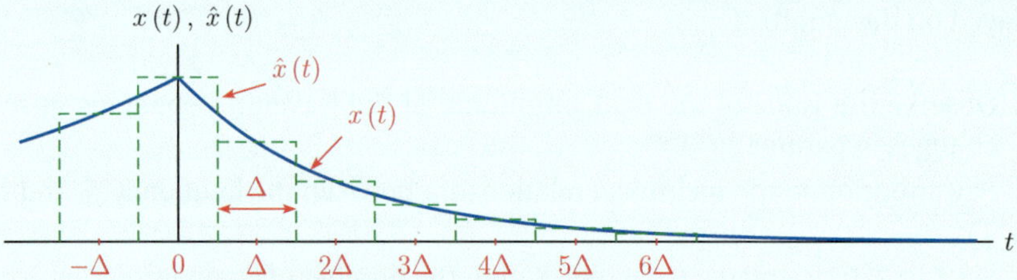

Figure 1.41 – Approximating a signal with rectangular pulses.

Clearly, $p(t)$ is a rectangular pulse of width Δ and height $(1/\Delta)$, and it is centered around the time instant $t=0$. It has unit area independent of the value of Δ. Substituting Eqn. (1.32) into Eqn. (1.31) leads to

$$\hat{x}(t) = \sum_{n=-\infty}^{\infty} x(n\Delta)\, p(t-n\Delta)\, \Delta \tag{1.33}$$

We know from Eqns. (1.5) and (1.6) and Fig. 1.23 that the unit impulse function $\delta(t)$ can be obtained as the limit of a rectangular pulse with unit area by making it narrower and taller at the same time. Therefore, the pulse $p(t)$ turns into an impulse if we force $\Delta \to 0$.

$$\lim_{\Delta\to 0} \left[p(t)\right] = \lim_{\Delta\to 0} \left[\frac{1}{\Delta}\, \Pi\left(\frac{t}{\Delta}\right)\right] = \delta(t) \tag{1.34}$$

Thus, if we take the limit of Eqn. (1.33) as $\Delta \to 0$, the approximation becomes an equality:

$$\begin{aligned} x(t) &= \lim_{\Delta \to 0} \left[\hat{x}(t) \right] \\ &= \lim_{\Delta \to 0} \left[\sum_{n=-\infty}^{\infty} x(n\Delta)\, p(t - n\Delta)\, \Delta \right] \end{aligned} \tag{1.35}$$

In the limit [2] we get $p(t) \to \delta(t)$, $n\Delta \to \lambda$, and $\Delta \to d\lambda$. Also, the summation in Eqn. (1.35) turns into an integral, and we obtain Eqn. (1.29). This result will be useful as we derive the convolution relationship in the next chapter.

Interactive App: Impulse decomposition

The interactive app in `appImpDecomp.m` illustrates the derivation of impulse decomposition as outlined in Eqns. (1.29) through (1.35). It shows the progression of $\hat{x}(t)$ as defined by Eqn. (1.30) into $x(t)$ as the value of Δ approaches zero.

a. Initially set the pulse width to $\Delta = 1$. Observe that each pulse is centered at $t = n\Delta$, and each pulse matches the height of the signal at its midpoint.

b. Using the slider control, gradually reduce the value of the pulse width Δ starting at $\Delta = 1$ and moving closer to 0. Keep paying attention to the width and the center point of each pulse.

Software resource: `appImpDecomp.m`

1.7 Signal Classifications

In this section we will summarize various methods and criteria for classifying the types of continuous-time signals that will be useful in our future discussions.

1.7.1 Real vs. complex signals

A real signal is one in which the amplitude is real-valued at all time instants. For example, if we model an electrical voltage or current with a mathematical function of time, the resulting signal $x(t)$ is obviously real-valued. In contrast, a complex signal is one in which the signal amplitude may also have an imaginary part. A complex signal may be written in *Cartesian form* using its real and imaginary parts as

$$x(t) = x_r(t) + j\, x_i(t) \tag{1.36}$$

or in *polar form* using its magnitude and phase as

$$x(t) = \left| x(t) \right| e^{j \measuredangle x(t)} \tag{1.37}$$

[2] It should be noted that we are not aiming for a mathematically rigorous proof here, but rather some insight. We are assuming that $x(t)$ is a well behaving function that allows the limit in Eqn. (1.35).

The two forms in Eqns. (1.36) and (1.37) can be related to each other through the following set of equations:

$$\left|x(t)\right| = \left[x_r^2(t) + x_i^2(t)\right]^{1/2} \tag{1.38}$$

$$\measuredangle x(t) = \tan^{-1}\left[\frac{x_i(t)}{x_r(t)}\right] \tag{1.39}$$

$$x_r(t) = \left|x(t)\right| \cos\left(\measuredangle x(t)\right) \tag{1.40}$$

$$x_i(t) = \left|x(t)\right| \sin\left(\measuredangle x(t)\right) \tag{1.41}$$

In deriving Eqns. (1.40) and (1.41) we have used Euler's formula.[3] Even though we will mostly focus on the use of real-valued signals in our discussion in the rest of this text, there will be times when the use of complex signal models will prove useful.

1.7.2 Periodic vs. non-periodic signals

A signal is said to be *periodic* if it satisfies

$$x(t+T_0) = x(t) \tag{1.42}$$

at all time instants t, and for a specific value of $T_0 \neq 0$. The value T_0 is referred to as the *period* of the signal. An example of a periodic signal is shown in Fig. 1.42.

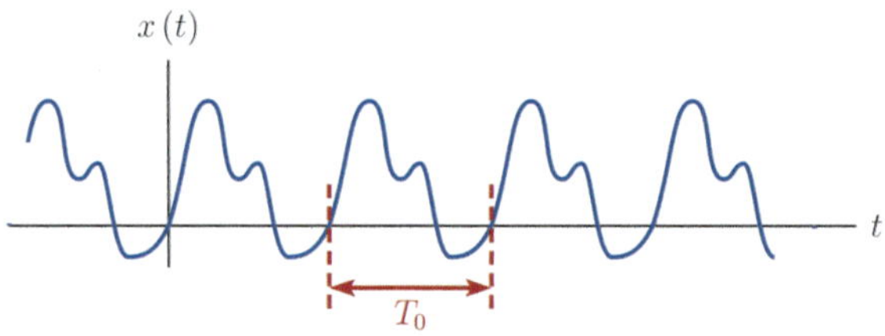

Figure 1.42 – Example of a signal that is periodic.

Applying the periodicity definition of Eqn. (1.42) to $x(t+T_0)$ instead of $x(t)$ we obtain

$$x(t+2T_0) = x(t+T_0)$$

which, when combined with Eqn. (1.42), implies that

$$x(t+2T_0) = x(t)$$

and, through repeated use of this process, we can show that

$$x(t+kT_0) = x(t)$$

for any integer value of k. Thus we conclude that, if a signal is periodic with period T_0, then it is also periodic with periods of $2T_0, 3T_0, \ldots, kT_0, \ldots$ where k is any integer. When we refer to the period of a periodic signal, we will imply the smallest positive value of T_0 that satisfies Eqn. (1.42). In cases when we discuss multiple periods for the same signal, we will refer to T_0 as the

[3] $e^{\pm ja} = \cos(a) \pm j\sin(a)$.

fundamental period in order to avoid ambiguity. The *fundamental frequency* of a periodic signal is defined as the reciprocal of its fundamental period:

$$f_0 = \frac{1}{T_0} \tag{1.43}$$

If the fundamental period T_0 is expressed in seconds, then the corresponding fundamental frequency f_0 is expressed in Hz.

Example 1.4: Working with a complex periodic signal

Consider a signal defined by

$$\begin{aligned} x(t) &= x_r(t) + j\,x_i(t) \\ &= A\cos\left(2\pi f_0 t + \theta\right) + j\,A\sin\left(2\pi f_0 t + \theta\right) \end{aligned}$$

Using Eqns. (1.38) and (1.39), polar complex form of this signal can be obtained as $x(t) = |x(t)|\, e^{j\measuredangle x(t)}$, with magnitude and phase given by

$$\left|x(t)\right| = \left[\left[A\cos\left(2\pi f_0 t + \theta\right)\right]^2 + \left[A\sin\left(2\pi f_0 t + \theta\right)\right]^2\right]^{1/2} = A$$

and

$$\measuredangle x(t) = \tan^{-1}\left[\frac{\sin\left(2\pi f_0 t + \theta\right)}{\cos\left(2\pi f_0 t + \theta\right)}\right] = \tan^{-1}\left[\tan\left(2\pi f_0 t + \theta\right)\right] = 2\pi f_0 t + \theta$$

respectively. In deriving these results we have relied on the appropriate trigonometric identities. Once the magnitude $|x(t)|$ and the phase $\measuredangle x(t)$ are obtained, we can express the signal $x(t)$ in polar complex form:

$$x(t) = \left|x(t)\right|\, e^{j\measuredangle x(t)} = A\, e^{j\,(2\pi f_0 t + \theta)}$$

The components of the Cartesian and polar complex forms of the signal are shown in Fig. 1.43. The real and imaginary parts of $x(t)$ have a 90 degree phase difference between them.

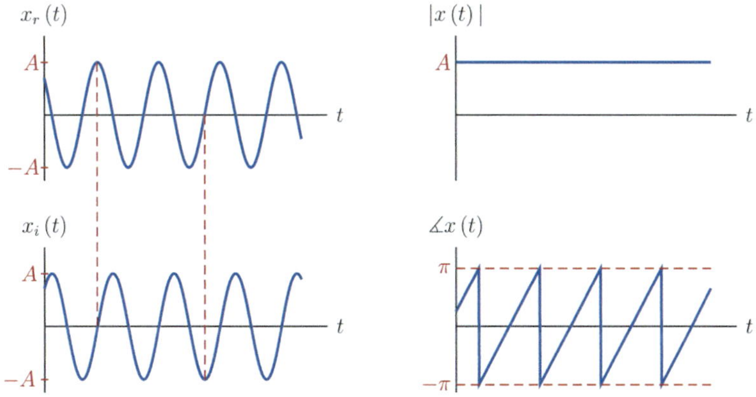

Figure 1.43 – Cartesian and polar representations of the signal $x(t)$ in Example 1.4.

When the real part of the signal goes through a peak, the imaginary part goes through zero and vice versa. In the above development, the phase of $x(t)$ was found to be a linear function of the time variable t. The broken-line appearance of the phase characteristic in Fig. 1.43 is due to normalization of the phase angles to keep them in the range $-\pi \le \measuredangle x(t) < \pi$.

Software resource: `exct_1_4.m`

Interactive App: Complex exponential signal in Cartesian and polar forms

The interactive app in `appCompExp.m` is based on Example 1.4 and Fig. 1.43. It allows experimentation with the four graphs in Fig. 1.43 while varying the parameters A, f_0, and θ. Slider controls can be used for varying the three parameters within their prescribed ranges $0.5 \le A \le 5$, $50 \text{ Hz} \le f_0 \le 500 \text{ Hz}$, and $-180° \le \theta \le 180°$ respectively. Vary the signal parameters and observe the following:

a. Observe that, as the amplitude A is increased, the graphs for $x_r(t)$, $x_i(t)$, and $|x(t)|$ scale accordingly, however, the phase $\measuredangle x(t)$ is not affected.

b. Changing the parameter θ causes a time delay in real- and imaginary-part signals $x_r(t)$ and $x_i(t)$. Compute this delay on paper for a few combinations of f_0 and θ, and compare it to the result obtained through the app.

Software resource: `appCompExp.m`

Software resource: See MATLAB Exercise 1.5.

Example 1.5: Periodicity of continuous-time sinusoidal signals

Consider two continuous-time sinusoidal signals

$$x_1(t) = A_1 \sin\left(2\pi f_1 t + \theta_1\right) \quad \text{and} \quad x_2(t) = A_2 \sin\left(2\pi f_2 t + \theta_2\right)$$

Determine the conditions under which their sum

$$x(t) = x_1(t) + x_2(t)$$

is also periodic. Also, determine the fundamental period of the signal $x(t)$ as a function of the relevant parameters of $x_1(t)$ and $x_2(t)$.

Solution: We know that the sinusoidal signals $x_1(t)$ and $x_2(t)$ are periodic with fundamental periods of T_1 and T_2 respectively, and we can write

$$x_1(t + m_1 T_1) = x_1(t)\,, \quad T_1 = 1/f_1 \tag{1.44}$$

$$x_2(t + m_2 T_2) = x_2(t)\,, \quad T_2 = 1/f_2 \tag{1.45}$$

where m_1 and m_2 are arbitrary integer values. In order for the sum $x(t)$ to be periodic with period T_0, we need the relationship $x(t + T_0) = x(t)$ to be valid at all time instants. In terms of the signals $x_1(t)$ and $x_2(t)$ this requires

$$x_1(t+T_0)+x_2(t+T_0)=x_1(t)+x_2(t) \tag{1.46}$$

Using Eqns. (1.44) and (1.45) with Eqn. (1.46) we reach the conclusion that Eqn. (1.46) would hold if two integers m_1 and m_2 can be found such that

$$T_0=m_1\,T_1=m_2\,T_2$$

or, in terms of the frequencies involved,

$$\frac{1}{f_0}=\frac{m_1}{f_1}=\frac{m_2}{f_2}$$

The sum of two sinusoidal signals $x_1(t)$ and $x_2(t)$ is periodic provided that their frequencies are integer multiples of a frequency f_0, i.e., $f_1=m_1 f_0$ and $f_2=m_2 f_0$. The period of the sum signal is $T_0=1/f_0$. The frequency f_0 is the fundamental frequency of $x(t)$. This reasoning can easily be extended to any number of sinusoidal signals added together. Consider a signal $x(t)$ defined as

$$x(t)=\sum_{n=1}^{N}A_n\,\sin\left(2\pi f_n t+\theta_n\right)$$

The signal $x(t)$ is periodic provided that the frequencies of all of its components are integer multiples of a fundamental frequency f_0, that is,

$$f_n=m_n f_0\,,\quad n=1,\ldots,N$$

This conclusion will be significant when we discuss the Fourier series representation of periodic signals in Chapter 3.

Example 1.6: More on the periodicity of sinusoidal signals

Discuss the periodicity of the signals

a. $x(t)=\sin(2\pi\,1.5\,t)+\sin(2\pi\,2.5\,t)$

b. $y(t)=\sin(2\pi\,1.5\,t)+\sin(2\pi\,2.75\,t)$

Solution:

a. For this signal, the fundamental frequency is $f_0=0.5$ Hz. The two signal frequencies can be expressed as

$$f_1=1.5\text{ Hz }=3f_0\ \text{ and }\ f_2=2.5\text{ Hz }=5f_0$$

The resulting fundamental period is $T_0=1/f_0=2$ seconds. Within one period of $x(t)$ there are $m_1=3$ full cycles of the first sinusoid and $m_2=5$ cycles of the second sinusoid. This is illustrated in Fig. 1.44.

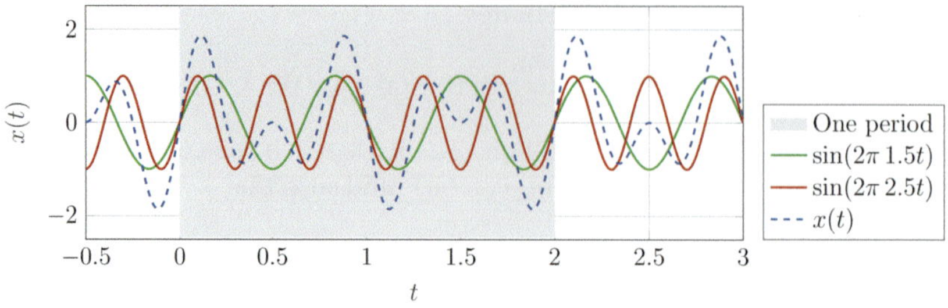

Figure 1.44 – Periodicity of $x(t)$ of Example 1.6.

b. In this case the fundamental frequency is $f_0 = 0.25$ Hz. The two signal frequencies can be expressed as

$$f_1 = 1.5\text{ Hz } = 6f_0 \quad \text{and} \quad f_2 = 2.75\text{ Hz } = 11f_0$$

The resulting fundamental period is $T_0 = 1/f_0 = 4$ seconds. Within one period of $x(t)$ there are $m_1 = 6$ full cycles of the first sinusoid and $m_2 = 11$ cycles of the second sinusoid. This is illustrated in Fig. 1.45.

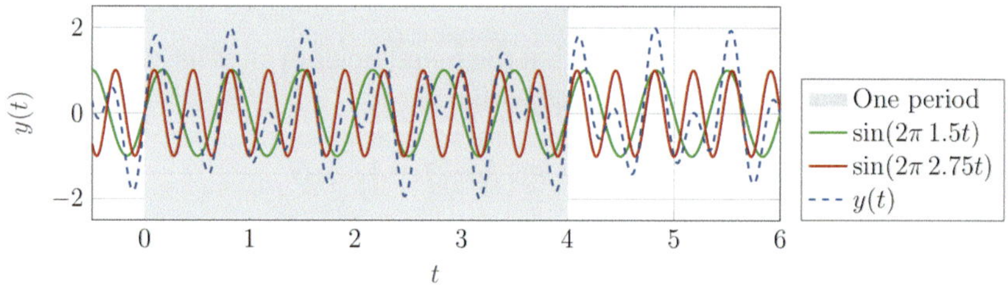

Figure 1.45 – Periodicity of $y(t)$ of Example 1.6.

Software resources: `exct_1_6a.m` , `exct_1_6b.m`

Interactive App: Periodicity of a signal with multiple sinusoids

The app in `appSin2.m` allows exploration of the periodicity of multi-tone signals. A signal in the form

$$x(t) = \sum_{n=1}^{N} A_n \sin\left(2\pi f_n t + \theta_n\right)$$

can be graphed using up to four frequencies specified by the user. The beginning and the end of one period of the signal is indicated on the graph with vertical dashed lines. Note that amplitudes of individual sinusoidal components have no effect on periodicity and are therefore set equal to unity in this app.

a. Create the conditions of part (a) of Example 1.6 by setting frequencies f_1 and f_2 equal to 1.5 Hz and 2.5 Hz, respectively. Check the box "Include in the sum" for the first two signal components, and uncheck it for the remaining two. Observe the resulting signal and its periodic nature. Does it agree with what was determined in Example 1.6?

b. Repeat the exercise with the conditions of part (b) of Example 1.6.

c. Using pencil and paper, determine if the signal

$$x(t) = \sin(3\pi t) + \sin(4.5\pi t) + \sin(6\pi t) + \sin(8.5\pi t)$$

is periodic. If it is, determine its fundamental frequency and period. Afterward use the app to check your result.

Software resource: `appSin2.m`

1.7.3 Deterministic vs. random signals

Deterministic signals are those that can be described completely in analytical form in the time domain. The signals that we worked with in this chapter have all been examples of deterministic signals. Random signals, on the other hand, are signals that occur due to random phenomena that cannot be modeled analytically. An example of a random signal is the speech signal converted to the form of an electrical voltage waveform by means of a microphone. A very short segment from a speech signal was shown in Fig. 1.2. Other examples of random signals are the vibration signal recorded during an earthquake by a seismograph, the noise signal generated by a resistor due to random movements and collisions of its electrons and the thermal energy those collisions produce. A common feature of these signals is the fact that the signal amplitude is not known at any given time instant, and cannot be expressed using a formula. Instead, we must try to model the statistical characteristics of the random signal rather than the signal itself. These statistical characteristics studied include average values of the signal amplitude or squared signal amplitude, distribution of various probabilities involving signal amplitude, and distribution of normalized signal energy or power among different frequencies. Study of random signals is beyond the scope of this text and the reader is referred to one of the many excellent texts available on the subject [9, 10, 14].

1.8 Energy and Power Definitions

1.8.1 Energy of a signal

With physical signals and systems, the concept of *energy* is associated with a signal applied to a load. The signal source delivers the energy which must be dissipated by the load. For example, consider a voltage source with voltage $v(t)$ connected to the terminals of a resistor with resistance R as shown in Fig. 1.46(a). Let $i(t)$ be the current that flows through the resistor. If we wanted to use the voltage $v(t)$ as our basis in energy calculations, the total energy dissipated in the resistor would be

$$E = \int_{-\infty}^{\infty} v(t)\, i(t)\, dt = \int_{-\infty}^{\infty} \frac{v^2(t)}{R}\, dt \tag{1.47}$$

Alternatively, consider the arrangement in Fig. 1.46(b) where a current source with a time varying current $i(t)$ is connected to the terminals of the same resistor. In this case, the energy dissipated in the resistor would be computed as

$$E = \int_{-\infty}^{\infty} v(t)\, i(t)\, dt = \int_{-\infty}^{\infty} R\, i^2(t)\, dt \tag{1.48}$$

In the context of this text we use mathematical signal models that are independent of the physical nature of signals that are being modeled. A signal such as $x(t)$ could represent a voltage, a current, or some other time-varying physical quantity. It would be desirable to have a definition of signal energy that is based on the mathematical model of the signal alone, without any reference to the load and to the physical quantity from which the mathematical model may have been derived.

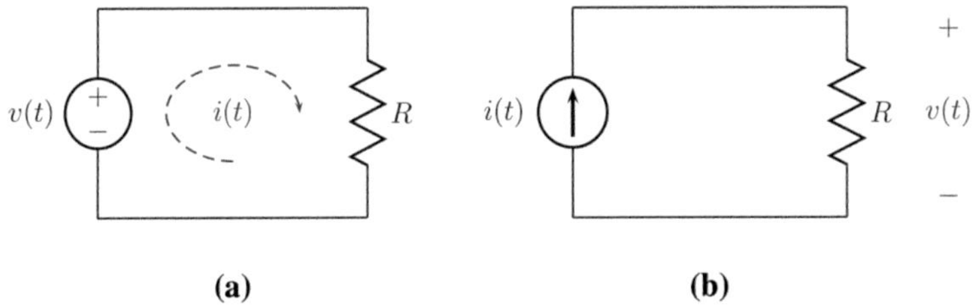

Figure 1.46 – Energy dissipation in a load when **(a)** a voltage source is used and **(b)** a current source is used.

If we don't know whether the function $x(t)$ represents a voltage or a current, which of the two equations discussed above should we use for computing the energy produced by the signal? Comparing Eqns. (1.47) and (1.48) it is obvious that, if the resistor value is chosen to be $R = 1$ Ω, then both equations would produce the same numerical value:

$$E = \int_{-\infty}^{\infty} \frac{v^2(t)}{(1)}\, dt = \int_{-\infty}^{\infty} (1)\, i^2(t)\, dt \tag{1.49}$$

This is precisely the approach we will take. In order to simplify our analysis and eliminate the need to always pay attention to the physical quantities that lead to our mathematical models, we will define the *normalized energy* of a real-valued signal $x(t)$ as

$$E_x = \int_{-\infty}^{\infty} x^2(t)\, dt \tag{1.50}$$

provided that the integral in Eqn. (1.50) can be computed. If the signal $x(t)$ is complex-valued, then its normalized energy is computed as

$$E_x = \int_{-\infty}^{\infty} |x(t)|^2\, dt \tag{1.51}$$

The term $|x(t)|$ is the norm of the complex signal. It is computed through

$$\left|x(t)\right|^2 = x(t)\, x^*(t)$$

where $x^*(t)$ is the complex conjugate of $x(t)$.

Note that, for a real-valued signal, the term $x^2(t)$ in Eqn. 1.50 is non-negative. Therefore, we could simply use Eqn. 1.51 for both real- and complex-valued signals.

Normalized energy of a signal:

$$E_x = \int_{-\infty}^{\infty} |x(t)|^2\, dt\,, \quad \text{if it can be computed.}$$

In situations where normalized signal energy is not computable, we may be able to compute the *normalized power* of the signal instead. This will be discussed in Section 1.8.3.

Example 1.7: Energy of a right-sided exponential signal

Compute the normalized energy of the right-sided exponential signal

$$x(t) = A\,e^{-\alpha t}\,u(t)$$

where $\alpha > 0$.

Solution: The signal $x(t)$ is shown in Fig. 1.47.

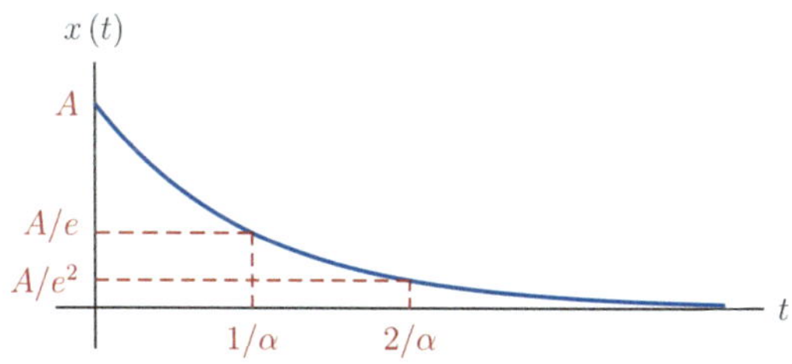

Figure 1.47 – The signal $x(t)$ for Example 1.7.

Notice how the signal amplitude starts with a value of $x(0) = A$ and declines to $x(1/\alpha) = A/e$ after $1/\alpha$ seconds elapse. At the time instant $t = 2/\alpha$ seconds, the amplitude is further down to $x(2/\alpha) = A/e^2$. Thus, the parameter α controls the rate at which the signal amplitude decays over time. The normalized energy of this signal can be computed by application of Eqn. (1.50) as

$$E_x = \int_{0}^{\infty} A^2\,e^{-2\alpha t}\,dt = A^2\,\frac{e^{-2\alpha t}}{-2\alpha}\bigg|_0^{\infty} = \frac{A^2}{2\alpha} \tag{1.52}$$

The restriction $\alpha > 0$ was necessary since, without it, we could not have evaluated the integral above.

Interactive App: Right-sided exponential signal

The interactive app in `appExp.m` is based on Example 1.7, Fig. 1.47, and Eqn. (1.52). Parameters A and α of the right-sided exponential signal $x(t)$ can be controlled with the slider controls available.

1. The shape of the right-sided exponential depends on the parameter α. Observe the key points with amplitudes A/e and A/e^2 as the parameter α is varied.

2. Observe how the displayed signal energy E_x relates to the shape of the exponential signal.

Software resource: `appExp.m`

1.8.2 Time averaging operator

In preparation for defining the power in a signal, we need to first define the time average of a signal. We will use the operator $\langle \ldots \rangle$ to indicate time average. If the signal $x(t)$ is periodic with period T_0, its time average can be computed as

$$\langle x(t) \rangle = \frac{1}{T_0} \int_{-T_0/2}^{T_0/2} x(t)\, dt \tag{1.53}$$

If the signal $x(t)$ is non-periodic, the definition of time average in Eqn. (1.53) can be generalized with the use of the limit operator as

$$\langle x(t) \rangle = \lim_{T \to \infty} \left[\frac{1}{T} \int_{-T/2}^{T/2} x(t)\, dt \right] \tag{1.54}$$

One way to make sense out of Eqn. (1.54) is to view the non-periodic signal $x(t)$ as though it is periodic with an infinitely large period so that we never get to see the signal pattern repeat itself.

Time average of a signal:

For a periodic signal: $$\langle x(t) \rangle = \frac{1}{T_0} \int_{-T_0/2}^{T_0/2} x(t)\, dt$$

For a non-periodic signal: $$\langle x(t) \rangle = \lim_{T \to \infty} \left[\frac{1}{T} \int_{-T/2}^{T/2} x(t)\, dt \right]$$

Example 1.8: Time average of a pulse train

Determine the time average of a periodic pulse train with amplitude A and period T_0, defined by the equations

$$x(t) = \begin{cases} A, & 0 < t < t_p \\ 0, & t_p < t < T_0 \end{cases}, \quad \text{and} \quad x(t + k\,T_0) = x(t)$$

for all t and all integer values of k. The signal $x(t)$ is shown in Fig. 1.48.

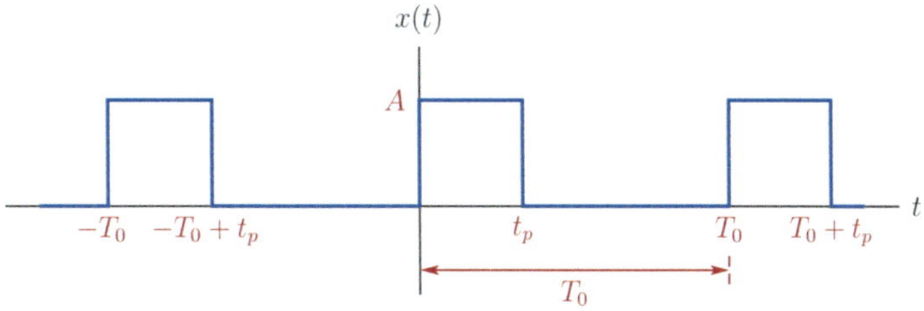

Figure 1.48 – Periodic pulse train signal $x(t)$ for Example 1.8.

Solution: Applying the definition for the time average of a periodic signal given by Eqn. (1.53) we obtain

$$\begin{aligned}\langle x(t)\rangle &= \frac{1}{T_0}\int_0^{T_0} x(t)\,dt \\ &= \frac{1}{T_0}\int_0^{t_p} (A)\,dt + \frac{1}{T_0}\int_{t_p}^{T_0} (0)\,dt \\ &= \frac{A\,t_p}{T_0}\end{aligned}$$

Recall that the parameter t_p is the width of the pulse within each period. Since the length of one period is T_0, the ratio t_p/T_0 in the final result represents the fraction of the period in which the pulse signal exists. In this context, we will refer to the ratio

$$d = \frac{t_p}{T_0}$$

as the *duty cycle* of the pulse train. Thus, the time average of the pulse train with amplitude A and duty cycle d is

$$\langle x(t)\rangle = A\,d$$

Interactive App: Time average of a pulse train

The MATLAB app in `appTimeAvg.m` illustrates the time averaging of pulse-train $x(t)$ as shown in Fig. 1.48 of Example 1.8 with amplitude A and duty cycle d. Values of both parameters can be adjusted using the slider controls provided. A dashed red line that shows the value of the time average $\langle x(t)\rangle$.

a. Start with the duty cycle equal to $d = 0.3$. Gradually increase the duty cycle toward $d = 1$ and observe the change in the time average.

b. With the duty cycle set at $d = 0.5$, vary the pulse amplitude A and observe the effect on the time average.

Software resource: `appTimeAvg.m`

1.8.3 Power of a signal

Consider again the simple circuits in Fig. 1.46 that were used in defining normalized signal energy. Using $v(t)$ as the voltage across the terminals of the load and $i(t)$ as the current flowing through the load, the *instantaneous power* dissipated in the load resistor would be

$$p_{\text{inst}}(t) = v(t)\, i(t) = \frac{v^2(t)}{R} = R\, i^2(t)$$

If the load is chosen to have a value of $R = 1\ \Omega$, the *normalized instantaneous power* can be defined as

$$p_{\text{norm}}(t) = x^2(t) \tag{1.55}$$

where $x(t)$ could represent either the voltage or the current. Often, a more useful concept is the *normalized average power* defined as the time average of $p_{\text{norm}}(t)$, that is

$$P_x = \langle x^2(t) \rangle \tag{1.56}$$

Note that this definition will work for both periodic and non-periodic signals provided that we use the appropriate form of the time averaging operator. For a periodic signal, Eqn. (1.56) can be used with Eqn. (1.53) to yield

$$P_x = \frac{1}{T_0} \int_{-T_0/2}^{T_0/2} x^2(t)\, dt \tag{1.57}$$

For a non-periodic signal, Eqn. (1.54) can be substituted into Eqn. (1.56) and we have

$$P_x = \lim_{T\to\infty} \left[\frac{1}{T} \int_{-T/2}^{T/2} x^2(t)\, dt \right] \tag{1.58}$$

In our discussion, we are assuming that the signal $x(t)$ is real-valued. If we need to consider complex-valued signals, then the definition of signal power can be generalized by using the squared norm of the signal in the time averaging operation:

$$P_x = \left\langle |x(t)|^2 \right\rangle \tag{1.59}$$

Thus, the power of a periodic complex signal is

$$P_x = \frac{1}{T_0} \int_{-T_0/2}^{T_0/2} |x(t)|^2\, dt \tag{1.60}$$

and the power of a non-periodic complex signal is

$$P_x = \lim_{T\to\infty} \left[\frac{1}{T} \int_{-T/2}^{T/2} |x(t)|^2\, dt \right] \tag{1.61}$$

Combining Eqns. (1.56) and (1.59), we can express the normalized average power of any signal, periodic or non-periodic, real- or complex-valued, as follows:

***Normalized average power* of a signal:**
$$P_x = \left\langle |x(t)|^2 \right\rangle$$

Example 1.9: Normalized average power of a pulse train

Determine the normalized average power of the pulse train of Example 1.8.

Solution: Recall that the period of the pulse train in Example 1.8 is T_0 and the width of each pulse is t_p. Applying the definition for the normalized average power of a periodic signal given by Eqn. (1.57) we obtain

$$\begin{aligned} P_x = \langle x^2(t) \rangle &= \frac{1}{T_0} \int_0^{T_0} x^2(t)\, dt \\ &= \frac{1}{T_0} \int_0^{t_p} (A^2)\, dt + \frac{1}{T_0} \int_{t_p}^{T_0} (0)\, dt \\ &= \frac{A^2\, t_p}{T_0} = A^2\, d \end{aligned}$$

Example 1.10: Power of a sinusoidal signal

Consider the signal

$$x(t) = A \sin(2\pi f_0 t + \theta)$$

which is not limited in time. It is obvious that the energy of this signal cannot be computed since the integral in Eqn. (1.50) would not yield a finite value. On the other hand, signal power can be computed through the use of Eqn. (1.57). The period of the signal is $T_0 = 1/f_0$, therefore

$$P_x = f_0 \int_{-1/2f_0}^{1/2f_0} A^2 \sin^2\left(2\pi f_0 t + \theta\right) dt$$

Using the appropriate trigonometric identities, we can write

$$P_x = f_0 \int_{-1/2f_0}^{1/2f_0} \frac{A^2}{2}\, dt - f_0 \int_{-1/2f_0}^{1/2f_0} \frac{A^2}{2} \cos\left(4\pi f_0 t + 2\theta\right) dt$$

The second integral in the above expression evaluates to zero, since we are integrating a cosine function with the frequency $2f_0$ over an interval of $T_0 = 1/f_0$. The cosine function has two full cycles within this interval. Thus, the normalized average power in a sinusoidal signal with amplitude A is

$$P_x = f_0 \int_{-1/2f_0}^{1/2f_0} \frac{A^2}{2}\, dt = \frac{A^2}{2} \tag{1.62}$$

This is a result that will be worth remembering.

Example 1.11: Right-sided exponential signal revisited

Let us consider the right-sided exponential signal of Example 1.7 again. Recall that the analytical expression for the signal was

$$x(t) = A e^{-\alpha t}\, u(t)$$

Using Eqn. (1.58) it can easily be shown that, for $\alpha > 0$, the power of the signal is $P_x = 0$. On the other hand, it is interesting to see what happens if we allow $\alpha = 0$. In this case, the signal $x(t)$ becomes a step function:

$$x(t) = A\, u(t)$$

The signal $x(t)$ non-periodic and real-valued. We will use Eqn. (1.58) to find the normalized average signal power as

$$\begin{aligned} P_x &= \lim_{T\to\infty} \left[\frac{1}{T} \int_{-T/2}^{T/2} x(t)\, dt \right] \\ &= \lim_{T\to\infty} \left[\frac{1}{T} \int_{-T/2}^{0} (0)\, dt + \frac{1}{T} \int_{0}^{T/2} A^2\, dt \right] \\ &= \lim_{T\to\infty} \left[\frac{A^2}{2} \right] = \frac{A^2}{2} \end{aligned}$$

and the normalized signal energy becomes infinitely large, i.e., $E_x \to \infty$.

1.8.4 Energy signals vs. power signals

In Examples 1.7 through 1.11 we have observed that the concept of signal energy is useful for some signals and not for others. Same can be said for signal power. Based on our observations, we can classify signals encountered in practice into two categories as *energy signals* and *power signals.* Defining characteristics of the two types of signals are as follows:

Energy signals vs. power signals:

- Energy signals are those that have finite energy and zero power, i.e., $E_x < \infty$ and $P_x = 0$.
- Power signals are those that have finite power and infinite energy, i.e., $E_x \to \infty$ and $P_x < \infty$.

Just as a side note, all voltage and current signals that can be generated in the laboratory or that occur in the electronic devices that we use in our daily lives are energy signals. A power signal is impossible to produce in any practical setting since doing so would require an infinite amount of energy. The concept of a power signal exists as a mathematical idealization only.

1.8.5 RMS value of a signal

The *root-mean-square (RMS)* value of a signal $x(t)$ is defined as

RMS value:

$$X_{RMS} = \left[\langle x^2(t)\rangle\right]^{1/2} \tag{1.63}$$

The time average in Eqn. (1.63) is computed through the use of either Eqn. (1.53) or Eqn. (1.54), depending on whether $x(t)$ is periodic or not. The RMS value concept is useful in many fields electrical engineering such as electronics, electromagnetics, power systems, and communications.

Example 1.12: RMS value of a sinusoidal signal

In Example 1.10, we have found the normalized average power of the sinusoidal signal $x(t) = A\sin(2\pi f_0 t + \theta)$ to be $P_x = \langle x^2(t)\rangle = A^2/2$. Based on Eqn. (1.63), the RMS value of this signal is

$$X_{RMS} = \frac{A}{\sqrt{2}} \tag{1.64}$$

as illustrated in Fig. 1.49.

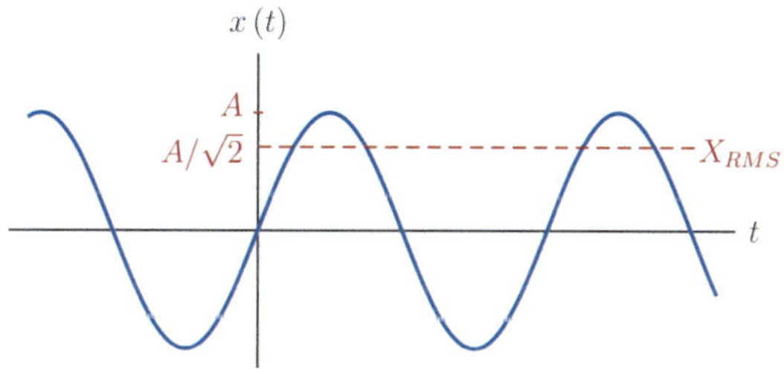

Figure 1.49 – RMS value of a sinusoidal signal.

Example 1.13: RMS value of a multi-tone signal

Consider the two-tone signal given by

$$x(t) = a_1\cos(2\pi f_1 t + \theta_1) + a_2\cos(2\pi f_2 t + \theta_2)$$

where the two frequencies are distinct, i.e., $f_1 \neq f_2$. Determine the RMS value and the normalized average power of this signal.

Solution: Computation of both the RMS value and the normalized average power for the signal $x(t)$ requires the computation of the time average of $x^2(t)$ given by

$$\begin{aligned} x^2(t) =& a_1^2\cos^2(2\pi f_1 t + \theta_1) + a_2^2\cos^2(2\pi f_2 t + \theta_2) \\ &+ 2\,a_1\,a_2\cos(2\pi f_1 t + \theta_1)\cos(2\pi f_2 t + \theta_2) \end{aligned}$$

Using the appropriate trigonometric identities, $x^2(t)$ can be written as

$$x^2(t) = \frac{a_1^2}{2} + \frac{a_2^2}{2} + \frac{a_1^2}{2}\cos(2\pi[2f_1]t + 2\theta_1) + \frac{a_2^2}{2}\cos(2\pi[2f_2]t + 2\theta_2)$$
$$+ a_1 a_2 \cos(2\pi[f_1 + f_2]t + \theta_1 + \theta_2) + a_1 a_2 \cos(2\pi[f_1 - f_2]t + \theta_1 - \theta_2)$$

We can now apply the time averaging operator to $x^2(t)$ to yield

$$\langle x^2(t)\rangle = \frac{a_1^2}{2} + \frac{a_2^2}{2} + \frac{a_1^2}{2}\langle\cos(2\pi[2f_1]t + 2\theta_1)\rangle + \frac{a_2^2}{2}\langle\cos(2\pi[2f_2]t + 2\theta_2)\rangle$$
$$+ a_1 a_2 \langle\cos(2\pi[f_1 + f_2]t + \theta_1 + \theta_2)\rangle + a_1 a_2 \langle\cos(2\pi[f_1 - f_2]t + \theta_1 - \theta_2)\rangle$$

The time average of each of the cosine terms is equal to zero, and we obtain

$$\langle x^2(t)\rangle = \frac{a_1^2}{2} + \frac{a_2^2}{2}$$

Thus, the RMS value of the signal $x(t)$ is

$$X_{RMS} = \sqrt{\frac{a_1^2}{2} + \frac{a_2^2}{2}}$$

and its normalized average power is

$$P_x = \frac{a_1^2}{2} + \frac{a_2^2}{2}$$

The generalization of these results to signals containing more than two sinusoidal components is straightforward. If $x(t)$ contains M sinusoidal components with amplitudes $a_1, a_2, \ldots, a_M$, and distinct frequencies $f_1, f_2, \ldots, f_M$, its RMS value and normalized average power are

$$X_{RMS} = \sqrt{\frac{a_1^2}{2} + \frac{a_2^2}{2} + \ldots + \frac{a_M^2}{2}} \tag{1.65}$$

and

$$P_x = \frac{a_1^2}{2} + \frac{a_2^2}{2} + \ldots + \frac{a_M^2}{2}, \tag{1.66}$$

respectively.

1.9 Symmetry properties

Some signals have certain symmetry properties that could be utilized in a variety of ways in the analysis. For example, we will see in the study of Fourier analysis that, signals with certain symmetry properties have simpler transforms. More importantly, a signal that may not have any symmetry properties can still be written as a linear combination of signals with certain symmetry properties, creating opportunities for simplified analysis and better understanding of signal and transform relationships.

1.9.1 Even and odd symmetry

A real-valued signal is said to have *even symmetry* if it has the property

$$x(-t) = x(t) \tag{1.67}$$

for all values of t. A signal with even symmetry remains unchanged when it is time reversed. Fig. 1.50(a) shows an example of a signal with even symmetry property. Similarly, a real-valued signal is said to have *odd symmetry* if it has the property

$$x(-t) = -x(t) \tag{1.68}$$

for all t. Time reversal has the same effect as negation on a signal with odd symmetry. This is illustrated in Fig. 1.50(b).

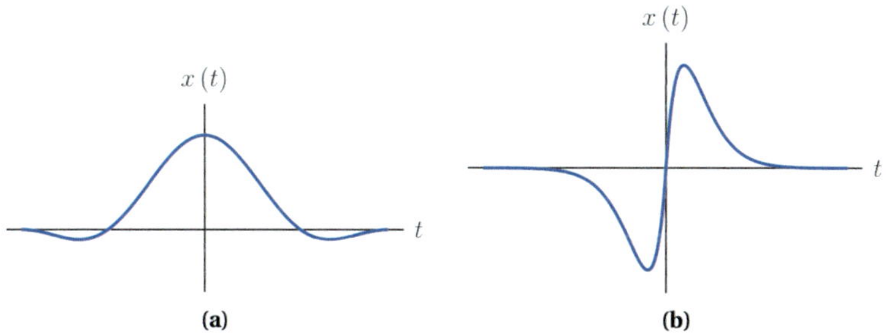

Figure 1.50 – **(a)** Even signal and **(b)** odd signal.

As a specific example the signal $x(t) = \cos(\omega t)$ has even symmetry since $\cos(-\omega t) = \cos(\omega t)$. Similarly, the signal $x(t) = \sin(\omega t)$ has odd symmetry since $\sin(-\omega t) = -\sin(\omega t)$.

1.9.2 Decomposition into even and odd components

It is always possible to split a real-valued signal into two components in such a way that one of the components is an even function of time, and the other is an odd function of time. Consider the following representation of the signal $x(t)$:

$$x(t) = x_e(t) + x_o(t) \tag{1.69}$$

where the two components $x_e(t)$ and $x_o(t)$ are defined as

$$x_e(t) = \frac{x(t) + x(-t)}{2} \tag{1.70}$$

and

$$x_o(t) = \frac{x(t) - x(-t)}{2} \tag{1.71}$$

The definitions of the two signals $x_e(t)$ and $x_o(t)$ guarantee that they always add up to $x(t)$. Furthermore, symmetry properties $x_e(-t) = x_e(t)$ and $x_o(-t) = -x_o(t)$ are guaranteed by the definitions of the signals $x_e(t)$ and $x_o(t)$ regardless of any symmetry properties the signal $x(t)$ may or may not possess.

Proof of Eqns. (1.70) and (1.71)

Symmetry properties of $x_e(t)$ and $x_o(t)$ can be easily verified by forming the expressions for $x_e(-t)$ and $x_o(-t)$ and comparing them to Eqns. (1.70) and (1.71). Let us use Eqn. (1.70) to write $x_e(-t)$:

$$x_e(-t) = \frac{x(-t) + x(t)}{2} = x_e(t) \tag{1.72}$$

Similarly, $x_o(-t)$ can be found from Eqn. (1.71) as

$$x_o(-t) = \frac{x(-t) - x(t)}{2} = -x_o(t) \tag{1.73}$$

Thus we have shown that $x_e(t)$ is always even, and $x_o(t)$ is always odd.

We will refer to $x_e(t)$ and $x_o(t)$ as the *even component* and the *odd component* of the signal $x(t)$ respectively.

Example 1.14: Even and odd components of a rectangular pulse

Determine even and odd components of the rectangular pulse signal

$$x(t) = \Pi\left(t - \tfrac{1}{2}\right) = \begin{cases} 1, & 0 < t < 1 \\ 0, & \text{elsewhere} \end{cases}$$

which is shown in Fig. 1.51.

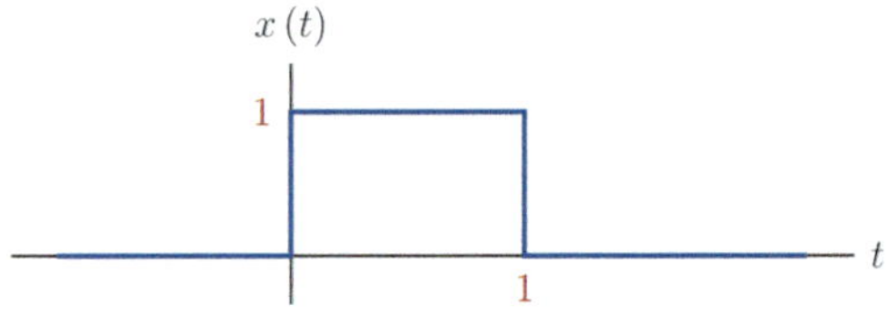

Figure 1.51 – The rectangular pulse signal of Example 1.14.

Solution: The even component of the signal $x(t)$ as computed from Eqn. (1.70) is

$$x_e(t) = \frac{\Pi\left(t - \frac{1}{2}\right) + \Pi\left(-t - \frac{1}{2}\right)}{2} = \frac{1}{2}\,\Pi(t/2)$$

Its odd component is computed using Eqn. (1.71) as

$$x_o(t) = \frac{1}{2}\left[\Pi\left(t - \tfrac{1}{2}\right) - \Pi\left(-t - \tfrac{1}{2}\right)\right]$$

The two components of $x(t)$ are graphed in Fig. 1.52.

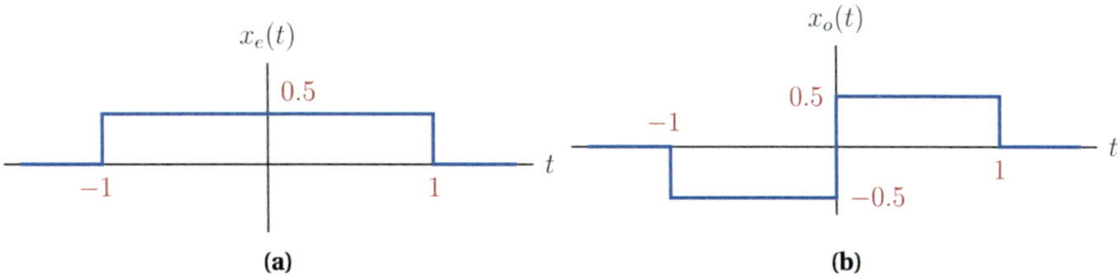

Figure 1.52 – **(a)** The even component and **(b)** the odd component of the rectangular pulse signal of Example 1.14.

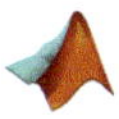

Software resource: exct_1_14.m

Example 1.15: Even and odd components of a sinusoidal signal

Consider the sinusoidal signal given by

$$x(t) = 5\cos(10t + \pi/3)$$

What symmetry properties does this signal have, if any? Determine its even and odd components.

Solution: The signal $x(t)$ is shown in Fig. 1.53. It can be observed that it has neither even symmetry nor odd symmetry.

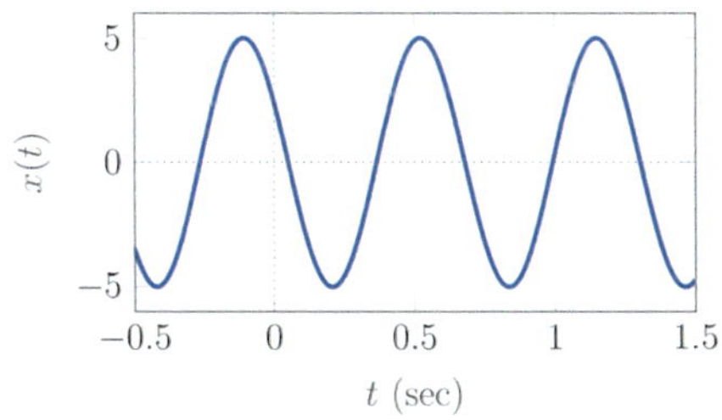

Figure 1.53 – The signal of Example 1.15.

We will use Eqns. (1.70) and (1.71) in decomposing this signal into its even and odd components. The even component is

$$x_e(t) = \frac{5}{2}\cos(10t + \pi/3) + \frac{5}{2}\cos(-10t + \pi/3)$$

Using the trigonometric identity

$$\cos(a + b) = \cos(a)\cos(b) - \sin(a)\sin(b)$$

$x_e(t)$ can be written as

$$x_e(t) = 5\cos(10t)\cos(\pi/3) = 2.5\cos(10t)$$

Similarly, the odd component of $x(t)$ is

$$x_o(t) = \frac{5}{2}\cos(10t + \pi/3) - \frac{5}{2}\cos(-10t + \pi/3)$$

which, through the use of the same trigonometric identity, can be written as

$$x_o(t) = -5\sin(10t)\sin(\pi/3) = -4.3301\sin(10t)$$

Even and odd components of $x(t)$ are shown in Fig. 1.54.

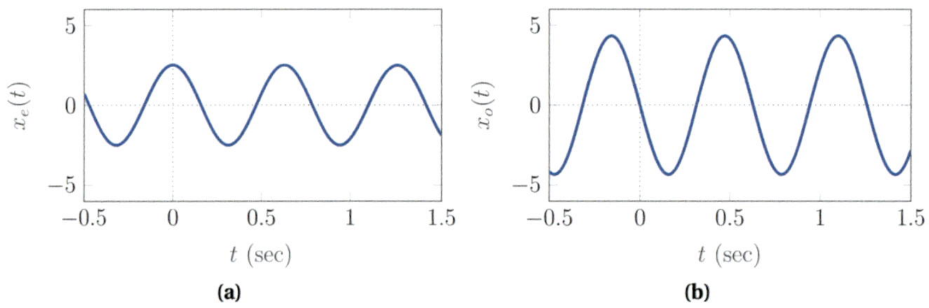

Figure 1.54 – **(a)** Even component and **(b)** odd component of the signal of Example 1.15.

Software resource: `exct_1_15.m`

Software resource: See MATLAB Exercise 1.6.

1.9.3 Symmetry properties for complex signals

Even and odd symmetry definitions given by Eqns. (1.67) and (1.68) for real-valued signals can be extended to work with complex-valued signals as well. A complex-valued signal is said to be *conjugate symmetric* if it satisfies

$$x(-t) = x^*(t) \tag{1.74}$$

for all t. For a signal with conjugate symmetry, time reversal has the same effect as complex conjugation. Similarly, a complex-valued signal is said to be *conjugate antisymmetric* if it satisfies

$$x(-t) = -x^*(t) \tag{1.75}$$

for all t. If a signal is conjugate antisymmetric, time reversing it causes the signal to be conjugated and negated simultaneously.

It is easy to see from Eqns. (1.67) and (1.74) that, if the signal is real-valued, that is, if $x(t) = x^*(t)$, the definition of conjugate symmetry reduces to that of even symmetry. Similarly, conjugate antisymmetry property reduces to odd symmetry for a real valued $x(t)$ as revealed by a comparison of Eqns. (1.68) and (1.75).

Example 1.16: Symmetry of a complex exponential signal

Consider the complex exponential signal

$$x(t) = A\,e^{j\omega t}, \quad A\text{: real}$$

What symmetry property does this signal have, if any?

Solution: Time reversal of $x(t)$ results in

$$x(-t) = A\,e^{-j\omega t}$$

Complex conjugate of the signal $x(t)$ is

$$x^*(t) = \left(A\,e^{j\omega t}\right)^* = A\,e^{-j\omega t}$$

Since $x(-t) = x^*(t)$, we conclude that the signal $x(t)$ is conjugate symmetric.

1.9.4 Decomposition of complex signals

It is possible to express any complex signal $x(t)$ as the sum of two signals

$$x(t) = x_E(t) + x_O(t) \tag{1.76}$$

such that $x_E(t)$ is conjugate symmetric and $x_O(t)$ is conjugate antisymmetric. The two components of the decomposition in Eqn. (1.76) are computed as

$$x_E(t) = \frac{x(t) + x^*(-t)}{2} \tag{1.77}$$

and

$$x_O(t) = \frac{x(t) - x^*(-t)}{2} \tag{1.78}$$

Definitions of $x_E(t)$ and $x_O(t)$ ensure that they always add up to $x(t)$. Furthermore, the two components are guaranteed to have the conjugate symmetry and conjugate antisymmetry properties, respectively.

Proof of Eqns. (1.77) and (1.78)

By conjugating the first component $x_E(t)$ we get

$$x_E^*(t) = \left[\frac{x(t) + x^*(-t)}{2}\right]^* = \frac{x^*(t) + x(-t)}{2} = x_E(-t)$$

proving that $x_E(t)$ is conjugate symmetric. Conjugation of the other component $x_O(t)$ yields

$$x_O^*(t) = \left[\frac{x(t) - x^*(-t)}{2}\right]^* = \frac{x^*(t) - x(-t)}{2} = -x_O(-t)$$

proving that $x_O(t)$ is conjugate antisymmetric.

1.10 Phasor Representation of Sinusoidal Signals

In certain situations we find it convenient to represent a sinusoidal signal with a complex quantity referred to as a *phasor.* Use of the phasor concept is beneficial in analyzing linear systems in which multiple sinusoidal signals may exist with the same frequency but with differing amplitude and phase values. Consider again a sinusoidal signal in the form

$$x(t) = A\cos\left(2\pi f_0 t + \theta\right)$$

with three adjustable parameters, namely the amplitude A, the phase θ, and the frequency f_0. Using Euler's formula, it is possible to express the sinusoidal signal $x(t)$ as the real part of a complex signal in the form

$$x(t) = \mathrm{Re}\left\{A\,e^{j(2\pi f_0 t+\theta)}\right\} \tag{1.79}$$

The exponential function in Eqn. (1.79) can be factored into two terms, yielding

$$x(t) = \mathrm{Re}\left\{A\,e^{j\theta}\,e^{j2\pi f_0 t}\right\} \tag{1.80}$$

Let the complex variable $\boldsymbol{X}$ be defined as

$$\boldsymbol{X} \triangleq A\,e^{j\theta} \tag{1.81}$$

The complex variable $\boldsymbol{X}$ can be viewed as a vector with norm A and angle θ. Vector interpretation of $\boldsymbol{X}$ is illustrated in Fig. 1.55.

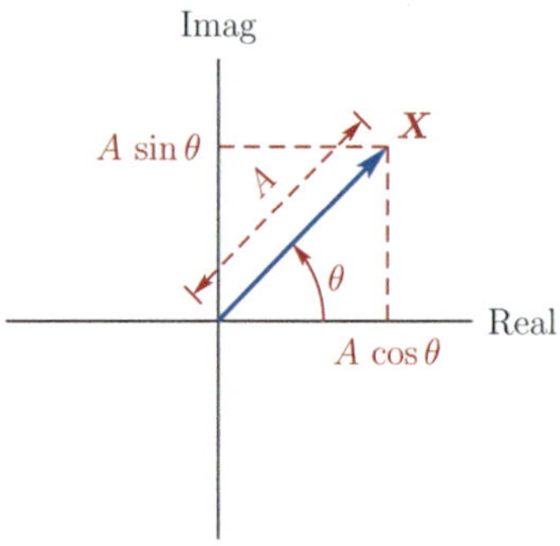

Figure 1.55 – Complex variable $\boldsymbol{X}$ shown as a vector.

Using $\boldsymbol{X}$ in Eqn. (1.80), we obtain

$$x(t) = \mathrm{Re}\left\{\boldsymbol{X}\,e^{j2\pi f_0 t}\right\} \tag{1.82}$$

In Eqn. (1.82) the exponential term $e^{j2\pi f_0 t}$ represents rotation. It is also complex-valued, and has unit norm. Its angle $2\pi f_0 t$ is a linear function of time. Over an interval of one second, the angle of the exponential term grows by $2\pi f_0$ radians, corresponding to f_0 revolutions around the origin. The expression in brackets in Eqn. (1.82) combines the vector $\boldsymbol{X}$ with this exponential term. The combination represents a vector with norm A and initial phase angle θ that rotates at the rate of f_0 revolutions per second. We will refer to this rotating vector as a *phasor.* At any time instant, the amplitude of the signal $x(t)$ equals the real part of the phasor. This relationship is shown in Fig. 1.56.

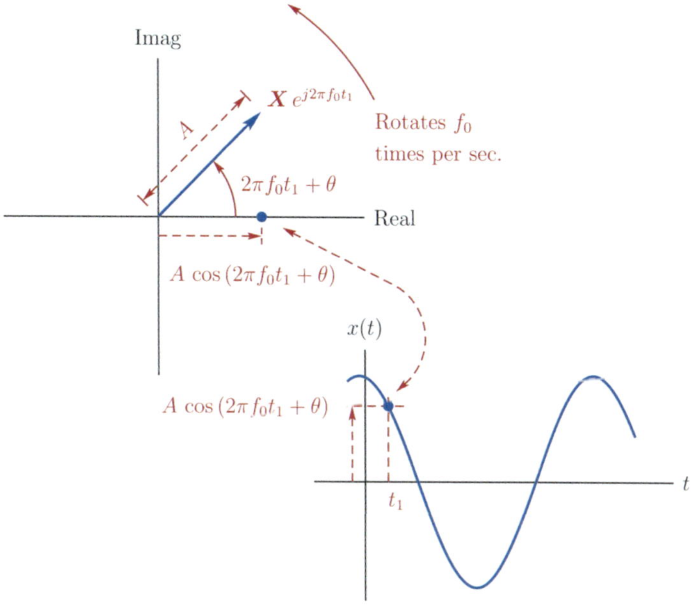

Figure 1.56 – Phasor representation of the signal $x(t)$.

Interactive App: Complex phasors

The interactive app in `appPhasor.m` provides a graphical user interface for experimenting with phasor representation of sinusoidal signals.

The graph window in the upper right section depicts the complex plane. One or two phasors can be visible; the first phasor is shown in blue and the second phasor is shown in red. The real part of each phasor is shown with thick dashed lines of the same color as the phasor.

The graph window at the bottom shows the time domain representation of the signal that corresponds to each phasor, using the same color scheme. The sum of the two sinusoidal signals is shown in black. The real part of each phasor, displayed with thick dashed lines on the complex plane, is duplicated on the time graph to show its relationship to the respective time-varying signal.

a. Check the box "Show phasor 1 & signal 1". Uncheck the two boxes "Show phasor 2 & signal 2" and "Show sum". Set the frequency and the phase of the signal to $f_1 = 2$ Hz and $\theta_1 = 0$ degrees respectively. Make sure the time value is at $t = 0$. Observe the position of the phasor and its projection on the real axis, which is indicated by a small circle. Locate the corresponding small circle on the graph for the signal $x(t)$ at $t = 0$.

b. Gradually advance the time variable starting at $t = 0$ and going toward $t = 2$. As the time variable is varied, pay attention to the real part of the phasor, and how it translates to the amplitude of $x(t)$ at that time instant.

c. Note that from $t = 0$ to $t = 2$ the phasor completes two full rotations, and the sinusoidal signal $x(t)$ completes two full cycles. Change the frequency to $f_1 = 3$ Hz, reset the time to $t = 0$, and repeat the steps (a) and (b).

d. Reset the time to $t = 0$. Check the boxes "Show phasor 2 & signal 2" and "Show sum". Set the frequencies as $f_1 = 2$ Hz and $f_2 = 3$ Hz, and repeat the steps (a) and (b) above. Pay attention to how the two phasors rotate at different speeds.

Software resource: `appPhasor.m`

Software resource: See MATLAB Exercise 1.6.

Summary of Key Points

- The mathematical model for a signal is in the form of a formula, function, algorithm, or a graph that approximately describes the time variations of the physical signal.
- A *continuous-time* signal is one in which the amplitude is defined for all values of the time variable t. Continuous-time signals are also referred to as *analog* signals.
- In contrast with a continuous-time signal, a *discrete-time* signal is one in which the amplitude is defined only at a discrete set of time instants, typically at integer multiples of some step size T_s, i.e., only at time instants $t = nT_s$. As a result, a discrete-time signal is expressed as a function of the index n rather than the time variable t.
- In some discrete-time signals, allowable amplitude values are also limited to a discrete set. The resulting signals are called *digital* signals.
- A constant offset can be applied to a continuous-time signal by adding it to the signal amplitude at each time instant.
- Applying a constant gain factor to a continuous-time signal is accomplished by multiplying the signal amplitude with the gain factor at each time instant.
- Addition of two signals is accomplished by adding amplitudes values at each time instant. Similarly, two signals are multiplied by multiplying amplitudes at each time instant.
- Time shifting, time scaling, and time reversal operations are described in sections 1.4.2, 1.4.3, and 1.4.4 respectively.
- Unit impulse function is defined by means of Eqns. (1.1) and (1.2). Its two important properties are the sampling property

  Eqn. (1.9): $$f(t)\,\delta(t-t_1) = f(t_1)\,\delta(t-t_1)$$

  and the sifting property

  Eqn. (1.10): $$\int_{-\infty}^{\infty} f(t)\,\delta(t-t_1)\,dt = f(t_1)$$

☞ The unit impulse function can be expressed as the limit of a rectangular pulse with unit area under it as the width of the pulse tends to zero. It can also be expressed as the limit of other functions. Two examples of this idea, namely the Gaussian pulse and the squared sinc pulse, are given in Eqns. (1.7) and (1.8).

☞ The unit step function is defined as

Eqn. (1.13): $$u(t) = \begin{cases} 1, & t > 0 \\ 0, & t < 0 \end{cases}$$

It is useful in describing signals that start or end at a certain time instant. Consequently, it can be used for constructing pulse type signals. The unit impulse function can be written as the derivative of the unit step function.

Eqn. (1.15): $$\delta(t) = \frac{du}{dt}$$

☞ The unit pulse function $\Pi(t)$ defined by Eqn. (1.16) represents a pulse with unit height and unit width, centered around the time origin. It can be constructed as the difference of two unit step signals.

Eqn. (1.17): $$\Pi(t) = u\left(t + \tfrac{1}{2}\right) - u\left(t - \tfrac{1}{2}\right)$$

Any pulse with arbitrary amplitude, width, and midpoint values can be obtained from a unit pulse signal by means of amplitude scaling, time scaling, and time shifting operations.

☞ The unit ramp function is defined in terms of the unit step function as $r(t) = t\,u(t)$. It can also be obtained as the running integral of the unit step function:

Eqn. (1.22): $$r(t) = \int_{-\infty}^{t} u(\lambda)\,d\lambda$$

☞ The unit triangle function Λ represents a triangle with two of its corners on the time axis at $t = -1$ and $t = 1$, and its third corner on the vertical axis at unit amplitude, as shown in Fig. 1.38. It can be constructed using scaled and shifted unit ramp functions:

Eqn. (1.24): $$\Lambda(t) = r(t+1) - 2\,r(t) + r(t-1)$$

☞ A sinusoidal signal is characterized by three parameters, namely the amplitude A, the frequency $\omega_0 = 2\pi f_0$, and the phase θ.

Eqn. (1.25): $$x(t) = A\cos(\omega_0 t + \theta)$$

☞ Impulse decomposition of a signal $x(t)$ is expressed as

Eqn. (1.29): $$x(t) = \int_{-\infty}^{\infty} x(\lambda)\,\delta(t-\lambda)\,d\lambda$$

☞ A real signal is one in which the amplitude is real-valued at all time instants. A complex signal is one in which the signal amplitude may have both real and imaginary parts. A complex signal may be written in *Cartesian form* using its real and imaginary parts as

Eqn. (1.36): $$x(t) = x_r(t) + j\,x_i(t)$$

or in *polar form* as

Eqn. (1.37): $$x(t) = |x(t)|\, e^{j\measuredangle x(t)}$$

The two forms can be related to each other through Eqns. (1.38) through Eqn. (1.41).

☞ A signal is said to be *periodic* if it satisfies

Eqn. (1.42): $$x(t+T_0) = x(t)$$

at all time instants t, and for a specific value of $T_0 \neq 0$. The value T_0 is referred to as the *period* of the signal. If a signal is periodic with period T_0, then it is also periodic with kT_0 where k is any integer. When we refer to the period of a periodic signal, we will imply the smallest positive value of T_0 that satisfies periodicity condition. When multiple periods are considered for the same signal, we will refer to T_0 as the *fundamental period* in order to avoid ambiguity. The *fundamental frequency* of a periodic signal is defined as the reciprocal of its fundamental period:

☞ A signal that is the sum of N sinusoidal components with different frequencies is periodic if a fundamental frequency f_0 can be found such that

$$f_n = m_n f_0\,, \quad n = 1,\ldots,N$$

☞ The *normalized energy* of a real-valued signal $x(t)$ is defined as

Eqn. (1.50): $$E_x = \int_{-\infty}^{\infty} x^2(t)\, dt$$

provided that the integral can be computed. If the signal under consideration $x(t)$ is complex-valued, then its normalized energy is

Eqn. (1.51): $$E_x = \int_{-\infty}^{\infty} |x(t)|^2\, dt$$

again assuming that the integral can be computed. The term $|x(t)|$ is the norm of the complex signal.

☞ Time averaging operator has multiple definitions depending on the type of signal involved. If the signal $x(t)$ in question is periodic with period T_0, its time average is computed as

Eqn. (1.53): $$\langle x(t)\rangle = \frac{1}{T_0}\int_{-T_0/2}^{T_0/2} x(t)\, dt$$

If the signal $x(t)$ is non-periodic, its time average is computed as

Eqn. (1.54): $$\langle x(t)\rangle = \lim_{T\to\infty}\left[\frac{1}{T}\int_{-T/2}^{T/2} x(t)\, dt\right]$$

☞ *Normalized average power* of a real-valued signal $x(t)$ is defined as

Eqn. (1.55): $$P_x = \langle x^2(t)\rangle$$

where $\langle\ldots\rangle$ represents the time averaging operator. This definition will work for both periodic and non-periodic signals provided that we use the appropriate form of the time averaging operator. If the signal $x(t)$ is complex-valued, its normalized average power is computed as

Eqn. (1.59): $$P_x = \left\langle |x(t)|^2 \right\rangle$$

☞ Signals can be classified as *energy signals* and *power signals.* Energy signals are those that have finite energy and zero power, i.e., $E_x < \infty$ and $P_x = 0$. Power signals are those that have finite power and infinite energy, i.e., $E_x \to \infty$ and $P_x < \infty$.

☞ The *root-mean-square (RMS)* value of a signal $x(t)$ is defined as

Eqn. (1.63): $$X_{RMS} = \left[\langle x^2(t)\rangle\right]^{1/2}$$

☞ A real-valued signal is said to have *even symmetry* if it has the property $x(-t) = x(t)$ for all values of t. Similarly, a real-valued signal is said to have *odd symmetry* if it has the property $x(-t) = -x(t)$ for all t.

☞ Any real-valued signal can be split into two components in such a way that one of the components is even and the other is odd.

Eqn. (1.69): $$x(t) = x_e(t) + x_o(t)$$

Even and odd components are

Eqns. (1.70) and (1.71): $$x_e(t) = \frac{x(t) + x(-t)}{2}, \quad x_o(t) = \frac{x(t) - x(-t)}{2}$$

☞ A complex-valued signal is said to be *conjugate symmetric* if it satisfies $x(-t) = x^*(t)$ for all t. Similarly, a complex-valued signal is said to be *conjugate antisymmetric* if it satisfies $x(-t) = -x^*(t)$ for all t.

☞ It is possible to express any complex signal $x(t)$ as the sum of two components such that one of the components is conjugate symmetric and the other is conjugate antisymmetric. The two components are

Eqns. (1.77) and (1.78): $$x_E(t) = \frac{x(t) + x^*(-t)}{2}, \quad x_O(t) = \frac{x(t) - x^*(-t)}{2}$$

☞ A sinusoidal signal in the form

$$x(t) = A\cos\left(2\pi f_0 t + \theta\right)$$

can be expressed as

Eqn. (1.82): $$x(t) = \mathrm{Re}\left\{\boldsymbol{X}\, e^{j2\pi f_0 t}\right\}$$

where the complex quantity $\boldsymbol{X} \triangleq A\, e^{j\theta}$ is referred to as a *phasor.* As the phasor rotates at the frequency f_0, its real part tracks the amplitude of the sinusoidal signal.

Further Reading

[1] A.B. Carlson and P.B. Crilly. *Communication Systems.* McGraw-Hill, 2009.

[2] L.W. Couch. *Digital and Analog Communication Systems.* Prentice Hall, 2013.

[3] Geoffrey Grimmett and Dominic J. Welsh. *Probability: An Introduction.* Oxford University Press, 2014.

[4] Misza Kalechman. *Practical MATLAB Basics for Engineers.* CRC Press, 2018.

[5] R.P. Kanwal. *Generalized Functions: Theory and Applications.* Birkhèauser, 2004.

[6] Kenneth Lange. *Applied Probability.* Springer, 2010.

[7] Mark Meerschaert. *Mathematical Modeling.* Academic Press, 2013.

[8] Peter V. O'Neil. *Advanced Engineering Mathematics.* Cengage Learning, 2017.

[9] A. Papoulis and S. Unnikrishna Pillai. *Probability, Random Variables, and Stochastic Processes.* 4th. McGraw-Hill, 2002.

[10] P.Z. Peebles. *Probability, Random Variables, and Random Signal Principles.* McGraw-Hill, 2001.

[11] P. Prandoni and M. Vetterli. *Signal Processing for Communications.* Communication and Information Sciences. Taylor & Francis, 2008.

[12] Sheldon M. Ross. *Introduction to Probability Models.* Academic Press, 2014.

[13] A. Spanias, T. Painter, and V. Atti. *Audio Signal Processing and Coding.* Wiley, 2006.

[14] Henry Stark and John W. Woods. *Probability, Statistics, and Random Processes for Engineers.* 4th. Prentice Hall, 2011.

[15] Kenneth Arthur Stroud and Dexter J. Booth. *Engineering Mathematics.* Bloomsbury Publishing, 2020.

[16] M. Tohyama. *Sound and Signals.* Springer, 2011.

[17] Daniel T. Valentine and Brian H. Hahn. *Essential MATLAB for Engineers and Scientists.* Academic Press, 2022.

[18] Bernard P. Zeigler, Herbert Praehofer, and Tag Gon Kim. *Theory of Modeling and Simulation.* Academic Press, 2000.

MATLAB Exercises with Solutions

MATLAB Exercise 1.1: Computing and graphing signals

Often we have the need to simulate mathematical models of signals on a computer. Sometimes this is done to check the results of hand calculations involving a signal, or to gain graphical insight into signal behavior. At other times we may need a numerical approximation to the mathematical model of a signal for use in a simulation study involving a signal-system interaction problem. In this exercise we will explore methods of computing and graphing continuous-time signal models in MATLAB.

Caveat: Since any row or column vector must have a finite-number of elements, a completely accurate representation of a continuous-time signal is not possible in MATLAB. The best we can do is to approximate a continuous-time signal by constructing a vector that holds amplitudes of the signal at equally-spaced time instants. We will refer to these amplitude values as *samples* of the signal. If the vector used for this purpose has a sufficient number of samples, and if the time spacing of these samples is chosen carefully, the result can still be a reasonably good approximation to a continuous-time signal. We will defer a more formal discussion of determining the appropriate time spacing for sample amplitudes to textbooks on discrete-time signals and systems, and specifically to the study of the subject of *sampling.*

In graphing a continuous-time signal represented by amplitude samples we simply connect those samples with straight lines to give the appearance of a continuous-time signal.

a. Compute the signal

$$x_1(t) = 5\sin(12t)$$

at 500 points in the time interval $0 \le t \le 5$, and graph the result.

Solution: Type the following three lines into MATLAB command window to obtain the graph shown in Fig. 1.57.

```
>> t = linspace(0,5,500);
>> x1 = 5*sin(12*t);
>> plot(t,x1);
```

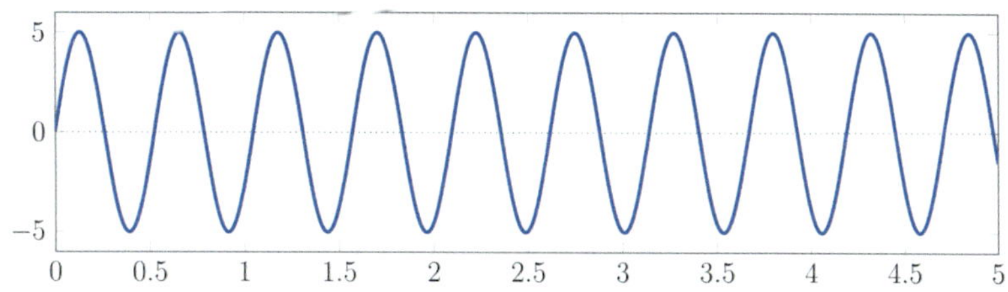

Figure 1.57 – Graph obtained in MATLAB Exercise 1.1 part (a).

The MATLAB function `linspace()` creates a vector `t` of time instants starting at $t = 0$ and going up to $t = 5$ using a total of 500 elements. In the second line we compute the vector `x1` containing amplitudes of the signal at time instants specified in vector `t`. The built-in function `plot()` takes two arguments, namely the independent variable and the dependent variable.

b. Compute and graph the signal

$$x_2(t) = \begin{cases} e^{-3t} - e^{-6t}, & t \ge 0 \\ 0, & t < 0 \end{cases}$$

in the time interval $0 \le t \le 5$ s, using a time increment of $\Delta t = 0.01$ s.

Solution: This signal can be computed and graphed with the following statements:

```
>> t = [0:0.01:5];
>> x2 = exp(-3*t)-exp(-6*t);
>> plot(t,x2);
```

The resulting graph is shown in Fig. 1.58.

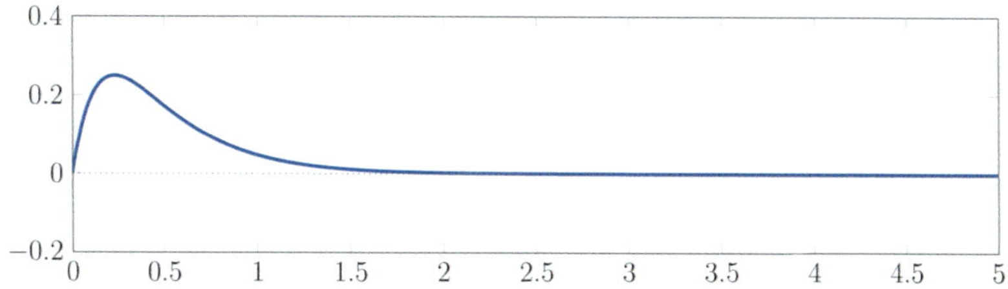

Figure 1.58 – Graph obtained in MATLAB Exercise 1.1 part (b).

The first line demonstrates an alternative method of constructing a time vector. We use three numbers separated by colons between a pair of square brackets. The notation

```
>> t = [a:b:c]
```

instructs MATLAB to create a vector with a starting element of a, an increment of b, and a final value of less than or equal to c.

c. Compute and graph the signal

$$x_3(t) = \begin{cases} e^{-3t} - e^{-6t}, & t \geq 0 \\ 0, & t < 0 \end{cases}$$

in the time interval $-2 \leq t \leq 3$ seconds, using a time increment of $\Delta t = 0.01$ s.

Solution: The signal we need to graph is the same as the signal in part (b). The slight complication presented by this exercise is due the fact that we need to graph the signal starting at a negative time instant, and signal amplitudes are equal to zero for negative values of t. As a result, two separate descriptions of the signal need to be utilized, one for $t < 0$ and the other for $t \geq 0$. In part (b) we avoided this problem by graphing the signal only for positive values of t and thus using a single analytical expression to compute all relevant signal amplitudes.

The simplest method of dealing with this problem is to use the logic operator ">=" which stands for *greater than or equal to.* The following lines produce the graph shown in Fig. 1.59.

```
>> t = [-2:0.01:3];
>> w = (t>=0);
>> x3 = (exp(-3*t)-exp(-6*t)).*w;
>> plot(t,x3);
```

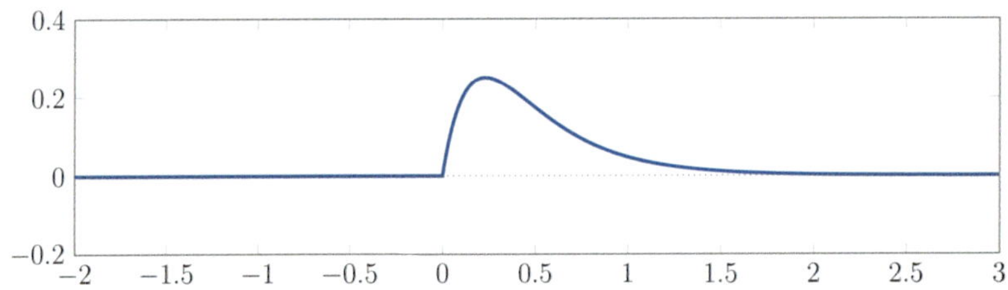

Figure 1.59 – Graph obtained in MATLAB Exercise 1.1 part (c).

Line 2 in the listing above does something interesting: It compares the variable t against the value of zero, but recall that t is a *vector.* When we compare a vector against a particular value, we ask the question for each element of that vector: Is it greater than or equal to zero? The result of the comparison is a logical vector in which an element is equal to true if the answer is yes, and false otherwise. MATLAB represents logical values true and false using 1 and 0, respectively. Furthermore, element by element multiplication of a floating point vector with a logical vector produces a floating point vector. We make use of this feature in line 3 of the listing when we multiply the vector exp(-3*t)-exp(-6*t) with the logical vector w to obtain the vector x3.

Common mistake: It is important to use the *element-by-element multiplication* operator ".*" and not the *vector multiplication* operator "*" in line 3 above. The expression for computing the vector `x3` involves two vectors of the same size, namely `(exp(-3*t) -exp(-6*t))` and `w`. The operator "`.*`" creates a vector of the same size by multiplying the corresponding elements of the two vectors. The operator "$*$", on the other hand, is used for computing the *scalar product* of two vectors, and would not work in this case.

d. Compute and graph the signal

$$x_4(t) = \begin{cases} e^{-3t} - e^{-6t}, & 0 \le t \le 1 \\ 0, & \text{otherwise} \end{cases}$$

in the time interval $-2 \le t \le 3$ second, using a time increment of $\Delta t = 0.01$ second.

Solution: This signal is similar to the one in part (c). The only difference is that $x_4(t) = 0$ for $t > 1$. The logical test needs to be modified to reflect this. Using the *logical and* operator "&" we can type the following lines into MATLAB command window to compute and graph $x_4(t)$:

```
>> t = [-2:0.01:3];
>> w = ((t>=0)&(t<=1));
>> x4 = (exp(-3*t)-exp(-6*t)).*w;
>> plot(t,x4);
```

The resulting graph is shown in Fig. 1.60.

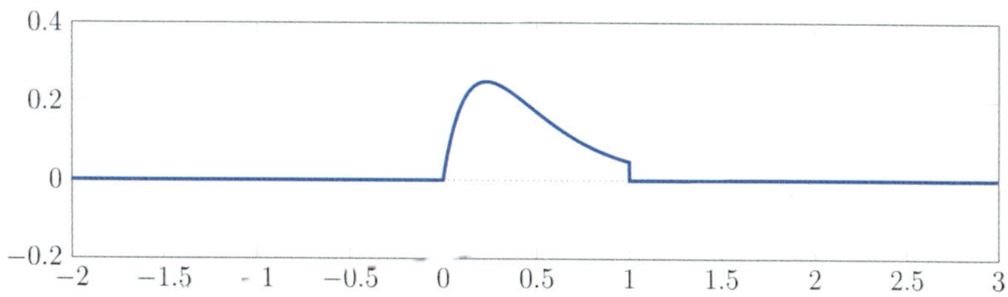

Figure 1.60 – Graph obtained in MATLAB Exercise 1.1 part (d).

Software resources: `mexct_1_1a.m`, `mexct_1_1b.m`, `mexct_1_1c.m` `mexct_1_1d.m`

MATLAB Exercise 1.2: Describing signals using piecewise linear segments

Consider the signal $x(t)$ obtained in Example 1.1 and shown in Fig. 1.8. A signal such as this one can also be described by specifying the endpoints of linear segments and interpolating between them. MATLAB function `interp1()` is used for this purpose. The endpoints of the signal under consideration are

$$(t_p, x_p) = \{\, (-1,0),\ (0,0),\ (1,1),\ (3,-1),\ (4,-1),\ (5,0),\ (6,0) \,\}$$

The code below computes the signal for $-1 \le t \le 6$ and graphs the result.

```
>> tp = [-1,0,1,3,4,5,6];
>> xp = [0,0,1,-1,-1,0,0];
>> t = [-1:0.01:6];
```

```
>> x = interp1(tp,xp,t,'linear');
>> plot(t,x,'b-',tp,xp,'ro'); grid;
```

The built-in MATLAB function `interp1()` takes four arguments in this case. The first two are vectors containing time and amplitude values of segment endpoints. The third argument is the vector containing the set of time instants at which we want the amplitudes to be computed using interpolation. The last argument indicates that *linear* interpolation is to be used.

Caution: The function `interp1()` requires that all t_p values be distinct, that is, no two points could have the same time value. As a result, signals that have discontinuities (vertical edges) cannot be graphed using this approach. We will find a way around this problem in MATLAB Exercise 1.3.

Software resource: mexct_1_2.m

MATLAB Exercise 1.3: Using piecewise linear segments – revisited

Consider the signal $x(t)$ shown in Fig. 1.61.

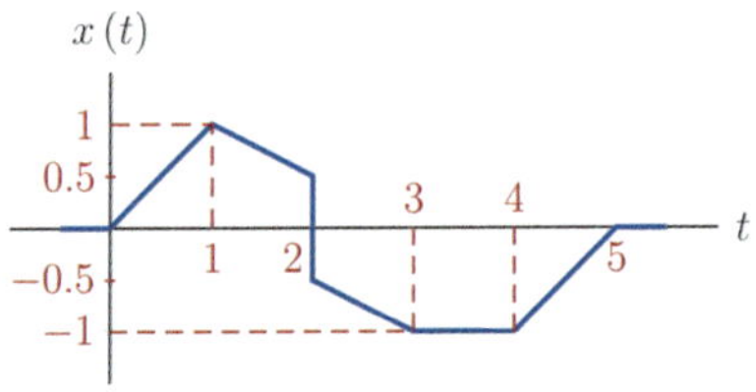

Figure 1.61 – Signal for MATLAB Exercise 1.3.

Compared to the signal of MATLAB Exercise 1.2, this $x(t)$ exhibits a discontinuity at $t = 2$. We will still attempt to use linear interpolation by means of the built-in function `interp1()` with the endpoints

$$\left(t_p, x_p\right) = \left\{ (-1,0),\ (0,0),\ (1,1),\ (2,0.5),\ (2,-0.5),\ (3,-1),\ (4,-1),\ (5,0),\ (6,0) \right\}$$

The code below, as our first attempt, produces the error message shown:

```
>> tp = [-1,0,1,2,2,3,4,5,6];
>> xp = [0,0,1,0.5,-0.5,-1,-1,0,0];
>> t = [-1:0.01:6];
>> x = interp1(tp,xp,t,'linear');

Error using matlab.internal.math.interp1
Sample points must be unique.
```

One method of tricking MATLAB into thinking that the two time instants at $t = 2$ are unique is to use the built-in variable `eps` which represents the smallest positive value distinguishable from zero that MATLAB can represent. We will modify the vector of time instants and try the following:

```
>> tp = [-1,0,1,2-eps,2+eps,3,4,5,6];
>> xp = [0,0,1,0.5,-0.5,-1,-1,0,0];
>> t = [-1:0.01:6];
>> x = interp1(tp,xp,t,'linear');
>> plot(t,x,'b-',tp,xp,'ro'); grid;
```

This works, and results in the graph shown in Fig. 1.62.

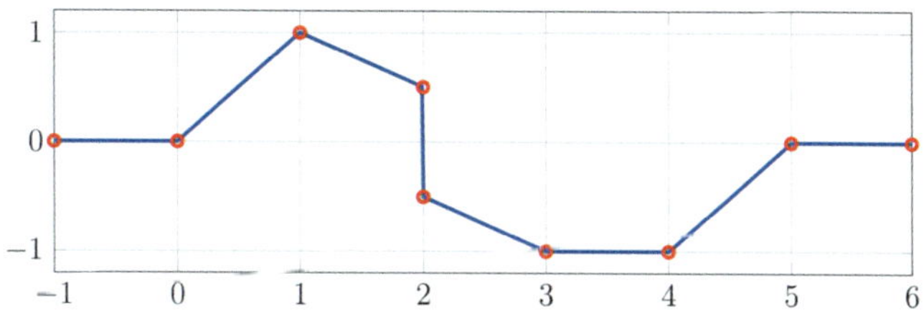

Figure 1.62 – Graph obtained in MATLAB Exercise 1.3.

Software resource: mexct_1_3.m

MATLAB Exercise 1.4: Exploring signal operations

In MATLAB Exercise 1.1 we have used vectors for representing mathematical models for continuous-time signals. This approach is often useful for simulating signal-system interaction problems and will be utilized extensively in later parts of this textbook. In this exercise, however, we will consider alternative approaches for representing signals in MATLAB for the purpose of experimenting with signal operations such as time shifting, time scaling, and time reversal.

Consider again the signal $x(t)$ used in Example 1.3 and shown in Fig. 1.16. Analytically $x(t)$ can be written in the form

$$x(t) = \begin{cases} 2\,, & -1 \le t < 2 \\ 1\,, & 2 \le t < 3 \\ -0.5\,, & 3 \le t \le 6 \end{cases}$$

The following six lines of code create a vector x to represent this signal and graph it.

```
>> t = [-10:0.01:10];
>> w1 = (t>=-1)&(t<2);
>> w2 = (t>=2)&(t<3);
>> w3 = (t>=3)&(t<=6);
>> x = 2*w1+1*w2-0.5*w3;
>> plot(t,x);
```

It is also possible to combine lines 2 through 5 into one compact line instead of explicitly defining logical vectors w1, w2 and w3.

```
>> t = [-10:0.01:10];
>> x = 2*((t>=-1)&(t<2))+1*((t>=2)&(t<3))-0.5*((t>=3)&(t<=6));
>> plot(t,x);
```

Since x is a vector, there is no easy way to compute and graph the signals $g(t) = x(2t-5)$ and $h(t) = x(-4t+2)$ that were graphed in Example 1.3. Instead of using a vector, we will change our strategy, and use a MATLAB function to represent the signal. Create an ASCII text file named "`sigx.m`" in the current directory and place the following three lines in it:

```
function x = sigx(t)
  x = 2*((t>=-1)&(t<2))+1*((t>=2)&(t<3))-0.5*((t>=3)&(t<=6));
end
```

The result is a MATLAB function `sigx()` that returns the value of the signal $x(t)$ at any specified time instant, or even at a set of specified time instants.

A vector g that holds the amplitude values for the signal $g(t) = x(2t-5)$ can now be computed and graphed with the following set of statements:

```
>> t = [-10:0.01:10];
>> g = sigx(2*t-5);
>> plot(t,g);
```

The graph that results should match Fig. 1.17(c). Similarly, a vector h with amplitude values of the signal $h(t) = x(-4t+2)$ can be computed and graphed with the statements

```
>> t = [-10:0.01:10];
>> h = sigx(-4*t+2);
>> plot(t,h);
```

The graph produced should match Fig. 1.18(c).

Use of the function `sigx()` allows us to write signal operations in a natural form. One slight drawback is that it requires the creation of an ASCII file "`sigx.m`" on disc with the function code in it. This may not always be desirable, especially if the signal $x(t)$ will only be used for one exercise and will not be needed again. An alternative method is to create an *anonymous function* that does not need to be saved into a file. An anonymous function is created for the current MATLAB session, and stays in memory as long as MATLAB remains active. It is discarded when MATLAB is closed, or when the session variables are cleared with the `clear` command. The following statement creates an anonymous function with the name sx that works the same way as the function `sigx()` used above:

```
>> sx = @(t) 2*((t>=-1)&(t<2))+1*((t>=2)&(t<3))-0.5*((t>=3)&(t<=6));
```

An easy way to remember the syntax is to just read the operator `@()` as "*anonymous function of ...*". Signals $g(t)$ and $h(t)$ can now be computed and graphed using

```
>> plot(t,sx(2*t-5));
>> plot(t,sx(-4*t+2));
```

Software resources: `mexct_1_4a.m` , `mexct_1_4b.m` , `mexct_1_4c.m` `mexct_1_4d.m` , `SignalOperations.mlx` , `sigx.m`

MATLAB Exercise 1.5: Creating periodic signals

MATLAB has a number of functions for creating specific periodic waveforms. In this exercise we will explore functions `square()` and `sawtooth()`. Afterward, we will present a simple technique for creating arbitrary periodic waveforms.

a. One period of a square wave signal is defined as

$$x_1(t) = \begin{cases} 1, & 0 < t < \dfrac{T}{2} \\ -1, & \dfrac{T}{2} < t < T \end{cases}$$

MATLAB function `square()` produces a square wave signal with period equal to $T = 1$ s by default, so we need to use time scaling operation to obtain it with any other value of T. The code below will compute and graph a square wave with period $T = 1$ s for the time interval $-1 < t < 10$ s. The resulting graph is shown in Fig. 1.63.

```
>> t = [-1:0.01:10];
>> x1 = square(2*pi*t);
>> plot(t,x1);
```

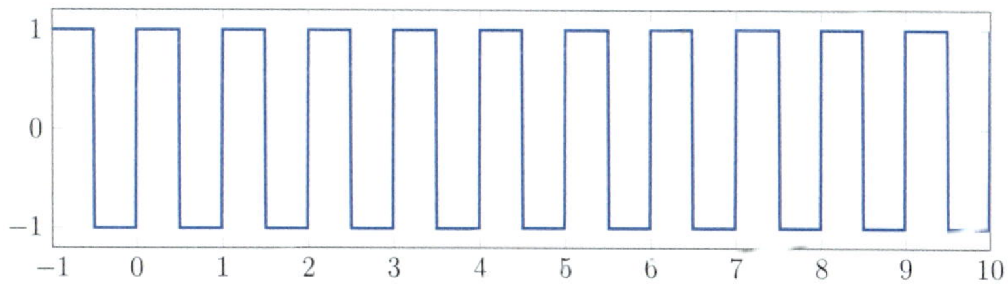

Figure 1.63 – Graph obtained in MATLAB Exercise 1.5 part (a).

b. A square wave signal with a duty cycle of $0 < d \leq 1$ is defined as

$$x_2(t) = \begin{cases} 1, & 0 < t < Td \\ -1, & Td < t < T \end{cases}$$

The following code will compute and graph a square wave signal with period $T = 1$ s and duty cycle $d = 0.2$ (or 20 percent). The resulting graph is shown in Fig. 1.64.

```
>> x2 = square(2*pi*t,20);
>> plot(t,x2);
```

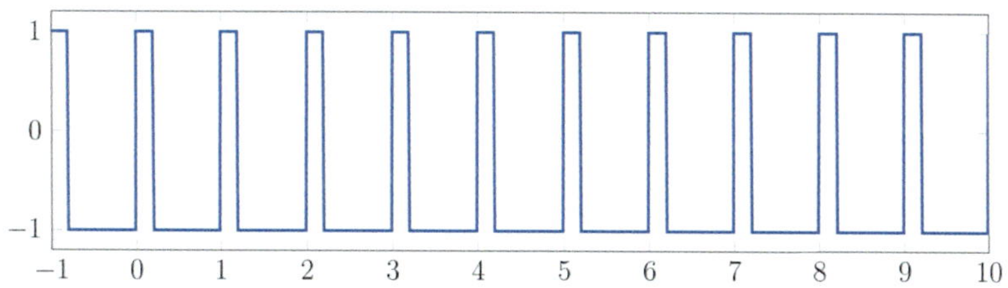

Figure 1.64 – Graph obtained in MATLAB Exercise 1.5 part (b).

c. The function `sawtooth()` generates a sawtooth waveform with period $T = 1$ s. Its definition for one period is

$$x_3(t) = \frac{t}{T} \quad \text{for } 0 < t < T$$

The code below can be used for obtaining a sawtooth signal with period $T = 1.5$ s. The resulting graph is shown in Fig. 1.65.

```
>> x3 = sawtooth(2*pi*t/1.5);
>> plot(t,x3);
```

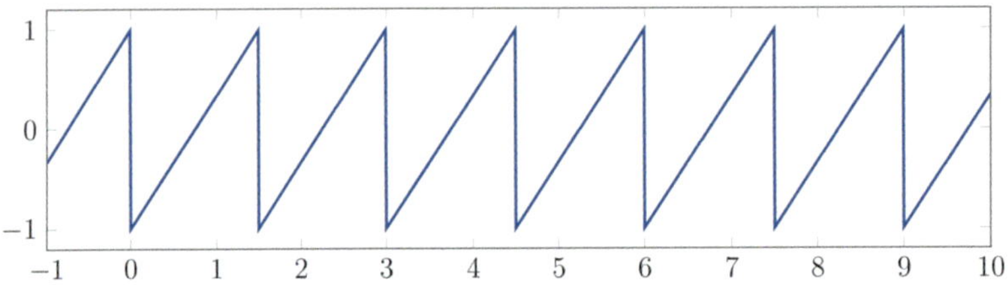

Figure 1.65 – Graph obtained in MATLAB Exercise 1.5 part (c).

d. Let a signal $x_4(t)$ that is periodic with period $T = 2.5$ s be defined through the following:

$$x_4(t) = \begin{cases} t, & 0 \le t < 1 \\ e^{-5(t-1)}, & 1 \le t < 2.5 \end{cases} \quad \text{and} \quad x_4(t + 2.5k) = x_4(t) \text{ for all integer } k$$

The following can be used for graphing this signal in the time interval $-2 < t < 12$ s.

```
>> t = [-2:0.01:12];
>> x4p = @(t) t.*((t>=0)&(t<1))+exp(-5*(t-1)).*((t>=1)&(t<2.5));
>> x4 = x4p(mod(t,2.5));
>> plot(t,x4); grid;
```

The strategy is straightforward: We first create an anonymous function `x4p()` to describe one period of the signal. In the next step we make use of the built-in function `mod()` which returns the remainder after dividing its first argument by its second argument. The value `tp=mod(t,2.5)` is in the interval $0 \le \text{tp} < 2.5$. When used with the anonymous function `x4p()`, it creates the effect of periodic repetition. The resulting graph is shown in Fig. 1.66.

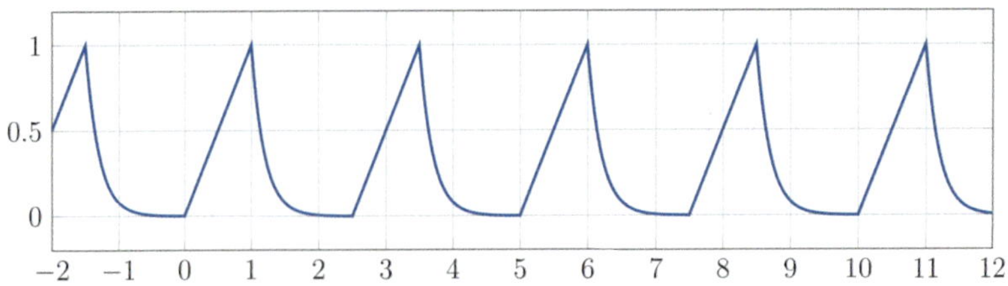

Figure 1.66 – Graph obtained in MATLAB Exercise 1.5 part (d).

Software resources: `mexct_1_5a.m`, `mexct_1_5b.m`, `mexct_1_5c.m` `mexct_1_5d.m`, `PeriodicSignals.mlx`

MATLAB Exercise 1.6: Functions for basic building blocks

It is possible to extend MATLAB by developing custom functions for performing tasks that are encountered often. In this exercise we will develop a few simple functions for the computation of basic signal building blocks described in Section 1.5. In order to avoid possible naming conflicts with MATLAB built-in functions and toolboxes, we will use signals and systems prefix "`ss_`" for the functions we write. The unit step function can be implemented by creating a text file with the name "`ss_step.m`" and the following three lines in it:

```
function x = ss_step(t)
  x = 1*(t>=0);
end
```

It returns the amplitude values for the unit step function evaluated at each time instant specified by the vector `t`. The unit ramp function can be implemented with the following three lines saved into a text file named "`ss_ramp.m`":

```
function x = ss_ramp(t)
  x = t.*(t>=0);
end
```

Functions for computing unit pulse and unit triangle signals can be developed based on the functions `ss_step()` and `ss_ramp()`. We will obtain the unit pulse by applying Eqn. (1.17) as follows:

```
function x = ss_pulse(t)
  x = ss_step(t+0.5)-ss_step(t-0.5);
end
```

Similarly, a unit triangle can be obtained by the application of Eqn. (1.24):

```
function x = ss_triangle(t)
  x = ss_ramp(t+1)-2*ss_ramp(t)+ss_ramp(t-1);
end
```

Software resources: `ss_step.m`, `ss_ramp.m`, `ss_pulse.m`, `ss_triangle.m`

MATLAB Exercise 1.7: Testing the functions for basic building blocks

We will now test the functions developed in MATLAB Exercise 1.6 for basic building blocks. Consider the signals $x_1(t)$ and $x_2(t)$ used in Example 1.2 and shown in Fig. 1.11.

a. The signal $x_1(t)$ can be expressed using unit step functions as

$$x_1(t) = 2u(t) - u(t-1) - 2u(t-2) + u(t-3)$$

Using the function `ss_step()`, the signal $x_1(t)$ can be computed and graphed over the time interval $-1 \le t \le 4$ as follows:

```
% Script: mexct_1_7a.m
t = linspace(-1,4,500);
x1 = 2*ss_step(t)-ss_step(t-1)-2*ss_step(t-2)+ss_step(t-3);
plot(t,x1); grid;
```

b. An alternative representation of the signal $x_1(t)$ can be written in terms of the unit pulse functions as

$$x_1(t) = 2\Pi(t-0.5) + \Pi(t-1.5) - \Pi(t-2.5)$$

Using the function `ss_pulse()`, the following script will compute and graph the same signal:

```
% Script: mexct_1_7b.m
t = linspace(-1,4,500);
x1 = 2*ss_pulse(t-0.5)+ss_pulse(t-1.5)-ss_pulse(t-2.5);
plot(t,x1); grid;
```

c. The signal $x_2(t)$ can be expressed using unit ramp functions as

$$x_2(t) = 0.5r(t) - 2.5r(t-2) + 3r(t-3) - r(t-4)$$

Below is a script for computing and graphing $x_2(t)$ using the function `ss_ramp()`.

```
% Script: mexct_1_7c.m
t = linspace(-1,5,500);
x2 = 0.5*ss_ramp(t)-2.5*ss_ramp(t-2)+3*ss_ramp(t-3)-ss_ramp(t-4);
plot(t,x2); grid;
```

Software resources: mexct_1_7a.m , mexct_1_7b.m , mexct_1_7c.m

MATLAB Exercise 1.8: Exploring symmetry properties of signals

Consider again the signal $x_2(t)$ of Example 1.2 shown in Fig. 1.11(b). In part (c) of MATLAB Exercise 1.7 we have expressed $x_2(t)$ in terms of scaled and time-shifted unit ramp functions, and used it for testing the function `ss_ramp()`. In this exercise we will compute and graph even and odd components of the signal. Recall that in Section 1.9.2 even and odd components of a signal were given by Eqns. (1.70) and (1.71) which are repeated here for convenience:

$$\text{Eqn. (1.70):} \qquad x_e(t) = \frac{x(t) + x(-t)}{2}$$

$$\text{Eqn. (1.71):} \qquad x_o(t) = \frac{x(t) - x(-t)}{2}$$

In the script `mexct_1_8a.m` listed below, we first take the MATLAB statement we used for computing `x2` in the previous MATLAB Exercise, and convert it to an anonymous function. Once that is done, Eqns. (1.70) and (1.71) can be applied directly for computing even and odd components of the signal.

```
% Script: mexct_1_8.m
t = linspace(-5,5,1000);
x2 = @(t) 0.5*ss_ramp(t)-2.5*ss_ramp(t-2)+3*ss_ramp(t-3)-ss_ramp(t-4);
xe = (x2(t)+x2(-t))/2;  % Eqn. (1.70)
xo = (x2(t)-x2(-t))/2;  % Eqn. (1.71)
tiledlayout(1,2);       % Plots in a 1 by 2 (horizontal) arrangement
nexttile;               % Plot the first tile
plot(t,xe); grid;
title('Even component');
xlabel('t (sec)');
ylabel('Amplitude');
nexttile;               % Plot the second tile
plot(t,xo); grid;
title('Odd component');
xlabel('t (sec)');
ylabel('Amplitude');
```

The resulting graph is shown in Fig. 1.67.

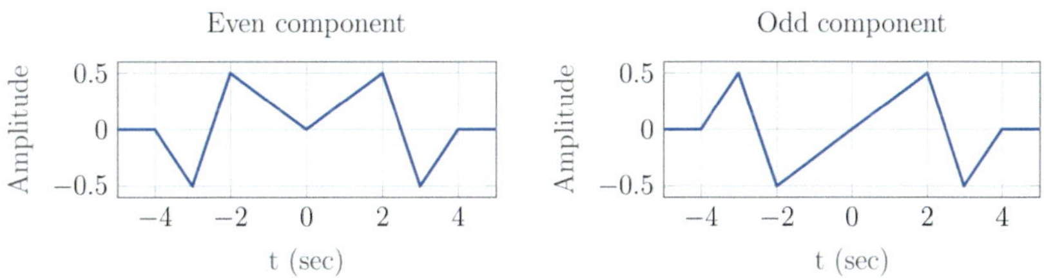

Figure 1.67 – Graph obtained in MATLAB Exercise 1.8.

Software resource: mexct_1_8.m

MATLAB Exercise 1.9: Case study – Pulse width modulation

In Example 1.8 of this Chapter we have determined the average value of a periodic pulse train with pulse amplitude of A and duty cycle d to be

$$\langle x(t) \rangle = A\,d$$

Similarly in Example 1.9 we have determined that the normalized average power of the same pulse train is

$$P_x = A^2\,d$$

The average value or the average power of a pulse train can be adjusted to a desired level by simply varying the duty cycle. This is referred to as *pulse with modulation (PWM)*. It finds practical applications in power electronics, specifically in converter circuits that are used for converting electrical power from one form to another.

Consider, for example, a laptop computer. Its external power adapter takes the sinusoidal line voltage and produces a DC voltage at the connection point, typically 19 V. Inside the laptop computer, a switched-mode power supply must convert this voltage to a multitude of other DC voltage levels needed for various components such as the screen, the CPU, the network interface, and so on. This is an example of DC to DC conversion and is illustrated by the diagram in Fig. 1.68.

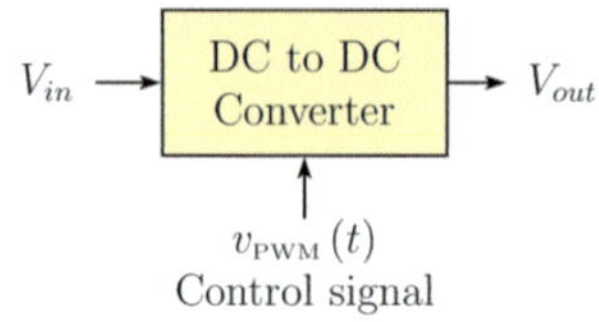

Figure 1.68 – Block diagram for DC to DC conversion.

The control signal for the converter is a pulse width modulated signal with the appropriate duty cycle. Similar circuits exist in cell phones, dimmers used for adjusting the intensity of LED lights and speed control systems for controlling the rotation speed of electric motors such as those used in electric vehicles.

A pulse width modulated signal with the desired duty cycle can be generated through the use of a comparator as shown in Fig. 1.69.

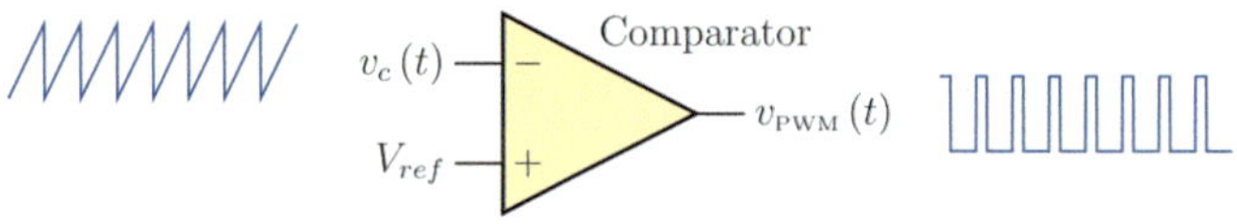

Figure 1.69 – Simple technique for obtaining a pulse width modulated signal.

The comparator circuit produces an output voltage depending on a comparison of its two input signals. Let the input signal $v_c(t)$ at the negative terminal be a periodic sawtooth waveform as shown in Fig. 1.70. This signal is referred to as the *carrier*. A reference voltage V_{ref} is applied to the positive input terminal. We will assume that the comparator circuit is set to produce an output signal as follows:

$$v_{\text{PWM}}(t) = \begin{cases} A, & \text{if } V_{ref} \geq v_c(t) \\ 0, & \text{if } V_{ref} < v_c(t) \end{cases} \tag{1.83}$$

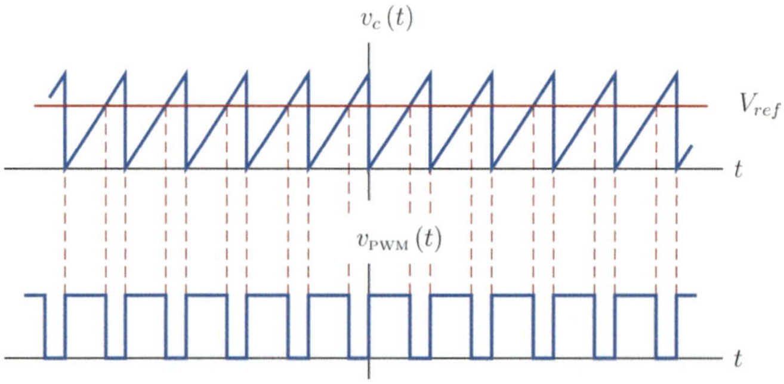

Figure 1.70 – Signals involved in the generation of $v_{\text{PWM}}(t)$.

The MATLAB script `mexct_1_9a.m` shown below implements the strategy outlined above:

```
1  % Script: mexct_1_9a.m
2  t = linspace(0,10,2000);
3  vc = 0.5*sawtooth(2*pi*2*t)+0.5;
4  Vref = 0.3;
5  vPWM = Vref>=vc;    % Eqn. (1.83)
6  tiledlayout(2,1);  % Plot in a 2 by 1 arrangement
7  nexttile;          % Plot the first tile
8  plot(t,vc,t,Vref*ones(size(t)));
9  axis([0,10,-0.2,1.2]);
10 title('Carrier and reference');
11 ylabel('Amplitude');
12 nexttile;          % Plot the next tile
13 plot(t,vPWM);
14 axis([0,10,-0.2,1.2]);
15 title('PWM signal');
16 xlabel('time (sec)');
17 ylabel('Amplitude');
```

In line 2 we use the `sawtooth()` function to generate a periodic sawtooth waveform with a fundamental frequency of 2 Hz, and then adjust it to fit into the amplitude range $[0, 1]$. Line 5 of the script is a direct implementation of Eqn. (1.83). The graphs produced by the script are shown in Fig. 1.71.

Another type of power converter is a DC to AC converter commonly used with solar panels. Electrical power generated by solar panels must be converted from a DC signal to an AC signal with proper amplitude and frequency so that it can be integrated into the existing power grid. For this type of applications, a time varying reference signal such as a sinusoid may be used instead of a DC voltage. This is illustrated in Fig. 1.72.

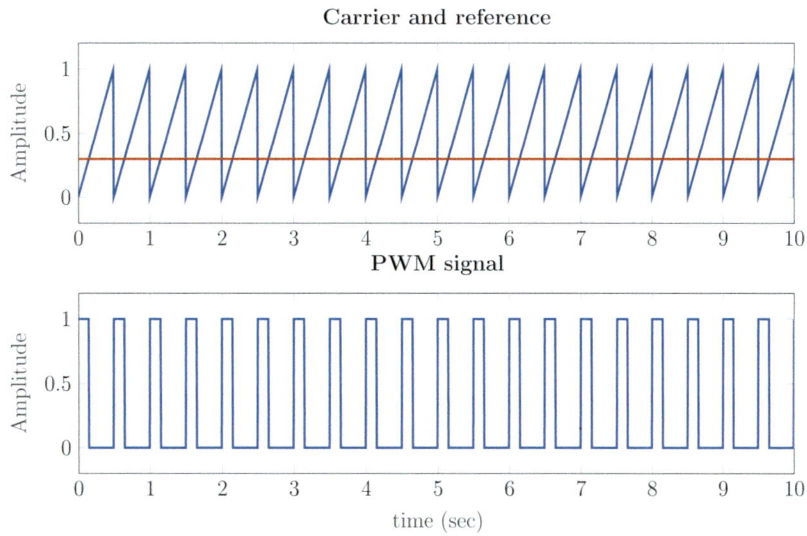

Figure 1.71 – Graph produced by MATLAB script `mexct_1_9a.m`.

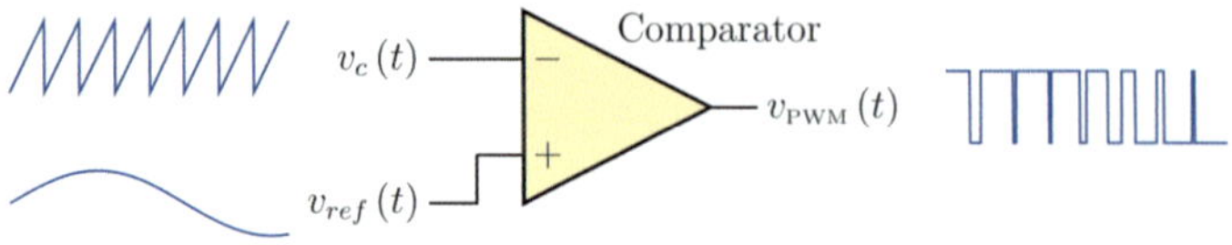

Figure 1.72 – Modulating a pulse train with a sinusoidal reference.

In this case, Eqn. (1.83) would need to be modified to reflect the time varying nature of the reference signal:

$$v_{\text{PWM}}(t) = \begin{cases} A, & \text{if } v_{ref}(t) \ge v_c(t) \\ 0, & \text{if } v_{ref}(t) < v_c(t) \end{cases} \tag{1.84}$$

The MATLAB script `mexct_1_9b.m` shown below implements this. The resulting graphs are shown in Fig.1.73.

```
% Script: mexct_1_9b.m
t = linspace(0,10,2000);
vc = 0.5*sawtooth(2*pi*2*t)+0.5;
vref = 0.45*sin(2*pi*0.15*t)+0.5;
vPWM = vref>=vc;    % Eqn. (1.83)
tiledlayout(2,1);   % Plot in a 2 by 1 arrangement
nexttile;           % Plot the first tile
plot(t,vc,t,vref);
axis([0,10,-0.2,1.2]);
title('Carrier and reference');
ylabel('Amplitude');
nexttile;           % Plot the next tile
plot(t,vPWM);
axis([0,10,-0.2,1.2]);
title('PWM signal');
xlabel('time (sec)');
ylabel('Amplitude');
```

The sinusoidal reference signal can be recovered from the pulse width modulated signal by time averaging each period of $v_{\text{PWM}}(t)$. (See Problem 1.46 at the end of this Chapter.)

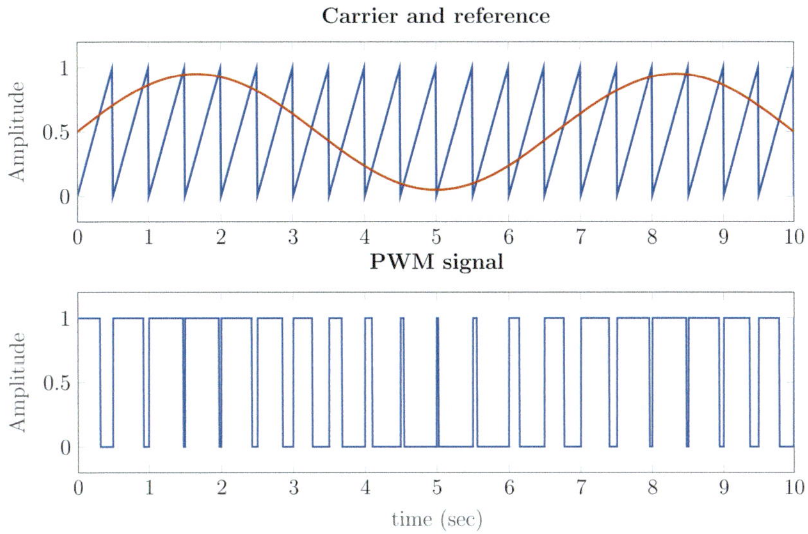

Figure 1.73 – Graph produced by MATLAB script `mexct_1_9b.m`.

Software resources: `mexct_1_9a.m` , `mexct_1_9b.m`

MATLAB Exercise 1.10: Case study – Synthesizing music

The history of electronic music production dates back to 1920's. Leon Theremin (1896-1993), a Russian engineer also known by the name Lev Termen , invented a device that came to be known as a *theremin* or an *etherophone*. The device relied on two high frequency oscillators built using vacuum tubes. Even though the operating frequencies of the two oscillators were outside the range of human hearing, their difference would be in the audible range, and that is what Theremin used for creating music (see Problems 1.6 and 1.47 at the end of this chapter). The instrument also had two external antennas. One antenna, known as the *pitch antenna*, controlled the frequency of one of the oscillators and, consequently, the audible frequency difference between the two. The other antenna was for controlling the volume. A trained theremin player could use coordinated hand movements close to the antennas to manipulate the electromagnetic fields around them and create music.

The first keyboard-based music synthesizer was built by Robert Moog (1934–2005), an American engineer, in 1964 while he was a doctoral student at Cornell University. Inspired by Theremin's work, Moog used voltage-controlled oscillators (VCO) in which the frequency of oscillations at the output of the circuit could be controlled by varying the voltage level at the input. He has also experimented with envelope generators, filters, and noise generators to further modify and enhance the sounds produced. A keyboard could be used to produce the voltage levels to control the oscillators and other circuitry.

In this exercise we will attempt to create MATLAB versions of some of the signals with which Robert Moog may have experimented in his quest to build the first keyboard-based music synthesizer. As a first step, we will generate a vector `t` of time instants going from $t = 0$ up to $t = 0.5$ s in 5000 steps. Note that this corresponds to a rate of 10,000 time steps per second.

Our next step will be to compute amplitudes of a sinusoidal signal with the frequency f = 261.63 Hz for the duration of 0.5 s. Type the following four lines into MATLAB command window.

```
>> t = linspace(0,0.5,5000);
>> f = 261.63;
>> x = 0.6*sin(2*pi*f*t);
>> sound(x,10000);
```

Upon running the code you should hear the sinusoidal signal being played as a sound. For those familiar with keyboard instruments, the frequency we used, namely f = 261.63 Hz, corresponds to the *middle* C sound also known as C_4 owing to the fact that it's part of the fourth octave on most keyboard instruments.

The `sound()` function of MATLAB plays the signal in vector `x` assuming all of its amplitude values are in the range $-1.0 \le x \le 1.0$. Amplitudes that fall outside this range are clipped. The second argument of the `sound()` function specifies the number of amplitude values contained in the vector `x` *for each second of audio.* In essence, the second argument specifies how fast MATLAB should play back the sound. In the study of discrete-time signals and systems this value is known as the *sampling rate.*

It is possible to play different notes by simply changing the frequency of the sinusoidal signal. Table 1.1 lists frequencies for musical notes C_4 through C_5, covering one full octave. For those familiar with piano keyboards, these correspond to the white keys.

Table 1.1 – Frequencies of some music notes in Hz.

C_4	D_4	E_4	F_4	G_4	A_4	B_4	C_5
261.63	293.66	329.63	349.23	392.00	440.00	493.88	523.25

Each note can be generated as a single tone with a peak amplitude of K and a duration of T_d.

$$x_i(t) = K \sin\left(2\pi f_i t\right), \quad 0 \le t \le T_d \tag{1.85}$$

On a keyboard T_d would depend on how long a key is pressed. In addition, if the keyboard has weight sensitive keys, the parameter K may be made a function of how hard a key is pressed. In our simulations we will set $K = 0.6$ and $T_d = 0.5$ s for simplicity. The following two lines can be used for creating an anonymous function called `tone()` that produces a vector with amplitudes of a sinusoidal signal for a duration of 0.5 s.

```
>> t = linspace(0,0.5,5000);
>> tone = @(f) 0.6*sin(2*pi*f*t);
```

Now we can create a vector for each of the notes in Table 1.1, and play the notes in the octave consecutively.

```
>> xC4 = tone(261.63);
>> xD4 = tone(293.66);
>> xE4 = tone(329.63);
>> xF4 = tone(349.23);
>> xG4 = tone(392.00);
>> xA4 = tone(440.00);
>> xB4 = tone(493.88);
```

```
>> xC5 = tone(523.25);
>> sound([xC4,xD4,xE4,xF4,xG4,xA4,xB4,xC5],10000)
```

Listening to the sequence makes it clear that, even though the notes seem to be accurate, the quality of the sound leaves a lot to be desired. There are noticeable clicks in transitions between the successive notes. Our synthesizer is not ready for prime time yet.

In order to understand the reason for the clicks in between successive notes, we need to take a close look at the signal at the transition point. Let us focus on the transition between the notes F_4 and G_4. At the end of the first note, the signal amplitude is

$$0.6 \sin(2\pi\,(349.23)\,(0.5)) = -0.3968$$

and at the beginning of the next note, the signal amplitude is

$$0.6 \sin(2\pi\,(293.66)\,(0.0)) = 0.0$$

The sudden jump from -0.3968 to 0.0 is a discontinuity that produces high-frequency content which is audible as a click. We will revisit this when we explore the Fourier transform in Chapter 3. Fig. 1.74 illustrates this discontinuity right around the transition time instant $t = 0.5$ s.

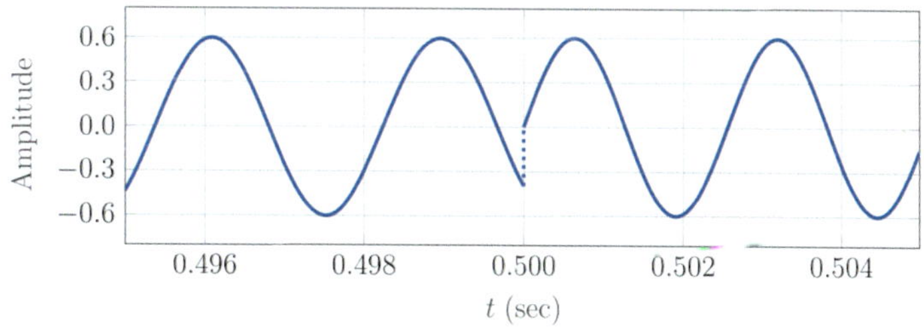

Figure 1.74 – Transition between F_4 and G_4.

In order to visualize this discontinuity in MATLAB, we can zoom in on a small number of amplitude values right around the transition using the following set of lines:

```
>> x = [xF4,xG4];
>> plot(x(4950:5050)); grid;
```

A possible solution for the clicking sounds between notes is to multiply each tone signal with a so-called *envelope* signal that tapers down at both ends. An example is shown in Fig. 1.75.

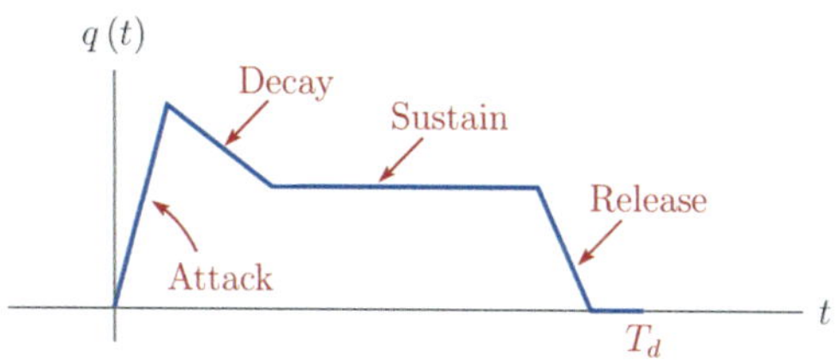

Figure 1.75 – The envelope signal $q(t)$ with various segments illustrated.

The signal $q(t)$ has specific segments that are often referred to as *attack, decay, sustain,* and *release.* Multiplication of $x_i(t)$ with $q(t)$ causes the peak amplitudes of the sinusoid to be *enveloped* by the shape of $q(t)$ as shown in Fig. 1.76. This is a particular form of *modulation* as we will see later in Chapter 7. The signal $q(t)$ is referred to as the *modulating envelope.*

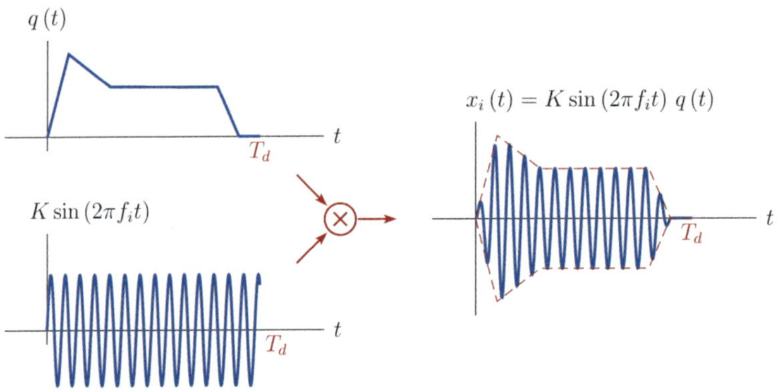

Figure 1.76 – Use of an envelope signal with a sinusoidal tone.

The modified waveform for each note is now in the form

$$x_i(t) = K\sin\left(2\pi f_i t\right) q(t)\,, \quad 0 \le t \le T_d$$

The envelope signal $q(t)$ shown in Fig. 1.75 can be implemented through piecewise linear segments using the MATLAB function `interp1()` we have encountered in Matlab Exercise 1.2:

```
>> tq = [0,0.05,0.15,0.4,0.45,0.5];
>> xq = [0,1.5,0.9,0.9,0,0];
>> q = interp1(tq,xq,t,'linear');
```

Putting all the code segments together, the completed MATLAB script is shown below:

```
% Script: mexct_1_10e.m
t = linspace(0,0.5,5000);            % Vector of time instants
% Set corner points for q(t)
tq = [0,0.05,0.15,0.4,0.45,0.5];
xq = [0,1.5,0.9,0.9,0,0];
% Obtain the signal q(t) through linear interpolation
q = interp1(tq,xq,t,'linear');
% Anonymous function for single tone shaped by q(t)
tone = @(f) 0.6*sin(2*pi*f*t).*q;  % Eqn. (1.85)
% Compute signals for each note
xC4 = tone(261.63);
xD4 = tone(293.66);
xE4 = tone(329.63);
xF4 = tone(349.23);
xG4 = tone(392.00);
xA4 = tone(440.00);
xB4 = tone(493.88);
xC5 = tone(523.25);
```

```
% Play the sound
sound([xC4,xD4,xE4,xF4,xG4,xA4,xB4,xC5],10000);
```

As an alternative to using a modulating envelope, it is also possible to apply a filter to the audio signal to remove high frequency artifacts such as clicks. Design and implementation of frequency selective filters will be the subject of Chapter 6.

Software resources: mexct_1_10a.m , mexct_1_10b.m , mexct_1_10c.m , mexct_1_10d.m , mexct_1_10e.m , mexct_1_10f.m , mexct_1_10g.m

MATLAB Exercise 1.11: Case study – Simulating Theremin's sounds

In MATLAB Exercise 1.10 we referred to the musical device invented by Leon Theremin. The sounds produced by Theremin's device, known in the music community as a *theremin* or an *etherophone*, are fundamentally different than those produced by a keyboard-based synthesizer such as the one built by Robert Moog. In the latter, a note is produced for as long as a particular key is pressed. When that key is released and a different key is subsequently pressed, a new and different note is produced. In between the two notes there is either an abrupt transition or a gap in the sound depending on how soon the second key is pressed after the first one is released.

In contrast, when gentle hand gestures are used for controlling oscillation frequencies in a theremin instrument, the transitions between notes must be gradual and continuous. This results in an almost eerie sound.

Consider the sequence of notes C_4, D_4, and E_4 first played on a keyboard-based synthesizer and then on a theremin instrument. Assume, for the sake of simplicity, that the resulting sound waveform is purely sinusoidal with no modulating envelope or filtering applied. Fig. 1.77 depicts the variation of frequency over time on each instrument.

Recall that, in the discussion of sinusoidal signals in Section 1.5, we focused our attention on signals with three parameters, namely amplitude, frequency, and phase. For a pure sinusoidal signal, these three parameters would have values that would be fixed for the duration of the signal. An example is

$$x(t) = K \sin\left(2\pi f_0 t + \alpha\right) \tag{1.86}$$

If we want to simulate the sounds of the theremin instrument, our focus needs to be on varying the frequency of the sinusoidal signal gradually over time. In signal processing terminology, such signals are referred to as *chirp signals*. Consider the generalized form of a sinusoidal signal expressed as

$$x(t) = \sin\left(\Theta(t)\right) \tag{1.87}$$

The *instantaneous frequency* of the signal is defined as

$$f_i(t) = \frac{1}{2\pi}\,\frac{d\Theta(t)}{dt} \tag{1.88}$$

and may or may not be a function of t depending on the phase function $\Theta(t)$. For the sinusoidal signal given by Eqn. (1.86) the instantaneous frequency is $f_i(t) = f_0$ (constant). If $f_i(t)$ is specified, the phase function $\Theta(t)$ can be found through integration of Eqn. (1.88).

$$\Theta(t) = 2\pi \int f_i(t)\, dt \tag{1.89}$$

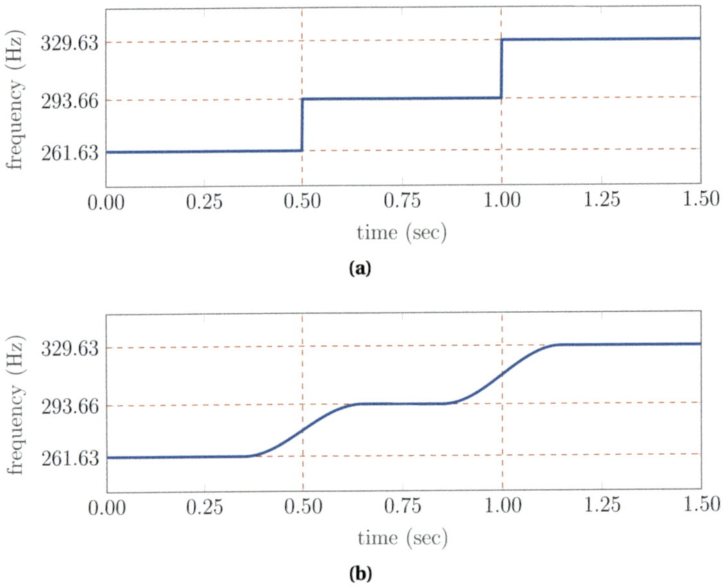

Figure 1.77 – Time variation of frequency for the sequence C_4, D_4, and E_4 **(a)** on a keyboard-based synthesizer and **(b)** on a theremin instrument.

Now suppose that we would like the instantaneous frequency to start from an initial value f_1 at time instant t_1, and and linearly increase or decrease to the value f_2 over a time interval of τ. This could be accomplished by selecting the instantaneous frequency as

$$f_i(t) = A + B\,t \tag{1.90}$$

where the coefficients A and B can easily be determined to satisfy the endpoint conditions $f_i(t_1) = f_1$ and $f_i(t_1+\tau) = f_2$ as follows:

$$B = \frac{f_2 - f_1}{\tau}, \qquad A = f_1 - \frac{f_2 - f_1}{\tau}\,t_1 = f_1 - B\,t_1 \tag{1.91}$$

The phase function $\Theta(t)$ can be obtained as

$$\begin{aligned}\Theta(t) =& 2\pi \int (A + B\,t)\,dt + \alpha \\ =& 2\pi A t + \pi B t^2 + \alpha\end{aligned} \tag{1.92}$$

where coefficients A and B are as given by Eqn. (1.91) and α is the desired initial phase at $t = t_1$. The chirp signal is given by

$$x(t) = \sin\left(2\pi A t + \pi B t^2 + \alpha\right) \tag{1.93}$$

In our MATLAB simulation of theremin instrument sounds we will take the approach outlined above. For simplicity, we will use instantaneous frequencies made up of piecewise linear segments as illustrated in Fig. 1.78. If we also assume that each linear segment starts at $t = 0$, then the coefficients found in Eqn. (1.91) are simplified to $A = f_1$ and $B = \left(f_2 - f_1\right)/\tau$. In this case the chirp signal of Eqn. (1.93) becomes

$$x(t) = \sin\left(2\pi f_1\,t + \frac{\pi\left(f_2 - f_1\right)}{\tau}\,t^2 + \alpha\right) \tag{1.94}$$

where α is the desired initial phase value at $t = 0$.

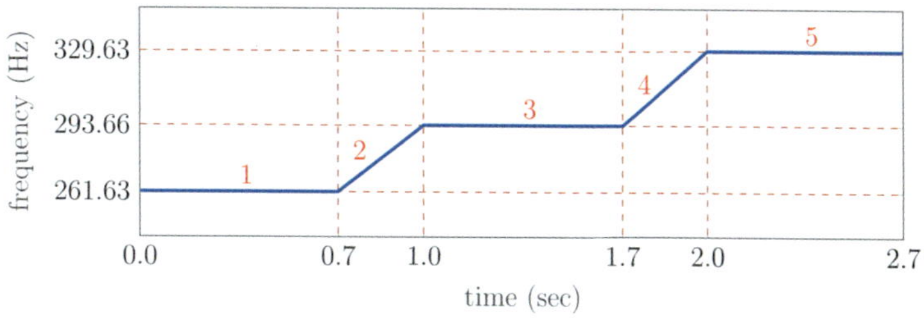

Figure 1.78 – Piecewise linear instantaneous frequency for MATLAB Exercise 1.11.

Consider the MATLAB script `mexct_1_11a.m` listed below. A *local function* `chirpsig()` is part of the script (lines 15 through 21), and computes the chirp signal at time instants contained in vector `t` using the development outlined in Eqns. (1.90) through (1.94). In MATLAB, a local function is a function that is appended to the end of a script file and is only visible within that script file. Multiple scripts may each have local functions with the same name without creating naming conflicts.

In lines 5 and 6 two time vectors are created; `t` for sinusoidal tones, and `tt` for chirp signals in transitions. Both vectors use a step size that corresponds to 10,000 time steps per second. Lines 7 through 11 compute each segment of the piecewise linear frequency characteristic as marked with numbers 1 through 5 on Fig. 1.78. Segments 1, 3, and 5 are computed using the `sin()` function. Segments 2 and 4 are computed using the `chirpsig()` function.

```
1  % Script: mexct_1_11a.m
2  C4 = 261.63;
3  D4 = 293.66;
4  E4 = 329.63;
5  t = linspace(0,0.7,7000);  % Time vector for each note
6  tt = linspace(0,0.3,3000); % Time vector for transitions
7  x1 = sin(2*pi*C4*t);       % Segment 1
8  x2 = chirpsig(tt,C4,D4,0); % Segment 2
9  x3 = sin(2*pi*D4*t);       % Segment 3
10 x4 = chirpsig(tt,D4,E4,0); % Segment 4
11 x5 = sin(2*pi*E4*t);       % Segment 5
12 x = 0.6*[x1,x2,x3,x4,x5];
13 sound(x,10000);
14
15 function x = chirpsig(t,f1,f2,alpha)
16   tau = t(end)-t(1);                  % Duration
17   B = (f2-f1)/tau;                    % Eqn. (1.91)
18   A = f1-B*t(1);
19   Theta = 2*pi*A*t+pi*B*t.^2+alpha;   % Eqn. (1.92)
20   x = sin(Theta);                     % Eqn. (1.93)
21 end
```

Listening to the resulting sound may be a bit disappointing for two reasons:

a. There are audible clicks between consecutive segments of the sound, that is, as we cross from one segment to the next.

b. There is an audible click at the very beginning of the waveform played back, and there is one at the very end.

The reason for both types of clicks is the same as what we have encountered in MATLAB Exercise 1.10, namely discontinuities in the signal. In this exercise, we will address the clicks in the first group, that is, the clicks between consecutive segments of the signal. Methods for addressing the clicks at the very beginning and the very end will be left to problem 1.48 at the end of this chapter.

In MATLAB Exercise 1.10 we solved the problems caused by discontinuities through the use of a modulating envelope. That solution would not be useful here since it would be against the physical nature of the theremin instrument. Instead we need to ensure that, for a smooth transition, the beginning phase value of each segment matches the ending phase value of the segment preceeding it. This highlights the usefulness of the parameter α we introduced in Eqn. (1.92).

Consider the modified script `mexct_1_11b.m` listed below. The function `chirpsig()` has been modified so that it now returns the ending phase of the segment as its second output argument. The built-in function `wrapToPi()` assures that the phase angle returned is within the range of $[-\pi, \pi]$ radians. We have also changed the code to use the function `chirpsig ()` for all five segments. The fixed-frequency segments 1, 3, and 5 are produced by using the same value for both arguments `f1` and `f2` in the function `chirpsig()`. This allows the ending phase to be computed and returned for each segment. Notice how, in lines 7 through 11, the `alpha` value returned by the function `chirpsig()` as the ending phase of each segment is used as the initial phase for the next segment.

```
1  % Script: mexct_1_11b.m
2  C4 = 261.63;
3  D4 = 293.66;
4  E4 = 329.63;
5  t = linspace(0,0.7,7000);    % Time vector for each note
6  tt = linspace(0,0.3,3000);   % Time vector for transitions
7  [x1,alpha] = chirpsig(t,C4,C4,0);        % Segment 1
8  [x2,alpha] = chirpsig(tt,C4,D4,alpha);   % Segment 2
9  [x3,alpha] = chirpsig(t,D4,D4,alpha);    % Segment 3
10 [x4,alpha] = chirpsig(tt,D4,E4,alpha);   % Segment 4
11 [x5,alpha] = chirpsig(t,E4,E4,alpha);    % Segment 5
12 x = 0.6*[x1,x2,x3,x4,x5];
13 sound(x,10000);
14
15 function [x,endphase] = chirpsig(t,f1,f2,alpha)
16   tau = t(end)-t(1);                     % Duration
17   B = (f2-f1)/tau;                       % Eqn. (1.91)
18   A = f1-B*t(1);
19   Theta = 2*pi*A*t+pi*B*t.^2+alpha;      % Eqn. (1.92)
20   x = sin(Theta);                        % Eqn. (1.93)
21   endphase = wrapToPi(Theta(end));       % Phase at the end
22 end
```

The sound that results from the script `mexct_1_11b.m` is significantly better compared to before; however, there are still subtle clicks between successive segments. The reason for this is the fact that the end data point of one segment and the beginning data point of the next segment have the same phase value, creating a narrow plateau that breaks the continuity of the phase. An easy solution is to discard the last data point of each segment. The script `mexct_1_11c.m` listed below is almost identical to the one before it, with the only change being the addition of line 22 in which the last data point of vector `x` is discarded.

```
1  % Script: mexct_1_11c.m
2  C4 = 261.63;
3  D4 = 293.66;
4  E4 = 329.63;
5  t = linspace(0,0.7,7000);    % Time vector for each note
6  tt = linspace(0,0.3,3000);   % Time vector for transitions
7  [x1,alpha] = chirpsig(t,C4,C4,0);         % Segment 1
8  [x2,alpha] = chirpsig(tt,C4,D4,alpha);    % Segment 2
9  [x3,alpha] = chirpsig(t,D4,D4,alpha);     % Segment 3
10 [x4,alpha] = chirpsig(tt,D4,E4,alpha);    % Segment 4
11 [x5,alpha] = chirpsig(t,E4,E4,alpha);     % Segment 5
12 x = 0.6*[x1,x2,x3,x4,x5];
13 sound(x,10000);
14
15 function [x,endphase] = chirpsig(t,f1,f2,alpha)
16   tau = t(end)-t(1);                  % Duration
17   B = (f2-f1)/tau;                    % Eqn. (1.91)
18   A = f1-B*t(1);
19   Theta = 2*pi*A*t+pi*B*t.^2+alpha;   % Eqn. (1.92)
20   x = sin(Theta);                     % Eqn. (1.93)
21   endphase = wrapToPi(Theta(end));    % Phase at the end
22   x(end) = [];                        % Discard the last data point
23 end
```

Software resources: mexct_1_11a.m , mexct_1_11b.m , mexct_1_11c.m

Problems

1.1. Sketch and label each of the signals defined below:

a. $$x_a(t) = \begin{cases} 0, & t<0 \text{ or } t>4 \\ 2, & 0<t<1 \\ 1, & 1<t<2 \\ t-1, & 2<t<3 \\ 2, & 3<t<4 \end{cases}$$

b. $$x_b(t) = \begin{cases} 0, & t<0 \text{ or } t>5 \\ t, & 0<t<2 \\ -t+4, & 2<t<3 \\ -2t+7, & 3<t<4 \\ t-5, & 4<t<5 \end{cases}$$

1.2. Consider the signals shown in Fig. P.1.2. For each signal write the analytical description in segmented form similar to the descriptions of the signals in Problem 1.1.

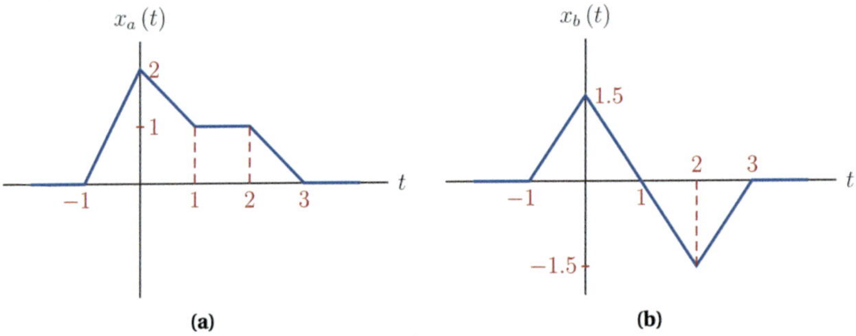

Figure P. 1.2

1.3. Using the two signals $x_a(t)$ and $x_b(t)$ given in Fig. P.1.2, compute and sketch the signals specified below:

a. $g_1(t) = x_a(t) + x_b(t)$

b. $g_2(t) = x_a(t)\, x_b(t)$

c. $g_3(t) = 2x_a(t) - x_b(t) + 3$

1.4. Consider the signal

$$x(t) = \left(e^{-t} - e^{-2t}\right) u(t)$$

Determine and sketch the following signals derived from $x(t)$ through signal operations:

a. $g_1(t) = x(2t-1)$

b. $g_2(t) = x(-t+2)$

c. $g_3(t) = x(-3t+5)$

d. $g_4(t) = x\left(\dfrac{t-1}{3}\right)$

1.5. Using the signal $x(t)$ shown in Fig. P.1.5 as the basis, compute and sketch each of the following signals:

a. $g_1(t) = x(-t)$

b. $g_2(t) = x(2t)$

c. $g_3(t) = x\left(\dfrac{t}{2}\right)$

d. $g_4(t) = x(-t+3)$

e. $g_5(t) = x\left(\dfrac{t-1}{3}\right)$

f. $g_6(t) = x(4t-3)$

g. $g_7(t) = x\left(1-\dfrac{t}{3}\right)$

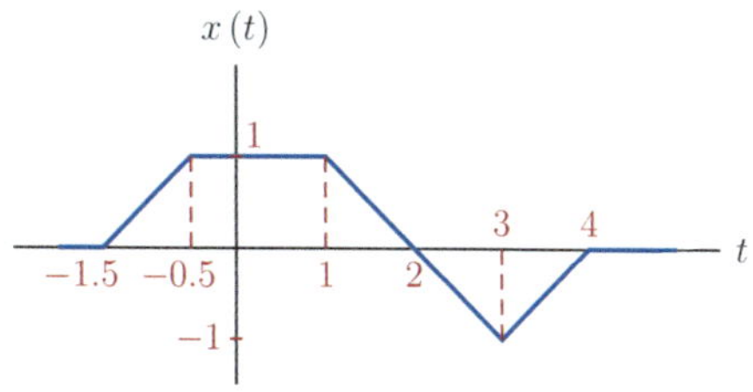

Figure P. 1.5

1.6. Consider two sinusoidal signals with frequencies f_1 and f_2:

$$x_1 = \sin(2\pi f_1 t)\ , \text{ and } x_2 = \sin(2\pi f_2 t)$$

a. Assume $f_1 \neq f_2$. Use the appropriate trigonometric identities to show that the product signal

$$x_3(t) = x_1(t)\, x_2(t)$$

contains sinusoids with frequencies $(f_1 - f_2)$ and $(f_1 + f_2)$.

b. Let $f_1 = f_2 = f$. Determine $x_3(t)$ for this case. Comment on the result.

1.7. Let b be a positive constant. Show that

$$\delta(bt) = \frac{1}{b}\,\delta(t)$$

Hint: Start with Eqn. (1.6) that expresses the impulse function as the limit of a pulse $q(t)$ with height $1/a$ and unit area. Apply time scaling to $q(t)$ and then take the limit as $a \to 0$.

1.8. Consider again Eqn. (1.6) that expresses the impulse function as the limit of a pulse $q(t)$ with height $1/a$ and area equal to unity. Show that, for small values of a, we have

$$\int_{-\infty}^{\infty} f(t)\, q(t-t_1)\, dt \approx f(t_1)$$

where $f(t)$ is any function that is continuous in the interval $t_1 - a/2 < t < t_1 + a/2$. Afterward, by taking the limit of this result, show that the sifting property of the impulse function holds.

1.9. Sketch each of the following functions.

a. $\delta(t) + \delta(t-1) + \delta(t-2)$

b. $\delta(t-1) + u(t)$

c. $e^{-t}\,\delta(t-1)$

d. $e^{-t}\,[u(t-1) - u(t-2)]$

e. $\displaystyle\sum_{n=0}^{\infty} e^{-t}\,\delta(t-0.1n)$

1.10. Sketch each of the following functions in the time interval $-1 \le t \le 5$. Afterward use the waveform explorer app `appWaveExplorerCT.m` to check your results.

a. $u(t)+u(t-1)-3u(t-2)+u(t-3)$

b. $r(t)-2r(t-2)+r(t-3)$

c. $u(t)+r(t-2)-u(t-3)-r(t-4)$

d. $2\Pi\left(\frac{t-1}{2}\right)-\Pi\left(\frac{t-2}{1.5}\right)+2\Pi(t-3)$

e. $\Lambda(t)+2\Lambda(t-1)+1.5\Lambda(t-3)-\Lambda(t-4)$

1.11. Express each of the signals shown in Fig. P.1.11 using scaled and time shifted unit step functions. Afterward use the waveform explorer app `appWaveExplorerCT.m` to check your results.

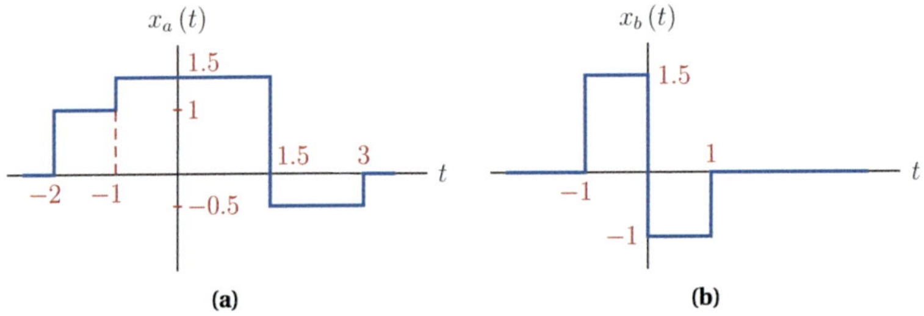

Figure P. 1.11

1.12. Consider again the signals in Fig. P.1.11. Express each signal using scaled and time-shifted unit pulse functions. Afterward use the waveform explorer app `appWaveExplorerCT.m` to check your results.

1.13. Express the signal $x(t)$ shown in Fig. P.1.13 using

a. Unit ramp functions

b. Unit triangle functions

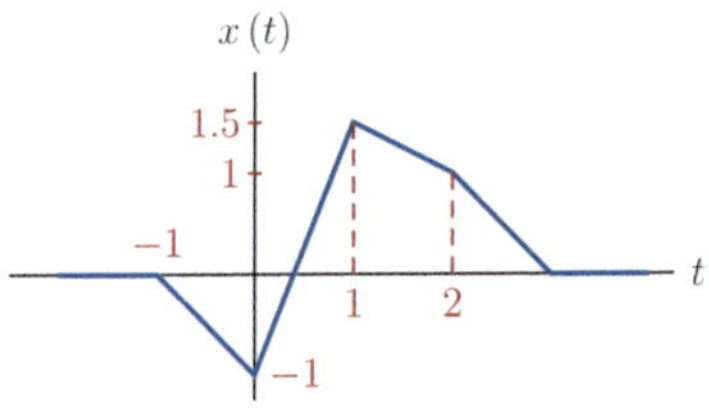

Figure P. 1.13

In each case, check your results using the waveform explorer app `appWaveExplorerCT.m`.

1.14. Suppose that in an experiment we collect data in the form of (t_i, x_i) pairs for $i = 1, \ldots, N$, and would like to construct a continuous-time signal $x(t)$ by linear interpolation, that is, by connecting the points with straight line segments as shown in Fig. P.1.14(a). The only restriction is that all t_i values are distinct; no two data points have the same time value. Let us define a *skewed* *triangle function* in the form

$$\Lambda_s(t,a,b)=\begin{cases}0\,, & t<-a \text{ or } t>b\\ \dfrac{t}{a}+1\,, & -a<t<0\\ -\dfrac{t}{b}+1\,, & 0<t<b\\ 0\,, & t>b\end{cases}$$

which is shown in Fig. 1.14(b). Devise a method of expressing the linearly interpolated signal $x(t)$ in terms of the skewed triangle functions.

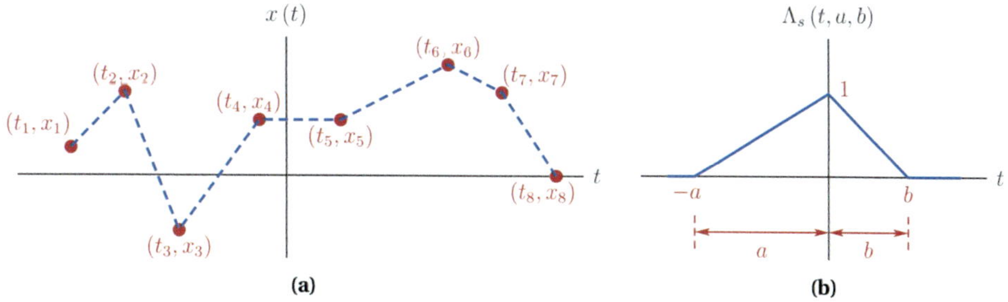

Figure P. 1.14

1.15. Time derivative of the unit impulse function $\delta(t)$ is called a *doublet*. Given a function $f(t)$ that is continuous at $t=0$ and also has continuous first derivative at $t=0$, show that

$$\int_{-\infty}^{\infty} f(t)\,\delta'(t)\,dt=-f'(0)=-\left.\frac{df(t)}{dt}\right|_{t=0}$$

Hint. Use integration by parts, and then apply the sifting property of the impulse function.

1.16. Signum function is defined as

$$\operatorname{sgn}(t)=\begin{cases}1\,, & t>0\\ -1\,, & t<0\end{cases}$$

Express the function $\operatorname{sgn}(t)$ in terms of the unit step function.

1.17. Using Euler's formula, prove the following identities:

a. $\cos(a)=\dfrac{1}{2}e^{ja}+\dfrac{1}{2}e^{-ja}$

b. $\sin(a)=\dfrac{1}{2j}e^{ja}-\dfrac{1}{2j}e^{-ja}$

c. $\dfrac{d}{da}[\cos(a)]=-\sin(a)$

d. $\cos(a+b)=\cos(a)\cos(b)-\sin(a)\sin(b)$

e. $\sin(a+b)=\sin(a)\cos(b)+\cos(a)\sin(b)$

f. $\cos^2(a)=\dfrac{1}{2}+\dfrac{1}{2}\cos(2a)$

1.18. Using the definition of periodicity, determine if each signal below is periodic or not. If the signal is periodic, determine the fundamental period and the fundamental frequency.

a. $x(t) = 3\cos(2t + \pi/10)$

b. $x(t) = 2\sin\left(\sqrt{20}t\right)$

c. $x(t) = 3\cos(2t + \pi/10)\, u(t)$

d. $x(t) = \cos^2(3t - \pi/3)$

e. $x(t) = e^{-|t|}\cos(2t)$

f. $x(t) = \sum_{k=-\infty}^{\infty} e^{-(t-kT_s)}\, u(t - kT_s)$

g. $x(t) = e^{j(2t+\pi/10)}$

h. $x(t) = e^{(-1+j2)t}$

1.19. Determine the periodicity of each of the multi-tone signals given below. For those that are periodic, determine the fundamental period and the fundamental frequency.

a. $x(t) = 2\cos(5t + \pi/10) + 3\sin(5\pi t)$

b. $x(t) = \cos(10\pi t) - 3\sin(30\pi t + \pi/4)$

c. $x(t) = \sin\left(\sqrt{2}t\right) + \sin(2t)$

d. $x(t) = \cos(45\pi t) + \cos(55\pi t)$

1.20. Consider the exponentially damped sinusoidal signal

$$x(t) = e^{-2t}\sin(2\pi t)\, u(t)$$

Develop a hand sketch of this signal using the following three steps. For each step, label the horizontal and vertical axes and mark a few critical values on each axis.

a. Sketch the signal $x_1(t) = e^{-2t}\, u(t)$.
b. Sketch the signal $x_2(t) = \sin(2\pi t)\, u(t)$.
c. Sketch the signal $x(t) = x_1(t)\, x_2(t)$

1.21. A signal $g(t)$ is defined in terms of another signal $x(t)$ as

$$g(t) = A\,x(t)\,, \quad A \neq 0$$

a. If $x(t)$ is an energy signal with normalized energy equal to E_x, show that $g(t)$ is also an energy signal. Determine the normalized energy of $g(t)$ in terms of E_x and A.
b. If $x(t)$ is a power signal with normalized average power equal to P_x, show that $g(t)$ is also a power signal. Determine the normalized average power of $g(t)$ in terms of P_x and A.

1.22. A signal $g(t)$ is defined in terms of another signal $x(t)$ as

$$g(t) = A\,x(t) + B\,, \quad A, B \neq 0$$

a. If $x(t)$ is an energy signal with normalized energy equal to E_x, is $g(t)$ is an energy signal or a power signal?
b. Determine either the normalized energy or the normalized average power of $g(t)$ as appropriate, in terms of E_x, A and B.

1.23. Determine the normalized energy of each of the signals given below.

a. $x(t) = e^{-2|t|}$

b. $x(t) = e^{-2t}\, u(t)$

c. $x(t) = e^{-2t} \cos(5t)\, u(t)$

Hint: Express the cosine term in $x(t)$ using Euler's formula prior to squaring and integrating.

1.24. Determine the normalized energy of each of the signals shown in Fig. P.1.2.

1.25. Determine the normalized average power of each of the periodic signals shown in Fig. P.1.25

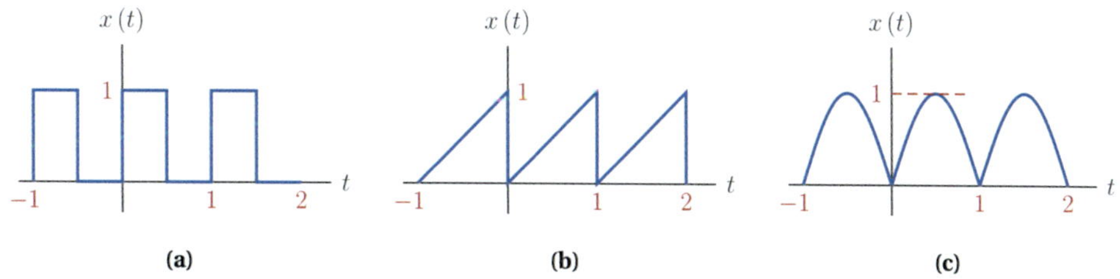

Figure P. 1.25

1.26. In Example 1.13 the RMS value of a multi-tone signal was derived for the case of distinct frequencies. Now consider the signal

$$x(t) = 2\cos(2\pi f_1 t) + 3\sin(2\pi f_1 t) - 6\cos(2\pi f_2 t)$$

where the first two terms have the same frequency. Determine the RMS value of the signal.

1.27. Identify which of the signals in Fig. P.1.27 are even, which ones are odd, and which signals are neither even nor odd.

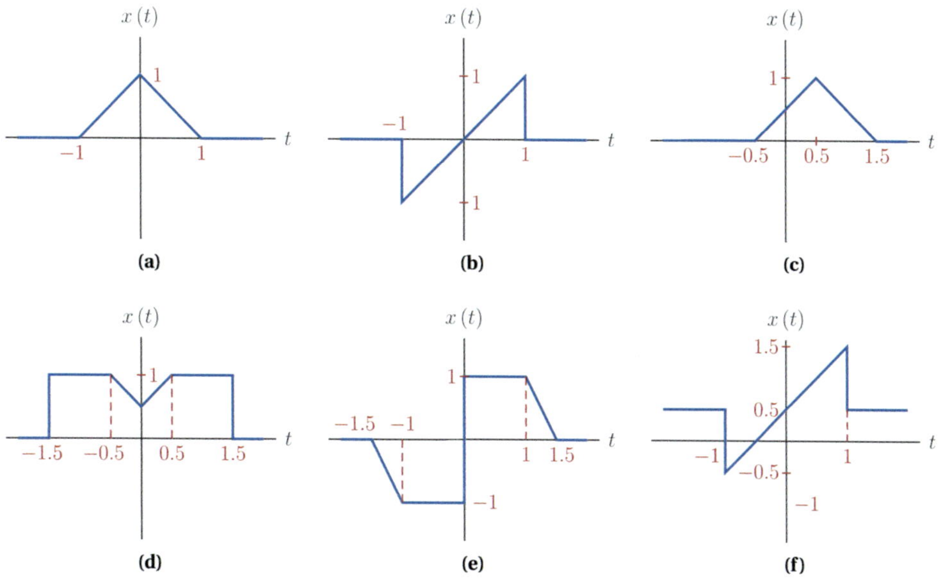

Figure P. 1.27

1.28. Consider a signal $x(t)$. Show that

a. If $x(t)$ is even, then

$$\int_{-\lambda}^{\lambda} x(t)\,dt = 2\int_{0}^{\lambda} x(t)\,dt$$

b. If $x(t)$ is odd, then

$$\int_{-\lambda}^{\lambda} x(t)\,dt = 0$$

1.29. Let $x(t) = x_1(t)\,x_2(t)$. Show that

a. If both $x_1(t)$ and $x_2(t)$ are even, then $x(t)$ is even.
b. If both $x_1(t)$ and $x_2(t)$ are odd, then $x(t)$ is even.
c. If $x_1(t)$ is even and $x_2(t)$ is odd, then $x(t)$ is odd.

1.30. Determine the conjugate symmetric and conjugate antisymmetric components of the signal $x(t) = e^{(-2+j10\pi)t}$. Sketch each component in polar form, that is, in terms of its magnitude and phase.

1.31. For each of the signals listed below, find the even and odd components $x_e(t)$ and $x_o(t)$. In each case, sketch the original signal and its two components.

a. $x(t) = e^{-5t}\sin(t)\,u(t)$

b. $x(t) = e^{-3|t|}\cos(t)$

c. $x(t) = e^{-3|t|}\sin(t)$

d. $x(t) = \left(t\,e^{-3t} + 2\right)u(t)$

e. $x(t) = e^{-2|t-1|}$

1.32. Determine and sketch even and odd components of the signal $x(t)$ used in Problem 1.13 and shown in Fig. P.1.13.

1.33. Transform each of the sinusoidal signals given below into phasor form.

a. $x(t) = 3\cos(200\pi t)$

b. $x(t) = 7\sin(100\pi t)$

c. $x(t) = 2\sin(10\pi t) - 5\cos(10\pi t)$

1.34. Using an operating frequency of $f_0 = 10$ Hz, express the phasors given below as time domain signals.

a. $\boldsymbol{X} = 5\,e^{j14^\circ}$

b. $\boldsymbol{X} = 2\,e^{j28^\circ} + 3\,e^{j18^\circ}$

c. $\boldsymbol{X} = 2\,e^{j28^\circ} - 3\,e^{j18^\circ}$

MATLAB Problems

1.35. Write a MATLAB script to compute and graph the signals $x_a(t)$ and $x_b(t)$ described in Problem 1.1. Use an anonymous function for each signal based on the analytical description given (see MATLAB Exercise 1.1). Graph each signal in the interval $-1 \le t \le 6$ s with a time increment of $\Delta t = 0.01$ s. Label the axes appropriately.

1.36.

a. Consider the signals $x_a(t)$ and $x_b(t)$ described in Problem 1.2. Using the `interp1()` function of MATLAB, compute each signal in the time interval $-2 \le t \le 4$ s with a time increment of $\Delta t = 0.01$ s (see MATLAB Exercise 1.2).

b. Afterward, compute the signals $g_1(t)$, $g_2(t)$, and $g_3(t)$ from the signals $x_a(t)$ and $x_b(t)$ as specified in Problem 1.3. Graph each signal in the same time interval as in part (a). Carefully label the axes of each graph.

1.37. Refer to the signals $g_1(t)$ and $g_2(t)$ found in Example 1.2 and shown in Fig. 1.12. Develop a MATLAB script to verify the analytical forms of these signals. Your script should

a. compute samples of each signal in the time interval $-1 \le t \le 4$ s using a time increment of $\Delta t = 0.01$ s.

b. graph the results using the functions `tiledlayout()` and `nexttile()`.

Ensure that the computations in part (a) correspond directly to the results found in Example 1.2.

1.38. Refer to the signal $x(t)$ used in Problem 1.5 and shown in Fig. P.1.5.

a. Express the signal $x(t)$ through the use of a MATLAB anonymous function that utilizes the function `interp1()`.

b. Using MATLAB, compute and graph each of the signals for parts (a) through (g) of Problem 1.5 in the time interval $-10 < t < 10$ s, using a time increment of $\Delta t = 0.02$ s. Compare MATLAB graphs to those obtained manually in Problem 1.4.

1.39. Refer to the signal $x(t)$ used in Problem 1.4.

a. Express the signal $x(t)$ through the use of a MATLAB anonymous function.

b. Using MATLAB, compute and graph each of the signals for parts (a) through (d) of Problem 1.4 in the time interval $-10 < t < 10$ s, using a time increment of $\Delta t = 0.02$ s. Compare MATLAB graphs to those obtained manually in Problem 1.4.

1.40. A *full-wave rectified* sinusoidal signal is defined as

$$x(t) = \left|\sin(2\pi f_0 t)\right|$$

a. Define an anonymous function to compute a full-wave rectified sinusoidal signal with a specified frequency f_0 at a set of time instants listed in a vector `t`.

b. Using the anonymous function defined in part (a), compute and graph the signal $x(t)$ with $f_0 = 10$ Hz for the time interval $0 \le t \le 1$ s using a time increment of $\Delta t = 0.001$ s.

1.41. A *half-wave rectified* sinusoidal signal is defined as

$$x(t) = \begin{cases} \sin(2\pi f_0 t), & \sin(2\pi f_0 t) \ge 0 \\ 0, & \sin(2\pi f_0 t) < 0 \end{cases}$$

An easy method of computing $x(t)$ would be by making use of a full-wave rectified sinusoidal signal as

$$x(t) = 0.5\sin(2\pi f_0 t) + 0.5\left|\sin(2\pi f_0 t)\right|$$

a. Define an anonymous function to compute a half-wave rectified sinusoidal signal with a specified frequency f_0 at a set of time instants listed in a vector `t`.

b. Using the anonymous function defined in part (a), compute and graph the signal $x(t)$ with $f_0 = 10$ Hz for the time interval $0 \le t \le 1$ s using a time increment of $\Delta t = 0.001$ s.

1.42. Consider the half-wave rectified sinusoidal signal we used in Problem 1.41. The signal can also be generated by starting with a full periodic sinusoid, and zeroing out its negative amplitudes through the use of logic operations similar to what was done in MATLAB Exercise 1.1 parts (c) and (d). Repeat the requirements of problem 1.41 using this method.

1.43. Using MATLAB functions `ss_step()`, `ss_ramp()`, `ss_pulse()` and `ss_tri()` developed in MATLAB Exercise 1.6, graph each of the signals specified in Problem 1.10 parts (a) through (e) in the time interval $-2 < t < 8$ s, using a time increment of $\Delta t = 0.01$ s. Compare MATLAB graphs to those obtained manually in Problem 1.10.

1.44. Consider the signal $x(t)$ used in Problem 1.13 and graphed in Fig. P.1.13.

a. Express this signal through an anonymous MATLAB function that utilizes the function `ss_ramp()`, and graph the result in an appropriate time interval.
b. Repeat part (a) using the function `ss_tri()` instead of the function `ss_ramp()`.
c. In MATLAB, compute and graph even and odd components of the signal $x(t)$. Code Eqns. (1.70) and (1.71) directly in MATLAB for this purpose. Compare your MATLAB graphs to those obtained manually in Problem 1.32.

1.45. Write a MATLAB script to perform the steps in Problem 1.20. Specifically, compute and graph each of the three signals specified in the time interval $-1 < t < 4$ s using a time increment of $\Delta t = 0.01$ s.

1.46. In the last part of MATLAB Exercise 1.9 we have generated a pulse width modulated signal from a sinusoidal reference. Start with the script file "`mexct_1_9b.m`" and modify it so that an approximation to the sinusoidal reference signal can be extracted from the signal $v_{\text{PWM}}(t)$ by time averaging each of its periods. Graph the original reference sinusoidal signal and the recovered approximation together on the same coordinate system.

1.47. In Problem 1.6 the issue of multiplying two sinusoidal signals with frequencies f_1 and f_2 was addressed. Write a MATLAB script to perform the following steps:

a. Generate samples of the two signals

$$x_1 = \sin\left(2\pi f_1 t\right), \text{ and } x_2 = \sin\left(2\pi f_2 t\right)$$

in the time interval $0 \le t \le 0.5$ s using 3000 time steps. Use the frequencies $f_1 = 100$ Hz and $f_2 = 104$ Hz.

b. Obtain the product

$$x_3(t) = x_1(t)\, x_2(t)$$

and graph it. Are the sum and difference frequencies clearly identifiable in the graph?

1.48. Recall that, in MATLAB Exercise 1.11, we were able to solve the problem of audible clicks between consecutive segments of the sound waveform, but the clicks occurring at the very beginning and the very end of the sound were not resolved. Those clicks are due to abrupt starting and ending of the waveform. One way to eliminate them is to use a *fade-in* envelope at the beginning, and a *fade-out* envelope at the end. Modify the MATLAB script "`mexct_1_11c.m`" developed in MATLAB Exercise 1.11 to incorporate fade-in and fade-out features into it using the envelope functions $q_{in}(t)$ and $q_{out}(t)$ shown in Fig. P.1.48. Use the function `interp1()` to create the envelope signals. Apply fade-in to `x1`, and apply fade-out to `x5`. Listen to the output sound and make sure that it is free from clicks.

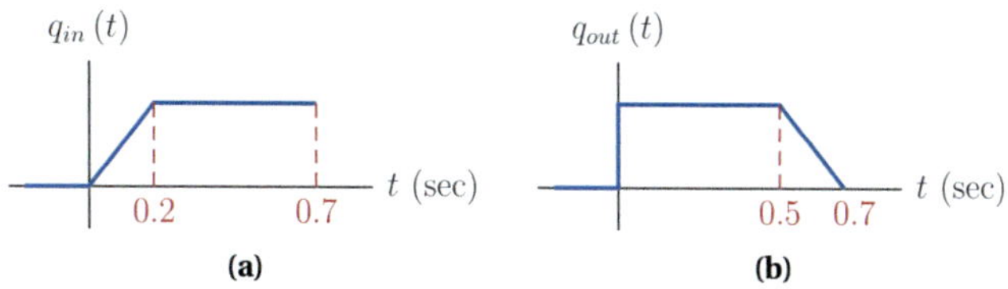

Figure P. 1.48

MATLAB Projects

1.49. In Problem 1.14 a method was developed for linearly interpolating between data points (t_i, x_i) for $i = 1, \ldots, N$ using the *skewed triangle function* shown in Fig. P.1.14b.

a. Write a MATLAB function `ss_trs()` to compute values of the skewed triangle function at time instants specified in vector `t`. The function should have the syntax

```
x = ss_trs(t,a,b)
```

where `t` is a row vector. Arguments `a` and `b` are both positive scalars. Test your function to compute and graph the functions $\Lambda_s(t,3,2)$, $\Lambda_s(t,1,4)$, and $\Lambda_s(t,2.7,1.3)$ in the time interval $-5 \le t \le 5$ s using a time increment of $\Delta t = 0.01$ s.

b. Next, write a MATLAB function `ss_interp()` to interpolate between a set of data points. The function should have the syntax

```
x = ss_interp(tvec,xvec,t)
```

The arguments `tvec` and `xvec` are row vectors of equal length. The vector `tvec` holds $t_1, \ldots, t_N$, and the vector `xvec` holds $x_1, \ldots, x_N$. The returned vector `x` holds values of the interpolated signal at time instants specified in the vector `t`.

c. Write a MATLAB script to test the function `ss_interp()` with the set of data points

$$(t_i, x_i) = (-4, 0.5), (-3, 1.5), (-2, -1), (-0.5, 1), (1, 1), (3, 2), (4, 1.5), (5, 0)$$

Graph the interpolated signal in the time interval $-4 \le t \le 5$ s using a time increment of $\Delta t = 0.01$ s.

1.50. The concept of instantaneous frequency was introduced in MATLAB Exercise 1.11 in the context of simulating the sounds of the Theremin instrument. In the script `mexct_1_11c.m`, a local function `chirpsig()` was utilized for generating sinusoidal signals with linearly varying instantaneous frequency.

a. Using the local function `chirpsig()` as the starting point, develop a MATLAB function `ss_chirp()` to generate a chirp signal with a specified set of parameters. Your function should have the syntax

```
[x,endphase] = ss_chirp(t,f1,f2,alpha)
```

where `f1` and `f2` are the initial and final instantaneous frequency values for the time interval represented by the vector `t`. The parameter `alpha` is the initial phase over which the instantaneous frequency linearly changes from f_1 to f_2. The computed signal and the final phase angle are returned in variables `x` and `endphase`, respectively.

b. Write a MATLAB script that uses the function `ss_chirp()` to produce a siren sound with the instantaneous frequency varying from $f_1 = 200$ Hz to $f_2 = 800$ Hz over a period of 2 seconds, and then coming back down to 200 Hz over the next 2 seconds. Hint: Use the function `ss_chirp()` twice; the first time for increasing the instantaneous frequency, and the second time for decreasing it. Make sure the phase angles are matched at the transition point.

1.51. In this project the concept of *dual-tone multi-frequency (DTMF)* signaling will be explored. As the name implies, DTMF signals are mixtures of two sinusoids at distinct frequencies. They are used in communications over analog telephone lines. A particular version of DTMF signaling is utilized in dialing a number with push-button telephone handsets, a scheme known as *touch tone dialing*. When the caller dials a number, the DTMF generator produces a dual-tone signal for each digit dialed. The synthesized signal is in the form

$$x_k(t) = \sin\left(2\pi f_1 t\right) + \sin\left(2\pi f_2 t\right), \qquad 0 \le t \le T_d$$

Frequency assignments for the digits on a telephone keypad are shown in Fig. P.1.50.

f_1 \ f_2	1209 Hz	1336 Hz	1477 Hz
697 Hz	1	2	3
770 Hz	4	5	6
852 Hz	7	8	9
941 Hz	*	0	#

Figure P. 1.50

The goal of this project is to develop a DTMF synthesizer function for MATLAB.

a. Develop a function named `ss_dtmf1()` to produce the signal for one digit. The syntax of the function should be

```
x = ss_dtmf1(n,t)
```

The first argument `n` is the digit for which the DTMF signal is to be generated. Let values $n = 0$ through $n = 9$ represent the corresponding keys on the keypad. Map the remaining two keys "$*$" and "#" to values $n = 10$ and $n = 11$, respectively. Finally, the value $n = 12$ should represent a pause, that is, a silent period. The vector `t` contains the time instants at which the DTMF signal $x(t)$ is evaluated and returned in vector `x`.

b. Develop a function named `ss_dtmf()` with the syntax

```
x = ss_dtmf(number,dt,nd,np)
```

The arguments for the function `ss_dtmf()` are defined as follows:

`number`: The phone number to be dialed, entered as a vector. For example, to dial the number 555-1212, the vector `number` would be entered as

```
number = [5,5,5,1,2,1,2]
```

`dt`: The time increment Δt to be used in computing the amplitudes of the DTMF signal.

`nd`: Parameter to control the duration of the DTMF signal for each digit. The duration of each digit should be

$$T_d = n_d \, \Delta t$$

`np`: Parameter to control the duration of pause between consecutive digits. The duration of pause should be

$$T_p = n_p \, \Delta t$$

The function `ss_dtmf()` should use the function `ss_dtmf1()` to produce the signals $x_k(t)$ for each digit (and the pauses between digits) and append them together to create the signal $x(t)$.

c. Write a script to test the function `ss_dtmf()` with the number 555-1212. Use a time increment of 125 ţs corresponding to 8000 values per second. The duration of each digit should be 200 ms with 80 ms pauses between digits.

d. Play back the resulting signal $x(t)$ using the `sound()` function.

CHAPTER 2

ANALYZING CONTINUOUS-TIME SYSTEMS IN THE TIME DOMAIN

Chapter Objectives

- Develop the notion of a *continuous-time system.* Learn simplifying assumptions made in the analysis of systems. Discuss the concepts of *linearity* and *time invariance* and their significance.
- Explore the use of differential equations for representing continuous-time systems. Develop methods for solving differential equations to compute the output signal of a system in response to a specified input signal.
- Learn to represent a differential equation in the form of a block diagram that can be used as the basis for simulating a system.
- Discuss the significance of the *impulse response* as an alternative description form for linear and time-invariant systems. Learn how to compute the output signal for a linear and time-invariant system using *convolution.* Understand the graphical interpretation of the steps involved in carrying out the convolution operation.
- Learn the concepts of causality and stability as they relate to physically realizable and usable systems.

2.1 Introduction

In Chapter 1 we have focused on understanding signals. In this chapter we will begin our examination of the system concept. An overly simplified and rather broad definition of a system may be given as follows:

DOI: 10.1201/9781032675251-2

What is a *system*?

In general, a system is any physical entity that takes in a set of one or more physical signals and, in response, produces a new set of one or more physical signals.

Consider a microphone that senses the variations in air pressure created by the voice of a singer and produces a small electrical signal in the form of a time-varying voltage. The microphone acts as a system that facilitates the conversion of an acoustic signal to an electrical signal. Next, consider an amplifier that is connected to the output terminals of the microphone. It takes the small-amplitude electrical signal from the microphone, and produces a larger-scale replica of it suitable for use with a loudspeaker. Finally a loudspeaker that is connected to the output terminals of the amplifier converts the electrical signal to sound. The microphone, the amplifier, and the loudspeaker can be viewed as individual systems. Alternatively, we can look at the combination of all three components as one system that consists of three subsystems working together.

Another example is the transmission of images from a television studio to the television sets in our homes. The system that achieves the goal of bringing sound and images from the studio to our homes is a rather complex one that consists of a large number of subsystems that carry out the tasks of 1) converting sound and images to electrical signals at the studio, 2) transforming electrical signals into different formats for the purposes of enhancement, encoding and transmission, 3) transmitting electrical or electromagnetic signals over the air or over a cable connection, 4) receiving electrical or electromagnetic signals at the destination point, 5) processing electrical signals to convert them to formats suitable for a television set, and 6) displaying the images and playing the sound on the television set. Within such a large-scale system some of the physical signals are continuous-time signals while others are discrete-time or digital signals.

Not all systems of interest are electrical. An automobile is an example of a large-scale system that consists of numerous mechanical, electrical, and electromechanical subsystems working together. Other examples are economic systems, mass transit systems, ecological systems and computer networks. In each example, a large-scale system is constructed as a collection of interconnected smaller-scale subsystems that work together to achieve a prescribed goal.

In this textbook we will not attempt a complete treatment of any particular large-scale system. Instead, we will focus our efforts on developing techniques for understanding, analyzing, and designing subsystems with accurate and practical models based on laws of physics and mathematical transforms. In general, a system can be viewed as any physical entity that defines the cause-effect relationships between a set of signals known as *inputs* and another set of signals known as *outputs*. The input signals are *excitations* that drive the system and the output signals are the *responses* of the system to those excitations.

In Chapter 1 we have discussed basic methods of mathematically modeling continuous-time and discrete-time signals as functions of time. In this chapter and the next, we will extend the discussion to mathematical modeling of systems that act on signals.

What is a *mathematical model* for a system?

The mathematical model of a system is a function, formula or algorithm (or a set of functions, formulas, algorithms) to approximately recreate the same cause-effect relationship between the mathematical models of the input and the output signals.

If we focus our attention on single-input/single-output systems, the interplay between the system and its input and output signals can be graphically illustrated as shown in Fig. 2.1.

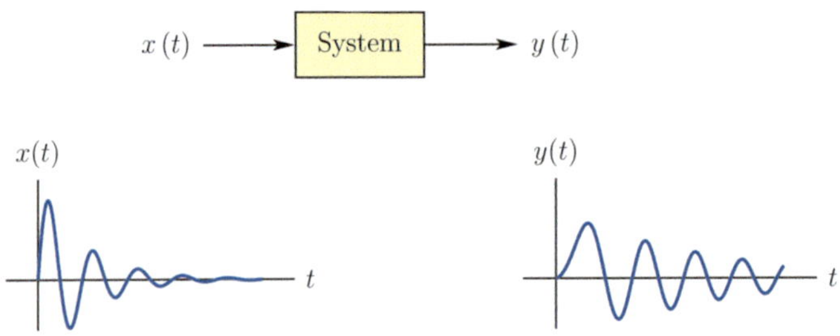

Figure 2.1 – Continuous-time signal-system interaction.

If the input and the output signals are modeled with mathematical functions $x(t)$ and $y(t)$, respectively, then the system itself needs to be modeled as a function, formula, or algorithm of some sort that converts the signal $x(t)$ into the signal $y(t)$. Sometimes we may have the need to design a system to achieve a particular desired effect in the conversion of the input signal $x(t)$ to the output signal $y(t)$. For example, an amplifier increases the power level of a signal that may be too weak to be heard or transmitted, while preserving its information content. A frequency-selective filter can be used for either removing or boosting certain frequencies in a signal. A speech recognition system can analyze the signal $x(t)$ in an attempt to recognize certain words, with the goal of performing certain tasks such as connecting the caller to the right person via an automated telephone menu system.

In contrast with systems that we design in order to achieve a desired outcome, sometimes the effect of a particular system on the input signal is not a desired one, but one that we are forced to accept, tolerate, or handle. For example, a wire connection is often used so that a signal that exists at a particular location could be duplicated at a distant remote location. In such a case, we expect the wire connection to be transparent within the scheme of transmitting the signal from point A to point B; however, that is usually not the case. The physical conductors used for the wire connection are far from ideal and cause changes in the signal as it travels from A to B. The characteristics of the connection itself represent a system that is more of a nuisance than anything else. We may need to analyze this system and find ways to compensate for the undesired effects of this system if we are to communicate successfully.

The relationship between the input and the output signals of a continuous-time system will be mathematically modeled as

$$y(t) = \text{Sys}\left\{x(t)\right\} \tag{2.1}$$

where the operator $\text{Sys}\{\ldots\}$ represents the transformation applied to $x(t)$. This transformation can be anything from a very simple one to a very complicated one. Consider, for example, a system that amplifies its input signal by a constant gain factor K to yield an output signal

$$y(t) = K\,x(t)$$

or one that delays its input signal by a constant time delay τ to produce

$$y(t) = x(t-\tau)$$

or a system that produces an output signal that is proportional to the square of the input signal as in

$$y(t) = K\left[x(t)\right]^2$$

In all three examples above, we have system definitions that could be expressed in the form of simple functions. This is not always the case. More interesting systems have more complicated definitions that cannot be reduced to a simple function of the input signal, but must rather be expressed by means of an algorithm, a number of functional relationships interconnected in some way, or a differential equation.

In deriving valid mathematical models for physical systems, we rely on established laws of physics that are applicable to individual components of a system. For example, consider a resistor in an electronic circuit. In mathematically modeling a resistor we use Ohm's law which states that the voltage between the terminals of a resistor is proportional to the current that flows through it. This leads to the well-known relationship $v = Ri$ between the resistor voltage and the resistor current. On the other hand, laboratory experiments indicate that the resistance of an actual carbon resistor varies as a function of temperature, a fact which is neglected in the mathematical model based on Ohm's law. Thus, Ohm's law provides a simplification of the physical relationship involved. Whether this simplification is acceptable or not depends on the specifics of the circuit in which the resistor is used. How significant are the deviations of the mathematical model of the resistor from the behavior of the actual resistor? How significant are the temperature changes that cause these deviations? How sensitive is the circuit to the variations in the value of R? Answers to these and similar questions are used for determining if the simple mathematical model is appropriate, or if a more sophisticated model should be used. Modeling of a system always involves some simplification of the physical relationships between input and output signals. This is necessary in order to obtain mathematical models that are practical for use in understanding system behavior. Care should be taken to avoid oversimplification and to ensure that the resulting model is a reasonably accurate approximation of reality.

Two commonly used simplifying assumptions for mathematical models of systems are *linearity* and *time invariance* which will be the subjects of Section 2.2. Section 2.3 focuses on the use of differential equations for representing continuous-time systems. Restrictions that must be placed on differential equations in order to model linear and time-invariant systems will be discussed in Section 2.4. Methods for solving linear constant-coefficient differential equations will be discussed in Section 2.5. Derivation of block diagrams for simulating continuous-time linear and time-invariant systems will be the subject of Section 2.6. In Section 2.7 we discuss the significance of the impulse response and its use in the context of the convolution operator for determining the output signal of a system. Concepts of causality and stability of systems are discussed in Sections 2.8 and 2.9, respectively.

2.2 Linearity and Time Invariance

In most of this textbook we will focus our attention on a particular class of systems referred to as *linear and time-invariant* systems. Linearity and time invariance will be two important properties which, when present in a system, will allow us to analyze the system using well-established techniques of the linear system theory. In contrast, the analysis of systems that are not linear and time-invariant tends to be more difficult and often relies on methods that are specific to the types of systems being analyzed.

2.2.1 Linearity in continuous-time systems

A system is said to be *linear* if the mathematical transformation $y(t) = \text{Sys}\{x(t)\}$ that governs the input-output relationship of the system satisfies the following two equations for any two input signals $x_1(t)$, $x_2(t)$ and any arbitrary constant gain factor α_1.

Conditions for linearity:

$$\mathrm{Sys}\left\{x_1(t)+x_2(t)\right\} = \mathrm{Sys}\left\{x_1(t)\right\} + \mathrm{Sys}\left\{x_2(t)\right\} \tag{2.2}$$

$$\mathrm{Sys}\left\{\alpha_1\, x_1(t)\right\} = \alpha_1\, \mathrm{Sys}\left\{x_1(t)\right\} \tag{2.3}$$

The condition in Eqn. (2.2) is the *additivity rule* which can be stated as follows: The response of a linear system to the sum of two signals is the same as the sum of individual responses to each of the two input signals. The condition in Eqn. (2.3) is the *homogeneity rule*. Verbally stated, scaling the input signal of a linear system by a constant gain factor causes the output signal to be scaled with the same gain factor. The two criteria given by Eqns. (2.2) and (2.3) can be combined into one equation which is referred to as the *superposition principle*.

Superposition principle:

$$\mathrm{Sys}\left\{\alpha_1\, x_1(t)+\alpha_2\, x_2(t)\right\} = \alpha_1\, \mathrm{Sys}\left\{x_1(t)\right\} + \alpha_2\, \mathrm{Sys}\left\{x_2(t)\right\} \tag{2.4}$$

A continuous-time system is linear if it satisfies the superposition principle stated in Eqn. (2.4) for any two arbitrary input signals $x_1(t)$, $x_2(t)$ and any two arbitrary constants α_1 and α_2. The superposition principle applied to two signals can be expressed verbally as follows: The response of the system to a weighted sum of two input signals is equal to the same weighted sum of the responses of the system to individual input signals. This concept is very important and is graphically illustrated in Fig. 2.2.

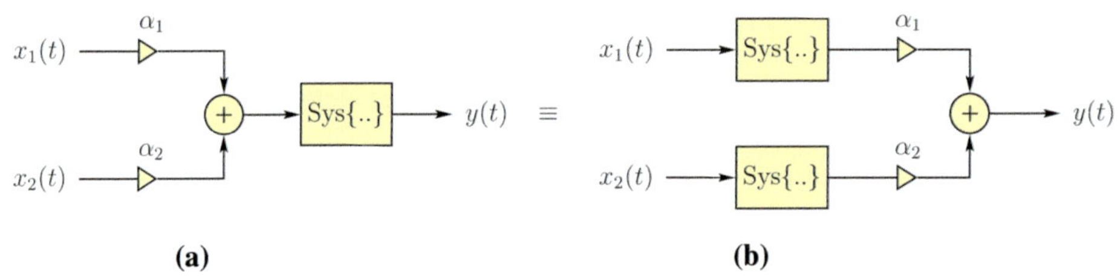

Figure 2.2 – Illustration of Eqn. (2.4). The two configurations shown are equivalent if the system under consideration is linear.

Furthermore, if superposition works for the weighted sum of any two input signals, it can be proven by induction that it also works for an arbitrary number of input signals. Mathematically we have

$$\mathrm{Sys}\left\{\sum_{i=1}^{N}\alpha_i\, x_i(t)\right\} = \sum_{i=1}^{N}\alpha_i\, \mathrm{Sys}\left\{x_i(t)\right\} \tag{2.5}$$

The generalized form of the superposition principle can be expressed verbally as follows: The response of a linear system to a weighted sum of N arbitrary signals is equal to the same weighted sum of the individual responses of the system to each of the N signals. Let $y_i(t)$ be the response to the input component $x_i(t)$ alone, that is $y_i(t) = \mathrm{Sys}\{x_i(t)\}$ for $i = 1,\ldots,N$. Superposition principle implies that

$$y(t) = \mathrm{Sys}\left\{\sum_{i=1}^{N}\alpha_i\, x_i(t)\right\} = \sum_{i=1}^{N}\alpha_i\, y_i(t) \tag{2.6}$$

This is graphically illustrated in Fig. 2.3.

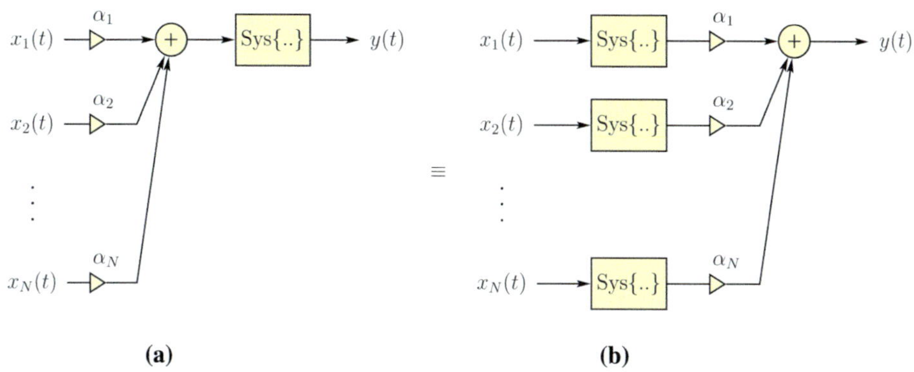

Figure 2.3 – Illustration of Eqn. (2.4). The two configurations shown are equivalent if the system under consideration is linear.

Testing a system for linearity involves checking whether the superposition principle holds true with two arbitrary input signals. Example 2.1 will demonstrate this procedure.

Example 2.1: Testing linearity of continuous-time systems

Four different systems are described below through their input-output relationships. For each, determine if the system is linear or not:

a. $y(t) = 5\,x(t)$

b. $y(t) = 5\,x(t) + 3$

c. $y(t) = 3\,[x(t)]^2$

d. $y(t) = \cos\left(x(t)\right)$

Solution:

a. If two input signals $x_1(t)$ and $x_2(t)$ are applied to the system individually, they produce the output signals $y_1(t) = 5\,x_1(t)$ and $y_2(t) = 5\,x_2(t)$, respectively. Let the input signal be $x(t) = \alpha_1\,x_1(t) + \alpha_2\,x_2(t)$. The corresponding output signal is found using the system definition:

$$y(t) = 5\,x(t) = \alpha_1\left[5\,x_1(t)\right] + \alpha_2\left[5\,x_2(t)\right] = \alpha_1\,y_1(t) + \alpha_2\,y_2(t)$$

Superposition principle holds; therefore this system is linear.

b. If two input signals $x_1(t)$ and $x_2(t)$ are applied to the system individually, they produce the output signals $y_1(t) = 5\,x_1(t) + 3$ and $y_2(t) = 5\,x_2(t) + 3$, respectively. We will again use the combined input signal $x(t) = \alpha_1\,x_1(t) + \alpha_2\,x_2(t)$ for testing. The corresponding output signal for this system is

$$y(t) = 5\,x(t) + 3 = 5\,\alpha_1\,x_1(t) + 5\,\alpha_2\,x_2(t) + 3$$

The output signal $y(t)$ cannot be expressed in the form $y(t) = \alpha_1\,y_1(t) + \alpha_2\,y_2(t)$. Superposition principle does not hold true in this case. The system is not linear.

c. Using two input signals $x_1(t)$ and $x_2(t)$ individually, the corresponding output signals produced by this system are $y_1(t) = 3\,[x_1(t)]^2$ and $y_2(t) = 3\,[x_2(t)]^2$, respectively. Applying the linear combination $x(t) = \alpha_1\,x_1(t) + \alpha_2\,x_2(t)$ to the system produces the output signal

$$\begin{aligned} y(t) =& 3\,\left[\alpha_1\,x_1(t) + \alpha_2\,x_2(t)\right]^2 \\ =& 3\,\alpha_1^2\,\left[x_1(t)\right]^2 + 6\,\alpha_1\alpha_2\,x_1(t)\,x_2(t) + 3\,\alpha_2^2\,\left[x_2(t)\right]^2 \end{aligned}$$

It is clear that the system is not linear.

d. The test signals $x_1(t)$ and $x_2(t)$ applied to the system individually produce the output signals $y_1(t) = \cos[x_1(t)]$ and $y_2(t) = \cos[x_2(t)]$, respectively. Their linear combination $x(t) = \alpha_1\,x_1(t) + \alpha_2\,x_2(t)$ produces the output signal

$$y(t) = \cos\left[\alpha_1\,x_1(t) + \alpha_2\,x_2(t)\right]$$

The system is not linear.

Software resource: See MATLAB Exercise 2.1.

2.2.2 Time invariance in continuous-time systems

Another important concept in the analysis of systems is time invariance. A system is said to be *time-invariant* if its behavior characteristics do not change in time. Consider a continuous-time system with the input-output relationship

$$\text{Sys}\left\{x(t)\right\} = y(t)$$

If the input signal applied to a time-invariant system is time-shifted by τ seconds, the only effect of this delay should be to cause an equal amount of time shift in the output signal, but to otherwise leave the shape of the output signal unchanged. If that is the case, we would expect the relationship

$$\text{Sys}\left\{x(t-\tau)\right\} = y(t-\tau)$$

to also be valid. This relationship is depicted in Fig. 2.4.

Condition for time invariance:

$$\text{Sys}\left\{x(t)\right\} = y(t) \quad \text{implies that} \quad \text{Sys}\left\{x(t-\tau)\right\} = y(t-\tau) \tag{2.7}$$

Alternatively, the relationship described by Eqn. (2.7) can be characterized by the equivalence of the two system configurations shown in Fig. 2.5. A system can be tested for time invariance by checking whether Eqn. (2.7) holds true for any arbitrary input signal. This procedure will be demonstrated in Example 2.2.

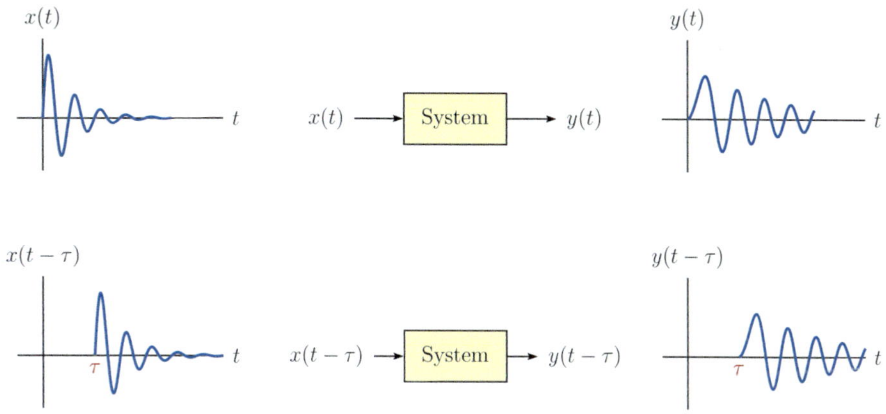

Figure 2.4 – Illustration of time invariance.

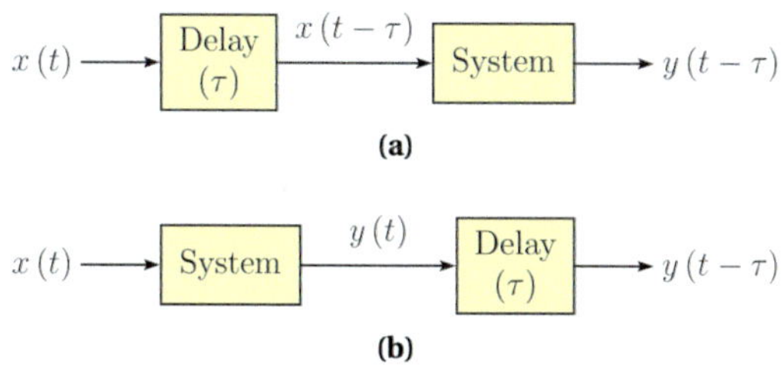

Figure 2.5 – Another interpretation of time invariance. The two configurations shown are equivalent for a time-invariant system.

Example 2.2: Testing time invariance of continuous-time systems

Three different systems are described below through their input-output relationships. For each, determine whether the system is time-invariant or not:

a. $y(t) = 5\,x(t)$

b. $y(t) = 3\cos\big(x(t)\big)$

c. $y(t) = 3\cos(t)\,x(t)$

Solution:

a. For this system, if the input signal $x(t)$ is delayed by τ seconds, the corresponding output signal would be

$$\text{Sys}\{x(t-\tau)\} = 5\,x(t-\tau) = y(t-\tau)$$

and therefore the system is time-invariant.

b. Let the input signal be $x(t-\tau)$. The output of the system is

$$\text{Sys}\{x(t-\tau)\} = 3\cos(x(t-\tau)) = y(t-\tau)$$

This system is time-invariant as well.

c. Again using the delayed input signal $x(t-\tau)$ we obtain the output

$$\text{Sys}\{x(t-\tau)\} = 3\cos(t)\, x(t-\tau) \neq y(t-\tau)$$

This system is not time-invariant since the time-shifted input signal leads to a response that is not the same as a similarly time-shifted version of the original output signal.

Before we leave this example we will use this last part of the problem as an opportunity to address a common source of confusion. The question may be raised as to whether the term t in the argument of the cosine function should be replaced with $t-\tau$ as well. In other words, should we have written

$$\text{Sys}\{x(t-\tau)\} \stackrel{?}{=} 3\cos(t-\tau)\, x(t-\tau)$$

which would have led to the conclusion that the system under consideration might be time-invariant? The answer is no. The term $\cos(t)$ is part of the system definition and not part of either the input or the output signal. Therefore we cannot include it in the process of time shifting input and output signals.

Software resource: See MATLAB Exercise 2.2.

Example 2.3: Using linearity property

A continuous-time system with input-output relationship $y(t) = \text{Sys}\{x(t)\}$ is known to be linear. Whether the system is time-invariant or not is not known. Assume that the responses of the system to four input signals $x_1(t)$, $x_2(t)$, $x_3(t)$, and $x_4(t)$ shown in Fig. 2.6 are known. Discuss how the information provided can be used for finding the response of this system to the signal $x(t)$ shown in Fig. 2.7.

Solution: Through a first glance at the four input signals given we realize that $x(t)$ is a time-shifted version of $x_1(t)$, that is, $x(t) = x_1(t-0.5)$. As a result, we may be tempted to conclude that $y(t) = y_1(t-0.5)$, however, this would be the wrong approach since the system is not necessarily time-invariant. We do not know for a fact that the input signal $x_1(t-0.5)$ produces the response $y_1(t-0.5)$.

For the same reason, we cannot base our solution on the idea of constructing the input signal from $x_2(t)$ and $x_3(t)$ as

$$x(t) = x_2(t) + x_3(t-1.5)$$

since it involves time shifting the signal $x_3(t)$, and we do not know the response of the system to $x_3(t-1.5)$. We need to find a way to construct the input signal from input signals given, using only scaling and addition operators, but without using any time shifting. A possible solution is to write $x(t)$ as

$$x(t) = 0.6\, x_2(t) + 0.8\, x_4(t)$$

Since the system is linear, the output signal is

$$y(t) = 0.6\, y_2(t) + 0.8\, y_4(t)$$

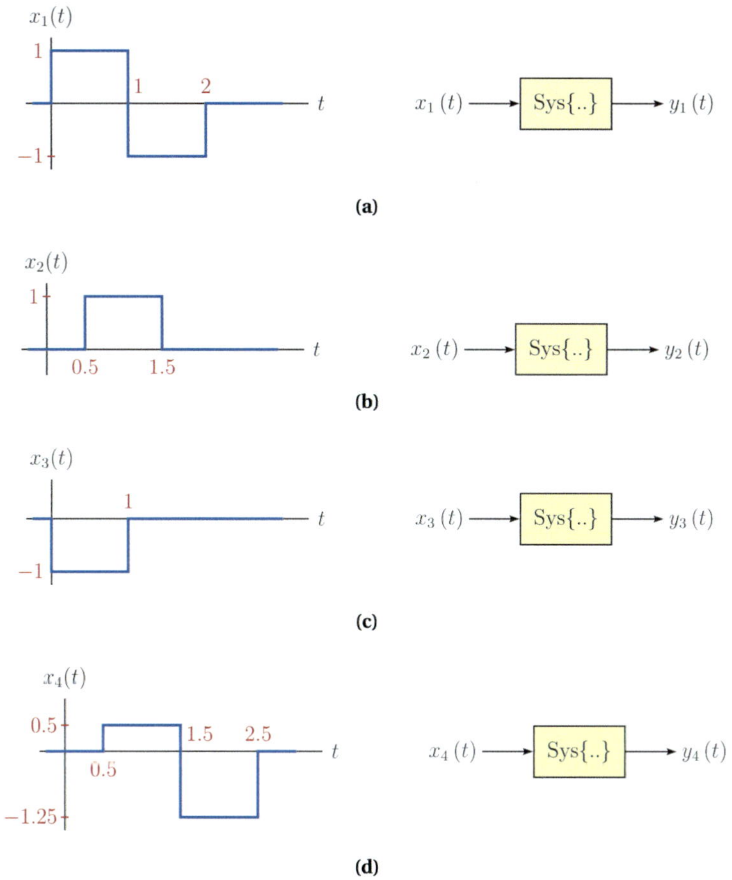

Figure 2.6 – Input-output pairs for Example 2.3.

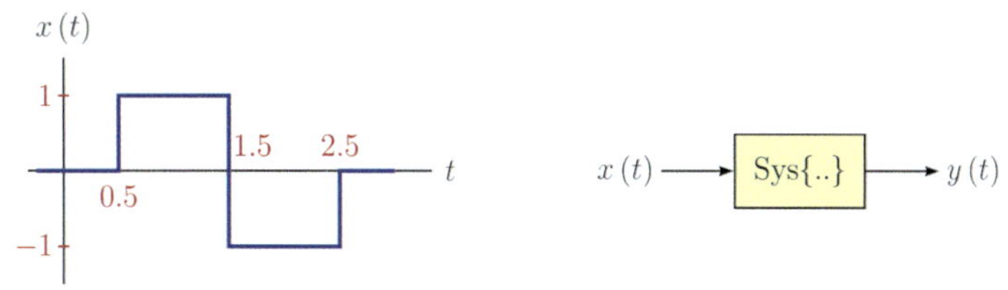

Figure 2.7 – Input signal for Example 2.3.

2.2.3 CTLTI systems

In the rest of this textbook, we will mostly work with continuous-time systems that are both linear and time-invariant. A number of time- and frequency-domain analysis and design techniques will be developed for such systems. For simplicity, we will use the acronym *CTLTI* to refer to *continuous-time linear and time-invariant* systems.

2.3 Differential Equations for Continuous-Time Systems

One method of representing the relationship established by a system between its input and output signals is a differential equation that approximately describes the interplay of the physical quantities within the system. Such a differential equation typically involves the input and the output signals as well as various derivatives of either or both. Following is an example:

$$\frac{d^2y}{dt^2}+3x(t)\,\frac{dy}{dt}+y(t)-2x(t)=0$$

If we need to model a mechanical system with a differential equation, physical relationships between quantities such as mass, force, torque, and acceleration may be used. In the case of an electrical circuit we use the physical relationships that exist between various voltages and currents in the circuit, as well as the properties of the circuit components that link those voltages and currents. For example, the voltage between the leads of an inductor is known to be approximately proportional to the time rate of change of the current flowing through the inductor. Using functions $v_L(t)$ and $i_L(t)$ as mathematical models for the inductor voltage and the inductor current respectively, the mathematical model for an *ideal inductor* is

$$v_L(t)=L\,\frac{di_L(t)}{dt}$$

Similarly, the current that flows through a capacitor is known to be proportional to the rate of change of the voltage between its terminals, leading to the mathematical model of an *ideal capacitor* as

$$i_C(t)=C\,\frac{dv_C(t)}{dt}$$

where $v_C(t)$ and $i_C(t)$ are the mathematical models for the capacitor voltage and the capacitor current, respectively. This is illustrated in Fig. 2.8.

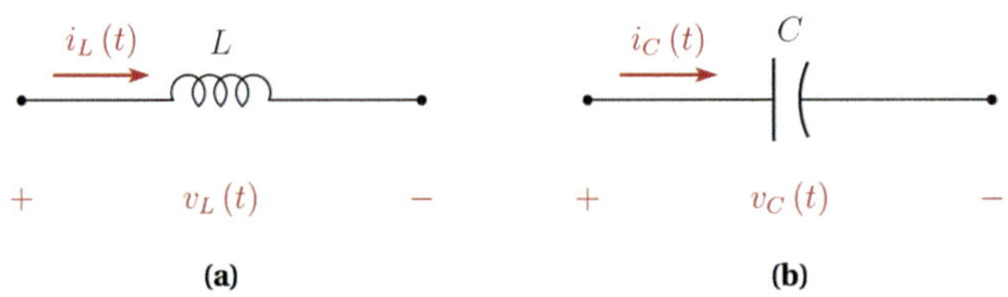

Figure 2.8 – Mathematical models for **(a)** ideal inductor and **(b)** ideal capacitor.

Ideal inductor and the ideal capacitor represent significant simplification of real-life versions of these devices with their physical voltage and current quantities. The following examples will demonstrate the process of obtaining a differential equation from the physical description of a system.

Example 2.4: Differential equation for simple RC circuit

Consider the simple first-order RC circuit shown in Fig. 2.9. The input signal $x(t)$ is the voltage applied to a series combination of a resistor and a capacitor and the output signal is the voltage $y(t)$ across the terminals of the capacitor.

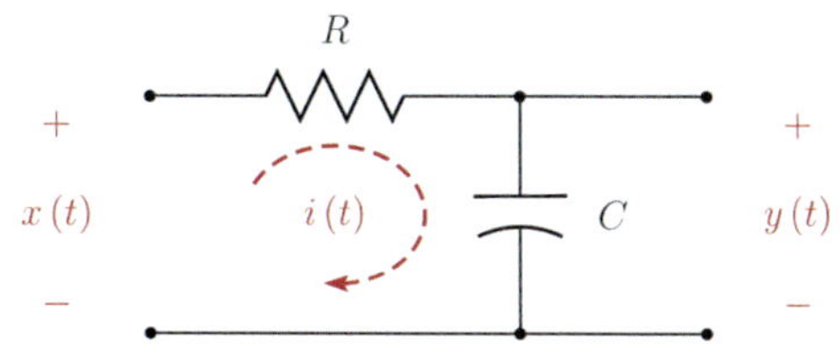

Figure 2.9 – RC circuit for Example 2.4.

Even though the RC circuit is a very simple example of a system, it will prove useful in discussing fundamental concepts of linear systems theory, and will be used as the basis of several examples that will follow. The techniques that we will develop using the simple RC circuit of Fig. 2.9 as a backdrop will be applicable to the solution of more complex problems involving a whole host of other linear systems.

The resistor and the capacitor carry the same current $i(t)$. Furthermore, the voltage drop across the terminals of the resistor is

$$v_R(t) = R\,i(t) \tag{2.8}$$

and the capacitor voltage $y(t)$ is governed by the equation

$$i(t) = C\,\frac{dy(t)}{dt} \tag{2.9}$$

Using Eqns. (2.8) and (2.9) the relationship between the input and the output signals is

$$RC\,\frac{dy(t)}{dt} + y(t) = x(t) \tag{2.10}$$

Multiplying both sides of Eqn. (2.10) by $1/RC$, we have

$$\frac{dy(t)}{dt} + \frac{1}{RC}\,y(t) = \frac{1}{RC}\,x(t) \tag{2.11}$$

Eqn. (2.11) describes the behavior of the system through its input-output relationship. If $x(t)$ is specified, $y(t)$ can be determined by solving Eqn. (2.11) using either analytical solution methods or numerical approximation techniques. The differential equation serves as a complete description of the system in this case. For example, the response of the system to the sinusoidal input signal $x(t) = \sin(2\pi f_0 t)$ can be found by solving the differential equation

$$\frac{dy(t)}{dt} + \frac{1}{RC}\,y(t) = \frac{1}{RC}\,\sin(2\pi f_0 t)$$

Example 2.5: Another RC circuit

Find a differential equation between the input and the output signals of the circuit shown in Fig. 2.10.

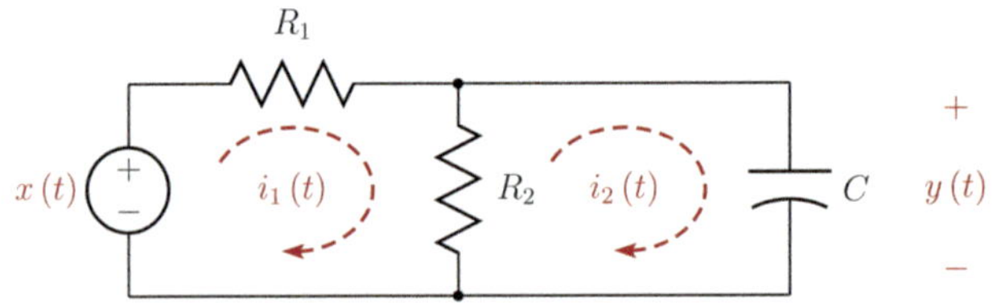

Figure 2.10 – Circuit for Example 2.5.

Solution: Using the two mesh currents $i_1(t)$ and $i_2(t)$, and applying Kirchhoff's voltage law (KVL) we obtain the following two equations:

$$-x(t) + R_1\, i_1(t) + R_2\left[i_1(t) - i_2(t)\right] = 0 \tag{2.12}$$

$$R_2\left[i_2(t) - i_1(t)\right] + y(t) = 0 \tag{2.13}$$

Capacitor current is proportional to the rate of change of capacitor voltage:

$$i_2(t) = C\,\frac{dy(t)}{dt} \tag{2.14}$$

Substituting Eqn. (2.14) into Eqn. (2.13) and solving for $i_1(t)$ yields

$$i_1(t) = C\,\frac{dy(t)}{dt} + \frac{1}{R_2}\,y(t) \tag{2.15}$$

Finally, using Eqns. (2.14) and (2.15) in Eqn. (2.12) results in

$$-x(t) + R_1 C\,\frac{dy(t)}{dt} - \frac{R_1 + R_2}{R_2}\,y(t) = 0$$

which can be rearranged to produce the differential equation we seek:

$$\frac{dy(t)}{dt} + \frac{R_1 + R_2}{R_1 R_2 C}\,y(t) = \frac{1}{R_1 C}\,x(t)$$

The order of a differential equation is determined by the highest order derivative that appears in it. In Examples 2.4 and 2.5 we obtained first-order differential equations; therefore, the systems they represent are also of first order. In the next example we will work with a circuit that will yield a second-order differential equation and, consequently, a second-order system.

Example 2.6: Differential equation for RLC circuit

Find a differential equation between the input signal $x(t)$ and the output signal $y(t)$ to serve as a mathematical model for the series RLC circuit shown in Fig. 2.11.

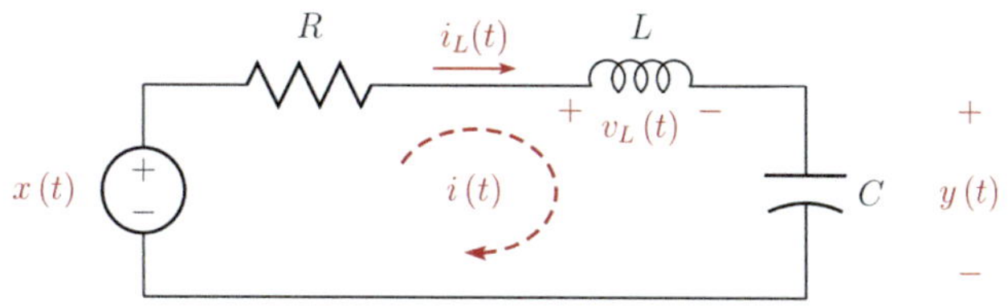

Figure 2.11 – Circuit for Example 2.6.

Solution: Applying KVL around the main loop we get

$$-x(t)+R\,i(t)+v_L(t)+y(t)=0 \tag{2.16}$$

with $x(t)$ and $y(t)$ representing the input and the output signals of the system, respectively. The inductor voltage is proportional to the time derivative of the current:

$$v_L(t)=L\,\frac{di(t)}{dt} \tag{2.17}$$

Similarly, realizing that the loop current $i(t)$ is proportional to the time derivative of the capacitor voltage $y(t)$, we can write

$$i(t)=C\,\frac{dy(t)}{dt} \tag{2.18}$$

Differentiating both sides of Eqn. (2.18) with respect to time yields

$$\frac{di(t)}{dt}=C\,\frac{d^2y(t)}{dt^2} \tag{2.19}$$

and substituting Eqn. (2.19) into Eqn. (2.17) we obtain

$$v_L(t)=LC\,\frac{d^2y(t)}{dt^2} \tag{2.20}$$

Finally, substituting Eqns. (2.18) and (2.20) into Eqn. (2.16) leads to the differential equation for the RLC circuit:

$$-x(t)+RC\,\frac{dy(t)}{dt}+LC\,\frac{d^2y(t)}{dt^2}+y(t)=0$$

Rearranging terms, we have

$$\frac{d^2y(t)}{dt^2}+\frac{R}{L}\,\frac{dy(t)}{dt}+\frac{1}{LC}\,y(t)=\frac{1}{LC}\,x(t) \tag{2.21}$$

Thus, the RLC circuit of Fig. 2.11 leads to a second-order differential equation.

2.4 Constant-Coefficient Ordinary Differential Equations

In Section 2.2 we have discussed the properties of a class of systems known as linear and time-invariant. Linearity and time invariance assumptions allow us to develop a robust set of methods and techniques for analyzing and designing systems. We will see in later parts of this text that, by limiting our focus to systems that are both linear and time-invariant (for which we use the acronym CTLTI), we will be able to use the convolution operation and the system function concept for the computation of the output signal.

In Examples 2.4, 2.5, and 2.6 we have explored methods of finding differential equations for several electrical circuits. In general, CTLTI systems can be modeled with *ordinary differential equations* that have *constant coefficients.* An ordinary differential equation is one that does not contain partial derivatives. The differential equation that represents a CTLTI system contains the input signal $x(t)$, the output signal $y(t)$ as well as simple time derivatives of the two, namely

$$\frac{d^k y(t)}{dt^k}, \quad k=0,\ldots,N$$

and

$$\frac{d^k x(t)}{dt^k}, \quad k=0,\ldots,M$$

A general constant-coefficient differential equation representing a CTLTI system is therefore in the form

$$a_N \frac{d^N y(t)}{dt^N} + a_{N-1} \frac{d^{N-1} y(t)}{dt^{N-1}} + \ldots + a_1 \frac{dy(t)}{dt} + a_0\, y(t) =$$

$$b_M \frac{d^M x(t)}{dt^M} + b_{M-1} \frac{d^{M-1} x(t)}{dt^{M-1}} + \ldots + b_1 \frac{dx(t)}{dt} + b_0\, x(t) \tag{2.22}$$

or it can be expressed in closed summation form.

Constant-coefficient ordinary differential equation:

$$\sum_{k=0}^{N} a_k \frac{d^k y(t)}{dt^k} = \sum_{k=0}^{M} b_k \frac{d^k x(t)}{dt^k} \tag{2.23}$$

The order of the differential equation (and therefore the order of the system) is the larger of N and M. As an example, the differential equation given by Eqn. (2.11) for the circuit of Fig. 2.9 fits the standard form of Eqn. (2.23) with $N=1$, $M=0$, $a_1=1$, $a_0=1/RC$, and $b_0=1/RC$. Similarly, the differential equation given by Eqn. (2.21) for the circuit of Fig. 2.11 also conforms to the standard format of Eqn. (2.23) with $N=2$, $M=0$, $a_2=1$, $a_1=R/L$, $a_0=1/LC$, and $b_0=1/LC$.

In general, a constant-coefficient ordinary differential equation in the form of Eqn. (2.23) has a family of solutions. In order to find a unique solution for $y(t)$, initial values of the output signal and its first $N-1$ derivatives need to be specified at a time instant $t=t_0$. We need to know

$$y(t_0)\,, \quad \left.\frac{dy(t)}{dt}\right|_{t=t_0}, \quad \ldots, \quad \left.\frac{d^{N-1} y(t)}{dt^{N-1}}\right|_{t=t_0}$$

to find the solution for $t>t_0$.

Example 2.7: Checking linearity and time invariance of a differential equation

Determine whether the first-order constant-coefficient differential equation

$$\frac{dy(t)}{dt} + a_0\, y(t) = b_0\, x(t) \tag{2.24}$$

represents a CTLTI system.

Solution: In order to check the linearity of the system described by the first-order differential equation given, we will assume that two input signals $x_1(t)$ and $x_2(t)$ produce the responses $y_1(t)$ and $y_2(t)$, respectively. The input-output pairs must satisfy the differential equation. Therefore we have

$$\frac{dy_1(t)}{dt} + a_0\, y_1(t) = b_0\, x_1(t) \tag{2.25}$$

and

$$\frac{dy_2(t)}{dt} + a_0\, y_2(t) = b_0\, x_2(t) \tag{2.26}$$

Now let a new input signal be constructed as a linear combination of $x_1(t)$ and $x_2(t)$ in the form

$$x_3(t) = \alpha_1\, x_1(t) + \alpha_2\, x_2(t)$$

where α_1 and α_2 are two arbitrary constants. If the system under consideration is linear, then with $x_3(t)$ as the input signal, the solution of the differential equation must be

$$y_3(t) = \alpha_1\, y_1(t) + \alpha_2\, y_2(t) \tag{2.27}$$

Does $y(t) = y_3(t)$ satisfy the differential equation when the input is equal to $x(t) = x_3(t)$? Using $y_3(t)$ in the differential equation we obtain

$$\frac{dy_3(t)}{dt} + a_0\, y_3(t) = \frac{d}{dt}\left[\alpha_1\, y_1(t) + \alpha_2\, y_2(t)\right] + a_0\left[\alpha_1\, y_1(t) + \alpha_2\, y_2(t)\right] \tag{2.28}$$

Rearranging terms on the right side of Eqn. (2.28) we have

$$\frac{dy_3(t)}{dt} + a_0\, y_3(t) = \alpha_1\left[\frac{dy_1(t)}{dt} + a_0\, y_1(t)\right] + \alpha_2\left[\frac{dy_2(t)}{dt} + a_0\, y_2(t)\right] \tag{2.29}$$

and substituting Eqns. (2.25) and (2.26) into Eqn. (2.29) results in

$$\begin{aligned}\frac{dy_3(t)}{dt} + a_0\, y_3(t) =& \alpha_1\left[b_0\, x_1(t)\right] + \alpha_2\left[b_0\, x_2(t)\right] \\ =& b_0\left[\alpha_1\, x_1(t) + \alpha_2\, x_2(t)\right] \\ =& b_0\, x_3(t)\end{aligned} \tag{2.30}$$

Signals $x_3(t)$ and $y_3(t)$ as a pair satisfy the differential equation. As a result, we may be tempted to conclude that the corresponding system is linear, however, we will take a cautious approach and investigate a bit further: Let us begin by recognizing that the differential equation we are considering is in the same form as the one obtained in Example 2.4 for the simple RC circuit. As a matter of fact it would be identical to it with $a_0 = b_0 = 1/RC$. In the simple RC circuit, the output signal is the voltage across the terminals of the capacitor.

What would happen if the capacitor is initially charged to V_0 volts? Would Eqn. (2.27) still be valid? Let us write Eqn. (2.27) at time $t = t_0$:

$$y_3(t_0) = \alpha_1 y_1(t_0) + \alpha_2 y_2(t_0) \tag{2.31}$$

On the other hand, because of the initial charge of the capacitor, any solution found must start with the same initial value $y(t_0) = V_0$ and continue from that point on. We must have

$$y_1(t_0) = V_0\,, \quad y_2(t_0) = V_0\,, \quad y_3(t_0) = V_0$$

It is clear that Eqn. (2.31) can only be satisfied if $V_0 = 0$, that is, if the capacitor has no initial charge. Therefore, the differential equation in Eqn. (2.24) represents a CTLTI system if and only if the initial value of the output signal is equal to zero.

Another argument to convince ourselves that a system with nonzero initial conditions cannot be linear is the following: Based on the second condition of linearity given by Eqn. (2.3), a zero input signal must produce a zero output signal (just set $\alpha_1 = 0$). A system with a nonzero initial state $y(t_0) = V_0$ produces a nonzero output signal even if the input signal is zero, and therefore cannot be linear.

Next we need to check the system described by the differential equation in Eqn. (2.24) for time invariance. If we replace the time variable t with $(t-\tau)$ we get

$$\frac{dy(t-\tau)}{dt} + a_0\, y(t-\tau) = b_0\, x(t-\tau)$$

Delaying the input signal by τ causes the output signal to also be delayed by τ. Therefore, the system is time-invariant.

In Example 2.7 we verified that the first-order constant-coefficient differential equation corresponds to a CTLTI system provided that the system is initially relaxed. It is also a straightforward task to prove that the general constant-coefficient differential equation given by Eqn. (2.23) corresponds to a CTLTI system if all initial conditions are equal to zero.

Proof that Eqn. (2.23) corresponds to a CTLTI system:

Assuming that two input signals $x_1(t)$ and $x_2(t)$ produce the output signals $y_1(t)$ and $y_2(t)$, respectively, we will check and see if the input signal $x_3(t) = \alpha_1 x_1(t) + \alpha_2 x_2(t)$ leads to the output signal $y_3(t) = \alpha_1 y_1(t) + \alpha_2 y_2(t)$. Through repeated differentiation it can be shown that

$$\begin{aligned} \frac{d^k x_3(t)}{dt^k} &= \frac{d^k}{dt^k}\left\{\alpha_1 x_1(t) + \alpha_2 x_2(t)\right\} \\ &= \alpha_1 \frac{d^k x_1(t)}{dt^k} + \alpha_2 \frac{d^k x_2(t)}{dt^k} \end{aligned} \tag{2.32}$$

for $k = 0, \ldots, M$, and similarly,

$$\begin{aligned} \frac{d^k y_3(t)}{dt^k} &= \frac{d^k}{dt^k}\left\{\alpha_1 y_1(t) + \alpha_2 y_2(t)\right\} \\ &= \alpha_1 \frac{d^k y_1(t)}{dt^k} + \alpha_2 \frac{d^k y_2(t)}{dt^k} \end{aligned} \tag{2.33}$$

for $k = 0, \ldots, N$. Substituting Eqn. (2.33) into the left side of Eqn. (2.23) yields

$$\begin{aligned} \sum_{k=0}^{N} a_k \frac{d^k y_3(t)}{dt^k} &= \sum_{k=0}^{N} a_k \left\{ \alpha_1 \frac{d^k y_1(t)}{dt^k} + \alpha_2 \frac{d^k y_2(t)}{dt^k} \right\} \\ &= \alpha_1 \sum_{k=0}^{N} a_k \frac{d^k y_1(t)}{dt^k} + \alpha_2 \sum_{k=0}^{N} a_k \frac{d^k y_2(t)}{dt^k} \end{aligned} \tag{2.34}$$

Since we have assumed that $[x_1(t), y_1(t)]$ and $[x_2(t), y_2(t)]$ are solution pairs and therefore satisfy the differential equation, we have

$$\sum_{k=0}^{N} a_k \frac{d^k y_1(t)}{dt^k} = \sum_{k=0}^{M} b_k \frac{d^k x_1(t)}{dt^k} \tag{2.35}$$

and

$$\sum_{k=0}^{N} a_k \frac{d^k y_2(t)}{dt^k} = \sum_{k=0}^{M} b_k \frac{d^k x_2(t)}{dt^k} \tag{2.36}$$

Using Eqns. (2.35) and (2.36) in Eqn. (2.34) it can be shown that

$$\begin{aligned} \sum_{k=0}^{N} a_k \frac{d^k y_3(t)}{dt^k} &= \sum_{k=0}^{M} b_k \left\{ \alpha_1 \frac{d^k x_1(t)}{dt^k} + \alpha_2 \frac{d^k x_2(t)}{dt^k} \right\} \\ &= \sum_{k=0}^{M} b_k \frac{d^k x_3(t)}{dt^k} \end{aligned} \tag{2.37}$$

where, in the last step, we have used the result found in Eqn. (2.32). As an added condition, Eqn. (2.33) must also be satisfied at $t = t_0$ for all derivatives in the differential equation, requiring

$$\left. \frac{d^k y(t)}{dt^k} \right|_{t=t_0} = 0 \tag{2.38}$$

for $k = 0, \ldots, N-1$.

The differential equation

$$\sum_{k=0}^{N} a_k \frac{d^k y(t)}{dt^k} = \sum_{k=0}^{M} b_k \frac{d^k x(t)}{dt^k}$$

represents a linear system provided that all initial conditions are equal to zero:

$$y(t_0) = 0, \quad \left. \frac{dy(t)}{dt} \right|_{t=t_0} = 0, \quad \ldots, \quad \left. \frac{d^{N-1} y(t)}{dt^{N-1}} \right|_{t=t_0} = 0$$

Time invariance of the corresponding system can also be proven easily by replacing T with $t - \tau$ in Eqn. (2.23):

$$\sum_{k=0}^{N} a_k \frac{d^k y(t-\tau)}{dt^k} = \sum_{k=0}^{M} b_k \frac{d^k x(t-\tau)}{dt^k} \tag{2.39}$$

2.5 Solving Differential Equations

One method of determining the output signal of a system in response to a specified input signal is to solve the corresponding differential equation. In later parts of this text we will study alternative methods of accomplishing the same task. The use of these alternative methods will also be linked to the solution of the differential equation to provide further insight into linear system behavior.

In this section we will present techniques for solving linear constant-coefficient differential equations. Two distinct methods will be presented; one that can only be used with a first-order differential equation, and one that can be used with any order differential equation in the standard form of Eqn. (2.23).

2.5.1 Solution of the first-order differential equation

The first-order differential equation

$$\frac{dy(t)}{dt}+\alpha\,y(t)=r(t) \tag{2.40}$$

represents a first-order CTLTI system. This could be the differential equation of the RC circuit considered in Example 2.4 with $\alpha = 1/RC$ and $r(t) = (1/RC)\,x(t)$. In this section we will formulate the general solution of this differential equation for a specified initial value $y(t_0)$.

Solving the differential equation in Eqn. (2.40):

Let the function $f(t)$ be defined as

$$f(t)=e^{\alpha t}\,y(t) \tag{2.41}$$

Differentiating $f(t)$ with respect to time yields

$$\begin{aligned}\frac{df(t)}{dt} &= e^{\alpha t}\,\frac{dy(t)}{dt}+\alpha\,e^{\alpha t}\,y(t)\\ &= e^{\alpha t}\left[\frac{dy(t)}{dt}+\alpha\,y(t)\right]\end{aligned} \tag{2.42}$$

The expression in square brackets is recognized as $r(t)$ from Eqn. (2.40), therefore

$$\frac{df(t)}{dt}=e^{\alpha t}\,r(t) \tag{2.43}$$

Integrating both sides of Eqn. (2.43) over time, we obtain $f(t)$ as

$$f(t)=f(t_0)+\int_{t_0}^{t} e^{\alpha\tau}\,r(\tau)\,d\tau \tag{2.44}$$

Using Eqn. (2.41)

$$f(t_0)=e^{\alpha t_0}\,y(t_0) \tag{2.45}$$

the solution for $f(t)$ becomes

$$f(t)=e^{\alpha t_0}\,y(t_0)+\int_{t_0}^{t} e^{\alpha\tau}\,r(\tau)\,d\tau \tag{2.46}$$

The solution for $y(t)$ is found through the use of this result in Eqn. (2.41):

$$\begin{aligned} y(t) =& e^{-\alpha t} f(t) \\ =& e^{-\alpha(t-t_0)} y(t_0) + e^{-\alpha t} \int_{t_0}^{t} e^{\alpha\tau} r(\tau)\, d\tau \\ =& e^{-\alpha(t-t_0)} y(t_0) + \int_{t_0}^{t} e^{-\alpha(t-\tau)} r(\tau)\, d\tau \end{aligned} \tag{2.47}$$

Thus, Eqn. (2.47) represents the complete solution of the first-order differential equation in Eqn. (2.40) and is practical to use as long as the right-side function $r(t)$ allows easy evaluation of the integral.

The differential equation

$$\frac{dy(t)}{dt} + \alpha\, y(t) = r(t)\,, \qquad y(t_0): \text{specified} \tag{2.48}$$

is solved as

$$y(t) = e^{-\alpha(t-t_0)} y(t_0) + \int_{t_0}^{t} e^{-\alpha(t-\tau)} r(\tau)\, d\tau \tag{2.49}$$

We may be tempted to question the significance of the solution found in Eqn. (2.47) especially since it is only applicable to a first-order differential equation and therefore a first-order system. The result we have found will be very useful for working with higher-order systems, however, when we study the state-space description of a CTLTI system in Chapter 5.

Example 2.8: Unit step response of the simple RC circuit

Consider the simple RC circuit that was first introduced in Example 2.4. Let the element values be $R = 1\ \Omega$ and $C = 1/4$ F. Assume the initial value of the output at time $t = 0$ is $y(0) = 0$. Determine the response of the system to an input signal in the form of a unit step function, i.e., $x(t) = u(t)$ as shown in Fig. 2.12.

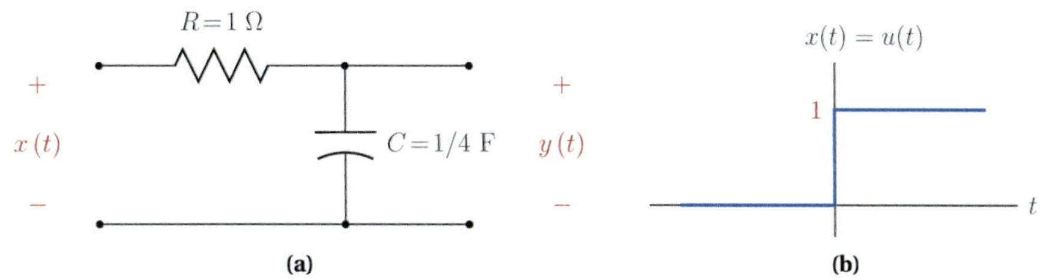

Figure 2.12 – **(a)** The circuit for Example 2.8 and **(b)** the input signal $x(t)$.

Solution: The differential equation of the circuit was obtained in Eqn. (2.11) of Example 2.4. Using the unit step input signal specified, it can be written as

$$\frac{dy(t)}{dt} + \frac{1}{RC}\,y(t) = \frac{1}{RC}\,u(t)$$

Applying Eqn. (2.49) with $\alpha = 1/RC$, $t_0 = 0$, and $r(t) = (1/RC)\,u(t)$ we have

$$y(t) = \int_0^t e^{-(t-\tau)/RC}\,\frac{1}{RC}\,u(t)\,d\tau = \frac{e^{-t/RC}}{RC}\int_0^t e^{\tau/RC}\,d\tau = 1 - e^{-t/RC} \tag{2.50}$$

for $t \ge 0$. In compact form, the result in Eqn. (2.50) is

$$y(t) = \left(1 - e^{-t/RC}\right)u(t) \tag{2.51}$$

If we now substitute the specified values of R and C into Eqn. (2.51), we obtain the output signal

$$y(t) = \left(1 - e^{-4t}\right)u(t) \tag{2.52}$$

which is graphed in Fig. 2.13.

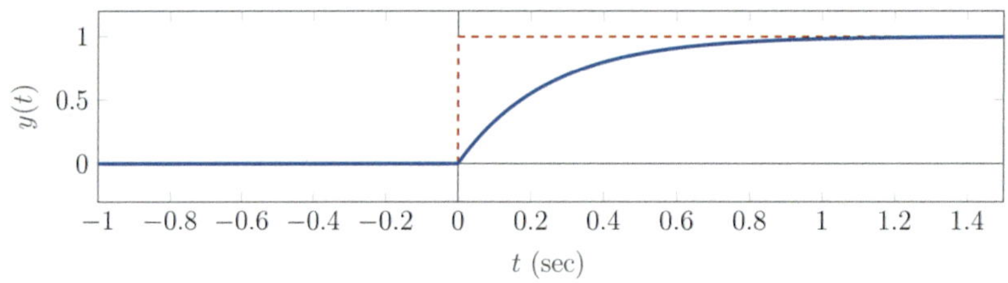

Figure 2.13 – The output signal $y(t)$ for Example 2.8.

Software resources: exct_2_8.m , exct_2_8_Live.mlx

Example 2.9: Pulse response of the simple RC circuit

Determine the response of the RC circuit of Example 2.4 to a rectangular pulse signal

$$x(t) = A\Pi\left(\frac{t}{\lambda}\right) \tag{2.53}$$

shown in Fig. 2.14. Element values for the circuit are $R = 1\ \Omega$ and $C = 1/4$ F. The initial value of the output signal at time $t = -\lambda/2$ is $y(-\lambda/2) = 0$.

Solution: The differential equation of the circuit is obtained from Eqn. (2.11) with the substitution of specified element values and the input signal:

$$\frac{dy(t)}{dt} + 4\,y(t) = 4A\Pi\left(\frac{t}{\lambda}\right)$$

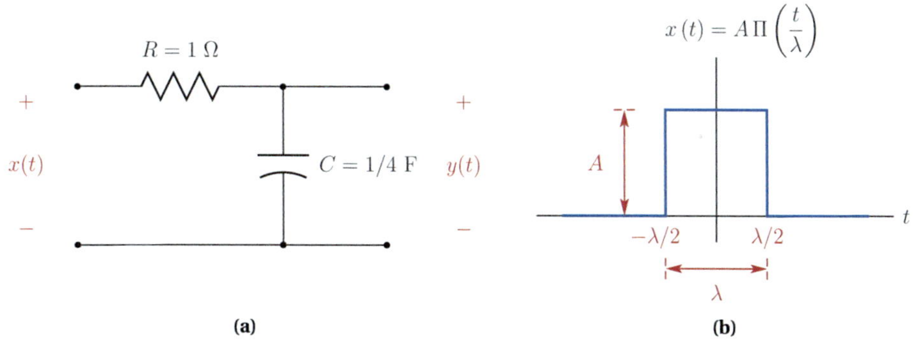

Figure 2.14 – **(a)** The circuit for Example 2.9 and **(b)** the input signal $x(t)$.

The solution is in the form of Eqn. (2.49) with $\alpha = 4$, $y(-\lambda/2) = 0$, and $r(t) = 4A\Pi(t/\lambda)$:

$$y(t) = \int_{-\lambda/2}^{t} e^{-4(t-\tau)}\, 4A\Pi\left(\frac{\tau}{\lambda}\right) d\tau \tag{2.54}$$

The integral in this result can be evaluated using two possible ranges of the variable t.

Case 1: $-\dfrac{\lambda}{2} < t \le \dfrac{\lambda}{2}$

In this range of t we have $\Pi(t/\lambda) = 1$, and the response in Eqn. (2.54) simplifies to

$$y(t) = 4A \int_{-\lambda/2}^{t} e^{-4(t-\tau)}\, d\tau$$

which can be evaluated as

$$y(t) = 4Ae^{-4t} \int_{-\lambda/2}^{t} e^{4\tau}\, d\tau = A\left[1 - e^{-2\lambda}\, e^{-4t}\right] \tag{2.55}$$

Case 2: $t > \dfrac{\lambda}{2}$

In this case we have $\Pi(t/\lambda) = 1$ for $-\lambda/2 \le t \le \lambda/2$ and $\Pi(t/\lambda) = 0$ for $t > \lambda/2$. The result in Eqn. (2.54) simplifies to

$$y(t) = 4A \int_{-\lambda/2}^{\lambda/2} e^{-4(t-\tau)}\, d\tau$$

which leads to the response

$$y(t) = 4Ae^{-4t} \int_{-\lambda/2}^{\lambda/2} e^{4\tau}\, d\tau = Ae^{-4t}\left[e^{2\lambda} - e^{-2\lambda}\right] \tag{2.56}$$

The complete response of the system can be expressed by combining the results found in Eqns. (2.55) and (2.56) as

$$y(t) = \begin{cases} A\left[1 - e^{-2\lambda}\, e^{-4t}\right], & -\dfrac{\lambda}{2} < t \le \dfrac{\lambda}{2} \\[2ex] Ae^{-4t}\left[e^{2\lambda} - e^{-2\lambda}\right], & t > \dfrac{\lambda}{2} \end{cases} \tag{2.57}$$

The signal $y(t)$ is shown in Fig. 2.15 for parameter values $A = 1$ and $\lambda = 1$.

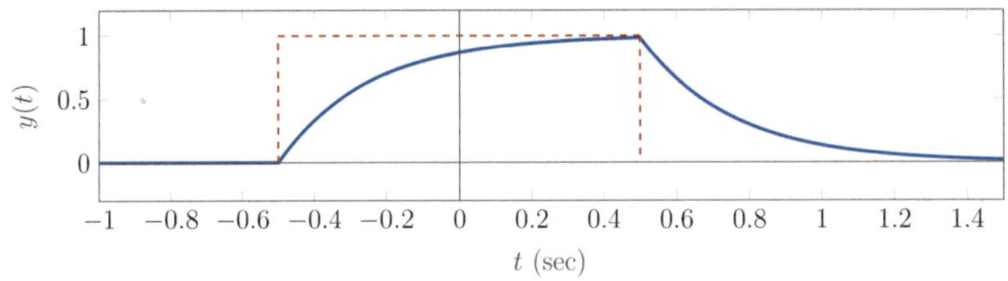

Figure 2.15 – The output signal $y(t)$ for Example 2.9.

Software resources: exct_2_9.m , exct_2_9_Live.mlx

Example 2.10: Pulse response of the simple RC circuit revisited

Rework the problem in Example 2.9 by making use of the unit step response found in Example 2.8 along with linearity and time invariance properties of the RC circuit.

Solution: The pulse signal used as input in Example 2.9 can be expressed as the difference of two unit step signals as illustrated in Fig. 2.16.

$$x(t) = A\Pi(t/w) = A\,u\left(t+\frac{w}{2}\right) - A\,u\left(t-\frac{w}{2}\right)$$

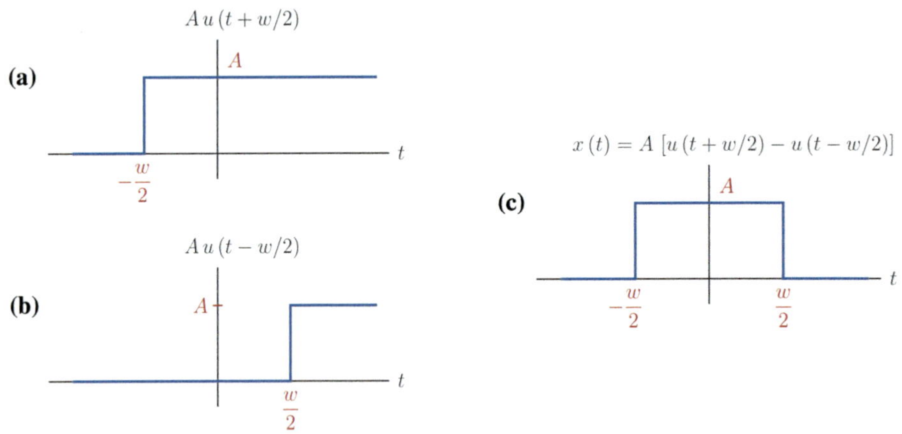

Figure 2.16 – Constructing a pulse from time-shifted step functions.

Since the system under consideration is linear, its response to $x(t)$ is

$$\begin{aligned}\text{Sys}\{x(t)\} &= \text{Sys}\left\{A\,u\left(t+\frac{w}{2}\right)\right\} - \text{Sys}\left\{A\,u\left(t-\frac{w}{2}\right)\right\}\\ &= A\,\text{Sys}\left\{u\left(t+\frac{w}{2}\right)\right\} - A\,\text{Sys}\left\{u\left(t-\frac{w}{2}\right)\right\}\end{aligned} \tag{2.58}$$

Furthermore, since the system is time-invariant, its responses to time-shifted unit step functions can be found from the unit step response that was computed in Example 2.8. It was determined that

$$\text{Sys}\left\{u(t)\right\} = \left(1 - e^{-4t}\right) u(t)$$

Using the time invariance property of the system we have

$$\text{Sys}\left\{u\left(t + \frac{w}{2}\right)\right\} = \left[1 - e^{-4(t+w/2)}\right] u\left(t + \frac{w}{2}\right) \tag{2.59}$$

and

$$\text{Sys}\left\{u\left(t - \frac{w}{2}\right)\right\} = \left[1 - e^{-4(t-w/2)}\right] u\left(t - \frac{w}{2}\right) \tag{2.60}$$

Substituting Eqns. (2.59) and (2.60) into Eqn. (2.58) we find the pulse response of the RC circuit as

$$\text{Sys}\left\{x(t)\right\} = A\left[1 - e^{-4(t+w/2)}\right] u\left(t + \frac{w}{2}\right) - A\left[1 - e^{-4(t-w/2)}\right] u\left(t - \frac{w}{2}\right)$$

which is in agreement with the result found in Example 2.9. The steps used in the solution are illustrated in Fig. 2.17.

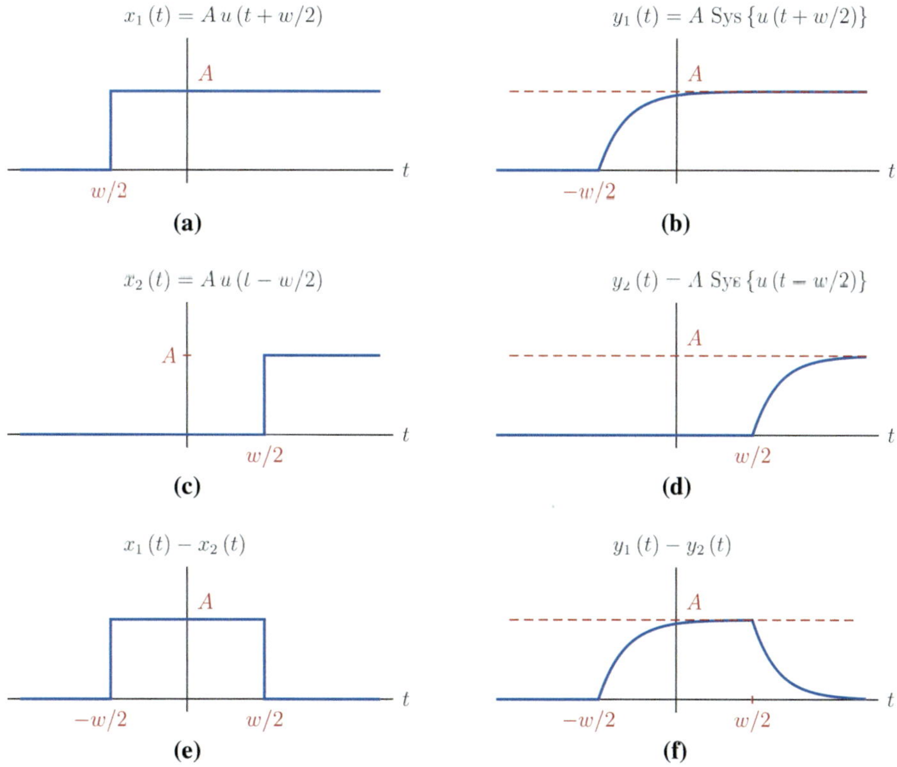

Figure 2.17 – The steps employed in the solution of Example 2.10.

Software resources: exct_2_10.m , exct_2_10_Live.mlx

Interactive App: Finding the pulse response of RC circuit using superposition

The interactive app in `appRC1.m` is based on Example 2.10, Eqn. (2.58), and Fig. 2.17. The response of the RC circuit to a pulse of width λ centered around the time origin is obtained from its step responses using superposition. The input pulse is expressed as the difference of two step signals, and the output signal is computed as the difference of the individual responses to step functions. Circuit parameters R and C as well as the pulse width w can be varied using slider controls.

a. Set the resistance to $R = 1\ \Omega$, the capacitance to $C = 0.25$ F, and the pulse width to $\lambda = 1$ s. Observe the responses to the two step signals and how the pulse response is formed as the difference of the two responses. Does the output signal look similar to the input signal? Pay attention to rising and falling edges of the input pulse and how the output signal is tracking them.

b. Gradually reduce the resistance to $R = 0.5\ \Omega$. How does this affect the shape of the output pulse? Think about what happens to the time constant of the RC circuit as R is reduced, and how that impacts the behavior of the output signal at the edges.

c. Set the circuit parameters back to $R = 1\ \Omega$ and $C = 0.25$ F. Gradually increase the pulse width from $\lambda = 1$ s to $\lambda = 2$ s. Does that help increase the similarity between the input and the output signals? Comment.

Software resource: `appRC1.m`

Software resource: See MATLAB Exercise 2.3.

2.5.2 Solution of the general differential equation

The solution method discussed in the previous section applies to a first-order differential equation only, although, in Chapter 5, we will discuss extension of this method to higher-order differential equations through the use of state variables. In our efforts to solve the general constant-coefficient differential equation in the form given by Eqn. (2.23) we will consider two separate components of the output signal $y(t)$ as follows:

$$y(t) = y_h(t) + y_p(t) \tag{2.61}$$

The first term, $y_h(t)$, is the solution of the *homogeneous differential equation* found by ignoring the input signal, that is, by setting $x(t)$ and all of its derivatives equal to zero in Eqn. (2.23) for all values of t:

$$\sum_{k=0}^{N} a_k \frac{d^k y(t)}{dt^k} = 0 \tag{2.62}$$

As a mathematical function, $y_h(t)$ is the *homogeneous solution* of the differential equation. From the perspective of the output signal of a system, $y_h(t)$ is called the *natural response* of the system. As one of the components of the output signal, the homogeneous solution of the differential equation or, equivalently, the natural response of the system to which it corresponds, $y_h(t)$ depends on the structure of the system as well as the initial state of the system. It does not depend, however, on the input signal applied to the system. It is the part of the response that is produced by the system due to a release of the energy stored within the system. Recall the circuits used in Examples 2.4 through 2.6. Some circuit elements such as capacitors and inductors are capable of storing energy which could later be released under certain circumstances. If, at some point in time, a capacitor or an inductor with stored energy is given a chance to release this energy, the circuit could produce a response through this release even when there is no external input signal being applied.

When we discuss the stability property of CTLTI systems in later sections of this chapter we will see that, for a stable system, $y_h(t)$ tends to gradually disappear in time. Because of this, it is also referred to as the *transient response* of the system.

In contrast, the second term $y_p(t)$ in Eqn. (2.61) is part of the solution that is due to the input signal $x(t)$ being applied to the system. It is referred to as the *particular solution* of the differential equation. It depends on the input signal $x(t)$ and the internal structure of the system, but it does not depend on the initial state of the system. It is the part of the response that remains active after the homogeneous solution $y_h(t)$ gradually becomes smaller and disappears. When we study the system function concept in later chapters of this text we will link the particular solution of the differential equation to the *steady-state response* of the system, that is, the response to an input signal that has been applied for a long enough time for the transient terms to die out.

2.5.3 Finding the natural response of a continuous-time system

Computation of the natural response of a CTLTI system requires solving the homogeneous differential equation. Before we tackle the problem of finding the homogeneous solution for the general constant-coefficient differential equation, we will consider the first-order case. The first-order homogeneous differential equation is in the form

$$\frac{dy(t)}{dt} + \alpha\, y(t) = 0 \tag{2.63}$$

where α is any arbitrary constant. The homogeneous differential equation given by Eqn. (2.63) has many possible solutions. In order to find a unique solution from among many, we also need to know the initial value of $y(t)$ at some time instant $t = t_0$. We will begin by writing Eqn. (2.63) in the alternative form

$$\frac{dy(t)}{dt} = -\alpha\, y(t) \tag{2.64}$$

It is apparent from Eqn. (2.64) that the solution we are after is a function $y(t)$ the time derivative of which is proportional to itself. An exponential function of time has that property, so we can make an educated guess for a solution in the form

$$y(t) = c\, e^{st} \tag{2.65}$$

Trying this guess[1] in the homogeneous differential equation we have

$$\frac{d}{dt}\left[c\, e^{st}\right] + \alpha\, c\, e^{st} = 0 \tag{2.66}$$

[1] Another way to justify the guess utilized in Eqn. (2.65) is to write Eqn. (2.63) in the form

$$\frac{dy}{y} = -\alpha\, dt$$

Integrating both sides leads to

$$\int \frac{dy}{y} = -\int \alpha\, dt \quad \Rightarrow \quad \ln(y) = -\alpha t + K \quad \Rightarrow \quad y(t) = e^K e^{-\alpha t} = c\, e^{-\alpha t}$$

which yields

$$s\,c\,e^{st} + \alpha\,c\,e^{st} = 0 \tag{2.67}$$

By factoring out the common term $c\,e^{st}$, Eqn. (2.67) becomes

$$c\,e^{st}\,(s+\alpha) = 0 \tag{2.68}$$

There are two ways to make the left side of Eqn. (2.68) equal zero:

1. Select $c\,e^{st} = 0$
2. Select $(s+\alpha) = 0$

The former choice leads to the trivial solution $y(t) = 0$ and is obviously not very useful. Also, as a solution it is only valid when the initial value of the output signal is $y(0) = 0$ as we cannot satisfy any other initial value with this solution. Therefore we must choose the latter and use $s = -\alpha$. Substituting this value of s into Eqn. (2.65) we get

$$y(t) = c\,e^{-\alpha t} \tag{2.69}$$

where the constant c must be determined based on the desired initial value of $y(t)$ at $t = t_0$. We will explore this in the next two examples.

Example 2.11: Natural response of the simple RC circuit

Consider again the RC circuit of Example 2.4 shown in Fig. 2.18 with the element values $R = 1\ \Omega$ and $C = 1/4$ F. Also, let the input terminals of the circuit be connected to a battery that supplies the circuit with an input voltage of 5 V up to the time instant $t = 0$.

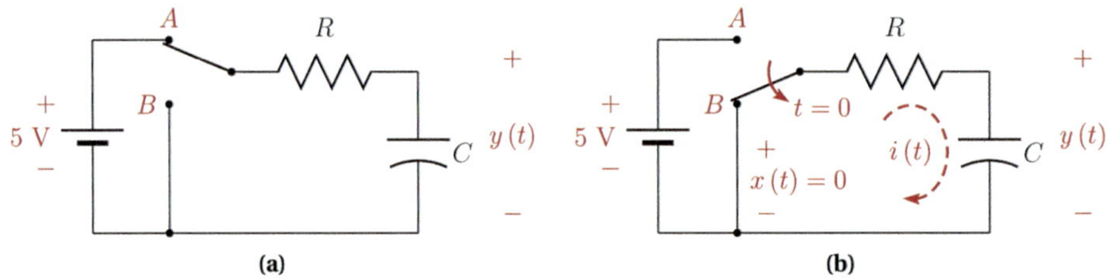

Figure 2.18 – The RC circuit for Example 2.11: **(a)** for $t < 0$ and **(b)** for $t \ge 0$.

Assuming the battery has been connected to the circuit for a long time before $t = 0$, the capacitor voltage has remained at a steady-state value of 5 V. Let the switch be moved from position A to position B at $t = 0$ ensuring that $x(t) = 0$ for $t \ge 0$. The initial value of capacitor voltage at time $t = 0$ is $y(0) = 5$ V. Find the output signal as a function of time.

Solution: Since $x(t) = 0$ for $t > 0$, the output signal is $y(t) = y_h(t)$, and we are trying to find the homogeneous solution. Substituting the specified parameter values, the homogeneous differential equation is found as

$$\frac{dy(t)}{dt} + 4\,y(t) = 0$$

We need $(s+4) = 0$, and the corresponding homogeneous solution is in the form

$$y_h(t) = c\,e^{-st} = c\,e^{-4t}$$

for $t \geq 0$. The initial condition $y_h(0) = 5$ must be satisfied. Substituting $t = 0$ into the homogeneous solution, we get

$$y_h(0) = c\,e^{-4(0)} = c = 5$$

Using the value found for the constant c, the natural response of the circuit is

$$y_h(t) = 5\,e^{-4t}\,, \quad \text{for } t \geq 0$$

The natural response can be expressed in a more compact form through the use of the unit step function:

$$y_h(t) = 5\,e^{-4t}\,u(t) \tag{2.70}$$

This solution is shown in Fig. 2.19.

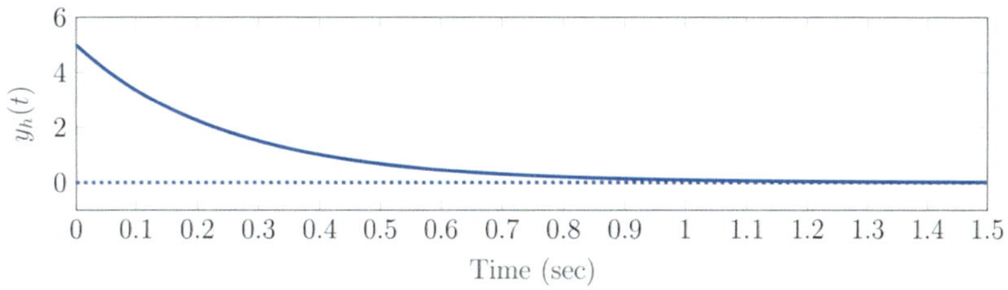

Figure 2.19 – The natural response of the circuit in Example 2.11.

Next we will check this solution against the homogeneous differential equation to verify its validity. The first derivative of the output signal is

$$\frac{dy_h(t)}{dt} = \frac{d}{dt}\left[5\,e^{-4t}\right] = -20\,e^{-4t}$$

Using $y_h(t)$ and $dy_h(t)/dt$ in the homogeneous differential equation we have

$$\frac{dy_h(t)}{dt} + 4\,y_h(t) = -20\,e^{-4t} + 4\left[5\,e^{-4t}\right] = 0$$

Software resources: `exct_2_11.m` , `exct_2_11_Live.mlx`

Example 2.12: Changing the start time in Example 2.11

Rework the problem in Example 2.11 with one minor change: The initial value of the output signal is specified at the time instant $t = -0.5$ seconds instead of at $t = 0$ and its value is $y(-0.5) = 10$. Physically this would correspond to using a 10 V battery instead of the 5 V battery shown in Fig. 2.18(a), and moving the switch from position A to position B at time instant $t = -0.5$ seconds instead of $t = 0$. These differences are illustrated in Fig. 2.20.

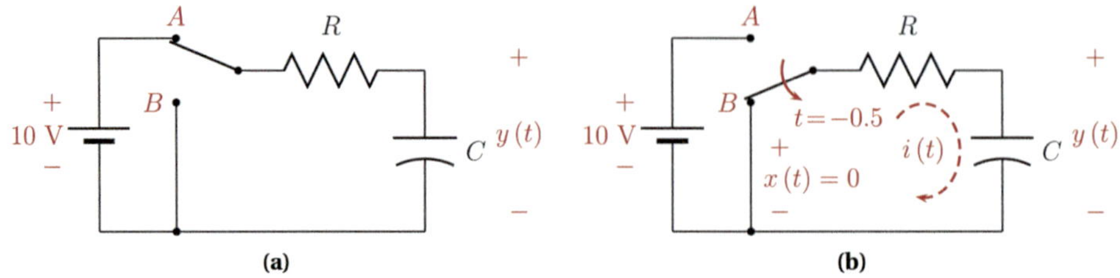

Figure 2.20 – The RC circuit for Example 2.12: **(a)** for $t < -0.5$ and **(b)** for $t \geq -0.5$.

Solution: The general form of the solution found in Example 2.11 is still valid, that is,

$$y_h(t) = c\,e^{-4t}$$

To satisfy $y_h(-0.5) = 10$ we need

$$y_h(-0.5) = c\,e^{-4(-0.5)} = c\,e^2 = 10$$

and therefore

$$c = \frac{10}{e^2} = 1.3534$$

The homogeneous solution is

$$y_h(t) = 1.3534\,e^{-4t}\,, \quad \text{for } t \geq -0.5$$

or, using the unit step function

$$y_h(t) = 1.3534\,e^{-4t}\,u(t+0.5)$$

If the initial condition is specified at a time instant other than $t = 0$, then the solution we find using the procedure outlined starts at that time instant as shown in Fig. 2.21.

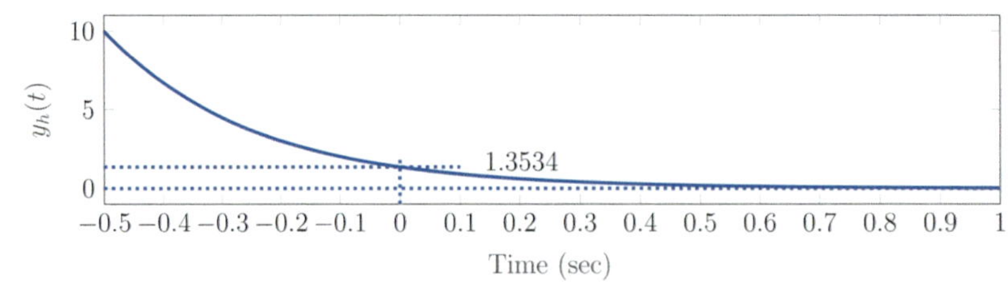

Figure 2.21 – The natural response of the circuit in Example 2.12.

Software resources: `exct_2_12.m`, `exct_2_12_Live.mlx`

We are now ready to solve the general homogeneous differential equation in the form

$$\sum_{k=0}^{N} a_k \frac{d^k y(t)}{dt^k} = 0 \tag{2.71}$$

If we use the same initial guess

$$y(t) = c\,e^{st} \tag{2.72}$$

for the solution, various derivatives of $y(t)$ will be

$$\frac{dy(t)}{dt} = s\,c\,e^{st}\,,\quad \frac{d^2y(t)}{dt^2} = s^2\,c\,e^{st}\,,\quad \frac{d^3y(t)}{dt^3} = s^3\,c\,e^{st}\,,\quad \ldots$$

Through repeated differentiation it can be shown that

$$\frac{d^ky(t)}{dt^k} = s^k\,c\,e^{st} \tag{2.73}$$

We need to determine which values of the parameter s would lead to valid solutions for the homogeneous differential equation. Substituting Eqn. (2.73) into Eqn. (2.71) results in

$$\sum_{k=0}^{N} a_k\,s^k c\,e^{st} = 0 \tag{2.74}$$

Since the term $c\,e^{st}$ is independent of the summation index k, it can be factored out to yield

$$c\,e^{st} \sum_{k=0}^{N} a_k\,s^k = 0 \tag{2.75}$$

requiring one of the following conditions to be satisfied for $y(t) = c\,e^{st}$ to be a solution of the differential equation:

1. $c\,e^{st} = 0$

   This leads to the trivial solution $y(t) = 0$ for the homogeneous equation, and is therefore not very interesting. Also, since this solution would leave us with no adjustable parameters, it cannot satisfy any non-zero initial conditions.

2. $\sum_{k=0}^{P} a_k\,s^k = 0$

   This is called the *characteristic equation* of the system. Values of s that are the solutions of the characteristic equation can be used in exponential functions as solutions of the homogeneous differential equation.

The characteristic equation:

$$\sum_{k=0}^{N} a_k\,s^k = 0 \tag{2.76}$$

The characteristic equation is found by starting with the homogeneous differential equation and replacing the k-th derivative of the output signal $y(t)$ with s^k.

To obtain the characteristic equation, substitute:

$$\frac{d^k y(t)}{dt^k} \quad \rightarrow \quad s^k$$

Let us write the characteristic equation in open form:

$$a_N s^N + a_{N-1} s^{N-1} + \ldots + a_1 s + a_0 = 0 \tag{2.77}$$

The polynomial of order N on the left side of Eqn. (2.77) is referred to as the *characteristic polynomial,* and is obtained by simply replacing each derivative in the homogeneous differential equation with the corresponding power of s. Let the roots of the characteristic polynomial be $s_1, s_2, \ldots, s_N$ so that Eqn. (2.77) can be written as

$$a_N (s - s_1)(s - s_2) \ldots (s - s_N) = 0 \tag{2.78}$$

Using any of the roots of the characteristic polynomial, we can construct a signal

$$y_k(t) = c_k e^{s_k t}, \quad k = 1, \ldots, N \tag{2.79}$$

that will satisfy the homogeneous differential equation. The general solution of the homogeneous differential equation is obtained as a linear combination of all valid terms in the form of Eqn. (2.79) as

$$y_h(t) = c_1 e^{s_1 t} + c_2 e^{s_2 t} + \ldots + c_N e^{s_N t} = \sum_{k=1}^{N} c_k e^{s_k t} \tag{2.80}$$

The unknown coefficients $c_1, c_2, \ldots, c_N$ are determined from the initial conditions. The exponential terms $e^{s_k t}$ in the homogeneous solution given by Eqn. (2.80) are called the *modes of the system.* In later chapters of this text we will see that the roots s_k of the characteristic polynomial of the system will be identical to the *poles* of the system function and to the *eigenvalues* of the state matrix.

Example 2.13: Time constant concept

We will again refer to the RC circuit introduced in Example 2.4 and shown in Fig. 2.9. The differential equation governing the behavior of the circuit was found in Eqn. (2.11). The characteristic equation of the system is found as

$$s + \frac{1}{RC} = 0$$

This is a first-order system, and its only mode is $e^{-t/RC}$. If the initial value of the output signal is $y(0) = V_0$, the homogeneous solution of the differential equation (or the natural response of the system) is

$$y_h(t) = V_0 e^{-t/RC} u(t) \tag{2.81}$$

Let parameter τ be defined as $\tau = RC$, and the natural response can be written as

$$y_h(t) = V_0 e^{-t/\tau} u(t) \tag{2.82}$$

For this type of a system with only one mode, the parameter τ is called the *time constant* of the system. Based on Eqn. (2.82) the time constant τ represents the amount it takes for the natural response to decay down to $1/e$ times or, equivalently, 36.8 percent of its initial value. This is illustrated in Fig. 2.22. It is evident from Fig. 2.22 that the line that is tangent to $y_h(t)$ at $t=0$ has a slope of $-V_0/\tau$, that is,

$$\left.\frac{dy_h(t)}{dt}\right|_{t=0} = -V_0/\tau$$

and it intersects the time axis at $t=\tau$. The amplitude of the output signal at $t=\tau$ is

$$y_h(\tau) = \frac{V_0}{e}$$

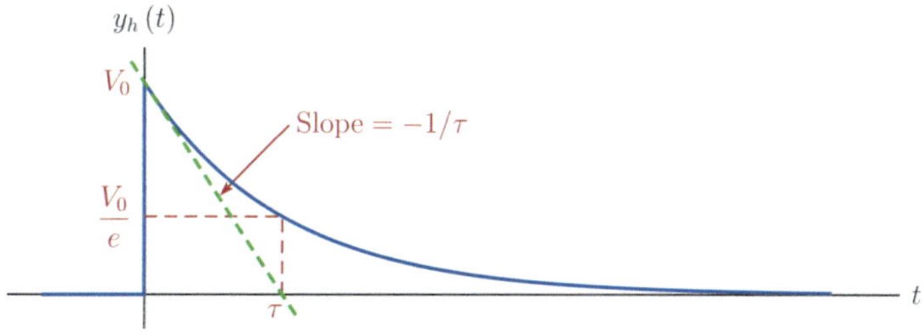

Figure 2.22 – Illustration of the time constant for a first-order system.

Software resources: exct_2_13.m , exct_2_13_Live.mlx

Interactive App: Time constant for first-order system

The interactive app in appRC2.m is based on Example 2.13, Eqn. (2.82), and Fig. 2.22. The natural response of the RC circuit is computed and graphed for a specified initial value. Circuit parameters R and C and the initial value V_0 of the output signal at time $t=0$ can be varied using slider controls.

a. Reduce the time constant τ by reducing either R or C and observe its effect on how long it takes the natural response to become negligibly small. Can you develop a rule-of-thumb on how many time constants it takes for the natural response to become negligible?

b. As the time constant is changed, observe the slope of the natural response at time $t=0$. The greater the slope, the faster the natural response disappears.

Software resource: appRC2.m

Example 2.14: Natural response of second-order system

The differential equation for the RLC circuit used in Example 2.6 and shown in Fig. 2.11 was given by Eqn. (2.21). Let the element values be $R = 5\ \Omega$, $L = 1$ H, and $C = 1/6$ F. At time $t = 0$, the initial inductor current is $i(0) = 2$ A and the initial capacitor voltage is $y(0) = 1.5$ V. No external input signal is applied to the circuit, therefore $x(t) = 0$. Determine the output voltage $y(t)$.

Solution: Without an external input signal, the particular solution is zero, and the total solution of the differential equation includes only the homogeneous solution, that is, $y(t) = y_h(t)$. The homogeneous differential equation is

$$\frac{d^2y(t)}{dt^2} + 5\,\frac{dy(t)}{dt} + 6\,y(t) = 0$$

The characteristic equation of the system is

$$s^2 + 5s + 6 = 0$$

with solutions $s_1 = -2$ and $s_2 = -3$. The homogeneous solution of the differential equation is in the form

$$y_h(t) = c_1\,e^{-2t} + c_2\,e^{-3t} \tag{2.83}$$

for $t \geq 0$. The unknown coefficients c_1 and c_2 need to be adjusted so that the specified initial conditions are satisfied. Evaluating $y_h(t)$ for $t = 0$ we obtain

$$y_h(0) = c_1 e^{-2(0)} + c_2 e^{-3(0)} = c_1 + c_2 = 1.5 \tag{2.84}$$

Using the initial value of the inductor current we have

$$i(0) = C\,\frac{dy_h(t)}{dt}\bigg|_{t=0} = 2$$

from which the initial value of the first derivative of the output signal can be obtained as

$$\frac{dy_h(t)}{dt}\bigg|_{t=0} = \frac{i(0)}{C} = \frac{2}{1/6} = 12$$

Differentiating the solution found in Eqn. (2.83) and imposing the initial value of $dy_h(t)/dt$ leads to

$$\begin{aligned}\frac{dy_h(t)}{dt}\bigg|_{t=0} &= \left[-2c_1 e^{-2t} - 3c_2 e^{-3t}\right]\bigg|_{t=0}\\ &= -2c_1 - 3c_2 = 12\end{aligned} \tag{2.85}$$

The coefficients c_1 and c_2 can now be determined by solving Eqns. (2.84) and (2.85) simultaneously, resulting in

$$c_1 = 16.5 \quad \text{and} \quad c_2 = -15$$

and the natural response of the system is

$$y_h(t) = 16.5\,e^{-2t} - 15\,e^{-3t}, \quad t \geq 0 \tag{2.86}$$

which is graphed in Fig. 2.23.

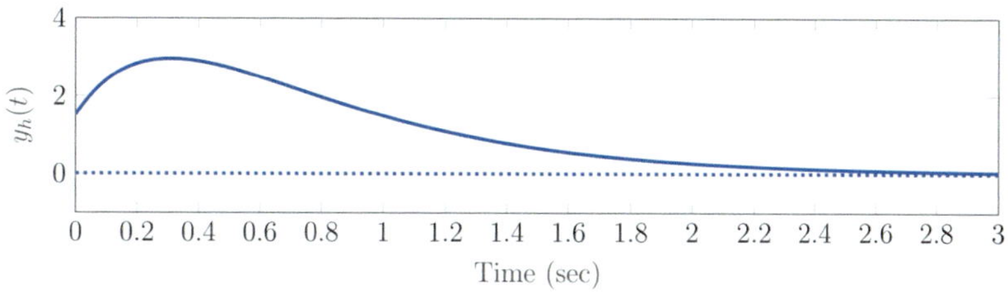

Figure 2.23 – The natural response of the second-order system in Example 2.14.

Software resources: exct_2_14.m , exct_2_14_Live.mlx

In Example 2.14 the two roots of the characteristic polynomial turned out to be both real-valued and different from each other. As a result we were able to express the solution of the homogeneous equation in the standard form of Eqn. (2.80). We will now review several possible scenarios for the types of roots obtained:

Case 1: All roots are distinct and real-valued

In this case the homogeneous solution is

$$y_h(t) = \sum_{k=1}^{N} c_k e^{s_k t} \tag{2.87}$$

as discussed above. If a real root s_k is negative, the corresponding term $c_k e^{s_k t}$ decays exponentially over time. Alternatively, if s_k is positive, the corresponding term grows exponentially over time, and may cause the output signal to become very large without bound. Fig. 2.24 illustrates these two possibilities.

Case 2: Characteristic polynomial has complex-valued roots

Since the coefficients of the differential equation, and consequently the coefficients of the characteristic polynomial, are all real-valued, any complex roots must appear in the form of conjugate pairs. Therefore, if $s_{1a} = \sigma_1 + j\omega_1$ is a complex root of the characteristic polynomial, then its complex conjugate $s_{1b} = s_{1a}^* = \sigma_1 - j\omega_1$ must also be a root. Let the part of the homogeneous solution that is due to these two roots be

$$\begin{aligned} y_{h1}(t) =& c_{1a}\, e^{s_{1a}t} + c_{1b}\, e^{s_{1b}t} \\ =& c_{1a}\, e^{(\sigma_1 + j\omega_1)t} + c_{1b}\, e^{(\sigma_1 - j\omega_1)t} \end{aligned} \tag{2.88}$$

where the coefficients c_{1a} and c_{1b} are to be determined from the initial conditions. Continuing with the reasoning we started above, since the coefficients of the differential equation are real-valued, the homogeneous solution $y_h(t)$ must also be real. Furthermore, $y_{h1}(t)$, the part of the solution that is due to the complex conjugate pair of roots we are considering, must also be real. This in turn implies that the coefficients c_{1a} and c_{1b} must form a complex conjugate pair as well.

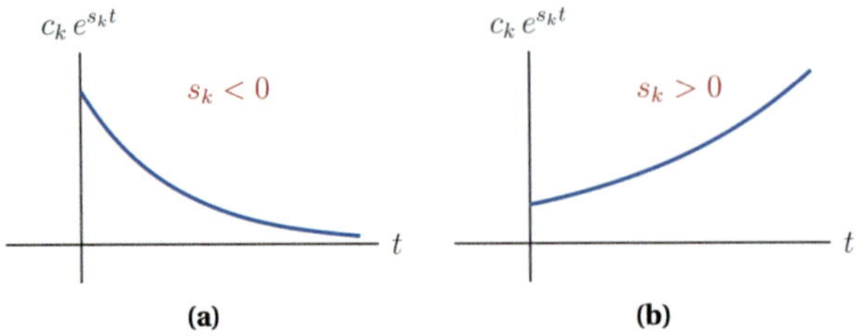

Figure 2.24 – Terms corresponding to real roots of the characteristic equation: **(a)** $s_k < 0$ and **(b)** $s_k > 0$.

Let us write the two coefficients in polar complex form as

$$c_{1a} = |c_1|\, e^{j\theta_1} \quad \text{and } c_{1b} = |c_1|\, e^{-j\theta_1} \tag{2.89}$$

Substituting Eqn. (2.89) into Eqn. (2.88) leads to the result

$$\begin{aligned} y_{h1}(t) =& |c_1|\, e^{\sigma_1 t}\, e^{j(\omega_1 t+\theta_1)} + |c_1|\, e^{\sigma_1 t}\, e^{-j(\omega_1 t+\theta_1)} \\ =& 2\,|c_1|\, e^{\sigma_1 t} \cos(\omega_1 t+\theta_1) \end{aligned} \tag{2.90}$$

a. A pair of complex conjugate roots for the characteristic polynomial leads to a solution component in the form of a cosine signal multiplied by an exponential signal.
b. The oscillation frequency of the cosine signal is determined by ω_1, the imaginary part of the complex roots.
c. The real part of the complex roots, σ_1, impacts the amplitude of the solution. If $\sigma_1 < 0$, then the amplitude of the cosine signal decays exponentially over time. In contrast, if $\sigma_1 > 0$, the amplitude of the cosine signal grows exponentially over time. These two possibilities are illustrated in Fig. 2.25.

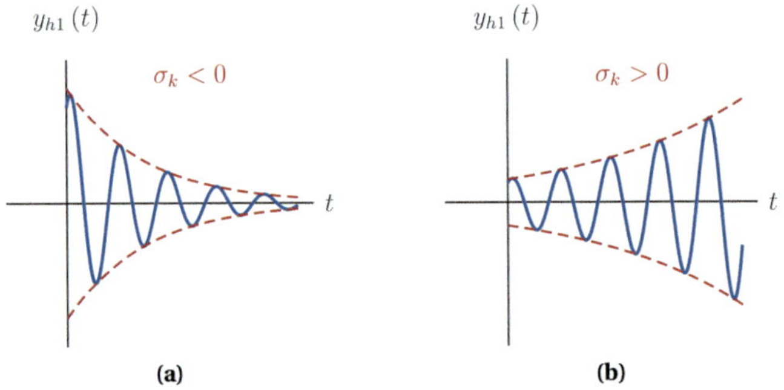

Figure 2.25 – Terms corresponding to pair of complex conjugate roots of the characteristic equation: **(a)** $\sigma_1 < 0$ and **(b)** $\sigma_1 > 0$.

Using the appropriate trigonometric identity[2] it is also possible to write Eqn. (2.90) as

$$\begin{aligned} y_{h1}(t) =& 2\,|c_1| \cos(\theta_1)\, e^{\sigma_1 t} \cos(\omega_1 t) - 2\,|c_1| \sin(\theta_1)\, e^{\sigma_1 t} \sin(\omega_1 t) \\ =& d_1\, e^{\sigma_1 t} \cos(\omega_1 t) + d_2\, e^{\sigma_1 t} \sin(\omega_1 t) \end{aligned} \tag{2.91}$$

2 $\cos(a+b) = \cos(a)\cos(b) - \sin(a)\sin(b)$.

Case 3: Characteristic polynomial has some repeated roots

Consider again the factored version of the characteristic equation first given by Eqn. (2.78):

$$a_N\,(s-s_1)\,(s-s_2)\ldots(s-s_N)=0$$

What if the first two roots are equal, that is, $s_2 = s_1$? If we used the approach we have employed so far, we would have a natural response in the form

$$\begin{aligned} y_h(t) &= c_1\,e^{s_1 t} + c_2\,e^{s_2 t} + \text{other terms} \\ &= c_1\,e^{s_1 t} + c_2\,e^{s_1 t} + \text{other terms} \\ &= (c_1+c_2)\,e^{s_1 t} + \text{other terms} \\ &= \bar{c}_1\,e^{s_1 t} + \text{other terms} \end{aligned} \tag{2.92}$$

The problem with the response in Eqn. (2.92) is that we have lost one of the coefficients. For a homogeneous differential equation of order N we need to satisfy N initial conditions at some time instant $t = t_0$, namely

$$y(t_0)\,, \quad \left.\frac{dy(t)}{dt}\right|_{t=t_0}\,, \quad \ldots, \quad \left.\frac{d^{N-1}y(t)}{dt^{N-1}}\right|_{t=t_0}$$

To satisfy N initial conditions we need as many adjustable parameters $c_1,\ldots,c_N$ sometimes referred to as N *degrees of freedom*. Losing one of the coefficients creates a problem for our ability to satisfy N initial conditions. In order to gain back the coefficient we have lost, we need an additional term for the two roots at $s = s_1$. A solution in the form

$$y_h(t) = c_{11}\,e^{s_1 t} + c_{12}\,t\,e^{s_1 t} + \text{other terms} \tag{2.93}$$

will work for this purpose. In general, a root of multiplicity r requires r terms in the homogeneous solution. If the characteristic polynomial has a factor $(s-s_1)^r$ then the resulting homogeneous solution will be

$$y_h(t) = c_{11}\,e^{s_1 t} + c_{12}\,t\,e^{s_1 t} + \ldots + c_{1r}\,t^{r-1}\,e^{s_1 t} + \text{other terms} \tag{2.94}$$

Example 2.15: Natural response of second-order system revisited

Consider again the RLC circuit which was first used in Example 2.6 and shown in Fig. 2.11. At time $t = 0$, the initial inductor current is $i(0) = 0.5$ A, and the initial capacitor voltage is $y(0) = 2$ V. No external input signal is applied to the circuit, therefore $x(t) = 0$. Determine the output voltage $y(t)$ if

a. the element values are $R = 2\ \Omega$, $L = 1$ H and $C = 1/26$ F,

b. the element values are $R = 6\ \Omega$, $L = 1$ H and $C = 1/9$ F.

Solution: Since no external input signal is applied to the circuit, the output signal is equal to the natural response, i.e., $y(t) = y_h(t)$, which we will obtain by solving the homogeneous differential equation. Two initial conditions are specified. The first one is that $y_h(0) = 2$. Using the specified initial value of the inductor current we have

$$i(0) = C\,\left.\frac{dy_h(t)}{dt}\right|_{t=0} = 0.5$$

which leads to the initial value for the first derivative of the output signal as

$$\left.\frac{dy_h(t)}{dt}\right|_{t=0} = \frac{0.5}{C}$$

Now we are ready to find the solutions for the two sets of component values given in parts (a) and (b):

a. Using the specified component values, the homogeneous differential equation is

$$\frac{d^2y(t)}{dt^2} + 2\frac{dy(t)}{dt} + 26\,y(t) = 0$$

and the characteristic equation is

$$s^2 + 2s + 26 = 0$$

The roots of the characteristic equation are $s_1 = -1 + j5$ and $s_2 = -1 - j5$. The natural response is therefore in the form given by Eqn. (2.91) with $\sigma_1 = -1$ and $\omega_1 = 5$ rad/s.

$$y_h(t) = d_1\,e^{-t}\cos(5t) + d_2\,e^{-t}\sin(5t)$$

Now we can impose the initial conditions. The first one is straightforward:

$$y_h(0) = d_1 = 2$$

For the specified capacitance value $C = 1/26$ F the initial value of $dy_h(t)/dt$ is

$$\left.\frac{dy_h(t)}{dt}\right|_{t=0} = 13$$

To impose the specified initial value of $dy_h(t)/dt$ we will first differentiate the homogeneous solution to obtain

$$\frac{dy_h(t)}{dt} = -d_1\,e^{-t}\cos(5t) - 5d_1\,e^{-t}\sin(5t) - d_2\,e^{-t}\sin(5t) + 5d_2\,e^{-t}\cos(5t)$$

Evaluating this derivative at $t = 0$ we have

$$\left.\frac{dy_h(t)}{dt}\right|_{t=0} = -d_1 + 5d_2 = 13$$

Since we previously found that $d_1 = 2$, we need $d_2 = 3$, and the natural response of the circuit is

$$y_h(t) = 2\,e^{-t}\cos(5t) + 3\,e^{-t}\sin(5t) \tag{2.95}$$

for $t \geq 0$. This solution is shown in Fig. 2.26.

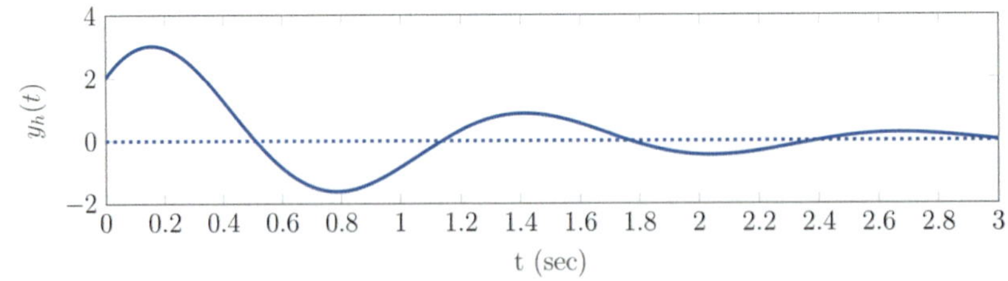

Figure 2.26 – Natural response of the RLC circuit in Example 2.15 for characteristic equation roots $s_{1,2} = -1 \pm j2$.

b. For this case the homogeneous differential equation becomes

$$\frac{d^2y(t)}{dt^2}+6\frac{dy(t)}{dt}+9\,y(t)=0$$

and the characteristic equation is

$$s^2+6s+9=0\,,\quad \text{or}\quad (s+3)^2=0$$

Since both roots of the characteristic equation are at $s_1=-3$ we will use a homogeneous solution in the form

$$y_h(t)=c_{11}\,e^{-3t}+c_{12}\,te^{-3t}$$

for $t\ge 0$. Imposing the initial value $y(0)=2$ yields

$$c_{11}=2$$

for the first coefficient. For the specified capacitance value of $C=1/9$ F the initial value of the derivative of the output signal is

$$\left.\frac{dy_h(t)}{dt}\right|_{t=0}=4.5$$

To satisfy the initial condition on $dy_h(t)/dt$ we will differentiate the homogeneous solution and evaluate the result at $t=0$ which yields

$$\begin{aligned}\left.\frac{dy_h(t)}{dt}\right|_{t=0} &= \frac{d}{dt}\left.\left[c_{11}\,e^{-3t}+c_{12}\,te^{-3t}\right]\right|_{t=0}\\ &= \left.\left[-3c_{11}\,e^{-3t}+c_{12}\,e^{-3t}-3c_{12}\,te^{-3t}\right]\right|_{t=0}\\ &= -3\,c_{11}+c_{12}=4.5\end{aligned}$$

and leads to the result

$$c_{12}=10.5$$

Therefore the natural response for this case is

$$y_h(t)=2\,e^{-3t}+10.5\,te^{-3t} \tag{2.96}$$

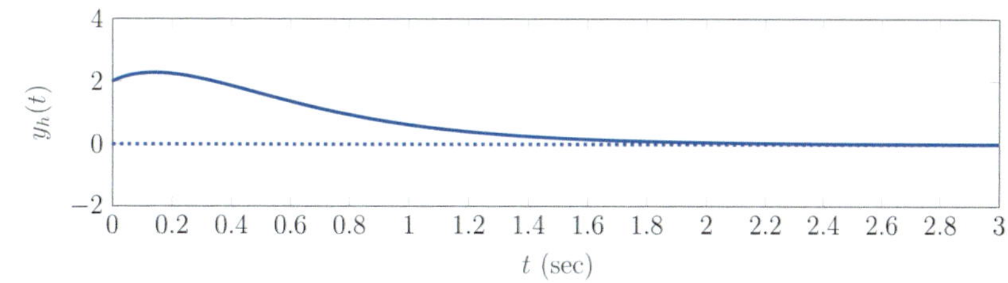

Figure 2.27 – Natural response of the RLC circuit in Example 2.15 for characteristic equation roots $s_1=s_2=-3$.

Software resources: exct_2_15a.m , exct_2_15b.m , exct_2_15_Live.mlx

Interactive App: Natural response of second-order continuous-time system

The interactive app in `appNatural.m` illustrates different types of homogeneous solutions for a second-order continuous-time system based on the roots of the characteristic polynomial. Recall that the three possibilities were explored above, namely distinct real roots, complex conjugate roots, and multiple roots.

In the app, program, the two roots can be specified using slider controls, and the corresponding natural response can be observed. If the roots are both real, then they can be controlled independently. If complex values are chosen, however, then the two roots move simultaneously to keep their complex conjugate relationship. The locations of the two roots s_1 and s_2 are marked on the complex plane. The differential equation, the characteristic equation and the analytical solution for the natural response are displayed and updated as the roots are moved.

a. Start with two real roots that are both negative, say $s_1 = -1$ and $s_2 = -0.5$. Set initial conditions as

$$y_h(t) = 0, \qquad \left.\frac{dy_h(t)}{dt}\right|_{t=0} = 1$$

Observe the shape of the natural response.

b. Gradually move s_2 to the right, bringing it closer to the vertical axis. Observe the changes in the shape of the natural response. What happens when the root reaches the vertical axis, that is, when $s_2 = 0$?

c. Gradually move s_2 to the right, bringing it closer to the vertical axis. Observe the changes in the shape of the natural response. What happens when the root reaches the vertical axis, that is, when $s_2 = 0$?

d. Keep moving s_2 to the right until $s_2 = 0.3$. How does the natural response change when a real root moves into the positive territory?

e. Set the two roots to be $s_{1,2} = -0.3 \pm j1.2$ and observe the natural response.

f. Gradually increase the imaginary parts of the two roots toward $\pm j4$ while keeping their real parts fixed at -0.3. How does the change in the imaginary parts affect the natural response?

g. Gradually move the real parts of the two roots to the right while keeping the imaginary parts equal to $\pm j4$. How does this affect the natural response? What happens when the roots cross over to the right side of the vertical axis?

Software resource: `appNatural.m`

2.5.4 Finding the forced response of a continuous-time system

In the preceding discussion we have concentrated our efforts on finding the homogeneous solution of the constant-coefficient differential equation which corresponds to the natural response $y_h(t)$ of the system under consideration when no external input signal is applied to it. To complete the solution process, we also need to determine the particular solution for a specified input signal $x(t)$ that is applied to the system. To find the particular solution, we start with an educated guess about the form of the solution we seek, and then adjust its parameters so that the result satisfies the differential equation. The form of the particular solution should include the input signal $x(t)$ and all of its derivatives that are linearly independent, assuming $x(t)$ has a finite number of linearly independent derivatives. For example, if the input signal is $x(t) = \cos(at)$, we need to construct a particular solution that includes the terms $\cos(at)$ and $\sin(at)$ in the form

$$y_p(t) = k_1 \cos(at) + k_2 \sin(at) \tag{2.97}$$

with parameters k_1 and k_2 to be determined from the differential equation. On the other hand, if the input signal is $x(t) = t^3$, then we need a particular solution that includes the terms t^3, t^2, t, and a constant term in the form

$$y_p(t) = k_3 t^3 + k_2 t^2 + k_1 t + k_0 \tag{2.98}$$

Table 2.1 lists some of the common types of input signals and the forms of particular solutions to be used for them.

Table 2.1 – Choosing a particular solution for various input signals.

Input signal	Particular solution
K (constant)	k_1
$K e^{at}$	$k_1 e^{at}$
$K \cos(at)$	$k_1 \cos(at) + k_2 \sin(at)$
$K \sin(at)$	$k_1 \cos(at) + k_2 \sin(at)$
$K t^n$	$k_n t^n + k_{n-1} t^{n-1} + \ldots + k_1 t + k_0$

The coefficients of the particular solution are determined from the differential equation by assuming all initial conditions are equal to zero (recall that the particular solution does not depend on the initial conditions of the differential equation or the initial state of the system). The specified initial conditions of the differential equation are imposed in the subsequent step for determining the unknown coefficients of the homogeneous solution, not the particular solution.

Now we have all the tools we need to determine the forced response of the system for a specified input signal. The following is a summary of the procedure to be used:

Finding the forced response of a system for a specified input signal:

1. Write the homogeneous differential equation. Find the characteristic equation by replacing derivatives of the output signal with corresponding powers of s.
2. Solve for the roots of the characteristic equation and write the homogeneous solution in the form of Eqn. (2.87). If some of the roots are complex conjugate pairs, then use the form in Eqn. (2.88) for them. If there are some repeated roots, use the procedure outlined in Eqn. (2.94) for them. Leave the homogeneous solution in parametric form with undetermined coefficients; do not attempt to compute the coefficients $c_1, c_2, \ldots$ of the homogeneous solution yet.
3. Find the form of the particular solution by either picking the appropriate form of it from Table 2.1, or by constructing it as a linear combination of the input signal and its time derivatives. (This latter approach requires that the input signal have a finite number of linearly independent derivatives.)

4. Try the particular solution in the non-homogeneous differential equation and determine the coefficients $k_1, k_2, \ldots$ of the particular solution. At this point the particular solution should be uniquely determined. However, the coefficients of the homogeneous solution are still undetermined.
5. Add the homogeneous solution and the particular solution together to obtain the total solution. Compute the necessary derivatives of the total solution. Impose the necessary initial conditions on the total solution and its derivatives. Solve the resulting set of equations to determine the coefficients $c_1, c_2, \ldots$ of the homogeneous solution.

These steps for finding the forced solution of a differential equation will be illustrated in the next example.

Example 2.16: Forced response of the first-order system for sinusoidal input

Consider once again the RC circuit of Fig. 2.9 with the element values of $R = 1\ \Omega$ and $C = 1/4$ F. The initial value of the output signal is $y(0) = 5$. Determine the output signal in response to a sinusoidal input signal in the form

$$x(t) = A\cos(\omega t)$$

with amplitude $A = 20$ and radian frequency $\omega = 8$ rad/s.

Solution: Using the specified component values, the non-homogeneous differential equation of the circuit under consideration is

$$\frac{dy(t)}{dt} + 4\,y(t) = 4\,x(t)$$

and we have found in Example 2.11 that the homogeneous solution is in the form

$$y_h(t) = c\,e^{-4t}$$

for $t \geq 0$. We will postpone the task of determining the value of the constant c until after we find the particular solution. Using Table 2.1, the form of the particular solution we seek is

$$y_p(t) = k_1\cos(\omega t) + k_2\sin(\omega t) \tag{2.99}$$

Differentiating Eqn. (2.99) with respect to time yields

$$\frac{dy_p(t)}{dt} = -\omega k_1\sin(\omega t) + \omega k_2\cos(\omega t) \tag{2.100}$$

The particular solution $y_p(t)$ must satisfy the non-homogeneous differential equation. Substituting Eqns. (2.99) and (2.100) along with the specified input signal $x(t)$ into the differential equation we obtain

$$-\omega k_1\sin(\omega t) + \omega k_2\cos(\omega t) + 4\,[k_1\cos(\omega t) + 4k_2\sin(\omega t)] = A\cos(\omega t)$$

which can be written in a more compact form as

$$(4k_1 + \omega k_2 - A)\cos(\omega t) + (4k_2 - \omega k_1)\sin(\omega t) = 0 \tag{2.101}$$

Eqn. (2.101) must be satisfied for all values of t, therefore we must have

$$4k_1 + \omega k_2 - A = 0 \tag{2.102}$$

and

$$4k_2 - \omega k_1 = 0 \tag{2.103}$$

Eqns. (2.102) and (2.103) can be solved simultaneously to yield

$$k_1 = \frac{4A}{16+\omega^2} \quad \text{and} \quad k_2 = \frac{A\omega}{16+\omega^2}$$

The forced solution is obtained by adding the homogeneous solution and the particular solution together:

$$y(t) = y_h(t) + y_f(t) = ce^{-4t} + \frac{4A}{16+\omega^2}\cos(\omega t) + \frac{A\omega}{16+\omega^2}\sin(\omega t) \tag{2.104}$$

Let us now substitute the numerical values $A = 20$ and $\omega = 8$ rad/s. The output signal becomes

$$y(t) = c\,e^{-4t} + \cos(8t) + 2\sin(8t)$$

Finally, we will impose the initial condition $y(0) = 5$ to obtain

$$y(0) = 5 = c + \cos(0) + 2\sin(0)$$

which yields $c = 4$. Therefore, the complete solution is

$$y(t) = 4\,e^{-4t} + \cos(8t) + 2\sin(8t) \tag{2.105}$$

for $t \ge 0$. The solution found in Eqn. (2.105) has two fundamentally different components, and can be written in the form

$$y(t) = y_t(t) + y_{ss}(t) \tag{2.106}$$

The first term

$$y_t(t) = 4\,e^{-4t} \tag{2.107}$$

represents the part of the output signal that disappears over time, that is,

$$\lim_{t\to\infty}\left\{y_t(t)\right\} = 0$$

It is called the *transient* component of the output signal. The second term

$$y_{ss}(t) = \cos(8t) + 2\sin(8t) \tag{2.108}$$

is the part of the output signal that remains after the transient term disappears. Therefore it is called the *steady-state response* of the system. Transient and steady-state components as well as the complete response are shown in Fig. 2.28.

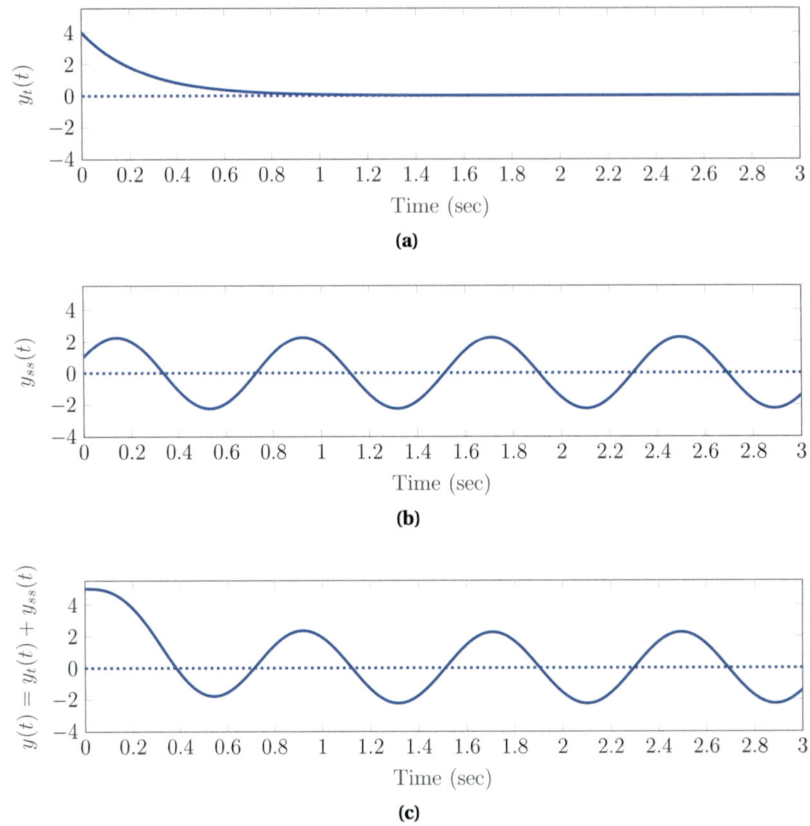

Figure 2.28 – Computation of the output signal of the circuit in Example 2.16: **(a)** transient component, **(b)** steady-state component, and **(c)** the complete output signal.

Before we leave this example, one final observation is in order: The homogeneous solution for the same circuit was found in Example 2.11 as

$$y_h(t) = 5\,e^{-4t}$$

with the input signal $x(t)$ set equal to zero. In this example we have used a sinusoidal input signal, and the transient part of the response obtained in Eqn. (2.107) does not match the homogeneous solution for the zero-input case; the two differ by a scale factor. This justifies our decision to postpone the computation of the constant c until the very end. The initial condition is specified for the total output signal $y(t)$ at time $t = 0$; therefore, we must first add the homogeneous solution and the particular solution, and only after doing that we can impose the specified value of $y(0)$ on the solution to determine the constant c. We will revisit this example when we study system function concepts in Chapter 4.

Software resources: `exct_2_16.m` , `exct_2_16_Live.mlx`

Interactive App: Forced response of RC circuit for sinusoidal input

The interactive app in `appForced.m` is based on Example 2.16. It allows experimentation with parameters of the problem. The amplitude is fixed at $A = 20$ so the input signal is

$$x(t) = 20\cos(\omega t)$$

The resistance R, the capacitance C, the radian frequency ω, and the initial output value $y(0)$ can be varied using slider controls. The effect of parameter changes on the transient response $y_t(t)$, the steady-state response $y_{ss}(t)$, and the total forced response $y_t(t) + y_{ss}(t)$ can be observed.

a. Start with the settings $R = 5\ \Omega$, $C = 10$ F, $\omega = 4$ rad/s, and $y(0) = 0$. Since the initial value of the output signal is zero, the resulting system is CTLTI. In addition, the initial value of the input signal is also zero, and therefore the output signal has no transient component. Observe the peak amplitude value of the steady-state component. Confirm that it matches with what was found in Example 2.16.

b. Now gradually increase the radian frequency ω up to $\omega = 12$ rad/s and observe the change in the peak amplitude of the steady-state component of the output. Compare with the result obtained in Eqn. (2.104).

c. Set parameter values as $R = 1\ \Omega$, $C = 0.3$ F, $\omega = 12$ rad/s, and $y(0) = 4$ V. The time constant is $\tau = RC = 0.3$ s. Comment on how long it takes for the transient component of the output to become negligibly small so that the output signal reaches its steady-state behavior. (As a rule of thumb, 4 to 5 time constants are sufficient for the output signal to be considered in steady state.)

d. Gradually increase the value of resistance to $R = 3\ \Omega$ and observe the changes in the transient behavior.

Software resource: `appForced.m`

2.6 Block Diagram Representation of Continuous-Time Systems

Up to this point we have studied methods for analyzing continuous-time systems represented in the time domain by means of constant-coefficient ordinary differential equations. Given a system description based on a differential equation, and an input signal $x(t)$, the output signal $y(t)$ can be determined by solving the differential equation.

In some cases we may need to realize a continuous-time system that has a particular differential equation. Alternately, it may be desired to simulate a continuous-time system on a digital computer in an approximate sense. In problems that involve the realization or the simulation of a continuous-time system, we start with a block diagram representation of the differential equation. Once a block diagram is obtained, it may either be realized using circuits that approximate the behavior of each component of the block diagram, or simulated on a computer using code segments that simulate the behavior of each component.

In general, the problem of converting a differential equation to a block diagram has multiple solutions that are functionally equivalent even though they may look different. In this section we will present one particular technique for obtaining a block diagram from a constant-coefficient ordinary differential equation. Alternative methods of constructing block diagrams will be presented in Chapter 4 in the context of obtaining a block diagram from a system function.

Block diagrams for continuous-time systems are constructed using three types of components, namely constant-gain amplifiers, signal adders, and integrators. These components are shown in Fig. 2.29.

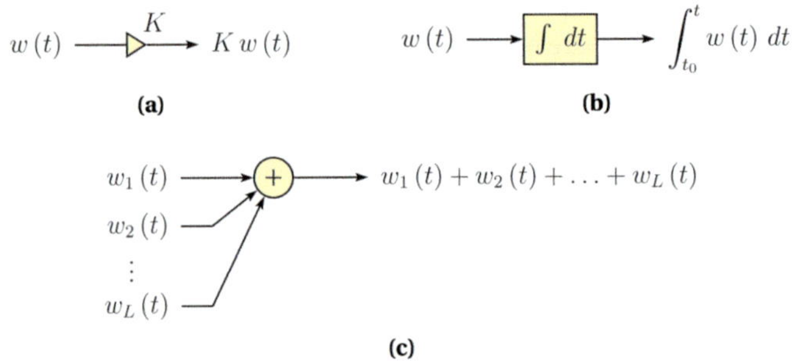

Figure 2.29 – Block diagram components for continuous-time systems: **(a)** constant-gain amplifier, **(b)** integrator, and **(c)** signal adder.

The technique for finding a block diagram from a differential equation is best explained with an example. Consider a third-order differential equation in the form

$$\frac{d^3y}{dt^3}+a_2\,\frac{d^2y}{dt^2}+a_1\,\frac{dy}{dt}+a_0\,y=b_2\,\frac{d^2x}{dt^2}+b_1\,\frac{dx}{dt}+b_0\,x \tag{2.109}$$

which is in the standard form of Eqn. (2.23) with $N=3$ and $M=2$. The differential equation is shown in compact notation with the understanding that $x(t)$, $y(t)$, and all of their derivatives are functions of time. Also, for convenience, we have chosen the coefficient of the highest derivative of $y(t)$ to be $a_N=1$ (which, in this case, happens to be a_3). In cases where $a_N\neq 1$ both sides of Eqn. (2.109) can be divided by a_N to satisfy this requirement.

As the first step for finding a block diagram for this differential equation, we will introduce an intermediate variable $w(t)$. This new variable will be used in place of $y(t)$ in the left side of the differential equation in Eqn. (2.109) and the result will be set equal to $x(t)$ to yield

$$\frac{d^3w}{dt^3}+a_2\,\frac{d^2w}{dt^2}+a_1\,\frac{dw}{dt}+a_0\,w=x \tag{2.110}$$

The differential equation in Eqn. (2.110) is relatively easy to implement in the form of a block diagram. Rearranging terms in Eqn. (2.110) we obtain

$$\frac{d^3w}{dt^3}=x-a_2\,\frac{d^2w}{dt^2}-a_1\,\frac{dw}{dt}-a_0\,w \tag{2.111}$$

One possible implementation is shown in Fig. 2.30.

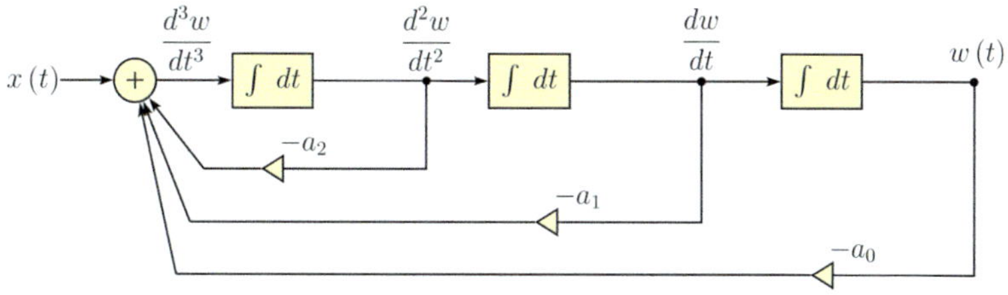

Figure 2.30 – The block diagram for Eqn. (2.111).

We will now show that the output signal $y(t)$ can be expressed in terms of the intermediate variable $w(t)$ as

$$y = b_2\,\frac{d^2w}{dt^2} + b_1\,\frac{dw}{dt} + b_0\,w \tag{2.112}$$

This is essentially the right side of the differential equation in Eqn. (2.109) in which $x(t)$ was replaced with $w(t)$. Together, Eqns. (2.110) and (2.112) are equivalent to Eqn. (2.109).

The proof for Eqn. (2.112):

The proof for Eqn. (2.112) will be given starting with the right side of Eqn. (2.109) and expressing the terms in it through the use of Eqn. (2.110). We have the following relationships:

$$b_0\,x = b_0\left(\frac{d^3w}{dt^3} + a_2\,\frac{d^2w}{dt^2} + a_1\,\frac{dw}{dt} + a_0\,w\right) \tag{2.113}$$

$$b_1\,\frac{dx}{dt} = b_1\left(\frac{d^4w}{dt^4} + a_2\,\frac{d^3w}{dt^3} + a_1\,\frac{d^2w}{dt^2} + a_0\,\frac{dw}{dt}\right) \tag{2.114}$$

$$b_2\,\frac{d^2x}{dt^2} = b_2\left(\frac{d^5w}{dt^5} + a_2\,\frac{d^4w}{dt^4} + a_1\,\frac{d^3w}{dt^3} + a_0\,\frac{d^2w}{dt^2}\right) \tag{2.115}$$

Adding Eqns. (2.113) through (2.115) we get

$$\begin{aligned}
b_0\,x + b_1\,\frac{dx}{dt} + b_2\,\frac{d^2x}{dt^2} =& \left(b_0\,\frac{d^3w}{dt^3} + b_1\,\frac{d^4w}{dt^4} + b_2\,\frac{d^5w}{dt^5}\right)\\
&+ a_2\left(b_0\,\frac{d^2w}{dt^2} + b_1\,\frac{d^3w}{dt^3} + b_2\,\frac{d^4w}{dt^4}\right)\\
&+ a_1\left(b_0\,\frac{dw}{dt} + b_1\,\frac{d^2w}{dt^2} + b_2\,\frac{d^3w}{dt^3}\right)\\
&+ a_0\left(b_0\,w + b_1\,\frac{dw}{dt} + b_2\,\frac{d^2w}{dt^2}\right)\\
=& \frac{d^3y}{dt^3} + a_2\,\frac{d^2y}{dt^2} + a_1\,\frac{dy}{dt} + a_0\,y
\end{aligned} \tag{2.116}$$

proving that Eqns. (2.110) and (2.112) are indeed equivalent to Eqn. (2.109). Thus, the output $y(t)$ can be obtained through the use of Eqn. (2.112).

Since the derivatives on the right-side of Eqn. (2.112) are readily available in the block diagram of Fig. 2.30, we will simply add the required connections to it to arrive at the completed block diagram shown in Fig. 2.31.

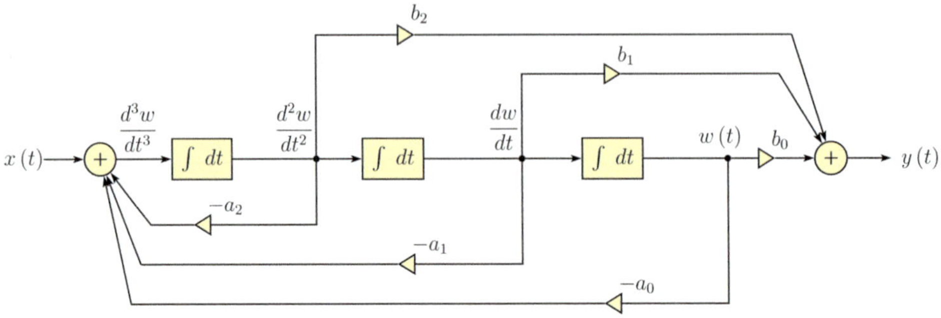

Figure 2.31 – The completed block diagram for Eqn. (2.109).

Even though we have used a third-order differential equation to demonstrate the process of constructing a block diagram, the extension of the technique to a general constant-coefficient differential equation is straightforward. In Fig. 2.31 the feed-forward gains of the block diagram are the right-side coefficients $b_0, b_1, \ldots, b_M$ of the differential equation. Feedback gains of the block diagram are the negated left-side coefficients $-a_0, -a_1, \ldots, -a_{N-1}$ of the differential equation. Recall that we must have $a_N = 1$ for this to work.

Imposing initial conditions

It is also possible to incorporate initial conditions into the block diagram. In the first step, initial values of $y(t)$ and its first $N-1$ derivatives need to be converted to corresponding initial values of $w(t)$ and its first $N-1$ derivatives. Afterward, appropriate initial value can be imposed on the output of each integrator as shown in Fig. 2.32.

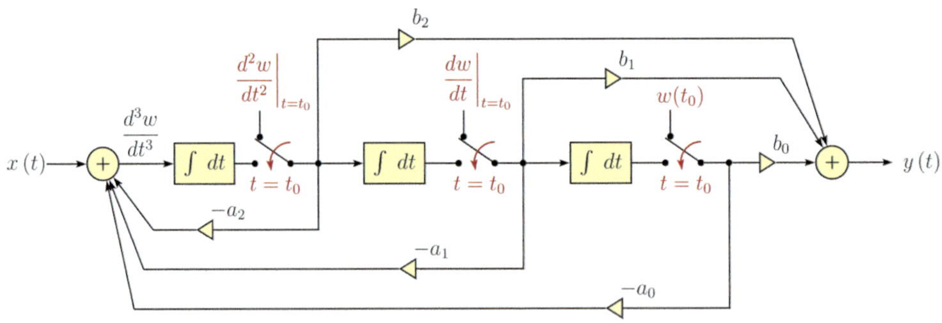

Figure 2.32 – Incorporating initial conditions into a block diagram.

Example 2.17: Block diagram for continuous-time system

Construct a block diagram to solve the differential equation

$$\frac{d^3y}{dt^3}+5\frac{d^2y}{dt^2}+17\frac{dy}{dt}+13\,y=x+2\frac{dx}{dt}$$

with the input signal $x(t)=\cos(20\pi t)$ and subject to initial conditions

$$y(0)=1\,,\qquad \left.\frac{dy}{dt}\right|_{t=0}=2\,,\qquad \left.\frac{d^2y}{dt^2}\right|_{t=0}=-4$$

Solution: Using the intermediate variable $w(t)$ as outlined in the preceding discussion we obtain the following pair of differential equations equivalent to the original differential equation:

$$\frac{d^3w}{dt^3}+5\frac{d^2w}{dt^2}+17\frac{dw}{dt}+13\,w=x \tag{2.117}$$

$$y=w+2\frac{dw}{dt} \tag{2.118}$$

Initial conditions specified in terms of the values of y, dy/dt, and d^2y/dt^2 at time instant $t=0$ need to be expressed in terms of the integrator outputs w, dw/dt, and d^2w/dt^2 at time instant $t=0$. For this purpose we will start by writing Eqn. (2.118) at time $t=0$:

$$y(0)=1=w(0)+2\left.\frac{dw}{dt}\right|_{t=0} \tag{2.119}$$

By differentiating both sides of Eqn. (2.118) and evaluating the result at $t=0$, we obtain

$$\left.\frac{dy}{dt}\right|_{t=0}=2=\left.\frac{dw}{dt}\right|_{t=0}+2\left.\frac{d^2w}{dt^2}\right|_{t=0} \tag{2.120}$$

Differentiating Eqn. (2.118) twice and evaluating the result at $t=0$ yields

$$\left.\frac{d^2y}{dt^2}\right|_{t=0}=-4=\left.\frac{d^2w}{dt^2}\right|_{t=0}+2\left.\frac{d^3w}{dt^3}\right|_{t=0} \tag{2.121}$$

Initial value of d^3w/dt^3 needed in Eqn. (2.121) is obtained from Eqn. (2.117) as

$$\left.\frac{d^3w}{dt^3}\right|_{t=0}=x(0)-5\left.\frac{d^2w}{dt^2}\right|_{t=0}-17\left.\frac{dw}{dt}\right|_{t=0}-13\,w(0) \tag{2.122}$$

Substitution of Eqn. (2.122) into Eqn. (2.121) results in

$$\left.\frac{d^2y}{dt^2}\right|_{t=0}=-4=-9\left.\frac{d^2w}{dt^2}\right|_{t=0}-34\left.\frac{dw}{dt}\right|_{t=0}-26\,w(0)+2\,x(0) \tag{2.123}$$

In addition we know that $x(0)=1$. Solving Eqns. (2.119), (2.120), and (2.123) the initial values of integrator outputs are obtained as

$$w(0)=\frac{-71}{45}\,,\qquad \left.\frac{dw}{dt}\right|_{t=0}=\frac{58}{45}\,,\qquad \left.\frac{d^2w}{dt^2}\right|_{t=0}=\frac{16}{45}$$

The block diagram can be constructed as shown in Fig. 2.33.

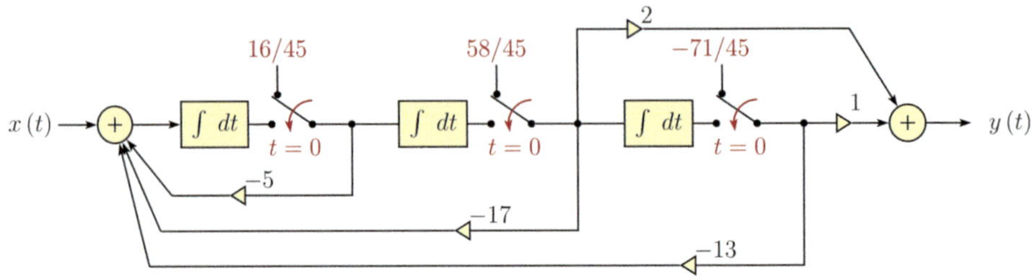

Figure 2.33 – Block diagram for Example 2.17.

2.7 Impulse Response and Convolution

In previous sections of this chapter we have explored the use of differential equations for describing the time-domain behavior of continuous-time systems, and have concluded that a CTLTI system can be completely described by means of a constant-coefficient ordinary differential equation. An alternative description of a CTLTI system can be given in terms of its *impulse response* $h(t)$ which is simply the forced response of the system under consideration when the input signal is a unit impulse. This is illustrated in Fig. 2.34.

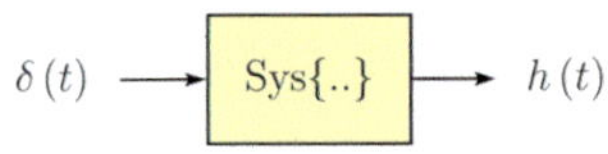

Figure 2.34 – Computation of the impulse response for a CTLTI system.

It will be shown later in this chapter that the impulse response also constitutes a complete description of a CTLTI system. Consequently, the response of such a system to any arbitrary input signal $x(t)$ can be uniquely determined from the knowledge of its impulse response.

It should be noted that our motivation for exploring additional description forms for CTLTI systems is not due to any deficiency or shortcoming of the differential equation of the system. We know that the differential equation is sufficient for the solution of any signal-system interaction problem. The impulse response $h(t)$ does not provide any new information or capability beyond that provided by the differential equation. What we gain through it is an alternative means of working with signal-system interaction problems which can sometimes be more convenient, and which can provide additional insight into system behavior. In the following sections we will discuss how the impulse response of a CTLTI system can be obtained from the underlying differential equation. The reverse operation is also possible, and will be discussed in later chapters.

2.7.1 Finding impulse response of a CTLTI system

When we worked on computing the forced response of a system from its differential equation, we relied on the entries in Table 2.1 to find a particular solution. First, the form of the particular solution appropriate for the type of input signal under consideration is obtained from the table. The particular solution is then tested against the differential equation and values of any unknown coefficients are computed. Afterward the particular solution is combined with the homogeneous solution to form the forced response of the system, and initial conditions are imposed to determine the values of any remaining coefficients.

In determining the impulse response of a system from its differential equation, we run into a roadblock: There is no entry in Table 2.1 that corresponds to an input signal in the form $x(t) = \delta(t)$. For a first-order differential equation we can use the technique outlined in Section 2.5.1 and obtain the impulse response through the use of Eqn. (2.47) with $t_0 = 0$, $y(t_0) = 0$, and $x(\tau) = \delta(\tau)$. However, this approach is not applicable to a higher-order differential equation. We will therefore take a slightly different approach in finding the impulse response of a CTLTI system:

Finding the impulse response of a CTLTI system from its differential equation:

1. Use a unit step function for the input signal and compute the forced response of the system using the techniques introduced in previous sections. This is the *unit step response* of the system.
2. Differentiate the unit step response of the system to obtain the impulse response, i.e.,

$$h(t) = \frac{dy(t)}{dt} \tag{2.124}$$

This idea relies on the fact that differentiation is a linear operator. Given that

$$y(t) = \text{Sys}\{x(t)\} \tag{2.125}$$

we have

$$\frac{dy(t)}{dt} = \frac{d}{dt}\left[\text{Sys}\{x(t)\}\right] = \text{Sys}\left\{\frac{dx(t)}{dt}\right\} \tag{2.126}$$

By choosing the input signal to be a unit step function, that is, $x(t) = u(t)$, and also recalling that $du(t)/dt = \delta(t)$, we obtain

$$\text{Sys}\{\delta(t)\} = \text{Sys}\left\{\frac{du(t)}{dt}\right\} = \frac{d}{dt}\left[\text{Sys}\{u(t)\}\right] \tag{2.127}$$

It is important to remember that the relationship expressed by Eqn. (2.127) is only valid for a CTLTI system. This is not a serious limitation at all, since we will rarely have a reason to compute an impulse response unless the system is linear and time-invariant.

Example 2.18: Impulse response of the simple RC circuit

Determine the impulse response of the first-order RC circuit of Example 2.4, shown in Fig. 2.9, first in parametric form, and then using element values $R = 1\ \Omega$ and $C = 1/4$ F. Assume the system is initially relaxed, that is, there is no initial energy stored in the system. (Recall that this is a necessary condition for the system to be CTLTI.)

Solution: We will solve this problem using two different methods. The differential equation for the circuit was given by Eqn. (2.11). Since we have a first-order differential equation, we can use the solution method that led to Eqn. (2.47) in Section 2.5.1. Letting $\alpha = 1/RC$, $t_0 = 0$, and $y(0) = 0$ we have

$$y(t) = \int_0^t e^{-(t-\tau)/RC}\,\frac{1}{RC}\,x(\tau)\,d\tau$$

If the input signal is chosen to be a unit impulse signal, then the output signal becomes the impulse response of the system. Setting $x(t) = \delta(t)$ leads to

$$h(t) = \int_0^t e^{-(t-\tau)/RC} \frac{1}{RC} \delta(\tau)\, d\tau$$

Using the sifting property of the unit impulse function, we obtain

$$h(t) = \frac{1}{RC} e^{-t/RC} u(t) \tag{2.128}$$

Now we will obtain the same result using the more general method developed in Section 2.7.1. Recall that the unit step response of the system was found in Example 2.8 Eqn. (2.51) as

$$y(t) = \left(1 - e^{-t/RC}\right) u(t) \tag{2.129}$$

Differentiating Eqn. (2.129) with respect to time we obtain the impulse response:

$$h(t) = \frac{dy(t)}{dt} = \frac{d}{dt}\left[\left(1 - e^{-t/RC}\right) u(t)\right] = \frac{1}{RC} e^{-t/RC} u(t)$$

which is the same result obtained in Eqn. (2.128). With the substitution of the specified element values, the impulse response becomes

$$h(t) = 4\, e^{-4t} u(t) \tag{2.130}$$

which is graphed in Fig. 2.35.

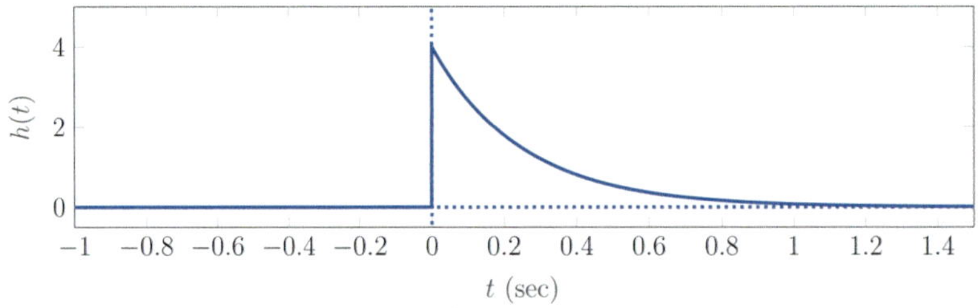

Figure 2.35 – Impulse response of the system in Example 2.18.

Software resources: exct_2_18.m , exct_2_18_Live.mlx

Example 2.19: Impulse response of a second-order system

Determine the impulse response of the RLC circuit that was used in Example 2.6 and shown in Fig. 2.11. Use element values $R = 2\ \Omega$, $L = 1$ H, and $C = 1/26$ F.

Solution: The homogeneous solution for the system was found in part (a) of Example 2.15 to be

$$y_h(t) = d_1\, e^{-t} \cos(5t) + d_2\, e^{-t} \sin(5t)$$

We will begin by finding the unit step response. The form of the particular solution is obtained from Table 2.1 as

$$y_p(t) = k_1$$

Testing the particular solution with the differential equation leads to the conclusion that $k_1 = 1$, and the complete solution is

$$\begin{aligned} y(t) =& y_h(t) + y_p(t) \\ =& d_1 e^{-t} \cos(5t) + d_2 e^{-t} \sin(5t) + 1 \end{aligned}$$

Since we are determining the impulse response, we will naturally assume that the system is CTLTI, and is therefore initially relaxed. This requires

$$y(0) = d_1 + 1 = 0 \tag{2.131}$$

and

$$\begin{aligned} \left.\frac{dy_h(t)}{dt}\right|_{t=0} =& \left.\left[(-d_1 + 5d_2)\, e^{-t} \cos(5t) + (-5d_1 - d_2)\, e^{-t} \sin(5t)\right]\right|_{t=0} \\ =& -d_1 + 5d_2 = 0 \end{aligned} \tag{2.132}$$

Solving Eqns. (2.131) and (2.132) yields $d_1 = -1$ and $d_2 = -0.2$. Therefore, the unit step response of the system is

$$y(t) = y_h(t) + y_p(t) = -e^{-t} \cos(5t) - 0.2\, e^{-t} \sin(5t) + 1$$

for $t \ge 0$. The impulse response is found by differentiating the unit step response with respect to time, and is

$$\begin{aligned} h(t) =& \frac{dy(t)}{dt} \\ =& e^{-t} \cos(5t) + 5\, e^{-t} \sin(5t) + 0.2\, e^{-t} \sin(5t) - e^{-t} \cos(5t) \\ =& 5.2\, e^{-t} \sin(5t) \end{aligned} \tag{2.133}$$

for $t \ge 0$. This result is graphed in Fig. 2.36.

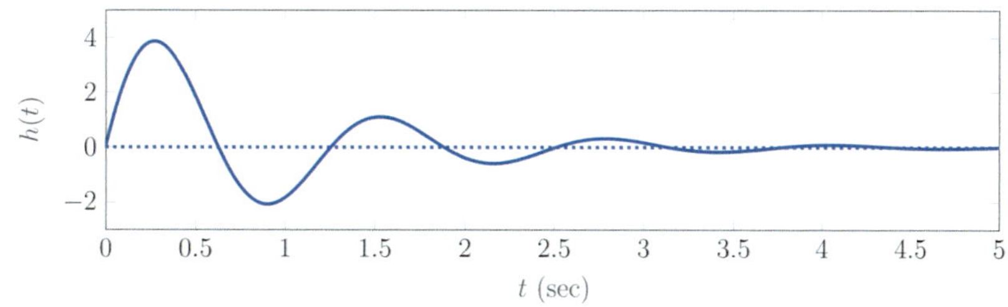

Figure 2.36 – Impulse response of the system in Example 2.19.

Software resources: `exct_2_19.m` , `exct_2_19_Live.mlx`

2.7.2 Convolution operation for CTLTI systems

In Section 1.6 of Chapter 1 we have formulated the decomposition of an arbitrary signal into impulse functions. Specifically, we have established that a signal $x(t)$ can be expressed through a decomposition in the form

$$x(t) = \int_{-\infty}^{\infty} x(\lambda)\, \delta(t-\lambda)\, d\lambda \tag{2.134}$$

Using Eqn. (2.134) the output signal $y(t)$ of a system can be written in terms of its input signal $x(t)$ as

$$y(t) = \text{Sys}\left\{x(t)\right\} = \text{Sys}\left\{\int_{-\infty}^{\infty} x(\lambda)\,\delta(t-\lambda)\,d\lambda\right\} \tag{2.135}$$

Assuming the system under consideration is linear, we will take liberty[3] to swap the order of integration and system transformation, and write the output signal as

$$y(t) = \int_{-\infty}^{\infty} \text{Sys}\left\{x(\lambda)\,\delta(t-\lambda)\right\} d\lambda \tag{2.136}$$

The justification for this action can be given by considering the integral of Eqn. (2.134) as the limit case of an infinite sum as demonstrated in Chapter 1, Eqn. (1.35).

For the sake of discussion let us assume that we know the system under consideration to be linear, but not necessarily time-invariant. The response of a linear system to an impulse signal scaled by $x(\lambda)$ and time-shifted by λ is

$$\text{Sys}\left\{x(\lambda)\,\delta(t-\lambda)\right\} = x(\lambda)\,\text{Sys}\left\{\delta(t-\lambda)\right\}$$

Eqn. (2.136) can therefore be written as

$$y(t) = \int_{-\infty}^{\infty} x(\lambda)\,\text{Sys}\left\{\delta(t-\lambda)\right\} d\lambda \tag{2.137}$$

Using Eqn. (2.137), the response of a linear system to any arbitrary input signal $x(t)$ can be computed, provided that we know the responses of the system to impulse signals time-shifted by all possible values. Thus, the knowledge necessary for determining the output of a linear system in response to an arbitrary signal $x(t)$ is

$$\left\{h_\lambda(t) = \text{Sys}\left\{\delta(t-\lambda)\right\}, \quad \text{all } \lambda\right\}$$

While Eqn. (2.137) provides us with a viable method of determining system output $y(t)$ for any input signal $x(t)$, the vast amount of knowledge that we need to possess about the system makes this highly impractical.

For the result in Eqn. (2.137) to be of practical use, the prerequisite knowledge about the system must be reduced to a manageable level. Let the *impulse response* of the system be defined as

$$h(t) = \text{Sys}\left\{\delta(t)\right\} \tag{2.138}$$

If, in addition to being linear, the system is also known to be time-invariant, then the response of the system to any shifted impulse signal can be derived from the knowledge of $h(t)$ alone, that is,

$$h_\lambda(t) = \text{Sys}\left\{\delta(t-\lambda)\right\} = h(t-\lambda) \tag{2.139}$$

[3] Eqn. (2.136) essentially represents the superposition principle applied to an infinite sum of terms $x(\lambda)\,\delta(t-\lambda)$. There are studies that conclude that one may find particular signals and circumstances under which the transition from Eqn. (2.135) to Eqn. (2.136) would be problematic. For the types of signals and systems encountered in engineering practice, however, Eqn. (2.136) will generally be valid.

consistent with the definition of time invariance given by Eqn. (2.7). Now the output signal can be written as

$$y(t) = \int_{-\infty}^{\infty} x(\lambda)\, h(t-\lambda)\, d\lambda \tag{2.140}$$

Eqn. (2.140) is known as the *convolution integral* for continuous-time signals. The output signal $y(t)$ of a CTLTI system is obtained by *convolving* the input signal $x(t)$ and the impulse response $h(t)$ of the system. This relationship is expressed in compact notation as

$$y(t) = x(t) * h(t) \tag{2.141}$$

where the symbol $*$ represents the *convolution operator.* We will show later in this section that the convolution operator is commutative, that is, the relationship in Eqn. (2.140) can also be written in the alternative form

$$y(t) = h(t) * x(t) = \int_{-\infty}^{\infty} h(\lambda)\, x(t-\lambda)\, d\lambda \tag{2.142}$$

The roles of $h(t)$ and $x(t)$ can be swapped without affecting the end result.

Continuous-time convolution summary:

$$y(t) = x(t) * h(t) = \int_{-\infty}^{\infty} x(\lambda)\, h(t-\lambda)\, d\lambda \tag{2.143}$$

$$= h(t) * x(t) = \int_{-\infty}^{\infty} h(\lambda)\, x(t-\lambda)\, d\lambda \tag{2.144}$$

In computing the convolution integral it is helpful to sketch the signals. The graphical steps involved in computing the convolution of two signals $x(t)$ and $h(t)$ at a specific time instant t can be summarized as follows:

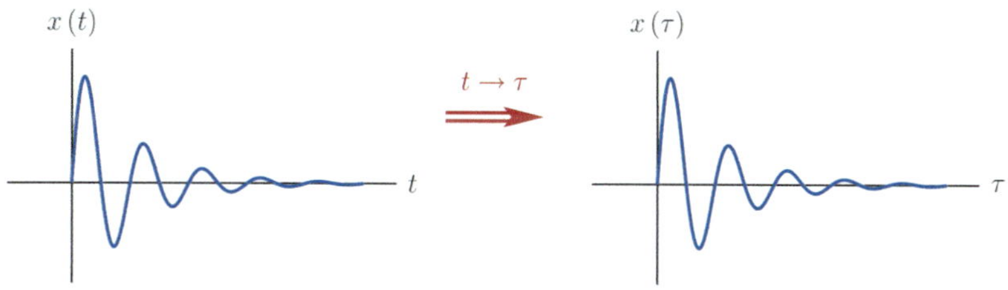

Figure 2.37 – Obtaining $x(\lambda)$ for the convolution integral.

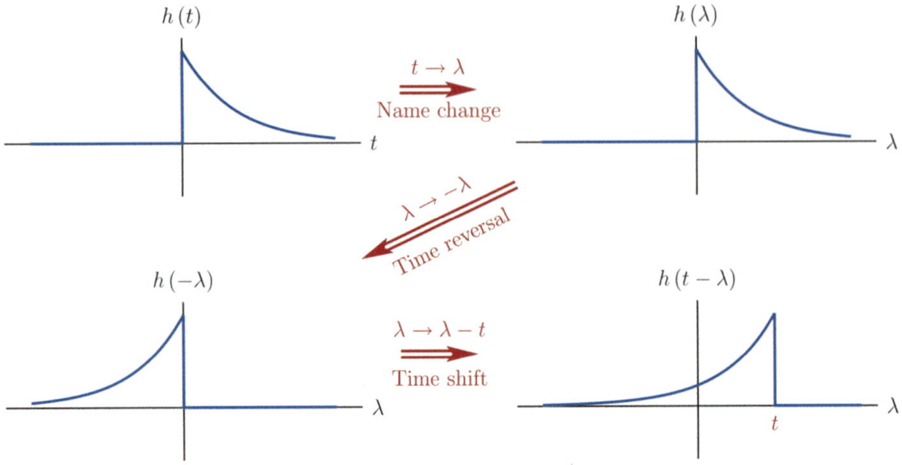

Figure 2.38 – Obtaining $h(t-\lambda)$ for the convolution integral.

Steps in continuous-time convolution:

1. Sketch the signal $x(\lambda)$ as a function of the independent variable λ. This corresponds to a simple name change on the independent variable, and the graph of the signal $x(\lambda)$ appears identical to the graph of the signal $x(t)$. (See Fig. 2.37.)
2. For one specific value of t, sketch the signal $h(t-\lambda)$ as a function of the independent variable λ. This task can be broken down into two steps as follows:

   a. Sketch $h(-\lambda)$ as a function of λ. This step amounts to time reversal of the signal $h(\lambda)$.

   b. In $h(\lambda)$ substitute $\lambda \rightarrow \lambda - t$. This step yields

   $$h(-\lambda)\Big|_{\lambda \rightarrow \lambda - t} = h(t-\lambda) \tag{2.145}$$

   and amounts to time shifting $h(-\lambda)$ by t.

   See Fig. 2.38 for an illustration of the steps for obtaining $h(t-\lambda)$.
3. Multiply the two signals sketched in 1 and 2 to obtain the product $f(\lambda) = x(\lambda)\, h(t-\lambda)$.
4. Compute the area under the product $f(\lambda) = x(\lambda)\, h(t-\lambda)$ by integrating it over the independent variable λ. The result is the value of the output signal at the specific time instant t.
5. Repeat steps 1 through 4 for all values of t that are of interest.

The next several examples will illustrate the details of the convolution operation for CTLTI systems.

Example 2.20: Unit step response of RC circuit revisited

The unit step response of the simple RC circuit was found in Example 2.8 using the direct solution method discussed in Section 2.5.1. Solve the same problem using the convolution operation.

Solution: In Example 2.18 we found the impulse response of the RC circuit to be

$$h(t) = \frac{1}{RC}\, e^{-t/RC}\, u(t)$$

Based on the convolution integral of Eqn. (2.140), the two functions we need are $x(\lambda)$ and $h(t-\lambda)$, both as functions of λ. Fig. 2.39 shows these functions for the two cases of $t \le 0$ and of $t > 0$. It is obvious that, for $t \le 0$ these two functions do not overlap anywhere, and therefore the integrand $x(\lambda)\, h(t-\lambda)$ of the convolution integral is zero for all values of λ. Thus, an easy conclusion is

$$y(t) = 0, \quad \text{for } t \le 0$$

If $t > 0$, the two functions $x(\lambda)$ and $h(t-\lambda)$ overlap for values of λ in the interval $(0, t)$. In this interval, we have

$$x(\lambda) = 1 \qquad \text{and} \qquad h(t-\lambda) = \frac{1}{RC}\, e^{-(t-\lambda)/RC}$$

Therefore, the output signal is

$$y(t) = \int_0^t \frac{1}{RC}\, e^{-(t-\lambda)/RC}\, d\lambda = 1 - e^{-t/RC}, \quad \text{for} \quad t > 0$$

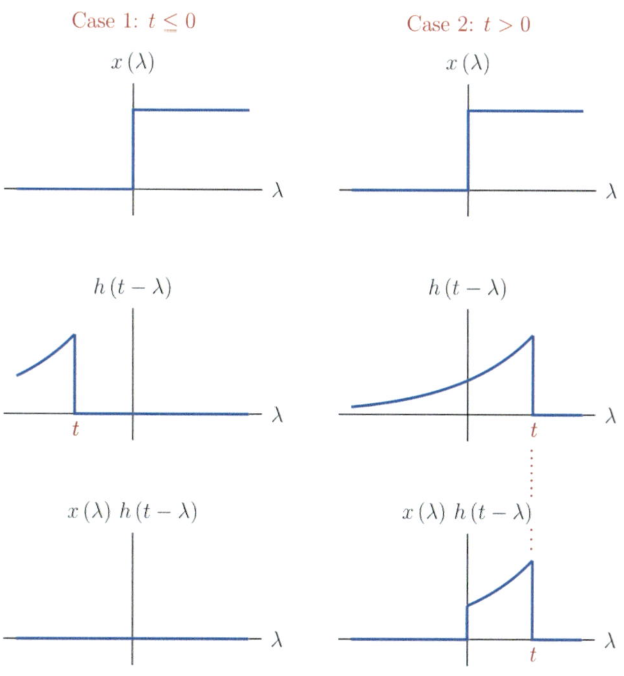

Figure 2.39 – Signals involved in the convolution integral of Example 2.20.

Using a unit step function to account for both segments of the solution found, the general solution valid for all t can be written as

$$y(t) = \left(1 - e^{-t/RC}\right) u(t)$$

which agrees with the result found in Example 2.8 by solving the differential equation directly.

Interactive App: Finding the unit step response of RC circuit using convolution

The interactive app in `appConv1.m` is based on Example 2.20, and allows visualization of the convolution operation. Signals $x(\lambda)$, $h(t-\lambda)$, and the product $[x(\lambda)\, h(t-\lambda)]$ are graphed on the left side of the app window. The time variable can be advanced by means of a slider control. As the time t is varied, graphs for the factor $h(t-\lambda)$ and the integrand $[x(\lambda)\, h(t-\lambda)]$ are updated. The area under the integrand $[x(\lambda)\, h(t-\lambda)]$ is shaded to correspond to the result of the integral for that particular value of t.

a. Start with the time slider set at $t = -0.5$ s. Set the time constant to $\tau = 1$ s. Gradually increment the time while observing the changes in $h(t-\lambda)$ and the product $x(\lambda)\, h(t-\lambda)$.

b. As the time variable crosses into the positive territory, pay special attention to the shaded area under the product $x(\lambda)\, h(t-\lambda)$ and how it translates to the amplitude of the output signal at that instant.

c. At $t = 0.5$ s, evaluate the integral

$$y(0.5) = \int_0^{0.5} e^{-(0.5-\lambda)/RC}\, d\lambda$$

using pencil and paper, and verify the value displayed in the app for the shaded area under the function. Recall that $\tau = RC$.

Software resource: `appConv1.m`

Example 2.21: Pulse response of RC circuit revisited

Using convolution, determine the response of the simple RC circuit of Example 2.4 to a unit pulse input signal $x(t) = \Pi(t)$.

Solution: As in the previous example, it will be useful to sketch the functions involved in the convolution integral, namely $x(\lambda)$ and $h(t-\lambda)$. In doing so, we will consider three distinctly different possibilities for the time variable t as shown in Fig. 2.40. The convolution integral will be evaluated separately for each of the three cases. It is suggested that the reader follow the development below while using the interactive app `appConv2.m` to observe the overlaps of $x(\lambda)$ and $h(t-\lambda)$ in each of the three cases.

Case 1: $t \le -\frac{1}{2}$

There is no overlap between the two functions in this case, therefore their product equals zero for all values of λ. As a result, the output signal is

$$y(t) = 0 \quad \text{for} \quad t \le -\frac{1}{2}$$

Case 2: $-\frac{1}{2} < t \le \frac{1}{2}$

For this case, the two functions overlap in the range $-\frac{1}{2} < \lambda \le t$. Setting integration limits accordingly, we can compute the output signal as

$$\begin{aligned} y(t) &= \int_{-1/2}^{t} \frac{1}{RC}\, e^{-(t-\lambda)/RC}\, d\lambda \\ &= \left(1 - e^{-(t+1/2)/RC}\right) \quad \text{for} \quad -\frac{1}{2} < t \le \frac{1}{2} \end{aligned}$$

Case 3: $t > \frac{1}{2}$

For this case, the overlap of the two functions will occur in the range $-\frac{1}{2} < \lambda \le \frac{1}{2}$. Setting integration limits to the endpoints of this new range, we have

$$\begin{aligned} y(t) &= \int_{-1/2}^{1/2} \frac{1}{RC}\, e^{-(t-\lambda)/RC}\, d\lambda \\ &= e^{-t/RC}\left(e^{1/2RC} - e^{-1/2RC}\right) \quad \text{for} \quad t > \frac{1}{2} \end{aligned}$$

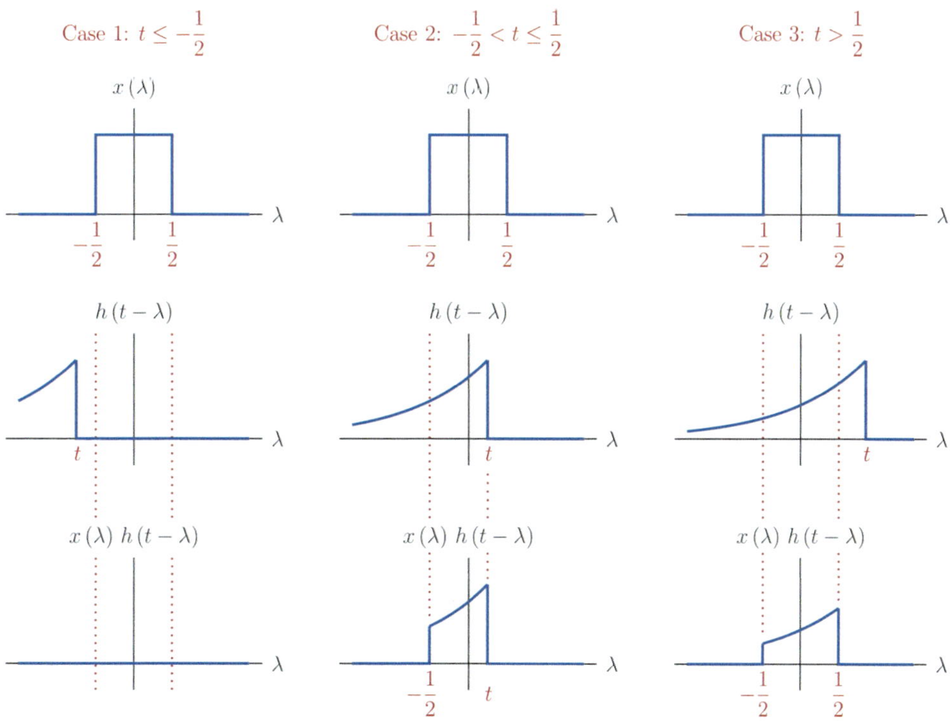

Figure 2.40 – Signals involved in the convolution integral of Example 2.21.

We have computed the output signal in each of the three distinct time intervals we have identified, mainly before the start of the pulse, during the pulse, and after the end of the pulse. Putting these three partial solutions together, the complete solution for the output signal is obtained as

$$y(t)=\begin{cases} 0\,, & t\le -\frac{1}{2} \\ \left(1-e^{-(t+1/2)/RC}\right), & -\frac{1}{2}<t\le\frac{1}{2} \\ e^{-t/RC}\left(e^{1/2RC}-e^{-1/2RC}\right), & t>\frac{1}{2} \end{cases} \tag{2.146}$$

and is shown in Fig. 2.41.

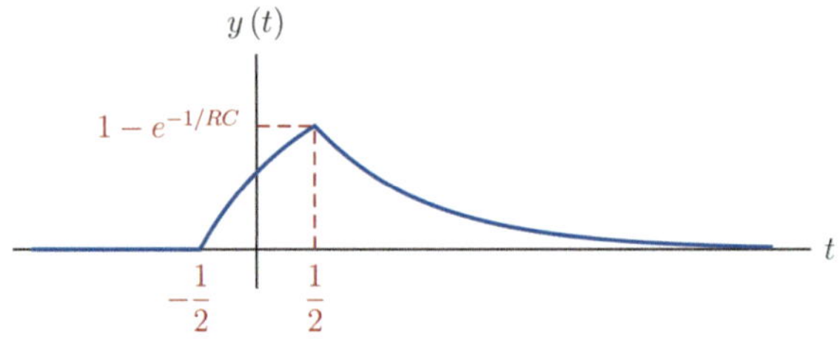

Figure 2.41 – Convolution result for Example 2.21.

It is easy to see that the output signal found in Eqn. (2.146) becomes identical to the result obtained in Eqn. (2.57) of Example 2.9 if we set $A=1$, $w=1$, $R=1\ \Omega$, and $C=1/4$ F.

Software resources: exct_2_21.m , exct_2_21_Live.mlx

Interactive App: Finding the pulse response of RC circuit using convolution

The interactive app in `appConv2.m` is based on Example 2.21, Fig. 2.40, and Fig. 2.41. It allows visualization of the convolution operation in finding the response of the RC circuit to a pulse centered at the time origin. The time t, time constant $\tau = RC$ and pulse width can be varied through the use of slider controllers. The graphical user interface of the app is similar to that of `appConv1.m` except it is adapted to the signals of Example 2.21.

a. Start with the time slider set at $t=-0.5$ s. Set the time constant to $\tau=1$ s and the pulse width to unity. Gradually increment the time while observing the changes in $h(t-\lambda)$ and the product $x(\lambda)\,h(t-\lambda)$.

b. As the time variable is advanced, pay special attention to the shaded area under the product $x(\lambda)\,h(t-\lambda)$ and how it translates to the amplitude of the output signal at that instant. Observe that the shaded area (and consequently the amplitude of the output signal) keeps increasing until $t=0.5$ s which is the right edge of the input pulse. At $t=0.3$ s, evaluate the integral

$$y(0.3)=\int_{-0.5}^{0.3} e^{-(0.3-\lambda)/RC}\,d\lambda$$

using pencil and paper, and verify the value displayed in the app for the shaded area under the function.

c. After you cross the threshold of $t = 0.5$ s, observe the reduction in the shaded area as the time variable is advanced. At $t = 0.7$ s, evaluate the integral

$$y(0.7) = \int_{-0.5}^{0.5} e^{-(0.7-\lambda)/RC}\, d\lambda$$

using pencil and paper, and verify the value displayed in the app for the shaded area under the function.

Software resource: `appConv2.m`

Example 2.22: A more involved convolution problem

Consider a system with the impulse response

$$h(t) = e^{-t}\,[u(t) - u(t-2)]$$

Let the input signal applied to this system be

$$x(t) = \Pi(t-0.5) - \Pi(t-1.5) = \begin{cases} 1\,, & 0 \le t < 1 \\ -1\,, & 1 \le t < 2 \\ 0\,, & \text{otherwise} \end{cases}$$

Determine the output signal $y(t)$ using convolution.

Solution: The functions $x(\lambda)$ and $h(t-\lambda)$ are shown in Fig. 2.42.

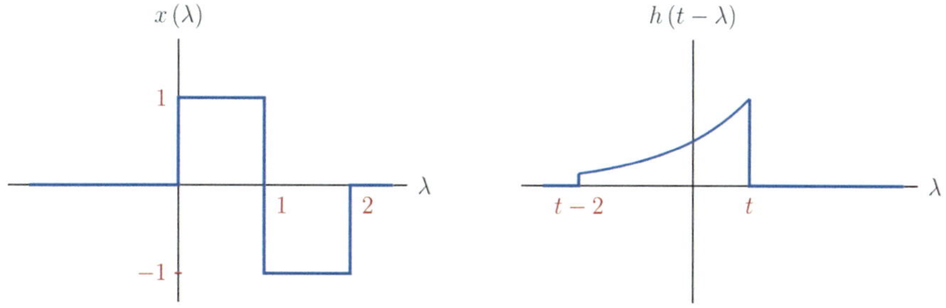

Figure 2.42 – The functions $x(\lambda)$ and $h(t-\lambda)$ for the convolution problem of Example 2.22.

For this problem we will need to distinguish between six different regions for the time variable t as shown in Figs. 2.43 and 2.44. The interactive app `appConv3.m` can be used along with the solution outlined below to observe the overlapped regions in time.

<u>Case 1: $t \le 0$</u>

The functions $x(\lambda)$ and $h(t-\lambda)$ do not overlap for $t \le 0$, and therefore their product equals zero for all values of λ. As a result, the output signal is

$$y(t) = 0 \quad \text{for} \quad t \le 0$$

Case 2: $0 < t \leq 1$

The two functions overlap for the range $0 < \lambda \leq t$. The output signal is

$$y(t) = \int_0^t (1)\, e^{-(t-\lambda)}\, d\lambda = 1 - e^{-t} \quad \text{for } 0 < t \leq 1$$

Case 3: $1 < t \leq 2$

The two functions overlap for the range $0 < \lambda \leq t$. In the interval $0 < \lambda \leq 1$ we have $x(\lambda) = 1$, and the resulting integrand is positive. On the other hand, in the interval $1 < \lambda \leq t$ we have $x(\lambda) = -1$, and the integrand is negative. As a result, two integrals will be needed for the solution:

$$y(t) = \int_0^1 (1)\, e^{-(t-\lambda)}\, d\lambda + \int_1^t (-1)\, e^{-(t-\lambda)}\, d\lambda$$
$$= -1 + 4.4366\, e^{-t} \quad \text{for } 1 < t \leq 2$$

Case 4: $2 < t \leq 3$

In this case the overlap region is $t-2 < \lambda \leq 2$. Furthermore, the left edge of $h(t-\lambda)$ is in the interval $0 < t-2 \leq 1$. In the interval $t-2 < \lambda \leq 1$ the input signal is $x(\lambda) = 1$ and the resulting integrand $[x(\lambda)\, h(t-\lambda)]$ is positive. In the remainder of the overlap interval, that is, for $1 < \lambda \leq 2$ we have $x(\lambda) = -1$ and the integrand is negative. Again, two integrals will be formed:

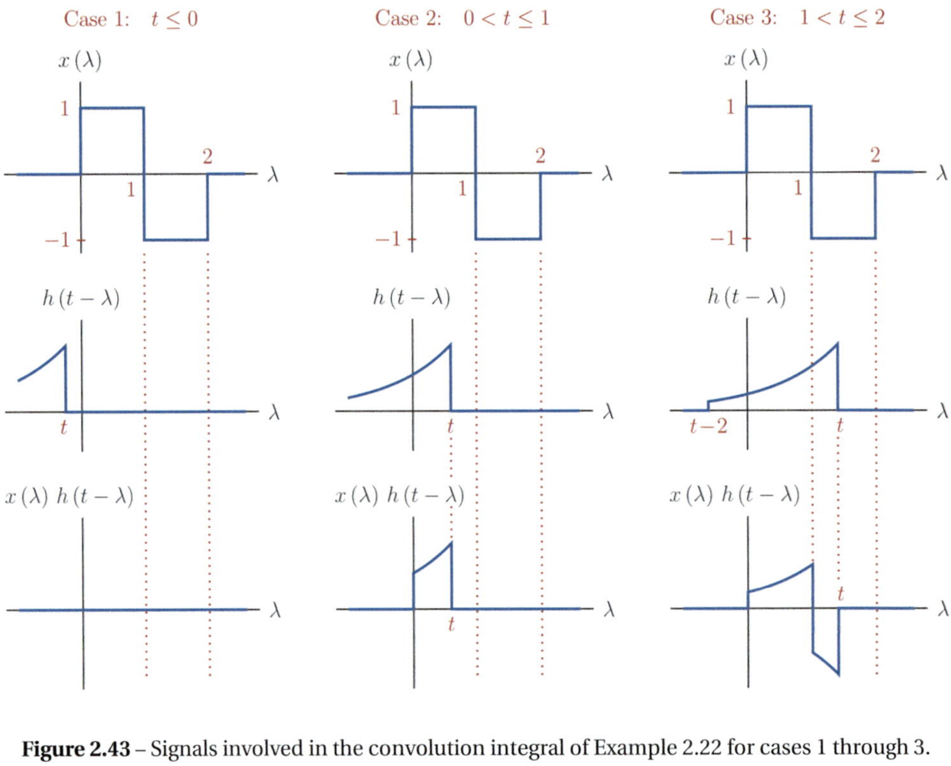

Figure 2.43 – Signals involved in the convolution integral of Example 2.22 for cases 1 through 3.

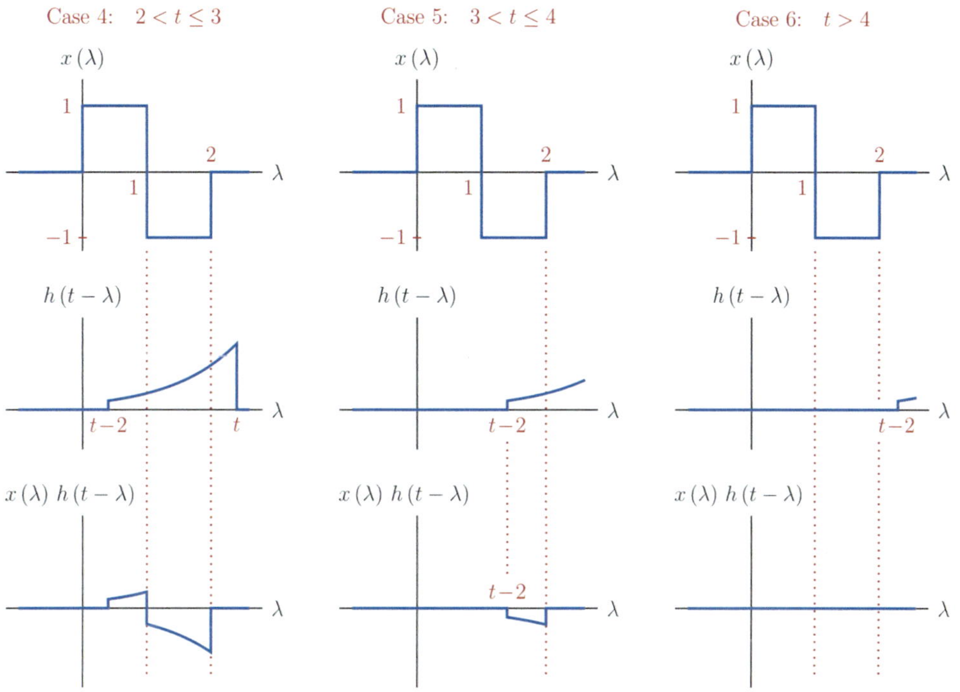

Figure 2.44 – Signals involved in the convolution integral of Example 2.22 for cases 4 through 6.

$$
\begin{aligned}
y(t) &= \int_{t-2}^{1} (1)\, e^{(t-\lambda)}\, d\lambda + \int_{1}^{2} (-1)\, e^{-(t-\lambda)}\, d\lambda \\
&= -0.1353 - 1.9525\, e^{-t} \quad \text{for } 2 < t \le 3
\end{aligned}
$$

Case 5: $3 < t \le 4$

In this case the overlap region is $t - 2 < \lambda \le 2$. The left edge of $h(t-\lambda)$ is in the interval $1 < t - 2 \le 2$. The output signal is

$$
y(t) = \int_{t-2}^{2} (-1)\, e^{-(t-\lambda)}\, d\lambda = 0.1353 - 7.3891\, e^{-t} \quad \text{for } 3 < t \le 4
$$

Case 6: $t > 4$

The functions $x(\lambda)$ and $h(t-\lambda)$ do not overlap for $t > 4$, and therefore the output signal is

$$
y(t) = 0 \quad \text{for } t > 4
$$

The complete output signal is graphed in Fig. 2.45 by combining the results found for all six regions of time.

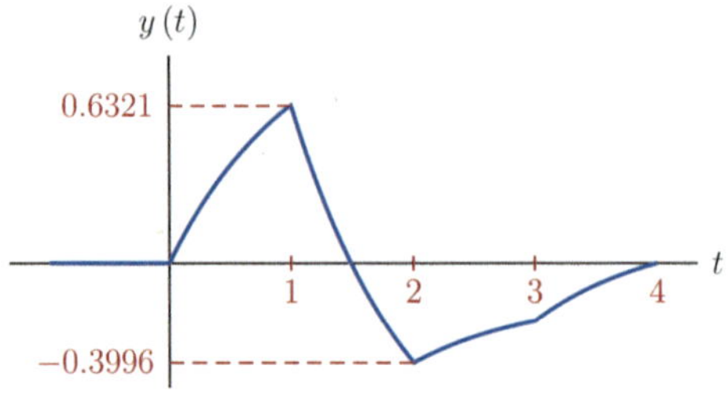

Figure 2.45 – Convolution result for Example 2.22.

Software resource: exct_2_22.m

Interactive App: A more involved convolution problem

The interactive app in appConv3.m is based on Example 2.22. It facilitates visualization of the overlapping regions between the functions $x(\lambda)$ and $h(t-\lambda)$ as the time variable t is varied. The graphical user interface is similar to the user interfaces of the apps appConv1.m and appConv2.m.

a. Start with the time slider set at $t = -0.2$ s. Gradually increment the time while observing the changes in $h(t-\lambda)$ and the product $x(\lambda)\,h(t-\lambda)$.

b. As the time variable is advanced, pay special attention to the shaded area under the product $x(\lambda)\,h(t-\lambda)$ and how it translates to the amplitude of the output signal at that instant. Observe the changes at critical time instants such as $t = 1,2,3,4$.

Software resource: appConv3.m

We have stated in Eqn. (2.142) that the convolution operator is commutative. The proof is straightforward and will now be given.

Proof that the convolution operator is commutative:

Let us apply the variable change $t-\lambda=\gamma$ to the integral in Eqn. (2.140), and write it as

$$\begin{aligned} y(t) =& x(t) * h(t) \\ =& \int_{\infty}^{-\infty} x(t-\gamma)\, h(\gamma)\,(-d\gamma) \end{aligned} \tag{2.147}$$

where we have also used the two consequences of the variable change employed, namely

1. $d\lambda = -d\gamma$
2. $\lambda \to \pm\infty$ means $\gamma \to \mp\infty$

Swapping the integration limits in Eqn. (2.147) and negating the integral to compensate for it, the output signal can now be written as

$$y(t) = \int_{-\infty}^{\infty} x(t-\gamma)\, h(\gamma)\, d\gamma$$
$$= h(t) * x(t) \tag{2.148}$$

thus proving the commutative property of the convolution operator.

The following example will illustrate the use of the alternative form of the convolution integral based on the commutative property.

Example 2.23: Using alternative form of convolution

Find the unit step response of the RC circuit with impulse response

$$h(t) = \frac{1}{RC}\, e^{-t/RC}\, u(t)$$

using the alternative form of the convolution integral given by Eqn. (2.148).

Solution: Using the convolution integral of Eqn. (2.148), the two functions we need are $h(\gamma)$ and $x(t-\gamma)$, both as functions of γ. Fig. 2.46 shows these functions for the two cases of $t \le 0$ and of $t > 0$.

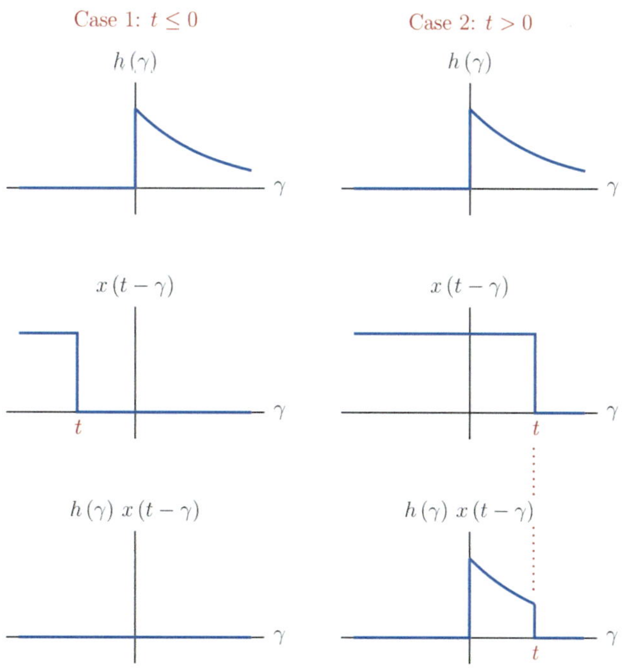

Figure 2.46 – Signals involved in the convolution integral of Example 2.23.

For $t \leq 0$ these two functions do not overlap anywhere, and therefore the integrand $h(\gamma)\, x(t-\gamma)$ of the convolution integral is zero for all values of γ. We therefore conclude that

$$y(t) = 0 \quad \text{for } t \leq 0$$

For values of $t > 0$, the two functions $h(\gamma)$ and $x(t-\gamma)$ overlap for values of γ in the interval $(0, t)$. In this interval, we have $x(t-\gamma) = 1$ and $h(\gamma) = (1/RC)\, e^{-\gamma/RC}$. Therefore, the output signal is

$$y(t) = \int_0^t \frac{1}{RC}\, e^{-\gamma/RC}\, d\gamma = 1 - e^{-t/RC} \quad \text{for } t > 0$$

Using a unit step function to account for both segments of the solution found, the general solution valid for all t is

$$y(t) = \left(1 - e^{-t/RC}\right) u(t)$$

which agrees with the result found in Example 2.20.

Interactive App: Alternative form of the convolution integral

The interactive app in `appConv4.m` is based on Example 2.23. It facilitates visualization of the overlapping regions between the functions $h(\lambda)$ and $x(t-\lambda)$ as the time variable t is varied. The graphical user interface is identical to the user interface of `appConv1.m`.

a. Start with the time slider set at $t = -0.5$ s. Set the time constant to $\tau = 1$ s. Gradually increment the time while observing the changes in $h(t-\lambda)$ and the product $x(\lambda)\, h(t-\lambda)$.

b. As the time variable crosses into the positive territory, pay special attention to the shaded area under the product $x(\lambda)\, h(t-\lambda)$ and how it translates to the amplitude of the output signal at that instant.

Software resource: `appConv4.m`

2.8 Causality in Continuous-Time Systems

Causality is an important feature of physically realizable systems. A system is said to be causal if the current value of the output signal depends only on current and past values of the input signal, but not on its future values. For example, a continuous-time system with input-output relationship

$$y(t) = x(t) + x(t-0.01) + x(t-0.02) \tag{2.149}$$

is causal since the output signal can be computed based on current and past values of the input signal. Conversely, the system given by

$$y(t) = x(t) + x(t-0.01) + x(t+0.01) \tag{2.150}$$

is non-causal since the computation of the output signal requires anticipation of a future value of the input signal.

Can a non-causal system be realized?

The answer depends on what type of realization we seek. Consider the system represented by Eqn. (2.149). For the sake of discussion, suppose that the signal $x(t)$ is the output of a microphone in a recording studio where a live band is playing. The signal is delayed by 0.01 seconds and 0.02 seconds in two separate systems, and the outputs of these delay systems are added to the original signal $x(t)$ to produce the signal $y(t)$ in *real time*, that is, while the input signal is being played. This type of signal processing is called *real-time processing*, and is illustrated in Fig. 2.47.

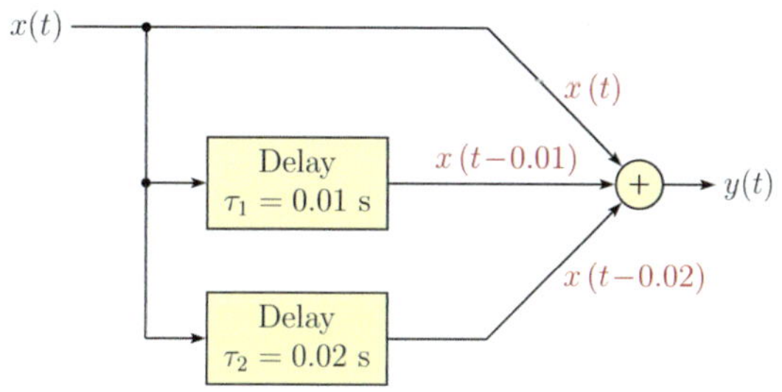

Figure 2.47 – Causal system given by Eqn. (2.149).

Now consider the system in Eqn. (2.150). Clearly, this system cannot be realized for real-time processing since the value $x(t+0.01)$ is not available at the time instant t when the output is computed. If we must realize this system (some non-causal systems may have desired properties in other respects) then we must realize it in *post processing mode*. The signals $x(t)$, $x(t-0.01)$, and $x(t+0.01)$ are each recorded. The output signal is then computed from these recordings after the fact, e.g., after the band finishes playing the song. Another way to look at this type of processing is to consider a new signal $w(t)$ defined as

$$w(t) = y(t-0.01) \tag{2.151}$$

and write the system description in terms of the new, delayed, output signal $w(t)$ in the form

$$w(t) = x(t-0.01) + x(t-0.02) + x(t) \tag{2.152}$$

The signal $w(t)$ can be computed in real time and we can listen to it while the band is playing. However, it represents a delayed version of the actual output signal $y(t)$ that is of interest to us. In essence we are doing post processing. Some non-causal systems can be made causal through the addition of delays as we have done in Eqn. (2.151). However, we will see that this is not always possible.

For CTLTI systems the causality property can be related to the impulse response of the system. Recall that the output signal $y(t)$ of the system is equal to the convolution of its input signal $x(t)$ and its impulse response $h(t)$. We will use the alternative form of the convolution integral given by

$$\begin{aligned} y(t) =& h(t) * x(t) \\ =& \int_{-\infty}^{\infty} h(\lambda)\, x(t-\lambda)\, d\lambda \end{aligned} \tag{2.153}$$

For the system under consideration to be causal, computation of $y(t)$ should not require any future values of the input signal. Consequently, the term $x(t-\lambda)$ in the integral should not contain references to any future values. We need

$$t-\lambda \le t \tag{2.154}$$

and equivalently $\lambda \ge 0$. Thus, the integrand in Eqn. (2.154) should not have any non-zero values for $\lambda < 0$. This in turn requires that the impulse response of the system be equal to zero for all negative values of its argument, i.e.,

$$h(t) = 0 \quad \text{for all} \quad t < 0 \tag{2.155}$$

This result intuitively makes sense. The impulse response of a causal CTLTI system cannot possibly start before $t = 0$ since doing so would require the system to anticipate the impulse to occur at $t = 0$. Using Eqn. (2.155), the convolution relationship in Eqn. (2.153) can be written in the right-sided form

$$y(t) = \int_0^{\infty} h(\lambda)\, x(t-\lambda)\, d\lambda \tag{2.156}$$

2.9 Stability in Continuous-Time Systems

Stable system:

A system is said to be *stable* in the *bounded-input bounded-output (BIBO)* sense if any bounded input signal is guaranteed to produce a bounded output signal.

An input signal $x(t)$ is said to be bounded if an upper bound B_x exists such that

$$|x(t)| < B_x < \infty \tag{2.157}$$

for all values of t. It doesn't matter how large the upper bound B_x may be, as long as it is finite. For the system to be stable a finite upper bound B_y must exist for the output signal in response to any input signal bounded as described by Eqn. (2.157).

For stability of a continuous-time system:

$$|x(t)| < B_x < \infty \quad \text{must imply that} \quad |y(t)| < B_y < \infty \tag{2.158}$$

If the system under consideration is CTLTI, then we would like to relate the stability condition given by Eqn. (2.158) to the impulse response of the system as well as its differential equation. Derivation of the necessary condition follows:

Required conditions for the impulse response:

The output signal of a CTLTI system is found from its input signal and impulse response through the use of the convolution integral

$$y(t) = \int_{-\infty}^{\infty} h(\lambda)\, x(t-\lambda)\, d\lambda$$

The absolute value of the output signal is

$$|y(t)| = \left| \int_{-\infty}^{\infty} h(\lambda)\, x(t-\lambda)\, d\lambda \right|$$

Absolute value of an integral is less than or equal to the integral of the absolute value of the integrand, so we can write the inequality

$$|y(t)| \le \int_{-\infty}^{\infty} |h(\lambda)\, x(t-\lambda)|\, d\lambda \tag{2.159}$$

The integrand in Eqn. (2.159) can be expressed as

$$|h(\lambda)\, x(t-\lambda)| = |h(\lambda)|\, |x(t-\lambda)|$$

and the inequality in Eqn. (2.159) becomes

$$|y(t)| \le \int_{-\infty}^{\infty} |h(\lambda)|\, |x(t-\lambda)|\, d\lambda \tag{2.160}$$

Since $|x(t-\tau)| < B_x$ we can write Eqn. (2.160) as

$$|y(t)| \le \int_{-\infty}^{\infty} |h(\lambda)|\, B_x\, d\lambda$$

or, equivalently

$$|y(t)| \le B_x \int_{-\infty}^{\infty} |h(\lambda)|\, d\lambda < B_y < \infty \tag{2.161}$$

Eqn. (2.161) leads to the condition for stability. The integral in Eqn. (2.161) must evaluate to a finite value for the output signal to be bounded.

Condition for stability of a CTLTI system:

$$\int_{-\infty}^{\infty} |h(\lambda)|\, d\lambda < \infty \tag{2.162}$$

For a CTLTI system to be stable, its impulse response must be *absolute integrable.*

Example 2.24: Stability of a first-order continuous-time system

Evaluate the stability of the first-order CTLTI system described by the differential equation

$$\frac{dy(t)}{dt} + a\,y(t) = x(t)$$

where a is a real-valued constant.

Solution: The impulse response of the system in question is

$$h(t) = e^{-at}\,u(t)$$

as can be easily verified through the use of Eqn. (2.47) with $r(t) = \delta(t)$, $t_0 = 0$, and $y(0) = 0$. (We take the initial value to be zero since the system is specified to be CTLTI.) Now we can check to see if the requirement given by Eqn. (2.162) is satisfied.

$$\int_{-\infty}^{\infty} \left|h(\lambda)\right|\,d\lambda = \int_{0}^{\infty} e^{-a\lambda}\,d\lambda = \frac{1}{a} \quad \text{provided that} \quad a > 0 \tag{2.163}$$

Thus the system is stable if $a > 0$. (Verify that the integral in Eqn. (2.163) cannot be evaluated if $a \le 0$.) It is also interesting to recognize that the characteristic equation that corresponds to the differential equation of the system is

$$s + a = 0$$

and its only solution is at $s_1 = -a$.

The stability of a CTLTI system can also be associated with the modes of the differential equation that governs the behavior of the system. In Example 2.24 the system proved to be stable for $a > 0$ which meant that the only root of the characteristic polynomial had to be negative. Furthermore, we have seen in Section 2.5.3 that the locations of the roots of the characteristic polynomial control the type of the transient behavior of the system. In the case of real roots, a negative real root leads to a decaying exponential signal while a positive real root produces a growing exponential signal. In the case of a pair of complex conjugate roots, if the real part of the pair is negative, the resulting signal is oscillatory with exponentially decaying amplitude. If the real part of the conjugate pair is positive, then the oscillatory response has a growing amplitude. Summarizing both cases we can conclude that, for a causal system, characteristic polynomial roots in the left half of the complex plane are associated with stable behavior, and those in the right half of the complex plane are associated with unstable behavior. For a causal CTLTI system to be stable, all roots of the characteristic polynomial must be in the left half of the complex plane.

A side note

In describing the associations between the roots of the characteristic polynomial and stability, we referred to a *causal* CTLTI system. If we were to consider an *anti-causal* system, the impulse response of which proceeds in the negative direction toward $t \to -\infty$, then the associations described above would have to be reversed. In that case roots in the right half of the complex plane would lead to stable behavior.

Why is stability important for a CTLTI system?

An unstable CTLTI system is capable of producing an unbounded output signal in response to at least some bounded input signals. What is the practical significance of this? Consider a battery-operated amplifier circuit that may be unstable due to a design flaw. We certainly don't expect the voltages in a battery-operated device to be infinitely large and it does not happen. What may happen is that the output voltage may go up to the maximum level it can physically attain (probably close to the power supply voltage) and stay there, causing circuit components to saturate, and causing the system to cease its useful linear operation. Another example would be an electromechanical system such as an electric motor. If it is unstable as a system, its speeds may reach levels that are too high for its normal operation and cause physical damage to the motor.

2.10 Approximate Numerical Solution of a Differential Equation

For a system modeled by means of a differential equation, the output signal can be determined by solving the differential equation with the specified input signal and initial conditions. In Section 2.5 we have explored time-domain methods for solving linear differential equations with constant coefficients. Transform-domain methods for accomplishing the same task will be presented in Chapter 4.

Sometimes we may run into differential equations for which analytical solutions may be difficult or impossible to obtain. The differential equation may have nonlinear terms, or the analytical definition of the input signal applied to the system may not be simple or practical enough for obtaining an analytical solution. In those cases, we may have to rely on approximate step-by-step solutions that are obtained on a computer by the use of an iterative numerical algorithm. A number of numerical algorithms for finding approximate solutions to differential equations are documented in the literature, ranging from very simple to very sophisticated. A thorough review of these algorithms is also well beyond the scope of this text, and can be found in a number of excellent textbooks on numerical analysis [4, 3, 19]. In this section we will briefly discuss the simplest of the numerical solution methods, the *Euler method*, which works reasonably well for our purposes as long as some basic precautions are taken.

Consider the RC circuit that was analyzed in Examples 2.8 and 2.9 using analytical solution techniques. The first-order differential equation that governs the operation of the circuit is

$$\frac{dy(t)}{dt}+\frac{1}{RC}\,y(t)=\frac{1}{RC}\,x(t) \tag{2.164}$$

where $x(t)$ and $y(t)$ represent the input and the output signals, respectively. The first step will be to rearrange the terms of the differential equation so that the derivative of the output signal is left by itself on the left side of the equal sign:

$$\frac{dy(t)}{dt}=-\frac{1}{RC}\,y(t)+\frac{1}{RC}\,x(t) \tag{2.165}$$

Eqn. (2.165) is in the general form

$$\frac{dy(t)}{dt}=g\left[t,y(t)\right] \tag{2.166}$$

where $g[\ldots]$ represents some function of t and $y(t)$. For this particular circuit we have

$$g\left[t,y(t)\right]=-\frac{1}{RC}\,y(t)+\frac{1}{RC}\,x(t) \tag{2.167}$$

We know from basic calculus that one way of approximating a derivative is through the use of finite differences. At the time instant $t = t_0$, the derivative on the left side of Eqn. (2.166) can be approximated using the *first forward difference* as

$$\left.\frac{dy(t)}{dt}\right|_{t=t_0} \approx \frac{y(t_0+T) - y(t_0)}{T} \tag{2.168}$$

provided that the time step T is sufficiently small. This is illustrated in Fig. 2.48. The slope of the tangent passing through the point $\left[t_0, y(t_0)\right]$ represents the true value of the derivative $dy(t)/dt$ at $t = t_0$. The slope of the chord passing through the points $\left[t_0, y(t_0)\right]$ and $\left[t_0 + T, y(t_0 + T)\right]$ is the approximate value of the same.

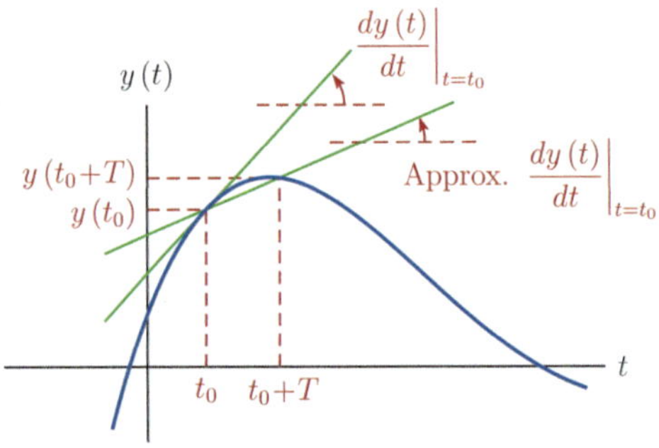

Figure 2.48 – Approximating the first derivative using a finite difference.

Substituting Eqn. (2.168) into Eqn. (2.166) and rearranging the terms we obtain

$$y(t_0 + T) \approx y(t_0) + T\, g\left[t_0, y(t_0)\right] \tag{2.169}$$

which gives us the iteration formula for the Euler method. The solution starts with a known initial value of the output signal at the time instant $t = t_0$, typically $t = 0$. Suppose $y(0)$ is given. We use it to approximate $y(T)$ as

$$y(T) \approx y(0) + T\, g\left[0, y(0)\right]$$

At the next step we have

$$y(2T) \approx y(T) + T\, g\left[T, y(T)\right]$$

and so on. For the RC circuit example in Eqn. (2.165) we can write

$$\begin{aligned} y((k+1)\,T) &\approx y(kT) + T\, g\left[kT, y(kT)\right] \\ &= y(kT) + T\left[-\frac{1}{RC}\, y(kT) + \frac{1}{RC}\, x(kT)\right] \end{aligned} \tag{2.170}$$

for $k = 0, 1, 2, \ldots$. For this approximation technique to work in a reasonably accurate manner, a sufficiently small step size T needs to be chosen. As far as the system is concerned, a good rule of thumb is for the step size T to be 10 percent of the time constant or smaller. In addition, time variations of the input signal must also be considered in selecting the step size. The Euler method may suffer from numerical instability problems if the step size T is not chosen sufficiently small. A variant known as the *backward Euler method* improves the stability of the solution, yet it is more difficult to use.

Although the Euler method as presented in this section only applies to a first-order differential equation, we will be able to generalize it to apply to higher-order systems when we discuss state-space models in Chapter 5. MATLAB Exercises 2.4 through 2.7 provide examples of finding numerical solutions in MATLAB using the Euler method or the more sophisticated *Runge-Kutta method* MATLAB implements.

Software resources: See MATLAB Exercises 2.4 through 2.7.

Summary of Key Points

- A system is a physical entity that takes in one or more physical signals as input and, in response, produces one or more new signals as output. For our purposes, the role of the system is to create a cause-effect relationship between input and output signals.

- The mathematical model of a system is a function, formula or algorithm (or a set of any of those) to approximately recreate the same cause-effect relationship between the mathematical models of input and output signals.

- A continuous-time system is one in which all input and output signals are continuous-time signals.

- A system is said to be linear if it satisfies both the additivity rule and the homogeneity rule. Additivity rule requires that the response of the system to the sum of any two input signals be the same as the sum of its individual responses to each of the input signals

  Eqn. (2.2): $$\text{Sys}\left\{x_1(t)+x_2(t)\right\}=\text{Sys}\left\{x_1(t)\right\}+\text{Sys}\left\{x_2(t)\right\}$$

  Homogeneity rule requires that the only effect of multiplying the input signal by a constant scale factor should be multiplying the output signal by the same scale factor.

  Eqn. (2.3): $$\text{Sys}\left\{\alpha_1\,x_1(t)\right\}=\alpha_1\,\text{Sys}\left\{x_1(t)\right\}$$

- In checking the linearity of a system, homogeneity and additivity rules can be combined into the superposition principle.

  Eqn. (2.4): $$\text{Sys}\left\{\alpha_1\,x_1(t)+\alpha_2\,x_2(t)\right\}=\alpha_1\,\text{Sys}\left\{x_1(t)\right\}+\alpha_2\,\text{Sys}\left\{x_2(t)\right\}$$

- A system is said to be time-invariant if the only effect of time shifting the input signal by a constant amount is a time shift of the output signal by the same amount.

  Eqn. (2.7): $$\text{Sys}\left\{x(t)\right\}=y(t)\quad\text{implies that}\quad\text{Sys}\left\{x(t-\tau)\right\}=y(t-\tau)$$

- The acronym *CTLTI* stands for *continuous-time linear and time-invariant.*

- One method of representing the relationship established by a system between its input and output signals is a differential equation. Differential equations are useful in building system models using the laws of physics that govern the behavior of various components of a system.

☞ CTLTI systems can be modeled using ordinary differential equations with constant coefficients in the form

Eqn. (2.23): $$\sum_{k=0}^{N} a_k \frac{d^k y(t)}{dt^k} = \sum_{k=0}^{M} b_k \frac{d^k x(t)}{dt^k}$$

☞ For an ordinary differential equation with constant coefficients to correspond to a CTLTI system, all initial conditions must be equal to zero, that is

Eqn. (2.38): $$\left.\frac{d^k y(t)}{dt^k}\right|_{t=t_0} = 0 \qquad \text{for } k = 0, \ldots, N-1$$

☞ In general, solving an ordinary differential equation with constant coefficients involves finding a *homogeneous solution* and a *particular solution.*

☞ The homogeneous solution is the solution of the *homogeneous differential equation* with the right side set equal to zero.

Eqn. (2.62): $$\sum_{k=0}^{N} a_k \frac{d^k y(t)}{dt^k} = 0$$

From a systems perspective, the homogeneous solution corresponds to the *natural response* of the system with no external input signals applied.

☞ The particular solution is the solution of the differential equation with the right side set for a specified input signal. From a systems perspective it corresponds to the *forced response* of the system, that is, the response to one or more external input signals.

☞ Block diagrams are useful for realizing systems described by differential equations. An ordinary differential equation can be converted to a block diagram by using three types of components, namely constant-gain amplifiers, signal adders, and integrators.

☞ A CTLTI system can be uniquely and completely described by its impulse response $h(t)$ which is simply its forced response when the input signal is a unit impulse.

☞ An easy method of finding the impulse response of a CTLTI system from its differential equation is to first find the unit step response of the system as a particular solution with a unit step input signal, and then differentiate it. Refer to Eqn. (2.124).

☞ The response of a CTLTI system to an input signal $x(t)$ can be found by computing the convolution of the input signal with the impulse response of the system.

Eqn. (2.143): $$y(t) = x(t) * h(t) = \int_{-\infty}^{\infty} x(\lambda)\, h(t-\lambda)\, d\lambda$$

Eqn. (2.144): $$y(t) = h(t) * x(t) = \int_{-\infty}^{\infty} h(\lambda)\, x(t-\lambda)\, d\lambda$$

☞ A system is said to be *causal* if, at any time instant, the current value of the output signal depends only on current and past values of the input signal, but not on its future values. A system must be causal if it is to be implemented for real-time processing.

- ☞ For a CTLTI system to be causal, the impulse response must be equal to zero for all negative values of the time variable t.
- ☞ A system is said to be *stable* in the *bounded-input bounded-output (BIBO)* sense if any bounded input signal is guaranteed to produce a bounded output signal.
- ☞ For a CTLTI system to be stable, its impulse response must be *absolute integrable*, that is,

  Eqn. (2.162): $$\int_{-\infty}^{\infty} \left|h(\lambda)\right| \, d\lambda < \infty$$

- ☞ Numerical methods exist for finding an approximate solution to a differential equation. The simplest of these methods is the Euler method which is based on approximating the first derivative using the first forward difference. Refer to Eqn. (2.169).
- ☞ Higher-order methods such as the fourth-order Runge-Kutta method provide better accuracy in finding an approximate solution to a differential equation.

Further Reading

[1] H.S. Bear. *Differential Equations: A Concise Course.* Addison-Wesley series in mathematics. Dover Publications, 2013.

[2] J.C. Butcher. *Numerical Methods for Ordinary Differential Equations.* Wiley, 2008.

[3] S.C. Chapra. *Applied Numerical Methods with MATLAB for Engineers and Scientists.* McGraw-Hill, 2008.

[4] S.C. Chapra and R.P. Canale. *Numerical Methods for Engineers.* McGraw-Hill, 2010.

[5] L. Edsberg. *Introduction to Computation and Modeling for Differential Equations.* Wiley, 2015.

[6] T.L. Harman, J.B. Dabney, and N. Richert. *Advanced Engineering Mathematics Using MATLAB V.4.* PWS BookWare Companion Series. PWS Publishing Company, 1997.

[7] A.Ü. Keskin. *Ordinary Differential Equations for Engineers: Problems with MATLAB Solutions.* Springer International Publishing, 2018.

[8] J.D. Logan. *A First Course in Differential Equations.* Undergraduate Texts in Mathematics. Springer New York, 2006.

[9] C. Lopez. *MATLAB Differential Equations.* MATLAB Solutions Series. Apress, 2014.

[10] C.B. Moler. *Numerical Computing with MATLAB: Revised Reprint.* Other Titles in Applied Mathematics. Society for Industrial and Applied Mathematics, 2010.

[11] P.V. O'Neil. *Advanced Engineering Mathematics.* Cengage Learning, 2011.

[12] B. Said-Houari. *Differential Equations: Methods and Applications.* Compact Textbooks in Mathematics. Springer International Publishing, 2016.

[13] D.A. Sanchez. *Ordinary Differential Equations and Stability Theory:* Dover Books on Mathematics. Dover Publications, 2019.

[14] Frank L. Severance. *System Modeling and Simulation: An Introduction.* John Wiley and Sons, 2001.

[15] L.F. Shampine, I. Gladwell, and S. Thompson. *Solving ODEs with MATLAB.* Cambridge University Press, 2003.

[16] A.H. Siddiqi and P. Manchanda. *A First Course in Differential Equations with Applications.* Macmillan Publishers India Limited, 2006.

[17] S.A. Wirkus and R.J. Swift. *A Course in Ordinary Differential Equations.* CRC Press, 2006.

[18] C.R. Wylie and L.C. Barrett. *Advanced Engineering Mathematics.* McGraw-Hill, 1995.

[19] W.Y. Yang et al. *Applied Numerical Methods Using MATLAB.* Wiley, 2020.

MATLAB Exercises with Solutions

MATLAB Exercise 2.1: Testing linearity of continuous-time systems

In Example 2.1 we have checked four different systems for linearity. In each case we started with two arbitrary signals $x_1(t)$ and $x_2(t)$, formulated the response of each system to the signal

$$x(t) = \alpha_1\, x_1(t) + \alpha_2\, x_2(t).$$

with arbitrary constants α_1 and α_2, and determined if the response to the combined signal matched the expected response

$$y(t) = \alpha_1\, y_1(t) + \alpha_2\, y_2(t).$$

for a linear system. In this exercise, we will simulate the four systems considered in Example 2.1 and test them using signals generated in MATLAB. The script listed below generates samples of two test signals

$$x_1(t) = \cos(2\pi 5 t) \qquad \text{and} \qquad x_2(t) = e^{-0.5t}$$

for $0 \le t \le 5$. The four systems to be tested are described through the use of four anonymous functions. Expected and actual responses of the first system `sys_a()` are computed and graphed. The graph produced by the script is shown in Fig. 2.49. The systems in parts (b) through (d) of Example 2.1 can be simulated by editing the code above and changing the anonymous function used for the system.

```
% Script: mexct_2_1.m
t = [0:0.01:4];                 % Create a time vector
x1 = cos(2*pi*5*t);             % Test signal 1
x2 = exp(-0.5*t);               % Test signal 2
alpha1 = 2;                     % Set parameters alpha1
alpha2 = 1.25;                  %   and alpha2
x = alpha1*x1+alpha2*x2;        % Combined signal
% Define anonymous functions for the systems in Example 2-1
sys_a = @(x) 5*x;
sys_b = @(x) 5*x+3;
sys_c = @(x) 3*x.*x;
sys_d = @(x) cos(x);
% Test the system in part (a) of Example 2-1
y1 = sys_a(x1);
```

```
y2 = sys_a(x2);
yExp = alpha1*y1+alpha2*y2;    % Expected response for a linear system
yAct = sys_a(x);               % Actual response
tiledlayout(1,2);
nexttile;
plot(t,yExp);                  % Graph expected response
title('y_{Exp} = \alpha_1 y_1 + \alpha_2 y_2')
xlabel('t (sec)');
ylabel('Amplitude');
nexttile;
plot(t,yAct);                  % Graph actual response
title('y_{Act} = Sys_a\{\alpha_1 x_1 + \alpha_2 x_2\}')
xlabel('t (sec)');
ylabel('Amplitude');
```

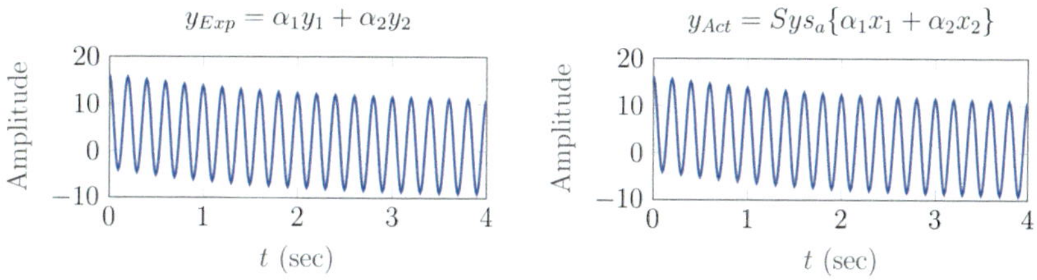

Figure 2.49 – Signals $y_{Exp}(t)$ and $y_{Act}(t)$ for MATLAB Exercise 2.1.

The two signals seem to be identical, supporting our earlier conclusion that the system in part (a) is linear. A word of caution is in order here: For a system to be linear, superposition principle must hold for *any two* signals $x_1(t)$ and $x_2(t)$ and not just the two signals we generated above. If we had only the graphs in Fig. 2.49 without our earlier analysis of Example 2.1, we would not be able to claim that the system is linear; the best we could do would be to say that the system *may be* linear. Nonlinearity of a system is easier to prove since we only need to show one case where superposition fails.

Software resource: mexct_2_1.m

MATLAB Exercise 2.2: Testing time invariance of continuous-time systems

In Example 2.2 we have checked three systems for time invariance. In each case an arbitrary test signal was used with the system in its original form $x(t)$ and in time-shifted form $x(t-\tau)$. If the system under consideration is time-invariant we need

$$\text{Sys}\{x(t-\tau)\} = y(t-\tau)$$

where $y(t)$ is the response to the original signal $x(t)$.

In this exercise, we will simulate the three systems considered in Example 2.2, and test them using signals generated in MATLAB. Let the test signal $x(t)$ be

$$x(t) = e^{-0.5t}\, u(t)$$

The script listed below tests the system in part (c) of Example 2.2 for time invariance. The graph produced by the script is shown in Fig. 2.50.

```
% Script: mexct_2_2.m
t = [0:0.01:10];                    % Create a time vector
x = @(t) exp(-0.5*t).*(t>=0);  % Anonymous function for test signal
% Define anonymous functions for the systems in Example 2-2
sys_a = @(x) 5*x;
sys_b = @(x) 3*cos(x);
sys_c = @(x) 3*cos(t).*x;
% Test the system in part (c) of Example 2-2
y1 = sys_c(x(t));
y2 = sys_c(x(t-2));
plot(t,y1,'b-',t,y2,'r-');    % Graph the two responses
title('Responses to x(t) and x(t-2)')
xlabel('t (sec)');
ylabel('Amplitude');
legend('Sys\{x(t)\}','Sys\{x(t-2)\}');
```

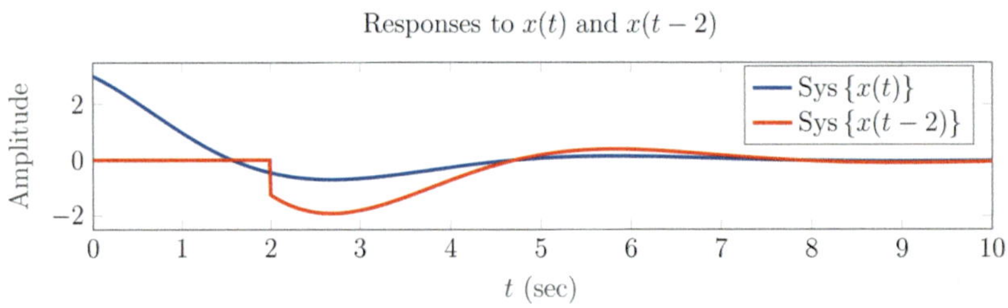

Figure 2.50 – Signals Sys $\{x(t)\}$ and Sys $\{x(t-2)\}$ for MATLAB Exercise 2.2.

The two signals seem to be different in shape, supporting our earlier conclusion that the system in part (c) is not time-invariant.

The systems in parts (a) and (b) of Example 2.2 can be simulated by editing the code above and changing the anonymous function used for the system.

Software resource: `mexct_2_2.m`

MATLAB Exercise 2.3: Using linearity to determine the response of the RC circuit

Refer to Examples 2.8, 2.9, and 2.10. In Example 2.8 the unit step response of the simple RC circuit was determined by solving the governing differential equation. The circuit was assumed to be initially relaxed, that is, $y(0) = 0$. In Example 2.9 we have determined the response of the same circuit to a pulse signal, again by solving the differential equation. Example 2.10 explored an alternative approach to finding the response to the pulse signal; namely using the response to a unit step and the superposition principle. (Superposition could be used since the circuit is initially relaxed and therefore linear and time-invariant.)

In this exercise we will explore this concept using MATLAB. The response of the circuit to a unit step signal was found in Example 2.8 to be

$$y_u(t) = \text{Sys}\{u(t)\} = \left(1 - e^{-4t}\right) u(t)$$

which can be expressed by means of an anonymous function as

```
yu = @(t) (1-exp(-4*t)).*(t>=0);
```

The script listed below uses the unit step response for determining and graphing the response of the system to the pulse signal

$$x_1(t) = \Pi(t) = u(t+0.5) - u(t-0.5)$$

```
% Script: mexct_2_3a.m
% Anonymous function for unit step response
yu = @(t) (1-exp(-4*t)).*(t>=0);
t = [-2:0.01:5];                % Vector of time instants
y1 = yu(t+0.5)-yu(t-0.5);     % Compute response to x1(t)
plot(t,y1); grid;
xlabel('Time (sec)');
ylabel('Amplitude');
```

This idea can be extended to input signals with higher complexity. Consider the signal

$$x_2(t) = u(t) - 2\,u(t-1) + u(t-2)$$

which is shown in Fig. 2.51.

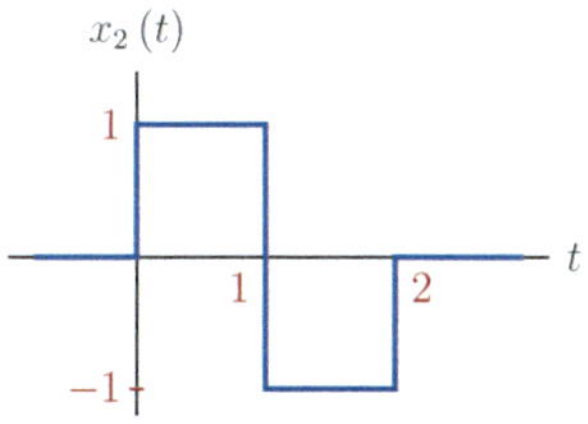

Figure 2.51 – Signal $x_2(t)$ for MATLAB Exercise 2.3.

The script listed below computes and graphs the response of the RC circuit to the signal $x_2(t)$. The resulting graph is shown in Fig. 2.52.

```
% Script: mexct_2_3b.m
% Anonymous function for unit step response
yu = @(t) (1-exp(-4*t)).*(t>=0);
t = [-2:0.01:5];                  % Vector of time instants
y2 = yu(t)-2*yu(t-1)+yu(t-2);   % Compute response to x2(t)
plot(t,y2); grid;
xlabel('Time (sec)');
ylabel('Amplitude');
```

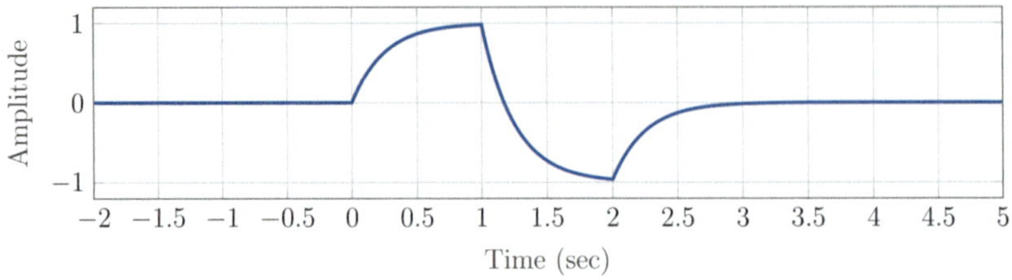

Figure 2.52 – Signal $y_2(t)$ for MATLAB Exercise 2.3.

Software resources: mexct_2_3a.m , mexct_2_3b.m

MATLAB Exercise 2.4: Numerical solution of the RC circuit using Euler method

In this exercise we will use the Euler method to find an approximate numerical solution for the RC circuit problem of Example 2.8 for a unit step input signal, and compare it to the exact solution that was found. The RC circuit and the input signal are shown in Fig. 2.53.

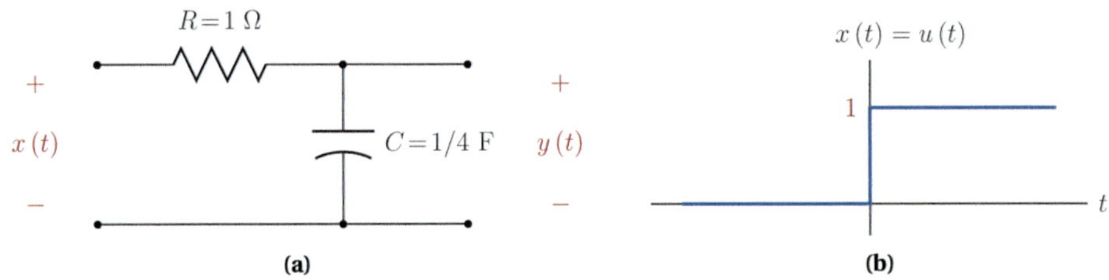

Figure 2.53 – **(a)** The circuit for MATLAB Exercise 2.4 and **(b)** the input signal $x(t)$.

For the specified input signal, the differential equation of the circuit is

$$\frac{dy(t)}{dt} + 4\,y(t) = 4\,u(t)$$

With the circuit initially relaxed, that is, with $y(0) = 0$, the exact solution for the output signal was found to be

$$y(t) = \left(1 - e^{-4t}\right)\,u(t) \tag{2.171}$$

In order to use the Euler method for approximate solution, we will write the differential equation in the form

$$\frac{dy(t)}{dt} = g\left(t, y(t)\right)$$

with the right-side function given by

$$g\left(t, y(t)\right) = -4\,y(t) + 4\,u(t)$$

The Euler method approximation $\hat{y}(t)$ to the solution is based on the iteration formula

$$\begin{aligned}\hat{y}\left((k+1)\,T_s\right) =& \hat{y}(kT_s) + T_s\,g\left(kT_s, \hat{y}(kT_s)\right)\\ =& \hat{y}(kT_s) + T_s\left(-4\,\hat{y}(kT_s) + 4\,u(kT_s)\right)\end{aligned}$$

The MATLAB script given below computes and graphs exact and approximate solutions, using a step size of $T_s = 0.1$ s. In addition, the percent error of the approximate solution is computed as

$$\varepsilon\,(kT_s) = \frac{\hat{y}\,(kT_s) - y\,(kT_s)}{y\,(kT_s)} \times 100$$

and graphed for $k = 1,2,\ldots,10$ to span a time interval of 1 s.

```
1  % Script: mexct_2_4a.m
2  Ts = 0.1;            % Time increment
3  t = [0:Ts:1];        % Vector of time instants
4  % Compute the exact solution.
5  y = 1-exp(-4*t);    % Eqn. (2.52)
6  % Compute the approximate solution using Euler method.
7  yHat = zeros(size(t));
8  yHat(1) = 0;        % Initial value.
9  for k = 1:length(yHat)-1,
10   g = -4*yHat(k)+4;              % Eqn. (2.167)
11   yHat(k+1) = yHat(k)+Ts*g;  % Eqn. (2.170)
12 end;
13 % Graph exact and approximate solutions.
14 tiledlayout(2,1);
15 nexttile;
16 plot(t,y,'-',t,yHat,'ro--'); grid;
17 axis([0,1,0,1]);
18 title('Exact and approximate solutions for RC circuit');
19 xlabel('Time (sec)');
20 ylabel('Amplitude');
21 legend('Exact solution','Approximate solution',..
22   'Location','SouthEast');
23 % Compute and graph the percent approximation error
24 nexttile;
25 err_pct = (yHat-y)./y*100;
26 plot(t(2:length(t)),err_pct(2:length(t))); grid;
27 axis([0,1,0,25])
28 title('Percent approximation error');
29 xlabel('Time (sec)');
30 ylabel('Error (%)');
```

The loop that runs from line 9 to 12 of the code computes the approximate solution step by step. The graphs produced by the MATLAB script above are given in Fig. 2.54. The largest approximation error seems to be around 21 percent and it goes down in time as the output signal settles close to its final value.

The time step that we used in the code above was $T_s = 0.1$ seconds which is 40 percent of the time constant of $\tau = 0.25$ s. For better accuracy a smaller time step is needed. Euler method is a first-order approximation method, and is the simplest of all numerical methods for solving differential equations. Fig. 2.55 shows the results obtained with the same script, by simply changing the step size to $T_s = 0.02$ seconds or 8 percent of the time constant. In this case, the largest error between the exact and approximate solutions is on the order or 4 percent.

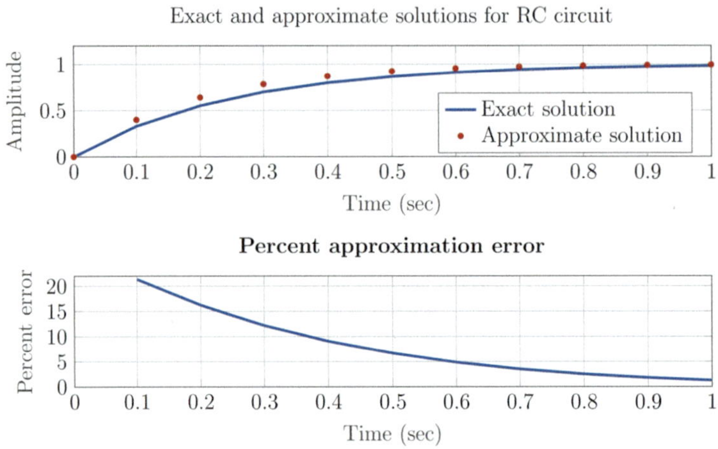

Figure 2.54 – Actual and approximate solutions for the RC circuit and the percent error for $\Delta t = 0.1$ s.

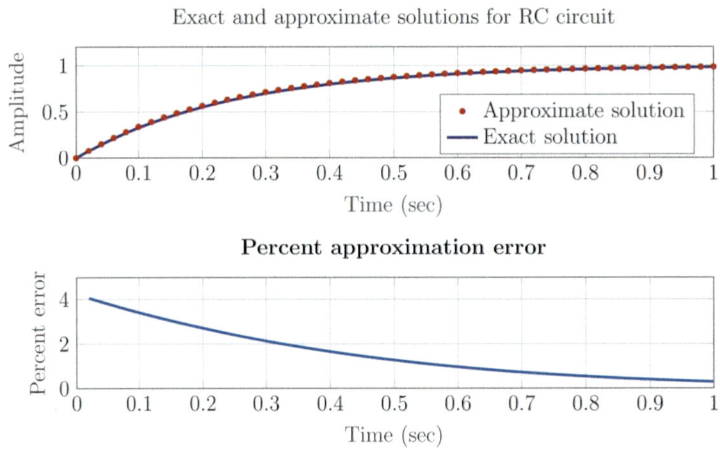

Figure 2.55 – Actual and approximate solutions for the RC circuit and the percent error for $\Delta t = 0.02$ s.

Another way to reduce the approximation error is to use a more sophisticated higher-order approximation method as we will see in the next exercise.

Software resources: `mexct_2_4a.m` , `mexct_2_4b.m`

MATLAB Exercise 2.5: Improved numerical solution of the RC circuit

For numerical solution of differential equations MATLAB has a number of functions that are based on more sophisticated algorithms than the Euler method we have used in MATLAB Exercise 2.4. In this exercise we will use the function `ode45()` which implements the fourth-order Runge-Kutta method, a more sophisticated numerical approximation than the Euler method of the previous example. For a given step size, approximation errors for the fourth-order Runge-Kutta method are generally smaller than those encountered with the Euler method.

The function `ode45()` utilizes the form of the first-order differential equation given by Eqn. (2.166). Before we can use `ode45()`, we need to write a function to compute the right side $g(t, y(t))$ of the differential equation. The code for such a function is given below:

```
function ydot = RCwithUnitStepInput(t,y)
  ydot = -4*y+4*(t>=0);
end
```

The function `RCwithUnitStepInput()` takes two arguments, t and y, and returns the corresponding value of $g(t, y(t))$. We will pass its handle as the first argument to the function `ode45()` in computing the approximate solution. The function name prefixed by the symbol @ denotes the handle. The script listed below computes an approximate solution in the time interval $0 \le t \le 1$ s, and compares it to the exact solution.

Often it is easier to just include the right side function for $g(t, y(t))$ at the end of the script file instead of saving it into a separate file, so that's what we have done. This makes the function `RCwithUnitStepInput()` local to the script and not generally accessible from outside.

```
% Script: mexct_2_5.m
% Compute the approximate solution using ode45()
tSpan = [0,1];           % Time span for approximate solution
[t,yApprox] = ode45(@RCwithUnitStepInput,tSpan,0);
% Compute the exact solution at the time instants in vector 't'
yExact = 1-exp(-4*t);  % Eqn. (2.171)
% Graph exact and approximate solutions
tiledlayout(2,1);
nexttile;
plot(t,yExact,'-',t(1:3:length(t)),yApprox(1:3:length(t)),'ro'); grid;
title('Exact and approximate solutions for RC circuit');
xlabel('Time (sec)');
ylabel('Amplitude');
legend('Exact solution','Approximate solution','Location','SouthEast');
% Compute and graph the percent approximation error
nexttile;
errPct = (yApprox-yExact)./yExact*100;
plot(t(2:length(t)),errPct(2:length(t))); grid;
title('Percent approximation error');
xlabel('Time (sec)');
ylabel('Percent error');

function ydot = RCwithUnitStepInput(t,y)
  ydot = -4*y+4*(t>=0);
end
```

Graphs produced by the script `mexct_2_5a.m` are shown in Fig. 2.56. It is easy to see that the accuracy of the solution obtained is much better than that obtained in the previous exercise (compare Fig. 2.56 to Fig. 2.54).

Note that the vector `tSpan` has two elements, namely the beginning and the end of the time interval in which the approximate solution is computed. The function `ode45()` internally determines the step size in computing each point in the solution.

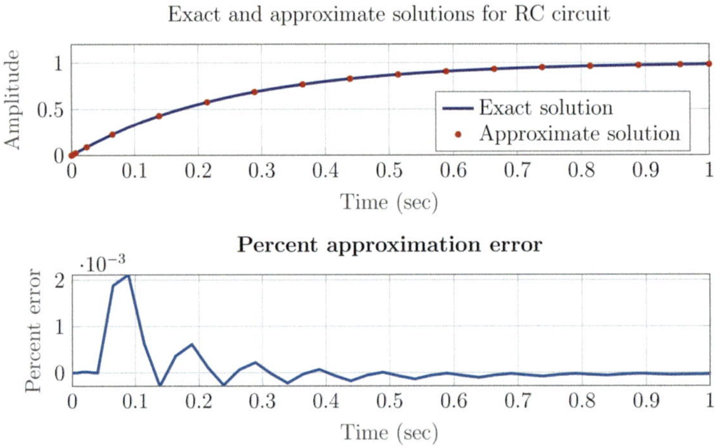

Figure 2.56 – Actual and approximate solutions for the RC circuit and the percent error with the `ode45()` function.

An examination of the vector `t` indicates that, in this particular case, it has 61 elements with variable spacing between them. An alternative is to provide the function `ode45()` with a vector of time instants instead of a range as shown below. This allows us to control the time instants for computing the approximate solution.

```
..
t = [0:0.01:1];
[t,yApprox] = ode45(@RCwithUnitStepInput,t,0);
..
```

Software resource: `mexct_2_5.m`

MATLAB Exercise 2.6: Finding natural and forced responses of a first-order system

Consider again the RC circuit of Fig. 2.53 with element values of $R = 1\ \Omega$ and $C = 1/4$ ţF. The initial value of the output signal is $y(0) = 5$.

a. Using the function `ode45()`, compute an approximation to the natural response of the system. Compare the result to that found in Example 2.11.

b. Approximate the forced response of the system to the input signal $x(t) = 20\cos(8t)$. Compare the result to what was found in Example 2.16.

Solution:

a. For this case the differential equation is

$$\frac{dy(t)}{dt}+4y(t)=0$$

The analytical solution with the specified initial condition was found in Example 2.11 to be

$$y_h(t)=5e^{-4t}u(t)$$

The script listed below computes both the exact solution and the approximate numerical solution. The two solutions are graphed as well as the percent error between them. The results are shown in Fig. 2.57.

```
% Script: mexct_2_6a.m
% Compute the approximate solution using ode45()
t = [0:0.01:3];            % Vector of time instants
[t,yApprox] = ode45(@RCnoInput,t,5);
% Compute the exact solution
yExact = 5*exp(-4*t);  % Eqn. (2.70)
% Graph exact and approximate solutions.
tiledlayout(2,1);
nexttile;
plot(t,yExact,'-',t(1:10:length(t)),yApprox(1:10:length(t)),'ro'); grid;
title('Exact and approximate solutions for RC circuit');
xlabel('Time (sec)');
ylabel('Amplitude');
legend('Exact solution','Approximate solution','Location','SouthEast');
% Compute and graph the percent approximation error
nexttile;
errPct = (yApprox-yExact)./yExact*100;
plot(t(2:length(t)),errPct(2:length(t))); grid;
title('Percent approximation error');
xlabel('Time (sec)');
ylabel('Percent error');

function ydot = RCnoInput(t,y)
  ydot = -4*y;
end
```

b. Recall that this problem was solved analytically in Example 2.16 and the forced response was determined to be

$$y(t)=4e^{-4t}+\cos(8t)+2\sin(8t)$$

for $t\geq 0$. The script listed below computes both the exact solution and the approximate numerical solution. Fig. 2.58 displays both solutions as well as the percent error between them.

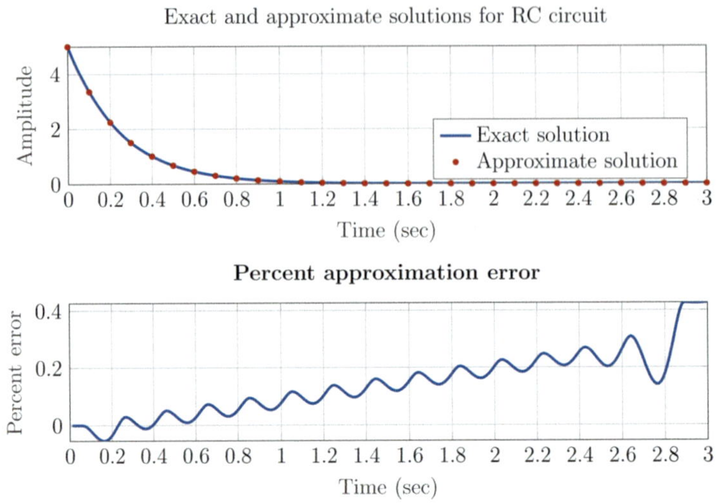

Figure 2.57 – Actual and approximate solutions for the natural response of the RC circuit and the percent error with the ode45() function.

```
% Script: mexct_2_6b.m
% Compute the approximate solution using ode45()
t = [0:0.01:3];          % Vector of time instants
[t,yApprox] = ode45(@RCwithCosineInput,t,5);
% Compute the exact solution
yExact = 4*exp(-4*t)+cos(8*t)+2*sin(8*t);  % Eqn. (2.105)
% Graph exact and approximate solutions.
tiledlayout(2,1);
nexttile;
plot(t,yExact,'-',t(1:5:length(t)),yApprox(1:5:length(t)),'ro'); grid;
title('Exact and approximate solutions for RC circuit');
xlabel('Time (sec)');
ylabel('Amplitude');
legend('Exact solution','Approximate solution','Location','NorthEast');
% Compute and graph the percent approximation error
nexttile;
errPct = (yApprox-yExact)./(yExact)*100;
plot(t(2:length(t)),errPct(2:length(t))); grid;
title('Percent approximation error');
xlabel('Time (sec)');
ylabel('Percent error');

function ydot = RCwithCosineInput(t,y)
  ydot = -4*y+20*cos(8*t);
end
```

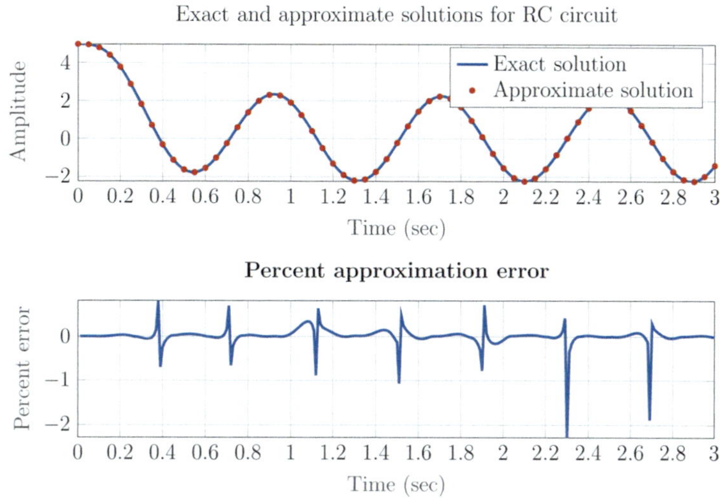

Figure 2.58 – Actual and approximate solutions for the forced response of the RC circuit and the percent error with the `ode45()` function.

Software resources: `mexct_2_6a.m`, `mexct_2_6b.m`

MATLAB Exercise 2.7: Finding natural and forced responses of a second-order system

Consider again the RLC circuit which was used in Examples 2.6 and 2.15. The circuit was shown in Fig. 2.11. Element values are $R = 2\ \Omega$, $L = 1$ H, and $C = 1/26$ F. Let the initial inductor current and the initial capacitor voltage at time $t = 0$ be $i(0) = 0.5$ A and $y(0) = 2$ V, respectively.

a. Using MATLAB function `ode45()` find and graph the approximate natural response of the system in the time interval $0 \le t \le 8$ s.

b. Find and graph the approximate forced response of the system in the same time interval for the input signal $x(t) = u(t-3)$.

Solution: The differential equation governing the behavior of the circuit was found in Example 2.6 to be

$$\frac{d^2y(t)}{dt^2} + \frac{R}{L}\frac{dy(t)}{dt} + \frac{1}{LC}y(t) = \frac{1}{LC}x(t)$$

Substituting the specified element values we obtain

$$\frac{d^2y(t)}{dt^2} + 2\frac{dy(t)}{dt} + 26\,y(t) = 26\,x(t) \tag{2.172}$$

We have a bit of a problem that we need to address here before we can use the function `ode45()`. Recall that `ode45()` expects a handle to a function that evaluates the right side of a *first-order* differential equation in the form

$$\frac{y(t)}{dt} = g\big(t, y(t)\big)$$

We need to write the second-order differential equation in Eqn. (2.172) as two first-order equations. Let two new variables be defined as

$$y_1(t) = y(t) \qquad \text{and} \qquad y_2(t) = \frac{dy(t)}{dt}$$

so that the differential equation in Eqn. (2.172) can be written as

$$\frac{dy_1(t)}{dt} = y_2(t)$$

$$\frac{dy_2(t)}{dt} = -26\, y_1(t) - 2\, y_2(t) + 26\, x(t)$$

With these changes, the function $g(t, y(t))$ becomes a vector function, that is, it receives a vector of two variables $y_1(t)$ and $y_2(t)$ at a specified time instant, and returns a vector of two derivatives $y_1'(t)$ and $y_2'(t)$ evaluated at the same time instant.

$$\begin{bmatrix} y_1'(t_1) \\ y_2'(t_1) \end{bmatrix} = g\left(t_1, \begin{bmatrix} y_1(t_1) \\ y_2(t_1) \end{bmatrix}\right)$$

Initial values of inductor current and capacitor voltage need to be translated to initial values of the two new variables $y_1(t)$ and $y_2(t)$. We have

$$y_1(0) = y(0) = v_C(0) = 2 \qquad \text{and} \qquad y_2(0) = \left.\frac{dy(t)}{dt}\right|_{t=0} = \frac{i_L(0)}{C} = 26$$

a. We are now ready to write the script for solving part (a) of the problem.

```
% Script: mexct_2_7a.m
% Compute the approximate solution using ode45()
t = [0:0.01:8];
[t,yApprox] = ode45(@RLCnoInput,t,[2,13]);
% Graph the approximate solution
plot(t,yApprox(:,1)); grid;
title('Approximate natural response of RLC circuit');
xlabel('Time (sec)');
ylabel('Amplitude');

function ydot = RLCnoInput(t,y)
  y1 = y(1);
  y2 = y(2);
  y1dot = y2;
  y2dot = -2*y2-26*y1;
  ydot = [y1dot;y2dot];
end
```

The response obtained is shown in Fig. 2.59.

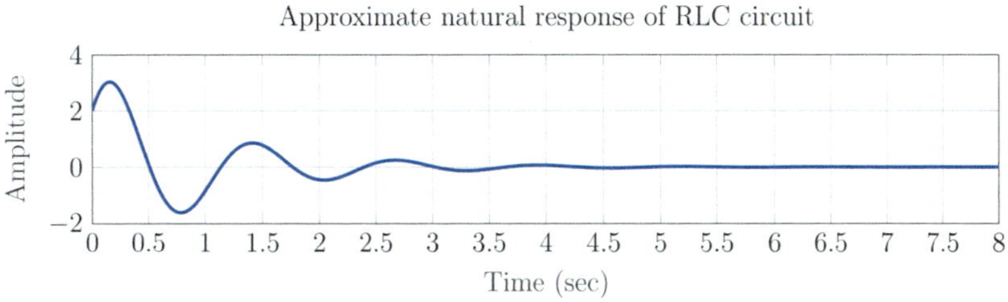

Figure 2.59 – Natural response of the RLC circuit in MATLAB Exercise 2.7.

b. The script for solving part (b) of the problem is easily obtained by replacing the function `RLCnoInput()` with the function `RLCwithUnitStepInput()` as shown below.

```
% Script: mexct_2_7b.m
% Compute the approximate solution using ode45()
t = [0:0.01:8];
[t,yApprox] = ode45(@RLCwithUnitStepInput,t,[2,13]);
% Graph the approximate solution
plot(t,yApprox(:,1)); grid;
title('Approximate forced response of RLC circuit');
xlabel('Time (sec)');
ylabel('Amplitude');

function ydot = RLCwithUnitStepInput(t,y)
  y1 = y(1);
  y2 = y(2);
  y1dot = y2;
  y2dot = -2*y2-26*y1+26*(t>=3);
  ydot = [y1dot;y2dot];
end
```

Fig. 2.60 shows the forced response of the circuit.

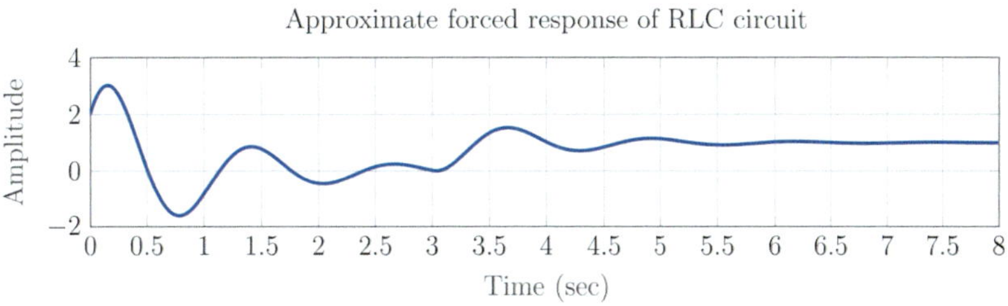

Figure 2.60 – Forced response of the RLC circuit in MATLAB Exercise 2.7.

Software resources: mexct_2_7a.m , mexct_2_7b.m

MATLAB Exercise 2.8: Using symbolic math for solving differential equations

Use symbolic math capabilities of MATLAB to find the natural response of the RLC circuit considered in MATLAB Exercise 2.7.

Solution: The script listed below accomplishes this task.

```
1 % Script: mexct_2_8.m
2 syms y(t);                                % Declare y(t) to be a symbolic variable
3 eqn = diff(y,t,2)==-2*diff(y,t)-26*y;    % Diff. equation
4 Dy = diff(y,t);                           % Dy is the first derivative
5 cond = [y(0)==2, Dy(0)==13];              % Initial conditions
6 solution = dsolve(eqn,cond)               % Solve using dsolve()
7 fplot(solution,[0,3]); grid;              % Graph the solution
```

Pay attention to how the differential equation is expressed in line 3. Terms `diff(y,t)` and `diff(y,t,2)` correspond to first and second derivatives, respectively. The symbolic variable `eqn` represents the differential equation to be solved, and `cond` represents the initial conditions. The solution is found by using the function `dsolve()` in line 6. MATLAB prints the solution in the following form:

```
solution =

exp(-t)*(2*cos(5*t) + 3*sin(5*t))
```

Compare this result to that found in part (a) of Example 2.15.

Software resource: mexct_2_8.m

Problems

2.1. A number of systems are specified below in terms of their input-output relationships. For each case, determine if the system is linear and/or time-invariant.

a. $y(t) = |x(t)| + x(t)$

b. $y(t) = t\, x(t)$

c. $y(t) = e^{-t}\, x(t)$

d. $y(t) = \int_{-\infty}^{t} x(\lambda)\, d\lambda$

e. $y(t) = \int_{t-1}^{t} x(\lambda)\, d\lambda$

f. $y(t) = (t+1) \int_{-\infty}^{t} x(\lambda)\, d\lambda$

2.2. Consider the cascade combination of two systems shown in Fig. P.2.2(a).

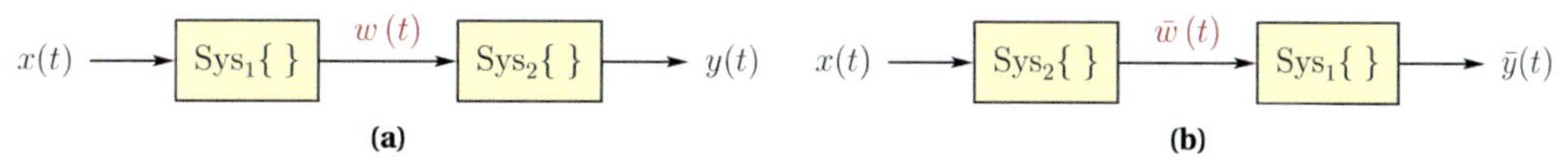

Figure P. 2.2

a. Let the input-output relationships of the two subsystems be given as

$$\text{Sys}_1\{x(t)\} = 3\,x(t) \quad \text{and} \quad \text{Sys}_2\{w(t)\} = w(t-2)$$

Write the relationship between $x(t)$ and $y(t)$.

b. Let the order of the two subsystems be changed as shown in Fig. P.2.2(b). Write the relationship between $x(t)$ and $\bar{y}(t)$. Does changing the order of two subsystems change the overall input-output relationship of the system?

2.3. Repeat Problem 2.2 with the following sets of subsystems:

a. $\text{Sys}_1\{x(t)\} = 3\,x(t)$ and $\text{Sys}_2\{w(t)\} = t\,w(t)$

b. $\text{Sys}_1\{x(t)\} = 3\,x(t)$ and $\text{Sys}_2\{w(t)\} = w(t) + 5$

2.4. Find a differential equation between the input voltage $x(t)$ and the output voltage $y(t)$ for the circuit shown in Fig. P.2.4. At $t = 0$ the initial values are

$$i_L(0) = 1 \text{ A}, \qquad v_C(0) = 2 \text{ V}$$

Express the initial conditions for $y(t)$ and $dy(t)/dt$.

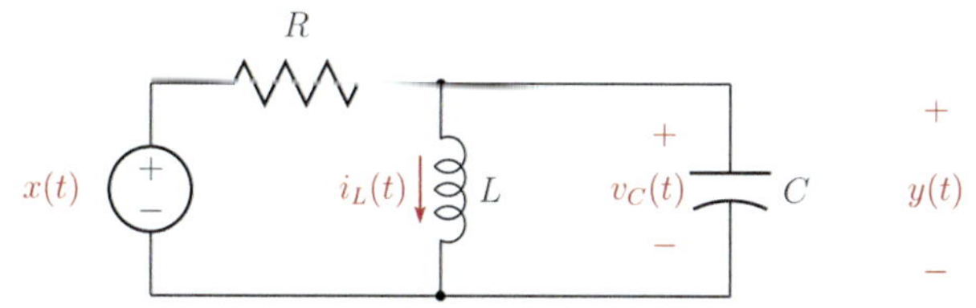

Figure P. 2.4

2.5. Find a differential equation between the input voltage $x(t)$ and the output voltage $y(t)$ for the circuit shown in Fig. P.2.5. At $t = 0$ the initial values are

$$v_1(0) = 2 \text{ V}, \qquad v_2(0) = -2 \text{ V}$$

Express the initial conditions for $y(t)$ and $dy(t)/dt$.

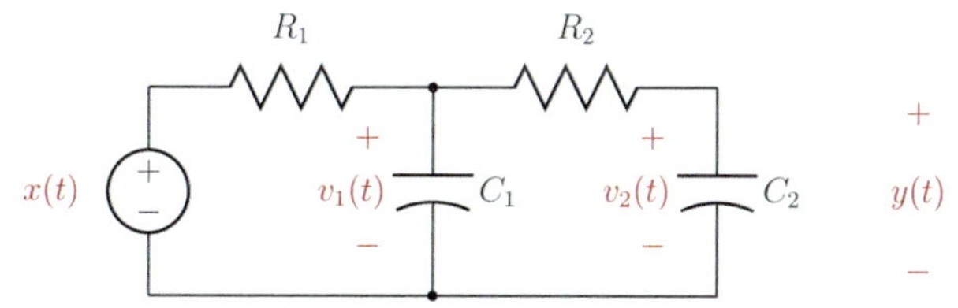

Figure P. 2.5

2.6. One of the commonly used examples for modeling dynamic systems with differential equations is the so-called *predator-prey* problem which is stated as follows:

Suppose that the following set of rules were experimentally determined to govern the populations of predators and prey that live in the same environment:

- The number of prey grows at a rate proportional to its current value if there are no predators.
- The number of predators decreases at a rate proportional to its current value if there is no prey.
- The number of encounters between predator and prey is proportional to the product of the current populations of each.
- Each encounter between the predator and the prey increases the number of predators and decreases the number of prey.

Designate $x(t)$ as the number of prey at time t and $y(t)$ as the number of predators at time t. Using parametric proportion constants as needed, write two differential equations to model the system. Are the equations linear or nonlinear? Justify your answer.

2.7. Consider the differential equation for the RC circuit in Fig. 2.12:

$$\frac{dy(t)}{dt}+4\,y(t)=4\,x(t)$$

Let the input signal be a unit step, that is, $x(t)=u(t)$. Using the first-order differential equation solution technique discussed in Section 2.5.1 find the solution $y(t)$ for $t\ge 0$ subject to each initial condition specified below:

a. $y(0)=0$

b. $y(0)=5$

c. $y(0)=1$

d. $y(0)=-1$

e. $y(0)=-3$

2.8. Solve each of the first-order differential equations given below for the specified input signal and subject to the specified initial condition. Use the first-order solution technique discussed in Section 2.5.1.

a. $\dfrac{dy(t)}{dt}+4\,y(t)=x(t)\,,\qquad x(t)=u(t)\,,\qquad y(0)=-1$

b. $\dfrac{dy(t)}{dt}+2\,y(t)=2\,x(t)\,,\qquad x(t)=u(t)-u(t-5)\,,\qquad y(0)=2$

c. $\dfrac{dy(t)}{dt}+5\,y(t)=3\,x(t)\,,\qquad x(t)=\delta(t)\,,\qquad y(0)=0.5$

d. $\dfrac{dy(t)}{dt}+5\,y(t)=3\,x(t)\,,\qquad x(t)=t\,u(t)\,,\qquad y(0)=-4$

e. $\dfrac{dy(t)}{dt}+y(t)=2\,x(t)\,,\qquad x(t)=e^{-2t}\,u(t)\,,\qquad y(0)=-1$

2.9. For each homogeneous differential equation given below, find the characteristic equation and show that it only has simple real roots. Find the homogeneous solution for $t\ge 0$ in each case subject to the initial conditions specified.

a. $$\frac{d^2y(t)}{dt^2}+3\frac{dy(t)}{dt}+2y(t)=0\,,\quad y(0)=3\,,\quad \left.\frac{dy(t)}{dt}\right|_{t=0}=0$$

b. $$\frac{d^2y(t)}{dt^2}+4\frac{dy(t)}{dt}+3y(t)=0\,,\quad y(0)=-2\,,\quad \left.\frac{dy(t)}{dt}\right|_{t=0}=1$$

c. $$\frac{d^2y(t)}{dt^2}-y(t)=0\,,\quad y(0)=1\,,\quad \left.\frac{dy(t)}{dt}\right|_{t=0}=-2$$

d. $$\frac{d^3y(t)}{dt^3}+6\frac{d^2y(t)}{dt^2}+11\frac{dy(t)}{dt}+6y(t)=0\,,\quad y(0)=2\,,\quad \left.\frac{dy(t)}{dt}\right|_{t=0}=-1\,,\quad \left.\frac{d^2y(t)}{dt^2}\right|_{t=0}=1$$

2.10. For each homogeneous differential equation given below, find the characteristic equation and show that at least some of its roots are complex. Find the homogeneous solution for $t \ge 0$ in each case subject to the initial conditions specified.

a. $$\frac{d^2y(t)}{dt^2}+3y(t)=0\,,\quad y(0)=2\,,\quad \left.\frac{dy(t)}{dt}\right|_{t=0}=0$$

b. $$\frac{d^2y(t)}{dt^2}+2\frac{dy(t)}{dt}+2y(t)=0\,,\quad y(0)=-2\,,\quad \left.\frac{dy(t)}{dt}\right|_{t=0}=-1$$

c. $$\frac{d^2y(t)}{dt^2}+4\frac{dy(t)}{dt}+13y(t)=0\,,\quad y(0)=5\,,\quad \left.\frac{dy(t)}{dt}\right|_{t=0}=0$$

d. $$\frac{d^3y(t)}{dt^3}+3\frac{d^2y(t)}{dt^2}+4\frac{dy(t)}{dt}+2y(t)=0\,,\quad y(0)=1\,,\quad \left.\frac{dy(t)}{dt}\right|_{t=0}=0\,,\quad \left.\frac{d^2y(t)}{dt^2}\right|_{t=0}=-2$$

2.11. For each homogeneous differential equation given below, find the characteristic equation and show that it has multiple-order roots. Find the homogeneous solution for $t \ge 0$ in each case subject to the initial conditions specified.

a. $$\frac{d^2y(t)}{dt^2}+2\frac{dy(t)}{dt}+y(t)=0\,,\quad y(0)=1\,,\quad \left.\frac{dy(t)}{dt}\right|_{t=0}=0$$

b. $$\frac{d^3y(t)}{dt^3}+7\frac{d^2y(t)}{dt^2}+16\frac{dy(t)}{dt}+12y(t)=0\,,\quad y(0)=0\,,\quad \left.\frac{dy(t)}{dt}\right|_{t=0}=-2\,,\quad \left.\frac{d^2y(t)}{dt^2}\right|_{t=0}=1$$

c. $$\frac{d^3y(t)}{dt^3}+6\frac{d^2y(t)}{dt^2}+12\frac{dy(t)}{dt}+8y(t)=0\,,\quad y(0)=-1\,,\quad \left.\frac{dy(t)}{dt}\right|_{t=0}=0\,,\quad \left.\frac{d^2y(t)}{dt^2}\right|_{t=0}=3$$

2.12. Consider the simple RC circuit used in Example 2.8 and shown in Fig. P.2.12(a).

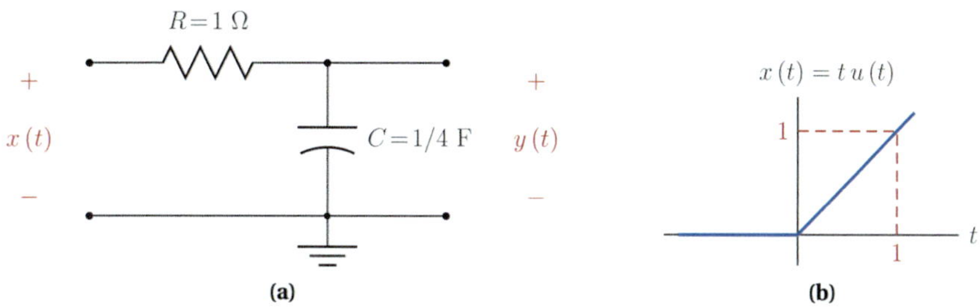

Figure P. 2.12

The differential equation between the input voltage and the output voltage was found to be

$$\frac{dy(t)}{dt}+4y(t)=4x(t)$$

Assuming the circuit is initially relaxed, compute and sketch the response to the unit ramp signal $x(t) = r(t) = t\,u(t)$ shown in Fig. P.2.12(b).

2.13. For each first-order differential equation given below, find the solution for $t \ge 0$ with the specified input signal and subject to the specified initial value. Use the general solution technique outlined in Section 2.5.4.

a. $$\frac{dy(t)}{dt} + 4\,y(t) = x(t)\,, \qquad x(t) = u(t)\,, \qquad y(0) = -1$$

b. $$\frac{dy(t)}{dt} + 2\,y(t) = 2\,x(t)\,, \qquad x(t) = \sin(2t) + 2\cos(t)\,, \qquad y(0) = 2$$

c. $$\frac{dy(t)}{dt} + 5\,y(t) = 3\,x(t)\,, \qquad x(t) = t\,u(t)\,, \qquad y(0) = -4$$

d. $$\frac{dy(t)}{dt} + y(t) = 2\,x(t)\,, \qquad x(t) = e^{-2t}\,u(t)\,, \qquad y(0) = -1$$

2.14. For each differential equation given below, find the solution for $t \ge 0$ with the specified input signal and subject to the specified initial value. Use the general solution technique outlined in Section 2.5.4.

a. $$\frac{d^2y(t)}{dt^2} + 3\frac{dy(t)}{dt} + 2\,y(t) = x(t)\,, \quad x(t) = u(t)\,, \quad y(0) = 3\,, \quad \left.\frac{dy(t)}{dt}\right|_{t=0} = 0$$

b. $$\frac{d^2y(t)}{dt^2} + 4\frac{dy(t)}{dt} + 3\,y(t) = x(t)\,, \quad x(t) = (t+1)\,u(t)\,, \quad y(0) = -2\,, \quad \left.\frac{dy(t)}{dt}\right|_{t=0} = 1$$

c. $$\frac{d^2y(t)}{dt^2} + 3\,y(t) = x(t)\,, \quad x(t) = u(t)\,, \quad y(0) = 2\,, \quad \left.\frac{dy(t)}{dt}\right|_{t=0} = 0$$

d. $$\frac{d^2y(t)}{dt^2} + 2\frac{dy(t)}{dt} + y(t) = x(t)\,, \quad x(t) = e^{-2t}\,u(t)\,, \quad y(0) = 1\,, \quad \left.\frac{dy(t)}{dt}\right|_{t=0} = 0$$

2.15. A system is described by the differential equation

$$\frac{d^2y(t)}{dt^2} + 4\frac{dy(t)}{dt} + 3\,y(t) = \frac{dx(t)}{dt} - 2\,x(t)$$

and has the initial conditions

$$y(0) = -2\,, \quad \left.\frac{dy(t)}{dt}\right|_{t=0} = 1$$

Using the technique outlined in Section 2.6 draw a block diagram to simulate this system. Incorporate the initial conditions into the block diagram.

2.16. Draw a block diagram for each DTLTI system described below by a differential equation.

a. $$\frac{d^2y(t)}{dt^2} + 4\frac{dy(t)}{dt} + 13\,y(t) = x(t)$$

b. $$\frac{d^3y(t)}{dt^3} + 3\frac{d^2y(t)}{dt^2} + 4\frac{dy(t)}{dt} + 2\,y(t) = \frac{d^2x(t)}{dt^2} + 6\frac{dx(t)}{dt} + 3\,x(t)$$

c. $$\frac{d^3y(t)}{dt^3} + 6\frac{d^2y(t)}{dt^2} + 12\frac{dy(t)}{dt} + 8\,y(t) = \frac{d^2x(t)}{dt^2} + 4\frac{dx(t)}{dt} + 2\,x(t)$$

2.17. A continuous-time system is described by the block diagram shown in Fig. P.2.17.

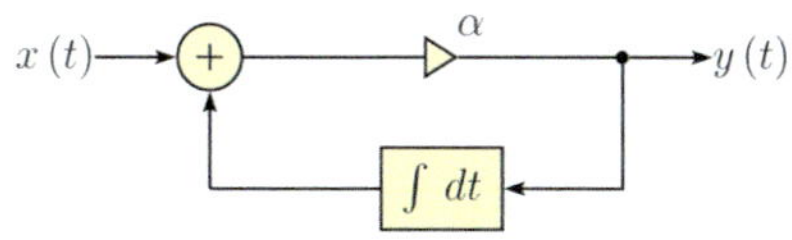

Figure P. 2.17

Write a differential equation between the signals $x(t)$ and $y(t)$.

2.18. Two CTLTI systems with impulse responses $h_1(t)$ and $h_2(t)$ are connected in cascade as shown in Fig. P.2.18(a).

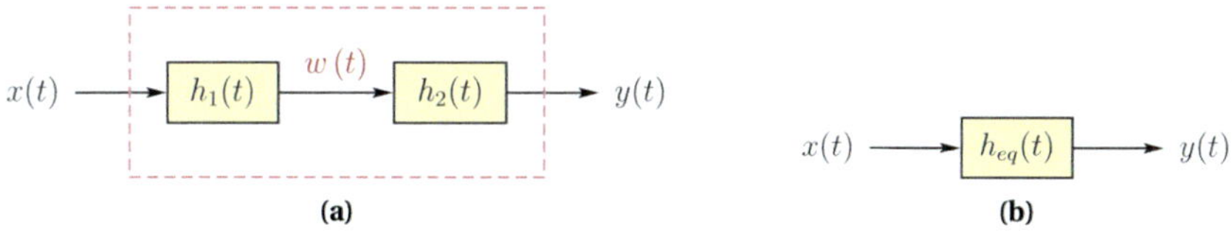

Figure P. 2.18

a. Determine the impulse response $h_{eq}(t)$ of the equivalent system, as shown in Fig. 2.18(b) in terms of $h_1(t)$ and $h_2(t)$.

Hint: Use convolution to express $w(t)$ in terms of $x(t)$. Afterward use convolution again to express $y(t)$ in terms of $w(t)$.

b. Let $h_1(t) = h_2(t) = \Pi(t-0.5)$ where $\Pi(t)$ is the unit pulse. Determine and sketch $h_{eq}(t)$ for the equivalent system.

c. With $h_1(t)$ and $h_2(t)$ as specified in part (b), let the input signal be a unit step, that is, $x(t) = u(t)$. Determine and sketch the signals $w(t)$ and $y(t)$.

2.19. Two CTLTI systems with impulse responses $h_1(t)$ and $h_2(t)$ are connected in parallel as shown in Fig. P.2.19(a).

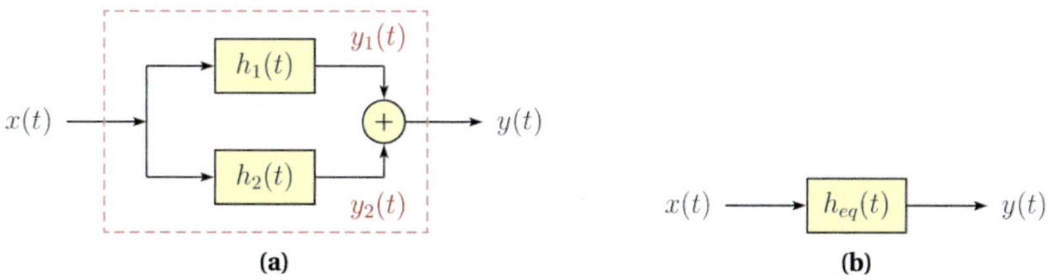

Figure P. 2.19

a. Determine the impulse response $h_{eq}(t)$ of the equivalent system, as shown in Fig. 2.19(b) in terms of $h_1(t)$ and $h_2(t)$.

Hint: Use convolution to express the signals $y_1(t)$ and $y_2(t)$ in terms of $x(t)$. Afterward express $y(t)$ in terms of $y_1(t)$ and $y_2(t)$.

b. Let $h_1(t) = e^{-t}\,u(t)$ and $h_2(t) = -e^{-3t}\,u(t)$. Determine and sketch $h_{eq}(t)$ for the equivalent system.

c. With $h_1(t)$ and $h_2(t)$ as specified in part (b), let the input signal be a unit step, that is, $x(t) = u(t)$. Determine and sketch the signals $y_1(t)$, $y_2(t)$, and $y(t)$.

2.20. Three CTLTI systems with impulse responses $h_1(t)$, $h_2(t)$, and $h_3(t)$ are connected as shown in Fig. P.2.20(a).

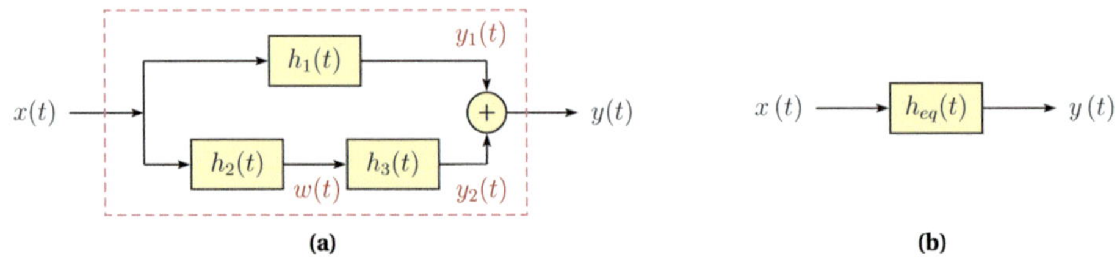

Figure P. 2.20

a. Determine the impulse response $h_{eq}(t)$ of the equivalent system, as shown in Fig. 2.20(b) in terms of $h_1(t)$, $h_2(t)$, and $h_3(t)$.

b. Let $h_1(t) = e^{-t}\,u(t)$, $h_2(t) = \delta(t-2)$, and $h_3(t) = e^{-2t}\,u(t)$. Determine and sketch $h_{eq}(t)$ for the equivalent system.

c. With $h_1(t)$, $h_2(t)$, and $h_3(t)$ as specified in part (b), let the input signal be a unit step, that is, $x(t) = u(t)$. Determine and sketch the signals $w(t)$, $y_1(t)$, $y_2(t)$, and $y(t)$.

2.21. Consider the CTLTI system shown in Fig. P.2.21(a).

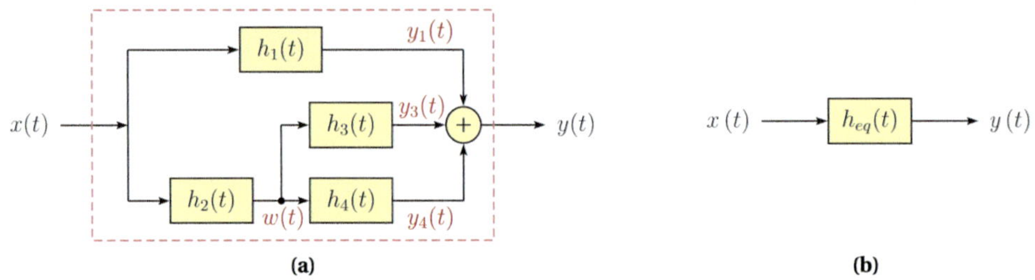

Figure P. 2.21

a. Express the impulse response of the system as a function of the impulse responses of the subsystems.

b. Let

$$h_1(t) = e^{-t}\,u(t)$$

$$h_2(t) = h_3(t) = u(t) - u(t-1)$$

and

$$h_4(t) = \delta(t-1)$$

Determine the impulse response $h_{eq}(t)$ of the equivalent system.

c. Let the input signal be a unit step, that is, $x(t) = u(t)$. Determine and sketch the signals $w(t)$, $y_1(t)$, $y_3(t)$, and $y_4(t)$.

2.22. Let $y(t)$ be the response of a CTLTI system to an input signal $x(t)$. Starting with the definition of the convolution integral, show that the response of the system to the derivative $dx(t)/dt$ is equal to $dy(t)/dt$.

2.23. Using the convolution integral given by Eqns. (2.143) and (2.144) prove each of the relationships below:

a. $x(t) * \delta(t) = x(t)$

b. $x(t) * \delta(t - t_0) = x(t - t_0)$

c. $x(t) * u(t-2) = \int_{-\infty}^{t-2} x(\lambda)\, d\lambda$

d. $x(t) * u(t - t_0) = \int_{-\infty}^{t-t_0} x(\lambda)\, d\lambda$

e. $x(t) * \Pi\left(\frac{t - t_0}{T}\right) = \int_{t-t_0-T/2}^{t-t_0+T/2} x(\lambda)\, d\lambda$

2.24. The impulse response of a CTLTI system is

$$h(t) = \delta(t) - \delta(t-1)$$

Determine sketch the response of this system to the triangular waveform shown in Fig. P.2.24.

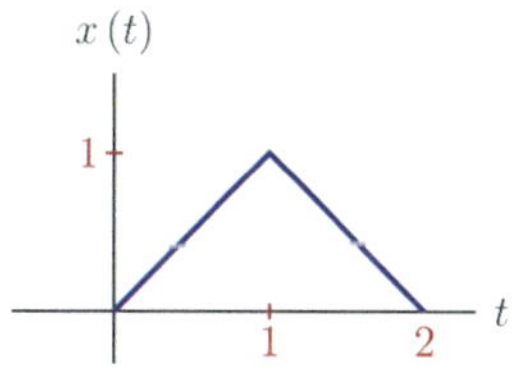

Figure P. 2.24

2.25. A CTLTI system has the impulse response

$$h(t) = \delta(t) + 0.5\,\delta(t-1) + 0.3\,\delta(t-2) + 0.2\,\delta(t-3)$$

Determine sketch the response of this system to the exponential input signal $x(t) = e^{-t}\,u(t)$ shown in Fig. P.2.25.

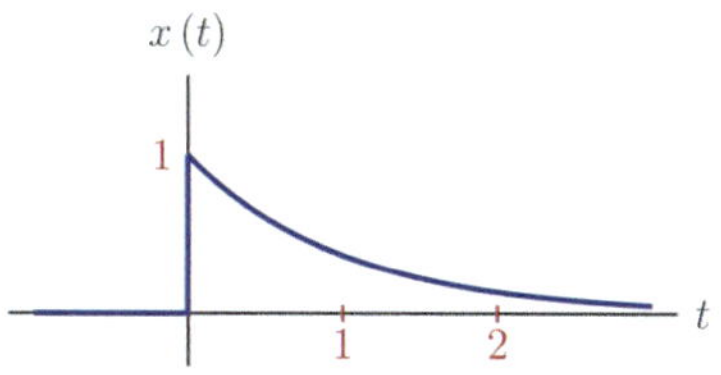

Figure P. 2.25

2.26. Let $x(t)$ be a signal limited in the range $(-\tau, \tau)$, that is

$$x(t) = 0 \quad \text{for } |t| > \tau$$

Show that a periodic extension of $x(t)$ can be obtained by convolving it with a periodic impulse train as follows:

$$\tilde{x}(t) = x(t) * \sum_{n=-\infty}^{\infty} \delta(t - nT_s)$$

where the period is $T_s \ge 2\tau$.

2.27. For each pair of signals $x(t)$ and $h(t)$ given below, find the convolution $y(t) = x(t) * h(t)$. In each case sketch the signals involved in the convolution integral and determine proper integration limits.

a. $x(t) = u(t)\,, \quad h(t) = e^{-2t}\, u(t)$

b. $x(t) = u(t)\,, \quad h(t) = \left[e^{-t} - e^{-2t}\right] u(t)$

c. $x(t) = u(t-2)\,, \quad h(t) = e^{-2t}\, u(t)$

d. $x(t) = u(t) - u(t-2)\,, \quad h(t) = e^{-2t}\, u(t)$

e. $x(t) = e^{-t}\, u(t)\,, \quad h(t) = e^{-2t}\, u(t)$

2.28. For each pair of signals $x(t)$ and $h(t)$ given below, find the convolution $y(t) = x(t) * h(t)$ first graphically and then analytically using the convolution integral.

a. $x(t) = \Pi\left(\dfrac{t-2}{4}\right), \quad h(t) = u(t)$

b. $x(t) = 3\Pi\left(\dfrac{t-2}{4}\right), \quad h(t) = e^{-t}\, u(t)$

c. $x(t) = \Pi\left(\dfrac{t-2}{4}\right), \quad h(t) = \Pi\left(\dfrac{t-2}{4}\right)$

d. $x(t) = \Pi\left(\dfrac{t-2}{4}\right), \quad h(t) = \Pi\left(\dfrac{t-3}{6}\right)$

2.29. Let $x(t)$ and $h(t)$ both be finite-duration signals such that, given parameter values $t_1 < t_2$ and $t_3 < t_4$, we have

$$x(t) = 0 \quad \text{for } t < t_1 \text{ or } t > t_2$$
$$h(t) = 0 \quad \text{for } t < t_3 \text{ or } t > t_4$$

Show that the convolution of $x(t)$ and $h(t)$ is also a finite duration signal:

$$y(t) = x(t) * h(t) = 0 \quad \text{for } t < t_5 \text{ or } t > t_6$$

Determine the limits t_5 and t_6 in terms of t_1, t_2, t_3, and t_4.

2.30. In Section 2.9 it was shown that, for a CTLTI system to be stable in BIBO sense, the impulse response must be absolute integrable, that is,

$$\int_{-\infty}^{\infty} |h(\lambda)|\, d\lambda < \infty$$

Suppose a particular system does not satisfy this condition. Prove that the system cannot be stable by finding a bounded input signal $x(t)$ that produces an unbounded output signal $y(t)$.

2.31. The system shown in Fig. P.2.31 represents addition of echoes to the signal $x(t)$:

$$y(t) = x(t) + \alpha_1\, x(t-\tau_1) + \alpha_2\, x(t-\tau_2)$$

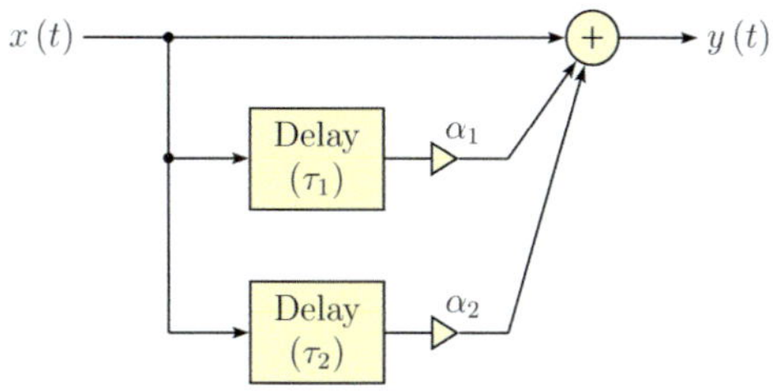

Figure P. 2.31

Comment on

a. linearity,
b. time invariance,
c. causality, and
d. stability

of the system.

2.32. For each system described below find the impulse response. Afterward determine if the system is causal and/or stable.

a. $$y(t) = \text{Sys}\{x(t)\} = \int_{-\infty}^{t} x(\lambda)\,d\lambda$$

b. $$y(t) = \text{Sys}\{x(t)\} = \int_{t-T}^{t} x(\lambda)\,d\lambda\,, \qquad T > 0$$

c. $$y(t) = \text{Sys}\{x(t)\} = \int_{t-T}^{t+T} x(\lambda)\,d\lambda\,, \qquad T > 0$$

MATLAB Problems

2.33. Refer to Problem 2.2. The cascade combination of the two subsystems will be tested using the input signal

$$x(t) = e^{-t}\cos(2t)\,u(t)$$

Write a MATLAB script to do the following:

a. Create an anonymous function `x()` to compute the input signal at all time instants specified in a vector `t`.

b. Implement the cascade system as shown in Fig. 2.2(a). Compute the signals $w(t)$ and $y(t)$ in the time interval $-1 \le t \le 5$ s using a time increment of 0.01 s.

c. Implement the cascade system in the alternative form shown in Fig. 2.2(b). Compute the signals $\bar{w}(t)$ and $\bar{y}(t)$ in the time interval $-1 \le t \le 5$ s using a time increment of 0.01 s. Compare $y(t)$ and $\bar{y}(t)$.

2.34. Repeat the steps of Problem 2.33 using the subsystems described in Problem 2.3.

2.35. Consider a *clipper* system with the input-output relationship $y(t) = \text{Sys}\{x(t)\}$ shown in Fig. P. 2.35.

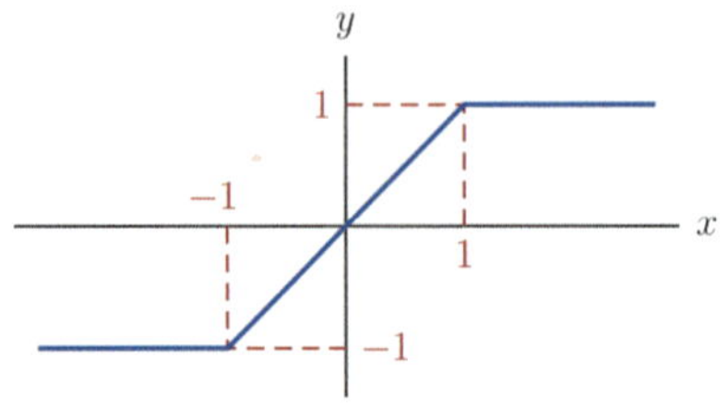

Figure P. 2.35

a. Develop a MATLAB function `ss_clip()` to implement this system. The syntax of your function should be

```
y = ss_clip(x)
```

where x is a vector containing the input signal and y is a vector of corresponding values of the output signal.

b. Demonstrate that a clipper is a nonlinear system by finding two appropriate signals $x_1(t)$ and $x_2(t)$ and computing the response of the clipper to the signal

$$x(t) = \alpha_1\, x_1(t) + \alpha_2\, x_2(t)$$

2.36. Consider a *dead zone* system with the input-output relationship $y(t) = \text{Sys}\{x(t)\}$ shown in Fig. P. 2.36.

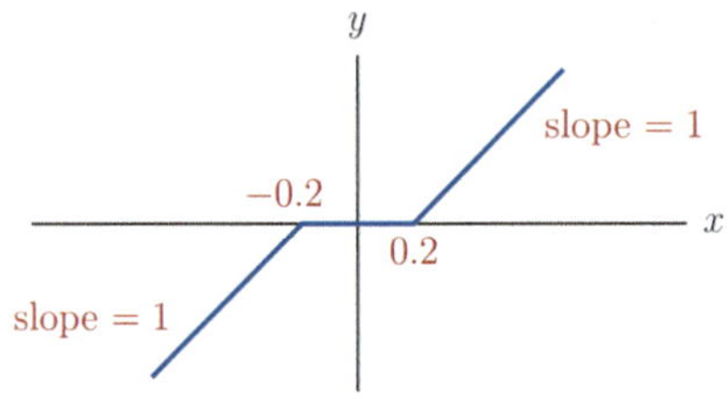

Figure P. 2.36

a. Develop a MATLAB function `ss_deadzone()` to implement this system. The syntax of your function should be

```
y = ss_deadzone(x)
```

where x is a vector containing the input signal and y is a vector of corresponding values of the output signal.

b. Demonstrate the nonlinearity of this system by finding two appropriate signals $x_1(t)$ and $x_2(t)$ and computing the response to the signal

$$x(t) = \alpha_1\, x_1(t) + \alpha_2\, x_2(t)$$

2.37. Consider the differential equations given in Problem 2.8. Using the Euler method, find a numerical solution for each differential equation with the specified input signal and initial condition. Refer to MATLAB Exercise 2.4 for an example of doing this. For each case use a step size T_s equal to 10 percent of the time constant, and obtain the numerical solution in the time interval $0 \le t \le 2$ s. Compare the numerical solution to the exact solution found in Problem 2.33.

Hint: Part (c) may be tricky due to the unit impulse input signal. Approximate $\delta(t)$ using a pulse with unit area and a width equal to the step size.

2.38. Repeat Problem 2.37 using MATLAB function `ode45()` instead of the Euler method.

2.39. Consider the system described by the block diagram in Fig. P.2.17. Let the input signal to the system be

$$x(t) = 3\sin(5t)$$

a. Write the differential equation of the system in the form

$$\frac{dy(t)}{dt} = g(t, y)$$

b. Let $\alpha = 0.7$. Using MATLAB function `ode45()`, compute the output signal $y(t)$ for $0 \le t \le 5$ s using a time increment $\Delta t = 0.01$ s. Use an anonymous function for evaluating $g(t, y)$. See MATLAB Exercise 2.5 for an example of how to do this.

c. Repeat part (c) for $\alpha = -0.7$. Comment on the effect of different values of α.

2.40. Refer to Problem 2.12 in which the unit ramp response of the simple RC circuit of Fig. P.2.12(a) was determined. Since the circuit is assumed to be initially relaxed, it is linear and time-invariant. As a result, the unit ramp response obtained in Problem 2.12 can be used with superposition for finding the response of the circuit to signals with higher complexity. Consider the trapezoidal input signal shown in Fig. P.2.40.

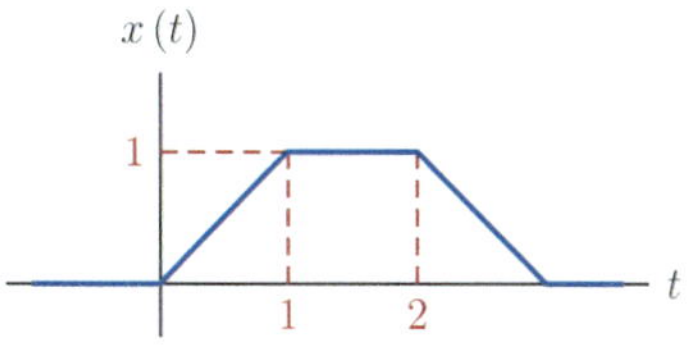

Figure P. 2.40

Develop a script to do the following:

a. Create an anonymous function `xr()` to compute the unit ramp function at time instants in a specified vector. You may wish to review the script developed in MATLAB Exercise 2.3 as an example.

b. Create an anonymous function `yr()` to compute the unit ramp response of the RC circuit at time instants in a specified vector.

c. Express the input signal $x(t)$ in terms of scaled and time-shifted unit ramp functions, and compute it in the time interval $-1 \le t \le 5$ s with a time increment of 0.01 s.

d. Express the response $y(t)$ of the circuit in terms of scaled and time-shifted versions of the unit ramp response, and compute it in the same time interval with the same increment.

e. Graph the input and the output signals on the same coordinate system.

2.41. Refer again to the simple RC circuit shown in Fig. 2.12(a). In MATLAB Exercise 2.3 an anonymous function was developed for computing the unit step response of the circuit. In Problem 2.40 a similar anonymous script was developed for computing its unit ramp response. Consider the input signal shown in Fig. P.2.41.

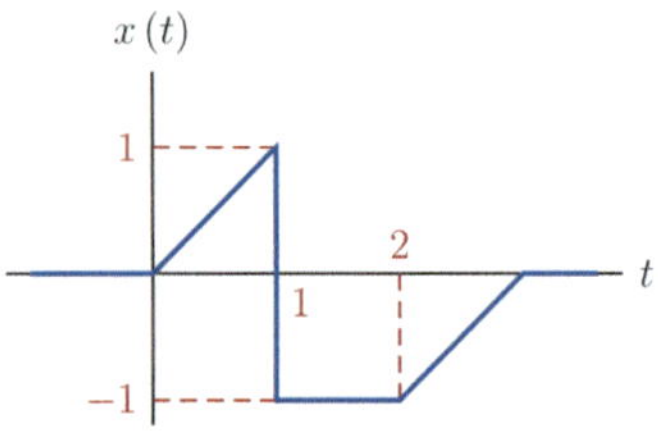

Figure P. 2.41

Develop a MATLAB script to do the following:

a. Express the input signal $x(t)$ using a combination of scaled and time-shifted unit step and unit ramp functions.

b. Use the anonymous function `yu()` from MATLAB Exercise 2.3 and the anonymous function `yr()` from Problem 2.40. Express the output signal $y(t)$ of the circuit in terms of scaled and time-shifted versions of the unit step response and the unit ramp response, and compute it in the time interval $-1 \le t \le 5$ s with a time increment of 0.01 s.

c. Graph the input and the output signals on the same coordinate system.

2.42. Write a script to verify the solution to Problem 2.24. Express the triangular waveform $x(t)$ using an anonymous function. Afterward compute and graph the output signal $y(t)$ in the time interval $-1 \le t \le 5$ s.

2.43. Write a script to verify the solution to Problem 2.25. Express the input signal $x(t) = e^{-t}\,u(t)$ using an anonymous function. Afterward compute and graph the output signal $y(t)$ in the time interval $-1 \le t \le 7$ s.

MATLAB Projects

2.44. Consider a system with the input-output relationship $y(t) = \text{Sys}\{x(t)\}$ shown in Fig. P. 2.44. It has the effect of mapping the amplitudes of the input signal to a finite number (in this case, eight) of predetermined amplitude values. It is referred to as a *quantizer* and its variants (typically with a larger number of amplitude levels) are used in communication systems.

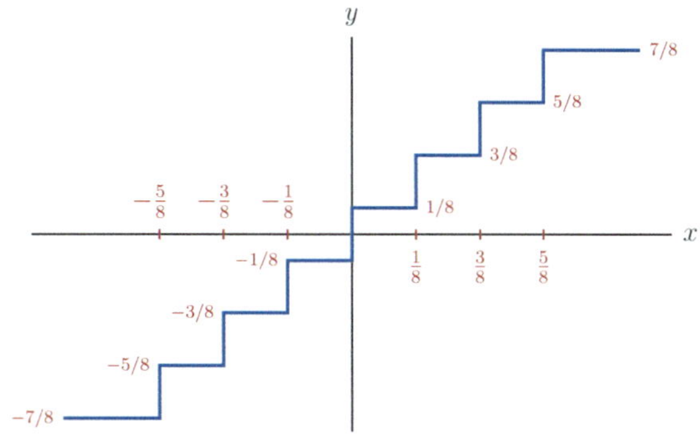

Figure P. 2.44

a. Develop a MATLAB function `ss_quant()` to implement this system. The syntax of your function should be

```
y = ss_quant(x)
```

where `x` is a vector containing the input signal, and `y` is a vector of corresponding values of the output signal.

b. Let the input signal to this quantizer be

$$x(t) = \sin(2\pi 10t)$$

Write a MATLAB script to perform the following steps:

b.1. Compute the signal $x(t)$ in the time interval $0 \le t \le 0.25$ s using a time increment of $\Delta t = 0.001$ s. The result should be the vector `x`.

b.2. Using the function `ss_quant()`, compute the output signal $y(t)$ of the quantizer. Graph the signals $x(t)$ and $y(t)$ on the same coordinate system.

b.3. The *quantization error* signal can be defined as

$$\varepsilon(t) = y(t) - x(t)$$

Compute and graph the quantization error $\varepsilon(t)$.

c. Is this system linear? Is it time-invariant? Comment.

2.45. Consider a general second-order homogeneous differential equation

$$\frac{d^2y(t)}{dt^2} + a_1\frac{dy(t)}{dt} + a_0\,y(t) = 0$$

with initial conditions

$$y(0) = p_1\,, \qquad \left.\frac{dy(t)}{dt}\right|_{t=0} = p_2$$

The coefficients a_1 and a_0 of the differential equation and the initial values p_1 and p_2 are all real-valued, and will be left as parameters.

a. Let s_1 and s_2 be the roots of the characteristic polynomial. On paper, solve the homogeneous differential equation for the three possibilities for the roots:

a.1. The roots are real and distinct.

a.2. The roots are a complex conjugate pair.

a.3. The two roots are real and equal.

Find the solution $y(t)$ as a function of the roots s_1 and s_2 as well as the initial values p_1 and p_2.

b. Develop a MATLAB function `ss_diff2solve()` with the syntax

```
y = ss_diff2solve(a1,a0,p1,p2,t)
```

The vector `t` contains the time instants at which the solution should be computed. The returned vector `y` holds the solution at the time instants in vector `t` so that the solution can be graphed with

```
plot(t,y)
```

Your function should perform the following steps:

b.1. Form the characteristic polynomial and find its roots.

b.2. Determine which of the three categories the roots fit (simple real roots, complex conjugate pair, or multiple roots).

b.3. Compute the solution accordingly.

c. Test the function `ss_diff2solve()` with the homogeneous differential equations in Problem 2.9a,b,c, Problem 2.10a,b,c, and Problem 2.11a.

2.46. Refer to the discussion in Section 2.10 regarding numerical solution of a differential equation using the Euler method. A modified version of this method is the backward Euler method which uses the approximation

$$\left.\frac{dy(t)}{dt}\right|_{t=t_0} \approx \frac{y(t_0)-y(t_0-T)}{T}$$

instead of Eqn. (2.168). With this modification Eqn. (2.169) becomes

$$y(t_0) \approx y(t_0-T) + T\,g\left[t_0, y(t_0)\right]$$

Consider the RC circuit problem the numerical solution of which was explored in MATLAB Exercise 2.4.

a. Start with the script developed in MATLAB Exercise 2.4. Modify it so that the step size is $T = 0.7$ s and the approximate solution is computed and graphed for $0 < t \leq 7$ s.

b. Develop a new script that uses the backward Euler method to solve the same problem. To accomplish this, start with the original differential equation and approximate the derivative on the left side with the backward difference. Simplify the resulting equation and use it in the code. Compute and graph the approximate solution using a step size of $T = 0.7$ s as in part (a).

c. Compare the results obtained in parts (a) and (b). Notice that the step size T used in both scripts is very large compared to the time constant. How does the performance of the backward Euler method compare to that of the Euler method?

2.47. Backward Euler method can be extended to apply to second-order differential equations using the approximation

$$\left.\frac{d^2y(t)}{dt^2}\right|_{t=t_0} \approx \frac{y(t_0)-2\,y(t_0-T)+y(t_0-2T)}{T^2}$$

for the second derivative. Using this approximation, develop a script to compute and graph the approximate solution of the homogeneous differential equation

$$\frac{d^2y(t)}{dt^2}+2\frac{dy(t)}{dt}+26\,y(t)=0$$

subject to the initial conditions

$$y(0)=2\,,\qquad \left.\frac{dy(t)}{dt}\right|_{t=0}=13$$

Experiment with the choice of the step size T. Compare your approximate solutions to the exact solution found in Example 2.15 part (a).

CHAPTER 3

FOURIER ANALYSIS FOR CONTINUOUS-TIME SIGNALS AND SYSTEMS

Chapter Objectives

- Learn techniques for representing continuous-time periodic signals using orthogonal sets of periodic basis functions.
- Study properties of exponential, trigonometric, and compact Fourier series, and conditions for their existence.
- Learn the Fourier transform for non-periodic signals as an extension of Fourier series for periodic signals.
- Study properties of the Fourier transform. Understand energy and power spectral density.
- Explore frequency-domain characteristics of continuous-time linear and time-invariant (CTLTI) systems. Understand the system function concept.
- Learn the use of frequency-domain analysis methods for solving signal-system interaction problems with periodic and non-periodic input signals.

3.1 Introduction

In Chapters 1 and 2 we have developed techniques for analyzing continuous-time signals and systems from a time-domain perspective. A continuous-time signal can be modeled as a function of time. A CTLTI system can be represented by means of a constant-coefficient linear differential

DOI: 10.1201/9781032675251-3

equation, or alternatively by means of an impulse response. The output signal of a CTLTI system can be determined by solving the corresponding differential equation or by using the convolution operation.

In Chapter 1 we have also discussed the idea of viewing a signal as a combination of simple signals acting as "building blocks". Examples of this were the use of unit step, unit ramp, unit pulse, and unit triangle functions for constructing signals (see Section 1.5) and the use of the unit impulse function for impulse decomposition of a signal (see Section 1.6).

Another especially useful set of building blocks is a set in which each member function has a unique frequency. Representing a signal as a linear combination of single-frequency building blocks allows us to develop a *frequency-domain* view of a signal that is particularly useful in understanding signal behavior and signal-system interaction problems. If a signal can be expressed as a superposition of single-frequency components, knowing how a linear and time-invariant system responds to each individual component helps us understand overall system behavior in response to the signal. This is the essence of frequency-domain analysis.

In Section 3.2 we discuss methods of analyzing periodic continuous-time signals in terms of their frequency content. Frequency-domain analysis methods for non-periodic signals are presented in Section 3.3. In Section 3.4 representation of signal energy and power in the frequency domain is discussed. System function concept is introduced in Section 3.5. The application of frequency-domain analysis methods to the analysis of CTLTI systems is discussed in Sections 3.6 and 3.7.

3.2 Analysis of Periodic Continuous-Time Signals

Most periodic continuous-time signals encountered in engineering problems can be expressed as linear combinations of sinusoidal *basis functions,* the more technical name for the so-called building blocks we referred to in the introductory section. The idea of representing periodic functions of time in terms of trigonometric basis functions was first realized by French mathematician and physicist Jean Baptiste Joseph Fourier (1768–1830) as he worked on problems related to heat transfer and propagation. The basis functions in question can either be individual sine and cosine functions, or they can be in the form of complex exponential functions that combine sine and cosine functions together.

Later in this section we will study methods of expressing periodic continuous-time signals in three different but equivalent formats, namely the *trigonometric Fourier series (TFS),* the *exponential Fourier series (EFS),* and the *compact Fourier series (CFS).* Before we start a detailed study of the mathematical theory of Fourier series, however, we will find it useful to consider the problem of *approximating* a periodic signal using a few trigonometric functions. This will help us build some intuitive understanding of frequency-domain methods for analyzing periodic signals.

3.2.1 Approximating a periodic signal with trigonometric functions

It was established in Section 1.7.2 of Chapter 1 that a signal $\tilde{x}(t)$ which is periodic with period T_0 has the property

$$\tilde{x}(t+T_0)=\tilde{x}(t) \tag{3.1}$$

for all t. Furthermore, it was shown through repeated use of Eqn. (3.1) that a signal that is periodic with period T_0 is also periodic with kT_0 for any integer k.

In working with periodic signals in this chapter, we will adopt the convention of using the tilde (~) character over the name of the signal as a reminder of its periodicity. For the sake of discussion let us consider the square wave signal $\tilde{x}(t)$ with a period of T_0 as shown in Fig. 3.1.

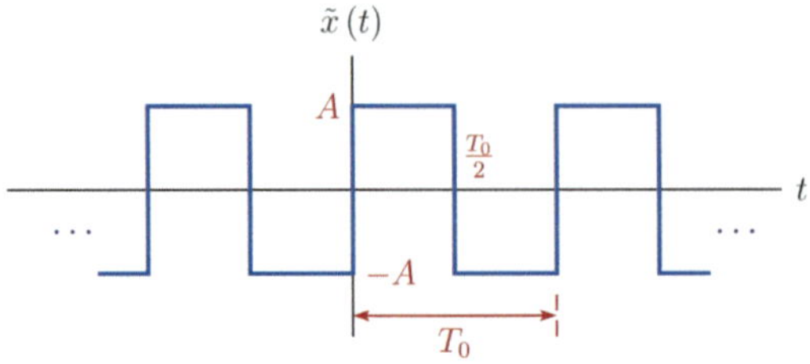

Figure 3.1 – Periodic square wave signal.

Suppose that we wish to approximate this signal using just one trigonometric function. Two questions that need to be answered are:

1. Should we use a sine or a cosine?
2. How should we adjust the parameters of the trigonometric function?

The first question is the easier one to answer. The square wave signal $\tilde{x}(t)$ shown in Fig. 3.1 has odd symmetry signified by $\tilde{x}(-t) = -\tilde{x}(t)$ (see Section 1.9 of Chapter 1). Among the two choices for a trigonometric function, the sine function also has odd symmetry, that is, $\sin(-a) = -\sin(a)$ for any real-valued parameter a. On the other hand, the cosine function has even symmetry since $\cos(-a) = \cos(a)$. Therefore it would intuitively make sense to choose the sine function over the cosine function. Our approximation would be in the form

$$\tilde{x}(t) \approx b_1 \sin(\omega t) \tag{3.2}$$

Since $\tilde{x}(t)$ has a fundamental period of T_0, it would make sense to pick a sine function with the same fundamental period, so that

$$\sin\big(\omega(t+T_0)\big) = \sin(\omega t) \tag{3.3}$$

For Eqn. (3.3) to work we need $\omega T_0 = 2\pi$ and consequently

$$\omega = \frac{2\pi}{T_0} = \omega_0 = 2\pi f_0 \tag{3.4}$$

where we have defined the frequency f_0 as the reciprocal of the period, that is, $f_0 = 1/T_0$. Let $\tilde{x}^{(1)}(t)$ represent an approximate version of the signal $\tilde{x}(t)$, so that

$$\tilde{x}^{(1)}(t) = b_1 \sin(\omega_0 t) \tag{3.5}$$

In Eqn. (3.5) we have used the superscripted signal name $\tilde{x}^{(1)}$ to signify the fact that we are using only one trigonometric function in this approximation. Our next task is to determine the value of the coefficient b_1. How should b_1 be chosen that Eqn. (3.5) represents the best approximation of the given type possible for the actual signal $\tilde{x}(t)$? There is a bit of subjectivity in this question since we have not yet defined what the "best approximation" means for our purposes.

Let us define the *approximation error* as the difference between the square wave signal and its approximation:

$$\tilde{\varepsilon}_1(t) = \tilde{x}(t) - \tilde{x}^{(1)}(t) = \tilde{x}(t) - b_1 \sin(\omega_0 t) \tag{3.6}$$

The subscript used on the error signal $\tilde{\varepsilon}_1(t)$ signifies the fact that it is the approximation error that results when only one trigonometric function is used. $\tilde{\varepsilon}_1(t)$ is also periodic with period T_0.

One possible method of choosing the best value for the coefficient b_1 would be to choose the value that makes the normalized average power of $\tilde{\varepsilon}_1(t)$ as small as possible.

Recall that the normalized average power in a periodic signal was defined in Chapter 1, Eqn. (1.57). Adapting it to the error signal $\tilde{\varepsilon}_1(t)$ we have

$$P_\epsilon = \frac{1}{T_0}\int_0^{T_0} \left[\tilde{\varepsilon}_1(t)\right]^2 dt \tag{3.7}$$

This is also referred to as the *mean-squared error (MSE)*. For simplicity we will drop the constant scale factor $1/T_0$ in front of the integral in Eqn. (3.7), and minimize the *cost function*

$$J = \int_0^{T_0} \left[\tilde{\varepsilon}_1(t)\right]^2 dt = \int_0^{T_0} \left[\tilde{x}(t) - b_1 \sin(\omega_0 t)\right]^2 dt \tag{3.8}$$

instead. The value of the coefficient b_1 that is optimum in the sense of producing the smallest possible value for MSE is found by differentiating the cost function with respect to b_1 and setting the result equal to zero.

$$\frac{dJ}{db_1} = 0 \quad \Longrightarrow \quad \text{Solve for } b_1 \tag{3.9}$$

Finding the optimum value of b_1 in Eqn. (3.5):

$$\frac{dJ}{db_1} = \frac{d}{db_1}\left[\int_0^{T_0} \left[\tilde{x}(t) - b_1 \sin(\omega_0 t)\right]^2 dt\right] = 0$$

Changing the order of integration and differentiation leads to

$$\frac{dJ}{db_1} = \int_0^{T_0} \frac{d}{db_1}\left[\tilde{x}(t) - b_1 \sin(\omega_0 t)\right]^2 dt = 0$$

Carrying out the differentiation we obtain

$$\int_0^{T_0} 2\left[\tilde{x}(t) - b_1 \sin(\omega_0 t)\right]\left[-\sin(\omega_0 t)\right] dt = 0$$

or equivalently

$$-\int_0^{T_0} \tilde{x}(t)\sin(\omega_0 t)\, dt + b_1 \int_0^{T_0} \sin^2(\omega_0 t)\, dt = 0 \tag{3.10}$$

It can be shown that the second integral in Eqn. (3.10) yields

$$\int_0^{T_0} \sin^2(\omega_0 t)\, dt = \frac{T_0}{2}$$

Substituting this result into Eqn. (3.10) yields the optimum choice for the coefficient b_1 as

$$b_1 = \frac{2}{T_0}\int_0^{T_0} \tilde{x}(t)\sin(\omega_0 t)\, dt \tag{3.11}$$

For the square wave signal $\tilde{x}(t)$ in Fig. 3.1 we have

$$b_1 = \frac{2}{T_0}\int_0^{T_0/2} (A)\sin(\omega_0 t)\, dt + \frac{2}{T_0}\int_{T_0/2}^{T_0} (-A)\sin(\omega_0 t)\, dt = \frac{4A}{\pi} \tag{3.12}$$

Thus, the optimum value of b_1 is $b_1 = 4A/\pi$. The best approximation to $\tilde{x}(t)$ using only one trigonometric function is

$$\tilde{x}^{(1)}(t) = \frac{4A}{\pi}\sin(\omega_0 t) \tag{3.13}$$

and the approximation error is

$$\tilde{\varepsilon}_1(t) = \tilde{x}(t) - \frac{4A}{\pi}\sin(\omega_0 t) \tag{3.14}$$

The signal $\tilde{x}(t)$, its single-frequency approximation $\tilde{x}^{(1)}(t)$ and the approximation error $\tilde{\varepsilon}_1(t)$ are shown in Fig. 3.2.

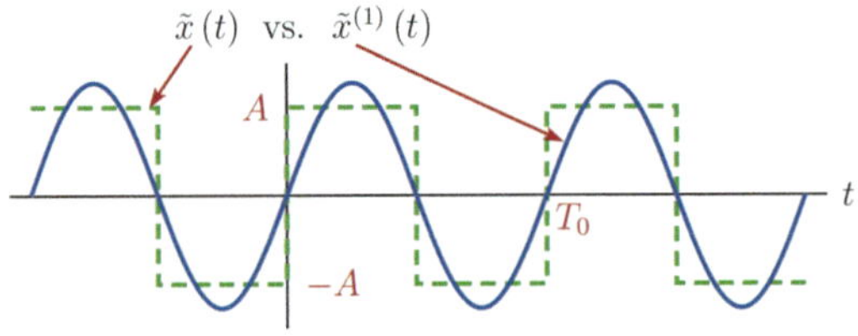

(a)

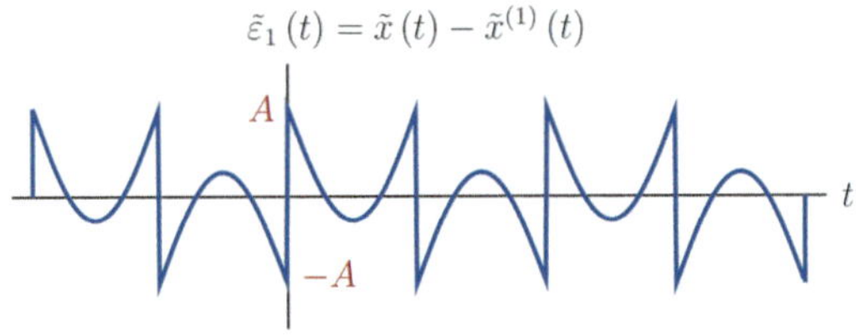

(b)

Figure 3.2 – **(a)** The square-wave signal $\tilde{x}(t)$ and its single-frequency approximation $\tilde{x}^{(1)}(t)$ and **(b)** the approximation error $\tilde{\varepsilon}_1(t)$.

In the next step we will try a two-frequency approximation to $\tilde{x}(t)$ and see if the approximation error can be reduced. We know that the basis function $\sin(2\omega_0 t) = \sin(4\pi f_0 t)$ is also periodic with the same period T_0; it just has two full cycles in the interval $(0, T_0)$ as shown in Fig. 3.3.

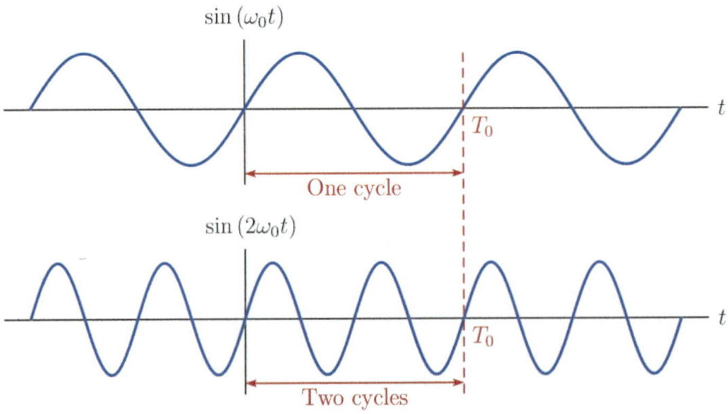

Figure 3.3 – Signals $\sin(\omega_0 t)$ and $\sin(2\omega_0 t)$.

Let the approximation using two frequencies be defined as

$$\tilde{x}^{(2)}(t) = b_1 \sin(\omega_0 t) + b_2 \sin(2\omega_0 t) \tag{3.15}$$

The corresponding approximation error is

$$\tilde{\varepsilon}_2(t) = \tilde{x}(t) - \tilde{x}^{(2)}(t) = \tilde{x}(t) - b_1 \sin(\omega_0 t) - b_2 \sin(2\omega_0 t) \tag{3.16}$$

The cost function will be set up in a similar manner:

$$J = \int_0^{T_0} \left[\tilde{x}(t) - b_1 \sin(\omega_0 t) - b_2 \sin(2\omega_0 t)\right]^2 dt \tag{3.17}$$

We now need to differentiate the cost function with respect to b_1 and b_2 and set both derivatives equal to zero to obtain two equations to solve.

$$\frac{\partial J}{\partial b_1} = 0\,, \quad \frac{\partial J}{\partial b_2} = 0 \quad \Longrightarrow \quad \text{Solve for } b_1 \text{ and } b_2 \tag{3.18}$$

Finding the optimum values for b_1 and b_2 in Eqn. (3.15):

Differentiating J with respect to first b_1 and then b_2 we obtain

$$\frac{\partial J}{\partial b_1} = \int_0^{T_0} 2\left[\tilde{x}(t) - b_1 \sin(\omega_0 t) - b_2 \sin(2\omega_0 t)\right]\left[-\sin(\omega_0 t)\right] dt$$

and

$$\frac{\partial J}{\partial b_2} = \int_0^{T_0} 2\left[\tilde{x}(t) - b_1 \sin(\omega_0 t) - b_2 \sin(2\omega_0 t)\right]\left[-\sin(2\omega_0 t)\right] dt$$

Setting the partial derivatives equal to zero leads to the following two equations:

$$\int_0^{T_0} \tilde{x}(t) \sin(\omega_0 t)\, dt = b_1 \int_0^{T_0} \sin^2(\omega_0 t)\, dt + b_2 \int_0^{T_0} \sin(2\omega_0 t) \sin(\omega_0 t)\, dt = 0 \tag{3.19}$$

$$\int_0^{T_0} \tilde{x}(t) \sin(2\omega_0 t)\, dt = b_1 \int_0^{T_0} \sin(\omega_0 t) \sin(2\omega_0 t)\, dt + b_2 \int_0^{T_0} \sin^2(2\omega_0 t)\, dt = 0 \tag{3.20}$$

We have already established that

$$\int_0^{T_0} \sin^2(\omega_0 t)\, dt = \frac{T_0}{2}$$

Similarly it can be shown that

$$\int_0^{T_0} \sin^2(2\omega_0 t)\, dt = \frac{T_0}{2} \quad \text{and} \quad \int_0^{T_0} \sin(\omega_0 t) \sin(2\omega_0 t)\, dt = 0$$

Eqns. (3.19) and (3.20) can now be solved for the optimum values of the coefficients b_1 and b_2:

$$b_1 = \frac{2}{T_0} \int_0^{T_0} \tilde{x}(t) \sin(\omega_0 t)\, dt \tag{3.21}$$

and

$$b_2 = \frac{2}{T_0} \int_0^{T_0} \tilde{x}(t) \sin(2\omega_0 t)\, dt \tag{3.22}$$

It is interesting to note that the expression for b_1 given in Eqn. (3.21) for the two-frequency approximation problem is the same as that found in Eqn. (3.11) for the single-frequency approximation problem. This is a result of the *orthogonality* of the two sine functions $\sin(\omega_0 t)$ and $\sin(2\omega_0 t)$ and will be discussed in more detail in the next section.

Using the square-wave signal $\tilde{x}(t)$ in Eqns. (3.21) and (3.22) we arrive at the following optimum values for the two coefficients:

$$b_1 = \frac{4A}{\pi} \quad \text{and} \quad b_2 = 0 \tag{3.23}$$

Interestingly, the optimum contribution from the sinusoidal term at radian frequency $2\omega_0$ turned out to be zero, resulting in

$$\tilde{x}^{(2)}(t) = \frac{4A}{\pi} \sin(\omega_0 t) + (0) \sin(2\omega_0 t) = \tilde{x}^{(1)}(t) \tag{3.24}$$

for this particular signal $\tilde{x}(t)$. It can be shown (see Problem 3.1 at the end of this chapter) that a three-frequency approximation

$$\tilde{x}^{(3)}(t) = b_1 \sin(\omega_0 t) + b_2 \sin(2\omega_0 t) + b_3 \sin(3\omega_0 t) \tag{3.25}$$

has the optimum coefficient values

$$b_1 = \frac{4A}{\pi}, \quad b_2 = 0, \quad \text{and} \quad b_3 = \frac{4A}{3\pi} \tag{3.26}$$

The signal $\tilde{x}(t)$, its three-frequency approximation $\tilde{x}^{(3)}(t)$ and the approximation error $\tilde{\varepsilon}_3(t)$ are shown in Fig. 3.4. Some observations are in order:

1. Based on a casual comparison of Figs. 3.2(a) and 3.4(b), the normalized average power of the error signal $\tilde{\varepsilon}_3(t)$ seems to be less than that of the error signal $\tilde{\varepsilon}_1(t)$. Consequently, $\tilde{x}^{(3)}(t)$ is a better approximation to the signal than $\tilde{x}^{(1)}(t)$.
2. On the other hand, the peak value of the approximation error seems to be $\pm A$ for both $\tilde{\varepsilon}_1(t)$ and $\tilde{\varepsilon}_3(t)$. If we were to try higher-order approximations using more trigonometric basis functions, the peak approximation error would still be $\pm A$ (see Problem 3.4 at the end of this chapter). This is due to fact that, at each discontinuity of $\tilde{x}(t)$, the value of the approximation is equal to zero independent of the number of trigonometric basis functions used. In general, the approximation at a discontinuity will yield the average of the signal amplitudes right before and right after the discontinuity. This leads to the well-known *Gibbs phenomenon* and will be explored further in Section 3.2.6.

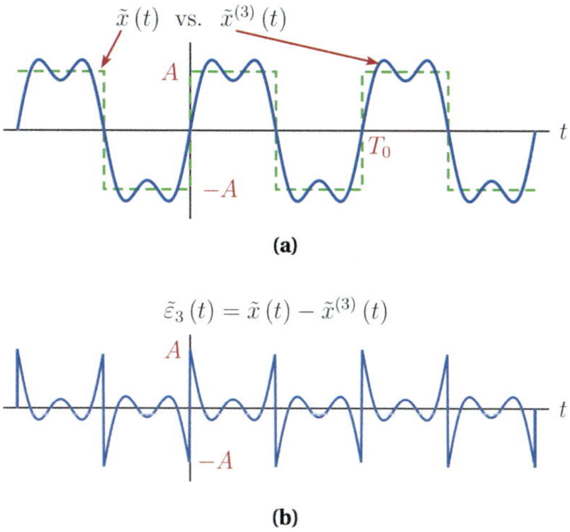

Figure 3.4 – **(a)** The square-wave signal $\tilde{x}(t)$ and its three-frequency approximation $\tilde{x}^{(3)}(t)$ and **(b)** the approximation error $\tilde{\varepsilon}_3(t)$.

Interactive App: Approximation of square wave using sinusoids

The interactive app in `appApprox.m` provides a graphical user interface for experimenting with approximations to the square wave signal of Fig. 3.1 using a specified number of trigonometric functions. Parameter values $A = 1$ and $T_0 = 1$ s are used. On the left side, slider controls allow parameters b_i to be adjusted freely for $i = 1, \ldots, 7$. The approximation

$$\tilde{x}^{(m)}(t) = \sum_{k=1}^{m} b_k \sin(k\omega_0 t)$$

and the corresponding approximation error

$$\tilde{\varepsilon}_m(t) = \tilde{x}(t) - \tilde{x}^{(m)}(t)$$

are computed and graphed. The value of m may be controlled by setting unneeded coefficients equal to zero. For example $\tilde{x}^{(5)}(t)$, the approximation with 5 trigonometric terms, may be explored by simply setting $b_6 = b_7 = 0$ and adjusting the remaining coefficients. In addition to graphing the signals, the program also computes and displays the value of the cost function

$$J = \int_0^{T_0} \left[\tilde{\varepsilon}_m(t)\right]^2 dt$$

a. With all other coefficients reset to zero, adjust the value of b_1 until J becomes as small as possible. How does the best value of b_1 correspond to the result found in Eqn. (3.13)? Make a note of the smallest value of J obtained.

b. Keeping b_1 at the best value obtained, start adjusting b_2. Move the slider control in the negative direction first; then move it in the positive direction. Can J be further reduced through the use of b_2? Can you confirm that the optimum value of b_2 is zero as determined in Eqn. (3.23)?

c. Keeping b_1 and b_2 at the optimum values obtained, adjust b_3 to obtain the smallest value possible for the cost function J. Once that point is reached, consider this question: Now that b_3 has been changed, do we need to go back and readjust b_1? Try and see if readjustment is needed? If not, why not?

d. Continue in this manner adjusting the remaining coefficients one at a time. Observe the shape of the approximation error signal and the value of J after each adjustment.

Software resource: `appApprox.m`

3.2.2 Trigonometric Fourier series (TFS)

We are now ready to generalize the results obtained in the foregoing discussion about approximating a signal using trigonometric functions. Consider a signal $\tilde{x}(t)$ that is periodic with fundamental period T_0 and associated fundamental frequency $f_0 = 1/T_0$. We may want to represent this signal using a linear combination of sinusoidal functions in the form

$$\begin{aligned}\tilde{x}(t) = & a_0 + a_1 \cos(\omega_0 t) + a_2 \cos(2\omega_0 t) + \ldots + a_k \cos(k\omega_0 t) \ldots \\ & + b_1 \sin(\omega_0 t) + b_2 \sin(2\omega_0 t) + \ldots + b_k \sin(k\omega_0 t) + \ldots\end{aligned} \tag{3.27}$$

or, using more compact notation

$$\tilde{x}(t) = a_0 + \sum_{k=1}^{\infty} a_k \cos(k\omega_0 t) + \sum_{k=1}^{\infty} b_k \sin(k\omega_0 t) \tag{3.28}$$

where $\omega_0 = 2\pi f_0$ is the fundamental frequency in rad/s. Eqn. (3.28) is referred to as the *trigonometric Fourier series (TFS)* representation of the periodic signal $\tilde{x}(t)$, and the sinusoidal functions with radian frequencies of $\omega_0, 2\omega_0, \ldots, k\omega_0$ are referred to as the *basis functions.* Thus, the set of basis functions includes

$$\phi_k(t) = \cos(k\omega_0 t)\ , \qquad k = 0, 1, 2, \ldots, \infty \tag{3.29}$$

and

$$\psi_k(t) = \sin(k\omega_0 t)\ , \qquad k = 1, 2, \ldots, \infty \tag{3.30}$$

Using the notation established in Eqns. (3.29) and (3.30), the series representation of the signal $\tilde{x}(t)$ given by Eqn. (3.27) can be written in a more generalized fashion as

$$\tilde{x}(t) = a_0 + \sum_{k=1}^{\infty} a_k \phi_k(t) + \sum_{k=1}^{\infty} b_k \psi_k(t) \tag{3.31}$$

We will call the frequencies that are integer multiples of the fundamental frequency the *harmonics.* The frequencies $2\omega_0, 3\omega_0, \ldots, k\omega_0$ are the second, the third, and the k-th harmonics of the fundamental frequency, respectively. The basis functions at harmonic frequencies are all periodic with a period of T_0. Therefore, considering our "building blocks" analogy, the signal $\tilde{x}(t)$ which is periodic with period T_0 is represented in terms of building blocks (basis functions) that are also periodic with the same period. Intuitively this makes sense.

Before we tackle the problem of determining the coefficients of the Fourier series representation, it is helpful to observe some properties of harmonically related sinusoids. Using trigonometric identities it can be shown that

$$\int_{t_0}^{t_0+T_0} \cos(m\omega_0 t) \cos(k\omega_0 t)\, dt = \begin{cases} T_0/2, & m = k \\ 0, & m \neq k \end{cases} \tag{3.32}$$

This is a very significant result. Two cosine basis functions at harmonic frequencies $m\omega_0$ and $k\omega_0$ are multiplied, and their product is integrated over one full period $(t_0, t_0 + T_0)$. When the integer multipliers m and k represent two different harmonics of the fundamental frequency, the result of the integral is zero. A non-zero result is obtained only when $m = k$, that is, when the two cosine functions in the integral are the same. A set of basis functions $\left\{\cos(k\omega_0 t),\ k = 0, \ldots, \infty\right\}$ that satisfies Eqn. (3.32) is said to be an *orthogonal* set. Similarly it is easy to show that

$$\int_{t_0}^{t_0+T_0} \sin(m\omega_0 t)\sin(k\omega_0 t)\,dt = \begin{cases} T_0/2, & m = k \\ 0, & m \neq k \end{cases} \tag{3.33}$$

meaning that the set of basis functions $\left\{\sin(k\omega_0 t),\ k = 1, \ldots, \infty\right\}$ is orthogonal as well. Furthermore it can be shown that the two sets are also orthogonal to each other, that is,

$$\int_{t_0}^{t_0+T_0} \cos(m\omega_0 t)\sin(k\omega_0 t)\,dt = 0 \tag{3.34}$$

for any combination of the two integers m and k (even when $m = k$). In Eqns. (3.32) through (3.34) the integral on the left side of the equal sign can be started at any arbitrary time instant t_0 without affecting the result. The only requirement is that integration be carried out over one full period of the signal. Detailed proofs of orthogonality conditions under a variety of circumstances are given in Appendix D. We are now ready to determine the unknown coefficients $\left\{a_k;\ k = 0, 1, \ldots, \infty\right\}$ and $\left\{b_k;\ k = 1, \ldots, \infty\right\}$ of Eqns. (3.27) and (3.28).

Finding TFS coefficients in Eqn. (3.28):

Let us first change summation indices in Eqn. (3.28) from k to m, then multiply both sides of it with $\cos(k\omega_0 t)$ and integrate over one full period:

$$\begin{aligned} \int_{t_0}^{t_0+T_0} \tilde{x}(t)\cos(k\omega_0 t)\,dt = & \int_{t_0}^{t_0+T_0} a_0 \cos(k\omega_0 t)\,dt \\ & + \int_{t_0}^{t_0+T_0} \left[\sum_{m=1}^{\infty} a_m \cos(m\omega_0 t)\right] \cos(k\omega_0 t)\,dt \\ & + \int_{t_0}^{t_0+T_0} \left[\sum_{m=1}^{\infty} b_m \sin(m\omega_0 t)\right] \cos(k\omega_0 t)\,dt \end{aligned} \tag{3.35}$$

Swapping the order of integration and summation in Eqn. (3.35) leads to

$$\begin{aligned} \int_{t_0}^{t_0+T_0} \tilde{x}(t)\cos(k\omega_0 t)\,dt = & \; a_0 \int_{t_0}^{t_0+T_0} \cos(k\omega_0 t)\,dt \\ & + \sum_{m=1}^{\infty} a_m \left[\int_{t_0}^{t_0+T_0} \cos(m\omega_0 t)\cos(k\omega_0 t)\,dt\right] \\ & + \sum_{m=1}^{\infty} b_m \left[\int_{t_0}^{t_0+T_0} \sin(m\omega_0 t)\cos(k\omega_0 t)\,dt\right] \end{aligned} \tag{3.36}$$

Let us consider the three terms on the right side of Eqn. (3.36) individually:

1. Let $k > 0$ (we will handle the case of $k = 0$ separately). The first term on the right side of Eqn. (3.36) evaluates to zero since it includes, as a factor, the integral of a cosine function over a full period.

2. Each term within the first summation has a factor which is the integral of the product of two cosines over a span of T_0. Using the orthogonality property observed in Eqn. (3.32), it is easy to see that all terms in this summation disappear with the exception of one term for which $m = k$.
3. Each term within the second summation has a factor which is the integral of the product of a sine function and a cosine function over a span of T_0. Using Eqn. (3.34) we conclude that all terms of this summation disappear.

Therefore, Eqn. (3.36) simplifies to

$$\int_{t_0}^{t_0+T_0} \tilde{x}(t)\cos(k\omega_0 t)\,dt = a_k \int_{t_0}^{t_0+T_0} \cos^2(k\omega_0 t)\,dt = a_k \frac{T_0}{2}$$

which can be solved for the only remaining coefficient a_k to yield

$$a_k = \frac{2}{T_0}\int_{t_0}^{t_0+T_0} \tilde{x}(t)\cos(k\omega_0 t)\,dt\,, \quad \text{for} \quad k = 1,\ldots,\infty \tag{3.37}$$

Similarly, by multiplying both sides of Eqn. (3.28) with $\sin(k\omega_0 t)$ and repeating the procedure used above, it can be shown that b_k coefficients can be computed as (see Problem 3.6 at the end of this chapter)

$$b_k = \frac{2}{T_0}\int_{t_0}^{t_0+T_0} \tilde{x}(t)\sin(k\omega_0 t)\,dt\,, \quad \text{for} \quad k = 1,\ldots,\infty \tag{3.38}$$

Finally, we need to compute the value of the constant coefficient a_0. Integrating both sides of Eqn. (3.28) over a full period, we obtain

$$\begin{aligned}\int_{t_0}^{t_0+T_0} \tilde{x}(t)\,dt = &\int_{t_0}^{t_0+T_0} a_0\,dt \\ &+ \int_{t_0}^{t_0+T_0}\left[\sum_{k=1}^{\infty} a_k\cos(k\omega_0 t)\right]dt + \int_{t_0}^{t_0+T_0}\left[\sum_{k=1}^{\infty} b_k\sin(k\omega_0 t)\right]dt\end{aligned} \tag{3.39}$$

Again changing the order of integration and summation operators, Eqn. (3.38) becomes

$$\begin{aligned}\int_{t_0}^{t_0+T_0} \tilde{x}(t)\,dt = &\int_{t_0}^{t_0+T_0} a_0\,dt \\ &+ \sum_{k=1}^{\infty} a_k\left[\int_{t_0}^{t_0+T_0}\cos(k\omega_0 t)\,dt\right] + \sum_{k=1}^{\infty} b_k\left[\int_{t_0}^{t_0+T_0}\sin(k\omega_0 t)\,dt\right]\end{aligned} \tag{3.40}$$

Every single term within each of the two summations will be equal to zero due to the periodicity of the sinusoidal functions being integrated. This allows us to simplify Eqn. (3.40) to

$$\int_{t_0}^{t_0+T_0} \tilde{x}(t)\,dt = \int_{t_0}^{t_0+T_0} a_0\,dt = a_0 T_0$$

which can be solved for a_0 to yield

$$a_0 = \frac{1}{T_0}\int_{t_0}^{t_0+T_0} \tilde{x}(t)\,dt \tag{3.41}$$

Key results in the above development are Eqns. (3.37), (3.38), and (3.41). A close examination of Eqn. (3.41) reveals that the coefficient a_0 represents the time average of the signal $\tilde{x}(t)$ as defined in Eqn. (1.53) of Chapter 1. Because of this, a_0 is also referred to as the *average value* or the *dc component* of the signal.

Combining the results obtained up to this point, the TFS expansion of a signal can be summarized as follows:

Trigonometric Fourier series (TFS):

1. Synthesis equation:

$$\tilde{x}(t) = a_0 + \sum_{k=1}^{\infty} a_k \cos(k\omega_0 t) + \sum_{k=1}^{\infty} b_k \sin(k\omega_0 t) \tag{3.42}$$

2. Analysis equations:

$$a_k = \frac{2}{T_0} \int_{t_0}^{t_0+T_0} \tilde{x}(t) \cos(k\omega_0 t)\, dt\,, \quad \text{for} \quad k = 1, \ldots, \infty \tag{3.43}$$

$$b_k = \frac{2}{T_0} \int_{t_0}^{t_0+T_0} \tilde{x}(t) \sin(k\omega_0 t)\, dt\,, \quad \text{for} \quad k = 1, \ldots, \infty \tag{3.44}$$

$$a_0 = \frac{1}{T_0} \int_{t_0}^{t_0+T_0} \tilde{x}(t)\, dt \qquad \text{(dc component)} \tag{3.45}$$

Example 3.1: Trigonometric Fourier series of a periodic pulse train

A pulse train signal $\tilde{x}(t)$ with a period of $T_0 = 3$ s is shown in Fig. 3.5. Determine the coefficients of the TFS representation of this signal.

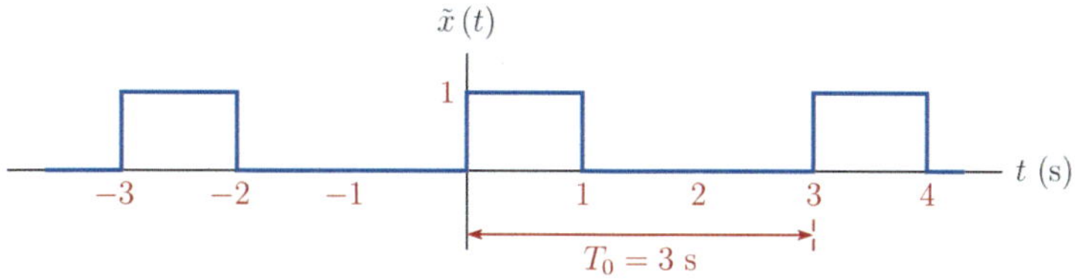

Figure 3.5 – The periodic pulse train used in Example 3.1.

Solution: In using the integrals given by Eqns. (3.43), (3.44), and (3.45), we can start at any arbitrary time instant t_0 and integrate over a span of 3 s. Applying Eqn. (3.45) with $t_0 = 0$ and $T_0 = 3$ s, we have

$$a_0 = \frac{1}{3}\left[\int_0^1 (1)\, dt + \int_1^3 (0)\, dt\right] = \frac{1}{3}$$

The fundamental frequency is $f_0 = 1/T_0 = 1/3$ Hz, and the corresponding value of ω_0 is

$$\omega_0 = 2\pi f_0 = \frac{2\pi}{3} \quad \text{rad/s}$$

Using Eqn. (3.43), we have

$$a_k = \frac{2}{3}\left[\int_0^1 (1)\cos(2\pi kt/3)\,dt + \int_1^3 (0)\cos(2\pi kt/3)\,dt\right] = \frac{\sin(2\pi k/3)}{\pi k}, \quad \text{for } k = 1,2,\ldots,\infty$$

Finally, using Eqn. (3.44), we get

$$b_k = \frac{2}{3}\left[\int_0^1 (1)\sin(2\pi kt/3)\,dt + \int_1^3 (0)\sin(2\pi kt/3)\,dt\right] = \frac{1-\cos(2\pi k/3)}{\pi k}, \quad \text{for } k = 1,2,\ldots,\infty$$

Using these coefficients in the synthesis equation given by Eqn. (3.42), the signal $x(t)$ can now be expressed in terms of the basis functions as

$$\tilde{x}(t) = \frac{1}{3} + \sum_{k=1}^{\infty}\left(\frac{\sin(2\pi k/3)}{\pi k}\right)\cos(2\pi kt/3) + \sum_{k=1}^{\infty}\left(\frac{1-\cos(2\pi k/3)}{\pi k}\right)\sin(2\pi kt/3) \tag{3.46}$$

Example 3.2: Approximation with a finite number of harmonics

Consider again the signal $\tilde{x}(t)$ of Example 3.1. Based on Eqns. (3.27) and (3.28), it would theoretically take an infinite number of cosine and sine terms to obtain an accurate representation of it. On the other hand, values of coefficients a_k and b_k are inversely proportional to k, indicating that the contributions from the higher-order terms in Eqn. (3.46) will decline in significance. As a result we may be able to neglect high-order terms and still obtain a reasonable approximation to the pulse train. Approximate $\tilde{x}(t)$ using a) the first 4 harmonics, and b) the first 10 harmonics.

Solution: Recall that we obtained the following in Example 3.1:

$$a_0 = \frac{1}{3}, \qquad a_k = \frac{\sin(2\pi k/3)}{\pi k}, \qquad b_k = \frac{1-\cos(2\pi k/3)}{\pi k}$$

These coefficients have been numerically evaluated for up to $k = 10$ and are shown in Table 3.1. Let $\tilde{x}^{(m)}(t)$ be an approximation to the signal $\tilde{x}(t)$ utilizing the first m harmonics of the fundamental frequency:

$$\tilde{x}^{(m)}(t) = a_0 + \sum_{k=1}^{m} a_k \cos(k\omega_0 t) + \sum_{k=1}^{m} b_k \sin(k\omega_0 t) \tag{3.47}$$

Using $m = 4$, we have

$$\begin{aligned}\tilde{x}^{(4)}(t) =& 0.3333 + 0.2757\cos(2\pi t/3) - 0.1378\cos(4\pi t/3) + 0.0689\cos(8\pi t/3) + \\ & 0.4775\sin(2\pi t/3) + 0.2387\sin(4\pi t/3) + 0.1194\sin(8\pi t/3)\end{aligned}$$

A similar but lengthier expression can be written for the case $m = 10$ which we will skip to save space. Fig. 3.6 shows the two approximations.

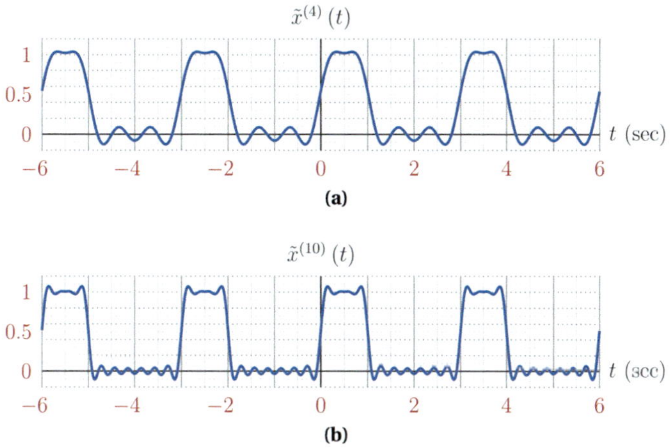

Figure 3.6 – Approximation to the pulse train of Example 3.1 using **(a)** 4 harmonics and **(b)** 10 harmonics.

Table 3.1 – TFS coefficients for the pulse train of Example 3.2.

k	a_k	b_k
0	0.3333	
1	0.2757	0.4775
2	−0.1378	0.2387
3	0.0	0.0
4	0.0689	0.1194
5	−0.0551	0.0955

k	a_k	b_k
6	0.0	0.0
7	0.0394	0.0682
8	−0.0345	0.0597
9	0.0	0.0
10	0.0276	0.0477
11	−0.0251	0.0434

Software resources: `exct_3_2.m` , `exct_3_2_Live.mlx`

Interactive App: TFS for periodic pulse train

The interactive app in `appTFS1.m` provides a graphical user interface for computing finite-harmonic approximations to the pulse train of Examples 3.1 and 3.2. Values of the TFS coefficients a_k and b_k for the pulse train are displayed in the spreadsheet-style table on the left.

Selecting an integer value m from the drop-down list labeled "Largest harmonic to include in approximation" causes the finite-harmonic approximation $\tilde{x}^{(m)}(t)$ to be computed and graphed. At the same time, checkboxes next to the coefficient pairs $\{a_k, b_k\}$ that are included in the approximation are automatically checked.

Alternatively coefficient sets $\{a_k, b_k\}$ may be included or excluded arbitrarily by checking the box to each set of coefficients on or off. When this is done, the graph displays the phrase "free format" as well as a list of the coefficient indices included.

a. Use the drop-down list to compute and graph finite-harmonic approximations for $m = 5, 10, 14$. Observe the similarity of the approximated signal to the original pulse train as larger values of m are used.

b. Pay attention to the overshoot that occurs at the discontinuities of the signal $\tilde{x}(t)$, that is, at time instants when the signal amplitude changes from -1 to $+1$ or vice versa. Does increasing the number of harmonics in $\tilde{x}^{(m)}(t)$ seem to help reduce the amount of overshoot?

c. Use the free format approach for observing the individual contributions of individual $\{a_k, b_k\}$ pairs. (This requires checking one box in the table with all others unchecked.)

Software resource: `appTFS1.m`

Software resource: See MATLAB Exercises 3.1 and 3.2.

Example 3.3: Periodic pulse train revisited

Determine the TFS coefficients for the periodic pulse train shown in Fig. 3.7.

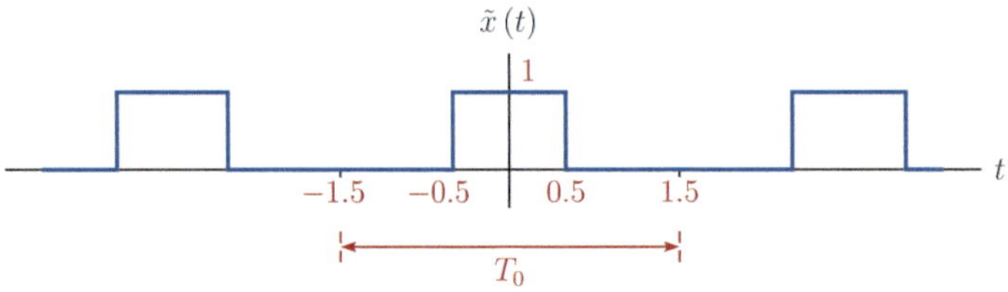

Figure 3.7 – The periodic pulse train used in Example 3.3.

Solution: This is essentially the same pulse train we have used in Example 3.1 with one minor difference: It is shifted in the time domain so that the main pulse is centered around the time origin $t = 0$. As a consequence, the resulting signal is an even function of time, that is, it has the property $\tilde{x}(-t) = \tilde{x}(t)$ for $-\infty < t < \infty$.

Let us take one period as the interval from $t_0 = -1.5$ to $t_0 + T_0 = 1.5$ s. Applying Eqn. (3.45) with these limits, we have

$$a_0 = \frac{1}{3}\int_{t=-0.5}^{0.5} (1)\,dt = \frac{1}{3}$$

Using Eqn. (3.43) yields

$$a_k = \frac{2}{3}\int_{-0.5}^{0.5} (1)\cos(2\pi k t/3)\,dt = \frac{2\sin(\pi k/3)}{\pi k}$$

and using Eqn. (3.44)

$$b_k = \frac{2}{3}\int_{-0.5}^{0.5} (1)\sin(2\pi k t/3)\,dt = 0$$

Thus, $\tilde{x}(t)$ can be written as

$$\tilde{x}(t) = \frac{1}{3} + \sum_{k=1}^{\infty} \left(\frac{2\sin(\pi k/3)}{\pi k} \right) \cos(k\omega_0 t)$$

where the fundamental frequency is $f_0 = 1/3$ Hz. In this case, the signal is expressed using only the $\cos(k\omega_0 t)$ terms of the TFS expansion. This result is intuitively satisfying since we have already recognized that $\tilde{x}(t)$ exhibits even symmetry, and therefore it can be represented using only the even basis functions $\{\cos(k\omega_0 t),\ k = 0, 1, \ldots, \infty\}$, omitting the odd basis functions $\{\sin(k\omega_0 t),\ k = 1, 2, \ldots, \infty\}$.

Software resource: `exct_3_3.m`

Interactive App: TFS for periodic pulse train revisited

The interactive app in `appTFS2.m` is based on Example 3.3, and computes finite-harmonic approximations to the periodic pulse train with even symmetry as shown in Fig. 3.7. It shares the same graphical user interface as in the program `appTFS1.m` with the only difference being the even symmetry of the pulse train used. Values of TFS coefficients a_k and b_k are displayed in the spreadsheet-style table on the left. Selecting an integer value m from the drop-down list labeled "Largest harmonic to include in approximation" causes the finite-harmonic approximation $\tilde{x}^{(m)}(t)$ to be computed and graphed. Observe that the signal $\tilde{x}(t)$ has even symmetry, and $b_k = 0$ for all k as we have determined in Example 3.3.

a. Use the drop-down list to compute and graph various finite-harmonic approximations. Observe the improvement in the approximation as larger values of m are used.

b. Pay attention to the subtle detail that $a_k = 0$ for every third harmonic, that is, when k is an integer multiple of 3. The reason for this will become clearer when we explore the concept of *duty cycle* in Example 3.7 in the next section.

Software resource: `appTFS2.m`

Example 3.4: Trigonometric Fourier series for a square wave

Determine the TFS coefficients for the periodic square wave shown in Fig 3.8.

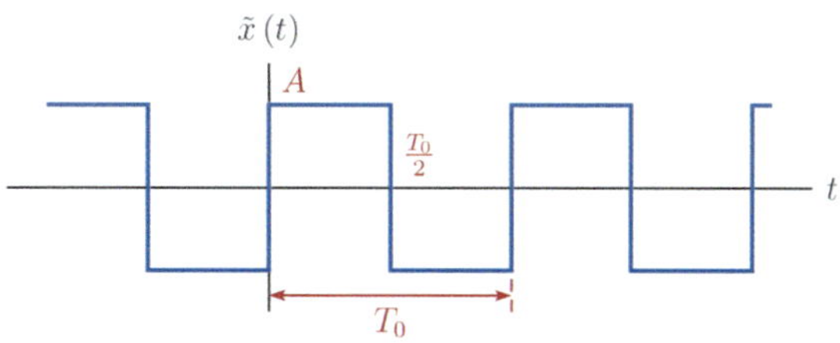

Figure 3.8 – The square wave signal used in Example 3.4.

Solution: This is a signal with odd symmetry, that is, $\tilde{x}(-t) = -\tilde{x}(t)$ for $-\infty < t < \infty$. Intuitively we can predict that the constant term a_0 and the coefficients a_k of the even basis functions $\{\cos(k\omega_0 t),\ k = 1,\ldots,\infty\}$ in the TFS representation of $\tilde{x}(t)$ should all be equal to zero and only the odd terms of the series should have significance. Applying Eqn. (3.43) with integration limits $t_0 = -T_0/2$ and $t_0 + T_0 = T_0/2$ yields

$$a_k = \frac{2}{T_0}\left[\int_{-T_0/2}^{0} (-A)\cos(2\pi k t/T_0)\,dt + \int_{t=0}^{T_0/2} (A)\cos(2\pi k t/T_0)\,dt\right] = 0$$

as we have anticipated. Next, we will use Eqn. (3.45) to determine a_0:

$$a_0 = \frac{1}{T_0}\left[\int_{-T_0/2}^{0} (-A)\,dt + \int_{0}^{T_0/2} (A)\,dt\right] = 0$$

Finally, using Eqn. (3.44), we get the coefficients of the sine terms:

$$\begin{aligned} b_k &= \frac{2}{T_0}\left[\int_{-T_0/2}^{0} (-A)\sin(2\pi k t/T_0)\,dt + \int_{0}^{T_0/2} (A)\sin(2\pi k t/T_0)\,dt\right] \\ &= \frac{2A}{\pi k}\left[1-\cos(\pi k)\right] \end{aligned}$$

Using the identity $\cos(\pi k) = (-1)^k$, this last result can be written as

$$b_k = \begin{cases} \dfrac{4A}{\pi k}, & k \text{ odd} \\ 0, & k \text{ even} \end{cases} \tag{3.48}$$

Compare this result with the coefficients b_1 through b_3 we have computed in Eqn. (3.26) in the process of finding the optimum approximation to the square wave signal $\tilde{x}(t)$ using three sine terms. Table 3.2 lists the TFS coefficients up to the 10-th harmonic. For simplicity, the amplitude A was set equal to unity.

Table 3.2 – TFS coefficients for the square-wave signal of Example 3.4 using $A = 1$.

k	a_k	b_k
0	0.0	
1	0.0	1.2732
2	0.0	0.0
3	0.0	0.4244
4	0.0	0.0
5	0.0	0.2546

k	a_k	b_k
6	0.0	0.0
7	0.0	0.1819
8	0.0	0.0
9	0.0	0.1415
10	0.0	0.0
11	0.0	0.1157

Finite-harmonic approximations $\tilde{x}^{(m)}(t)$ of the signal $\tilde{x}(t)$ for $m = 3$ and $m = 9$ are shown in Fig. 3.9.

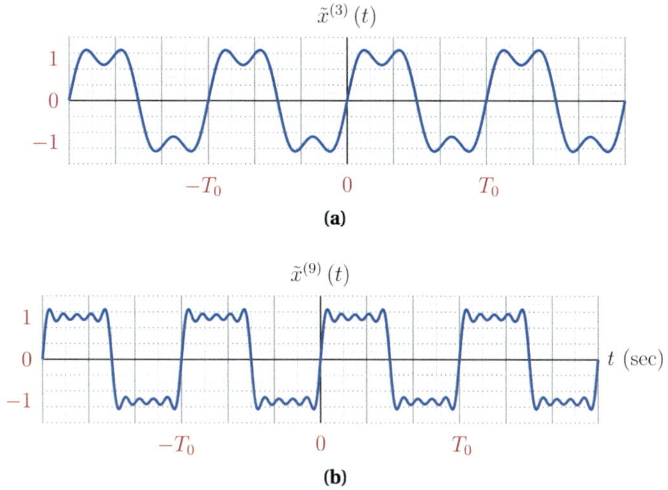

Figure 3.9 – Approximation to the square wave signal of Example 3.4 **(a)** using $m = 3$ and **(b)** using $m = 9$.

Software resources: exct_3_4a.m , exct_3_4b.m

Interactive App: TFS for periodic square wave signal

The interactive app in `appTFS3.m` is based on the TFS representation of a square wave signal as discussed in Example 3.4. It computes finite-harmonic approximations to the square wave signal with odd symmetry as shown in Fig. 3.9. Recall from Example 3.4 that, since the square wave signal in question has odd symmetry, the cosine terms are not needed; therefore $a_k = 0$ for $k = 1, \ldots, \infty$. The app extends the user interface of the two apps `appTFS1.m` and `appTFS2.m` by allowing the amplitude A and the period T_0 to be varied through the use of slider controls.

a. Set the amplitude to $A = 1$, the period to $T_0 = 1$ s, and the number of coefficients to $m = 3$. Observe the coefficient values b_1 and b_3.

b. Gradually increase the amplitude to $A = 2$ and then to $A = 3$. Observe the changes in TFS coefficients. Do they change proportionally? Recall that TFS coefficients for this signal were given by Eqn. (3.48).

c. Set the amplitude back to $A = 1$, and gradually increase the pulse period T_0. Observe the change in the fundamental frequency and, consequently, the frequencies of the sinusoidal basis functions. Do the TFS coefficients change when T_0 is varied?

d. Pay attention to the fact that even indexed coefficients b_k are equal to zero, that is, $b_k = 0$ for $k = 2, 4, 6, \ldots$. This will make more sense when we explore the concept of duty cycle.

Software resource: `appTFS3.m`

3.2.3 Exponential Fourier series (EFS)

Fourier series representation of the periodic signal $\tilde{x}(t)$ in Eqn. (3.28) can also be written in alternative forms. Consider the use of complex exponentials $\left\{e^{jk\omega_0 t},\ n=-\infty,\ldots,\infty\right\}$ as basis functions so that the signal $\tilde{x}(t)$ is expressed as a linear combination of them in the form

$$\tilde{x}(t)=\sum_{k=-\infty}^{\infty} c_k\, e^{jk\omega_0 t} \tag{3.49}$$

where the coefficients c_k are allowed to be complex-valued even though the signal $\tilde{x}(t)$ is real. This is referred to as the *exponential Fourier series (EFS)* representation of the signal $\tilde{x}(t)$. Before we consider this idea for an arbitrary periodic signal $\tilde{x}(t)$ we will study a special case.

Single-tone signals

Let us first consider the simplest of periodic signals: a single-tone signal in the form of a cosine or a sine waveform. We know that Euler's formula can be used for expressing such a signal in terms of two complex exponential functions:

$$\begin{aligned}\tilde{x}(t)=&A\cos(\omega_0 t+\theta)\\ =&\frac{A}{2}\,e^{j(\omega_0 t+\theta)}+\frac{A}{2}\,e^{-j(\omega_0 t+\theta)}\\ =&\frac{A}{2}\,e^{j\theta}\,e^{j\omega_0 t}+\frac{A}{2}\,e^{-j\theta}\,e^{-j\omega_0 t}\end{aligned} \tag{3.50}$$

Comparing Eqn. (3.50) with Eqn. (3.49) we conclude that the cosine waveform can be written in the EFS form of Eqn. (3.49) with coefficients

$$c_1=\frac{A}{2}\,e^{j\theta}\,,\quad c_{-1}=\frac{A}{2}\,e^{-j\theta},\quad \text{and}\quad c_k=0 \text{ for all other } k \tag{3.51}$$

If the signal under consideration is $\tilde{x}(t)=A\sin(\omega_0 t+\theta)$, a similar representation can be obtained using Euler's formula:

$$\begin{aligned}\tilde{x}(t)=&A\sin(\omega_0 t+\theta)\\ =&\frac{A}{2j}\,e^{j(\omega_0 t+\theta)}-\frac{A}{2j}\,e^{-j(\omega_0 t+\theta)}\end{aligned} \tag{3.52}$$

Using the substitutions

$$\frac{1}{j}=-j=e^{-j\pi/2}\quad\text{and}\quad -\frac{1}{j}=j=e^{j\pi/2}$$

Eqn. (3.52) can be written as

$$\tilde{x}(t)=\frac{A}{2}\,e^{j(\theta-\pi/2)}\,e^{j\omega_0 t}+\frac{A}{2}\,e^{-j(\theta-\pi/2)}\,e^{-j\omega_0 t} \tag{3.53}$$

Comparison of Eqn. (3.53) with Eqn. (3.49) leads us to the conclusion that the sine waveform in Eqn. (3.53) can be written in the EFS form of Eqn. (3.49) with coefficients

$$c_1=\frac{A}{2}\,e^{j(\theta-\pi/2)}\,,\quad c_{-1}=\frac{A}{2}\,e^{-j(\theta-\pi/2)},\quad \text{and}\quad c_k=0 \text{ for all other } k \tag{3.54}$$

The EFS representations of the two signals are shown graphically, in the form of a *line spectrum,* in Fig. 3.10(a) and (b).

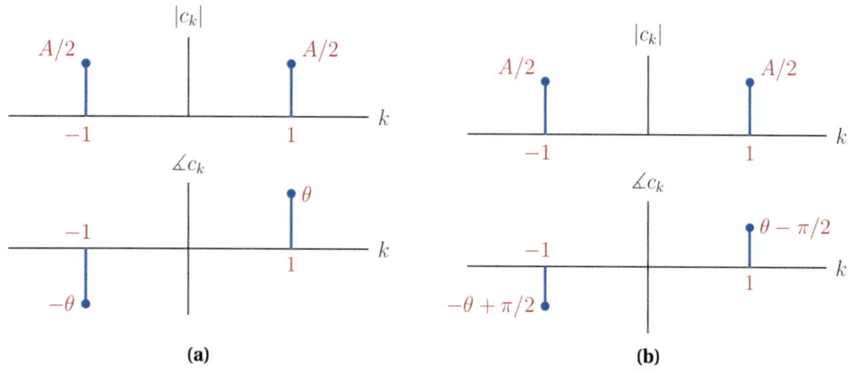

Figure 3.10 – The EFS representation of **(a)** $\tilde{x}(t) = A\cos(\omega_0 t + \theta)$ and **(b)** $\tilde{x}(t) = A\sin(\omega_0 t + \theta)$.

The magnitudes of the coefficients are identical for the two signals, and the phases differ by $\pi/2$. This is consistent with the relationship

$$A\sin(\omega_0 t + \theta) = A\cos(\omega_0 t + \theta - \pi/2) \tag{3.55}$$

The general case

We are now ready to consider the EFS representation of an arbitrary periodic signal $\tilde{x}(t)$. In order for the series representation given by Eqn. (3.49) to be equivalent to that of Eqn. (3.28) we need

$$c_0 = a_0 \tag{3.56}$$

and

$$c_{-k}\,e^{-jk\omega_0 t} + c_k\,e^{jk\omega_0 t} = a_k\cos(k\omega_0 t) + b_k\sin(k\omega_0 t) \tag{3.57}$$

for each value of the integer index k. Using Euler's formula in Eqn. (3.57) yields

$$[c_{-k} + c_k]\cos(k\omega_0 t) + \left[-jc_{-k} + jc_k\right]\sin(k\omega_0 t) = a_k\cos(k\omega_0 t) + b_k\sin(k\omega_0 t) \tag{3.58}$$

Since Eqn. (3.58) must be satisfied for all t, we require that coefficients of sine and cosine terms on both sides be identical. Therefore

$$c_k + c_{-k} = a_k \tag{3.59}$$

and

$$j\,(c_k - c_{-k}) = b_k \tag{3.60}$$

Solving Eqns. (3.59) and (3.60) for c_k and c_{-k} we obtain

$$c_k = \frac{1}{2}\left(a_k - jb_k\right) \tag{3.61}$$

and

$$c_{-k} = \frac{1}{2}\left(a_k + jb_k\right) \tag{3.62}$$

for $k = 1, \ldots, \infty$. Thus, EFS coefficients can be computed from the knowledge of TFS coefficients through the use of Eqns. (3.61) and (3.62).

Comparison of Eqns. (3.61) and (3.62) reveals another interesting result: For a real-valued signal $\tilde{x}(t)$, positive and negative indexed coefficients are complex conjugates of each other, that is,

$$c_k = c_{-k}^* \tag{3.63}$$

What if we would like to compute the EFS coefficients of a signal without first having to obtain the TFS coefficients? It can be shown (see Appendix D) that the exponential basis functions also form an orthogonal set:

$$\int_{t_0}^{t_0+T_0} e^{jm\omega_0 t}\, e^{-jk\omega_0 t}\, dt = \begin{cases} T_0\,, & m=k \\ 0\,, & m \neq k \end{cases} \tag{3.64}$$

EFS coefficients can be determined by making use of the orthogonality of the basis function set.

Derivation of EFS coefficients:

Let us first change the summation index in Eqn. (3.49) from k to m. Afterward we will multiply both sides with $e^{-jk\omega_0 t}$ to get

$$\begin{aligned} \tilde{x}(t)\, e^{-jk\omega_0 t} =& e^{-jk\omega_0 t} \sum_{m=-\infty}^{\infty} c_m\, e^{jm\omega_0 t} \\ =& \sum_{m=-\infty}^{\infty} c_m\, e^{jm\omega_0 t}\, e^{-jk\omega_0 t} \end{aligned} \tag{3.65}$$

Integrating both sides of Eqn. (3.65) over one period of $\tilde{x}(t)$ and making use of the orthogonality property in Eqn. (3.64), we obtain

$$\begin{aligned} \int_{t_0}^{t_0+T_0} \tilde{x}(t)\, e^{-jk\omega_0 t}\, dt =& \int_{t_0}^{t_0+T_0} \left[\sum_{m=-\infty}^{\infty} c_m\, e^{jm\omega_0 t}\, e^{-jk\omega_0 t} \right] dt \\ =& \sum_{m=-\infty}^{\infty} c_m \left[\int_{t_0}^{t_0+T_0} e^{jm\omega_0 t}\, e^{-jk\omega_0 t}\, dt \right] \\ =& c_k\, T_0 \end{aligned} \tag{3.66}$$

Eqn. (3.66) can be solved for the coefficient c_k to yield

$$c_k = \frac{1}{T_0} \int_{t_0}^{t_0+T_0} \tilde{x}(t)\, e^{-jk\omega_0 t}\, dt \tag{3.67}$$

If we evaluate Eqn. (3.67) for $k=0$, we obtain

$$c_0 = \frac{1}{T_0} \int_{t_0}^{t_0+T_0} \tilde{x}(t)\, dt \tag{3.68}$$

The right side of Eqn. (3.68) is essentially the definition of the time average operator given by Eqn. (1.53) in Chapter 1. Therefore, c_0 is the dc value of the signal $\tilde{x}(t)$.

In general, the coefficients of the EFS representation of a periodic signal $\tilde{x}(t)$ are complex-valued. They can be graphed in the form of a line spectrum if each coefficient is expressed in polar complex form with its magnitude and phase:

$$c_k = \left| c_k \right| e^{j\theta_k} \tag{3.69}$$

Magnitude and phase values in Eqn. (3.69) are computed by

$$\left| c_k \right| = \left[\left(\mathrm{Re}\left\{ c_k \right\} \right)^2 + \left(\mathrm{Im}\left\{ c_k \right\} \right)^2 \right]^{1/2} \tag{3.70}$$

and

$$\theta_k = \measuredangle c_k = \tan^{-1}\left(\frac{\text{Im}\{c_k\}}{\text{Re}\{c_k\}}\right), \tag{3.71}$$

respectively. Example 3.5 will illustrate this.

Exponential Fourier series (EFS)

1. Synthesis equation:

$$\tilde{x}(t) = \sum_{k=-\infty}^{\infty} c_k\, e^{jk\omega_0 t} \tag{3.72}$$

2. Analysis equation:

$$c_k = \frac{1}{T_0}\int_{t_0}^{t_0+T_0} \tilde{x}(t)\, e^{-jk\omega_0 t}\, dt \tag{3.73}$$

Example 3.5: Exponential Fourier series for periodic pulse train

Determine the EFS coefficients of the signal $\tilde{x}(t)$ of Example 3.3, shown in Fig. 3.7, through direct application of Eqn. (3.73).

Solution: Using Eqn. (3.73) with $t_0 = -1.5$ s and $T_0 = 3$ s, we obtain

$$c_k = \frac{1}{3}\int_{-0.5}^{0.5} (1)\, e^{-j2\pi kt/3}\, dt = \frac{\sin(\pi k/3)}{\pi k} \tag{3.74}$$

For this particular signal $\tilde{x}(t)$, the EFS coefficients c_k are real-valued. This will not always be the case, and is due to the even symmetry property of the signal $\tilde{x}(t)$. (Remember that, in working with the same signal in Example 3.3, we found $b_k = 0$ for all k. As a result, we have $c_k = c_{-k} = a_k/2$ for all k.)

Before the coefficients $\{c_k\}$ can be graphed or used for reconstructing the signal $\tilde{x}(t)$ the center coefficient c_0 needs special attention. Both the numerator and the denominator of the expression for c_k become zero for $k = 0$. For this case we need to use L'Hospital's rule which yields

$$c_0 = \left.\frac{\dfrac{d}{dk}\left[\sin(\pi k/3)\right]}{\dfrac{d}{dk}\left[\pi k\right]}\right|_{k=0} = \left.\frac{(\pi/3)\cos(\pi k/3)}{\pi}\right|_{k=0} = \frac{1}{3}$$

The signal $\tilde{x}(t)$ can be expressed in terms of complex exponential basis functions as

$$\tilde{x}(t) = \sum_{k=-\infty}^{\infty} \left(\frac{\sin(\pi k/3)}{\pi k}\right) e^{j2\pi kt/3}$$

A line graph of the set of coefficients c_k is useful for illustrating the make-up of the signal $\tilde{x}(t)$ in terms of its harmonics, and is shown in Fig. 3.11.

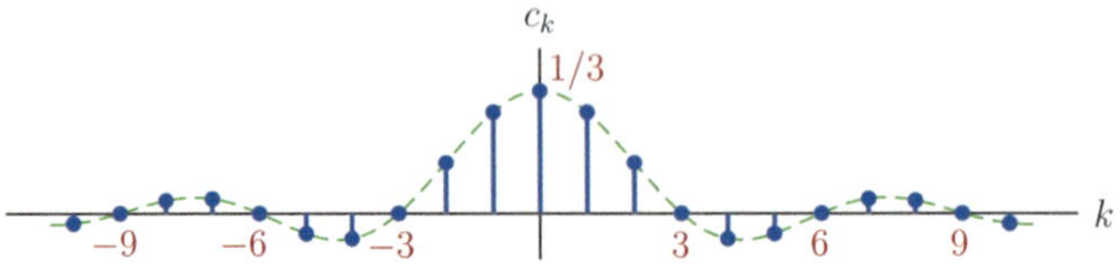

Figure 3.11 – The line spectrum for the pulse train of Example 3.5.

Note that this is not quite in the magnitude-and-phase form of the line spectrum discussed earlier. Even though the coefficients $\{c_k\}$ are real-valued, some of the coefficients are negative. Consequently, the graph in Fig. 3.11 does not qualify to be the magnitude of the spectrum. Realizing that $e^{j\pi} = -1$, any negative-valued coefficient $c_k < 0$ can be expressed as

$$c_k = |c_k|\, e^{j\pi}$$

using a phase angle of π radians to account for the negative multiplier. The line spectrum is shown in Fig. 3.12 in its proper form.

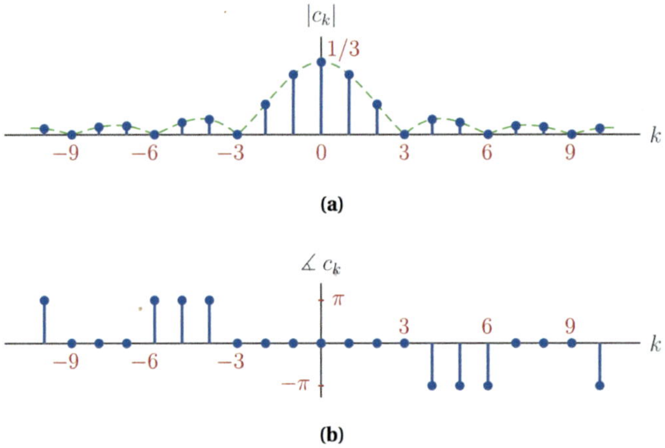

Figure 3.12 – The line spectrum for Example 3.5 in proper magnitude and phase format.

Interactive App: EFS for periodic pulse train with even symmetry

The interactive app in `appEFS1.m` is based on Example 3.5, and provides a graphical user interface for computing finite-harmonic approximations for the pulse train of Fig. 3.7. Values of the EFS coefficients c_k are displayed in Cartesian format in the spreadsheet-style table on the left. Selecting an integer value m from the drop-down list labeled "Largest harmonic to include in approximation" causes the finite-harmonic approximation $\tilde{x}^{(m)}(t)$ to be computed and graphed. At the same time, checkboxes next to the coefficients c_k included in the finite-harmonic approximation are automatically checked. Alternatively, arbitrary coefficients may be included or excluded by checking or unchecking the box next to each coefficient. When this is done, the graph displays the phrase "free format" as well as a list of indices of the coefficients included in computation.

a. Using the drop-down list, set $m = 1$. Note that the middle three lines in the line spectrum are displayed in red. These correspond to EFS coefficients $|c_k|$ for $k = -1, \ldots, 1$ so that the finite-harmonic approximation displayed is

$$\tilde{x}^{(2)}(t) = c_{-1}\, e^{-j\omega_0 t} + c_0 + c_1\, e^{j\omega_0 t}$$

b. Set $m = 2$. Now the middle five lines of the line spectrum should be displayed in red, corresponding to EFS coefficients $|c_k|$ for $k = -2, \ldots, 2$. Thus the new finite-harmonic approximation is

$$\tilde{x}^{(1)}(t) = c_{-2}\, e^{-j2\omega_0 t} + c_{-1}\, e^{-j\omega_0 t} + c_0 + c_1\, e^{j\omega_0 t} + c_2\, e^{j2\omega_0 t}$$

c. Pay attention to the fact that the line spectrum is purely real, that is, $\text{Im}\{c_k\} = 0$ for all k. This has to do with the even symmetry of the signal $\tilde{x}(t)$. We will explore properties of Fourier series for even and odd signals in Section 3.2.7.

d. Use the free format approach for observing the individual contributions of individual c_k coefficients by checking only one box in the table with all others unchecked.

Software resource: `appEFS1.m`

Software resource: See MATLAB Exercise 3.3.

Example 3.6: Periodic pulse train revisited

In Example 3.5 the EFS coefficients of the pulse train in Fig. 3.7 were found to be purely real. As discussed, this is due to the even symmetry of the signal. In this example we will remove this symmetry to see how it impacts the EFS coefficients. Consider the pulse train shown in Fig. 3.13.

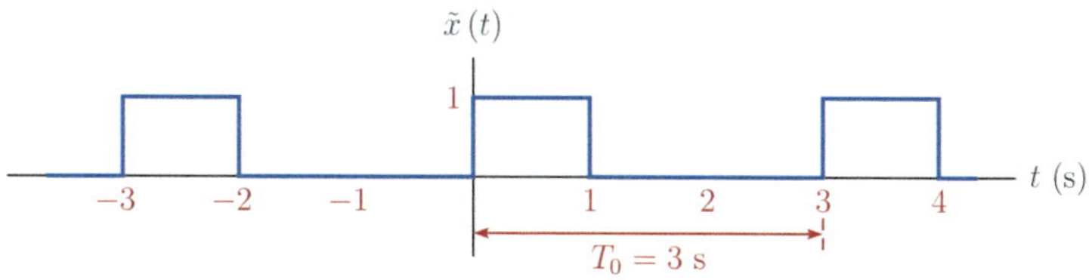

Figure 3.13 – The periodic pulse train for Example 3.6.

EFS coefficients can be found through the use of Eqn. (3.73). Setting $t_0 = 0$ and $T_0 = 3$ s and integrating over one period we obtain

$$c_k = \frac{1}{3}\int_0^1 (1)\, e^{-j2\pi k t/3}\, dt = \frac{-1}{j2\pi k}\left[e^{\,j2\pi k/3} - 1\right]$$

After some simplification, it can be shown that real and imaginary parts of the coefficients can be expressed as

$$\text{Re}\left\{c_k\right\} = \frac{\sin\left(2\pi k/3\right)}{2\pi k} \quad \text{and} \quad \text{Im}\left\{c_k\right\} = \frac{\cos\left(2\pi k/3\right)-1}{2\pi k}$$

Contrast these results with the TFS coefficients determined in Example 3.1 for the same signal. TFS representation of the signal $\tilde{x}(t)$ was given in Eqn. (3.46). Recall that the EFS coefficients are related to TFS coefficients by Eqns. (3.61) and (3.62). Using Eqns. (3.70) and (3.71), magnitude and phase of c_k can be computed as

$$\left|c_k\right| = \frac{1}{\sqrt{2}\pi\left|k\right|}\sqrt{1-\cos\left(2\pi k/3\right)} \quad \text{and} \quad \theta_k = \tan^{-1}\left(\frac{\cos\left(2\pi k/3\right)-1}{\sin\left(2\pi k/3\right)}\right) \tag{3.75}$$

The expression for magnitude can be further simplified. Using the appropriate trigonometric identity we can write

$$\cos\left(2\pi k/3\right) = 2\cos^2\left(\pi k/3\right) - 1 = 1 - 2\sin^2\left(\pi k/3\right)$$

Substituting this into the result for $\left|c_k\right|$ yields

$$\left|c_k\right| = \frac{1}{3}\left|\text{sinc}\left(k/3\right)\right|$$

where we have used the sinc function defined as

$$\text{sinc}\left(\alpha\right) \triangleq \frac{\sin\left(\pi\alpha\right)}{\pi\alpha} \tag{3.76}$$

The line spectrum is graphed in Fig. (3.14).

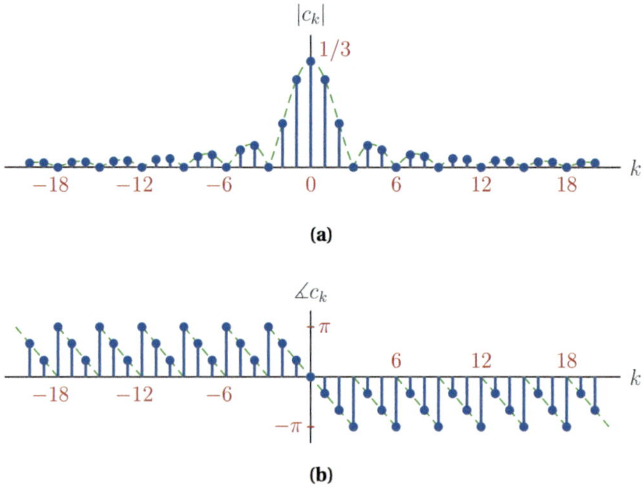

Figure 3.14 – Line spectrum for the pulse train of Example 3.6: **(a)** magnitude and **(b)** phase.

Software resource: exct_3_6.m

Interactive App: Effects of time shifting the signal on EFS

The interactive app in `appEFS2.m` is based on Example 3.6 in which we have removed the even symmetry of the pulse train of Example 3.5. If the periodic waveforms used in Examples 3.5 and 3.6 are denoted as $\tilde{x}_1(t)$ and $\tilde{x}_2(t)$, respectively, the relationship between them may be written as

$$\tilde{x}_2(t) = \tilde{x}_1(t - t_d) = \tilde{x}_1(t - 0.5)$$

The app provides a graphical user interface for adjusting the time delay and observing its effects on the exponential Fourier series coefficients as well as the corresponding finite-harmonic approximation. In contrast with the app `appEFS1.m`, the EFS coefficients are not purely real. Therefore, both the magnitude and the phase of the line spectrum is displayed.

a. Set the time delay to $t_d = 0.5$ s. Using the drop-down list, set $m = 1$. Note that the middle three lines in both the magnitude and the phase of the line spectrum are displayed in red. These correspond to EFS coefficients $|c_k|$ for $k = -1, \ldots, 1$ so that the finite-harmonic approximation displayed is

$$\tilde{x}^{(2)}(t) = c_{-1}\, e^{-j\omega_0 t} + c_0 + c_1\, e^{j\omega_0 t}$$

b. Set $m = 8$. Change the time delay to $t_d = 0$, and confirm that we now have the even symmetric signal in Fig. 3.7 of Example 3.3. Observe the magnitude and the phase of the line spectrum. We know that a zero time delay should give us the same result as in `appEFS1.m`. The magnitude of the line spectrum is the same as before, but the phase of some of the EFS coefficients is 180 degrees (π radians). How would you explain this?

c. Gradually change the time delay from $t_d = 0$ toward $t_d = 1$ s. Observe the magnitude and the phase of the line spectrum.

Software resource: `appEFS2.m`

Software resource: See MATLAB Exercise 3.4.

Example 3.7: Effects of duty cycle on the spectrum

Find the EFS representation of the pulse train depicted in Fig. 3.15.

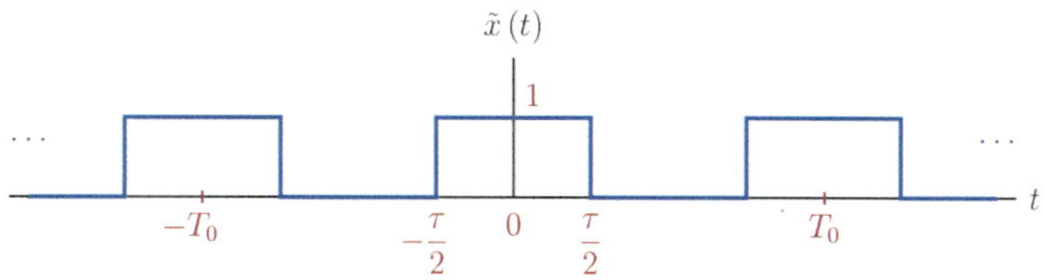

Figure 3.15 – Periodic pulse train for Example 3.7.

Solution: Each pulse occupies a time duration of τ within a period length of T_0. The *duty cycle* of a pulse train is defined as the ratio of the pulse width to the period, that is,

$$d \triangleq \frac{\tau}{T_0} \tag{3.77}$$

Using Eqn. (3.73) with $t_0 = -\tau/2$ we obtain the EFS coefficients as

$$c_k = \frac{1}{T_0}\left[\int_{-\tau/2}^{\tau/2} (1)\, e^{-j2\pi k t/T_0}\, dt\right] = \frac{\sin(\pi k d)}{\pi k}$$

This result can be written in a more compact form as

$$c_k = d\,\mathrm{sinc}(kd) \tag{3.78}$$

Magnitudes of coefficients c_k are shown in Figs. 3.16, 3.17, and 3.18 for duty cycle values of $d = 0.1$, $d = 0.2$, and $d = 0.3$, respectively.

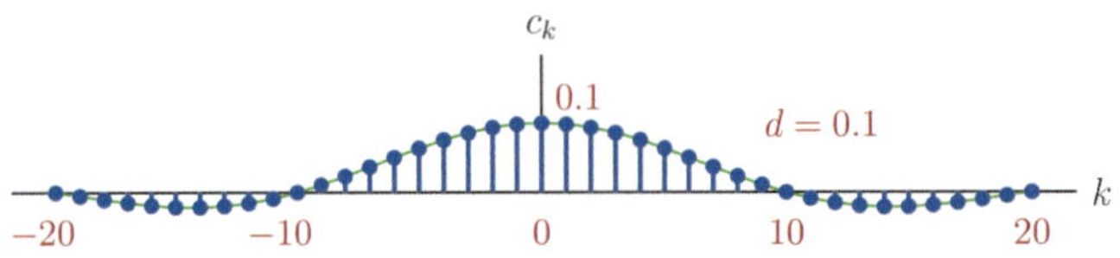

Figure 3.16 – Line spectrum for the pulse train of Example 3.7 with duty cycle $d = 0.1$.

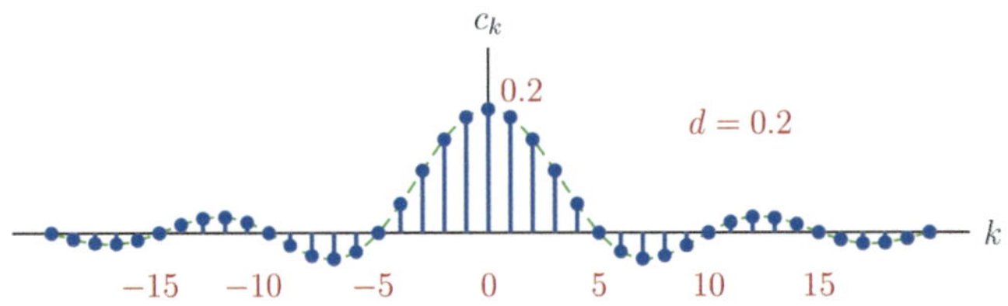

Figure 3.17 – Line spectrum for the pulse train of Example 3.7 with duty cycle $d = 0.2$.

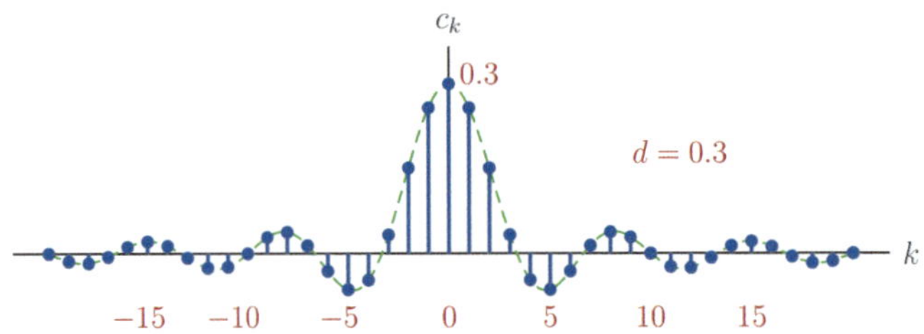

Figure 3.18 – Line spectrum for the pulse train of Example 3.7 with duty cycle $d = 0.3$.

In Eqn. (3.78) values of coefficients c_k depend only on the duty cycle and not on the period T_0.

On the other hand, the period T_0 impacts actual locations of the coefficients on the frequency axis, and consequently the frequency spacing between successive coefficients. Since the fundamental frequency is $f_0 = 1/T_0$, the coefficient c_k occurs at the frequency $kf_0 = k/T_0$ in the line spectrum.

The duty cycle plays an important role in pulse train type waveforms used in electronics and communication systems. Having a small duty cycle means increased blank space between individual pulses, and this can be beneficial in certain circumstances. For example, we may want to take advantage of the large gap between pulses, and utilize the target system to process other signals using a strategy known as *time-division multiplexing*. On the other hand, a small duty cycle does not come without cost. Smaller values of the duty cycle produce increased high-frequency content as the coefficients of large harmonics seem to be stronger for $d = 0.1$ compared to the other two cases.

Software resources: exct_3_7.m , exct_3_7_Live.mlx

Example 3.8: Spectrum of periodic sawtooth waveform

Consider a sawtooth signal $\tilde{x}(t)$, with a period of $T_0 = 1$ s, defined by

$$\tilde{x}(t) = at\,, \quad 0 < t < 1 \quad \text{and} \quad \tilde{x}(t+1) = \tilde{x}(t)$$

and shown in Fig. 3.19. The parameter a is a real-valued constant. Determine the EFS coefficients for this signal.

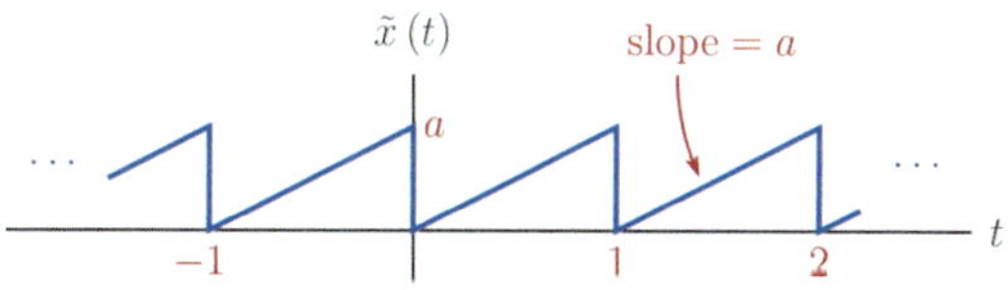

Figure 3.19 – Periodic sawtooth waveform of Example 3.8.

Solution: The fundamental frequency of the signal is

$$f_0 = \frac{1}{T_0} = 1\,\text{Hz}, \qquad \omega_0 = \frac{2\pi}{T_0} = 2\pi\,\text{rad/s}$$

Using Eqn. (3.73) we obtain

$$\begin{aligned} c_k = \int_0^1 a t\, e^{-j2\pi k t}\, dt &= a \left[\left. \frac{t\, e^{-j2\pi k t}}{-j2\pi k} \right|_0^1 + \int_0^1 \frac{e^{-j2\pi k t}}{j2\pi k}\, dt \right] \\ &= a \left[-\frac{e^{-j2\pi k}}{j2\pi k} + \frac{e^{-j2\pi k} - 1}{4\pi^2 k^2} \right] \end{aligned}$$

where we have used *integration by parts*. Applying Euler's formula to the exponential terms in this result, real and imaginary parts of the EFS coefficients are obtained as

$$\text{Re}\{c_k\} = a \left[\frac{\sin(2\pi k)}{2\pi k} + \frac{\cos(2\pi k) - 1}{4\pi^2 k^2} \right]$$

and

$$\text{Im}\{c_k\} = a \left[\frac{\cos(2\pi k)}{2\pi k} - \frac{\sin(2\pi k)}{4\pi^2 k^2} \right]$$

respectively. These two expressions can be greatly simplified by recognizing that $\sin(2\pi k) = 0$ and $\cos(2\pi k) = 1$ for all integer k. For $k = 0$, the values of $\text{Re}\{c_k\}$ and $\text{Im}\{c_k\}$ need to be resolved by using L'Hospital's rule. Thus, the real and imaginary parts of c_k are

$$\text{Re}\{c_k\} = \begin{cases} \dfrac{a}{2}, & k = 0 \\ 0, & k \neq 0 \end{cases} \quad \text{and} \quad \text{Im}\{c_k\} = \begin{cases} 0, & k = 0 \\ \dfrac{a}{2\pi k}, & k \neq 0 \end{cases} \tag{3.79}$$

Magnitudes of the EFS coefficients are obtained as

$$|c_k| = \begin{cases} \dfrac{|a|}{2}, & k = 0 \\ \left|\dfrac{a}{2\pi k}\right|, & k \neq 0 \end{cases}$$

For the phase angle we need to pay attention to the sign of the parameter a. If $a \geq 0$, we have

$$\theta_k = \begin{cases} 0\,, & k = 0 \\ \pi/2\,, & k > 0 \\ -\pi/2\,, & k < 0 \end{cases}$$

If $a < 0$, then θ_k needs to be modified as

$$\theta_k = \begin{cases} \pi\,, & k = 0 \\ -\pi/2\,, & k > 0 \\ \pi/2\,, & k < 0 \end{cases}$$

Magnitude and phase spectra for $\tilde{x}(t)$ are graphed in Fig. 3.20(**a**) and (**b**) with the assumption that $a > 0$.

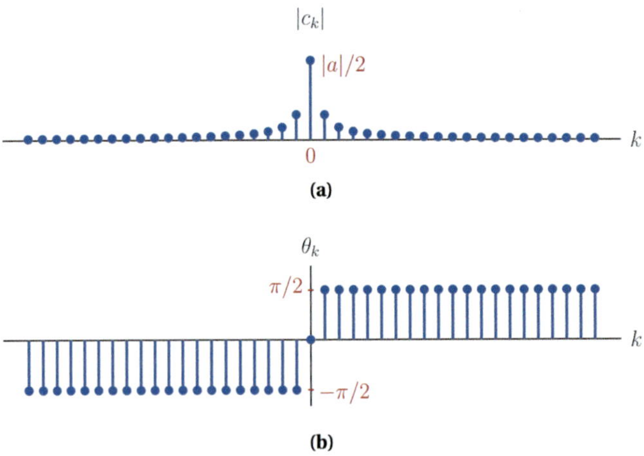

Figure 3.20 – Line spectrum for the periodic sawtooth waveform of Example 3.8: **(a)** magnitude and **(b)** phase.

Software resources: `exct_3_8a.m` , `exct_3_8b.m` , `exct_3_8_Live.mlx`

Example 3.9: Spectrum of multi-tone signal

Determine the EFS coefficients and graph the line spectrum for the multi-tone signal

$$\tilde{x}(t) = \cos\left(2\pi\left[10f_0\right]t\right) + 0.8\cos\left(2\pi f_0 t\right)\cos\left(2\pi\left[10f_0\right]t\right)$$

shown in Fig. 3.21.

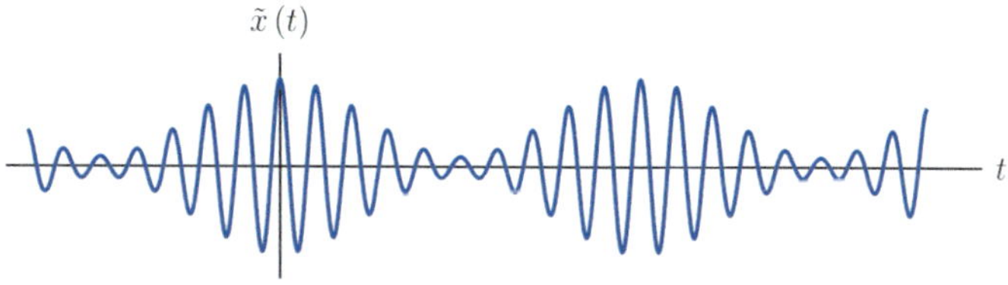

Figure 3.21 – Multi-tone signal of Example 3.9.

Solution: Using the trigonometric identity

$$cos(a)\cos(b) = \frac{1}{2}\cos(a+b) + \frac{1}{2}\cos(a-b)$$

the signal $\tilde{x}(t)$ can be written as

$$\tilde{x}(t) = \cos\left(2\pi\left[10f_0\right]t\right) + 0.4\cos\left(2\pi\left[11f_0\right]t\right) + 0.4\cos\left(2\pi\left[9f_0\right]t\right)$$

Applying Euler's formula, we have

$$\begin{aligned}\tilde{x}(t) =& 0.5\,e^{j2\pi(10f_0)t} + 0.5\,e^{-j2\pi(10f_0)t}\\ &+ 0.2\,e^{j2\pi(11f_0)t} + 0.2\,e^{-j2\pi(11f_0)t} + 0.2\,e^{j2\pi(9f_0)t} + 0.2\,e^{-j2\pi(9f_0)t}\end{aligned}$$

Inspecting this result, we find the significant EFS coefficients for the signal $\tilde{x}(t)$ to be

$$\begin{aligned} c_9 &= c_{-9} = 0.2\\ c_{10} &= c_{-10} = 0.5\\ c_{11} &= c_{-11} = 0.2\end{aligned}$$

All other EFS coefficients are equal to zero. The resulting line spectrum is shown in Fig. 3.22.

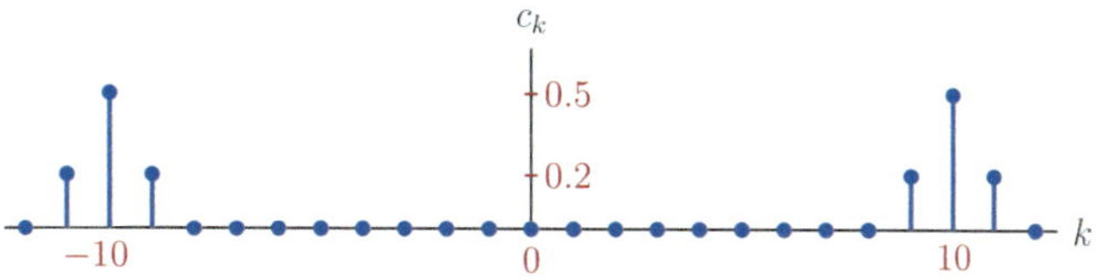

Figure 3.22 – Line spectrum for the multi-tone signal of Example 3.9.

Software resource: exct_3_9.m

Example 3.10: Spectrum of half-wave rectified sinusoidal signal

Determine the EFS coefficients and graph the line spectrum for the half-wave periodic signal $\tilde{x}(t)$ defined by

$$\tilde{x}(t) = \begin{cases} \sin(\omega_0 t)\,, & 0 \le t < T_0/2 \\ 0\,, & T_0/2 \le t < T_0 \end{cases} \qquad \text{and} \quad \tilde{x}(t+T_0) = \tilde{x}(t)$$

and shown in Fig. 3.23.

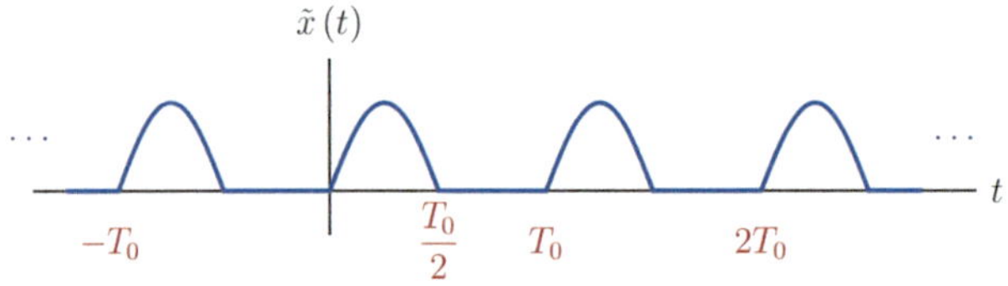

Figure 3.23 – Half-wave rectified sinusoidal signal of Example 3.10.

Solution: Using the analysis equation given by Eqn. (3.73), the EFS coefficients are

$$c_k = \frac{1}{T_0} \int_0^{T_0/2} \sin(\omega_0 t)\, e^{-jk\omega_0 t}\, dt$$

In order to simplify the evaluation of the integral we will write the sine function using Euler's formula and obtain

$$c_k = \frac{1}{j2T_0} \int_0^{T_0/2} \left[e^{j\omega_0 t} - e^{-j\omega_0 t} \right] e^{-jk\omega_0 t}\, dt$$

which can be evaluated to yield

$$\begin{aligned} c_k &= \frac{1}{j2T_0} \left(\frac{e^{-j\omega_0 (k-1)}}{-j\omega_0 (k-1)} \right) \Bigg|_{t=0}^{T_0/2} - \frac{1}{j2T_0} \left(\frac{e^{-j\omega_0 (k+1)}}{-j\omega_0 (k+1)} \right) \Bigg|_{t=0}^{T_0/2} \\ &= \frac{1}{4\pi (k-1)} \left[e^{-j\pi (k-1)} - 1 \right] - \frac{1}{4\pi (k+1)} \left[e^{-j\pi (k+1)} - 1 \right] \end{aligned} \tag{3.80}$$

We need to consider even and odd values of k separately.

Case 1: k odd and $k \ne \mp 1$

In this case both $k-1$ and $k+1$ are non-zero and even. We have $e^{-j\pi(k-1)} = e^{-j\pi(k+1)} = 1$, and therefore $c_k = 0$.

Case 2: $k = 1$

If $k = 1$, c_1 has an indeterminate form which can be resolved through the use of L'Hospital's rule:

$$c_1 = \frac{-j\pi\, e^{-j\pi(k-1)}}{4\pi} \Bigg|_{k=1} = -\frac{j}{4}$$

Case 3: $k = -1$

Similar to the case of $k = 1$ an indeterminate form is obtained for $k = -1$. Through the use of L'Hospital's rule the coefficient c_{-1} is found to be

$$c_{-1} = \frac{j\pi\, e^{-j\pi(k+1)}}{4\pi} \Bigg|_{k=-1} = \frac{j}{4}$$

Case 4: k even

In this case both $k-1$ and $k+1$ are odd, and we have $e^{-j\pi(k-1)} = e^{-j\pi(k+1)} = -1$.

$$e^{-j\pi(k-1)} - 1 = e^{-j\pi(k+1)} - 1 = -2$$

Using this result in Eqn. (3.80) we get

$$c_k = -\frac{1}{2\pi(k-1)} + \frac{1}{2\pi(k+1)} = \frac{-1}{\pi\left(k^2-1\right)}$$

Combining the results of the four cases, the EFS coefficients are

$$c_k = \begin{cases} 0\,, & k \text{ odd and } k \neq \mp 1 \\ -j/4\,, & k = 1 \\ j/4\,, & k = -1 \\ \dfrac{-1}{\pi\left(k^2-1\right)}\,, & k \text{ even} \end{cases}$$

The resulting line spectrum is shown in Fig. 3.24.

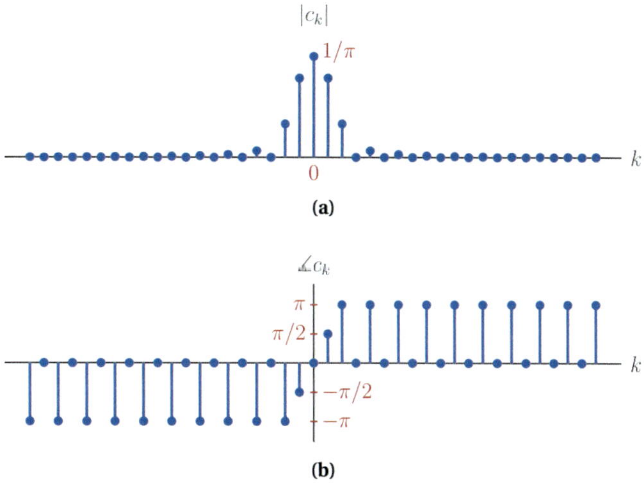

Figure 3.24 – Line spectrum for the half-wave sinusoidal signal of Example 3.10.

Software resources: exct_3_10a.m , exct_3_10b.m , exct_3_10_Live.mlx

3.2.4 Compact Fourier series (CFS)

Yet another form of the Fourier series representation of a periodic signal is the *compact Fourier series (CFS)* expressed as

$$\tilde{x}(t) = d_0 + \sum_{k=1}^{\infty} d_k \cos\left(k\omega_0 t + \phi_k\right) \tag{3.81}$$

Using the appropriate trigonometric identity, Eqn. (3.81) can be written as

$$\begin{aligned}\tilde{x}(t) &= d_0 + \sum_{k=1}^{\infty}\Big[d_k \cos(k\omega_0 t)\cos(\phi_k) - d_k \sin(k\omega_0 t)\sin(\phi_k)\Big] \\ &= d_0 + \sum_{k=1}^{\infty} d_k \cos(k\omega_0 t)\cos(\phi_k) - \sum_{k=1}^{\infty} d_k \sin(k\omega_0 t)\sin(\phi_k)\end{aligned}$$

Recognizing that this last equation for $\tilde{x}(t)$ is in a format identical to the TFS expansion of the same signal given by Eqn. (3.42), the coefficients of the corresponding terms must be equal. Therefore the following must be true:

$$a_0 = d_0 \tag{3.82}$$

$$a_k = d_k \cos(\phi_k)\,, \quad k = 1, \ldots, \infty \tag{3.83}$$

$$b_k = -d_k \sin(\phi_k)\,, \quad k = 1, \ldots, \infty \tag{3.84}$$

Solving for d_k and ϕ_k from Eqns. (3.82), (3.83), and (3.84) we obtain

$$d_k = \sqrt{\left\{a_k^2 + b_k^2\right\}}\,, \qquad k = 1, \ldots, \infty \tag{3.85}$$

and

$$\phi_k = -\tan^{-1}\left(\frac{b_k}{a_k}\right), \qquad k = 1, \ldots, \infty \tag{3.86}$$

with $d_0 = a_0$, and $\phi_0 = 0$. CFS coefficients can also be obtained from the EFS coefficients by using Eqns. (3.83) and (3.84) in conjunction with Eqns. (3.61) and (3.62), and remembering that $c_{-k} = c_k^*$ for real-valued signals:

$$d_k = 2\left[(\mathrm{Re}\{c_k\})^2 + (\mathrm{Im}\{c_k\})^2\right]^{1/2} = 2\,|c_k|\,, \qquad k = 1, \ldots, \infty \tag{3.87}$$

and

$$\phi_k = \tan^{-1}\left(\frac{\mathrm{Im}\{c_k\}}{\mathrm{Re}\{c_k\}}\right) = \theta_k\,, \qquad k = 1, \ldots, \infty \tag{3.88}$$

In the use of Eqn. (3.88), attention must be paid to the quadrant of the complex plane in which the coefficient c_k resides.

Example 3.11: Compact Fourier series for a periodic pulse train

Determine the CFS coefficients for the periodic pulse train that was used in Example 3.6. Afterward, using the CFS coefficients, find an approximation to $\tilde{x}(t)$ using $m = 4$ harmonics.

Solution: CFS coefficients can be obtained from the EFS coefficients found in Example 3.6 along with Eqns. (3.87) and (3.88):

$$d_0 = c_0\,, \quad d_k = 2\,|c_k| = \frac{2}{3}\,|\mathrm{sinc}(k/3)|, \quad \text{and} \quad \phi_k = \tan^{-1}\left(\frac{\cos(2\pi k/3) - 1}{\sin(2\pi k/3)}\right), \qquad k = 1, \ldots, \infty$$

Table 3.3 lists values of some of the compact Fourier series coefficients for the pulse train $\tilde{x}(t)$. Let $\tilde{x}^{(m)}(t)$ be an approximation to the signal $\tilde{x}(t)$ utilizing the first m harmonics of the fundamental frequency, that is,

$$\tilde{x}^{(m)}(t) = d_0 + \sum_{k=1}^{m} d_k \cos\left(k\omega_0 t + \phi_k\right) \tag{3.89}$$

Using $m = 4$ and $f_0 = 1/3$ Hz we have

$$\begin{aligned}\tilde{x}^{(4)}(t) =& \; 0.3333 + 0.5513\cos\left(2\pi\left(\tfrac{1}{3}\right)t - 1.0472\right)\\ &+ 0.2757\cos\left(2\pi\left(\tfrac{2}{3}\right)t - 2.0944\right) + 0.1378\cos\left(2\pi\left(\tfrac{4}{3}\right)t - 1.0472\right)\end{aligned} \tag{3.90}$$

The individual terms in Eqn. (3.90) as well as the resulting finite-harmonic approximation are shown in Fig. 3.25.

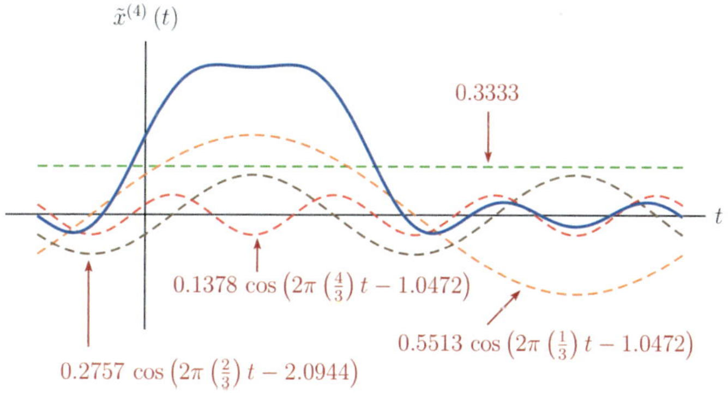

Figure 3.25 – The contributing terms in Eqn. (3.90) and the resulting finite-harmonic approximation.

Table 3.3 – Compact Fourier series coefficients for the waveform in Example 3.11.

k	d_k	ϕ_k (rad)
0	0.3333	0.0000
1	0.5513	−1.0472
2	0.2757	−2.0944
3	0.0000	0.0000
4	0.1378	−1.0472
5	0.1103	−2.0944
6	0.0000	0.0000
7	0.0788	−1.0472
8	0.0689	−2.0944
9	0.0000	0.0000
10	0.0551	−1.0472

3.2.5 Existence of Fourier series

For a specified periodic signal $\tilde{x}(t)$, one question that needs to be considered is the existence of the Fourier series: Is it always possible to determine the Fourier series coefficients? The answer is well-known in mathematics. A periodic signal can be uniquely expressed using sinusoidal basis functions at harmonic frequencies provided that the signal satisfies a set of conditions known as *Dirichlet conditions* named after the German mathematician Johann Peter Gustav Lejeune Dirichlet (1805–1859). A thorough mathematical treatment of Dirichlet conditions is beyond the scope of this text. For the purpose of the types of signals we will encounter in this text, however, it will suffice to summarize the three conditions as follows:

Existence conditions for Fourier series:

1. The signal $\tilde{x}(t)$ must be integrable over one period in an absolute sense, that is

$$\int_0^{T_0} |\tilde{x}(t)|\,dt < \infty \tag{3.91}$$

   Any periodic signal in which the amplitude values are bounded will satisfy Eqn. (3.91). In addition, periodic repetitions of singularity functions such as a train of impulses repeated every T_0 seconds will satisfy it as well.
2. If the signal $\tilde{x}(t)$ has discontinuities, it must have at most a finite number of them in one period. Signals with an infinite number of discontinuities in one period cannot be expanded into Fourier series.
3. The signal $\tilde{x}(t)$ must have at most a finite number of minima and maxima in one period. Signals with an infinite number of minima and maxima in one period cannot be expanded into Fourier series.

Most signals we encounter in engineering applications satisfy the existence criteria listed above. Consequently, they can be represented in series expansion forms given by Eqns. (3.28), (3.49), or (3.81). At points where the signal $\tilde{x}(t)$ is continuous, its Fourier series expansion converges perfectly. At a point of discontinuity, however, the Fourier series expansion (TFS, EFS, or CFS representation of the signal) yields the average value obtained by approaching the discontinuity from opposite directions. If the signal $\tilde{x}(t)$ has a discontinuity at $t = t_0$, and if $\{c_k\}$ are the EFS coefficients for it, we have

$$\sum_{k=-\infty}^{\infty} c_k\, e^{jk\omega_0 t_0} = \frac{1}{2} \lim_{t \to t_0^-} \left[\tilde{x}(t)\right] + \frac{1}{2} \lim_{t \to t_0^+} \left[\tilde{x}(t)\right] \tag{3.92}$$

This is illustrated in Fig. 3.26.

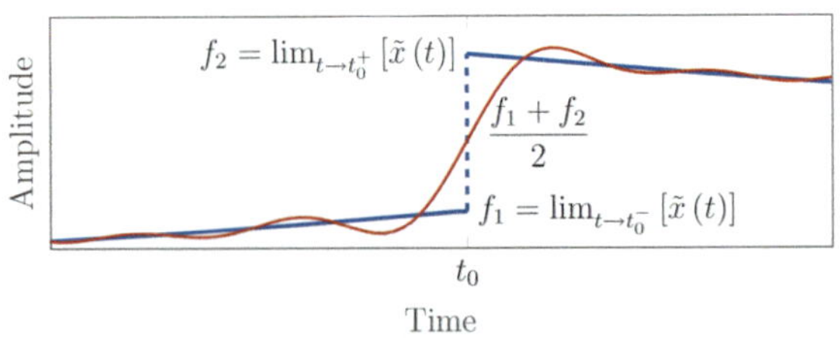

Figure 3.26 – Convergence of Fourier series at a discontinuity.

3.2.6 Gibbs phenomenon

Let us further explore the issue of convergence of the Fourier series at a discontinuity. Consider the periodic square wave shown in Fig. 3.27. The TFS coefficients of this signal were found in Example 3.4, and are repeated below:

$$a_k = 0 \quad \text{and} \quad b_k = \begin{cases} \dfrac{4}{\pi k}, & k \text{ odd} \\ 0, & k \text{ even} \end{cases}$$

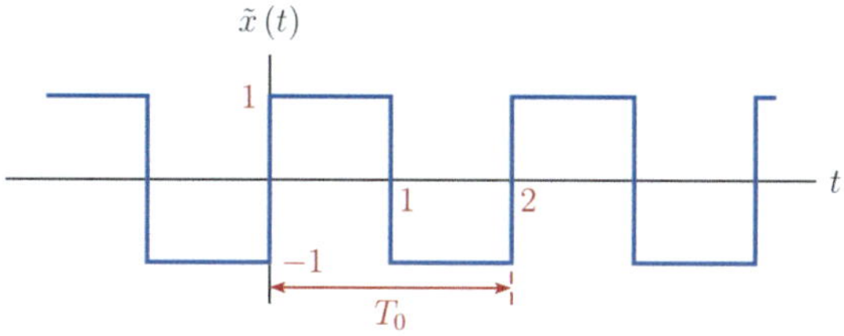

Figure 3.27 – The square wave with period $T_0 = 2$ s.

Finite-harmonic approximation to the signal $\tilde{x}(t)$ using m harmonics is

$$\tilde{x}^{(m)}(t) = \sum_{\substack{k=1 \\ k \text{ odd}}}^{m} \left(\frac{4}{\pi k}\right) \sin(k\pi t)$$

and the approximation error that occurs when m harmonics is used is

$$\varepsilon^{(m)}(t) = \tilde{x}(t) - \tilde{x}^{(m)}(t)$$

Several finite-harmonic approximations to $\tilde{x}(t)$ are shown in Fig. 3.28 for $m = 1,3,9,$ and 25. The approximation error is also shown for each case. We observe that the approximation error is non-uniform. It seems to be relatively large right before and right after a discontinuity, and it gets smaller as we move away from discontinuities. The Fourier series approximation overshoots the actual signal value at one side of each discontinuity and undershoots it at the other side.

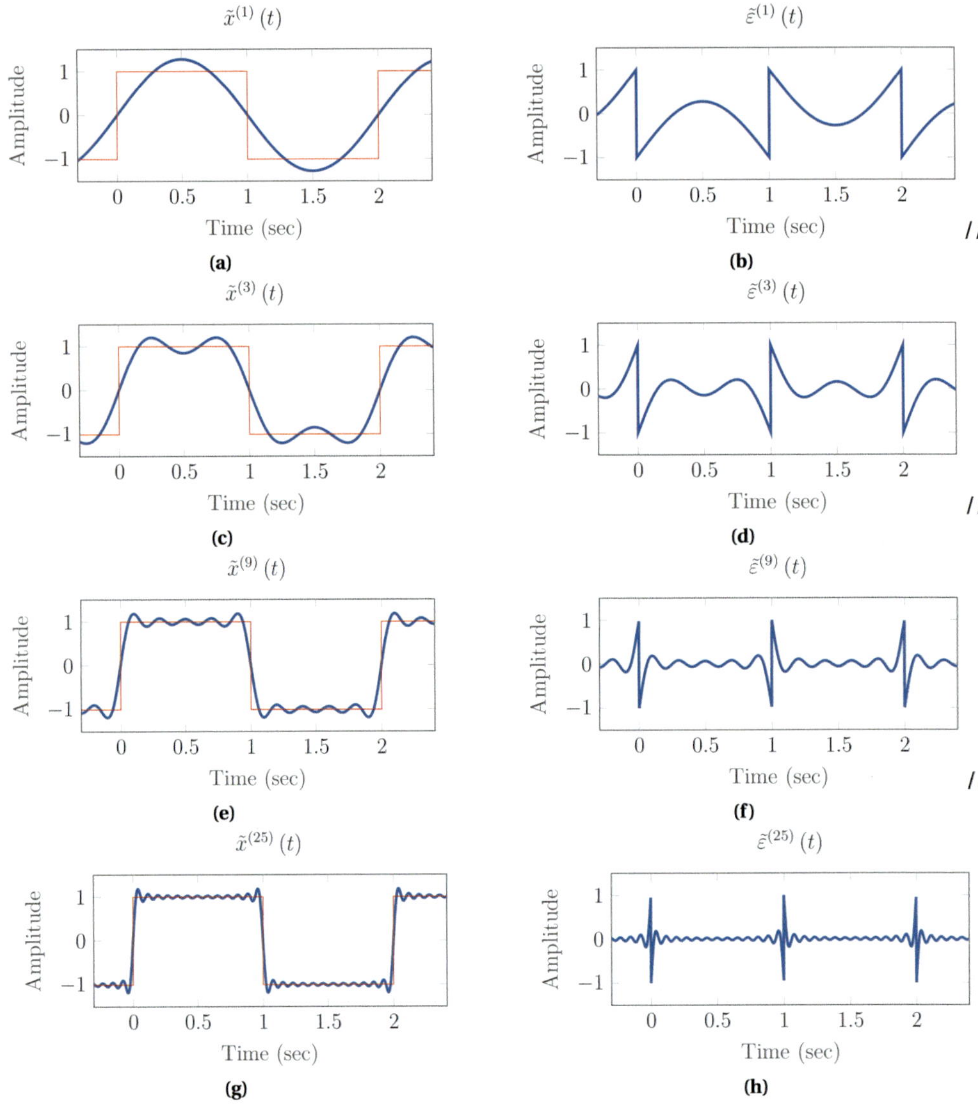

Figure 3.28 – Finite-harmonic approximations to a square wave signal with period $T_0 = 2$ s and the corresponding approximation errors.

The amount of overshoot or undershoot is about 9 percent of the height of the discontinuity, and cannot be reduced by increasing m. This is a form of the *Gibbs phenomenon* named after American scientist Josiah Willard Gibbs (1839–1903) who explained it in 1899. An enlarged view of the approximation and the error right around the discontinuity at $t = 1$ s are shown in Fig. 3.29 for $m = 25$. A further enlarged view of the approximation error corresponding to the shaded area in Fig. 3.29(b) is shown in Fig. 3.30. Notice that the positive and negative lobes around the discontinuity have amplitudes of about ± 0.18 which is 9 percent of the amplitude of the discontinuity. One way to explain the reason for the Gibbs phenomenon would be to link it to the inability of sinusoidal basis functions that are continuous at every point to approximate a discontinuity in the signal.

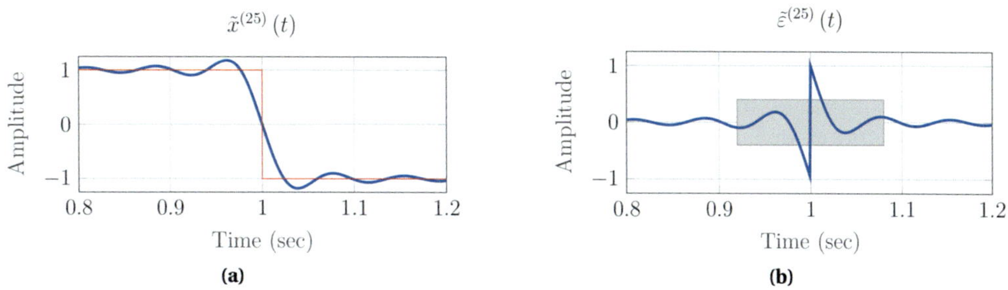

Figure 3.29 – Enlarged view of the finite-harmonic approximation and the approximation error for $m = 25$.

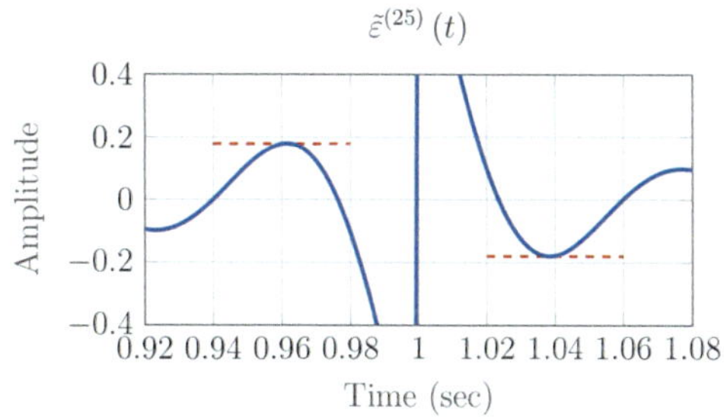

Figure 3.30 – Further enlarged view of the approximation error for $m = 25$.

Software resource: See live script `GibbsEffect.mlx`.

3.2.7 Properties of Fourier series

Some fundamental properties of the Fourier series will be briefly explored in this section using the exponential (EFS) form of the series. A more thorough discussion of these properties will be given in Section 3.3.5 for the Fourier transform.

Linearity

Linearity of Fourier series:

For any two signals $\tilde{x}(t)$ and $\tilde{y}(t)$ periodic with $T_0 = 2\pi/\omega_0$ and with their respective series expansions

$$\tilde{x}(t) = \sum_{k=-\infty}^{\infty} c_k\, e^{jk\omega_0 t} \quad \text{and} \quad \tilde{y}(t) = \sum_{k=-\infty}^{\infty} d_k\, e^{jk\omega_0 t}$$

and any two constants α_1 and α_2, it can be shown that the following relationship holds:

$$\alpha_1\,\tilde{x}(t) + \alpha_2\,\tilde{y}(t) = \sum_{k=-\infty}^{\infty} \left[\alpha_1\, c_k + \alpha_2\, d_k\right] e^{jk\omega_0 t} \tag{3.93}$$

Symmetry of Fourier series

Conjugate symmetry and conjugate antisymmetry properties were defined for complex-valued continuous-time signals in Section 1.9.3 of Chapter 1. Similar definitions can be used for Fourier series coefficients as well. EFS coefficients c_k are said to be *conjugate symmetric* if they satisfy

$$c_{-k} = c_k^* \tag{3.94}$$

for all k. Similarly, the coefficients form a *conjugate antisymmetric* set if they satisfy

$$c_{-k} = -c_k^* \tag{3.95}$$

for all k.

If the signal $\tilde{x}(t)$ is real-valued, it can be shown that its EFS coefficients form a conjugate symmetric set. Conversely, if $\tilde{x}(t)$ is purely imaginary, its EFS coefficients form a conjugate antisymmetric set.

Purely real and purely imaginary signals:

$$\tilde{x}(t)\text{: Real,} \quad \text{Im}\left\{\tilde{x}(t)\right\} = 0 \quad \text{implies that} \quad c_{-k} = c_k^* \tag{3.96}$$

$$\tilde{x}(t)\text{: Imag,} \quad \text{Re}\left\{\tilde{x}(t)\right\} = 0 \quad \text{implies that} \quad c_{-k} = -c_k^* \tag{3.97}$$

Polar form of EFS coefficients

Recall that the EFS coefficients can be written in polar form as

$$c_k = \left|c_k\right| e^{j\theta_k} \tag{3.98}$$

If the set $\left\{c_k\right\}$ is conjugate symmetric, the relationship in Eqn. (3.94) leads to

$$\left|c_{-k}\right| e^{j\theta_{-k}} = \left|c_k\right| e^{-j\theta_k} \tag{3.99}$$

using the polar form of the coefficients. The consequences of Eqn. (3.99) are obtained by equating the magnitudes and the phases on both sides.

Conjugate symmetric coefficients: $c_{-k} = c_k^*$

$$\text{Magnitude:} \quad \left|c_{-k}\right| = \left|c_k\right| \tag{3.100}$$

$$\text{Phase:} \quad \theta_{-k} = -\theta_k \tag{3.101}$$

It was established in Eqn. (3.96) that the EFS coefficients of a real-valued $\tilde{x}(t)$ are conjugate symmetric. Based on the results in Eqns. (3.100) and (3.101) the magnitude spectrum is an even function of k and the phase spectrum is an odd function.

Similarly, if the set $\left\{c_k\right\}$ is conjugate antisymmetric, the relationship in Eqn. (3.95) reflects on polar form of c_k as

$$\left|c_{-k}\right| e^{j\theta_{-k}} = -\left|c_k\right| e^{-j\theta_k} \tag{3.102}$$

The negative sign on the right side of Eqn. (3.102) needs to be incorporated into the phase since we could not write $\left|c_{-k}\right| = -\left|c_k\right|$ (recall that magnitude must to be non-negative). Using $e^{\mp j\pi} = -1$, Eqn. (3.102) becomes

$$\left|c_{-k}\right| e^{j\theta_k} = \left|c_k\right| e^{-j\theta_k} e^{\mp j\pi} = \left|c_k\right| e^{-j(\theta_k \pm \pi)} \tag{3.103}$$

The consequences of Eqn. (3.103) are summarized below.

Conjugate antisymmetric coefficients: $c_{-k} = -c_k^*$

$$\text{Magnitude:} \quad \left|c_{-k}\right| = \left|c_k\right| \tag{3.104}$$

$$\text{Phase:} \quad \theta_{-k} = -\theta_k \pm \pi \tag{3.105}$$

A purely imaginary signal $\tilde{x}(t)$ leads to a set of EFS coefficients with conjugate antisymmetry. The corresponding magnitude spectrum is an even function of k as suggested by Eqn. (3.104). The phase is neither even nor odd.

Fourier series for even and odd signals

If the real-valued signal $\tilde{x}(t)$ is an even function of time, the resulting EFS spectrum c_k is real-valued for all k. Conversely it can be shown that, if the real-valued signal $\tilde{x}(t)$ has odd-symmetry, the resulting EFS spectrum is purely imaginary.

For real-valued periodic signals:

$$\tilde{x}(-t) = \tilde{x}(t), \text{ all } t \quad \text{implies that} \quad \text{Im}\left\{c_k\right\} = 0, \text{ all } k \tag{3.106}$$

$$\tilde{x}(-t) = -\tilde{x}(t), \text{ all } t \quad \text{implies that} \quad \text{Re}\left\{c_k\right\} = 0, \text{ all } k \tag{3.107}$$

Time shifting

The relationship between the EFS coefficients of a signal and those of its time-shifted version can be useful in certain types of problems. The consequence of time shifting $\tilde{x}(t)$ is multiplication of its EFS coefficients by a complex exponential function of frequency.

Time shifting the signal $\tilde{x}(t)$:

For a signal $\tilde{x}(t)$ with EFS expansion

$$\tilde{x}(t) = \sum_{k=-\infty}^{\infty} c_k e^{jk\omega_0 t}$$

it can be shown that

$$\tilde{x}(t-\tau) = \sum_{k=-\infty}^{\infty} \left[c_k e^{-jk\omega_0 \tau}\right] e^{jk\omega_0 t} \tag{3.108}$$

3.3 Analysis of Non-Periodic Continuous-Time Signals

In Section 3.2 of this chapter we have discussed methods of representing periodic signals by means of harmonically related basis functions. The ability to express a periodic signal as a linear combination of standard basis functions allows us to use the superposition principle when such a signal is used as input to a linear and time-invariant system. We must also realize, however, that we often work with signals that are not necessarily periodic. We would like to have similar capability when we use non-periodic signals in conjunction with linear and time-invariant systems. In this section we will work on generalizing the results of the previous section to apply to signals that are not periodic. These efforts will lead us to the *Fourier transform* for continuous-time signals.

3.3.1 Fourier transform

In deriving the Fourier transform for non-periodic signals we will opt to take an intuitive approach at the expense of mathematical rigor. Our approach is to view a non-periodic signal as the limit case of a periodic one, and make use of the exponential Fourier series discussion of Section 3.2. Consider the non-periodic signal $x(t)$ shown in Fig. 3.31.

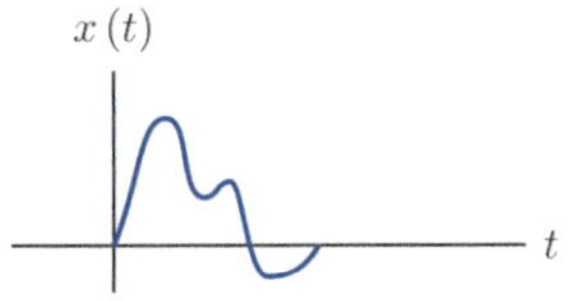

Figure 3.31 – A non-periodic signal $x(t)$.

What frequencies are contained in this signal? What kind of a specific mixture of various frequencies needs to be assembled in order to construct this signal from a standard set of basis functions?

We already know how to represent periodic signals in the frequency domain. Let us construct a periodic extension $\tilde{x}(t)$ of the signal $x(t)$ by repeating it at intervals of T_0. This is illustrated in Fig. 3.32.

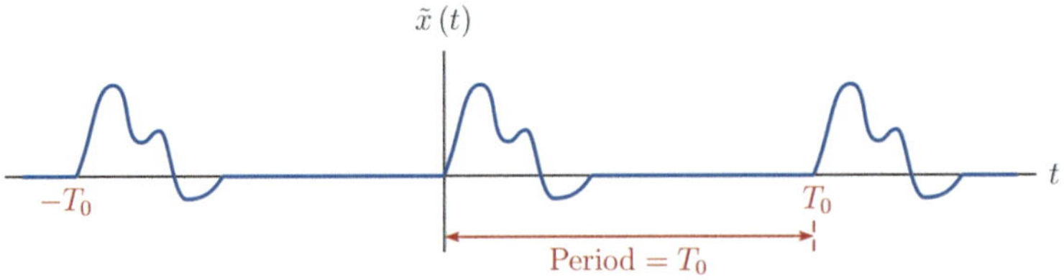

Figure 3.32 – Periodic extension $\tilde{x}(t)$ of the signal $x(t)$.

The periodic extension $\tilde{x}(t)$ can be expressed as a sum of time-shifted versions of $x(t)$ shifted by all integer multiples of T_0 to yield

$$\tilde{x}(t) = \ldots + x(t+T_0) + x(t) + x(t-T_0) + x(t-2T_0) + \ldots \tag{3.109}$$

The selected period T_0 should be sufficiently large so as not to change the shape of the signal by causing overlaps. Putting Eqn. (3.109) in summation form we obtain

$$\tilde{x}(t) = \sum_{k=-\infty}^{\infty} x(t - kT_0) \tag{3.110}$$

Since $\tilde{x}(t)$ is periodic, it can be analyzed in the frequency domain by using the methods developed in the previous section. The EFS representation of a periodic signal was given by Eqn. (3.72) which is repeated here:

$$\text{Eqn. (3.72):} \qquad \tilde{x}(t) = \sum_{k=-\infty}^{\infty} c_k \, e^{jk\omega_0 t}$$

The coefficients c_k of the EFS expansion of the signal $\tilde{x}(t)$ are computed as

$$\text{Eqn. (3.73):} \qquad c_k = \frac{1}{T_0} \int_{-T_0/2}^{T_0/2} \tilde{x}(t) \, e^{-jk\omega_0 t} \, dt$$

We have used the starting point $t_0 = -T_0/2$ for the integral in Eqn. (3.73) so that the lower and upper integration limits are symmetric. Fundamental frequency is the reciprocal of the period, that is,

$$f_0 = \frac{1}{T_0}$$

and the fundamental radian frequency is

$$\omega_0 = 2\pi f_0 = \frac{2\pi}{T_0}$$

Realizing that $\tilde{x}(t) = x(t)$ within the span $-T_0/2 < t < T_0/2$ of the integral, let us write Eqn. (3.73) as

$$c_k = \frac{1}{T_0} \int_{-T_0/2}^{T_0/2} x(t) \, e^{-jk\omega_0 t} \, dt \tag{3.111}$$

If the period T_0 is allowed to become very large, the periodic signal $\tilde{x}(t)$ would start to look more and more similar to $x(t)$. In the limit we would have

$$\lim_{T_0 \to \infty} \left[\tilde{x}(t) \right] = x(t) \tag{3.112}$$

As the period T_0 becomes very large, the fundamental frequency f_0 becomes very small. In the limit, as we force $T_0 \to \infty$, the fundamental frequency becomes infinitesimal:

$$T_0 \to \infty \quad \text{implies that} \quad \Delta f = \frac{1}{T_0} \to 0 \text{ and } \Delta\omega = \frac{2\pi}{T_0} \to 0 \tag{3.113}$$

In Eqn. (3.113) we have switched to the notation Δf and $\Delta\omega$ instead of f_0 and ω_0 to emphasize the infinitesimal nature of the fundamental frequency. Applying this change to the result in Eqn. (3.111) leads to

$$c_k = \frac{1}{T_0} \int_{-T_0/2}^{T_0/2} x(t) \, e^{-jk\Delta\omega t} \, dt \tag{3.114}$$

where c_k is the contribution of the complex exponential at the frequency $\omega = k\Delta\omega$. Because of the large T_0 term appearing in the denominator of the right side of Eqn. (3.114), each individual coefficient c_k is very small in magnitude, and in the limit we have $c_k \to 0$. In addition, successive harmonics $k\Delta\omega$ are very close to each other due to infinitesimally small $\Delta\omega$. Let us multiply both sides of Eqn. (3.114) by T_0 to obtain

$$c_k\, T_0 = \int_{-T_0/2}^{T_0/2} x(t)\, e^{-jk\Delta\omega t}\, dt \tag{3.115}$$

If we now take the limit as $T_0 \to \infty$ and substitute $\omega = k\Delta\omega$ we obtain

$$\begin{aligned} X(\omega) &= \lim_{T_0\to\infty} \left[c_k\, T_0\right] \\ &= \lim_{T_0\to\infty} \left[\int_{-T_0/2}^{T_0/2} x(t)\, e^{-jk\Delta\omega t}\, dt\right] \\ &= \int_{-\infty}^{\infty} x(t)\, e^{-j\omega t}\, dt \end{aligned} \tag{3.116}$$

The function $X(\omega)$ is the *Fourier transform* of the non-periodic signal $x(t)$. It is a continuous function of frequency as opposed to the EFS line spectrum for a periodic signal that utilizes only integer multiples of a fundamental frequency. We can visualize this by imagining that each harmonic of the fundamental frequency comes closer and closer to its neighbors, finally closing the gaps and turning into a continuous function of ω.

At the start of the discussion it was assumed that the signal $x(t)$ has finite duration. What if this is not the case? Would we be able to use the transform defined by Eqn. (3.116) with infinite-duration signals? The answer is yes, as long as the integral in Eqn. (3.116) converges.

Before discussing the conditions for the existence of the Fourier transform, we will address the following question: What does the transform $X(\omega)$ mean? In other words, how can we use the new function $X(\omega)$ for representing the signal $x(t)$? Recall that the Fourier series coefficients c_k for a periodic signal $\tilde{x}_p(t)$ were useful because we could construct the signal from them using Eqn. (3.72). Let us apply the limit operator to Eqn. (3.72):

$$x(t) = \lim_{T_0\to\infty} \left[\tilde{x}(t)\right] = \lim_{T_0\to\infty} \left[\sum_{k=-\infty}^{\infty} c_k\, e^{jk\omega_0 t}\right] \tag{3.117}$$

If we multiply and divide the term inside the summation by T_0, we obtain

$$x(t) = \lim_{T_0\to\infty} \left[\tilde{x}(t)\right] = \lim_{T_0\to\infty} \left[\sum_{k=-\infty}^{\infty} c_k\, T_0\, e^{jk\omega_0 t} \left(\frac{1}{T_0}\right)\right] \tag{3.118}$$

We know that

$$\frac{1}{T_0} = \frac{\omega_0}{2\pi}$$

and for large values of T_0 we can write

$$\omega_0 \to \Delta\omega \quad \text{and} \quad \frac{1}{T_0} \to \frac{\Delta\omega}{2\pi}$$

In the limit, Eqn. (3.118) becomes

$$x(t) = \lim_{T_0\to\infty} \left[\tilde{x}(t)\right] = \lim_{T_0\to\infty} \left[\sum_{k=-\infty}^{\infty} c_k\, T_0\, e^{jk\Delta\omega t} \left(\frac{\Delta\omega}{2\pi}\right)\right] \tag{3.119}$$

Also realizing that

$$\lim_{T_0\to\infty} \left[c_k\, T_0\right] = X(k\Delta\omega) = X(\omega)$$

and changing the summation to an integral in the limit we have

$$x(t) = \frac{1}{2\pi} \int_{-\infty}^{\infty} X(\omega)\, e^{j\omega t}\, d\omega \tag{3.120}$$

Eqn. (3.120) explains how the transform $X(\omega)$ can be used for constructing the signal $x(t)$. We can interpret the integral in Eqn. (3.120) as a continuous sum of complex exponentials at harmonic frequencies that are infinitesimally close to each other. Thus, $x(t)$ and $X(\omega)$ represent two different ways of looking at the same signal, one by observing the amplitude of the signal at each time instant and the other by considering the contribution of each frequency component to the signal.

In summary, we have derived a transform relationship between $x(t)$ and $X(\omega)$ through the following equations:

Fourier transform for continuous-time signals:

1. Synthesis equation: (Inverse transform)

$$x(t) = \frac{1}{2\pi} \int_{-\infty}^{\infty} X(\omega)\, e^{j\omega t}\, d\omega \tag{3.121}$$

2. Analysis equation: (Forward transform)

$$X(\omega) = \int_{-\infty}^{\infty} x(t)\, e^{-j\omega t}\, dt \tag{3.122}$$

Often we will use the Fourier transform operator $\mathscr{F}$ and its inverse $\mathscr{F}^{-1}$ in a shorthand notation as

$$X(\omega) = \mathscr{F}\left\{x(t)\right\}$$

for the *forward transform* and

$$x(t) = \mathscr{F}^{-1}\left\{X(\omega)\right\}$$

for the *inverse transform*. An even more compact notation for expressing the relationship between $x(t)$ and $X(\omega)$ is in the form

$$x(t) \overset{\mathscr{F}}{\longleftrightarrow} X(\omega)$$

In general, the Fourier transform, as computed by Eqn. (3.122), is a complex function of ω. It can be written in Cartesian complex form as

$$X(\omega) = X_r(\omega) + j\, X_i(\omega)$$

or in polar complex form as

$$X(\omega) = \left|X(\omega)\right|\, e^{j\measuredangle X(\omega)}$$

Sometimes it will be more convenient to express the Fourier transform of a signal in terms of the frequency f in Hz rather than the radian frequency ω in rad/s. The conversion is straightforward by substituting $\omega = 2\pi f$ and $d\omega = 2\pi\, df$ in Eqns. (3.121) and (3.122) which leads to the following equations:

Fourier transform for continuous-time signals using f instead of ω):

1. Synthesis equation: (Inverse transform)

$$x(t) = \int_{-\infty}^{\infty} X(f)\, e^{j2\pi f t}\, df \tag{3.123}$$

2. Analysis equation: (Forward transform)

$$X(f) = \int_{-\infty}^{\infty} x(t)\, e^{-j2\pi f t}\, dt \tag{3.124}$$

Note the lack of the scale factor $1/2\pi$ in front of the integral of the inverse transform when f is used. This is consistent with the relationship $d\omega = 2\pi\, df$.

3.3.2 Existence of Fourier transform

The Fourier transform integral given by Eqn. (3.122) may or may not converge for a given signal $x(t)$. A complete theoretical study of convergence conditions is beyond the scope of this text, but we will provide a summary of the practical results that will be sufficient for our purposes. Let $\hat{x}(t)$ be defined as

$$\hat{x}(t) = \frac{1}{2\pi} \int_{-\infty}^{\infty} X(\omega)\, e^{j\omega t}\, d\omega \tag{3.125}$$

and let $\varepsilon(t)$ be the error defined as the difference between $x(t)$ and $\hat{x}(t)$:

$$\varepsilon(t) = x(t) - \hat{x}(t) = x(t) - \frac{1}{2\pi} \int_{-\infty}^{\infty} X(\omega)\, e^{j\omega t}\, d\omega \tag{3.126}$$

For perfect convergence of the transform at all time instants, we would naturally want $\varepsilon(t) = 0$ for all t. However, this is not possible at time instants for which $x(t)$ exhibits discontinuities. German mathematician Johann Peter Gustav Lejeune Dirichlet showed that the following set of conditions, referred to as *Dirichlet conditions,* is sufficient for the convergence error $\varepsilon(t)$ to be zero at all time instants except those that correspond to discontinuities of the signal $x(t)$:

Conditions for existence of the Fourier transform:

1. The signal $x(t)$ must be integrable in an absolute sense, that is

$$\int_{-\infty}^{\infty} |x(t)|\, dt < \infty \tag{3.127}$$

2. If the signal $x(t)$ has discontinuities, it must have at most a finite number of them in any finite time interval.
3. The signal $x(t)$ must have at most a finite number of minima and maxima in any finite time interval.

From a practical perspective, all signals we will encounter in our study of signals and systems will satisfy the second and third conditions regarding the number of discontinuities and the number of extrema, respectively. The absolute integrability condition given by Eqn. (3.127) ensures that the result of the integral in Eqn. (3.121) is equal to the signal $x(t)$ at all time instants except discontinuities. At points of discontinuities, Eqn. (3.121) yields the average value obtained by approaching the discontinuity from opposite directions. If the signal $x(t)$ has a discontinuity at $t = t_0$, we have

$$\hat{x}(t_0) = \frac{1}{2} \lim_{t \to t_0^-} \left[x(t) \right] + \frac{1}{2} \lim_{t \to t_0^+} \left[x(t) \right] \tag{3.128}$$

An alternative approach to the question of convergence is to require

$$\int_{-\infty}^{\infty} |\varepsilon(t)|^2 \, dt = \int_{-\infty}^{\infty} |x(t) - \hat{x}(t)|^2 \, dt = 0 \tag{3.129}$$

which ensures that the normalized energy in the error signal $\varepsilon(t)$ is zero even if the error signal itself is not equal to zero at all times. This condition ensures that the transform $X(\omega)$ as defined by Eqn. (3.122) is finite. It can be shown that, the condition stated by Eqn. (3.129) is satisfied provided that the signal $x(t)$ is square integrable, that is

$$\int_{-\infty}^{\infty} |x(t)|^2 \, dt < \infty \tag{3.130}$$

We have seen in Chapter 1 (see Eqn. (1.51)) that a signal that satisfies Eqn. (3.130) is referred to as an *energy signal*. Therefore, all energy signals have Fourier transforms. Furthermore, we can find Fourier transforms for some signals that do not satisfy Eqn. (3.130), such as periodic signals, if we are willing to accept the use of the impulse function in the transform. Another example of a signal that is not an energy signal is the unit step function. It is neither absolute integrable nor square integrable. We will show, however, that a Fourier transform can be found for the unit step function as well if we allow the use of the impulse function in the transform.

3.3.3 Developing further insight

The interpretation of the Fourier transform for a non-periodic signal as a generalization of the EFS representation of a periodic signal is fundamental. We will take the time to apply this idea step by step to an isolated pulse in order to develop further insight into the Fourier transform. Consider the signal $x(t)$ shown in Fig. 3.33, an isolated rectangular pulse centered at $t = 0$ with amplitude A and width τ.

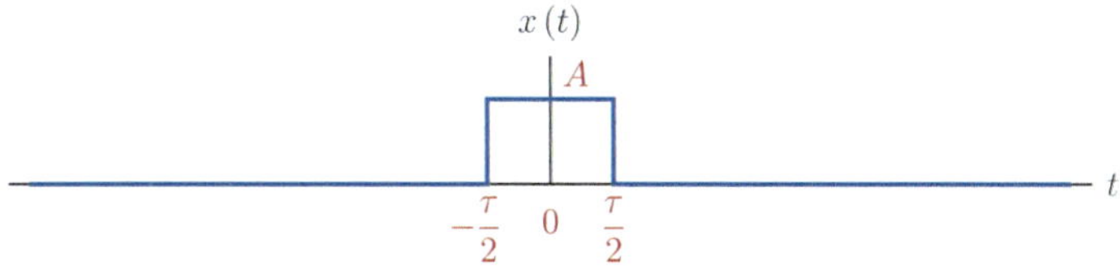

Figure 3.33 – An isolated rectangular pulse.

The analytical definition of $x(t)$ is

$$x(t) = A\Pi\left(\frac{t}{\tau}\right) = \begin{cases} A, & |t| < \tau/2 \\ 0, & |t| > \tau/2 \end{cases}$$

Let the signal $x(t)$ be extended into a pulse train $\tilde{x}(t)$ with a period T_0 as shown in Fig. 3.34.

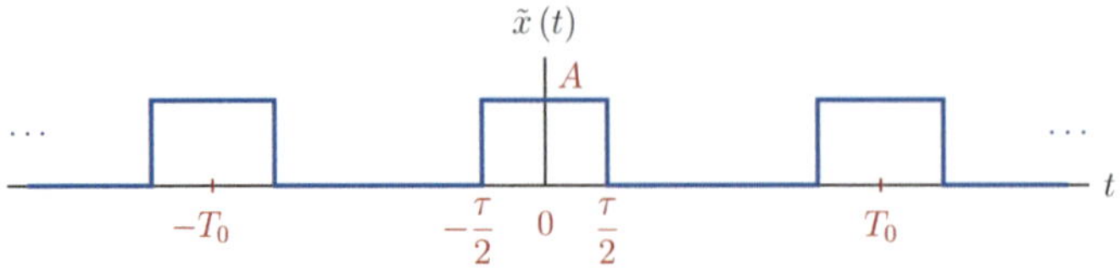

Figure 3.34 – Periodic extension of the isolated rectangular pulse into a pulse train.

By adapting the result found in Example 3.7 to the problem at hand, the EFS coefficients of the signal $\tilde{x}(t)$ can be written as

$$c_k = Ad\,\text{sinc}(kd) \tag{3.131}$$

where d is the duty cycle of the pulse train, and is given by

$$d = \frac{\tau}{T_0} \tag{3.132}$$

Substituting Eqn. (3.132) into Eqn. (3.131)

$$c_k = \frac{A\tau}{T_0}\,\text{sinc}(k\tau/T_0) \tag{3.133}$$

Let us multiply both sides of Eqn. (3.133) by T_0 and write the scaled EFS coefficients as

$$c_k T_0 = A\tau\,\text{sinc}\left(kf_0\tau\right) \tag{3.134}$$

where we have also substituted $1/T_0 = f_0$ in the argument of the sinc function. The outline (or the envelope) of the scaled EFS coefficients $c_k T_0$ is a sinc function with a peak value of $A\tau$ at $k = 0$. The first zero crossing of the sinc-shaped envelope occurs at

$$kf_0\tau = 1 \quad \Longrightarrow \quad k = \frac{1}{f_0\tau} \tag{3.135}$$

which may or may not yield an integer result. Subsequent zero crossings occur for

$$kf_0\tau = 2,3,4,\ldots$$

or equivalently for

$$k = \frac{2}{f_0\tau},\ \frac{3}{f_0\tau},\ \frac{4}{f_0\tau},\ \ldots$$

The coefficients $c_k T_0$ are graphed in Fig. 3.35(a) for $A = 1$, $\tau = 0.1$ s, and $T_0 = 0.25$ s corresponding to a fundamental frequency of $f_0 = 4$ Hz. Spectral lines for $k = 1$, $k = 2$, and $k = 3$ represent the strength of the frequency components at $f_0 = 4$ Hz, $2f_0 = 8$ Hz, and $3f_0 = 12$ Hz, respectively. For the example values given, the first zero crossing of the sinc envelope occurs at $k = 2.5$, between the spectral lines for $k = 2$ and $k = 3$.

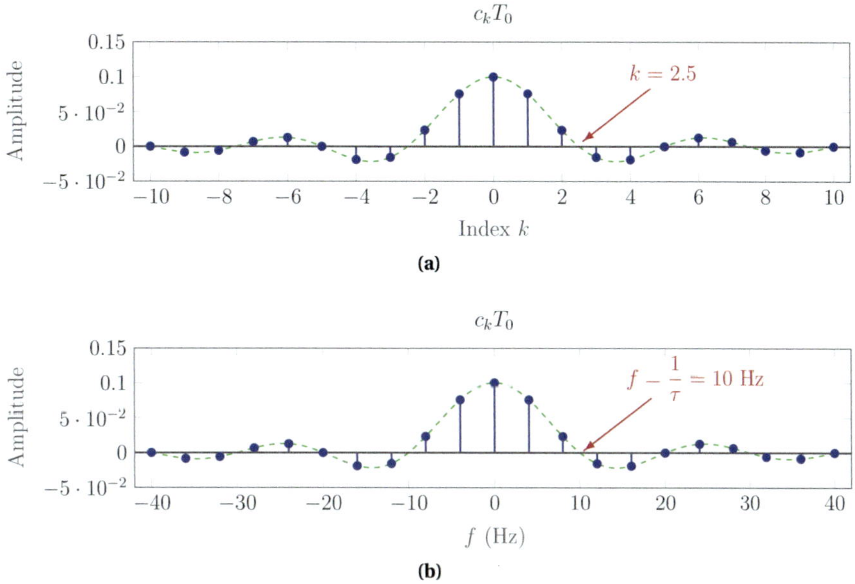

Figure 3.35 – **(a)** The scaled line spectrum $c_k T_0$ for $A = 1$, $\tau = 0.1$ s and $T_0 = 0.25$ s and **(b)** the scaled line spectrum as a function of frequency f.

In Fig. 3.35(b) the same line spectrum is shown with actual frequencies f on the horizontal axis instead of the integer index k. The spectral line for index k now appears at the frequency $f = kf_0 = 4k$ Hz. For example, the spectral lines for $k = 1, 2, 3$ appear at frequencies $f = 4, 8, 12$ Hz, respectively. The first zero crossing of the sinc envelope is at the frequency

$$f = \frac{1}{\tau} = 10 \text{ Hz}$$

and subsequent zero crossings appear at frequencies

$$f = \frac{2}{\tau}, \frac{3}{\tau}, \frac{4}{\tau}, \ldots = 20, 30, 40, \ldots \text{ Hz}$$

What would happen if the period T_0 is gradually increased while keeping the pulse amplitude A and the pulse width τ unchanged? The spectral lines in 3.35(b) are at integer multiples of the fundamental frequency f_0. Since $f_0 = 1/T_0$, increasing the period T_0 would cause f_0 to become smaller, and the spectral lines to move inward, closer to each other. The zero crossings of the sinc envelope would not change, however, since the locations of the zero crossings depend on the pulse width τ only. Figs. 3.36(a) through (c) show the line spectra for $T_0 = 5, 10, 20$ s, respectively. In Fig. 3.36(a), the first zero-crossing of the sinc envelope is still at $1/\tau = 10$ Hz, yet the frequency spacing of the spectral lines is now reduced to 2 Hz, therefore the first zero-crossing coincides with the fifth harmonic. In Fig. 3.36(b), the frequency spacing of the spectral lines is further reduced to 1 Hz and, consequently, the first zero-crossing of the sinc envelope coincides with the 10-th harmonic, and is still at 10 Hz.

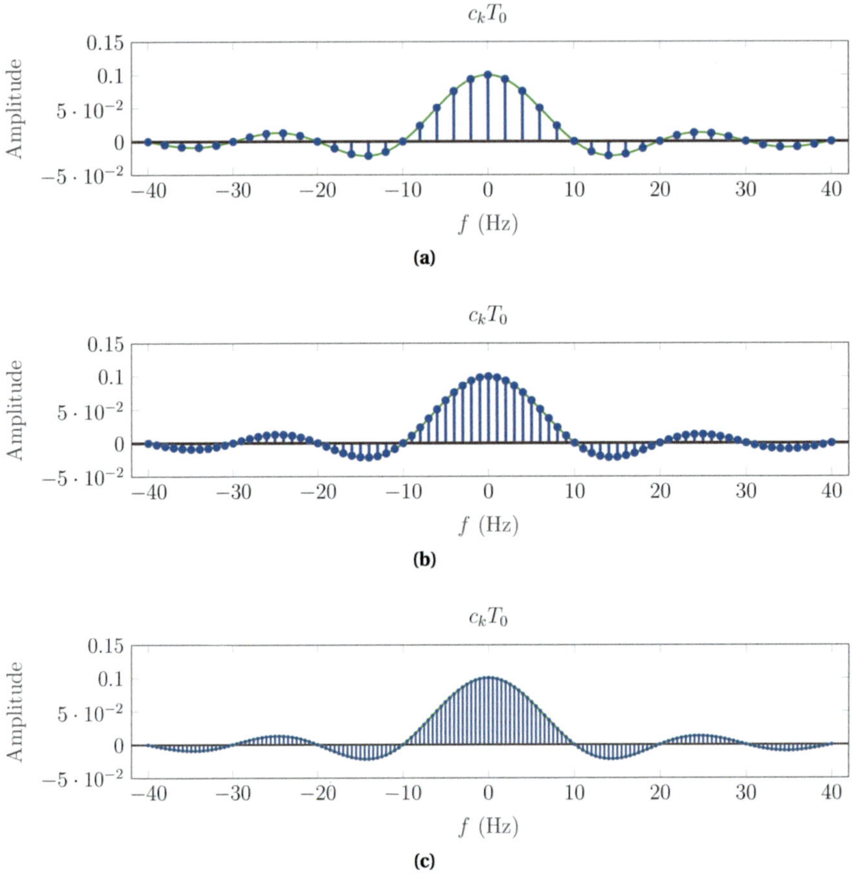

Figure 3.36 – **(a)** The scaled line spectra for $T_0 = 0.5$ s and **(b)** for $T_0 = 1$ s, c) for $T_0 = 2$ s.

We can conclude that, as the period becomes infinitely large, the distances between adjacent spectral lines become infinitesimally small and they eventually converge to a continuous function in the shape of the sinc envelope. It appears that the Fourier transform of the isolated pulse with amplitude A and width τ is

$$X\left(f\right) = A\tau \,\text{sinc}\left(f\tau\right) \tag{3.136}$$

or, using the radian frequency variable ω,

$$X\left(\omega\right) = A\tau \,\text{sinc}\left(\frac{\omega\tau}{2\pi}\right) \tag{3.137}$$

Interactive App: Developing further insight into Fourier transform

The interactive app in `appFT1.m` is based on the concepts introduced in the discussion above along with Figs. 3.33 through 3.36. The periodic pulse train with amplitude A, pulse width τ and period T_0 is graphed. Its line spectrum based on the EFS coefficients is also shown. The spectrum graph includes the outline (or the envelope) of the EFS coefficients. Compare these graphs to Figs. 3.34 and 3.36. The amplitude is fixed at $A = 5$ since it is not a significant parameter for this app. The period T_0 and the pulse width τ may be varied by adjusting the slider controls, allowing us to duplicate the cases in Fig. 3.36.

a. Set pulse width to $\tau = 0.1$ s, and the pulse period to $T_0 = 0.25$ s. Note the envelope, or the outline, of the EFS line spectrum, and specifically the frequency of its first zero crossing. To what value of the index k does the zero crossing correspond?

b. Gradually increase the pulse width τ toward $\tau = 0.2$ s. Does the frequency spacing of the spectral lines change? Does the envelope change? For $\tau = 0.2$m s, where is the first zero crossing of the envelope? To what value of the index k does the zero crossing correspond?

c. Set the pulse width to $\tau = 0.1$ s, and the pulse period to $T_0 = 0.25$ s. Gradually increase the pulse period while observing its effects on the line spectrum. Observe how the spectral lines move in closer with increasing pulse period, and how the envelope remains unchanged. Remember that the fundamental frequency is inversely proportional to the period, however, the envelope is only a function of the pulse width.

d. For large values of the signal period observe how the envelope approaches the Fourier transform of a single isolated pulse.

Software resource: `appFT1.m`

3.3.4 Fourier transforms of some signals

In this section we will work on examples of determining the Fourier transforms of some fundamental signals.

Example 3.12: Fourier transform of a rectangular pulse

Using the forward Fourier transform integral in Eqn. (3.122), find the Fourier transform of the isolated rectangular pulse signal

$$x(t) = A\,\Pi\left(\frac{t}{\tau}\right)$$

shown in Fig. 3.37.

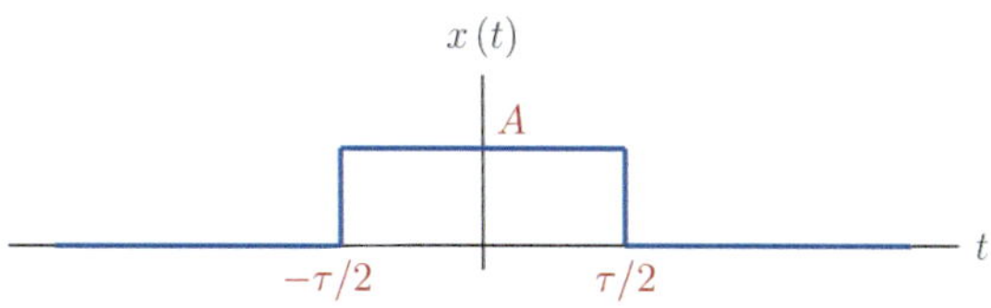

Figure 3.37 – Isolated pulse with amplitude A and width τ for Example 3.12.

Solution: Recall that this is the same isolated pulse the Fourier transform of which was determined in Section 3.3.3 as the limit case of the EFS representation of a periodic pulse train. In this example we will take a more direct approach to obtain the same result through the use of the Fourier transform integral in Eqn. (3.122):

$$X(\omega) = \int_{-\tau/2}^{\tau/2} (A)\, e^{-j\omega t}\, dt = A\, \frac{e^{-j\omega t}}{-j\omega}\Bigg|_{-\tau/2}^{\tau/2} = \frac{2A}{\omega}\, \sin\left(\frac{\omega\tau}{2}\right)$$

The sinc function can be used for simplifying this result to

$$X(\omega) = A\tau\, \frac{\sin(\omega\tau/2)}{(\omega\tau/2)} = A\tau\, \text{sinc}\left(\frac{\omega\tau}{2\pi}\right) \tag{3.138}$$

In this case it will be easier to graph the transform in terms of the independent variable f instead of ω. Substituting $\omega = 2\pi f$ into Eqn. (3.138) we obtain

$$X(f) = A\tau\, \text{sinc}(f\tau) \tag{3.139}$$

The spectrum $X(f)$, shown in Fig. 3.38, is purely real owing to the fact that the signal $x(t)$ exhibits even symmetry. The peak value of the spectrum is $A\tau$ and occurs at the frequency $f = 0$. The zero crossings of the spectrum occur at frequencies that satisfy $f\tau = k$ where k is any non-zero integer.

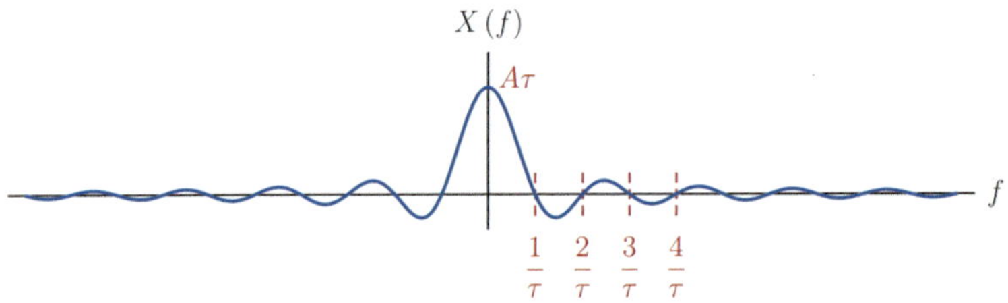

Figure 3.38 – Transform of the pulse in Example 3.12.

Software resources: exct_3_12.m, exct_3_12_Live.mlx

The particular spectrum obtained in Example 3.12 is of special interest especially in digital systems. For example, in digital communications, rectangular pulses such as the one considered in Example 3.12 may be used for representing binary 0's or 1's. Let us observe the relationship between the pulse width and the shape of the spectrum:

1. Largest values of the spectrum occur at frequencies close to $f = 0$. Thus, low frequencies seem to be the more significant ones in the spectrum, and the significance of frequency components seems to decrease as we move further away from $f = 0$ in either direction.
2. The zero-crossings of the spectrum occur for values of f that are integer multiples of $1/\tau$ for multiplier values $k = \pm 1, \pm 2, \ldots, \pm\infty$. As a result, if the pulse width is decreased, these zero-crossings move further away from the frequency $f = 0$ resulting in the spectrum being *stretched out* in both directions. This increases the relative significance of large frequencies. Narrower pulses have frequency spectra that expand to higher frequencies.

3. If the pulse width is increased, zero-crossings of the spectrum move inward, that is, closer to the frequency $f = 0$ resulting in the spectrum being squeezed in from both directions. This decreases the significance of large frequencies, and causes the spectrum to be concentrated more heavily around the frequency $f = 0$. Wider pulses have frequency spectra that are more concentrated at low frequencies.

Fig. 3.39 illustrates the effects of changing the pulse width on the frequency spectrum. In all three cases, the product of the pulse width and the pulse amplitude is fixed, i.e., $A\tau = 1$. Consequently, the peak of the frequency spectrum is equal to unity in each case. The placements of the zero-crossings of the spectrum depend on the pulse width. In Fig. 3.39(a), the zero-crossings are 1 Hz apart, consistent with the pulse width of $\tau = 1$ s. In parts (b) and (c) of the figure, the spacing between adjacent zero crossings is 2 Hz and $\frac{1}{2}$ Hz, corresponding to pulse widths of $\tau = 0.5$ s and $\tau = 2$ s, respectively.

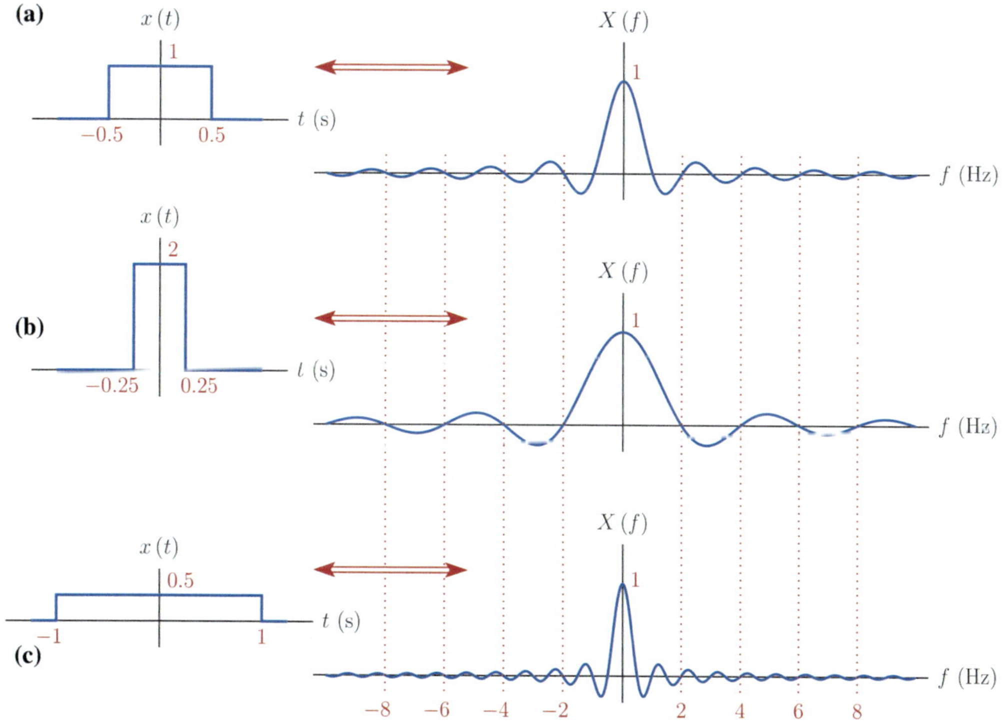

Figure 3.39 – Effects of changing the pulse width on the frequency spectrum: **(a)** $A = 1$ and $\tau = 1$, **(b)** $A = 2$ and $\tau = 0.5$, and **(c)** $A = 0.5$ and $\tau = 2$.

Interactive App: Fourier transform of a rectangular pulse

The interactive app in `appFT2.m` is based on the concepts introduced in Example 3.12 and Figs. 3.37, 3.38, 3.39. The amplitude A and the width τ of the rectangular pulse signal $x(t)$ may be varied by using the two slider controls or by typing values into the corresponding edit fields. The Fourier transform $X(f)$ is computed and graphed as the pulse parameters are varied.

a. Set the pulse width at $\tau = 0.2$ s.Make a note of the zero crossings of the transform $X(f)$. How far apart are the zero crossings?

b. Change the pulse width to $\tau = 0.4$ s.What is the relationship between the pulse width and the zero crossings of the spectrum?

c. Slowly increase the pulse width while observing the changes in the spectrum. Pay attention to how the spectrum becomes more concentrated around $f = 0$ as the pulse is made wider. Also observe how the height of the spectrum at $f = 0$ changes with pulse amplitude and pulse width.

Software resource: `appFT2.m`

Example 3.13: Fourier transform of a rectangular pulse revisited

Using the forward Fourier transform integral in Eqn. (3.122), find the Fourier transform of the isolated rectangular pulse given by

$$x(t) = A\Pi\left(\frac{t-\tau/2}{\tau}\right)$$

which is shown in Fig. 3.40.

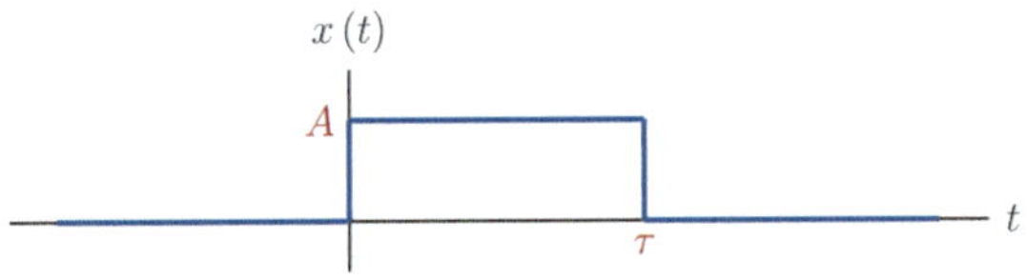

Figure 3.40 – Isolated pulse with amplitude A and width τ for Example 3.13.

Solution: The signal $x(t)$ is essentially a time-shifted version of the pulse signal used in Example 3.12. Using the Fourier transform integral in Eqn. (3.122) we obtain

$$X(\omega) = \int_0^{\tau} (A)\, e^{-j\omega t}\, dt = A\, \frac{e^{-j\omega t}}{-j\omega}\bigg|_0^{\tau} = \frac{A}{-j\omega}\left[e^{-j\omega\tau} - 1\right] \tag{3.140}$$

At this point we will use a trick which will come in handy in similar situations in the rest of this text as well. Let us write the part of Eqn. (3.140) in square brackets as follows:

$$\left[1 - e^{j\omega\tau}\right] = \left[e^{-j\omega\tau/2} - e^{j\omega\tau/2}\right]\, e^{-j\omega\tau/2} \tag{3.141}$$

The use of Euler's formula on the result of Eqn. (3.141) yields

$$\left[1 - e^{j\omega\tau}\right] = -j2\sin(\omega\tau/2)\, e^{-j\omega\tau/2} \tag{3.142}$$

Substituting Eqn. (3.142) into Eqn. (3.140), the Fourier transform of $x(t)$ is found as

$$X(\omega) = \frac{2A}{\omega}\sin\left(\frac{\omega\tau}{2}\right) e^{-j\omega\tau/2} = A\tau\, \text{sinc}\left(\frac{\omega\tau}{2\pi}\right) e^{-j\omega\tau/2} \tag{3.143}$$

The transform is complex-valued and is best expressed in polar form. In preparation for that, we will find it convenient to write the transform in terms of the frequency variable f through the substitution $\omega = 2\pi f$:

$$X(f) = A\tau \,\text{sinc}(f\tau)\, e^{-j\pi f\tau} \tag{3.144}$$

We would ultimately like to express the transform of Eqn. (3.144) in the form

$$X(f) = |X(f)|\, e^{j\theta(f)}$$

Let the functions $B(f)$ and $\beta(f)$ be defined as

$$B(f) = A\tau \,\text{sinc}(f\tau)$$

and

$$\beta(f) = -\pi f\tau$$

so that

$$X(f) = B(f)\, e^{j\beta(f)}$$

The functions $B(f)$ and $\beta(f)$ are shown in Fig. 3.41.

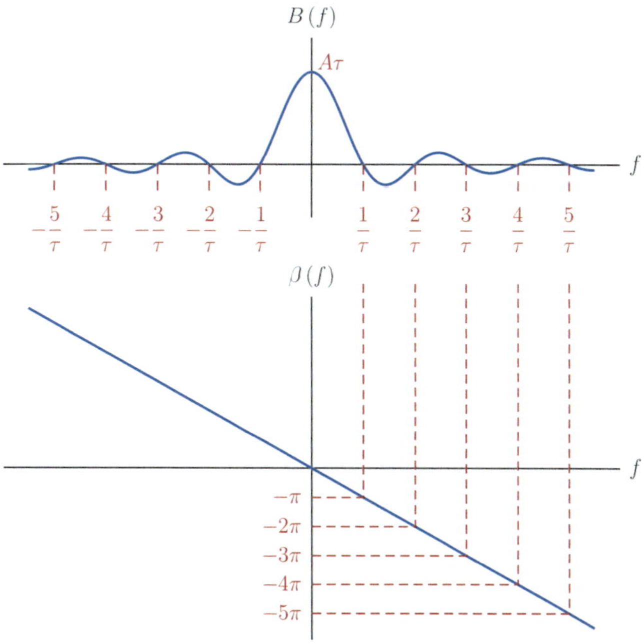

Figure 3.41 – The spectrum $X(f)$ in Example 3.13 expressed in the form $X(f) = B(f)\, e^{j\beta(f)}$.

A quick glance at Fig. 3.41 reveals the reason that keeps us from declaring the functions $B(f)$ and $\beta(f)$ as the magnitude and the phase of the transform $X(f)$: The function $B(f)$ could be negative for some values of f, and therefore cannot be the magnitude of the transform. However, $|X(\omega)|$ can be written in terms of $B(f)$ by paying attention to its sign. In order to outline the procedure to be used, we will consider two separate example intervals of the frequency variable f:

Case 1: $-1/\tau < f < 1/\tau$

This is an example of a frequency interval in which $B(f) \ge 0$ and

$$B(f) = |B(f)|$$

Therefore, we can write the transform as

$$X(f) = B(f)\, e^{j\beta(f)} = |B(f)|\, e^{j\beta(f)}$$

resulting in the magnitude

$$|X(f)| = |B(f)| = B(f)$$

and the phase

$$\theta(f) = \beta(f)$$

for the transform.

Case 2: $1/\tau < f < 2/\tau$

For this interval we have $B(f) < 0$. The function $B(f)$ can be expressed as

$$B(f) = -|B(f)| = |B(f)|\, e^{\pm j\pi}$$

Using this form of $B(f)$ in the transform leads to

$$X(f) = |B(f)|\, e^{\pm j\pi}\, e^{j\beta(f)}$$

which results in the magnitude

$$|X(f)| = |B(f)| = -B(f)$$

and the phase

$$\theta(f) = \beta(f) \pm \pi$$

for the transform $X(f)$.

Comparison of the results in the two cases above leads us to the following generalized expressions for the magnitude and the phase of the transform $X(f)$ as

$$|X(f)| = |B(f)|$$

and

$$\theta(f) = \begin{cases} \beta(f) & \text{if } B(f) \ge 0 \\ \beta(f) \pm \pi & \text{if } B(f) < 0 \end{cases}$$

In summary, for frequencies where $B(f)$ is negative, we add $\pm\pi$ radians to the phase term to account for the factor (-1). This is illustrated in Fig. 3.42. Phase angles outside the interval $(-\pi, \pi)$ radians are indistinguishable from the corresponding angles within the interval since we can freely add any integer multiple of 2π radians to the phase. Because of this, it is customary to fit the phase values inside the interval $(-\pi, \pi)$, a practice referred to as *phase wrapping*. Fig. 3.43 shows the phase characteristic after the application of phase wrapping.

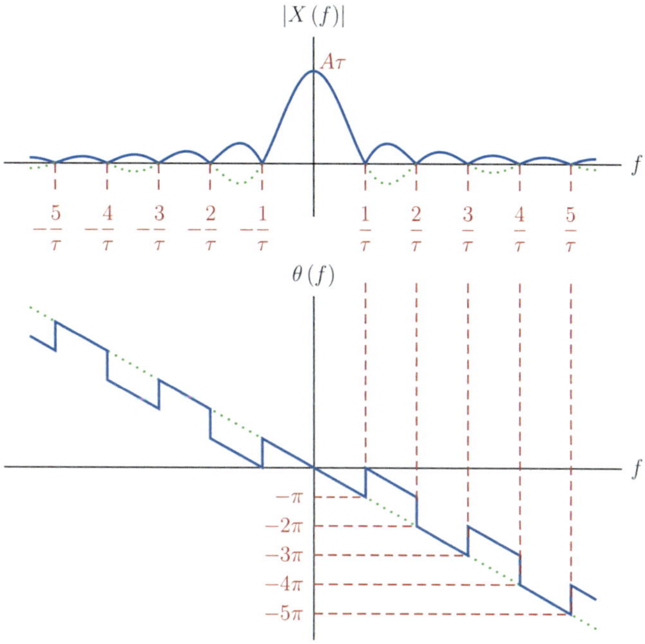

Figure 3.42 – Adjustment of magnitude and phase of the transform in Example 3.13.

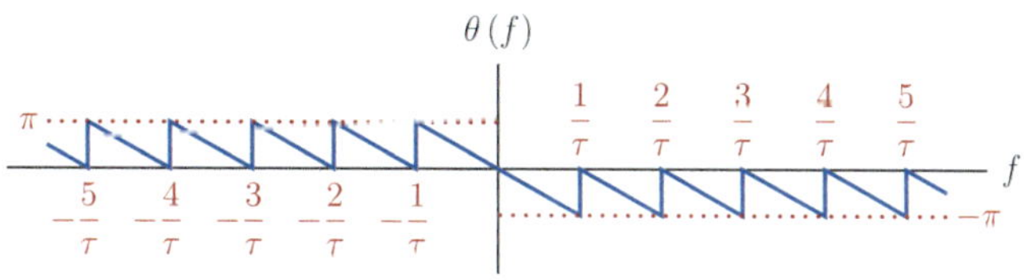

Figure 3.43 – Phase $\theta(f)$ of the transform in Example 3.13 after phase wrapping.

Software resources: exct_3_13.m , exct_3_13_Live.mlx

Example 3.14: Transform of the unit impulse function

The unit impulse function was defined in Section 1.5.1 of Chapter 1. The Fourier transform of the unit impulse signal can be found by direct application of the Fourier transform integral along with the sifting property of the unit impulse function.

$$\mathscr{F}\{\delta(t)\} = \int_{-\infty}^{\infty} \delta(t)\, e^{-j\omega t}\, dt = e^{-j\omega t}\Big|_{t=0} = 1 \qquad (3.145)$$

In an effort to gain further insight, we will also take an alternative approach to determining the Fourier transform of the unit impulse signal. Recall that in Section 1.5.1 we expressed the unit impulse function as the limit case of a rectangular pulse with unit area. Given the pulse signal

$$q(t) = \frac{1}{a}\,\Pi\left(\frac{t}{a}\right)$$

the unit impulse function can be expressed as

$$\delta(t) = \lim_{a\to 0}\left\{q(t)\right\}$$

If the parameter a is gradually made smaller, the pulse $q(t)$ becomes narrower and taller while still retaining unit area under it. In the limit, the pulse $q(t)$ becomes the unit impulse function $\delta(t)$. Using the result obtained in Example 3.12 with $A = 1/a$ and $\tau = a$, the Fourier transform of $q(t)$ is

$$Q(f) = \mathscr{F}\left\{q(t)\right\} = \text{sinc}\left(fa\right) \tag{3.146}$$

Using an intuitive approach, we may also conclude that the Fourier transform of the unit impulse function is an expanded, or stretched-out, version of the transform $Q(f)$ in Eqn. (3.146):

$$\mathscr{F}\left\{\delta(t)\right\} = \lim_{a\to 0}\left\{Q(f)\right\} = \lim_{a\to 0}\left\{\text{sinc}\left(fa\right)\right\} = 1$$

This conclusion is easy to justify from the general behavior of the function $\text{sinc}(fa)$. Zero crossings on the frequency axis appear at values of f that are integer multiples of $1/a$. As the value of the parameter a is reduced, the zero crossings of the sinc function move further apart and the function stretches out or flattens around its peak at $f = 0$. In the limit, it approaches a constant of unity. This behavior is illustrated in Fig. 3.44.

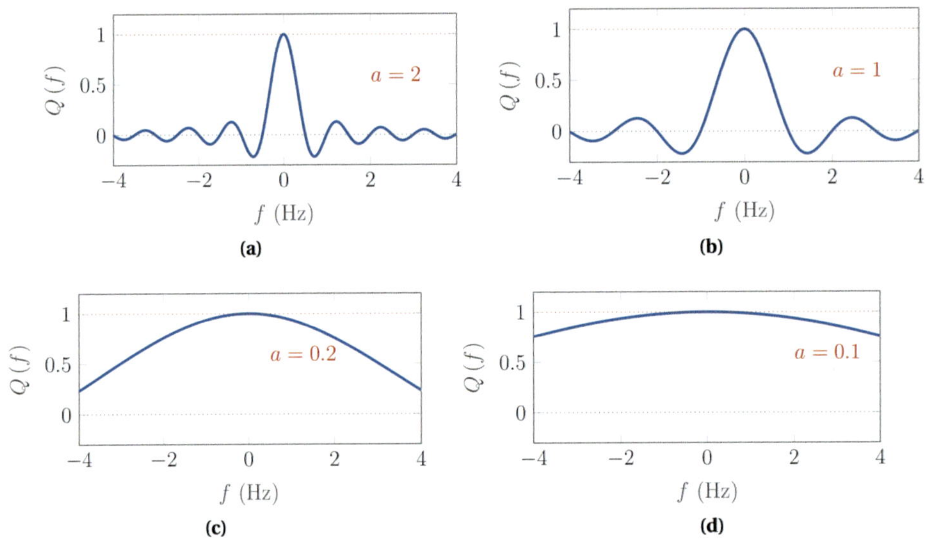

Figure 3.44 – Obtaining the Fourier transform of the unit impulse signal from the transform of a rectangular pulse: **(a)** $a = 2$, **(b)** $a = 1$, **(c)** $a = 0.2$, and **(d)** $a = 0.1$.

Software resources: exct_3_14.m , exct_3_14_Live.mlx

Interactive App: From rectangular pulse to unit impulse

The interactive app in `appFT3.m` illustrates the relationship between the Fourier transforms of the unit impulse and the rectangular pulse with unit area as discussed in the preceding section.

A rectangular pulse signal $q(t)$ with width equal to a and height equal to $1/a$ is shown along with its sinc-shaped Fourier transform $Q(f)$. The pulse width parameter a may be varied through the use of a slider control. If the parameter a is gradually reduced, $q(t)$ starts to look more and more like a unit impulse signal, and it becomes a unit impulse signal in the limit. Intuitively it would make sense for the transform of the pulse to look more and more like the transform of the unit impulse.

a. Start with $a = 2$ as in Fig. 3.44(a) and observe the spectrum of the pulse.
b. Gradually reduce the pulse width and compare the resulting spectrum to parts (b) through (d) of Fig. 3.44.
c. Now gradually increase the pulse width and observe the changes in the spectrum.

Software resource: `appFT3.m`

Example 3.15: Fourier transform of a right-sided exponential signal

Determine the Fourier transform of the right-sided exponential signal

$$x(t) = e^{-at}\,u(t)$$

with $a > 0$ as shown in Fig. 3.45.

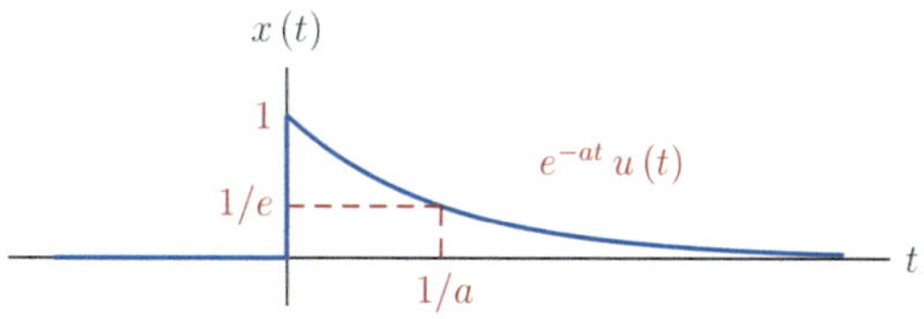

Figure 3.45 – Right-sided exponential signal for Example 3.15.

Solution: Application of the Fourier transform integral of Eqn. (3.122) to $x(t)$ yields

$$X(\omega) = \int_{-\infty}^{\infty} e^{-at}\,u(t)\,e^{-j\omega t}\,dt$$

Changing the lower limit of integral to $t = 0$ and dropping the factor $u(t)$ results in

$$X(\omega) = \int_{0}^{\infty} e^{-at}\,e^{-j\omega t}\,dt = \int_{0}^{\infty} e^{-(a+j\omega)t}\,dt = \frac{1}{a+j\omega} \tag{3.147}$$

This result is only valid for $a > 0$ since the integral could not have been evaluated otherwise. The magnitude and the phase of the transform are

$$|X(\omega)| = \left|\frac{1}{a+j\omega}\right| = \frac{1}{\sqrt{a^2+\omega^2}} \quad \text{and} \quad \theta(\omega) = \measuredangle X(\omega) = -\tan^{-1}\left(\frac{\omega}{a}\right)$$

Magnitude and phase characteristics are graphed in Fig. 3.46(a),(b).

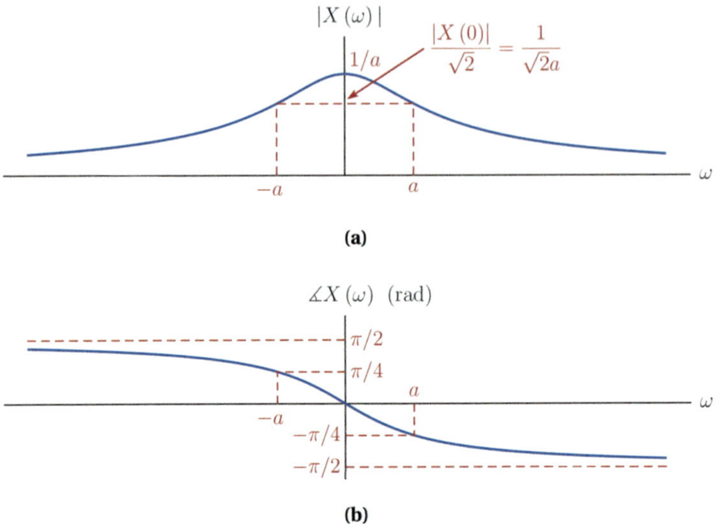

Figure 3.46 – **(a)** The magnitude and **(b)** the phase of the transform of the right-sided exponential signal in Example 3.15.

Some observations are in order: The magnitude of the transform is an even function of ω since

$$\left|X(-\omega)\right| = \frac{1}{\sqrt{a^2+(-\omega)^2}} = \frac{1}{\sqrt{a^2+\omega^2}} = X(\omega)$$

The peak magnitude occurs at $\omega = 0$ and its value is $\left|X(0)\right| = 1/a$. At the radian frequency $\omega = a$ the magnitude is equal to $\left|X(a)\right| = 1/\left(\sqrt{2}a\right)$ which is $1/\sqrt{2}$ times its peak value. On a logarithmic scale this corresponds to a drop of 3 decibels (dB). In contrast with the magnitude, the phase of the transform is an odd function of ω since

$$\theta(-\omega) = -\tan^{-1}\left(\frac{-\omega}{a}\right) = \tan^{-1}\left(\frac{\omega}{a}\right) = -\theta(\omega)$$

At frequencies $\omega \to \pm\infty$ the phase angle approaches

$$\lim_{\omega\to\pm\infty}\left[\theta(\omega)\right] = \mp\frac{\pi}{2}$$

Furthermore, at frequencies $\omega = \pm a$ the phase angle is

$$\theta(\pm a) = \mp\frac{\pi}{4}$$

We will also compute real and imaginary parts of the transform to write its cartesian form representation. Multiplying both the numerator and the denominator of the result in Eqn. (3.147) by $(a - j\omega)$ we obtain

$$X(\omega) = \frac{1}{(a+j\omega)}\,\frac{(a-j\omega)}{(a-j\omega)} = \frac{a-j\omega}{a^2+\omega^2}$$

from which real and imaginary parts of the transform can be extracted as

$$X_r(\omega) = \frac{a}{a^2+\omega^2} \quad \text{and} \quad X_i(\omega) = \frac{-\omega}{a^2+\omega^2}$$

respectively. $X_r(\omega)$ and $X_i(\omega)$ are shown in Fig. 3.47. In this case the real part is an even function of ω and the imaginary part is an odd function.

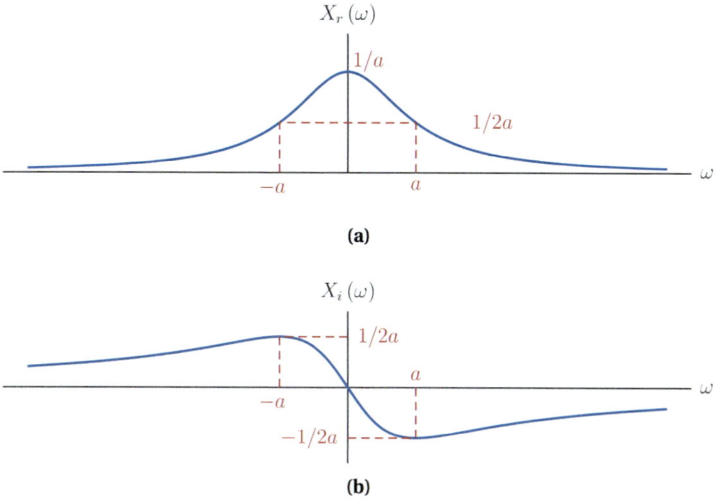

Figure 3.47 – **(a)** Real part and **(b)** imaginary part of the transform of the right-sided exponential signal in Example 3.15.

Software resources: `exct_3_15.m` , `exct_3_15_Live.mlx`

Interactive App: Fourier transform of right-sided exponential signal

The interactive app in `appFT4.m` is based on Example 3.15, Figs. 3.45, and 3.46. The right-sided exponential signal

$$x(t) = e^{-at}\,u(t)$$

and its Fourier transform are graphed. The parameter a may be modified through the use of a slider control. The Fourier transform may be displayed in either polar or Cartesian form.

a. Using the slider control increase the value of the parameter a. Observe how changes in a affect the signal $x(t)$ and the transform $X(\omega)$. Pay attention to the correlation between the width of the signal and the width of the transform. Does the fundamental relationship resemble the one observed earlier between the rectangular pulse and its Fourier transform?

b. Using the drop-down list, change the display form of the spectrum to Cartesian. What type of symmetry properties do the real and imaginary parts of the transform $X(\omega)$ exhibit?

Software resource: `appFT4.m`

Example 3.16: Fourier transform of a two-sided exponential signal

Determine the Fourier transform of the two-sided exponential signal given by

$$x(t) = e^{-a|t|}$$

where a is any non-negative real-valued constant. The signal $x(t)$ is shown in Fig. 3.48.

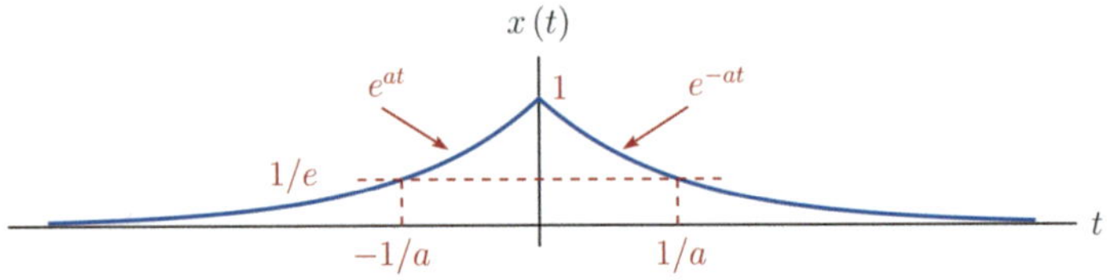

Figure 3.48 – Two-sided exponential signal $x(t)$ for Example 3.16.

Solution: Applying the Fourier transform integral of Eqn. (3.122) to our signal we get

$$X(\omega) = \int_{-\infty}^{\infty} e^{-a|t|}\, e^{-j\omega t}\, dt$$

Splitting the integral into two halves yields

$$X(\omega) = \int_{-\infty}^{0} e^{-a|t|}\, e^{-j\omega t}\, dt + \int_{0}^{\infty} e^{-a|t|}\, e^{-j\omega t}\, dt$$

Recognizing that

$$\text{for } t < 0 \quad \Rightarrow \quad e^{-a|t|} = e^{at}$$

$$\text{for } t > 0 \quad \Rightarrow \quad e^{-a|t|} = e^{-at}$$

the transform is

$$X(\omega) = \int_{-\infty}^{0} e^{at}\, e^{-j\omega t}\, dt + \int_{0}^{\infty} e^{-at} e^{-j\omega t}\, dt = \frac{1}{a-j\omega} + \frac{1}{a+j\omega} = \frac{2a}{a^2+\omega^2}$$

$X(\omega)$ is shown in Fig. 3.49.

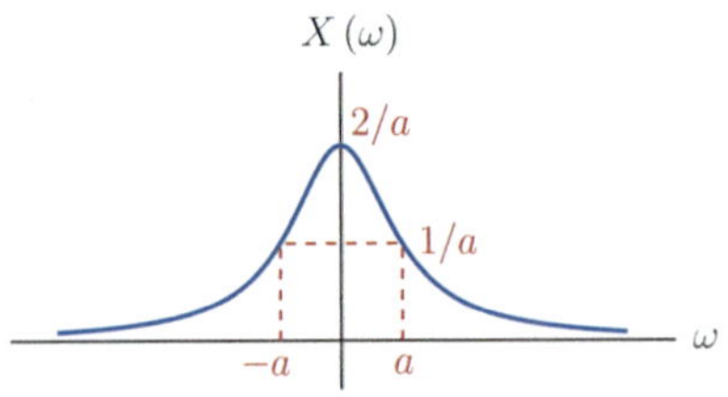

Figure 3.49 – Transform of the two-sided exponential signal of Example 3.16.

The transform $X(\omega)$ is purely real for all values of ω. This is a consequence of the signal $x(t)$ having even symmetry and will be explored further as we look at the symmetry properties of the Fourier transform in the next section. In addition to being purely real, the transform also happens to be non-negative in this case, resulting in a phase characteristic that is equal to zero for all frequencies.

Software resources: exct_3_16.m , exct_3_16_Live.mlx

Interactive App: Fourier transform of a two-sided exponential signal

The interactive app in appFT5.m is based on Example 3.16. The two-sided exponential signal

$$x(t) = e^{-a|t|}$$

and its Fourier transform are graphed. The transform is purely real for all ω owing to the even symmetry of the signal $x(t)$. The parameter a may be modified through the use of a slider control.

a. Set $a = 1$. Observe the decay rate of the exponential signal in both directions. Pay attention to the bandwidth of the frequency spectrum.

b. Gradually increase the value of a and observe the changes in the frequency spectrum. How does the bandwidth change?

c. Now gradually decrease the value of a. Notice how the signal $x(t)$ starts resembling a constant. What happens to the transform? Does it start to resemble an impulse function?

Software resource: appFT5.m

Example 3.17: Fourier transform of a triangular pulse

Find the Fourier transform of the triangular pulse signal given by

$$x(t) = A\Lambda\left(\frac{t}{\tau}\right) = \begin{cases} A + At/\tau\,, & -\tau < t < 0 \\ A - At/\tau\,, & 0 \le t < \tau \\ 0\,, & |t| \ge \tau \end{cases}$$

where $\Lambda(t)$ is the unit triangle function defined in Section 1.5.5 of Chapter 1. The signal $x(t)$ is shown in Fig. 3.50.

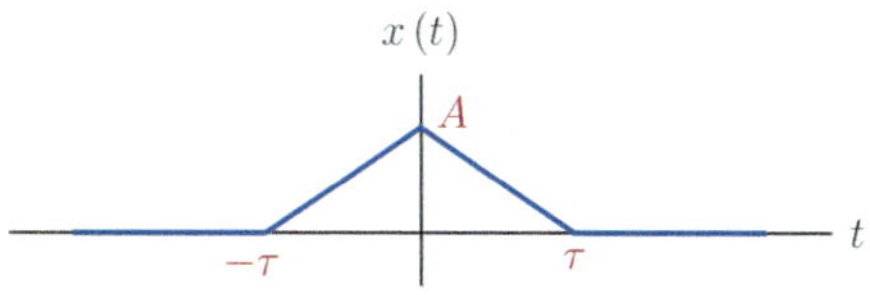

Figure 3.50 – Triangular pulse signal of Example 3.17.

Solution: Using the Fourier transform integral on the signal $x(t)$, we obtain

$$X(\omega) = \int_{-\tau}^{0} \left(A + \frac{At}{\tau}\right) e^{-j\omega t}\, dt + \int_{0}^{\tau} \left(A - \frac{At}{\tau}\right) e^{-j\omega t}\, dt$$

which can be evaluated to yield

$$\begin{aligned} X(\omega) =& A \int_{-\tau}^{0} e^{-j\omega t}\, dt + \frac{A}{\tau} \int_{-\tau}^{0} t\, e^{-j\omega t}\, dt + A \int_{0}^{\tau} e^{-j\omega t}\, dt - \frac{A}{\tau} \int_{0}^{\tau} t\, e^{-j\omega t}\, dt \\ =& A \int_{-\tau}^{\tau} e^{-j\omega t}\, dt + \frac{A}{\tau} \int_{-\tau}^{0} t\, e^{-j\omega t}\, dt - \frac{A}{\tau} \int_{0}^{\tau} t\, e^{-j\omega t}\, dt \end{aligned} \tag{3.148}$$

The last two integrals in Eqn. (3.148) are similar in form to entry (B.16) of the table of indefinite integrals in Appendix B, repeated here for convenience:

$$\text{Eqn. (B.16):} \qquad \int t\, e^{at}\, dt = \frac{1}{a^2}\, e^{at}\, [at - 1]$$

Setting $a = -j\omega$ results in

$$\int t\, e^{-j\omega t}\, dt = \frac{1}{\omega^2}\, e^{-j\omega t}\, [j\omega t + 1]$$

and the Fourier transform is

$$\begin{aligned} X(\omega) =& A \left(\frac{e^{-j\omega t}}{-j\omega} \right) \Bigg|_{-\tau}^{\tau} + \frac{A}{\omega^2 \tau} \left(e^{-j\omega t}\, [j\omega t + 1] \right) \Bigg|_{-\tau}^{0} - \frac{A}{\omega^2 \tau} \left(e^{-j\omega t}\, [j\omega t + 1] \right) \Bigg|_{0}^{\tau} \\ =& \frac{2A \sin(\omega\tau)}{\omega} + \frac{2A}{\omega^2 \tau} \left[1 - \omega\tau \sin(\omega\tau) - \cos(\omega\tau) \right] \\ =& \frac{A}{\omega^2 \tau} \left[1 - \cos(\omega\tau) \right] \end{aligned}$$

Using the appropriate trigonometric identity, it can be shown that

$$X(\omega) = \frac{4A}{\omega^2 \tau} \sin^2\left(\frac{\omega\tau}{2} \right) \tag{3.149}$$

It would be convenient to use the sinc function with the result in Eqn. (3.149). Recognizing that

$$\text{sinc}\left(\frac{\omega\tau}{2\pi} \right) = \frac{\sin(\omega\tau/2)}{(\omega\tau/2)} = \frac{2}{\omega\tau} \sin\left(\frac{\omega\tau}{2} \right)$$

the transform in Eqn. (3.149) can be written as

$$X(\omega) = A\tau\, \text{sinc}^2\left(\frac{\omega\tau}{2\pi} \right)$$

If the transform is written in terms of f instead of ω, we obtain

$$X(f) = A\tau\, \text{sinc}^2(f\tau)$$

Thus, the Fourier transform of a triangular pulse is proportional the square of the sinc function. Contrast this with the Fourier transform of a rectangular pulse which is proportional to the sinc function as we have seen in Example 3.12. The transform of the triangular pulse, shown in Fig. 3.51, seems to be concentrated more heavily around low frequencies compared to the transform of the rectangular pulse.

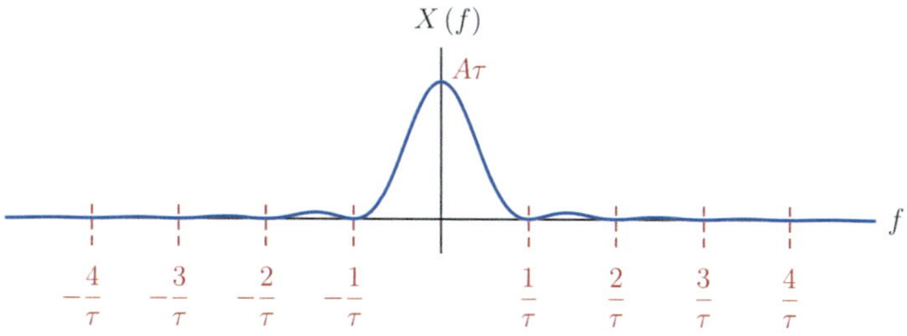

Figure 3.51 – Frequency spectrum of the triangular pulse signal of Example 3.17.

We will have occasion to look at the transform of a triangular pulse in more depth when we study the properties of the Fourier transform in the next section. Specifically, a much simpler method of obtaining $X(\omega)$ for a triangular pulse will be presented after studying the convolution property of the Fourier transform.

Software resources: exct_3_17.m , exct_3_17_Live.mlx

Interactive App: Fourier transform of a triangular pulse

The interactive app in `appFT6.m` is based on Example 3.17. The triangular pulse signal with peak amplitude A and width 2τ is shown along with its Fourier transform as determined in Example 3.17. The transform is purely real for all f owing to the even symmetry of the triangular pulse signal $x(t)$. The parameters A and τ may be modified through the use of slider controls.

a. Start with $A = 1$ and $\tau = 1$. Pay attention to the frequencies where $X(f) = 0$ (you may need to zoom in to see the zeros of the transform).

b. Gradually increase the value of τ. Observe how parameter changes affect the signal $x(t)$ and the transform $X(\omega)$. Again, pay attention to frequencies where $X(f) = 0$. How do they relate to the value of τ?

c. Now gradually reduce the value of τ and repeat.

Software resource: `appFT6.m`

Example 3.18: Fourier transform of the signum function

Determine the Fourier transform of the signum function defined as

$$x(t) = \text{sgn}(t) = \begin{cases} -1, & t<0 \\ 1, & t>0 \end{cases} \tag{3.150}$$

and shown graphically in Fig. 3.52.

Solution: If we were to apply the Fourier transform integral directly to this signal, we would obtain

$$X(\omega) = \int_{-\infty}^{0} (-1)\, e^{-j\omega t}\, dt + \int_{0}^{\infty} (1)\, e^{-j\omega t}\, dt$$

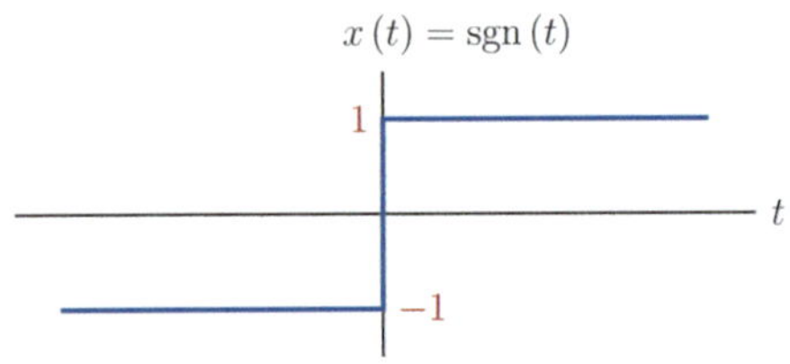

Figure 3.52 – The signum function.

in which the two integrals cannot be evaluated. Instead, we will define an intermediate signal $p(t)$ as

$$p(t) = \begin{cases} -e^{at}, & t<0 \\ e^{-at}, & t>0 \end{cases} \tag{3.151}$$

where $a \ge 0$. We will first determine the Fourier transform of this intermediate signal, and then obtain the Fourier transform of the signum function from this intermediate result. The Fourier transform of $p(t)$ is found as

$$\begin{aligned} P(\omega) &= \int_{-\infty}^{0} \left(-e^{at}\right) e^{-j\omega t}\, dt + \int_{0}^{\infty} \left(e^{-at}\right) e^{-j\omega t}\, dt \\ &= \left(\frac{-e^{at}\, e^{-j\omega t}}{a - j\omega}\right)\Bigg|_{-\infty}^{0} + \left(\frac{e^{-at}\, e^{-j\omega t}}{-a - j\omega}\right)\Bigg|_{0}^{\infty} \\ &= \frac{-j2\omega}{a^2 + \omega^2} \end{aligned}$$

The result is purely imaginary for all ω due to the odd symmetry of $p(t)$. As the value of a is reduced, the intermediate signal $p(t)$ starts to look more and more similar to the signum function, and becomes the signum function in the limit as $a \to 0$. The transition of $p(t)$ to $\text{sgn}(t)$ is illustrated in Fig. 3.53 for $a = 0.4, 0.15$, and 0.04.

Setting $a = 0$ in the transform $P(\omega)$, we obtain the Fourier transform of the original signal $x(t)$:

$$X(\omega) = \mathscr{F}\left\{\text{sgn}(t)\right\} = \lim_{a \to 0} \left[\frac{-j2\omega}{a^2 + \omega^2}\right] = \frac{2}{j\omega}$$

The magnitude of the transform is

$$\left|X(\omega)\right| = \frac{2}{|\omega|}$$

and the phase is

$$\theta(\omega) = \measuredangle X(\omega) = \begin{cases} \pi/2\,, & \omega < 0 \\ -\pi/2\,, & \omega > 0 \end{cases}$$

Magnitude and phase characteristics are shown in Fig. 3.54.

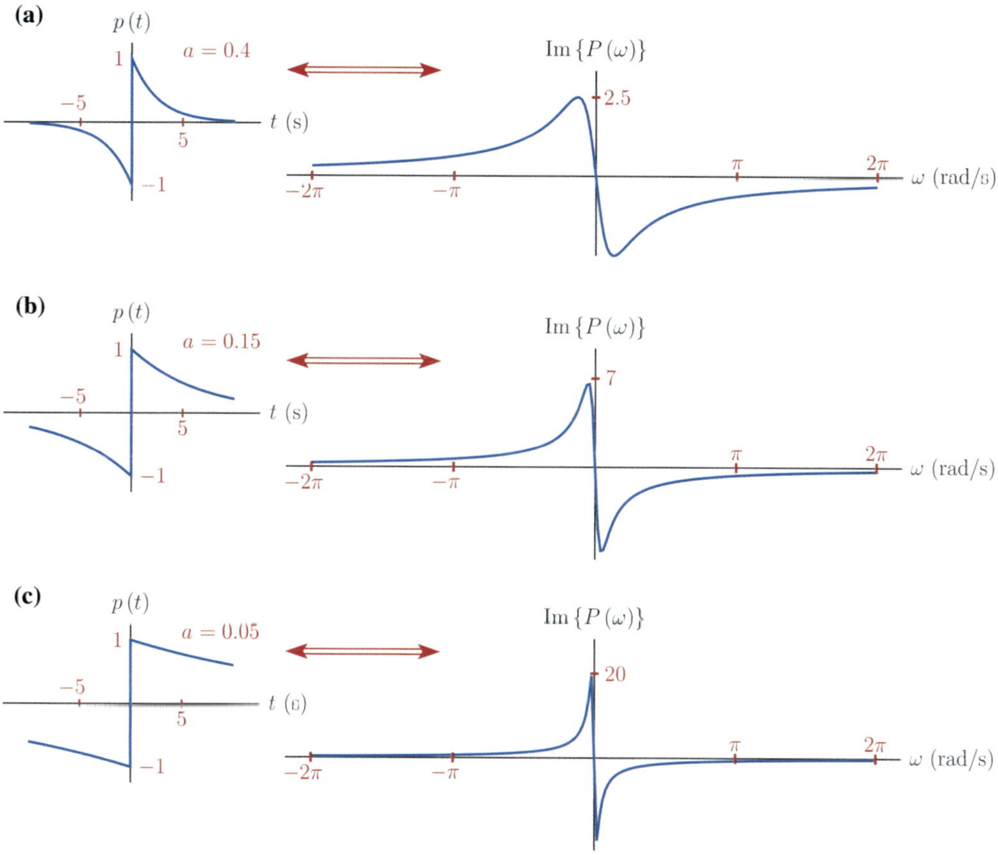

Figure 3.53 – Signum function as a limit case of $p(t)$ of Eqn. (3.151): **(a)** $p(t)$ for $a = 0.4$ and its transform, **(b)** $p(t)$ for $a = 0.15$ and its transform, and **(c)** $p(t)$ for $a = 0.05$ and its transform.

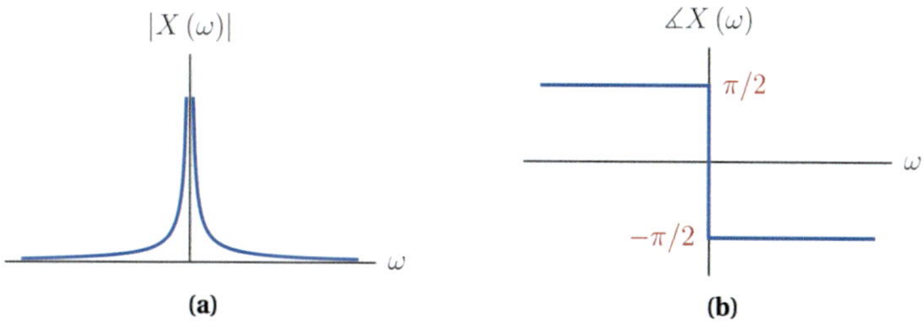

Figure 3.54 – Magnitude and phase of the Fourier transform of sgn(t).

Software resources: `exct_3_18.m` , `exct_3_18_Live.mlx`

Interactive App: Fourier transform of the signum function

The interactive app in `appFT7.m` is based on Example 3.18, and allows experimentation with the progression of the signal

$$p(t)=\begin{cases} -e^{at}\,, & t<0 \\ e^{-at}\,, & t>0 \end{cases}$$

and its corresponding Fourier transform $P(\omega)$ into the signal

$$x(t)=\text{sgn}(t)$$

and its associated transform $X(\omega)$ as the parameter a is made smaller. The parameter a may be adjusted through the use of a slider control. The program displays a graph of the signal $p(t)$ and a graph of the imaginary part of transform $P(\omega)$. The real part of the transform is always equal to zero for this particular signal and is therefore not shown.

a. Set $a=1$. Observe the signal $p(t)$ and its transform $P(\omega)$.

b. Gradually decrease the value of a. Observe how the signal $p(t)$ becomes more and more similar to the signum function. Pay attention to the changes in the shape of the spectrum. Compare to the transform $X(\omega)$ of the signum function which was found in Example 3.18 and graphed in Fig. 3.54.

Software resource: `appFT7.m`

3.3.5 Properties of the Fourier transform

In this section we will explore some of the fundamental properties of the Fourier transform that provide further insight into its use for understanding and analyzing characteristics of signals. We will also see that some of these properties greatly simplify the computation of the Fourier transform for certain types of signals.

Linearity

Fourier transform is a linear operator. For any two signals $x_1(t)$ and $x_2(t)$ with their respective transforms

$$x_1(t) \overset{\mathcal{F}}{\longleftrightarrow} X_1(\omega) \quad \text{and} \quad x_2(t) \overset{\mathcal{F}}{\longleftrightarrow} X_2(\omega)$$

and any two constants α_1 and α_2, the following relationship holds:

Linearity of the Fourier transform:

$$\alpha_1\,x_1(t)+\alpha_2\,x_2(t) \overset{\mathcal{F}}{\longleftrightarrow} \alpha_1\,X_1(\omega)+\alpha_2\,X_2(\omega) \tag{3.152}$$

Proof of Eqn. (3.152):

Using the forward transform equation given by Eqn. (3.122) with the time domain signal $\left[\alpha_1\, x_1(t) + \alpha_2\, x_2(t)\right]$ leads to:

$$\mathcal{F}\left\{\alpha_1\, x_1(t) + \alpha_2\, x_2(t)\right\} = \int_{-\infty}^{\infty} \left[\alpha_1\, x_1(t) + \alpha_2\, x_2(t)\right] e^{-j\omega t}\, dt$$

which can be simplified as

$$\begin{aligned}\mathcal{F}\left\{\alpha_1\, x_1(t) + \alpha_2\, x_2(t)\right\} =& \int_{-\infty}^{\infty} \alpha_1\, x_1(t)\, e^{-j\omega t}\, dt + \int_{-\infty}^{\infty} \alpha_2\, x_2(t)\, e^{-j\omega t}\, dt \\ =& \alpha_1 \int_{-\infty}^{\infty} x_1(t)\, e^{-j\omega t}\, dt + \alpha_2 \int_{-\infty}^{\infty} x_2(t)\, e^{-j\omega t}\, dt \\ =& \alpha_1\, \mathcal{F}\left\{x_1(t)\right\} + \alpha_2\, \mathcal{F}\left\{x_2(t)\right\} \end{aligned} \tag{3.153}$$

Duality

Consider the transform pair

$$x(t) \overset{\mathcal{F}}{\longleftrightarrow} X(\omega)$$

The transform relationship between $x(t)$ and $X(\omega)$ is defined by the inverse and forward Fourier transform integrals given by Eqns. (3.121) and (3.122) and repeated here for convenience:

$$\text{Eqn. (3.121):} \qquad x(t) = \frac{1}{2\pi} \int_{-\infty}^{\infty} X(\omega)\, e^{j\omega t}\, d\omega$$

$$\text{Eqn. (3.122):} \qquad X(\omega) = \int_{-\infty}^{\infty} x(t)\, e^{-j\omega t}\, dt$$

Suppose that we swap the roles of the independent variables to construct a new time-domain signal $X(t)$ and a new frequency-domain function $x(\omega)$. What would be the relationship between $X(t)$ and $x(\omega)$?

Derivation of Fourier transform of $X(t)$:

The expressions for the inverse transform in Eqn. (3.121) and the forward transform in Eqn. (3.122) are interestingly similar. The main differences between the two integrals are the sign of the exponential term and the scale factor $1/2\pi$ for the analysis equation. This similarity leads to the *duality principle* of the Fourier transform. Let us change the integration variable of Eqn. (3.121) from ω to λ, and write it as

$$x(t) = \frac{1}{2\pi} \int_{-\infty}^{\infty} X(\lambda)\, e^{j\lambda t}\, d\lambda \tag{3.154}$$

If $\omega \to \pm\infty$ then $\lambda \to \pm\infty$, so we don't need to change the limits of integration. Now, from a purely mathematical perspective, we will change the name of the independent variable from t to $-\omega$, that is, we will evaluate the function on the left side of the integral in Eqn. (3.154) for $t = -\omega$, and write the result as

$$x(-\omega) = \frac{1}{2\pi} \int_{-\infty}^{\infty} X(\lambda)\, e^{-j\lambda\omega}\, d\lambda \tag{3.155}$$

Changing the integration variable from λ to t yields

$$x(-\omega) = \frac{1}{2\pi} \int_{-\infty}^{\infty} X(t)\, e^{-jt\omega}\, dt \tag{3.156}$$

and by multiplying both sides of Eqn. (3.156) by 2π we obtain

$$2\pi\, x(-\omega) = \int_{-\infty}^{\infty} X(t)\, e^{-jt\omega}\, dt \tag{3.157}$$

Eqn. (3.157) has the same general form as Eqn. (3.122) with substitutions $x(t) \rightarrow X(t)$ and $X(\omega) \rightarrow 2\pi\, x(-\omega)$. Thus, the following relationship must also be true:

$$X(t) \overset{\mathcal{F}}{\longleftrightarrow} 2\pi\, x(-\omega) \tag{3.158}$$

The time-domain function $X(t)$ and the frequency-domain function $2\pi\, x(-\omega)$ form a Fourier transform pair. Thus, if $X(\omega)$ is the Fourier transform of $x(t)$, then $2\pi\, x(-\omega)$ is the Fourier transform of $X(t)$.

Duality property:

$$x(t) \overset{\mathcal{F}}{\longleftrightarrow} X(\omega) \quad \text{implies that} \quad X(t) \overset{\mathcal{F}}{\longleftrightarrow} 2\pi\, x(-\omega) \tag{3.159}$$

We may find it more convenient to express the duality property using the frequency f instead of the radian frequency ω since this approach eliminates the 2π factor from the equations:

Duality property (using f instead of ω):

$$x(t) \overset{\mathcal{F}}{\longleftrightarrow} X(f) \quad \text{implies that} \quad X(t) \overset{\mathcal{F}}{\longleftrightarrow} x(-f) \tag{3.160}$$

Duality property will be useful in deriving some of the other properties of the Fourier transform.

Example 3.19: Fourier transform of the sinc function

Find the Fourier transform of the signal

$$x(t) = \mathrm{sinc}(t)$$

Solution: The Fourier transform of a rectangular pulse was found in Example 3.12 as a sinc function. Specifically we obtained

$$\mathcal{F}\left\{ A\Pi\left(\frac{t}{\tau}\right)\right\} = A\tau\, \mathrm{sinc}\left(\frac{\omega\tau}{2\pi}\right) \tag{3.161}$$

This relationship can be used as a starting point in finding the Fourier transform of $x(t) = \mathrm{sinc}(t)$. Let $\tau = 2\pi$ so that the argument of the sinc function in Eqn. (3.161) becomes ω:

$$\mathcal{F}\left\{ A\Pi\left(\frac{t}{2\pi}\right)\right\} = 2\pi\, A\, \mathrm{sinc}(\omega)$$

In the next step, let $A = 1/2\pi$ to obtain

$$\mathcal{F}\left\{\frac{1}{2\pi}\Pi\left(\frac{t}{2\pi}\right)\right\} = \text{sinc}\,(\omega) \tag{3.162}$$

Applying the duality property given by Eqn. (3.160) to the transform pair in Eqn. (3.162), and remembering that the unit pulse function $\Pi(t)$ has even symmetry, we conclude that

$$\mathcal{F}\left\{\text{sinc}\,(t)\right\} = \Pi\left(\frac{-\omega}{2\pi}\right) = \Pi\left(\frac{\omega}{2\pi}\right) \tag{3.163}$$

Using f instead of ω in Eqn. (3.163) yields a relationship that is easier to remember:

$$\mathcal{F}\left\{\text{sinc}\,(t)\right\} = \Pi\left(f\right) \tag{3.164}$$

The Fourier transform of $x(t) = \text{sinc}\,(t)$ is a unit pulse in the frequency domain. The signal $x(t)$ and its transform are graphed in Fig. 3.55.

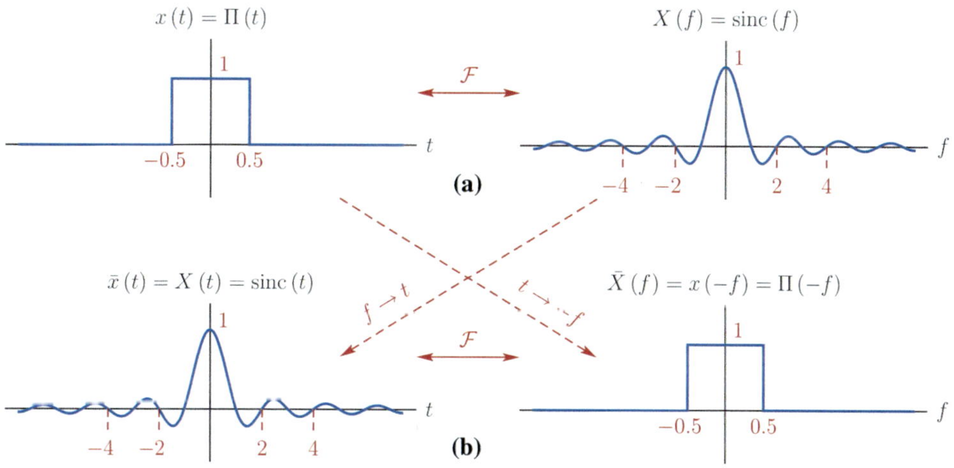

Figure 3.55 – **(a)** The function $x(t) = \Pi(t)$ and its Fourier transform $X(f)$ and **(b)** the function $\bar{x}(t) = \text{sinc}\,(t)$ and its Fourier transform $\bar{X}(f)$.

Example 3.20: Transform of a constant-amplitude signal

Find the Fourier transform of the constant-amplitude signal

$$x(t) = 1, \quad \text{all } t$$

Solution: Finding the Fourier transform of $x(t)$ by direct application of Eqn. (3.122) is not possible since the resulting integral

$$X(\omega) = \int_{-\infty}^{\infty} x(t)\, e^{-j\omega t}\, dt = \int_{-\infty}^{\infty} (1)\, e^{-j\omega t}\, dt$$

could not be evaluated. On the other hand, in Example 3.14 we have concluded that the Fourier transform of the unit impulse signal is a constant for all frequencies, that is

$$\mathcal{F}\left\{\delta(t)\right\} = 1, \quad \text{all } \omega \tag{3.165}$$

Applying the duality property to the transform pair in Eqn. (3.165) we obtain

$$\mathcal{F}\{1\} = 2\pi\,\delta(-\omega) = 2\pi\,\delta(\omega) \tag{3.166}$$

If f is used instead of ω in the transform, then we have

$$\mathcal{F}\{1\} = \delta(f) \tag{3.167}$$

This example also presents an opportunity for two important observations:

1. The signal $x(t) = 1$ does not satisfy the existence conditions discussed in Section 3.3.2; it is neither absolute integrable nor square integrable. Therefore, in a strict sense, its Fourier transform does not converge. By allowing the impulse function to be used in the transform, we obtain a function $X(\omega)$ that has the characteristics of a Fourier transform and that can be used in solving problems in the frequency domain.
2. Sometimes we express the Fourier transform in terms of the radian frequency ω; at other times we use the frequency f. In general, the conversion from one format to another is just a matter of using the relationship $\omega = 2\pi f$, so that

$$X(\omega) = X(f)\Big|_{f=\omega/2\pi}\,, \qquad X(f) = X(\omega)\Big|_{\omega=2\pi f}$$

   One exception to this rule is seen when the transform contains a singularity function. Compare Eqns. (3.166) and (3.167) in the example above. $X(\omega)$ has a 2π factor that $X(f)$ does not have. This behavior is consistent with scaling of the impulse function (see Problem 1.7 of Chapter 1).

Example 3.21: Another example of using the duality property

Using the duality property, find the Fourier transform of the signal

$$x(t) = \frac{1}{3+2t^2}$$

which is graphed in Fig. 3.56.

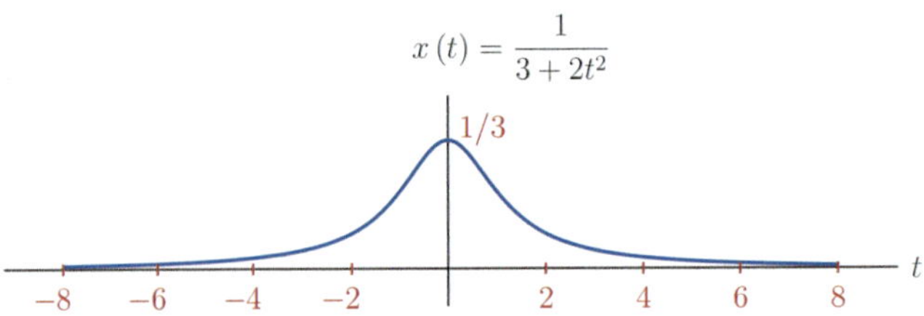

Figure 3.56 – The signal $x(t) = 1/(3+2t^2)$ of Example 3.21.

Solution: While it is possible to solve this problem by a direct application of the Fourier transform integral given by Eqn. (3.122), it would be difficult to evaluate the integral unless we resort to the use of integration tables. Instead, we will simplify the solution by using another transform pair that we already know, and adapting that transform pair to our needs.

In Example 3.16 we have found that

$$e^{-a|t|} \overset{\mathcal{F}}{\longleftrightarrow} \frac{2a}{a^2+\omega^2}$$

where the form of the transform bears a similarity to the time-domain expression for the signal $x(t)$. In both functions the numerator is constant, and the denominator has a second-order term next to a constant term. We will make use of this similarity through the use of the duality property of the Fourier transform. The duality property allows us to write

$$\frac{2a}{a^2+t^2} \overset{\mathcal{F}}{\longleftrightarrow} 2\pi\, e^{-a|\omega|} \tag{3.168}$$

as another valid transform pair. By taking steps that will preserve the validity of this transform pair, we will work toward making the time-domain component of the relationship in Eqn. (3.168) match the signal $x(t)$ given in the problem statement. The signal $x(t)$ has a $2t^2$ term in its denominator. Let us multiply both the numerator and the denominator of the time-domain component of Eqn. (3.168) by 2:

$$\frac{4a}{2a^2+2t^2} \overset{\mathcal{F}}{\longleftrightarrow} 2\pi\, e^{-a|\omega|}$$

To match the constant term in the denominator of $x(t)$ we will choose

$$2a^2 = 3 \quad \Longrightarrow \quad a = \sqrt{\frac{3}{2}}$$

and obtain

$$\frac{2\sqrt{6}}{3+2t^2} \overset{\mathcal{F}}{\longleftrightarrow} 2\pi\, e^{-\sqrt{\frac{3}{2}}|\omega|} \tag{3.169}$$

Afterward, by scaling both sides of Eqn. (3.169) with $2\sqrt{6}$ we get

$$\frac{1}{3+2t^2} \overset{\mathcal{F}}{\longleftrightarrow} \frac{\pi}{\sqrt{6}}\, e^{-\sqrt{\frac{3}{2}}|\omega|} \tag{3.170}$$

The transform found in Eqn. (3.170) is graphed in Fig. 3.57 using f as the independent variable instead of ω.

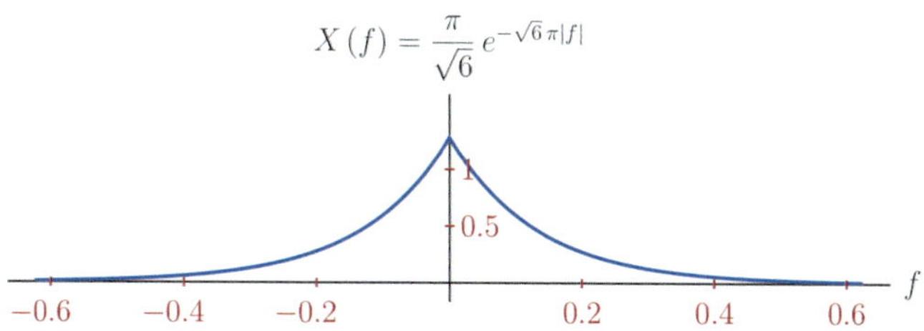

Figure 3.57 – Fourier transform of the function $x(t)$ of Example 3.21.

Software resource: exct_3_21.m

Example 3.22: Signum function revisited

In Example 3.18 the Fourier transform of the signal $x(t) = \operatorname{sgn}(t)$ was found to be

$$X(\omega) = \mathcal{F}\{\operatorname{sgn}(t)\} = \frac{2}{j\omega}$$

Determine the time-domain signal, the Fourier transform of which is

$$X(\omega) = -j\,\operatorname{sgn}(\omega)$$

Solution: This is another problem that could be simplified by the use of the duality property. Starting with the already known transform pair

$$\operatorname{sgn}(t) \overset{\mathcal{F}}{\longleftrightarrow} \frac{2}{j\omega}$$

and applying the duality property as expressed by Eqn. (3.159) we conclude that the following must also be a valid transform pair:

$$\frac{2}{jt} \overset{\mathcal{F}}{\longleftrightarrow} 2\pi\,\operatorname{sgn}(-\omega)$$

Recognizing that the sgn function has odd symmetry, that is, $\operatorname{sgn}(-\omega) = -\operatorname{sgn}(\omega)$,

$$\frac{2}{jt} \overset{\mathcal{F}}{\longleftrightarrow} -2\pi\,\operatorname{sgn}(\omega) \tag{3.171}$$

Multiplying both sides of this by $j/2\pi$, we have

$$\frac{1}{\pi t} \overset{\mathcal{F}}{\longleftrightarrow} -j\,\operatorname{sgn}(\omega)$$

The signal we are seeking is

$$x(t) = \frac{1}{\pi t}$$

which is graphed in Fig. 3.58.

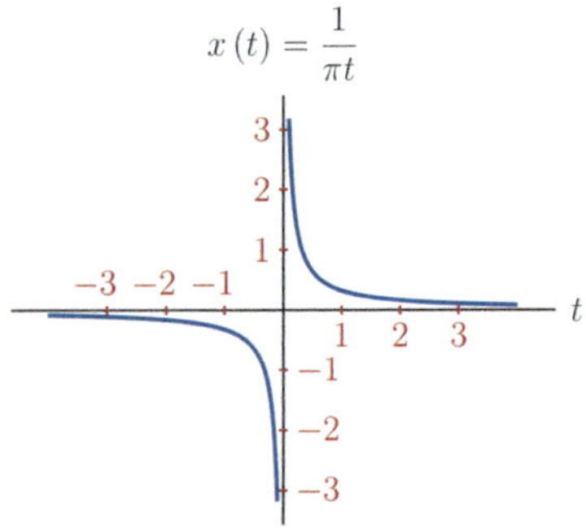

Figure 3.58 – The signal $x(t) = \dfrac{1}{\pi t}$ found in Example 3.22.

It is interesting to look at the magnitude and the phase of the transform $X(\omega)$ in Eqn. (3.171). The transform can be written as

$$X(\omega) = -j\,\mathrm{sgn}(\omega) = \begin{cases} -j\,, & \omega > 0 \\ j\,, & \omega < 0 \end{cases}$$

Realizing that $j = e^{j\pi/2}$ we can write

$$X(\omega) = \begin{cases} e^{-j\pi/2}\,, & \omega > 0 \\ e^{j\pi/2}\,, & \omega < 0 \end{cases}$$

The magnitude and the phase of $X(\omega)$ are shown in Fig. 3.59.

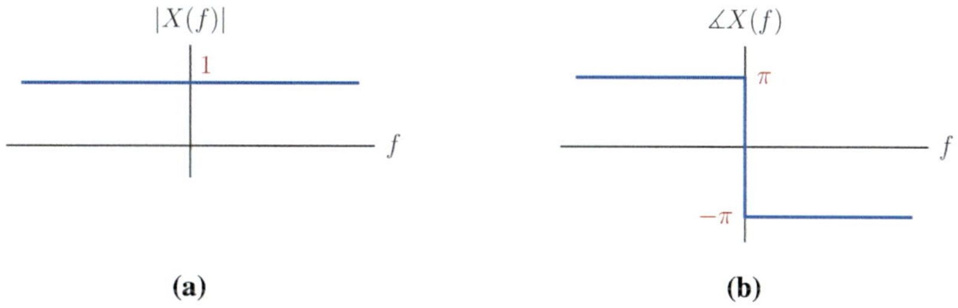

Figure 3.59 – **(a)** The magnitude and **(b)** the phase of $X(\omega)$ given by Eqn. (3.171).

The magnitude of the transform is equal to unity for all frequencies. The phase of the transform is $-\pi/2$ radians for positive frequencies and $+\pi/2$ radians for negative frequencies.

Example 3.23: Fourier transform of the unit step function

Determine the Fourier transform of the unit step function $x(t) = u(t)$.

Solution: The unit step function does not satisfy the conditions for the existence of the Fourier transform. An attempt to find the transform by direct application of Eqn. (3.122) would result in

$$\mathcal{F}\{u(t)\} = \int_{-\infty}^{\infty} u(t)\, e^{-j\omega t}\, dt = \int_{0}^{\infty} e^{-j\omega t}\, dt$$

which does not converge. Instead, we will try to make use of some of the earlier results found. We will also employ the linearity property of the Fourier transform. The unit step function can be expressed in terms of the signum function and a constant offset term as

$$u(t) = \frac{1}{2} + \frac{1}{2}\,\mathrm{sgn}(t)$$

This is illustrated in Fig. 3.60. Using the linearity of the Fourier transform we obtain

$$\mathcal{F}\{u(t)\} = \mathcal{F}\left\{\frac{1}{2} + \frac{1}{2}\,\mathrm{sgn}(t)\right\} = \frac{1}{2}\,\mathcal{F}\{1\} + \frac{1}{2}\,\mathcal{F}\{\mathrm{sgn}(t)\} \tag{3.172}$$

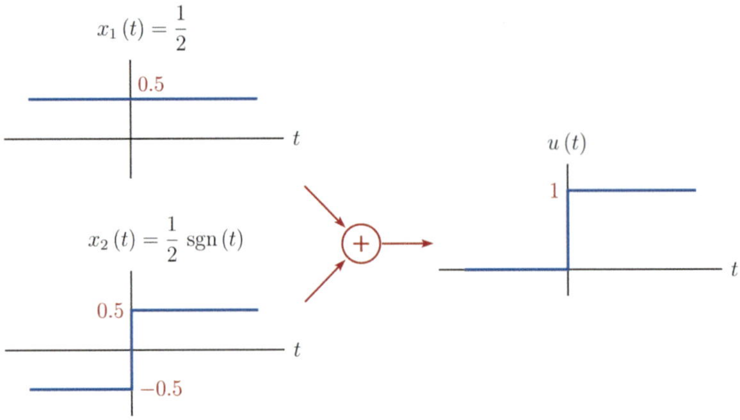

Figure 3.60 – Obtaining the unit step function as the sum of the signum function and a constant offset.

The transform of a constant signal was found in Example 3.20:

$$\mathcal{F}\{1\} = 2\pi\,\delta(\omega)$$

Furthermore, in Example 3.18 we found

$$\mathcal{F}\{\mathrm{sgn}(t)\} = \frac{2}{j\omega}$$

Using these two results in Eqn. (3.172), Fourier transform of the unit step function is

$$\mathcal{F}\{u(t)\} = \pi\,\delta(\omega) + \frac{1}{j\omega} \tag{3.173}$$

Fig. 3.61 shows the unit step signal and the magnitude of its transform.

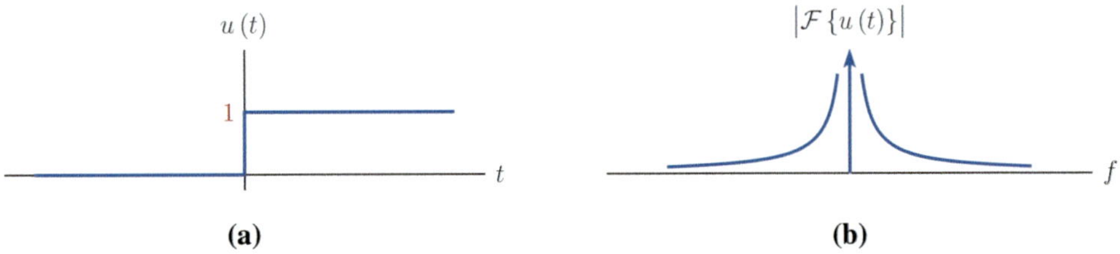

Figure 3.61 – **(a)** The unit step function and **(b)** the magnitude of its Fourier transform.

If the transform of the unit step function is needed in terms of f instead of ω, the result in Eqn. (3.173) needs to be modified as

$$\mathcal{F}\{u(t)\} = \frac{1}{2}\,\delta(f) + \frac{1}{j2\pi f} \tag{3.174}$$

Symmetry of the Fourier Transform

Conjugate symmetry and conjugate antisymmetry properties were defined for time-domain signals in Section 1.9 of Chapter 1. Same definitions apply to a frequency-domain transform as well: A transform $X(\omega)$ is said to be *conjugate symmetric* if it satisfies

$$X^*(\omega) = X(-\omega) \tag{3.175}$$

for all ω. Similarly, a transform $X(\omega)$ is said to be *conjugate antisymmetric* if it satisfies

$$X^*(\omega) = -X(-\omega) \tag{3.176}$$

for all ω. A transform does not have to be conjugate symmetric or conjugate antisymmetric. If it has any symmetry properties, however, they can be explored for simplifying the analysis in the frequency domain.

If the signal $x(t)$ is real-valued, it can be shown that its Fourier transform $X(\omega)$ is conjugate symmetric. Conversely, if the signal $x(t)$ is purely imaginary, its transform is conjugate antisymmetric.

Symmetry of the Fourier transform:

$$x(t)\text{: Real, } \operatorname{Im}\{x(t)\} = 0 \quad \text{implies that} \quad X^*(\omega) = X(-\omega) \tag{3.177}$$

$$x(t)\text{: Imag, } \operatorname{Re}\{x(t)\} = 0 \quad \text{implies that} \quad X^*(\omega) = -X(-\omega) \tag{3.178}$$

Derivation of symmetry properties of the Fourier transform:

1. Real $x(t)$:

   Let us begin by conjugating both sides of the forward transform integral given by Eqn. (3.122):

   $$X^*(\omega) = \left[\int_{-\infty}^{\infty} x(t)\, e^{-j\omega t}\, dt\right]^* = \int_{-\infty}^{\infty} x^*(t)\, e^{j\omega t}\, dt \tag{3.179}$$

   Since $x(t)$ is real valued, $x^*(t) = x(t)$, and we have

   $$X^*(\omega) = \int_{-\infty}^{\infty} x(t)\, e^{j\omega t}\, dt \tag{3.180}$$

   In the next step we will derive $X(-\omega)$ from Eqn. (3.122) with the substitution $\omega \to -\omega$:

   $$X(-\omega) = \int_{-\infty}^{\infty} x(t)\, e^{j\omega t}\, dt \tag{3.181}$$

   Comparing Eqns. (3.180) and (3.181) we conclude that $X^*(\omega) = X(-\omega)$ when $x(t)$ is real-valued.

2. Imaginary $x(t)$:

   In this case the complex conjugate of the transform is

   $$X^*(\omega) = \int_{-\infty}^{\infty} x^*(t)\, e^{j\omega t}\, dt = -\int_{-\infty}^{\infty} x(t)\, e^{j\omega t}\, dt \tag{3.182}$$

since $x^*(t) = -x(t)$ for an imaginary signal. Comparison of Eqns. (3.181) and (3.182) leads to the conclusion that $X^*(\omega) = -X(-\omega)$ when $x(t)$ is purely imaginary.

Cartesian and polar forms of the transform

A complex transform $X(\omega)$ can be written in polar form as

$$X(\omega) = \left|X(\omega)\right| e^{j\Theta(\omega)} \tag{3.183}$$

and in Cartesian form as

$$X(\omega) = X_r(\omega) + jX_i(\omega) \tag{3.184}$$

If $X(\omega)$ is conjugate symmetric, the relationship in Eqn. (3.175) can be written as

$$\left|X(\omega)\right| e^{-j\Theta(\omega)} = \left|X(-\omega)\right| e^{j\Theta(-\omega)} \tag{3.185}$$

using the polar form of the transform, and

$$X_r(\omega) - jX_i(\omega) = X_r(-\omega) + jX_i(-\omega) \tag{3.186}$$

using its Cartesian form. The consequences of Eqns. (3.185) and (3.186) can be obtained by equating the magnitudes and the phases on both sides of Eqn. (3.185) and by equating real and imaginary parts on both sides of Eqn. (3.186). The results can be summarized as follows:

Conjugate symmetric transform: $X^*(\omega) = X(-\omega)$

Magnitude:	$\left|X(-\omega)\right| = \left|X(\omega)\right|$	(3.187)
Phase:	$\Theta(-\omega) = -\Theta(\omega)$	(3.188)
Real part:	$X_r(-\omega) = X_r(\omega)$	(3.189)
Imag. part:	$X_i(-\omega) = -X_i(\omega)$	(3.190)

We have already established the fact that the transform of a real-valued $x(t)$ is conjugate symmetric. The results in Eqns. (3.187) and (3.188) suggest that, for such a transform, the magnitude is an even function of ω, and the phase is an odd function. Furthermore, based on Eqns. (3.189) and (3.190), the real part of the transform exhibits even symmetry while its imaginary part exhibits odd symmetry.

Similarly, if $X(\omega)$ is conjugate antisymmetric, the relationship in Eqn. (3.175) reflects on polar form of $X(\omega)$ as

$$\left|X(\omega)\right| e^{-j\Theta(\omega)} = -\left|X(-\omega)\right| e^{j\Theta(-\omega)} \tag{3.191}$$

The negative sign on the right side of Eqn. (3.191) needs to be incorporated into the phase since we could not write $\left|X(\omega)\right| = -\left|X(-\omega)\right|$ (recall that magnitude needs to be a non-negative function for all ω). Using $e^{\mp j\pi} = -1$, Eqn. (3.191) can be written as

$$\left|X(\omega)\right| e^{-j\Theta(\omega)} e^{j\mp\pi} = \left|X(-\omega)\right| e^{j\Theta(-\omega)}$$

Conjugate antisymmetry property of the transform can also be expressed in Cartesian form as

$$X_r(\omega) - jX_i(\omega) = -X_r(-\omega) - jX_i(-\omega) \tag{3.192}$$

The consequences of Eqns. (3.191) and (3.192) are given below.

Conjugate antisymmetric transform: $X^*(\omega) = -X(-\omega)$

Magnitude:	$\left\|X(-\omega)\right\| = \left\|X(\omega)\right\|$	(3.193)
Phase:	$\Theta(-\omega) = -\Theta(\omega) \pm \pi$	(3.194)
Real part:	$X_r(-\omega) = -X_r(\omega)$	(3.195)
Imag. part:	$X_i(-\omega) = X_i(\omega)$	(3.196)

A purely imaginary signal leads to a Fourier transform with conjugate antisymmetry. For such a transform the magnitude is still an even function of ω as suggested by Eqn. (3.193). The phase is neither even nor odd. The real part is an odd function of ω and the imaginary part is an even function.

Example 3.24: Symmetry properties for the transform of right-sided exponential signal

The Fourier transform of the right-sided exponential signal $x(t) = e^{-at}\,u(t)$ was found in Example 3.15 to be

$$X(\omega) = \mathcal{F}\left\{e^{-at}\,u(t)\right\} = \frac{1}{a+j\omega} \tag{3.197}$$

Elaborate on the symmetry properties of the transform.

Solution: Since the signal $x(t)$ is real-valued, its transform must be conjugate-symmetric. The complex conjugate of the Fourier transform is

$$X^*(\omega) = \left(\frac{1}{a+j\omega}\right)^* = \frac{1}{a-j\omega}$$

and the folded version of the transform, obtained by replacing ω with $-\omega$ in Eqn. (3.197), is

$$X(-\omega) = \left(\frac{1}{a+j\omega}\right)\Bigg|_{\omega\to-\omega} = \frac{1}{a-j\omega}$$

Thus we verify that $X^*(\omega) = X(-\omega)$ and the transform is conjugate symmetric. The magnitude and the phase of the transform are

$$|X(\omega)| = \left|\frac{1}{a+j\omega}\right| = \frac{1}{\sqrt{a^2+\omega^2}}$$

and

$$\Theta(\omega) = \measuredangle X(\omega) = -\tan^{-1}\left(\frac{\omega}{a}\right)$$

respectively. We observe that the magnitude is an even function of frequency, that is,

$$\left|X(-\omega)\right| = \frac{1}{\sqrt{a^2+(-\omega)^2}} = \frac{1}{\sqrt{a^2+\omega^2}} = \left|X(\omega)\right|$$

consistent with the conclusion reached in Eqn. (3.187). On the other hand, the phase has odd symmetry since

$$\Theta(-\omega) = -\tan^{-1}\left(\frac{-\omega}{a}\right) = \tan^{-1}\left(\frac{\omega}{a}\right) = -\Theta(\omega)$$

consistent with Eqn. (3.188). By multiplying both the numerator and the denominator of the transform in Eqn. (3.197) by $(a - j\omega)$ we have

$$X(\omega) = \frac{(1)}{(a + j\omega)} \frac{(a - j\omega)}{(a - j\omega)} = \frac{a}{a^2 + \omega^2} - j \frac{\omega}{a^2 + \omega^2}$$

Real and imaginary parts of the transform can be found as

$$X_r(\omega) = \text{Re}\left\{X(\omega)\right\} = \frac{a}{a^2 + \omega^2}$$

and

$$X_i(\omega) = \text{Im}\left\{X(\omega)\right\} = \frac{-\omega}{a^2 + \omega^2}$$

It is a trivial matter to show that $X_r(\omega)$ is an even function of ω, and $X_i(\omega)$ is an odd function.

Software resources: `exct_3_24a.m`, `exct_3_24b.m`

Transforms of even and odd signals

If the real-valued signal $x(t)$ is an even function of time, the resulting transform $X(\omega)$ is real-valued for all ω.

Transform of a signal with even symmetry:

$$x(-t) = x(t), \text{ all } t \quad \text{implies that} \quad \text{Im}\left\{X(\omega)\right\} = 0, \text{ all } \omega \tag{3.198}$$

Proof of Eqn. (3.198):

If the imaginary part of the transform $X(\omega)$ is zero, then the transform must remain unchanged when it is conjugated. Thus, another way to express Eqn. (3.198) is by

$$X^*(\omega) = X(\omega) \tag{3.199}$$

This is the relationship we will prove. Let us conjugate both sides of the Fourier transform integral in Eqn. (3.122) to obtain

$$X^*(\omega) = \left[\int_{-\infty}^{\infty} x(t)\, e^{-j\omega t}\, dt\right]^* = \int_{-\infty}^{\infty} x^*(t)\, e^{j\omega t}\, dt = \int_{-\infty}^{\infty} x(t)\, e^{j\omega t}\, dt \tag{3.200}$$

where we have used the fact that $x^*(t) = x(t)$ for a real-valued signal. Changing the integration variable from t to $-\lambda$, recognizing that $dt = -d\lambda$, and adjusting the integration limits accordingly we obtain

$$X^*(\omega) = -\int_{\infty}^{-\infty} x(-\lambda)\, e^{-j\omega\lambda}\, d\lambda \tag{3.201}$$

For a signal with even symmetry we have $x(-\lambda) = x(\lambda)$, and the integral in Eqn. (3.201) becomes

$$X^*(\omega) = -\int_{\infty}^{-\infty} x(\lambda)\, e^{-j\omega\lambda}\, d\lambda \tag{3.202}$$

Finally, swapping the two integration limits and negating the integral to compensate for it results in

$$X^*(\omega) = \int_{-\infty}^{\infty} x(\lambda)\, e^{-j\omega\lambda}\, d\lambda = X(\omega) \tag{3.203}$$

Conversely it can also be proven that, if the real-valued signal $x(t)$ has odd-symmetry, the resulting Fourier transform is purely imaginary. This can be mathematically stated as follows:

Transform of a signal with odd symmetry:

$$x(-t) = -x(t),\ \text{all } t \quad \text{implies that} \quad \text{Re}\left\{X(\omega)\right\} = 0,\ \text{all } \omega \tag{3.204}$$

Conjugating a purely imaginary transform is equivalent to negating it, so an alternative method of expressing the relationship in Eqn. (3.204) is

$$X^*(\omega) = -X(\omega) \tag{3.205}$$

The procedure for proving Eqn. (3.205) is very similar to the one given above for the transform of an even symmetric signal. (See Problem 3.27 at the end of this chapter.)

Example 3.25: Transform of a pulse with odd symmetry

Determine the Fourier transform of the signal

$$x(t) = \begin{cases} -1, & -1 < t < 0 \\ 1, & 0 < t < 1 \\ 0, & t < -1 \quad \text{or} \quad t > 1 \end{cases}$$

graphed in Fig. 3.62. Show that the transform is purely imaginary.

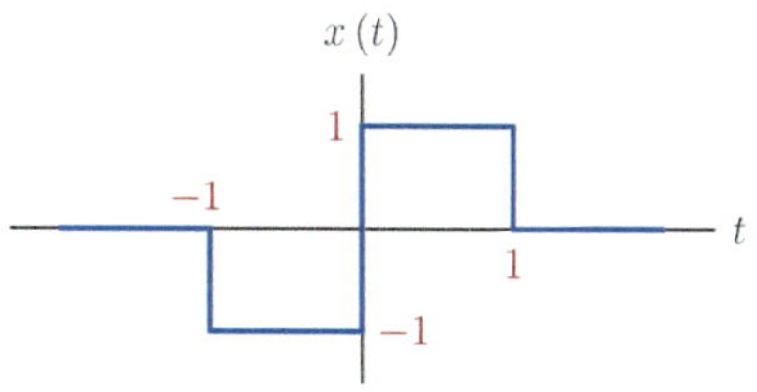

Figure 3.62 – The pulse of Example 3.25 with odd symmetry.

Solution: The transform can be found by direct use of the forward transform integral.

$$X(\omega) = \int_{-1}^{0} (-1)\, e^{-j\omega t}\, dt + \int_{0}^{1} (1)\, e^{-j\omega t}\, dt = \frac{j2}{\omega}\left[\cos(\omega) - 1\right]$$

As expected, the transform is purely imaginary. Its magnitude, phase and imaginary part are graphed in Fig. 3.63 using the frequency variable f instead of ω.

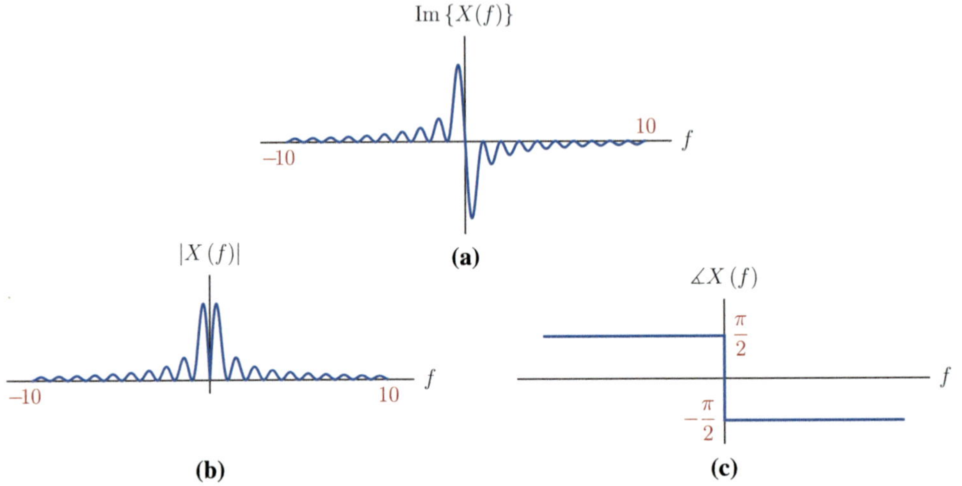

Figure 3.63 – The Fourier transform of the odd symmetric pulse of Example 3.25: **(a)** Imaginary part, **(b)** magnitude, and **(c)** phase.

Software resource: exct_3_25.m

Example 3.26: Transform of a two-sided exponential signal

The two-sided exponential signal $x(t) = e^{-a|t|}$ was shown in Fig. 3.48, and has even symmetry. Its Fourier transform was found in Example 3.16 as

$$X(\omega) = \frac{2a}{a^2 + \omega^2}$$

and was graphed in Fig. 3.49. As expected, the transform is real-valued for all ω.

Time shifting

Shifting the signal in the time domain is equivalent to multiplying the transform with a complex exponential factor. This relationship can be useful in certain types of problems where a time-domain signal is expressed in terms of time-shifted basis functions.

Time shifting the signal $x(t)$:

For a transform pair

$$x(t) \stackrel{\mathscr{F}}{\longleftrightarrow} X(\omega)$$

it can be shown that

$$x(t-\tau) \stackrel{\mathscr{F}}{\longleftrightarrow} X(\omega)\, e^{-j\omega\tau} \tag{3.206}$$

Proof of time shifting property:

Applying the Fourier transform integral in Eqn. (3.122) to $x(t-\tau)$ we obtain

$$\mathcal{F}\left\{x(t-\tau)\right\} = \int_{-\infty}^{\infty} x(t-\tau)\, e^{-j\omega t}\, dt \tag{3.207}$$

Let $\lambda = t - \tau$ in the integral of Eqn. (3.207) so that

$$\mathcal{F}\left\{x(t-\tau)\right\} = \int_{-\infty}^{\infty} x(\lambda)\, e^{-j\omega(\lambda+\tau)}\, d\lambda \tag{3.208}$$

The exponential function in the integral of Eqn. (3.208) can be written as a product of two exponential functions to obtain

$$\begin{aligned}\mathcal{F}\left\{x(t-\tau)\right\} &= \int_{-\infty}^{\infty} x(\lambda)\, e^{-j\omega\lambda}\, e^{-j\omega\tau}\, d\lambda \\ &= e^{-j\omega\tau} \int_{-\infty}^{\infty} x(\lambda)\, e^{-j\omega\lambda}\, d\lambda \\ &= e^{-j\omega\tau}\, X(\omega)\end{aligned} \tag{3.209}$$

Example 3.27: Time shifting a rectangular pulse

The transform of a rectangular pulse with amplitude A, width τ, and center at $t = 0$ was found in Example 3.12 as

$$A\Pi\left(\frac{t}{\tau}\right) \overset{\mathcal{F}}{\longleftrightarrow} A\tau\,\text{sinc}\left(\frac{\omega\tau}{2\pi}\right)$$

In Example 3.13 the transform of a time shifted version of this pulse was determined by direct application of the Fourier transform integral as

$$A\Pi\left(\frac{t-\tau/2}{\tau}\right) \overset{\mathcal{F}}{\longleftrightarrow} A\tau\,\text{sinc}\left(\frac{\omega\tau}{2\pi}\right) e^{-j\omega\tau/2}$$

The result in Eqn. (3.143) could easily have been obtained from the result in Eqn. (3.138) through the use of the time shifting property.

Example 3.28: Time shifting a two-sided exponential signal

Consider again the two-sided exponential signal used in earlier examples. A time shifted version of it is

$$x(t) = e^{-a|t-\tau|}$$

The signal $x(t)$ is shown in Fig 3.64 with the assumption that $a > 0$. In Example 3.16 its Fourier transform was determined to be

$$\mathcal{F}\left\{e^{-a|t|}\right\} = \frac{2a}{a^2+\omega^2}$$

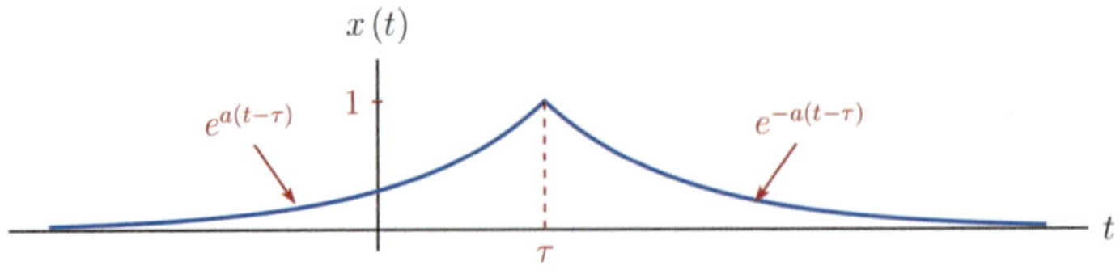

Figure 3.64 – The time shifted two-sided exponential signal of Example 3.28.

Using the time shifting property of the Fourier transform we obtain

$$X(\omega) = \mathscr{F}\left\{e^{-a|t-\tau|}\right\} = \frac{2a\, e^{-j\omega\tau}}{a^2+\omega^2}$$

The transform $X(\omega)$ is shown in Fig. 3.65.

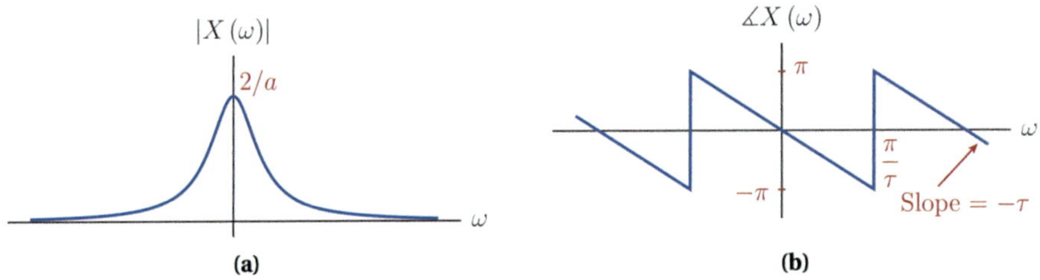

Figure 3.65 – Transform of the time shifted two-sided exponential signal of Example 3.28: **(a)** magnitude and **(b)** phase.

Compare this with the transform of the signal of Example 3.16. Note that time shifting the signal affects only the phase of the transform, and not its magnitude.

Software resource: exct_3_28.m

Frequency shifting

Shifting the Fourier transform in the frequency domain is equivalent to multiplication of the time-domain signal by a complex exponential factor. We will see that this property will also be practical to use in solving certain types of problems.

Frequency shifting $X(\omega)$:

For a transform pair

$$x(t) \stackrel{\mathscr{F}}{\longleftrightarrow} X(\omega)$$

it can be shown that

$$x(t)\, e^{j\omega_0 t} \stackrel{\mathscr{F}}{\longleftrightarrow} X(\omega-\omega_0) \tag{3.210}$$

Proof of frequency shifting property:

By directly applying the Fourier transform integral in Eqn. (3.122) to the signal $x(t)\, e^{j\omega_0 t}$ we obtain

$$\begin{aligned}\mathcal{F}\left\{x(t)\, e^{j\omega_0 t}\right\} &= \int_{-\infty}^{\infty} x(t)\, e^{j\omega_0 t}\, e^{-j\omega t}\, dt \\ &= \int_{-\infty}^{\infty} x(t)\, e^{-j(\omega-\omega_0)t}\, dt \\ &= X(\omega-\omega_0) \end{aligned} \tag{3.211}$$

An alternative proof can be obtained by using the duality principle in conjunction with the time shifting property. While it isn't necessarily shorter or simpler than the proof above, it nevertheless serves as another demonstration of the use of duality principle, and will therefore be given here. We start with the following two relationships:

$$x(t) \overset{\mathcal{F}}{\longleftrightarrow} X(\omega) \tag{3.212}$$

$$x(t-\tau) \overset{\mathcal{F}}{\longleftrightarrow} X(\omega)\, e^{-j\omega\tau} \tag{3.213}$$

If $x(t)$ and $X(\omega)$ are a valid transform pair in Eqn. (3.212), then it follows through the use of the time-shifting property that Eqn. (3.213) is also valid. Applying the duality principle to each of the transform pairs in Eqns. (3.212) and (3.213) we obtain the following two relationships:

$$X(t) \overset{\mathcal{F}}{\longleftrightarrow} 2\pi\, x(-\omega) \tag{3.214}$$

$$X(t)\, e^{-j\tau t} \overset{\mathcal{F}}{\longleftrightarrow} 2\pi\, x(-\omega-\tau) \tag{3.215}$$

Let $\tilde{x}(t) = X(t)$ and $\tilde{X}(\omega) = 2\pi\, x(-\omega)$. Also using the substitution $\omega_0 = -\tau$, the two transforms in Eqns. (3.214) and (3.215) can be rewritten as

$$\tilde{x}(t) \overset{\mathcal{F}}{\longleftrightarrow} \tilde{X}(\omega) \tag{3.216}$$

$$\tilde{x}(t)\, e^{j\omega_0 t} \overset{\mathcal{F}}{\longleftrightarrow} \tilde{X}(\omega-\omega_0) \tag{3.217}$$

Modulation property

Modulation property is an interesting consequence of the frequency shifting property combined with the linearity of the Fourier transform. As its name implies, it finds significant use in modulation techniques utilized in analog and digital communications. We will have the occasion to use it when we study amplitude modulation in Chapter 7.

Modulation property of the Fourier transform:

For a transform pair

$$x(t) \stackrel{\mathcal{F}}{\longleftrightarrow} X(\omega)$$

it can be shown that

$$x(t)\cos(\omega_0 t) \stackrel{\mathcal{F}}{\longleftrightarrow} \frac{1}{2}\left[X(\omega-\omega_0)+X(\omega+\omega_0)\right] \tag{3.218}$$

and

$$x(t)\sin(\omega_0 t) \stackrel{\mathcal{F}}{\longleftrightarrow} \frac{1}{2}\left[X(\omega-\omega_0)\,e^{-j\pi/2}+X(\omega+\omega_0)\,e^{j\pi/2}\right] \tag{3.219}$$

Multiplication of a signal by a cosine waveform causes its spectrum to be shifted in both directions by the frequency of the cosine waveform, and to be scaled by $\frac{1}{2}$. Multiplication of the signal by a sine waveform causes a similar effect with an added phase shift of $-\pi/2$ radians for positive frequencies and $\pi/2$ radians for negative frequencies.

Proof of modulation property:

Using Euler's formula, the left side of the relationship in Eqn. (3.218) can be written as

$$x(t)\cos(\omega_0 t) = \frac{1}{2}\,x(t)\,e^{\omega_0 t}+\frac{1}{2}\,x(t)\,e^{-\omega_0 t} \tag{3.220}$$

The desired proof is obtained by applying the frequency shifting property to the terms on the right side of Eqn. (3.220). Using Eqn. (3.210) we obtain

$$x(t)\,e^{j\omega_0 t} \stackrel{\mathcal{F}}{\longleftrightarrow} X(\omega-\omega_0) \tag{3.221}$$

and

$$x(t)\,e^{-j\omega_0 t} \stackrel{\mathcal{F}}{\longleftrightarrow} X(\omega+\omega_0) \tag{3.222}$$

which could be used together in Eqn. (3.220) to arrive at the result in Eqn. (3.218).

The proof of Eqn. (3.219) is similar, but requires one additional step. Again using Euler's formula, let us write the left side of Eqn. (3.219) as

$$x(t)\sin(\omega_0 t) = \frac{1}{2j}\,x(t)\,e^{j\omega_0 t}-\frac{1}{2j}\,x(t)\,e^{-j\omega_0 t} \tag{3.223}$$

Realizing that

$$\frac{1}{j} = -j = e^{-j\pi/2} \quad \text{and} \quad \frac{-1}{j} = j = e^{j\pi/2}$$

Eqn. (3.223) can be rewritten as

$$x(t)\sin(\omega_0 t) = \frac{1}{2}\,x(t)\,e^{j\omega_0 t}\,e^{-j\pi/2}+\frac{1}{2}\,x(t)\,e^{-j\omega_0 t}\,e^{j\pi/2} \tag{3.224}$$

The proof of Eqn. (3.219) can be completed by using Eqns. (3.221) and (3.222) on the right side of Eqn. (3.224).

Example 3.29: Modulated pulse

Find the Fourier transform of the modulated pulse given by

$$x(t) = \begin{cases} \cos(2\pi f_0 t), & |t| < \tau \\ 0, & |t| > \tau \end{cases}$$

Solution: Let $p(t)$ be defined as a rectangular pulse centered around the time origin with a pulse width of 2τ and a unit amplitude.

$$p(t) = \Pi\left(\frac{t}{2\tau}\right)$$

Using $p(t)$, the signal $x(t)$ can be expressed in a more compact form as

$$x(t) = p(t)\cos(2\pi f_0 t)$$

This is illustrated in Fig. 3.66.

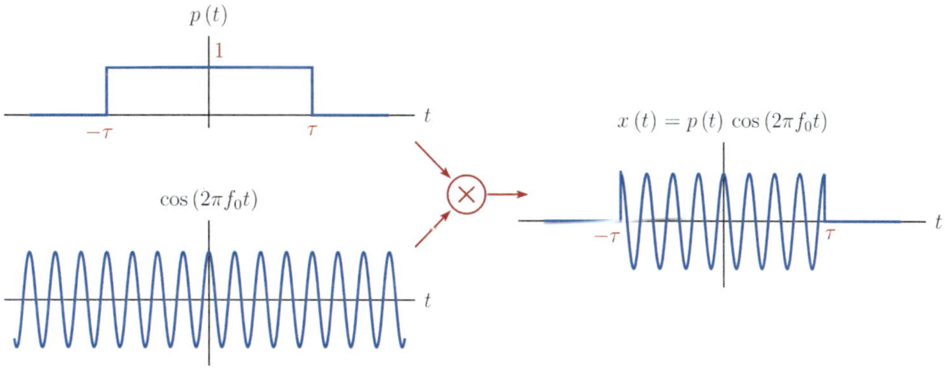

Figure 3.66 – Expressing the modulated pulse signal of Example 3.29 as the product of a rectangular pulse and a sinusoidal signal.

The transform of the pulse $p(t)$ is

$$P(f) = \mathscr{F}\{p(t)\} = 2\tau\,\text{sinc}(2\tau f)$$

We will apply the modulation property to this result to obtain the transform we seek:

$$\begin{aligned} X(f) &= \frac{1}{2}P(f-f_0) + \frac{1}{2}P(f+f_0) \\ &= \tau\,\text{sinc}\big(2\tau(f+f_0)\big) + \tau\,\text{sinc}\big(2\tau(f-f_0)\big) \end{aligned}$$

Construction of this spectrum is illustrated in Fig. 3.67.

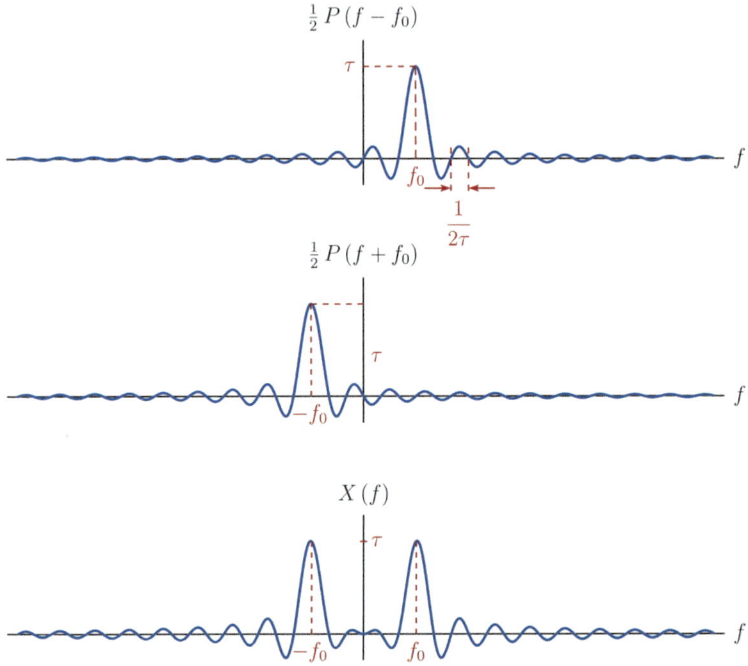

Figure 3.67 – Transform of the modulated pulse of Example 3.29 obtained by adding frequency-shifted versions of $P(\omega)$.

Software resources: exct_3_29.m , exct_3_29_Live.mlx

Interactive App: Modulated pulse

The interactive app in appFT8.m is based on Example 3.29. It allows experimentation with modulated pulse signal and its Fourier transform. Pulse width parameter τ and the frequency f_0 may be varied using slider controls. Both the modulated pulse signal and its transform are graphed.

- **a.** Start with $\tau = 1$ s and $f_0 = 5$ Hz. Observe the locations of the peaks of the transform $X(f)$.
- **b.** Gradually increase the frequency f_0. Pay attention to the peaks of the transform. How does the frequency f_0 affect the shape of the spectrum?
- **c.** Set $f_0 = 5$ Hz again. Gradually increase the pulse width parameter τ. What is the effect of changing the pulse width on the spectrum?

Software resource: appFT8.m

Time and frequency scaling

This property is significant in signal processing. If the time variable is scaled by a factor a, the frequency variable is scaled by the same factor, but in the opposite direction.

Time and frequency scaling property of the Fourier transform:

For a transform pair

$$x(t) \overset{\mathcal{F}}{\longleftrightarrow} X(\omega)$$

and any non-zero and real-valued constant a, it can be shown that

$$x(at) \overset{\mathcal{F}}{\longleftrightarrow} \frac{1}{|a|} X\left(\frac{\omega}{a}\right) \tag{3.225}$$

Proof of time and frequency scaling property:

The Fourier transform of $x(at)$ is

$$\mathcal{F}\{x(at)\} = \int_{-\infty}^{\infty} x(at)\, e^{-j\omega t}\, dt \tag{3.226}$$

A new independent variable λ will be introduced through the variable change $at = \lambda$. Let us substitute

$$t = \frac{\lambda}{a} \quad \text{and} \quad dt = \frac{d\lambda}{a} \tag{3.227}$$

in the integral of Eqn. (3.227). We need to consider the cases of $a > 0$ and $a < 0$ separately. If $a > 0$, then $t \to \pm\infty$ implies $\lambda \to \pm\infty$. Therefore, the limits of the integral remain unchanged under the variable change and we have

$$\begin{aligned}\mathcal{F}\{x(at)\} &= \frac{1}{a} \int_{-\infty}^{\infty} x(\lambda)\, e^{-j\omega\lambda/a}\, d\lambda \\ &= \frac{1}{a} X\left(\frac{\omega}{a}\right), \quad a > 0\end{aligned} \tag{3.228}$$

If $a < 0$, however, the same substitution leads to

$$\mathcal{F}\{x(at)\} = \frac{1}{a} \int_{\infty}^{-\infty} x(\lambda)\, e^{-j\omega\lambda/a}\, d\lambda \tag{3.229}$$

since $t \to \pm\infty$ now leads to $\lambda \to \mp\infty$. Swapping the lower and upper limits of the integral and negating the result to compensate for it yields

$$\begin{aligned}\mathcal{F}\{x(\lambda)\} &= -\frac{1}{a} \int_{-\infty}^{\infty} x(\lambda)\, e^{-j\omega\lambda/a}\, d\lambda \\ &= -\frac{1}{a} X\left(\frac{\omega}{a}\right), \quad a < 0\end{aligned} \tag{3.230}$$

It is possible to combine Eqns. (3.228) and (3.230) into one compact expression by letting the scale factor in front of the integral be expressed as $1/|a|$ so that it is valid for both

$a > 0$ and $a < 0$. Thus we obtain the desired proof:

$$\mathscr{F}\{x(at)\} = \frac{1}{|a|} X\left(\frac{\omega}{a}\right) \tag{3.231}$$

Example 3.30: Working with scaled and shifted pulses

Determine the Fourier transform of the signal

$$x(t) = \begin{cases} -1, & -0.5 < t < 1.5 \\ 0.75, & 1.5 < t < 5.5 \end{cases}$$

which is shown in Fig 3.68.

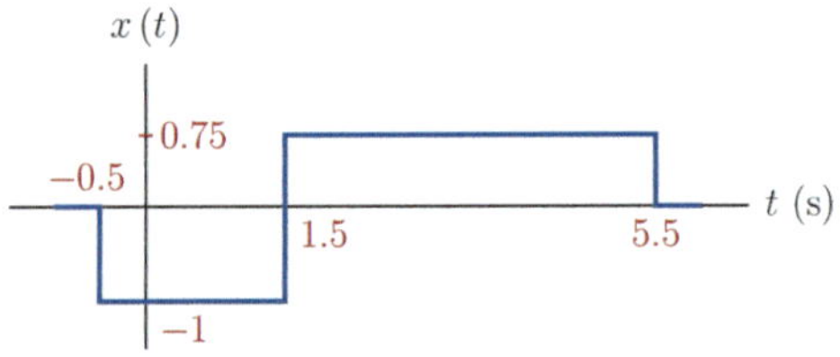

Figure 3.68 – The signal $x(t)$ of Example 3.30.

Solution: It certainly is possible to use the Fourier transform integral of Eqn. (3.122) directly on the signal $x(t)$. It would be more convenient, however, to make use of the properties of the Fourier transform. The signal $x(t)$ can be written as the sum of two rectangular pulses:

1. One with amplitude of $A_1 = -1$, width $\tau_1 = 2$ s, center at $t_{d1} = 0.5$ s
2. One with amplitude of $A_2 = 0.75$, width $\tau_2 = 4$ s, center at $t_{d2} = 3.5$ s

Using the unit pulse function $\Pi(t)$ we have

$$\begin{aligned} x(t) =& A_1 \Pi\left(\frac{t - t_{d1}}{\tau_1}\right) + A_2 \Pi\left(\frac{t - t_{d2}}{\tau_2}\right) \\ =& -\Pi\left(\frac{t - 0.5}{2}\right) + 0.75\,\Pi\left(\frac{t - 3.5}{4}\right) \end{aligned}$$

The Fourier transform of the unit pulse function is

$$\mathscr{F}\{\Pi(t)\} = \text{sinc}\left(\frac{\omega}{2\pi}\right)$$

Using the scaling property of the Fourier transform, the following two relationships can be written:

$$\Pi\left(\frac{t}{2}\right) \overset{\mathscr{F}}{\longleftrightarrow} 2\,\text{sinc}\left(\frac{\omega}{\pi}\right)$$

$$\Pi\left(\frac{t}{4}\right) \overset{\mathscr{F}}{\longleftrightarrow} 4\,\text{sinc}\left(\frac{2\omega}{\pi}\right)$$

In addition, the use of the time shifting property leads to

$$\Pi\left(\frac{t-0.5}{2}\right) \overset{\mathscr{F}}{\longleftrightarrow} 2\,\text{sinc}\left(\frac{\omega}{\pi}\right) e^{-j0.5\omega} \tag{3.232}$$

$$\Pi\left(\frac{t-3.5}{4}\right) \overset{\mathscr{F}}{\longleftrightarrow} 4\,\text{sinc}\left(\frac{2\omega}{\pi}\right) e^{-j3.5\omega} \tag{3.233}$$

Using Eqns. (3.232) and (3.233) the Fourier transform of $x(t)$ is

$$X(\omega) = -2\,\text{sinc}\left(\frac{\omega}{\pi}\right) e^{-j0.5\omega} + 3\,\text{sinc}\left(\frac{2\omega}{\pi}\right) e^{-j3.5\omega}$$

which is graphed in Fig. 3.69.

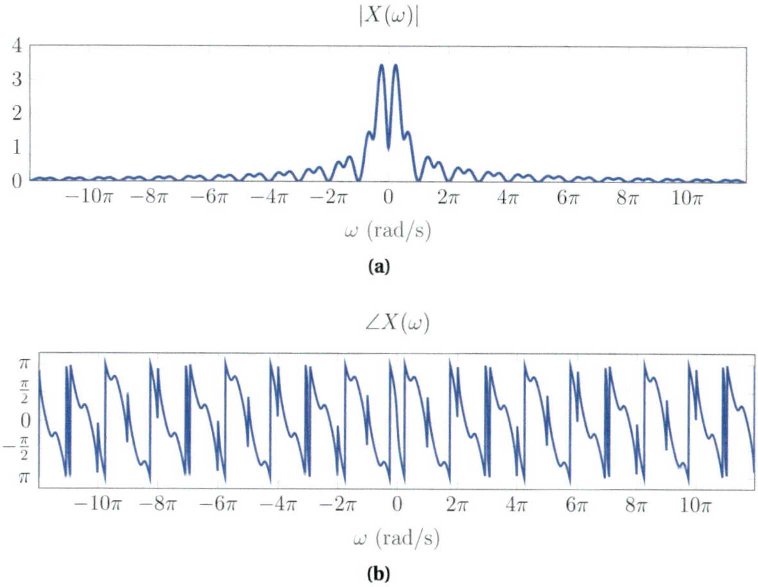

Figure 3.69 – Magnitude and phase of the spectrum of the signal $x(t)$ of Example 3.30.

Software resource: exct_3_30.m

Interactive App: Working with scaled and shifted pulses

The interactive app in appFT9.m is based on Example 3.30. It illustrates the use of time scaling and time shifting properties of the Fourier transform for computing the transform of a signal that can be expressed as the sum of two rectangular pulses. The signal to be constructed is in the form

$$x(t) = A_1\,\Pi\left(\frac{t-t_{d1}}{\tau_1}\right) + A_2\,\Pi\left(\frac{t-t_{d2}}{\tau_2}\right)$$

Parameters A_1, A_2, τ_1 τ_2, t_{d1}, and t_{d2} can be varied using slider controls. The individual spectrum of each pulse is displayed along with the spectrum of the signal $x(t)$.

a. Adjust parameters of the app to create the signal $x(t)$ of Example 3.30. Observe the magnitude and the phase of the spectrum.

b. Adjust the parameters to create the signal of Example 3.25 shown in Fig. 3.62. Compare the spectrum to that found and graphed in Fig. 3.63.

c. Shift the signal set in part (b) by $t_d = 1$ Hz. This can be accomplished by modifying parameters τ_1 and τ_2. Observe the frequency spectrum. Does it make sense in light of time shifting property of the Fourier transform?

d. Starting with the signal $x(t)$ created in part (b) above, construct the signal $x(t/2)$ by adjusting τ_1, τ_2, t_{d1}, and t_{d2} as needed. Observe the spectrum, and explain what is happening in the context of time and frequency scaling property of the Fourier transform.

Software resource: `appFT9.m`

Differentiation in the time domain

Differentiating the signal in the time domain is equivalent to multiplying its Fourier transform by a factor of $j\omega$. This property will be useful in determining the Fourier transform of each term in a differential equation.

Differentiation in time property of the Fourier transform:

For a given transform pair

$$x(t) \overset{\mathcal{F}}{\longleftrightarrow} X(\omega)$$

it can be shown that

$$\frac{d^n}{dt^n}\left[x(t)\right] \overset{\mathcal{F}}{\longleftrightarrow} \left(j\omega\right)^n X(\omega) \tag{3.234}$$

If we choose to use f instead or ω, then

$$\frac{d^n}{dt^n}\left[x(t)\right] \overset{\mathcal{F}}{\longleftrightarrow} \left(j2\pi f\right)^n X(f) \tag{3.235}$$

Proof of differentiation in the time domain property:

We will first prove the relationship in Eqn. (3.234) for $n = 1$. Once this is done, proof for $n > 1$ is easily obtained by repeating the process. The proof for Eqn. (3.235) is very similar, and will be left as an exercise (see Problem 3.30 at the end of this chapter).

Consider the inverse Fourier transform relationship given by Eqn. (3.121). Differentiating both sides of it with respect to t yields

$$\frac{d}{dt}\left[x(t)\right] = \frac{d}{dt}\left[\frac{1}{2\pi}\int_{-\infty}^{\infty} X(\omega)\, e^{j\omega t}\, d\omega\right] \tag{3.236}$$

Swapping the order of differentiation and integration on the right side of Eqn. (3.236) we can write

$$\begin{aligned}\frac{d}{dt}[x(t)] =& \frac{1}{2\pi}\int_{-\infty}^{\infty}\frac{d}{dt}\left[X(\omega)\,e^{j\omega t}\right]d\omega\\ =& \frac{1}{2\pi}\int_{-\infty}^{\infty}\left[j\omega\,X(\omega)\right]e^{j\omega t}\,d\omega\\ =& \mathscr{F}^{-1}\left\{j\omega\,X(\omega)\right\}\end{aligned} \tag{3.237}$$

which leads us to the conclusion

$$\frac{d}{dt}\left[x(t)\right] \overset{\mathscr{F}}{\longleftrightarrow} j\omega\,X(\omega) \tag{3.238}$$

Thus, differentiation of the signal with respect to the time variable corresponds to multiplication of the transform with $j\omega$. Differentiating both sides of Eqn. (3.237) one more time yields

$$\begin{aligned}\frac{d}{dt}\left[\frac{d}{dt}[x(t)]\right] =& \frac{d}{dt}\left[\frac{1}{2\pi}\int_{-\infty}^{\infty}j\omega\,X(\omega)\,e^{j\omega t}\,df\right]\\ =& \frac{1}{2\pi}\int_{-\infty}^{\infty}\frac{d}{dt}\left[j\omega\,X(\omega)\,e^{j\omega t}\right]df\\ =& \frac{1}{2\pi}\int_{-\infty}^{\infty}\left[(j\omega)^2\,X(\omega)\right]e^{j\omega t}\,df\\ =& \mathscr{F}^{-1}\left\{(j\omega)^2\,X(\omega)\right\}\end{aligned} \tag{3.239}$$

or in compact form

$$\frac{d^2}{dt^2}[x(t)] \overset{\mathscr{F}}{\longleftrightarrow} (j\omega)^2\,X(\omega) \tag{3.240}$$

It is obvious that each differentiation of the signal $x(t)$ will bring up another $(j\omega)$ factor in the corresponding transform. Differentiating the signal $x(t)$ for n times corresponds to multiplying the transform by $(j\omega)^n$.

Example 3.31: Triangular pulse revisited

Recall that the Fourier transform of the triangular pulse signal

$$x(t) = A\,\Lambda\left(\frac{t}{\tau}\right)$$

was found in Example 3.17 through direct application of the Fourier transform integral of Eqn. (3.122). Rework the problem using the differentiation-in-time property.

Solution: Instead of working with the signal $x(t)$ directly, we will find it convenient to work with its derivative

$$w(t) = \frac{dx(t)}{dt}$$

shown in Fig. 3.70.

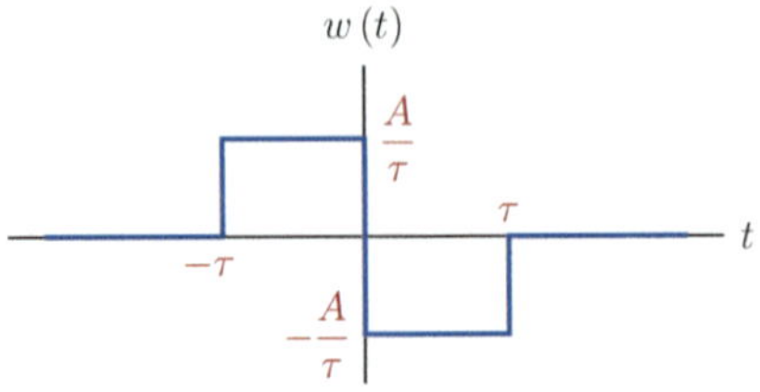

Figure 3.70 – The intermediate signal $w(t)$ for Example 3.31.

Using the unit pulse function along with time scaling and time shifting operations, $w(t)$ can be expressed as

$$w(t) = \frac{A}{\tau}\left[\Pi\left(\frac{t+\tau/2}{\tau}\right) - \Pi\left(\frac{t-\tau/2}{\tau}\right)\right] \tag{3.241}$$

Using the variable f instead of ω for convenience, the transform $W(f)$ can be computed from Eqn. (3.241) as

$$\begin{aligned}
W(f) =& A\,\mathrm{sinc}(f\tau)\, e^{j2\pi f(\tau/2)} - A\,\mathrm{sinc}(f\tau)\, e^{-j2\pi f(\tau/2)} \\
=& A\,\mathrm{sinc}(f\tau)\left[e^{j2\pi f(\tau/2)} - e^{-j2\pi f(\tau/2)}\right] \\
=& 2j\,A\,\mathrm{sinc}(f\tau)\,\sin(\pi f\tau)
\end{aligned} \tag{3.242}$$

The real part of $W(f)$ is equal to zero; its imaginary part is shown in Fig. 3.71. Frequency-domain equivalent of the relationship in Eqn. (3.242) is

$$W(f) = (j2\pi f)\, X(f) \tag{3.243}$$

Solving Eqn. (3.243) for $X(f)$ yields

$$X(f) = \frac{W(f)}{j2\pi f} = A\tau\,\mathrm{sinc}^2(f\tau)$$

which is in agreement with the answer found in Example 3.17.

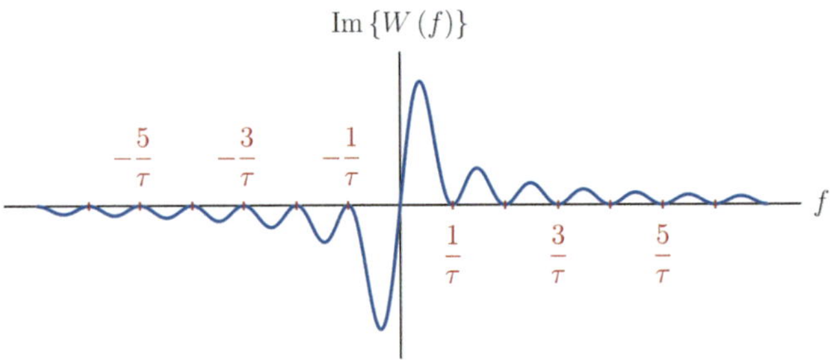

Figure 3.71 – Spectrum of the intermediate signal $w(t)$ for Example 3.31.

Example 3.32: Fourier transform of a trapezoidal pulse

Determine the Fourier transform of the trapezoidal pulse $x(t)$ shown in Fig. 3.72.

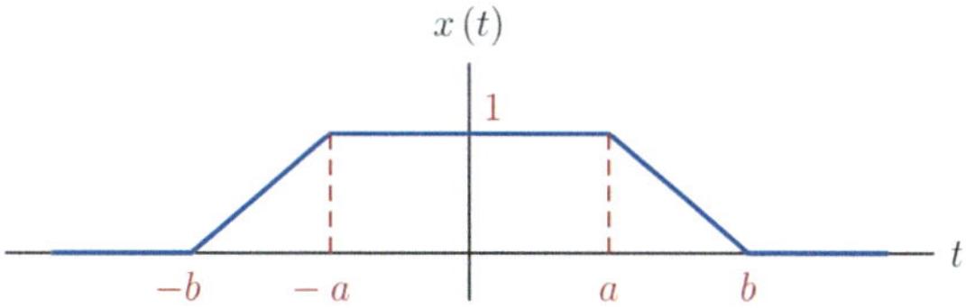

Figure 3.72 – Trapezoidal pulse of Example 3.32.

Solution: Analytical description of this signal could be given in the form

$$x(t) = \begin{cases} 0\,, & t < -b \\ (t+b)\,/\,(b-a)\,, & -b \le t < -a \\ 1\,, & -a \le t < a \\ (-t+b)\,/\,(b-a)\,, & a \le t < b \\ 0\,, & t > b \end{cases}$$

Once again we will use the intermediate signal $w(t)$ obtained by differentiating $x(t)$ with respect to t:

$$w(t) = \frac{dx(t)}{dt} = \begin{cases} 1/\,(b-a)\,, & -b \le t < -a \\ -1/\,(b-a)\,, & a \le t < b \\ 0\,, & \text{otherwise} \end{cases}$$

The signal $w(t)$ is shown in Fig. 3.73. Finding the Fourier transform of $w(t)$ is much easier compared to a direct evaluation of the Fourier transform integral for $x(t)$. For notational convenience, let two new parameters be defined as

$$\tau = b - a \quad \text{and} \quad \lambda = \frac{a+b}{2} \tag{3.244}$$

The parameter τ is the width of each of the two pulses that make up $w(t)$, and these pulses are centered at $\pm\lambda$. The derivative $w(t)$ can be written using the unit pulse function as

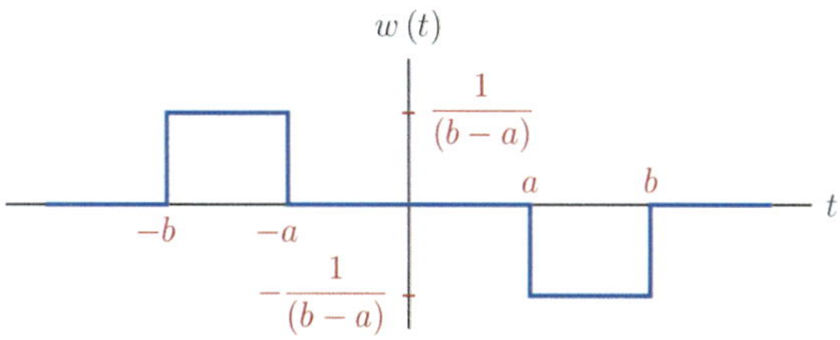

Figure 3.73 – The intermediate signal $w(t)$ for Example 3.32.

$$w(t) = \frac{1}{\tau}\,\Pi\left(\frac{t+\lambda}{\tau}\right) - \frac{1}{\tau}\,\Pi\left(\frac{t-\lambda}{\tau}\right)$$

and the corresponding transform $W(f)$ is

$$\begin{aligned} W(f) &= \mathrm{sinc}(f\tau)\, e^{j2\pi f\lambda} - \mathrm{sinc}(f\tau)\, e^{-j2\pi f\lambda} \\ &= \mathrm{sinc}(f\tau)\left[e^{j2\pi f\lambda} - e^{-j2\pi f\lambda}\right] \\ &= 2j\, \mathrm{sinc}(f\tau)\, \sin(2\pi f\lambda) \end{aligned} \tag{3.245}$$

The transform of $x(t)$ can now be found by scaling this result:

$$X(f) = \frac{W(f)}{j2\pi f} = 2\lambda\, \mathrm{sinc}(f\tau)\, \mathrm{sinc}(2f\lambda) \tag{3.246}$$

The transform $X(f)$ is graphed in Fig. 3.74 for various values of the parameters τ and λ.

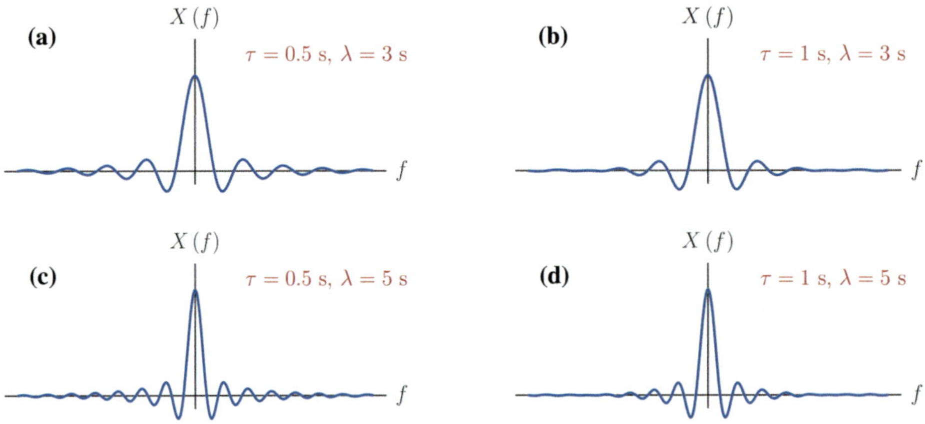

Figure 3.74 – The transform $X(f)$ for Example 3.32.

Having found the solution for $X(f)$ we will speculate on the differentiation in time property a bit further: What if we had differentiated $w(t)$ one more time to obtain a new signal $v(t)$?

$$v(t) = \frac{dw(t)}{dt} = \frac{d^2x(t)}{dt^2}$$

The signal $v(t)$ consists of some shifted and scaled impulses, and is shown in Fig. 3.75. The transform of $v(t)$ would be related to the transform of $x(t)$ by

$$X(f) = \frac{V(f)}{(j2\pi f)^2} = -\frac{V(f)}{4\pi^2 f^2} \tag{3.247}$$

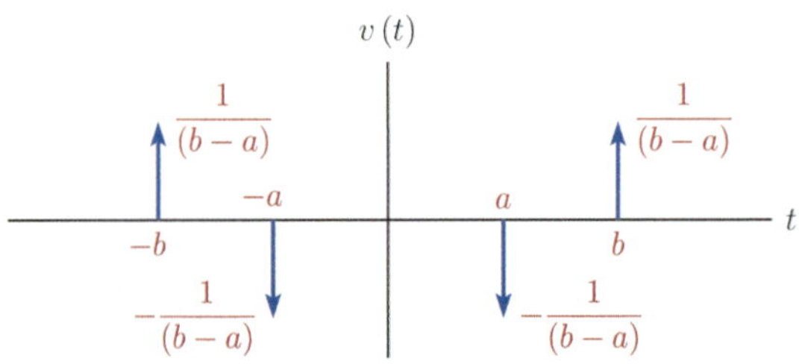

Figure 3.75 – The intermediate signal $v(t)$ for Example 3.32.

Analytically, $v(t)$ can be expressed as

$$v(t) = \frac{1}{b-a}\,\delta(t+b) - \frac{1}{b-a}\,\delta(t+a) - \frac{1}{b-a}\,\delta(t-a) + \frac{1}{b-a}\,\delta(t-b)$$

Knowing that $\mathcal{F}\{\delta(t)\} = 1$, and utilizing the time-shifting property of the Fourier transform, $V(f)$ is found to be

$$\begin{aligned} V(f) =& \frac{1}{b-a}\,e^{j2\pi f b} - \frac{1}{b-a}\,e^{j2\pi f a} - \frac{1}{b-a}\,e^{-j2\pi f a} + \frac{1}{b-a}\,e^{-j2\pi f b} \\ =& \frac{2}{b-a}\left[\cos(2\pi f b) - \cos(2\pi f a)\right] \end{aligned} \tag{3.248}$$

Based on Eqn. (3.244), we have

$$a = \lambda - \frac{\tau}{2} \quad \text{and} \quad b = \lambda + \frac{\tau}{2}$$

Substituting a and b into Eqn. (3.248) we obtain

$$V(f) = \frac{2}{\tau}\left[\cos\left(2\pi f\left(\lambda + \frac{\tau}{2}\right)\right) - \cos\left(2\pi f\left(\lambda - \frac{\tau}{2}\right)\right)\right] \tag{3.249}$$

Using the appropriate trigonometric identity, Eqn. (3.249) becomes

$$V(f) = -\frac{4}{\tau}\sin(\pi f \tau)\sin(2\pi f \lambda)$$

and the transform $X(f)$ is obtained by using Eqn. (3.247) as

$$X(f) = \frac{1}{\pi^2 f^2 \tau}\sin(\pi f \tau)\sin(2\pi f \lambda) = 2\lambda\,\mathrm{sinc}(f\tau)\,\mathrm{sinc}(2f\lambda)$$

which matches the result found earlier.

Software resources: exct_3_32.m , exct_3_32_Live.mlx

Interactive App: Fourier transform of a trapezoidal pulse

The interactive app in `appFT10.m` illustrates the steps involved in Example 3.32 for finding the Fourier transform of a trapezoidal pulse through the use of the differentiation in time property. The signals $x(t)$ and $w(t)$ are displayed on the left side of the graphical user interface window. On the right, the transforms are displayed. The parameters τ and λ may be adjusted using the two slider controls provided. Based on Eqn. (3.245) the real part of $W(f)$ is zero, and therefore only its imaginary part is shown. On the other hand, the transform $X(f)$ is purely real as given by Eqn. (3.246). We graph the scaled spectrum $X(f)/\lambda$ so that the peak magnitude of the resulting graph is fixed, and we can concentrate on the placement of zero crossings as the parameters τ and λ are varied.

a. Adjust the parameters τ and λ to create each of the four spectra shown in Fig. 3.74. Observe the frequency spectrum and specifically the locations of zeros. Do they seem to be evenly spaced?

b. Slightly change the parameter τ and observe the changes in the frequency spectrum. Do the locations of zero crossings seem to be dependent on τ?

c. Now slightly change the parameter λ and observe the changes in the frequency spectrum. Do the locations of zero crossings seem to be dependent on λ? Can you see a relationship between the value of λ and the spacing of the zeros?

d. It is also interesting to note that, for $\lambda = \tau/2$, the trapezoidal pulse $x(t)$ becomes identical to the triangular pulse of Example 3.31. Set $\lambda = 0.5$ and $\tau = 1$. Compare the transforms $W\left(f\right)$ and $X\left(f\right)/\lambda$ with the results of Example 3.31.

Software resource: `appFT10.m`

Differentiation in the frequency domain

Differentiation in frequency property of the Fourier transform:

For a transform pair

$$x(t) \overset{\mathcal{F}}{\longleftrightarrow} X(\omega)$$

it can be shown that

$$\left(-jt\right)^n x(t) \overset{\mathcal{F}}{\longleftrightarrow} \frac{d^n}{d\omega^n}\left[X(\omega)\right] \tag{3.250}$$

If we choose to use f instead or ω, then

$$\left(-j2\pi t\right)^n x(t) \overset{\mathcal{F}}{\longleftrightarrow} \frac{d^n}{df^n}\left[X\left(f\right)\right] \tag{3.251}$$

Proof of differentiation in the frequency domain property:

The proof is straightforward and is similar to the proof of Eqn. (3.234). It will be carried out for Eqn. (3.250). The very similar proof of Eqn. (3.251) will be left as an exercise (see Problems 3.31 and 3.32 at the end of this chapter). Let us differentiate both sides of the Fourier transform integral in Eqn. (3.122) with respect to ω:

$$\frac{d}{d\omega}\left[X(\omega)\right] = \frac{d}{d\omega}\left[\int_{-\infty}^{\infty} x(t)\, e^{-j\omega t}\, dt\right] \tag{3.252}$$

Changing the order of integration and differentiation on the right side of Eqn. (3.252) yields

$$\begin{aligned}\frac{d}{d\omega}\left[X(\omega)\right] &= \int_{-\infty}^{\infty} \frac{d}{d\omega}\left[x(t)\, e^{-j\omega t}\right] dt \\ &= \int_{-\infty}^{\infty} \left(-jt\right) x(t)\, e^{-j\omega t}\, dt \\ &= \mathcal{F}\left\{\left(-jt\right) x(t)\right\} \end{aligned} \tag{3.253}$$

to prove Eqn. (3.250) for $n = 1$. Proof for $n > 1$ can easily be obtained by repeated use of this procedure.

Alternatively, we might see the similarity between the expressions of the differentiation properties in time and frequency domains, and prove the latter from the knowledge of the former through the use of the duality principle. Let's start with the following two relationships:

$$x(t) \overset{\mathscr{F}}{\longleftrightarrow} X(\omega) \tag{3.254}$$

$$\frac{d^n}{dt^n}\left[x(t)\right] \overset{\mathscr{F}}{\longleftrightarrow} (j\omega)^n\, X(\omega) \tag{3.255}$$

If $x(t)$ and $X(f)$ are a valid transform pair in Eqn. (3.254), then Eqn. (3.255) is also valid by the differentiation in time property. Application of the duality principle to each of the transform pairs in Eqns. (3.254) and (3.255) leads to the following two relationships:

$$X(t) \overset{\mathscr{F}}{\longleftrightarrow} 2\pi\, x(-\omega) \tag{3.256}$$

$$(jt)^n\, X(t) \overset{\mathscr{F}}{\longleftrightarrow} 2\pi \frac{d^n}{d(-\omega)^n}\left[x(-\omega)\right] \tag{3.257}$$

Let $\bar{x}(t) = X(t)$ and $\bar{X}(\omega) = 2\pi\, x(-\omega)$. The two transforms in Eqns. (3.256) and (3.257) can be rewritten as

$$\bar{x}(t) \overset{\mathscr{F}}{\longleftrightarrow} \bar{X}(\omega) \tag{3.258}$$

$$(jt)^n\, \bar{x}(t) \overset{\mathscr{F}}{\longleftrightarrow} \frac{d^n}{d(-\omega)^n}\left[\bar{X}(\omega)\right] \tag{3.259}$$

Finally, multiplication of both sides of the transform pair in Eqn. (3.259) by $(-1)^n$ yields

$$(-jt)^n\, \bar{x}(t) \overset{\mathscr{F}}{\longleftrightarrow} \frac{d^n}{d\omega^n}\left[\bar{X}(\omega)\right] \tag{3.260}$$

which completes the proof. If Eqn. (3.258) represents a valid transform pair, then so does Eqn. (3.260).

Example 3.33: Pulse with trapezoidal spectrum

Find the signal $x(t)$ the Fourier transform of which is the trapezoidal function given by

$$X(f) = \begin{cases} 1\,, & |f| \le f_0(1-r) \\ \dfrac{1}{2r}\left[-\dfrac{|f|}{f_0} + 1 + r\right], & f_0(1-r) < |f| \le f_0(1+r) \\ 0\,, & |f| > f_0(1+r) \end{cases}$$

and shown in Fig. 3.76. Real-valued parameter r is adjustable in the range $0 < r < 1$.

Solution: The answer can easily be found by applying the duality property to the result obtained in Eqn. (3.245) of Example 3.32 with the adjustments $a = f_0(1-r)$, $b = f_0(1+r)$, $\lambda = f_0$, and $\tau = 2f_0 r$. (See Problem 3.34 at the end of this chapter.) Instead of using that approach, however, we will use this example as an opportunity to apply the differentiation in frequency property. Let

$$W(f) = \frac{dX(f)}{df}$$

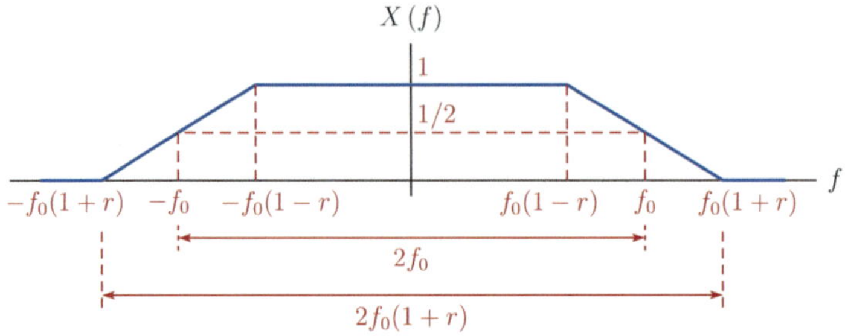

Figure 3.76 – The trapezoidal spectrum $X(f)$ for Example 3.33.

By carrying out the differentiation we obtain $W(f)$ as

$$W(f) = \frac{1}{2f_0\,r}\,\Pi\left(\frac{f+f_0}{2f_0\,r}\right) - \frac{1}{2f_0\,r}\,\Pi\left(\frac{f-f_0}{2f_0\,r}\right)$$

which is shown in Fig. 3.77. The Fourier transform of the sinc function is (see Example 3.19)

$$\mathcal{F}\left\{\text{sinc}(t)\right\} = \Pi(f)$$

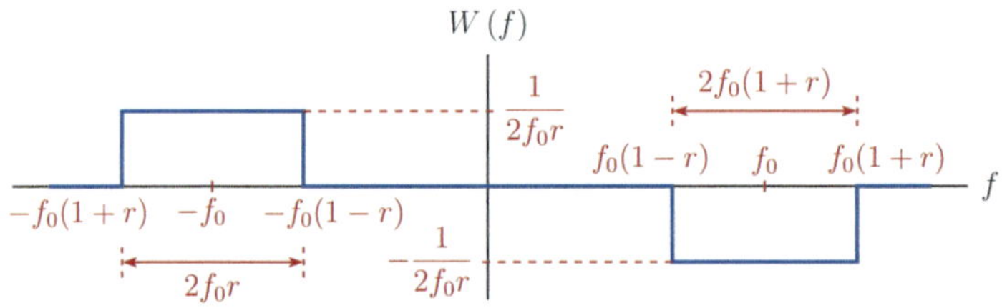

Figure 3.77 – The intermediate spectrum $W(f)$ for Example 3.33.

Using the time and frequency scaling property of the Fourier transform given by Eqn. (3.225) we obtain

$$\mathcal{F}\left\{\text{sinc}(2f_0rt)\right\} = \frac{1}{2f_0r}\,\Pi\left(\frac{f}{2f_0r}\right) \tag{3.261}$$

The two frequency-domain pulses that make up the transform $W(f)$ are simply frequency shifted and frequency scaled versions of the right side of Eqn. (3.261). Therefore, the inverse transform is

$$\begin{aligned} w(t) &= \text{sinc}\left(2f_0\,r\,t\right)\,e^{-j2\pi f_0 t} - \text{sinc}\left(2f_0\,r\,t\right)\,e^{j2\pi f_0 t} \\ &= \text{sinc}\left(2f_0\,r\,t\right)\left[e^{-j2\pi f_0 t} - e^{j2\pi f_0 t}\right] \\ &= -2j\,\sin\left(2\pi f_0 t\right)\,\text{sinc}\left(2f_0\,r\,t\right) \end{aligned}$$

Since $w(t) = -j2\pi t\,x(t)$, we can write $x(t)$ as

$$x(t) = 2f_0\,\text{sinc}\left(2f_0 t\right)\,\text{sinc}\left(2f_0\,r\,t\right) \tag{3.262}$$

The signal $x(t)$, shown in Fig. 3.78 for various values of the parameter r, exhibits some interesting properties:

1. Zero crossings of the signal occur at uniform time intervals of $1/(2f_0)$ regardless of the value of the parameter r which is referred to as the *roll-off factor*.
2. Larger values of r result in a reduction in the oscillatory behavior of the signal.

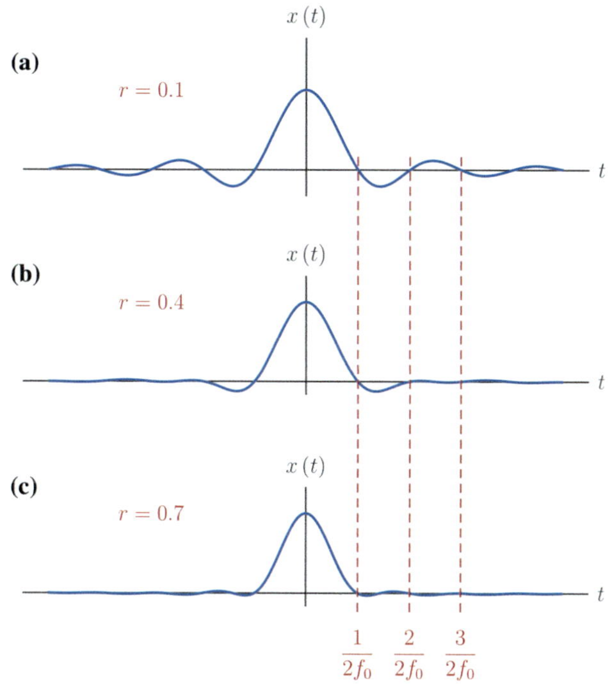

Figure 3.78 – The time-domain signal $x(t)$ with the trapezoidal spectrum $X(f)$ used in Example 3.33: **(a)** $r = 0.1$, **(b)** $r = 0.4$, and **(c)** $r = 0.7$.

These properties are very significant in communication systems in the context of *pulse shaping*. A signal with the properties listed above can be used for representing binary or M-ary symbols while eliminating the interference of those symbols with each other, a phenomenon known as *intersymbol interference (ISI)*. Harry Nyquist has shown that signals the frequency spectra of which have certain symmetry properties will exhibit uniformly spaced zero crossings. The trapezoidal spectrum shown in Fig. 3.76 is one example.

Software resources: `exct_3_33.m` , `exct_3_33_Live.mlx`

Interactive App: Pulse with trapezoidal spectrum

The interactive app in `appFT11.m` expands on the concepts explored in Example 3.33. The trapezoidal spectrum $X(f)$ with parameters f_0 and r and its corresponding time-domain signal $x(t)$ are shown. Values of parameters f_0 and r may be adjusted using slider controls. The zero crossings of the signal $x(t)$ are shown with marker lines. Specifically observe the following:

a. As the parameter f_0 is changed, the zero crossings of the signal coincide with integer multiples of $1/(2f_0)$. (You may need to zoom in to observe that zero crossings are always uniformly spaced.)

b. Vary the parameter r and observe its effect on the locations of zero crossings. Does it seem to have an effect?

c. The one-sided bandwidth of the trapezoidal spectrum is $f_0(1+r)$. If f_0 is fixed, bandwidth can be controlled by the parameter r. Observe the effect of increasing the bandwidth of the spectrum on the shape of the signal $x(t)$.

Software resource: `appFT11.m`

Convolution property

Convolution property of the Fourier transform will prove to be fundamental in establishing the significance of the *system function* concept in Section 3.5.

Convolution property of the Fourier transform:

For two transform pairs

$$x_1(t) \overset{\mathcal{F}}{\longleftrightarrow} X_1(\omega) \quad \text{and} \quad x_2(t) \overset{\mathcal{F}}{\longleftrightarrow} X_2(\omega)$$

it can be shown that

$$x_1(t) * x_2(t) \overset{\mathcal{F}}{\longleftrightarrow} X_1(\omega)\, X_2(\omega) \tag{3.263}$$

Proof of convolution property:

The proof is obtained by direct application of the Fourier transform integral in Eqn. (3.122) to the convolution of signals $x_1(t)$ and $x_2(t)$:

$$\mathcal{F}\left\{x_1(t) * x_2(t)\right\} = \int_{-\infty}^{\infty} \left[\int_{-\infty}^{\infty} x_1(\lambda)\, x_2(t-\lambda)\, d\lambda\right] e^{-j\omega t}\, dt \tag{3.264}$$

Interchanging the order of two integrals and rearranging terms we obtain

$$\begin{aligned}\mathcal{F}\left\{x_1(t) * x_2(t)\right\} &= \int_{-\infty}^{\infty} \left[\int_{-\infty}^{\infty} x_1(\lambda)\, x_2(t-\lambda)\, e^{-j\omega t}\, dt\right] d\lambda \\ &= \int_{-\infty}^{\infty} x_1(\lambda) \left[\int_{-\infty}^{\infty} x_2(t-\lambda)\, e^{-j\omega t}\, dt\right] d\lambda \end{aligned} \tag{3.265}$$

In Eqn. (3.265), we will recognize the expression in square brackets as the Fourier transform of the time shifted signal $x_2(t-\lambda)$. The use of time shifting property of the Fourier transform given by Eqn. (3.206) yields

$$\int_{-\infty}^{\infty} x_2(t-\lambda)\, e^{-j\omega t}\, dt = \mathcal{F}\left\{x_2(t-\lambda)\right\} = X_2(\omega)\, e^{-j\omega\lambda} \tag{3.266}$$

Substituting Eqn. (3.266) into Eqn. (3.265) we get

$$\begin{aligned}\mathcal{F}\left\{x_1(t) * x_2(t)\right\} &= \int_{-\infty}^{\infty} x_1(\lambda)\left[X_2(\omega)\, e^{-j\omega\lambda}\right] d\lambda \\ &= \left[\int_{-\infty}^{\infty} x_1(\lambda)\, e^{-j\omega\lambda}\, d\lambda\right] X_2(\omega) \\ &= X_1(\omega)\, X_2(\omega)\end{aligned} \tag{3.267}$$

Multiplication of two signals

Multiplication property of the Fourier transform:

For two transform pairs

$$x_1(t) \overset{\mathcal{F}}{\longleftrightarrow} X_1(\omega) \quad \text{and} \quad x_2(t) \overset{\mathcal{F}}{\longleftrightarrow} X_2(\omega)$$

it can be shown that

$$x_1(t)\, x_2(t) \overset{\mathcal{F}}{\longleftrightarrow} \frac{1}{2\pi} X_1(\omega) * X_2(\omega) \tag{3.268}$$

If we choose to use f instead of ω, then

$$x_1(t)\, x_2(t) \overset{\mathcal{F}}{\longleftrightarrow} X_1(f) * X_2(f) \tag{3.269}$$

Proof of multiplication property:

The right side of Eqn. (3.268) can be written as

$$\frac{1}{2\pi} X_1(\omega) * X_2(\omega) = \frac{1}{2\pi}\int_{-\infty}^{\infty} X_1(\lambda)\, X_2(\omega-\lambda)\, d\lambda \tag{3.270}$$

Applying the inverse Fourier transform to both sides of Eqn. (3.270) yields

$$\mathcal{F}^{-1}\left\{\frac{1}{2\pi} X_1(\omega) * X_2(\omega)\right\} = \frac{1}{2\pi}\int_{-\infty}^{\infty}\left[\frac{1}{2\pi}\int_{-\infty}^{\infty} X_1(\lambda)\, X_2(\omega-\lambda)\, d\lambda\right] e^{j\omega t}\, d\omega \tag{3.271}$$

Changing the order of the two integrals on the right side of Eqn. (3.271), and arranging terms, we obtain

$$\mathcal{F}^{-1}\left\{\frac{1}{2\pi} X_1(\omega) * X_2(\omega)\right\} = \frac{1}{2\pi}\int_{-\infty}^{\infty} X_1(\lambda)\left[\frac{1}{2\pi}\int_{-\infty}^{\infty} X_2(\omega-\lambda)\, e^{j\omega t}\, d\omega\right] d\lambda \tag{3.272}$$

Recall from the frequency shifting property of the Fourier transform, given by Eqn. (3.210), that $X_2(\omega-\lambda)$ is the Fourier transform of the signal $x_2(t)$ multiplied by a complex exponential. The expression in square brackets on the right side of Eqn. (3.272) represents the inverse Fourier transform of $X_2(\omega-\lambda)$, and evaluates to

$$\frac{1}{2\pi}\int_{-\infty}^{\infty} X_2(\omega-\lambda)\, e^{j\omega t}\, d\omega = x_2(t)\, e^{j\lambda t} \tag{3.273}$$

Substituting Eqn. (3.273) into Eqn. (3.272) and rearranging the terms on the right side, we have

$$\begin{aligned}\mathscr{F}^{-1}\left\{\frac{1}{2\pi}X_1(\omega)*X_2(\omega)\right\} =& \frac{1}{2\pi}\int_{-\infty}^{\infty} X_1(\lambda)\left[x_2(t)\,e^{j\lambda t}\right]d\lambda \\ =& x_2(t)\left[\frac{1}{2\pi}\int_{-\infty}^{\infty} X_1(\lambda)\,e^{j\lambda t}\,d\lambda\right] \\ =& x_1(t)\,x_2(t) \qquad (3.274)\end{aligned}$$

Thus we have proved the relationship in Eqn. (3.268).

Example 3.34: Transform of a truncated sinusoidal signal

A sinusoidal signal that is time-limited in the interval $-\tau < t < \tau$ is given by

$$x(t) = \begin{cases} \cos(2\pi f_0 t)\,, & -\tau < t < \tau \\ 0\,, & \text{otherwise} \end{cases}$$

Determine the Fourier transform of this signal using the multiplication property.

Solution: The signal $x(t)$ is the same as the modulated pulse signal used in Example 3.29 and shown in Fig. 3.66. In Example 3.29 the transform was determined using the modulation property. Let $x_1(t)$ and $x_2(t)$ be defined as

$$x_1(t) = \cos(2\pi f_0 t) \quad \text{and} \quad x_2(t) = \Pi\left(\frac{t}{2\tau}\right)$$

The signal $x(t)$ can be written as the product of these two signals:

$$x(t) = x_1(t)\,x_2(t) = \cos(2\pi f_0 t)\,\Pi\left(\frac{t}{2\tau}\right)$$

It will be more convenient to use f rather than ω in this case. Fourier transforms of the signals $x_1(t)$ and $x_2(t)$ are

$$X_1(f) = \frac{1}{2}\delta(f+f_0) + \frac{1}{2}\delta(f-f_0)$$

and

$$X_2(f) = 2\tau\,\text{sinc}(2\tau f)\,,$$

respectively. Using Eqn. (3.269), the Fourier transform of the product $x(t) = x_1(t)\,x_2(t)$ is

$$X(f) = X_1(f) * X_2(f) = \tau\,\text{sinc}\Big(2\tau(f+f_0)\Big) + \tau\,\text{sinc}\Big(2\tau(f-f_0)\Big)$$

which matches the result found in Example 3.29.

Software resources: exct_3_34.m , exct_3_34_Live.mlx

Interactive App: Transform of a truncated sinusoidal signal

The interactive app in `appFT12.m` is based on Example 3.34 and illustrates the multiplication property of the Fourier transform. On the left side of the screen, the signals $x_1(t)$ and $x_2(t)$ of Example 3.34 as well as their product $[x_1(t)\,x_2(t)]$ are graphed. The right side of the screen displays the corresponding spectra $X_1(f)$, $X_2(f)$, and $\left[X_1(f)*X_2(f)\right]$. Parameters τ and f_0 may be adjusted using slider controls.

a. Set $f_0=3$ Hz and $\tau=1$ Hz. Observe the spectra $X_1(f)$, $X_2(f)$, and $X(f)$. Pay attention to shapes of spectra and locations of peaks.

b. Gradually increase the frequency toward $f_0=5$ Hz. Observe the changes in the spectra $X_1(f)$ and $X(f)$. Explain the changes using the multiplication property of the Fourier transform. Afterward explain the same using the modulation property.

c. With f_0 set at 5 Hz, gradually increase the pulse width toward $\tau=2$ s. Observe the changes in the spectra $X_2(f)$ and $X(f)$. Explain using the multiplication property of the Fourier transform. Afterward explain the modulation property.

Software resource: `appFT12.m`

Integration

Integration property of the Fourier transform:

For a transform pair

$$x(t) \overset{\mathcal{F}}{\longleftrightarrow} X(\omega)$$

it can be shown that

$$\int_{-\infty}^{t} x(\lambda)\,d\lambda \overset{\mathcal{F}}{\longleftrightarrow} \frac{X(\omega)}{j\omega}+\pi\,X(0)\,\delta(\omega) \tag{3.275}$$

Proof of integration property:

Recall that in Example 3.23 we have found the Fourier transform of a unit step function to be

$$U(\omega)=\mathcal{F}\left\{u(t)\right\}=\pi\,\delta(\omega)+\frac{1}{j\omega}$$

Now consider the convolution of the signal $x(t)$ with a unit step signal:

$$x(t)*u(t)=\int_{-\infty}^{\infty} x(\lambda)\,u(t-\lambda)\,d\lambda \tag{3.276}$$

We know that

$$u(t-\lambda)=\begin{cases} 1, & \lambda<t \\ 0, & \lambda>t \end{cases} \tag{3.277}$$

Using Eqn. (3.277) to modify the integration limits in Eqn. (3.276) we obtain

$$x(t) * u(t) = \int_{-\infty}^{t} x(\lambda)\, d\lambda \tag{3.278}$$

The right side of Eqn. (3.278) is the function the Fourier transform of which we are seeking. We also know from Eqn. (3.264) that

$$\mathscr{F}\left\{x(t) * u(t)\right\} = X(\omega)\, U(\omega) \tag{3.279}$$

Thus we have

$$\mathscr{F}\left\{\int_{-\infty}^{t} x(\lambda)\, d\lambda\right\} = \pi\, X(\omega)\, \delta(\omega) + \frac{X(\omega)}{j\omega} \tag{3.280}$$

Using the sampling property of the impulse function, Eqn. (3.280) becomes

$$\mathscr{F}\left\{\int_{-\infty}^{t} x(\lambda)\, d\lambda\right\} = \pi\, X(0)\, \delta(\omega) + \frac{X(\omega)}{j\omega} \tag{3.281}$$

Table 3.4 contains a summary of key properties of the Fourier transform. Table 3.5 lists some of the fundamental Fourier transform pairs.

3.3.6 Applying Fourier transform to periodic signals

In developing frequency-domain analysis methods in the previous sections of this chapter we have distinguished between periodic and non-periodic continuous-time signals. Periodic signals were analyzed using various forms of the Fourier series such as TFS, EFS, or CFS. In contrast, Fourier transform was used for analyzing non-periodic signals. While this distinction will be appropriate when we work with one type of signal or the other, there may be times when we need to mix periodic and non-periodic signals within the same system. An example of this occurs in amplitude modulation where a non-periodic *message signal* may be multiplied with a periodic *carrier signal.* Another example is the use of a periodic signal as input to a system the impulse response of which is non-periodic. In such situations it would be convenient to use the Fourier transform for periodic signals as well.

In general, a periodic signal is a power signal that does not satisfy the existence conditions for the Fourier transform. It is neither absolute integrable nor square integrable. On the other hand, we have seen in Example 3.20 that a Fourier transform can be found for a constant-amplitude signal that does not satisfy the existence conditions, as long as we are willing to accept singularity functions in the transform. The next two examples will expand on this idea. Afterward we will develop a technique for converting the EFS representation of a periodic continuous-time signal to a Fourier transform that is suitable for use in solving problems.

Example 3.35: Fourier transform of complex exponential signal

Determine the transform of the complex exponential signal $x(t) = e^{j\omega_0 t}$.

Solution: The transform of the constant unit-amplitude signal was found in Example 3.20 to be

$$\mathscr{F}\left\{1\right\} = 2\pi\, \delta(\omega)$$

Using this result along with the frequency shifting property of the Fourier transform given by Eqn. (3.210), the transform of the complex exponential signal is

$$\mathscr{F}\left\{e^{j\omega_0 t}\right\} = 2\pi\,\delta(\omega-\omega_0) \tag{3.282}$$

This is illustrated in Fig. 3.79.

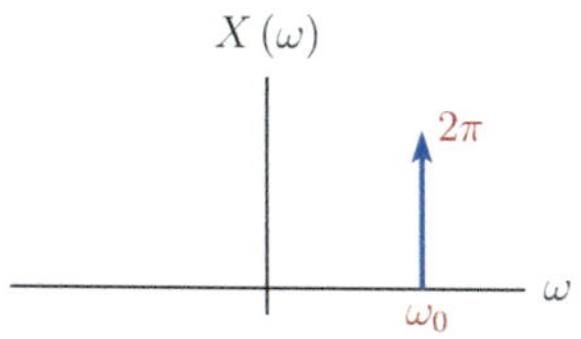

Figure 3.79 – The transform of complex exponential signal $x(t) = e^{j\omega_0 t}$.

Table 3.4 – Fourier transform properties.

Property	Signal	Transform	
Linearity	$\alpha\,x_1(t)+\beta\,x_2(t)$	$\alpha\,X_1(\omega)+\beta\,X_2(\omega)$	
Duality	$X(t)$	$2\pi\,x(-\omega)$	
Conjugate symmetry	$x(t)$ real	$X^*(\omega) = X(-\omega)$	
		Magnitude:	$\lvert X(-\omega)\rvert = \lvert X(\omega)\rvert$
		Phase:	$\Theta(-\omega) = -\Theta(\omega)$
		Real part:	$X_r(-\omega) = X_r(\omega)$
		Imaginary part:	$X_i(-\omega) = -X_i(\omega)$
Conjugate antisymmetry	$x(t)$ imaginary	$X^*(\omega) = -X(-\omega)$	
		Magnitude:	$\lvert X(-\omega)\rvert = \lvert X(\omega)\rvert$
		Phase:	$\Theta(-\omega) = -\Theta(\omega) \mp \pi$
		Real part:	$X_r(-\omega) = -X_r(\omega)$
		Imaginary part:	$X_i(-\omega) = X_i(\omega)$
Even signal	$x(-t) = x(t)$	$\mathrm{Im}\left\{X(\omega)\right\} = 0$	
Odd signal	$x(-t) = -x(t)$	$\mathrm{Re}\left\{X(\omega)\right\} = 0$	
Time shifting	$x(t-\tau)$	$X(\omega)\,e^{-j\omega\tau}$	
Frequency shifting	$x(t)\,e^{j\omega_0 t}$	$X(\omega-\omega_0)$	
Modulation property	$x(t)\cos(\omega_0 t)$	$\frac{1}{2}\left[X(\omega-\omega_0)+X(\omega+\omega_0)\right]$	
Time and frequency scaling	$x(at)$	$\dfrac{1}{\lvert a\rvert}\,X\left(\dfrac{\omega}{a}\right)$	
Differentiation in time	$\dfrac{d^n}{dt^n}\left[x(t)\right]$	$(j\omega)^n\,X(\omega)$	
Differentiation in frequency	$(-jt)^n\,x(t)$	$\dfrac{d^n}{d\omega^n}\left[X(\omega)\right]$	
Convolution	$x_1(t) * x_2(t)$	$X_1(\omega)\,X_2(\omega)$	
Multiplication	$x_1(t)\,x_2(t)$	$\dfrac{1}{2\pi}\,X_1(\omega) * X_2(\omega)$	
Integration	$\displaystyle\int_{-\infty}^{t} x(\lambda)\,d\lambda$	$\dfrac{X(\omega)}{j\omega}+\pi\,X(0)\,\delta(\omega)$	
Parseval's theorem	$\displaystyle\int_{-\infty}^{\infty}\lvert x(t)\rvert^2\,dt = \frac{1}{2\pi}\int_{-\infty}^{\infty}\lvert X(\omega)\rvert^2\,d\omega$		

Table 3.5 – Some Fourier transform pairs.

Name	Signal	Transform
Rectangular pulse	$x(t) = A\,\Pi(t/\tau)$	$X(\omega) = A\tau\,\text{sinc}\left(\frac{\omega\tau}{2\pi}\right)$
Triangular pulse	$x(t) = A\,\Lambda(t/\tau)$	$X(\omega) = A\tau\,\text{sinc}^2\left(\frac{\omega\tau}{2\pi}\right)$
Right-sided exponential	$x(t) = e^{-at}\,u(t)$	$X(\omega) = \frac{1}{a+j\omega}$
Two-sided exponential	$x(t) = e^{-a\|t\|}$	$X(\omega) = \frac{2a}{a^2+\omega^2}$
Signum function	$x(t) = \text{sgn}(t)$	$X(\omega) = \frac{2}{j\omega}$
Unit impulse	$x(t) = \delta(t)$	$X(\omega) = 1$
sinc function	$x(t) = \text{sinc}(t)$	$X(\omega) = \Pi\left(\frac{\omega}{2\pi}\right)$
Constant-amplitude signal	$x(t) = 1$, all t	$X(\omega) = 2\pi\,\delta(\omega)$
	$x(t) = \frac{1}{\pi t}$	$X(\omega) = -j\,\text{sgn}(\omega)$
Unit step function	$x(t) = u(t)$	$X(\omega) = \pi\,\delta(\omega) + \frac{1}{j\omega}$
Modulated pulse	$x(t) = \Pi\left(\frac{t}{\tau}\right)\cos(\omega_0 t)$	$X(\omega) = \frac{\tau}{2}\,\text{sinc}\left(\frac{(\omega-\omega_0)\tau}{2\pi}\right) + \frac{\tau}{2}\,\text{sinc}\left(\frac{(\omega+\omega_0)\tau}{2\pi}\right)$

Example 3.36: Fourier transform of sinusoidal signal

Determine the transform of the sinusoidal signal $x(t) = \cos(\omega_0 t)$.

Solution: The transform of the constant unit-amplitude signal was found in Example 3.20 to be

$$\mathcal{F}\{1\} = 2\pi\,\delta(\omega)$$

Using this result along with the modulation property of the Fourier transform given by Eqn. (3.218), the transform of the sinusoidal signal is

$$\mathcal{F}\left\{\cos(\omega_0 t)\right\} = \pi\,\delta(\omega-\omega_0) + \pi\,\delta(\omega+\omega_0) \tag{3.283}$$

This transform is shown in Fig. 3.80.

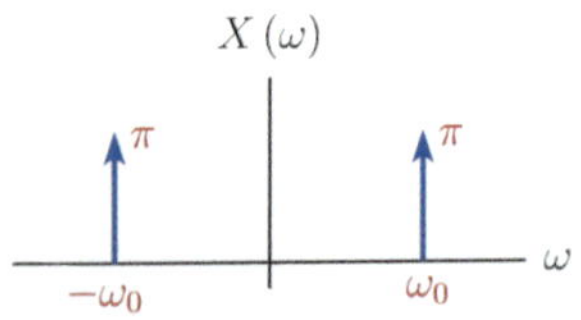

Figure 3.80 – The transform of sinusoidal signal $x(t) = \cos(\omega_0 t)$.

In Examples 3.35 and 3.36 we were able to obtain the Fourier transforms of two periodic signals, namely a complex exponential signal and a sinusoidal signal. Both transforms were possible through the use of impulse functions in the transform. The idea can be generalized to apply to any periodic continuous-time signal that has an EFS representation.

Generalization of the Fourier transform to periodic signals:

The EFS synthesis equation for a periodic signal $\tilde{x}(t)$ was given by Eqn. (3.72) which is repeated here for convenience:

$$\text{Eqn. (3.72):} \qquad \tilde{x}(t) = \sum_{k=-\infty}^{\infty} c_k \, e^{jk\omega_0 t}$$

Using the Fourier transform analysis equation given by Eqn. (3.122), the transform of $\tilde{x}(t)$ is

$$X(\omega) = \int_{-\infty}^{\infty} \tilde{x}(t)\, e^{-j\omega t}\, dt = \int_{-\infty}^{\infty} \left[\sum_{k=-\infty}^{\infty} c_k \, e^{jk\omega_0 t} \right] e^{-j\omega t}\, dt \tag{3.284}$$

Interchanging the order of the integral and the summation in Eqn. (3.284) and rearranging terms we obtain

$$X(\omega) = \sum_{k=-\infty}^{\infty} c_k \left[\int_{-\infty}^{\infty} e^{jk\omega_0 t}\, e^{-j\omega t}\, dt \right] \tag{3.285}$$

In Eqn. (3.285) the expression in square brackets is the Fourier transform of the signal $e^{jk\omega_0 t}$ which, using the result obtained in Example 3.35, can be written as

$$\int_{-\infty}^{\infty} e^{jk\omega_0 t}\, e^{-j\omega t}\, dt = 2\pi\, \delta(\omega - k\omega_0) \tag{3.286}$$

Using Eqn. (3.286) in Eqn. (3.285), the transform of the periodic signal $\tilde{x}(t)$ with EFS coefficients $\{c_k\}$ is obtained as

$$X(\omega) = \sum_{k=-\infty}^{\infty} 2\pi c_k \, \delta(\omega - k\omega_0) \tag{3.287}$$

The Fourier transform for a periodic continuous-time signal is obtained by converting each EFS coefficient c_k to an impulse with area equal to $2\pi c_k$ and placing it at the radian frequency $\omega = k\omega_0$. This process is illustrated in Fig. 3.81.

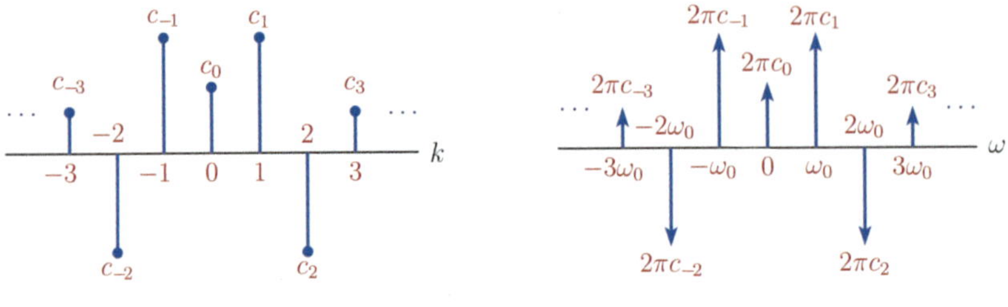

Figure 3.81 – **(a)** EFS coefficients for a signal $\tilde{x}(t)$ and **(b)** Fourier transform obtained through Eqn. (3.287).

Example 3.37: Fourier transform of periodic pulse train

Determine the Fourier transform of the periodic pulse train with duty cycle $d = \tau / T_0$ as shown in Fig. 3.82.

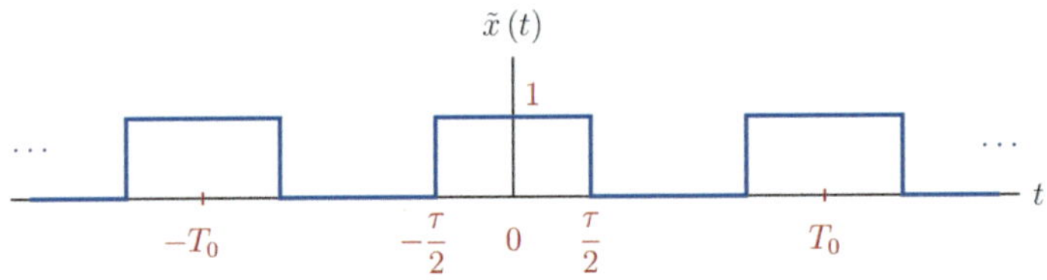

Figure 3.82 – Periodic pulse train for Example 3.37.

Solution: The EFS coefficients for $\tilde{x}(t)$ were determined in Example 3.7 as

$$c_k = d \operatorname{sinc}(kd)$$

Using Eqn. 3.287 the Fourier transform is

$$X(\omega) = \sum_{k=-\infty}^{\infty} 2\pi d \operatorname{sinc}(kd)\, \delta(\omega - k\omega_0)$$

where $\omega_0 = 1/T_0$ is the fundamental radian frequency.

3.4 Energy and Power in the Frequency Domain

In this section we will discuss a very important theorem of Fourier series and transform known as *Parseval's theorem* which can be used as the basis of computing energy or power of a signal from its frequency domain representation. Afterward energy and power spectral density concepts will be introduced.

3.4.1 Parseval's theorem

In Section 1.8 we have defined normalized energy and normalized average power for both periodic and non-periodic signals. Those definitions were based on time-domain descriptions of the signals involved. We know that a signal can be described either in the time domain (as a function of time) or in the frequency domain (through Fourier analysis). Knowing that time- and frequency-domain descriptions of a signal are equivalent, an interesting question that may be raised is the following: How can the normalized energy and power definitions be expressed in the frequency domain? How can we compute signal energy or power from the frequency representation (Fourier series or transform) of the signal?

Parseval's theorem:

For a periodic power signal $\tilde{x}(t)$ with period of T_0 and EFS coefficients $\{c_k\}$ it can be shown that

$$\frac{1}{T_0}\int_{t_0}^{t_0+T_0} |\tilde{x}(t)|^2\,dt = \sum_{k=-\infty}^{\infty} |c_k|^2 \tag{3.288}$$

For a non-periodic energy signal $x(t)$ with a Fourier transform $X(f)$, the following holds true:

$$\int_{-\infty}^{\infty} |x(t)|^2\,dt = \int_{-\infty}^{\infty} |X(f)|^2\,df \tag{3.289}$$

The left side of Eqn. (3.288) represents the normalized average power in a periodic signal as we have derived in Eqn. (1.60). The left side of Eqn. (3.289) represents the normalized signal energy as derived in Eqn. (1.51). The relationships given by Eqns. (3.288) and (3.289) relate signal energy or signal power to the frequency-domain representation of the signal. They are two forms of *Parseval's theorem.*

Proof of Parseval's theorem:

We will begin with the proof of Eqn. (3.288). We know from Eqn. (3.49) that

$$\tilde{x}(t) = \sum_{k=-\infty}^{\infty} c_k\, e^{j\omega_0 t}$$

The left side of Eqn. (3.288) can be written as

$$\begin{aligned}\frac{1}{T_0}\int_{t_0}^{t_0+T_0} |\tilde{x}(t)|^2\,dt &= \frac{1}{T_0}\int_{t_0}^{t_0+T_0}\left[\sum_{k=-\infty}^{\infty} c_k\, e^{jk\omega_0 t}\right]\left[\sum_{m=-\infty}^{\infty} c_m\, e^{jm\omega_0 t}\right]^* dt\\ &= \frac{1}{T_0}\int_{t_0}^{t_0+T_0}\left[\sum_{k=-\infty}^{\infty} c_k\, e^{jk\omega_0 t}\right]\left[\sum_{m=-\infty}^{\infty} c_m^*\, e^{-jm\omega_0 t}\right] dt\end{aligned} \tag{3.290}$$

By rearranging the order of the two summations and the integral in Eqn. (3.290), we have

$$\frac{1}{T_0}\int_{t_0}^{t_0+T_0} |\tilde{x}(t)|^2\,dt = \frac{1}{T_0}\sum_{k=-\infty}^{\infty} c_k \sum_{m=-\infty}^{\infty} c_m^* \int_{t_0}^{t_0+T_0} e^{j(k-m)\omega_0 t}\,dt \tag{3.291}$$

Using orthogonality of exponential basis functions (see Eqn. (3.64)), we have

$$\int_{t_0}^{t_0+T_0} e^{j(k-m)\omega_0 t}\,dt = \begin{cases} T_0, & k=m \\ 0, & k\neq m \end{cases} \tag{3.292}$$

As a result, the inner summation in Eqn. (3.291) evaluates to $c_k^*\,T_0$, and we have

$$\frac{1}{T_0}\int_{t_0}^{t_0+T_0} |\tilde{x}(t)|^2\,dt = \frac{1}{T_0}\sum_{k=-\infty}^{\infty} c_k\,c_k^*\,T_0 = \sum_{k=-\infty}^{\infty} |c_k|^2$$

which proves Eqn. (3.288). The proof for Eqn. (3.289) is quite similar. We know that for an energy signal $x(t)$ the inverse Fourier transform equation is

$$x(t) = \int_{-\infty}^{\infty} X(f)\,e^{j2\pi f t}\,dt$$

The normalized energy of $x(t)$ can be written as

$$\begin{aligned}\int_{-\infty}^{\infty} |x(t)|^2\,dt &= \int_{-\infty}^{\infty} x(t)\,x^*(t)\,dt \\ &= \int_{-\infty}^{\infty} x(t)\left[\int_{-\infty}^{\infty} X(f)\,e^{j2\pi f t}\,df\right]^*\,dt \\ &= \int_{-\infty}^{\infty} x(t)\left[\int_{-\infty}^{\infty} X^*(f)\,e^{-j2\pi f t}\,df\right]\,dt \end{aligned} \tag{3.293}$$

By interchanging the order of the two integrals in Eqn. (3.293) and rearranging terms we get

$$\begin{aligned}\int_{-\infty}^{\infty} |x(t)|^2\,dt &= \int_{-\infty}^{\infty} X^*(f)\left[\int_{-\infty}^{\infty} x(t)\,e^{-j2\pi f t}\,dt\right]\,df \\ &= \int_{-\infty}^{\infty} X^*(f)\,X(f)\,df \\ &= \int_{-\infty}^{\infty} |X(f)|^2\,df \end{aligned} \tag{3.294}$$

which proves Eqn. (3.289).

3.4.2 Energy and power spectral density

Examining the two statements of Parseval's theorem expressed by Eqns. (3.288) and (3.289), we reach the following conclusions:

1. In Eqn. (3.288) the left side corresponds to the normalized average power of the signal $\tilde{x}(t)$, and therefore the summation on the right side must also represent power. The term $|c_k|^2$ corresponds to the power of the frequency component at $f = kf_0$, that is, the power in the k-th harmonic only. Suppose we construct a new function $S_x(f)$ by starting with the line spectrum c_k and replacing each spectral line with an impulse function scaled by $|c_k|^2$ and

time shifted to the frequency kf_0.

$$S_x(f)=\sum_{k=-\infty}^{\infty}|c_k|^2\,\delta(f-kf_0) \tag{3.295}$$

In the next step we will integrate $S_x(f)$ to obtain

$$\begin{aligned}\int_{-\infty}^{\infty}S_x(f)\,df&=\int_{-\infty}^{\infty}\left[\sum_{k=-\infty}^{\infty}|c_k|^2\,\delta(f-kf_0)\right]df\\&=\sum_{k=-\infty}^{\infty}\left[\int_{-\infty}^{\infty}|c_k|^2\,\delta(f-kf_0)\,df\right]\end{aligned} \tag{3.296}$$

Using the sifting property of the unit impulse function Eqn. (3.296) becomes

$$\int_{-\infty}^{\infty}S_x(f)\,df=\sum_{k=-\infty}^{\infty}|c_k|^2 \tag{3.297}$$

Finally, substituting Eqn. (3.297) into Eqn. (3.288) we have

$$\frac{1}{T_0}\int_{t_0}^{t_0+T_0}|\tilde{x}(t)|^2\,dt=\int_{-\infty}^{\infty}S_x(f)\,df \tag{3.298}$$

In Eqn. (3.298) the normalized average power of the signal $x(t)$ is computed by integrating the function $S_x(f)$ over all frequencies. Consequently, the function $S_x(f)$ is the *power spectral density* of the signal $x(t)$.

If we use the radian frequency variable ω instead of f, Eqn. (3.298) becomes

$$\frac{1}{T_0}\int_{t_0}^{t_0+T_0}|\tilde{x}(t)|^2\,dt=\frac{1}{2\pi}\int_{-\infty}^{\infty}S_x(\omega)\,d\omega \tag{3.299}$$

As an interesting by-product of Eqn. (3.298), the power of $\tilde{x}(t)$ that is within a specific frequency range can be determined by integrating $S_x(f)$ over that frequency range. For example, the power contained at frequencies in the range $(-f_0,f_0)$ can be computed as

$$P_x \text{ in } (-f_0,f_0)=\int_{-f_0}^{f_0}S_x(f)\,df \tag{3.300}$$

2. In Eqn. (3.289) the left side is the normalized signal energy for the signal $x(t)$, and the right side must be the same. The integrand $|X(f)|^2$ is therefore the *energy spectral density* of the signal $x(t)$. Let the function $G_x(f)$ be defined as

$$G_x(f)=|X(f)|^2 \tag{3.301}$$

Substituting Eqn. (3.301) into Eqn. (3.289), the normalized energy in the signal $x(t)$ can be expressed as

$$\int_{-\infty}^{\infty}|x(t)|^2\,dt=\int_{-\infty}^{\infty}G_x(f)\,df \tag{3.302}$$

If ω is used in place of f, Eqn. (3.302) becomes

$$\int_{-\infty}^{\infty}|x(t)|^2\,dt=\frac{1}{2\pi}\int_{-\infty}^{\infty}G_x(\omega)\,d\omega \tag{3.303}$$

3. In Eqn. (3.289) we have expressed Parseval's theorem for an energy signal, and used it to lead to the derivation of the energy spectral density in Eqn. (3.301). We know that some non-periodic signals are power signals, therefore their energy cannot be computed. The example of one such signal is the unit step function. The power of a non-periodic signal was defined in Eqn. (1.61) which will be repeated here:

$$\text{Eqn. (1.61):} \qquad P_x = \lim_{T\to\infty}\left[\frac{1}{T}\int_{-T/2}^{T/2}\left|x(t)\right|^2\,dt\right]$$

In order to write the counterpart of Eqn. (3.289) for a power signal, we will first define a *truncated* version of $x(t)$ as

$$x_T(t) = \begin{cases} x(t)\,, & -T/2 < t < T/2 \\ 0\,, & \text{otherwise} \end{cases} \tag{3.304}$$

Let $X_T(f)$ be the Fourier transform of the truncated signal $x_T(t)$:

$$X_T(f) = \mathscr{F}\left\{x_T(t)\right\} = \int_{-T/2}^{T/2} x_T(t)\,e^{-j2\pi f t}\,dt \tag{3.305}$$

Now Eqn. (3.289) can be written in terms of the truncated signal and its transform as

$$\int_{-T/2}^{T/2}\left|x_T(t)\right|^2\,dt = \int_{-\infty}^{\infty}\left|X_T(f)\right|^2\,df \tag{3.306}$$

Scaling both sides of Eqn. (3.306) by $(1/T)$ and taking the limit as T becomes infinitely large, we obtain

$$\lim_{T\to\infty}\left[\frac{1}{T}\int_{-T/2}^{T/2}\left|x_T(t)\right|^2\,dt\right] = \lim_{T\to\infty}\left[\frac{1}{T}\int_{-\infty}^{\infty}\left|X_T(f)\right|^2\,df\right] \tag{3.307}$$

The left side of Eqn. (3.307) is the average normalized power in a non-periodic signal as we have established in Eqn. (1.61). Therefore, the power spectral density of $x(t)$ is

$$S_x(f) = \lim_{T\to\infty}\left[\frac{1}{T}\left|X_T(f)\right|^2\right] \tag{3.308}$$

Example 3.38: Power spectral density of a periodic pulse train

The EFS coefficients for the periodic pulse train $\tilde{x}(t)$ shown in Fig. 3.83 were found in Example 3.5 to be

$$c_k = \frac{1}{3}\,\text{sinc}\,(k/3)$$

Determine the power spectral density for $x(t)$. Also find the total power, the dc power, the power in the first three harmonics, and the power above 1 Hz.

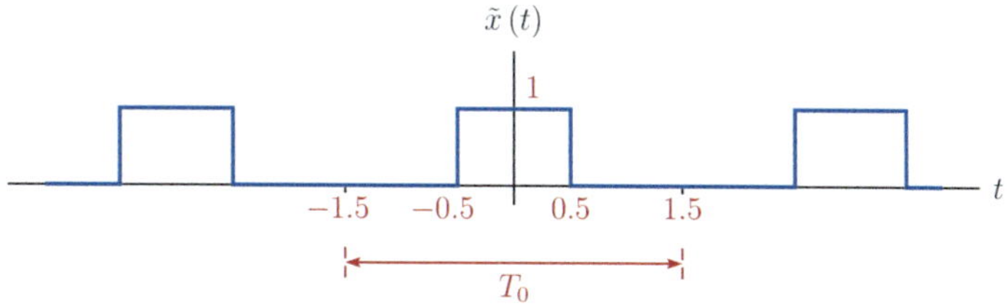

Figure 3.83 – Periodic pulse train used in Example 3.38.

Solution: The period of the signal is $T_0 = 3$ s, and therefore the fundamental frequency is $f_0 = \frac{1}{3}$ Hz. The power spectral density $S_x(f)$ can be written using Eqn. (3.295) as

$$S_x(f) = \sum_{k=-\infty}^{\infty} \left| \frac{1}{3} \operatorname{sinc}(k/3) \right|^2 \delta(f - k/3)$$

which is graphed in Fig. 3.84.

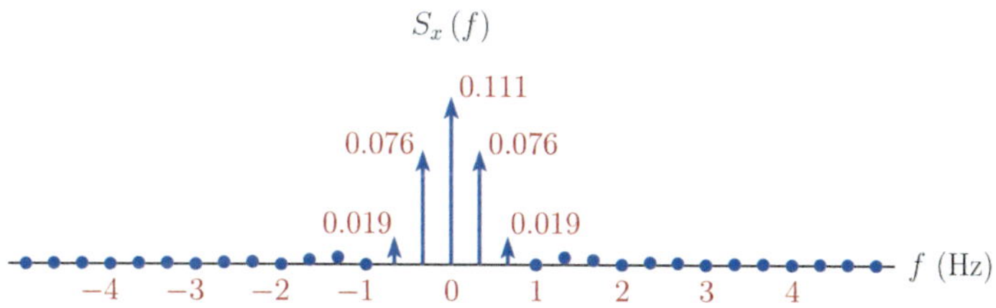

Figure 3.84 – Power spectral density for Example 3.38.

The total power in the signal $x(t)$ can be easily determined using the left side of Eqn. (3.288), i.e.,

$$P_x = \frac{1}{T_0} \int_{t_0}^{t_0+T_0} |\tilde{x}(t)|^2 \, dt = \frac{1}{3} \int_{-0.5}^{0.5} (1)^2 \, dt = 0.3333$$

The dc power in the signal $x(t)$ is

$$P_{dc} = |c_0|^2 = (0.3333)^2 = 0.1111$$

The power in the fundamental frequency is

$$P_1 = |c_{-1}|^2 + |c_1|^2 = 2\,|c_1|^2 = 2\,(0.0760) = 0.1520$$

Power values for the second and third harmonics can be found similarly as

$$P_2 = |c_{-2}|^2 + |c_2|^2 = 2\,|c_2|^2 = 2\,(0.0190)^2 = 0.0380$$

and

$$P_3 = |c_{-3}|^2 + |c_3|^2 = 2\,|c_3|^2 = 2\,(0)^2 = 0$$

The third harmonic is at frequency $f = 1$ Hz. Thus, the power above 1 Hz can be determined by subtracting the power values for the dc and the first three harmonics from the total, that is,

$$\begin{aligned} P_{hf} =& P_x - P_{dc} - P_1 - P_2 - P_3 \\ =& 0.3333 - 0.1111 - 0.1520 - 0.0380 - 0 = 0.0322 \end{aligned}$$

Example 3.39: Power spectral density of a sinusoidal signal

Find the power spectral density of the signal $\tilde{x}(t) = 5\cos(200\pi t)$.

Solution: Using Euler's formula, the signal in question can be written as

$$x(t) = \frac{5}{2}\,e^{-j200\pi t} + \frac{5}{2}\,e^{j200\pi t}$$

from an inspection of which we conclude that the only significant coefficients in the EFS representation of the signal are

$$c_{-1} = c_1 = \frac{5}{2}$$

with all other coefficients equal to zero. The fundamental frequency is $f_0 = 100$ Hz. Using Eqn. (3.295), the power spectral density is

$$\begin{aligned} S_x(f) &= \sum_{n=-\infty}^{\infty} |c_n|^2\,\delta(f-100n) \\ &= |c_{-1}|^2\,\delta(f+100) + |c_1|^2\,\delta(f-100) \\ &= \frac{25}{4}\,\delta(f+100) + \frac{25}{4}\,\delta(f-100) \end{aligned}$$

which is shown in Fig. 3.85

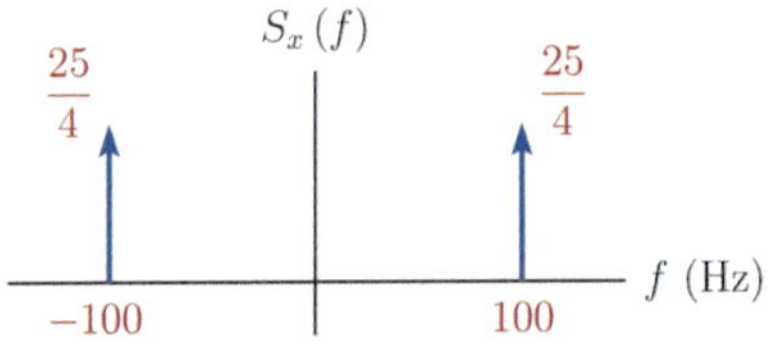

Figure 3.85 – Power spectral density for Example 3.39.

Example 3.40: Energy spectral density of a rectangular pulse

Determine the energy spectral density of the rectangular pulse

$$x(t) = 20\,\Pi(100t)$$

Solution: It was determined in Example 3.12 that

$$A\,\Pi(t/\tau) \overset{\mathcal{F}}{\longleftrightarrow} A\tau\,\mathrm{sinc}(f\tau)$$

Thus the Fourier transform of the pulse given above is

$$X(f) = 0.2\,\mathrm{sinc}(0.01f)$$

and the energy spectral density is

$$G_x(f) = |X(f)|^2 = 0.04\,\mathrm{sinc}^2(0.01f)$$

This result is shown in Fig. 3.86.

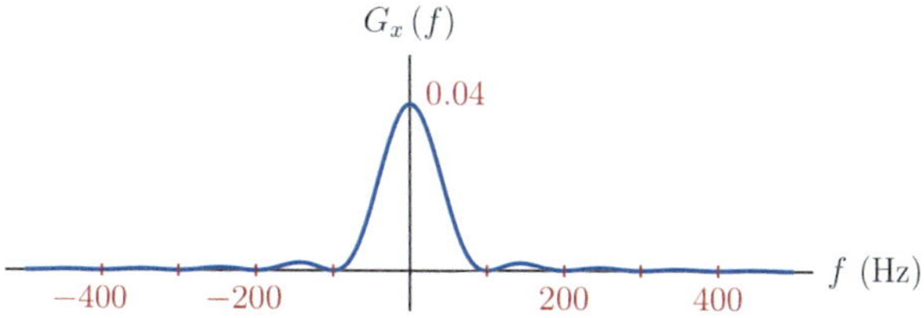

Figure 3.86 – Energy spectral density for Example 3.40.

Example 3.41: Energy spectral density of the sinc function

Determine the energy spectral density of the signal

$$x(t) = \text{sinc}(10t)$$

Afterward, compute a) the total energy in the sinc pulse and b) the energy in the sinc pulse at frequencies up to 3 Hz.

Solution: The spectrum of a sinc pulse is a rectangle. Through the use of the duality property of the Fourier transform, $X(f)$ is found to be

$$X(f) = \frac{1}{10}\,\Pi\left(\frac{f}{10}\right)$$

The energy spectral density is

$$G_x(f) = \frac{1}{100}\,\Pi\left(\frac{f}{10}\right)$$

which is shown in Fig. 3.87.

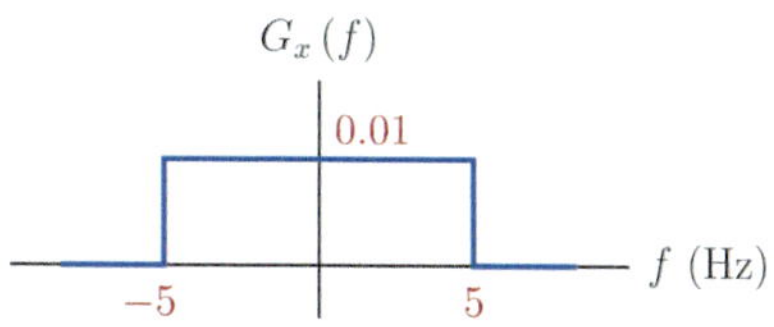

Figure 3.87 – Energy spectral density for Example 3.41.

The total energy in the signal $x(t)$ is

$$E_x = \int_{-\infty}^{\infty} G_x(f)\,df = \int_{-5}^{5}\left(\frac{1}{100}\right)df = 0.1$$

The energy at frequencies up to 3 Hz is found by integrating $G_x(f)$ in the interval -3 Hz $< f < 3$ Hz:

$$E_x \text{ in } (-3, 3\text{ Hz}) = \int_{-3}^{3} G_x(f)\,df = \int_{-3}^{3}\left(\frac{1}{100}\right)df = 0.06$$

Fig. 3.88 illustrates the use of $G_x(f)$ for computing the signal energy.

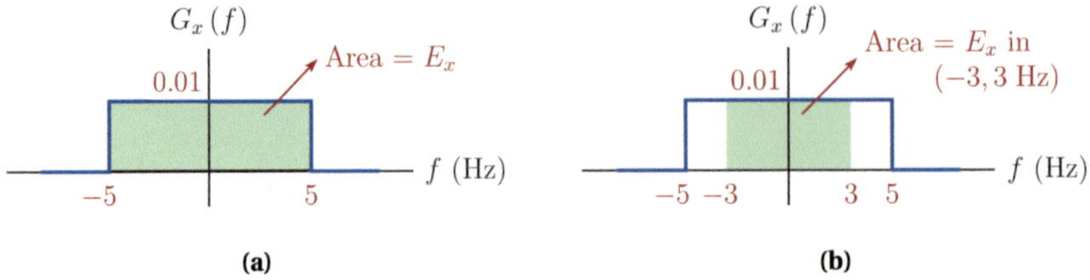

Figure 3.88 – Use of energy spectral density for computing signal energy: **(a)** total area under $G_x(f)$ corresponding to total energy of the signal and **(b)** partial area under $G_x(f)$ corresponding to signal energy in frequencies up to 3 Hz.

3.4.3 Autocorrelation

The energy spectral density $G_x(f)$ or the power spectral $S_x(f)$ density for a signal can be computed through direct application of the corresponding equations derived in Section 3.4.2; namely Eqn. (3.301) for an energy signal, and either Eqn. (3.295) or Eqn. (3.308) for a power signal, depending on its type. In some circumstances it is also possible to compute $G_x(f)$ or $S_x(f)$ from the knowledge of the *autocorrelation function* which will be defined in this section. Let $x(t)$ be a real-valued signal.

Autocorrelation function:

For an energy signal $x(t)$ the *autocorrelation function* is defined as

$$r_{xx}(\tau) = \int_{-\infty}^{\infty} x(t)\, x(t+\tau)\, dt \tag{3.309}$$

For a periodic power signal $\tilde{x}(t)$ with period T_0, the corresponding definition of the autocorrelation function is

$$\begin{aligned}\tilde{r}_{xx}(\tau) &= \langle \tilde{x}(t)\, \tilde{x}(t+\tau) \rangle \\ &= \frac{1}{T_0} \int_{-T_0/2}^{T_0/2} \tilde{x}(t)\, \tilde{x}(t+\tau)\, dt \end{aligned} \tag{3.310}$$

The triangle brackets indicate time average. The autocorrelation function for a periodic signal is also periodic as signified by the tilde (~) character used over the symbol r_{xx} in Eqn. (3.310). Finally, for a non-periodic power signal, the corresponding definition is

$$\begin{aligned} r_{xx}(\tau) &= \langle x(t)\, x(t+\tau) \rangle \\ &= \lim_{T_0 \to \infty} \left[\frac{1}{T_0} \int_{-T_0/2}^{T_0/2} x(t)\, x(t+\tau)\, dt \right] \end{aligned} \tag{3.311}$$

Even though we refer to $r_{xx}(\tau)$ as a *function*, we will often treat it as if it is a continuous-time signal, albeit one that uses a different independent variable, τ, than the signal $x(t)$ for which it is computed. The variable τ simply corresponds to the time shift between the two copies of $x(t)$ used in the definitions of Eqns. (3.309) through (3.311).

Energy spectral density:

It can be shown that, for an energy signal, the energy spectral density is the Fourier transform of the autocorrelation function, that is,

$$\mathcal{F}\left\{r_{xx}(\tau)\right\} = G_x(f) \tag{3.312}$$

Proof of Eqn. (3.311):

Let us begin by applying the Fourier transform definition to the autocorrelation function $r_{xx}(\tau)$ treated as a continuous-time signal:

$$\mathcal{F}\left\{r_{xx}(\tau)\right\} = \int_{-\infty}^{\infty} r_{xx}(\tau)\, e^{-j2\pi f\tau}\, d\tau \tag{3.313}$$

$$= \int_{-\infty}^{\infty} \left[\int_{-\infty}^{\infty} x(t)\, x(t+\tau)\, dt\right] e^{-j2\pi f\tau}\, d\tau \tag{3.314}$$

The two integrals in Eqn. (3.313) can be rearranged to yield

$$\mathcal{F}\left\{r_{xx}(\tau)\right\} = \int_{-\infty}^{\infty} x(t) \left[\int_{-\infty}^{\infty} x(t+\tau)\, e^{-j2\pi f\tau}\, d\tau\right] dt \tag{3.315}$$

Realizing that the inner integral in Eqn. (3.315) is equal to

$$\int_{-\infty}^{\infty} x(t+\tau)\, e^{-j2\pi f\tau}\, d\tau = \mathcal{F}\left\{x(t+\tau)\right\} = e^{j2\pi ft}\, X(f) \tag{3.316}$$

Eqn. (3.315) becomes

$$\mathcal{F}\left\{r_{xx}(\tau)\right\} = X(f) \int_{-\infty}^{\infty} x(t)\, e^{j2\pi ft}\, dt \tag{3.317}$$

Since $x(t)$ is real

$$\int_{-\infty}^{\infty} x(t)\, e^{j2\pi ft}\, dt = X^*(f) \tag{3.318}$$

and we obtain

$$\mathcal{F}\left\{r_{xx}(\tau)\right\} = X(f)\, X^*(f) = \left|X(f)\right|^2 \tag{3.319}$$

which completes the proof.

The definition of the autocorrelation function for a periodic signal $\tilde{x}(t)$ can be used in a similar manner in determining the power spectral density of the signal.

Power spectral density:

Let the signal $\tilde{x}(t)$ and the autocorrelation function $\tilde{r}_{xx}(\tau)$, both periodic with period T_0, have the EFS representations given by

$$\tilde{x}(t) = \sum_{k=-\infty}^{\infty} c_k \, e^{j2\pi k f_0 t} \tag{3.320}$$

and

$$\tilde{r}_{xx}(\tau) = \sum_{k=-\infty}^{\infty} d_k \, e^{j2\pi k f_0 \tau}, \tag{3.321}$$

respectively. It can be shown that the EFS coefficients $\{c_k\}$ and $\{d_k\}$ are related by

$$d_k = |c_k|^2 \tag{3.322}$$

and the power spectral density is the Fourier transform of the autocorrelation function, that is,

$$S_x(f) = \mathscr{F}\{\tilde{r}_{xx}(\tau)\} \tag{3.323}$$

Proof of Eqn. (3.323):

Using the EFS analysis equation given by Eqn. (3.73) with the definition of the autocorrelation function in Eqn. (3.310) leads to

$$\begin{aligned} d_k =& \frac{1}{T_0} \int_{-T_0/2}^{T_0/2} \tilde{r}_{xx}(\tau) \, e^{-j2\pi k f_0 \tau} \, d\tau \\ =& \frac{1}{T_0} \int_{-T_0/2}^{T_0/2} \left[\frac{1}{T_0} \int_{-T_0/2}^{T_0/2} \tilde{x}(t) \, \tilde{x}(t+\tau) \, dt \right] e^{-j2\pi k f_0 \tau} \, d\tau \end{aligned} \tag{3.324}$$

Rearranging the order of the two integrals Eqn. (3.324) can be written as

$$d_k = \frac{1}{T_0} \int_{-T_0/2}^{T_0/2} \tilde{x}(t) \left[\frac{1}{T_0} \int_{-T_0/2}^{T_0/2} \tilde{x}(t+\tau) \, e^{-j2\pi k f_0 \tau} \, d\tau \right] dt \tag{3.325}$$

Using the time shifting property of the EFS given by Eqn. (3.108) the expression in square brackets in Eqn. (3.325) is

$$\frac{1}{T_0} \int_{-T_0/2}^{T_0/2} \tilde{x}(t+\tau) \, e^{-j2\pi k f_0 \tau} \, d\tau = e^{j2\pi k f_0 t} \, c_k \tag{3.326}$$

Substituting Eqn. (3.326) into Eqn. (3.325) yields

$$\begin{aligned} d_k =& \frac{1}{T_0} \int_{-T_0/2}^{T_0/2} \tilde{x}(t) \, e^{j2\pi k f_0 t} \, c_k \, dt \\ =& c_k \frac{1}{T_0} \int_{-T_0/2}^{T_0/2} \tilde{x}(t) \, e^{j2\pi k f_0 t} \, dt \end{aligned} \tag{3.327}$$

Recognizing that

$$\frac{1}{T_0}\int_{-T_0/2}^{T_0/2} \tilde{x}(t)\, e^{j2\pi k f_0 t}\, dt = c_k^*$$

since $\tilde{x}(t)$ is real, we conclude

$$d_k = c_k\, c_k^* = \left|c_k\right|^2 \tag{3.328}$$

Using the technique developed in Section 3.3.6 the Fourier transform of $\tilde{r}_{xx}(\tau)$ is

$$\mathscr{F}\left\{\tilde{r}_{xx}(\tau)\right\} = \sum_{k=-\infty}^{\infty} d_k\, \delta\left(f - kf_0\right) \tag{3.329}$$

Using Eqn. (3.327) in Eqn. (3.329) results in

$$\mathscr{F}\left\{\tilde{r}_{xx}(\tau)\right\} = 2\pi \sum_{k=-\infty}^{\infty} \left|c_k\right|^2 \delta\left(f - kf_0\right) \tag{3.330}$$

which is identical to the expression given by Eqn. (3.295) for the power spectral density of a periodic signal.

A relationship similar to the one expressed by Eqn. (3.323) applies to random processes that are *wide-sense stationary*, and is known as the Wiener-Khinchin theorem. It is one of the fundamental theorems of random signal processing.

Example 3.42: Power spectral density of a sinusoidal signal revisited

Find the autocorrelation function for the signal

$$\tilde{x}(t) = 5\cos(200\pi t)$$

Afterward determine the power spectral density from the autocorrelation function.

Solution: Using Eqn. (3.310) with $T_0 = 0.01$ s we have

$$\tilde{r}_{xx}(\tau) = \frac{1}{0.01}\int_{-0.005}^{0.005} 25\cos(200\pi t)\cos(200\pi\,[t+\tau])\, dt$$

Using the appropriate trigonometric identity, the integral is evaluated as

$$\tilde{r}_{xx}(\tau) = \frac{25}{2}\cos(200\pi\tau)$$

Power spectral density is the Fourier transform of the autocorrelation function:

$$S_x\left(f\right) = \mathscr{F}\left\{\tilde{r}_{xx}(\tau)\right\} = \frac{25}{4}\,\delta\left(f+100\right) + \frac{25}{4}\,\delta\left(f-100\right)$$

which is in agreement with the earlier result found in Example 3.39.

Properties of the autocorrelation function

The autocorrelation function as defined by Eqns. (3.309), (3.310), and (3.311) has a number of important properties that will be listed here:

1. $r_{xx}(0) \ge |r_{xx}(\tau)|$ for all τ.

   To see why this is the case, we will consider the non-negative function $[x(t) \mp x(t+\tau)]^2$. The time average of this function must also be non-negative, so we can write

$$\left\langle [x(t) \mp x(t+\tau)]^2 \right\rangle \ge 0$$

   or equivalently

$$\left\langle x^2(t) \right\rangle \mp \left\langle 2\,x(t)\,x(t+\tau) \right\rangle + \left\langle x^2(t+\tau) \right\rangle \ge 0$$

   which implies that

$$r_{xx}(0) \mp 2\,r_{xx}(\tau) + r_{xx}(0) \ge 0$$

   which is the same as property 1.

2. $r_{xx}(-\tau) = r_{xx}(\tau)$ for all τ, that is, the autocorrelation function has even symmetry.

   Recall that the autocorrelation function is the inverse Fourier transform of the power spectral density, i.e., $r_{xx}(\tau) = \mathcal{F}^{-1}\{S_x(f)\}$. We know from the symmetry properties of the Fourier transform that, since $S_x(f)$ is purely real, $r_{xx}(\tau)$ must be an even function of τ.

3. If the signal $x(t)$ is periodic with period T, then its autocorrelation function $\tilde{r}_{xx}(\tau)$ is also periodic with the same period, that is, if $x(t) = x(t+kT)$ for all integer k, then $\tilde{r}_{xx}(\tau) = \tilde{r}_{xx}(\tau+kT)$ for all integer k. This property easily follows from the time-average based definition of the autocorrelation function given by Eqn. (3.310).

3.5 System Function Concept

In time-domain analysis of systems in Chapter 2 we have relied on two distinct description forms for CTLTI systems:

1. A linear constant-coefficient differential equation that describes the relationship between the input and the output signals
2. An impulse response which can be used with the convolution operation for determining the response of the system to an arbitrary input signal

In this section, the concept of *system function* will be introduced as the third method for describing the characteristics of a system.

System function $H(\omega)$:

The system function is simply the Fourier transform of the impulse response:

$$H(\omega) = \mathcal{F}\{h(t)\} = \int_{-\infty}^{\infty} h(t)\, e^{-j\omega t}\, dt \tag{3.331}$$

Recall that the impulse response is only meaningful for a system that is both linear and time-invariant, since the convolution operator could not be used otherwise. Consequently, the system function concept is valid for linear and time-invariant systems only.

In general, $H(\omega)$ is a complex function of ω, and can be written in polar form as

$$H(\omega) = |H(\omega)|\, e^{j\Theta(\omega)} \tag{3.332}$$

Example 3.43: System function for the simple RC circuit

The impulse response of the RC circuit shown in Fig. 3.89 was found in Example 2.18 to be

$$h(t) = \frac{1}{RC}\, e^{-t/RC}\, u(t)$$

Determine the system function.

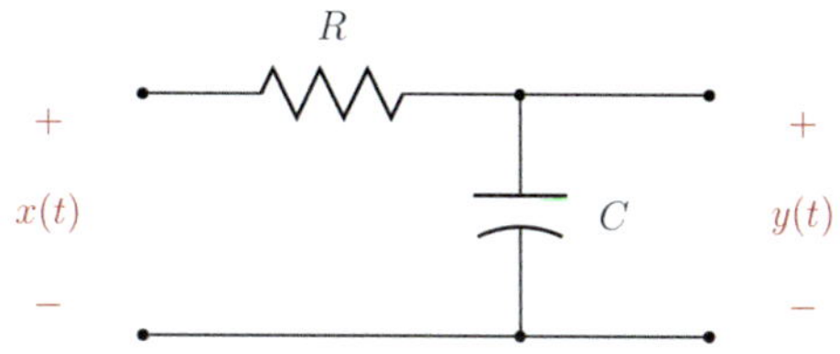

Figure 3.89 – The RC circuit for Example 3.43.

Solution: Taking the Fourier transform of the impulse response we obtain

$$H(\omega) = \int_0^{\infty} \frac{1}{RC}\, e^{-t/RC}\, e^{-j\omega t}\, dt = \frac{1}{1 + j\omega RC}$$

To simplify notation we will define $\omega_c = 1/RC$ and write the system function as

$$H(\omega) = \frac{1}{1 + j(\omega/\omega_c)} \tag{3.333}$$

The magnitude and the phase of the system function are

$$|H(\omega)| = \frac{1}{\sqrt{1 + (\omega/\omega_c)^2}} \tag{3.334}$$

and

$$\Theta(\omega) = \measuredangle H(\omega) = -\tan^{-1}(\omega/\omega_c), \tag{3.335}$$

respectively, and are shown in Fig. 3.90. The value of the system function at the frequency ω_c is $H(\omega_c) = 1/(1+j)$, and the corresponding magnitude is $|H(\omega_c)| = 1/\sqrt{2}$. Thus, ω_c represents the frequency at which the magnitude of the system function is 3 decibels below its peak value at $\omega = 0$, that is,

$$20\log_{10}\left[\frac{|H(\omega_c)|}{|H(0)|}\right] = 20\log_{10}\left[\frac{1}{\sqrt{2}}\right] \approx -3\text{ dB}$$

The frequency ω_c is often referred to as the *3-dB cutoff frequency* of the system.

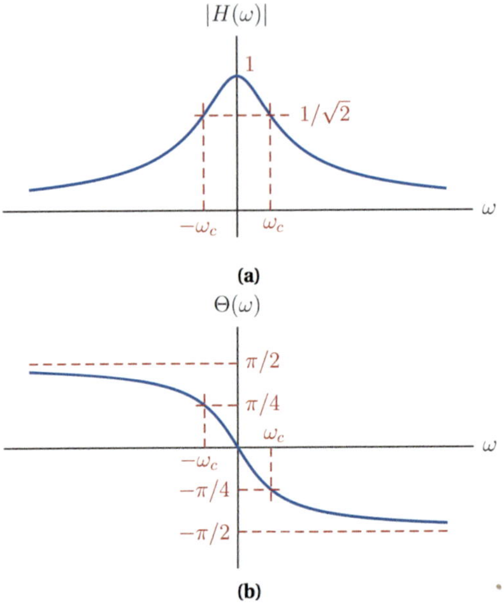

Figure 3.90 – The system function for the RC circuit in Example 3.22: **(a)** magnitude and **(b)** phase.

Software resources: `exct_3_43.m` , `exct_3_43_Live.mlx`

Interactive App: System function for the simple RC circuit

The interactive app in `appSysFun1.m` illustrates the system function concept for the simple RC circuit of Example 3.43. The demo program allows circuit parameters R and C to be varied using slider controls, and displays the effect of parameter changes on the magnitude and the phase of the system function.

a. Set parameter values to $R = 20\ \Omega$ and $C = 130$ ţF. What is the 3-dB cutoff frequency f_c of the system? Calculate it on paper, and verify the value shown in the app.

b. One particular bandwidth measure called the 3-*dB bandwidth* is defined as the frequency range in which the magnitude of the system function remains within 3-dB of its peak value, or greater than $1/\sqrt{2}$ in this case. What is the 3-dB bandwidth of the system?

c. Gradually increase the value of R and observe how it is related to the 3-dB cutoff frequency f_c and consequently the *bandwidth* of the system function.

d. What is the value of the phase of the system function at the frequency $f = f_c$. Does that value change as f_c is changed? If not, why not?

Software resource: `appSysFun1.m`

Software resource: See MATLAB Exercise 3.7.

Obtaining the system function from the differential equation

Sometimes we need to find the system function for a CTLTI system described by means of a differential equation. Two properties of the Fourier transform will be useful for this purpose; the convolution property and the time differentiation property.

1. Since the output signal is computed as the convolution of the impulse response and the input signal, that is, $y(t) = h(t) * x(t)$, the corresponding relationship in the frequency domain is $Y(\omega) = H(\omega)\, X(\omega)$. Consequently, the system function is equal to the ratio of the output transform to the input transform:

$$H(\omega) = \frac{Y(\omega)}{X(\omega)} \tag{3.336}$$

2. Using the time differentiation property, the transforms of the individual terms in the differential equation are

$$\frac{d^k y(t)}{dt^k} \stackrel{\mathcal{F}}{\longleftrightarrow} \left(j\omega\right)^k Y(\omega) \qquad k = 0, 1, \ldots \tag{3.337}$$

$$\frac{d^k x(t)}{dt^k} \stackrel{\mathcal{F}}{\longleftrightarrow} \left(j\omega\right)^k X(\omega) \qquad k = 0, 1, \ldots \tag{3.338}$$

The system function is obtained from the differential equation by first transforming both sides of the differential equation through the use of Eqns. (3.337) and (3.338), and then using Eqn. (3.336).

Example 3.44: Finding the system function from the differential equation

Determine the system function for a CTLTI system described by the differential equation

$$\frac{d^2 y(t)}{dt^2} + 2\,\frac{dy(t)}{dt} + 26\, y(t) = 10\, x(t)$$

Solution: Taking the Fourier transform of both sides of the differential equation we obtain

$$\left(j\omega\right)^2 Y(\omega) + 2\left(j\omega\right) Y(\omega) + 26\, Y(\omega) = 10\, X(\omega)$$

or

$$\left[\left(26 - \omega^2\right) + j2\omega\right] Y(\omega) = 10\, X(\omega)$$

The system function is

$$H(\omega) = \frac{Y(\omega)}{X(\omega)} = \frac{10}{\left(26 - \omega^2\right) + j2\omega}$$

3.6 CTLTI Systems with Periodic Input Signals

In earlier sections of this chapter we have developed methods for representing periodic signals as linear combinations of sinusoidal or complex exponential basis functions. Using the TFS representation we have

$$\text{Eqn. (3.28):} \qquad \tilde{x}(t) = a_0 + \sum_{k=1}^{\infty} a_k \cos(k\omega_0 t) + \sum_{k=1}^{\infty} b_k \sin(k\omega_0 t)$$

Alternatively, using the EFS representation

$$\text{Eqn. (3.49):} \qquad \tilde{x}(t) = \sum_{k=-\infty}^{\infty} c_k \, e^{jk\omega_0 t}$$

If a periodic signal is used as input to a CTLTI system, the superposition property can be utilized for finding the output signal. The response of the CTLTI system to the periodic signal can be determined as a linear combination of its responses to the individual basis functions in the TFS or EFS representation. We will analyze this concept in two steps; first for a complex exponential input signal, and then for a sinusoidal signal.

3.6.1 Response of a CTLTI system to complex exponential signal

Consider a CTLTI system with impulse response $h(t)$, driven by an input signal in the form of a complex exponential function of time

$$\tilde{x}(t) = e^{j\omega_0 t} \tag{3.339}$$

The response $y(t)$ of the system can be found through the use of the convolution relationship as we have derived in Section 2.7.2 of Chapter 2.

$$\begin{aligned} y(t) =& h(t) * \tilde{x}(t) \\ =& \int_{-\infty}^{\infty} h(\lambda)\, \tilde{x}(t-\lambda)\, d\lambda \end{aligned} \tag{3.340}$$

Derivation of the response of CTLTI system to $\tilde{x}(t) = e^{j\omega_0 t}$:

Using the signal $\tilde{x}(t)$ given by Eqn. (3.339) in Eqn. (3.340) we get

$$y(t) = \int_{-\infty}^{\infty} h(\lambda)\, e^{j\omega_0 (t-\lambda)}\, d\lambda \tag{3.341}$$

or, equivalently

$$y(t) = e^{j\omega_0 t} \int_{-\infty}^{\infty} h(\lambda)\, e^{-j\omega_0 \lambda}\, d\lambda \tag{3.342}$$

The integral in Eqn. (3.342) should be recognized as the system function evaluated at the specific radian frequency $\omega = \omega_0$. Therefore

$$y(t) = e^{j\omega_0 t}\, H(\omega_0) \tag{3.343}$$

The development in Eqns. (3.341) through (3.343) is based on the inherent assumption that the Fourier transform of $h(t)$ exists. This in turn requires the corresponding

CTLTI system to be stable. Any natural response the system may have exhibited at one point would have disappeared a long time ago. Consequently, the response found in Eqn. (3.343) is the *steady-state response* of the system.

For a CTLTI system driven by a complex exponential input signal, we have the following important relationship:

Response of CTLTI system to complex exponential input:

$$y(t) = \text{Sys}\left\{e^{j\omega_0 t}\right\} = e^{j\omega_0 t}\, H(\omega_0)$$
$$= \left|H(\omega_0)\right| e^{j[\omega_0 t + \Theta(\omega_0)]} \tag{3.344}$$

1. The response of a CTLTI system to a complex exponential input signal is a complex exponential output signal with the same frequency ω_0.
2. The effect of the system on the complex exponential input signal is to
   (a) Scale its amplitude by an amount equal to the magnitude of the system function at $\omega = \omega_0$.
   (b) Shift its phase by an amount equal to the phase of the system function at $\omega = \omega_0$.

3.6.2 Response of a CTLTI system to sinusoidal signal

Let the input signal to the CTLTI system under consideration be a sinusoidal signal in the form

$$\tilde{x}(t) = \cos(\omega_0 t) \tag{3.345}$$

The response of the system in this case will be determined by making use of the results of the previous section.

Derivation of the response of CTLTI system to $\tilde{x}(t) = \cos(\Omega_0 t)$:

We will use Euler's formula to write the input signal using two complex exponential functions as

$$\tilde{x}(t) = \cos(\omega_0 t) = \frac{1}{2}\, e^{j\omega_0 t} + \frac{1}{2}\, e^{-j\omega_0 t} \tag{3.346}$$

This representation of $\tilde{x}(t)$ allows the results of the previous section to be used. The output signal can be written using superposition:

$$y(t) = \frac{1}{2}\,\text{Sys}\left\{e^{j\omega_0 t}\right\} + \frac{1}{2}\,\text{Sys}\left\{e^{-j\omega_0 t}\right\}$$
$$= \frac{1}{2}\, e^{j\omega_0 t}\, H(\omega_0) + \frac{1}{2}\, e^{-j\omega_0 t}\, H(-\omega_0)$$
$$= \frac{1}{2}\, e^{j\omega_0 t} \left|H(\omega_0)\right| e^{j\Theta(\omega_0)} + \frac{1}{2}\, e^{-j\omega_0 t} \left|H(-\omega_0)\right| e^{j\Theta(-\omega_0)} \tag{3.347}$$

If the impulse response $h(t)$ is real-valued, the result in Eqn. (3.347) can be further simplified. Recall from Section 3.3.5 that, for real-valued $h(t)$, the transform $H(\omega)$ is conjugate symmetric, resulting in

$$\left|H(-\omega_0)\right| = \left|H(\omega_0)\right| \quad \text{and} \quad \Theta(-\omega_0) = -\Theta(\omega_0) \tag{3.348}$$

Using these relationships, Eqn. (3.347) becomes

$$\begin{aligned} y(t) &= \frac{1}{2}\left|H(\omega_0)\right| e^{j[\omega_0 t+\Theta(\omega_0)]} + \frac{1}{2}\left|H(\omega_0)\right| e^{-j[\omega_0 t+\Theta(\omega_0)]} \\ &= \left|H(\omega_0)\right| \cos\left(\omega_0 t + \Theta(\omega_0)\right) \end{aligned} \tag{3.349}$$

For a CTLTI system driven by a cosine input signal, we have the following important relationship:

Response of CTLTI system to cosine input:

$$y(t) = \text{Sys}\left\{\cos(\omega_0 t)\right\} = \left|H(\omega_0)\right| \cos\left(\omega_0 t + \Theta(\omega_0)\right) \tag{3.350}$$

1. When a CTLTI system is driven by single-tone input signal at frequency ω_0, its output signal is also a single-tone signal at the same frequency.
2. The effect of the system on the input signal is to
   (a) Scale its amplitude by an amount equal to the magnitude of the system function at $\omega = \omega_0$.
   (b) Shift its phase by an amount equal to the phase of the system function at $\omega = \omega_0$.

Example 3.45: Steady-state response of RC circuit for single-tone input

Consider the RC circuit of Example 3.43. Let the component values be chosen to yield a 3-dB cutoff frequency of $\omega_c = 160\pi$ rad/s, or equivalently $f_c = 80$ Hz. Let the input signal be in the form

$$\tilde{x}(t) = 5\cos\left(2\pi f t\right)$$

Compute the steady-state output signal for the cases $f_1 = 20$ Hz, $f_2 = 100$ Hz, $f_3 = 200$ Hz, and $f_4 = 500$ Hz.

Solution: For the system under consideration the system function is

$$H\left(f\right) = \frac{1}{1 + j\left(f/80\right)} \tag{3.351}$$

The magnitude and the phase of the system function were determined in Example 3.43 Eqns. (3.334) and (3.335). Using f instead of ω we have

$$\left|H\left(f\right)\right| = \frac{1}{\sqrt{1 + \left(f/80\right)^2}} \quad \text{and} \quad \Theta\left(f\right) = -\tan^{-1}\left(f/80\right)$$

as shown in Fig. 3.91.

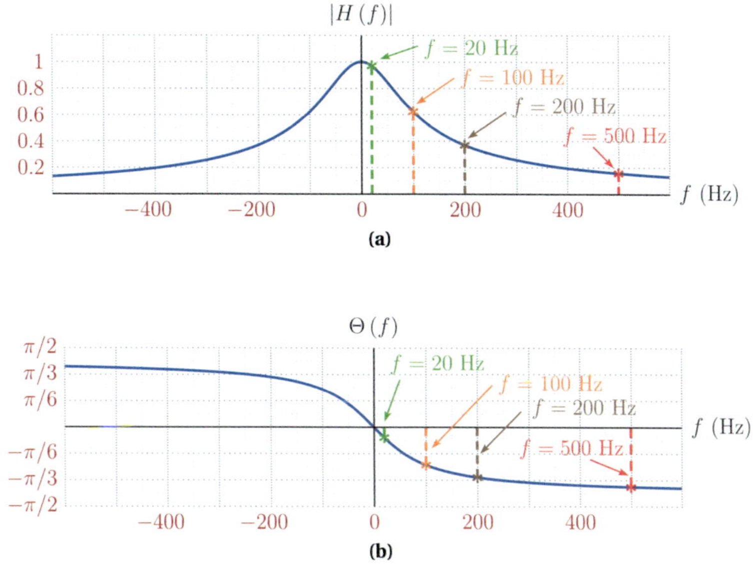

Figure 3.91 – The system function for the system in Example 3.45: **(a)** magnitude and **(b)** phase.

For the input frequency of $f_1 = 20$ Hz, the magnitude and the phase of the system function are

$$\left|H(20)\right| = \frac{1}{\sqrt{1+(20/80)^2}} = 0.9701 \tag{3.352}$$

and

$$\Theta(20) = -\tan^{-1}(20/80) = -0.245 \text{ radians} \tag{3.353}$$

The impact of the system on the 20 Hz input signal is amplitude scaling by a factor of 0.9701 and phase shift by -0.245 radians. The corresponding steady-state output signal is

$$\begin{aligned} y_1(t) &= (5)(0.9701)\cos(40\pi t - 0.245) \\ &= 4.8507\cos\big(40\pi(t-0.0019)\big) \end{aligned} \tag{3.354}$$

The phase shift of -0.245 radians translates to a time-delay of about $t_d = 1.9$ ms. The steady-state output signal $y_1(t)$ is shown in Fig. 3.92 along with the input signal for comparison.

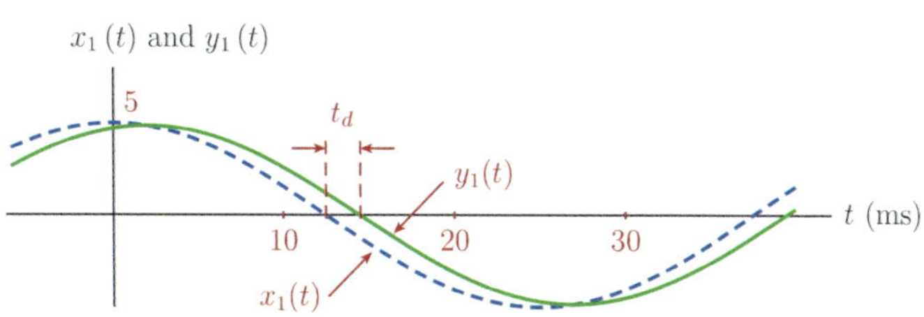

Figure 3.92 – The 20 Hz input signal and the corresponding output signal for Example 3.45.

Calculations in Eqns. (3.352) through 3.354) can easily be repeated for the other three frequency choices, but will be skipped here to save space. The results are summarized in Table 3.6 for all four frequencies. The frequency-discriminating nature of the system is evident.

Table 3.6 – Amplitude scaling, phase offset, and time delay values for Example 3.45.

f (Hz)	$\lvert H(f) \rvert$	$\Theta(f)$ (rad)	t_d (ms)
20	0.9701	−0.2450	1.95
100	0.6247	−0.8961	1.43
200	0.3714	−1.1903	0.94
500	0.1580	−1.4121	0.45

Input-output signal pairs for $f_2 = 100$ Hz, $f_3 = 200$ Hz, and $f_4 = 500$ Hz are shown in Fig. 3.93.

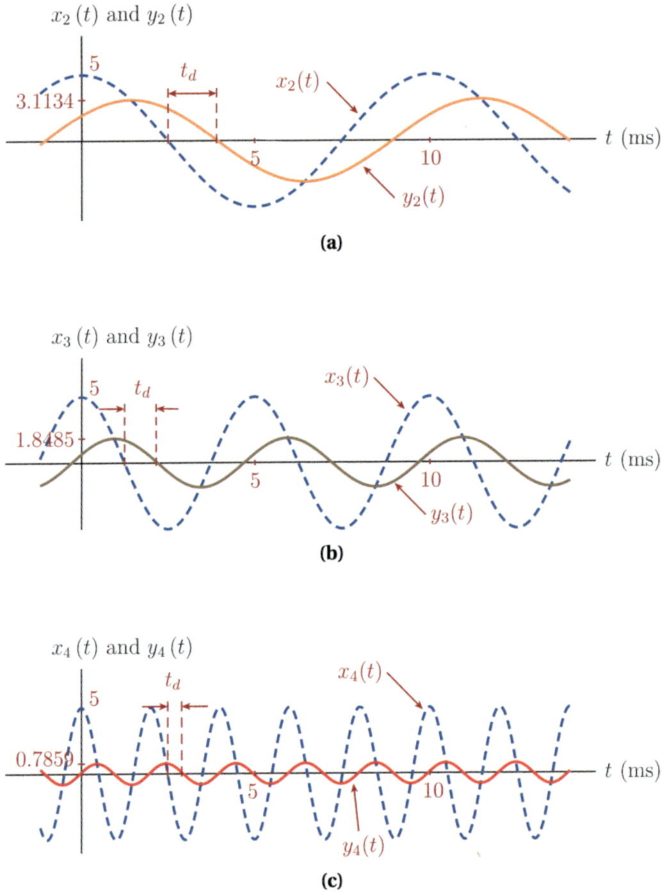

Figure 3.93 – The input-output signal pairs for Example 3.45: **(a)** $f_2 = 100$ Hz, **(b)** $f_3 = 200$ Hz, and **(c)** $f_4 = 500$ Hz.

Software resources: exct_3_45a.m , exct_3_45b.m , exct_3_45c.m , exct_3_45_Live.mlx

Interactive App: Steady-state response of RC circuit for single-tone input

The interactive app in `appSysFun2.m` parallels the development of Example 3.45. Magnitude, phase and time-delay characteristics of the system function under consideration are shown. The first slider control allows the value of f_c to be adjusted. With the choice of $f_c = 80$ Hz the transfer function of Example 3.45 is obtained. The second slider control specifies the frequency of the input signal which is marked on the magnitude, phase, and time-delay graphs. At the bottom of the screen, input and output signals are shown along with delay markers that illustrate the delay caused by the system.

a. Set the cutoff frequency to $f_c = 80$ Hz. Set the signal frequency to $f_1 = 20$ Hz. Compare the graph of input and output signals to Fig. 3.92.

b. Pay attention to magnitude and phase of the system function and how they impact the output signal.

c. Verify the numerical results of Example 3.45 by using $f_2 = 100$ Hz, $f_3 = 200$ Hz, and $f_4 = 500$ Hz for the input signal frequency. Compare input and output signals to those in Fig. 3.93.

d. With the signal frequency fixed at $f_2 = 100$ Hz, change the 3-dB bandwidth of the system and observe its effect on the output signal in terms of peak amplitude and time delay.

Software resource: `appSysFun2.m`

3.6.3 Response of a CTLTI system to periodic input signal

Using the development in the previous sections, we are now ready to consider the use of a general periodic signal $\tilde{x}(t)$ as input to a CTLTI system. Let $\tilde{x}(t)$ be a signal that satisfies the existence conditions for the Fourier series. Using the EFS representation of the signal, the response of the system is

$$\text{Sys}\{\tilde{x}(t)\} = \text{Sys}\left\{\sum_{k=-\infty}^{\infty} c_k\, e^{jk\omega_0 t}\right\} \tag{3.355}$$

Let us use the linearity of the system to write the response as

$$\text{Sys}\{\tilde{x}(t)\} = \sum_{k=-\infty}^{\infty} \text{Sys}\left\{c_k\, e^{jk\omega_0 t}\right\} = \sum_{k=-\infty}^{\infty} c_k\, \text{Sys}\left\{e^{jk\omega_0 t}\right\} \tag{3.356}$$

The response of the system to an exponential basis function at frequency $\omega = k\omega_0$ is given by

$$\text{Sys}\left\{e^{jk\omega_0 t}\right\} = e^{jk\omega_0 t}\, H(k\omega_0) \tag{3.357}$$

Using Eqn. (3.357) in Eqn. (3.356), the response of the system to $\tilde{x}(t)$ is found as

$$\text{Sys}\{\tilde{x}(t)\} = \sum_{k=-\infty}^{\infty} c_k\, H(k\omega_0)\, e^{jk\omega_0 t} \tag{3.358}$$

Two important observations should be made based on Eqn. (3.358):

1. For a CTLTI system driven by a periodic input signal, the output signal is also periodic with the same period.

2. If the EFS coefficients of the input signal are $\{c_k\,;\, k=1,\ldots,\infty\}$, then the EFS coefficients of the output signal are $\{c_k\, H(k\omega_0)\,;\, k=1,\ldots,\infty\}$.

Example 3.46: RC circuit with pulse train input

Consider again the RC circuit used in Example 3.45 with system function given by Eqn. (3.351). Let the input signal be a pulse train with period $T_0 = 50$ ms and duty cycle $d = 0.2$ as shown in Fig. 3.94. Determine the output signal in steady state.

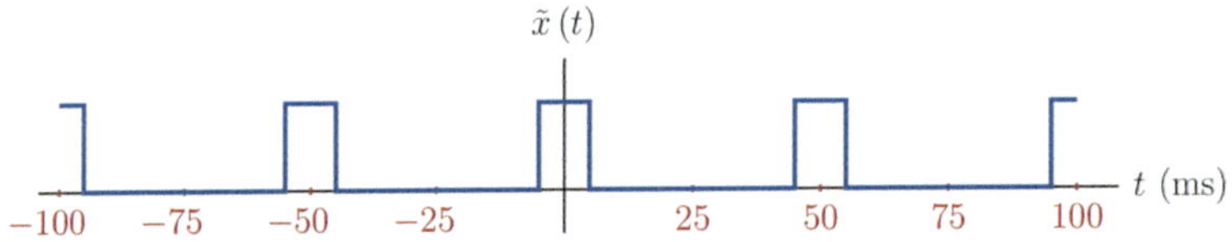

Figure 3.94 – The input signal for Example 3.46.

Solution: The EFS coefficients were determined in Example 3.7 for the type of periodic pulse train shown in Fig. 3.94. Using the specified duty cycle the coefficients are

$$c_k = 0.2\,\text{sinc}\,(0.2k) \tag{3.359}$$

The fundamental frequency is $f_0 = 1/T_0 = 20$ Hz, and the EFS representation of the signal $\tilde{x}(t)$ is

$$\tilde{x}(t) = \sum_{k=-\infty}^{\infty} 0.2\,\text{sinc}\,(0.2k)\, e^{j40\pi kt} \tag{3.360}$$

Let $\{d_k\,;\, k=-\infty,\ldots,\infty\}$ be the EFS coefficients of the output signal $\tilde{y}(t)$. Using Eqn. (3.358) we have

$$d_k = c_k\, H\big(kf_0\big) = \frac{c_k}{1+j\,(20k/80)}\,, \qquad k=-\infty,\ldots,\infty \tag{3.361}$$

The EFS representation of the output signal is

$$\tilde{y}(t) = \sum_{k=-\infty}^{\infty} \left(\frac{c_k}{1+j\,(20k/80)}\right) e^{j40\pi kt} \tag{3.362}$$

Using the expressions for the magnitude and the phase of the system function, Eqn. (3.361) can be written as

$$\left|d_k\right| = \left|c_k\right|\,\left|H\big(kf_0\big)\right| = \frac{\left|c_k\right|}{\sqrt{1+(20k/80)^2}} \tag{3.363}$$

and

$$\measuredangle d_k = \measuredangle c_k + \Theta\big(kf_0\big) = \measuredangle c_k - \tan^{-1}(20k/80) \tag{3.364}$$

The relationship between the system function $H\big(f\big)$ and the line spectra c_k and d_k is illustrated in Fig. 3.95.

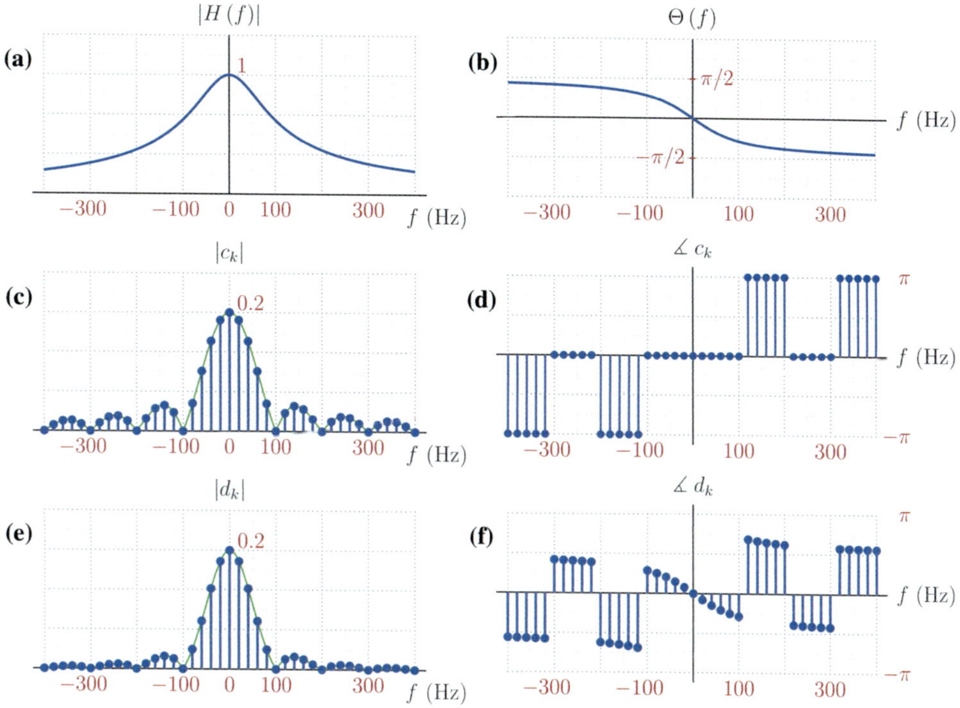

Figure 3.95 – System function and line spectra for Example 3.46: **(a)** the magnitude of the system function, **(b)** the phase of the system function, **(c)** the magnitude of the input spectrum, **(d)** the phase of the input spectrum, **(e)** the magnitude of the output spectrum, and **(f)** the phase of the output spectrum.

An approximation to the output signal $\tilde{y}(t)$ using terms up to the 75th harmonic is shown in Fig. 3.96.

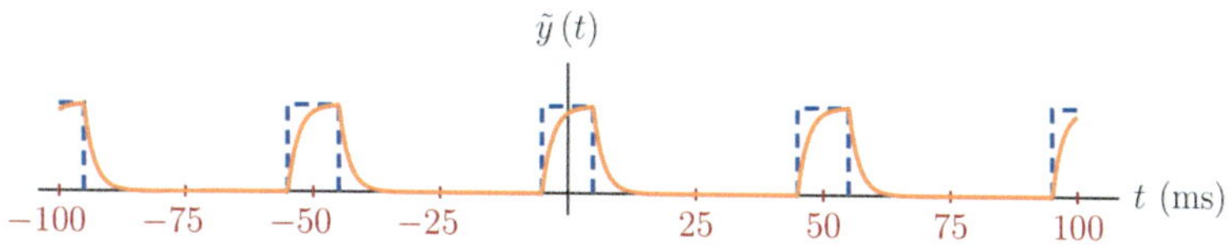

Figure 3.96 – The response of the system in Example 3.5 to pulse train with period $T_0 = 20$ ms and duty cycle $d = 0.2$.

Software resources: exct_3_46a.m , exct_3_46b.m , exct_3_46_Live.mlx

Interactive App: RC circuit with pulse train input

The interactive app in `appSysFun3.m` is based on Example 3.46, Figs. 3.94 and 3.95. It illustrates the use of the system function with the EFS coefficients of a periodic signal.

The system under consideration is the RC circuit that we have used extensively in this chapter. Its 3-dB cutoff frequency f_c may be adjusted by using a slider control.

The input signal is a periodic pulse train the period. The period and the duty cycle of the pulse train may be adjusted using slider controls. Continuing with the notation established in Example 3.46, we will let c_k and d_k represent the line spectra of the input and the output signals, respectively. In the app only the magnitude spectra $|c_k|$ and $|d_k|$ are shown due to screen size constraints.

a. Set the 3-dB cutoff frequency of the RC circuit to $f_c = 80$ Hz. Also, set the pulse period to $T_0 = 50$ ms and the pulse duty cycle to 25 percent. Observe the output line spectrum $|d_k|$ in relation to the input line spectrum $|c_k|$, keeping in mind that

$$|d_k| = |H(kf_0)|\,|c_k|$$

What is the frequency spacing of the line spectra $|c_k|$ and $|d_k|$? Pay attention to how the RC circuit modifies the line spectrum and how those modifications translate to the distortion in the output signal.

b. Gradually increase the pulse period to $T_0 = 100$ ms, and observe the line spectra. Does a larger pulse period help reduce the distortion in the output signal? Now reduce the pulse period to $T_0 = 20$ ms. What happens and why?

c. Set the pulse period back to $T_0 = 50$ ms. Increase the duty cycle from 25 percent to 50 percent. How does this affect the output signal?

d. With pulse period set to $T_0 = 50$ ms and the duty cycle set to 50 percent, gradually increase the 3-dB cutoff frequency of the RC circuit from $f_c = 80$ Hz to $f_c = 150$ Hz. Comment on the changes to the output signal.

Software resource: `appSysFun3.m`

3.7 CTLTI Systems with Non-Periodic Input Signals

Let us consider the case of using a non-periodic signal $x(t)$ as input to a CTLTI system. We have established in Section 2.7.2 of Chapter 2 that the output of a CTLTI system is equal to the convolution of the input signal with the impulse response, that is

$$y(t) = h(t) * x(t) = \int_{-\infty}^{\infty} h(\lambda)\, x(t-\lambda)\, d\lambda$$

Let us assume that

1. The system is stable ensuring that $H(\omega)$ converges.
2. The input signal has a Fourier transform.

We have seen in Section 3.3.5 of this chapter that the Fourier transform of the convolution of two signals is equal to the product of individual transforms. Thus, the output transform is the product of the input transform and the system function.

$$Y(\omega) = H(\omega)\, X(\omega) \tag{3.365}$$

Writing each transform involved in Eqn. (3.365) in polar form using its magnitude and phase we obtain the corresponding relationships:

Magnitude and phase of $Y(\omega)$:

$$|Y(\omega)| = |H(\omega)|\, |X(\omega)| \tag{3.366}$$

$$\measuredangle Y(\omega) = \measuredangle X(\omega) + \Theta(\omega) \tag{3.367}$$

1. The magnitude of the output spectrum is equal to the product of the magnitudes of the input spectrum and the system function.
2. The phase of the output spectrum is found by adding the phase characteristics of the input spectrum and the system function.

Example 3.47: Pulse response of RC circuit revisited

Consider again the RC circuit used Example 3.43. Let $f_c = 1/RC = 80$ Hz. Determine the Fourier transform of the response of the system to the unit pulse input signal $x(t) = \Pi(t)$.

Solution: In Example 2.21 of Chapter 2 the output signal $y(t)$ was found through the use of convolution operation. In this example we will approach the same problem from a system function perspective. The system function of the RC circuit under consideration was found in Example 3.43 to be

$$\text{Eqn. (3.333):} \qquad H(f) = \frac{1}{1 + j\,(f/f_c)}$$

The transform of the input signal is

$$X(f) = \text{sinc}(f)$$

Using $f_c = 80$ Hz, the transform of the output signal is

$$Y(f) = H(f)\, X(f) = \frac{1}{1 + j\left(\dfrac{f}{80}\right)}\, \text{sinc}(f) \tag{3.368}$$

Magnitude and phase of the transform $Y(\omega)$ can be found using Eqns. (3.366) and (3.367). The magnitude is

$$|Y(f)| = \frac{1}{\sqrt{1 + \left(\dfrac{f}{80}\right)^2}}\, |\text{sinc}(f)| \tag{3.369}$$

and the phase is

$$\measuredangle Y(f) = -\tan^{-1}\left(\frac{f}{80}\right) + \measuredangle\left[\text{sinc}(f)\right] \tag{3.370}$$

as illustrated in Fig. 3.97.

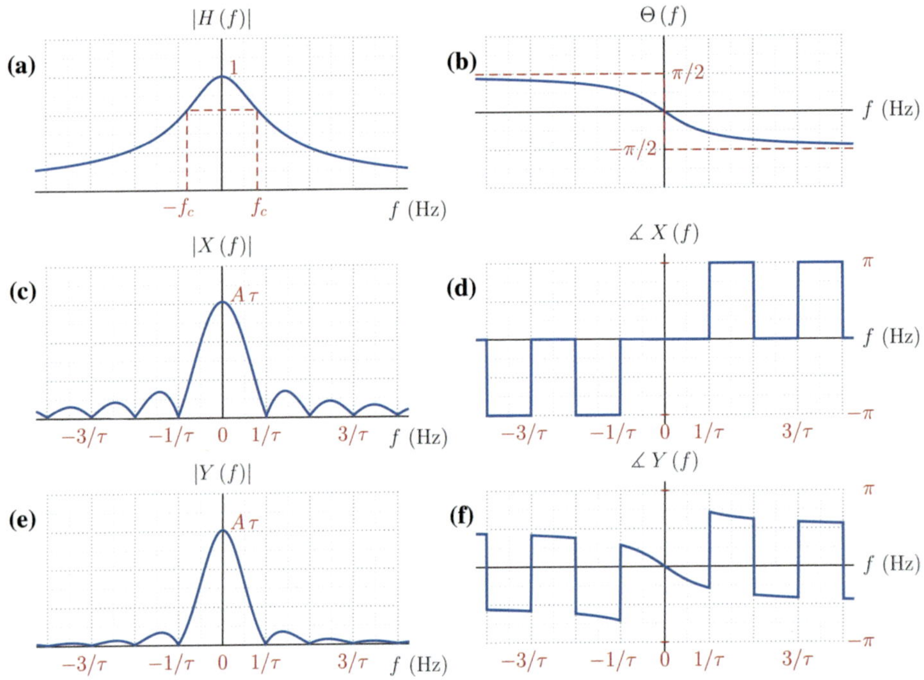

Figure 3.97 – Determining the Fourier transform of the output signal of a linear system: **(a)** the magnitude of the system function, **(b)** the phase of the system function, **(c)** the magnitude of the input spectrum, **(d)** the phase of the input spectrum, **(e)** the magnitude of the output spectrum, and **(f)** the phase of the output spectrum.

With a bit of work it can be shown that $Y(f)$ found above is indeed the Fourier transform of the signal $y(t)$ found in Example 2.21 using the convolution integral.

Software resources: `exct_3_47.m` , `exct_3_47_Live.mlx`

Interactive App: Pulse response of RC circuit revisited

The interactive app in `appSysFun4.m` is based on Example 3.47 and illustrates the use of the system function in conjunction with the Fourier transform of a pulse. The transform of the output signal is the product of the system function with the transform of the input signal as we have established in Eqns. (3.368) through (3.370).

The input signal $x(t)$ is a single pulse with unit amplitude and width equal to τ, and it is centered around $t=0$.

The user interface of the app displays magnitudes of the three transforms involved, namely $X(f)$, $H(f)$, and $Y(f)$. Time domain representations of the input and the output signals are also shown. Phase characteristics are not displayed in the app due to screen space constraints. System parameter f_c and input signal parameter τ may be varied using slider controls, and the effect of these variations may be observed on magnitude spectra as well as the output signal.

a. Set the 3-dB cutoff frequency of the RC circuit to $f_c = 80$ Hz and the pulse width to $\tau = 20$ ms. Observe the output magnitude spectrum $|Y(f)|$ in relation to the input spectrum $|X(f)|$, keeping in mind that

$$|Y(f)| = |H(f)|\,|X(f)|$$

Pay attention to how the RC circuit modifies the line spectrum, and how those modifications translate to the distortion in the output signal.

b. Gradually increase the pulse width to $\tau = 100$ ms, and observe the magnitude spectra. Does a wider input pulse help reduce the distortion in the output signal? Now reduce the pulse width down to $\tau = 10$ ms. What happens and why?

c. Set the pulse width back to $\tau = 20$ ms. Gradually increase the 3-dB cutoff frequency of the RC circuit from $f_c = 80$ Hz to $f_c = 150$ Hz. Comment on the changes to the output signal.

Software resource: `appSysFun4.m`

Summary of Key Points

☞ A signal $\tilde{x}(t)$ that is periodic with fundamental period T_0 and associated fundamental frequency $f_0 = 1/T_0$ can be represented with a trigonometric Fourier series (TFS) expansion in the form

Eqn. (3.42):
$$\tilde{x}(t) = a_0 + \sum_{k=1}^{\infty} a_k \cos(k\omega_0 t) + \sum_{k=1}^{\infty} b_k \sin(k\omega_0 t)$$

with series coefficients computed as

Eqn. (3.43):
$$a_k = \frac{2}{T_0}\int_{t_0}^{t_0+T_0} \tilde{x}(t)\cos(k\omega_0 t)\,dt\,, \quad \text{for} \quad k = 1,\ldots,\infty$$

Eqn. (3.44):
$$b_k = \frac{2}{T_0}\int_{t_0}^{t_0+T_0} \tilde{x}(t)\sin(k\omega_0 t)\,dt\,, \quad \text{for} \quad k = 1,\ldots,\infty$$

and

Eqn. (3.45):
$$a_0 = \frac{1}{T_0}\int_{t_0}^{t_0+T_0} \tilde{x}(t)\,dt \qquad \text{(dc component)}$$

☞ The basis functions used in the TFS expansion form an orthogonal set. It can be shown that

Eqn. (3.32):
$$\int_{t_0}^{t_0+T_0} \cos(m\omega_0 t)\cos(k\omega_0 t)\,dt = \begin{cases} T_0/2, & m = k \\ 0, & m \neq k \end{cases}$$

Eqn. (3.33): $$\int_{t_0}^{t_0+T_0} \sin(m\omega_0 t)\,\sin(k\omega_0 t)\,dt = \begin{cases} T_0/2, & m=k \\ 0, & m\neq k \end{cases}$$

and

Eqn. (3.34): $$\int_{t_0}^{t_0+T_0} \cos(m\omega_0 t)\,\sin(k\omega_0 t)\,dt = 0$$

☞ A signal $\tilde{x}(t)$ that is periodic with fundamental period T_0 and associated fundamental frequency $f_0 = 1/T_0$ can alternatively be represented with an exponential Fourier series (EFS) expansion in the form

Eqn. (3.72): $$\tilde{x}(t) = \sum_{k=-\infty}^{\infty} c_k\, e^{jk\omega_0 t}$$

with series coefficients computed as

Eqn. (3.73): $$c_k = \frac{1}{T_0}\int_{t_0}^{t_0+T_0} \tilde{x}(t)\, e^{-jk\omega_0 t}\,dt$$

☞ The basis functions used in the EFS expansion form an orthogonal set. It can be shown that

Eqn. (3.64): $$\int_{t_0}^{t_0+T_0} e^{jm\omega_0 t}\, e^{-jk\omega_0 t}\,dt = \begin{cases} T_0, & m=k \\ 0, & m\neq k \end{cases}$$

☞ Coefficients of TFS and EFS expansions can be easily related to each other. EFS coefficients can be computed from TFS coefficients using $c_0 = a_0$ and

Eqn. (3.61): $$c_k = \frac{1}{2}\left(a_k - j b_k\right), \qquad k = 1,\ldots,\infty$$

and

Eqn. (3.62): $$c_{-k} = \frac{1}{2}\left(a_k + j b_k\right), \qquad k = 1,\ldots,\infty$$

Alternatively, TFS coefficients can be computed from EFS coefficients using

Eqn. (3.59): $$c_k + c_{-k} = a_k, \qquad k = 1,\ldots,\infty$$

and

Eqn. (3.60): $$j\,(c_k - c_{-k}) = b_k, \qquad k = 1,\ldots,\infty$$

☞ For a real-valued signal $\tilde{x}(t)$, positive and negative-indexed EFS coefficients are complex conjugates of each other, that is,

Eqn. (3.63): $$c_k = c_{-k}^{*}$$

☞ Yet another form of the Fourier series representation of a periodic signal is the compact Fourier series (CFS) expressed as

Eqn. (3.81): $$\tilde{x}(t) = d_0 + \sum_{k=1}^{\infty} d_k \cos\left(k\omega_0 t + \phi_k\right)$$

CFS coefficients can be computed from EFS coefficients using

Eqn. (3.85): $$d_k = \sqrt{\{a_k^2 + b_k^2\}}\,, \qquad k = 1, \ldots, \infty$$

and

Eqn. (3.86): $$\phi_k = -\tan^{-1}\left(\frac{b_k}{a_k}\right), \qquad k = 1, \ldots, \infty$$

with $d_0 = a_0$ and $\phi_0 = 0$.

☞ For the Fourier series representation of a periodic signal $\tilde{x}(t)$ to exist, it must satisfy a set of conditions known as Dirichtlet conditions. Most importantly, the signal must be integrable over one period in an absolute sense, that is

Eqn. (3.91): $$\int_0^{T_0} \left|\tilde{x}(t)\right| dt < \infty$$

Any periodic signal in which the amplitude values are bounded will satisfy this condition. In addition, periodic repetitions of singularity functions such as a train of impulses repeated every T_0 seconds will satisfy it as well. If $\tilde{x}(t)$ has discontinuities, it must have a finite number of them in one period. The signal must also have at most a finite number of minima and maxima in one period.

☞ At a point of discontinuity in a periodic signal, the Fourier series expansion (TFS, EFS, or CFS representation of the signal) yields the average value obtained by approaching the discontinuity from opposite directions. If the signal $\tilde{x}(t)$ has a discontinuity at $t = t_0$, evaluating the Fourier series at that time instant yields

Eqn. (3.92): $$\sum_{k=-\infty}^{\infty} c_k\, e^{jk\omega_0 t_0} = \frac{1}{2} \lim_{t \to t_0^-} \left[\tilde{x}(t)\right] + \frac{1}{2} \lim_{t \to t_0^+} \left[\tilde{x}(t)\right]$$

☞ In finite-harmonic approximation of a periodic signal using m harmonics, the approximation error $\tilde{\varepsilon}^{(m)}(t)$ tends to be largest around discontinuities. In addition, Fourier series approximation overshoots the actual signal value at one side of each discontinuity and undershoots it at the other side. The amount of overshoot or undershoot is about 9 percent of the height of the discontinuity, and cannot be reduced by increasing m. This is known as the Gibbs phenomenon.

☞ Fourier series operator is linear. Let two signals $\tilde{x}(t)$ and $\tilde{y}(t)$ periodic with $T_0 = 2\pi/\omega_0$ have EFS coefficients c_k and d_k, respectively. For any two scalars α_1 and α_2, the new periodic signal $\tilde{w}(t) = \alpha_1\,\tilde{x}(t) + \alpha_2\,\tilde{y}(t)$ has the EFS coefficients $\alpha_1\,c_k + \alpha_2\,d_k$.

☞ For a periodic signal $\tilde{x}(t)$ that is purely real, the EFS coefficients are conjugate symmetric, that is, $c_{-k} = c_k^*$ for all k. For a $\tilde{x}(t)$ that is purely imaginary, the EFS coefficients are conjugate antisymmetric, that is, $c_{-k} = -c_k^*$ for all k.

☞ If the periodic signal $\tilde{x}(t)$ is even, EFS coefficients c_k are real-valued for all k. If $\tilde{x}(t)$ is odd, EFS coefficients c_k are purely imaginary for all k.

☞ If the periodic signal $\tilde{x}(t)$ has EFS coefficients c_k, then the EFS coefficients for $\tilde{x}(t-\tau)$ are $c_k\,e^{-jk\omega_0\tau}$.

☞ Fourier transform of the signal $x(t)$ is

Eqn. (3.122): $$X(\omega) = \int_{-\infty}^{\infty} x(t)\, e^{-j\omega t}\, dt$$

The signal $x(t)$ can be obtained from the transform $X(\omega)$ through

Eqn. (3.121): $$x(t) = \frac{1}{2\pi} \int_{-\infty}^{\infty} X(\omega)\, e^{j\omega t}\, d\omega$$

☞ For the Fourier transform $X(\omega)$ to exist, the signal $x(t)$ must satisfy a set of conditions known as Dirichlet conditions. First, it must be integrable in an absolute sense, that is

Eqn. (3.127): $$\int_{-\infty}^{\infty} \left|x(t)\right| dt < \infty$$

Furthermore, if it has discontinuities, it must have at most a finite number of them in any finite time interval. Also, it must have at most a finite number of minima and maxima in any finite time interval.

☞ An alternative condition for the existence of the Fourier transform is that the signal be square integrable, that is

Eqn. (3.130): $$\int_{-\infty}^{\infty} \left|x(t)\right|^2 dt < \infty$$

As a consequence, all energy signals have Fourier transforms.

☞ The Fourier transform of an isolated pulse centered around time origin with amplitude A and width τ is

Eqn. (3.136): $$X(f) = A\tau \,\text{sinc}\left(f\tau\right)$$

☞ Fourier transform is a linear operator. For any two signals $x_1(t)$ and $x_2(t)$ with their respective transforms $X_1(\omega)$ and $X_2(\omega)$ and arbitrary constants α_1, α_2, the superposition principle holds:

Eqn. (3.152): $$\alpha_1\, x_1(t) + \alpha_2\, x_2(t) \stackrel{\mathcal{F}}{\longleftrightarrow} \alpha_1\, X_1(\omega) + \alpha_2\, X_2(\omega)$$

☞ If the signal $x(t)$ is real-valued, its Fourier transform $X(\omega)$ is conjugate symmetric. $\text{Im}\left\{x(t)\right\} = 0$ implies that $X^*(\omega) = X(-\omega)$. Conversely, if the signal $x(t)$ is purely imaginary, its transform is conjugate antisymmetric. $\text{Re}\left\{x(t)\right\} = 0$ implies that $X^*(\omega) = -X(-\omega)$.

☞ For the Fourier transform of a real-valued signal $x(t)$, the magnitude $\left|X(\omega)\right|$ is even and the phase $\Theta(\omega)$ is odd.

☞ If the real-valued signal $x(t)$ is an even function of time, the resulting transform $X(\omega)$ is real-valued for all ω. Conversely, if the real-valued signal $x(t)$ is an odd function of time, its transform $X(\omega)$ is purely imaginary for all ω.

☞ Given a transform pair

$$x(t) \stackrel{\mathcal{F}}{\longleftrightarrow} X(\omega)$$

the following are also valid transform pairs:

Duality:

Eqn. (3.160): $$X(t) \overset{\mathcal{F}}{\longleftrightarrow} x(-f)$$

Time shifting:

Eqn. (3.206): $$x(t-\tau) \overset{\mathcal{F}}{\longleftrightarrow} X(\omega)\, e^{-j\omega\tau}$$

Frequency shifting:

Eqn. (3.210): $$x(t)\, e^{j\omega_0 t} \overset{\mathcal{F}}{\longleftrightarrow} X(\omega-\omega_0)$$

Modulation property:

Eqn. (3.218): $$x(t)\cos(\omega_0 t) \overset{\mathcal{F}}{\longleftrightarrow} \frac{1}{2}\left[X(\omega-\omega_0)+X(\omega+\omega_0)\right]$$

Eqn. (3.219): $$x(t)\sin(\omega_0 t) \overset{\mathcal{F}}{\longleftrightarrow} \frac{1}{2}\left[X(\omega-\omega_0)\, e^{-j\pi/2}+X(\omega+\omega_0)\, e^{j\pi/2}\right]$$

Time and frequency scaling:

Eqn. (3.225): $$x(at) \overset{\mathcal{F}}{\longleftrightarrow} \frac{1}{|a|}X\left(\frac{\omega}{a}\right)$$

Differentiation in the time domain:

Eqn. (3.234): $$\frac{d^n}{dt^n}\left[x(t)\right] \overset{\mathcal{F}}{\longleftrightarrow} (j\omega)^n\, X(\omega)$$

Differentiation in the frequency domain:

Eqn. (3.250): $$(-jt)^n\, x(t) \overset{\mathcal{F}}{\longleftrightarrow} \frac{d^n}{d\omega^n}\left[X(\omega)\right]$$

Integration:

Eqn. (3.275): $$\int_{-\infty}^{t} x(\lambda)\, d\lambda \overset{\mathcal{F}}{\longleftrightarrow} \frac{X(\omega)}{j\omega}+\pi\, X(0)\,\delta(\omega)$$

☞ Given two transform pairs

$$x_1(t) \overset{\mathcal{F}}{\longleftrightarrow} X_1(\omega) \quad \text{and} \quad x_2(t) \overset{\mathcal{F}}{\longleftrightarrow} X_2(\omega)$$

the following are also valid transform pairs:

Convolution property:

Eqn. (3.263): $$x_1(t) * x_2(t) \overset{\mathcal{F}}{\longleftrightarrow} X_1(\omega)\, X_2(\omega)$$

Multiplication of two signals:

Eqn. (3.268): $$x_1(t)\, x_2(t) \overset{\mathcal{F}}{\longleftrightarrow} \frac{1}{2\pi} X_1(\omega) * X_2(\omega)$$

☞ Fourier transform can be extended to periodic signals through the use of EFS coefficients in conjunction with singularity functions.

Eqn. (3.287):
$$X(\omega) = \sum_{k=-\infty}^{\infty} 2\pi c_k \, \delta(\omega - k\omega_0)$$

☞ Parseval's theorem:

For a periodic power signal $\tilde{x}(t)$ with period of T_0 and EFS coefficients $\{c_k\}$

Eqn. (3.288):
$$\frac{1}{T_0}\int_{t_0}^{t_0+T_0} |\tilde{x}(t)|^2 \, dt = \sum_{k=-\infty}^{\infty} |c_k|^2$$

For a non-periodic energy signal $x(t)$ with a Fourier transform $X(f)$

Eqn. (3.289):
$$\int_{-\infty}^{\infty} |x(t)|^2 \, dt = \int_{-\infty}^{\infty} |X(f)|^2 \, df$$

☞ For a periodic power signal $\tilde{x}(t)$ with fundamental frequency f_0 and EFS coefficients $\{c_k\}$, the *power spectral density* is

Eqn. (3.295):
$$S_x(f) = \sum_{k=-\infty}^{\infty} |c_k|^2 \, \delta(f - kf_0)$$

The power contained at frequencies in the range $(-f_0, f_0)$ can be computed as

Eqn. (3.300):
$$P_x \text{ in } (-f_0, f_0) = \int_{-f_0}^{f_0} S_x(f) \, df$$

☞ For a non-periodic energy signal $x(t)$ with Fourier transform $X(f)$, the *energy spectral density* is

Eqn. (3.301):
$$G_x(f) = |X(f)|^2$$

☞ For a non-periodic power signal, the power spectral density is computed as

Eqn. (3.308):
$$S_x(f) = \lim_{T\to\infty}\left[\frac{1}{T}|X_T(f)|^2\right]$$

☞ Autocorrelation function: For an energy signal $x(t)$ the autocorrelation function is defined as

Eqn. (3.309):
$$r_{xx}(\tau) = \int_{-\infty}^{\infty} x(t)\, x(t+\tau)\, dt$$

For a periodic power signal $\tilde{x}(t)$ with period T_0, the autocorrelation function is

Eqn. (3.310):
$$\tilde{r}_{xx}(\tau) = \frac{1}{T_0}\int_{-T_0/2}^{T_0/2} \tilde{x}(t)\, \tilde{x}(t+\tau)\, dt$$

For a non-periodic power signal, the autocorrelation function is

Eqn. (3.311):
$$r_{xx}(\tau) = \lim_{T_0\to\infty}\left[\frac{1}{T_0}\int_{-T_0/2}^{T_0/2} x(t)\, x(t+\tau)\, dt\right]$$

☞ For an energy signal $x(t)$, the energy spectral density is the Fourier transform of the autocorrelation function.

Eqn. (3.312): $$G_x(f) = \mathcal{F}\{r_{xx}(\tau)\}$$

☞ For a periodic power signal $\tilde{x}(t)$, the power spectral density is the Fourier transform of the autocorrelation function.

Eqn. (3.323): $$S_x(f) = \mathcal{F}\{\tilde{r}_{xx}(\tau)\}$$

☞ The autocorrelation function has the following properties:

1. $r_{xx}(0) \geq |r_{xx}(\tau)|$ for all τ.
2. $r_{xx}(-\tau) = r_{xx}(\tau)$ for all τ, that is, the autocorrelation function has even symmetry.
3. If the signal $x(t)$ is periodic with period T, then its autocorrelation function $\tilde{r}_{xx}(\tau)$ is also periodic with the same period.

☞ The system function is simply the Fourier transform of the impulse response.

Eqn. (3.331): $$H(\omega) = \mathcal{F}\{h(t)\} = \int_{-\infty}^{\infty} h(t)\, e^{-j\omega t}\, dt$$

☞ System function for a CTLTI system can be obtained from the differential equation in two steps:

1. Form the Fourier transform of each side of the differential equation. Use the time differentiation property of the Fourier transform for this step.

Eqn. (3.337): $$\frac{d^k y(t)}{dt^k} \overset{\mathcal{F}}{\longleftrightarrow} (j\omega)^k\, Y(\omega) \qquad k = 0, 1, \ldots$$

Eqn. (3.338): $$\frac{d^k x(t)}{dt^k} \overset{\mathcal{F}}{\longleftrightarrow} (j\omega)^k\, X(\omega) \qquad k = 0, 1, \ldots$$

2. Find the ratio of the output transform $Y(\omega)$ to the input transform $Y(\omega)$.

Eqn. (3.336): $$H(\omega) = \frac{Y(\omega)}{X(\omega)}$$

☞ The response of CTLTI system to complex exponential input is found as

Eqn. (3.344): $$y(t) = \text{Sys}\left\{e^{j\omega_0 t}\right\} = e^{j\omega_0 t}\, H(\omega_0)$$

☞ The response of a CTLTI system to a sinusoidal signal is found as

Eqn. (3.350): $$y(t) = \text{Sys}\{\cos(\omega_0 t)\} = |H(\omega_0)| \cos\left(\omega_0 t + \Theta(\omega_0)\right)$$

where $|H(\omega_0)|$ and $\Theta(\omega_0)$ are magnitude and phase of the system function, respectively.

☞ The response of a CTLTI system to periodic input signal $\tilde{x}(t)$ with EFS coefficients c_k is found as

Eqn. (3.358): $$\text{Sys}\{\tilde{x}(t)\} = \sum_{k=-\infty}^{\infty} c_k\, H(k\omega_0)\, e^{jk\omega_0 t}$$

☞ For a CTLTI system with system function $H(\omega)$ and non-periodic input and output signals $x(t)$ and $y(t)$, we have the following relationship between input and output transforms:

Eqn. (3.365): $$Y(\omega) = H(\omega)\, X(\omega)$$

Further Reading

[1] R.J. Beerends. *Fourier and Laplace Transforms.* Cambridge University Press, 2003.

[2] R.N. Bracewell. *The Fourier Transform and Its Applications.* Electrical Engineering Series. McGraw-Hill, 2000.

[3] Phil Dyke. *An Introduction to Laplace Transforms and Fourier Series.* Springer Science & Business Media, 2014.

[4] E.A. Gonzalez-Velasco. *Fourier Analysis and Boundary Value Problems.* Elsevier Science, 1996.

[5] Loukas Grafakos et al. *Classical Fourier Analysis.* Vol. 2. Springer, 2008.

[6] R. Haberman. *Applied Partial Differential Equations: With Fourier Series and Boundary Value Problems.* Pearson Education, 2012.

[7] Kenneth B Howell. *Principles of Fourier analysis.* CRC Press, 2016.

[8] Elias M. Stein and Rami Shakarchi. *Fourier Analysis: An Introduction.* Vol. 1. Princeton University Press, 2011.

MATLAB Exercises with Solutions

MATLAB Exercise 3.1: Computing finite-harmonic approximation to pulse train

Develop a MATLAB script to compute and graph the finite-harmonic approximation to the periodic pulse train $\tilde{x}(t)$ of Example 3.1 for $m = 4$.

Solution: Consider the script `mexct_3_1.m` listed below:

```
1  % Script: mexct_3_1.m
2  m = 4;                  % Number of harmonics to be used
3  t = [-7:0.01:7];        % Create a vector of time instants
4  f0 = 1/3;               % Fundamental frequency
5  omega0 = 2*pi*f0;
6  x = 1/3;                % Recall that a0=1/3.
7                          % We will start by setting x(t)=a0.
8  % Start the loop to compute the contribution of each harmonic.
9  for k = 1:m
10   % See Eqn. (3.46)
11   ak = 1/(pi*k)*sin(k*omega0);
12   bk = 1/(pi*k)*(1-cos(k*omega0));
13   x = x+ak*cos(k*omega0*t)+bk*sin(k*omega0*t);
14 end
15 % Graph the resulting approximation to x(t).
16 plot(t,x); grid;
17 title('Approximation using m=4 harmonics');
18 xlabel('Time (sec)');
19 ylabel('Amplitude');
```

The main part of the code is the loop construct between lines 9 and 14. Before entering the loop, the vector x holds a zero-order approximation to the signal, that is, $\tilde{x}^{(0)}(t) = a_0 = 1/3$.

After the first pass through the loop, the vector x holds samples of the signal

$$\tilde{x}^{(1)}(t) = a_0 + a_1 \cos\left(2\pi f_0 t\right) + b_1 \sin\left(2\pi f_0 t\right)$$

After the second pass through the loop, the vector x now holds

$$\tilde{x}^{(2)}(t) = a_0 + a_1 \cos\left(2\pi f_0 t\right) + b_1 \sin\left(2\pi f_0 t\right) + a_2 \cos\left(4\pi f_0 t\right) + b_2 \sin\left(4\pi f_0 t\right)$$

and so on. The graph produced by this listing is shown in Fig. 3.98.

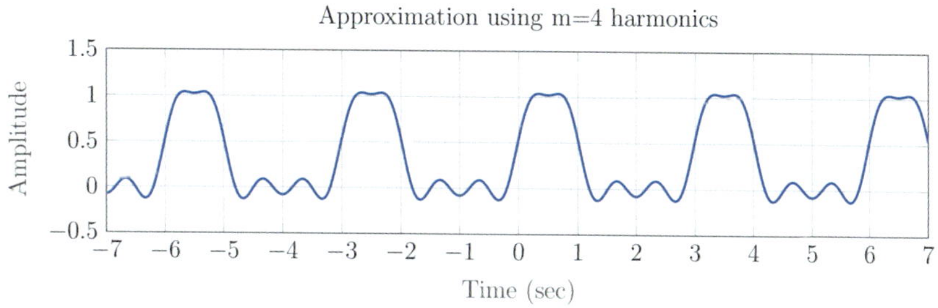

Figure 3.98 – Graph produced by the code in MATLAB Exercise 3.1.

Software resource: mexct_3_1.m

MATLAB Exercise 3.2: Computing multiple approximations to pulse train

MATLAB script mexct_3_2.m given below computes all approximations to the periodic pulse train $\tilde{x}(t)$ of Example 3.1 for $m = 1,\ldots,5$. The approximations $\tilde{x}^{(1)}(t)$, $\tilde{x}^{(2)}(t)$, $\tilde{x}^{(4)}(t)$, $\tilde{x}^{(5)}(t)$, are graphed on the same coordinate system for comparison. The approximation $\tilde{x}^{(3)}(t)$ is not graphed since $a_3 = b_3 = 0$ based on Table 3.1, and therefore $\tilde{x}^{(3)}(t) = \tilde{x}^{(2)}(t)$.

```
% Script: mexct_3_2.m
m = 5;                 % Maximum number of harmonics to be used
t = [-3:0.005:3];      % Create a vector of time instants
f0 = 1/3;              % Fundamental frequency
omega0 = 2*pi*f0;
x = 1/3*ones(size(t));   % Recall that a0=1/3
                         % We will start by setting x(t)=a0
xmat = x;              % Row vector containing dc value of 1/3
% Start the loop to compute the contribution of each harmonic
for k = 1:m
  ak = 1/(pi*k)*sin(k*omega0);
  bk = 1/(pi*k)*(1-cos(k*omega0));
  x = x+ak*cos(k*omega0*t)+bk*sin(k*omega0*t);
  xmat = [xmat;x];   % Append another row to matrix 'xmat'
end
% Graph the results
```

```
17 plot(t,xmat(2,:),'b--',t,xmat(3,:),'g-',t,xmat(5,:),'r-',..
18    t,xmat(6,:),'k-'); grid;
19 title('Approximation to x(t) using m=1,...,5');
20 xlabel('Time (sec)');
21 ylabel('Amplitude');
22 legend('x^{(1)}(t)','x^{(2)}(t)','x^{(4)}(t)','x^{(5)}(t)');
```

On line 8 the matrix xmat is initialized as a row vector holding the constant (zero-order) approximation to $\tilde{x}(t)$. The loop construct between lines 10 and 15 is a slightly modified version of the corresponding construct in MATLAB Exercise 3.1. At each pass through the loop a new row is added to the matrix xmat as the next approximation to the periodic signal. When the script is finished, the rows of the matrix xmat correspond to finite-harmonic approximations as follows:

$$\begin{aligned} \text{row } 1 &\Rightarrow \tilde{x}^{(0)}(t) \\ \text{row } 2 &\Rightarrow \tilde{x}^{(1)}(t) \\ \text{row } 3 &\Rightarrow \tilde{x}^{(2)}(t) \\ &\vdots \end{aligned}$$

The graph produced by this script is shown in Fig. 3.99.

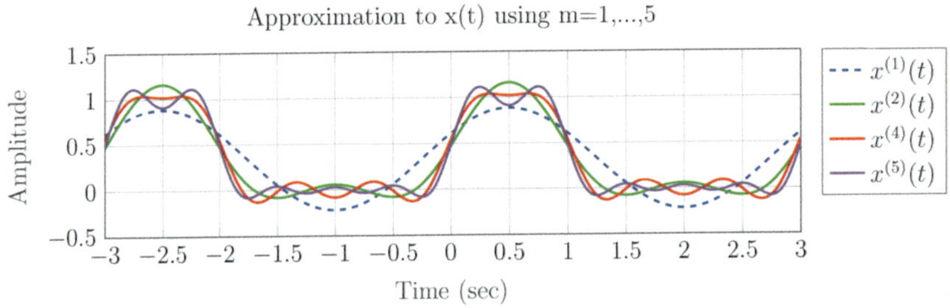

Figure 3.99 – Graph produced by the code in MATLAB Exercise 3.2.

Software resource: mexct_3_2.m

MATLAB Exercise 3.3: Graphing the line spectrum in Example 3.5

The line spectrum found in Example 3.5 can be graphed in MATLAB using a direct implementation of the general expression found for the EFS coefficients. The code given below computes and graphs the spectrum in the index range $-25 \le k \le 25$.

```
% Script: mexct_3_3a.m
k = [-25:25];          % Create a vector of index values
k = k+eps;             % Avoid division by zero
ck = sin(pi*k/3)./(pi*k);   % Eqn. (3.371)
% Graph the line spectrum
stem(k,ck);
```

```
7 axis([-25.5,25.5,-0.1,0.4]);
8 xlabel('Index k');
9 ylabel('Coefficient');
```

The resulting line spectrum graph is shown in Fig. 3.100.

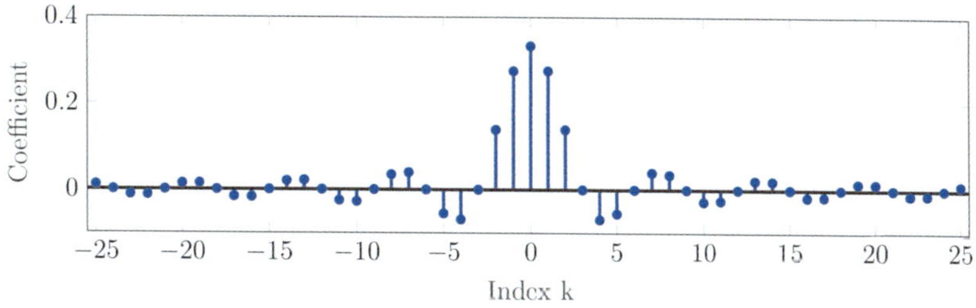

Figure 3.100 – The line spectrum for MATLAB Exercise 3.3.

Recall that the general expression for c_k as given by Eqn. (3.74) produced an indefinite value at $k = 0$ necessitating the use of L'Hospital's rule. Without any precautions, line 4 of the code given above would produce a division by zero error for $k = 0$. To avoid this, we add a very small positive value to all indices. This is done in line 3 of the code with the statement

```
3 k = k+eps;
```

MATLAB has a built-in variable `eps` for this purpose. It is the smallest positive floating-point number MATLAB can represent and therefore its effect on the end result should be negligible. An alternative and perhaps tidier method of avoiding the division-by-zero error is to use the sinc function defined as

$$\text{sinc}\,(\alpha) \triangleq \frac{\sin\,(\pi\alpha)}{\pi\alpha}$$

which allows the result of Eqn. (3.74) to be written in the form

$$c_k = \tfrac{1}{3}\,\text{sinc}\,(k/3) \tag{3.371}$$

MATLAB function `sinc()` computes the value at $k = 0$ correctly without requiring any special precautions on our part. A modified script file that uses the function `sinc()` and produces the same graph is given below:

```
1 % Script: mexct_3_3b.m
2 k = [-25:25];           % Create a vector of index values
3 ck = 1/3*sinc(k/3);     % Eqn. (3.371)
4 % Graph the line spectrum
5 stem(k,ck);
6 axis([-25.5,25.5,-0.1,0.4]);
7 xlabel('Index k');
8 ylabel('Coefficient');
```

If the line spectrum is desired in polar form with magnitude and phase, lines 5 through 8 of the script listed above may be replaced with the following:

```
5 tiledlayout(2,1);
6 nexttile;
7 stem(k,abs(ck));
```

```
axis([-25.5,25.5,-0.1,0.4]);
xlabel('Index');
ylabel('Magnitude');
nexttile;
stem(k,angle(ck));
axis([-25.5,25.5,-4,4]);
xlabel('Index');
ylabel('Phase (rad)');
```

Magnitude and phase of the EFS coefficients are graphed through the use of MATLAB functions abs() and angle(), respectively. The resulting graph is shown in Fig. 3.101.

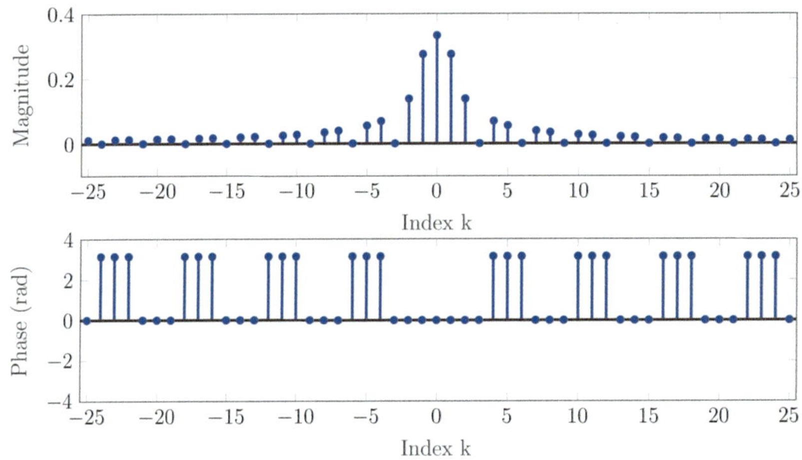

Figure 3.101 – The line spectrum in polar form for MATLAB Exercise 3.3.

Software resources: mexct_3_3a.m , mexct_3_3b.m , mexct_3_3c.m

MATLAB Exercise 3.4: Line spectrum for Example 3.6

MATLAB listing given below can be used for computing and displaying the EFS coefficients for the signal $\tilde{x}(t)$ of Example 3.6 for $k = -25, \ldots, 25$.

```
% Script: mexct_3_4a.m
k = [-25:25];       % Create a vector of index values
k = k+eps;          % Avoid division by zero
cReal = 1./(2*pi*k).*sin(2*pi*k/3);      % EFS coefficients (real part)
cImag = 1./(2*pi*k).*(cos(2*pi*k/3)-1);  % EFS coefficients (imag part)
cMag = abs(cReal+j*cImag);               % Magnitude
cPhs = angle(cReal+j*cImag);             % Phase
% Graph the line spectrum
tiledlayout(2,1);
nexttile;
stem(k,cMag);
axis([-25.5,25.5,-0.1,0.4]);
```

```
title('|c_{k}|');
xlabel('Index k');
ylabel('Magnitude');
nexttile;
stem(k,cPhs);
axis([-25.5,25.5,-4,4]);
title('\angle c_{k}');
xlabel('Index k');
ylabel('Phase (rad)');
```

A finite-harmonic approximation to the signal $\tilde{x}(t)$ can be computed and graphed using the script below:

```
1  % Script: mexct_3_4b.m
2  f0 = 1/3;             % Fundamental frequency
3  omega0 = 2*pi*f0;
4  t = [-3:0.01:6];      % Create a vector of time instants
5  x=zeros(size(t));     % Start with a vector of all zeros
6  for k=-25:25
7      k = k+eps;        % Avoid division by zero
8      cReal = 1/(2*pi*k)*sin(k*omega0);
9      cImag = 1/(2*pi*k)*(cos(k*omega0)-1);
10     x = x+(cReal+j*cImag)*exp(j*k*omega0*t);
11 end
12 % Graph the signal
13 plot(t,real(x)); grid;
14 title('Approximation to x(t)');
15 xlabel('Time (sec)');
16 ylabel('Amplitude');
```

On line 13 of the code we use the real() function on the vector x to remove any residual imaginary part that may be introduced as a result of round-off error.

Software resources: mexct_3_4a.m , mexct_3_4b.m

MATLAB Exercise 3.5: Developing a function to approximate EFS coefficients

It is possible to approximate EFS coefficients of a periodic signal using the built-in function fft() and modifying the result. The acronym *FFT* stands for *fast Fourier transform,* and represents a way of computing the *DFT (discrete Fourier transform)* in an efficient manner. Both FFT and DFT are subjects of discrete-time signals and systems. They are covered in the companion textbook *Discrete-Time Signals and Systems: A MATLAB Integated Approach.* We will, however, use the function fft() in the implementation of the function ss_efsapprox () shown below.

```
function c = ss_efsapprox(x,k)
  Nx = length(x);   % Size of vector 'x'
  Nk = length(k);   % Size of vector 'k'
  % Create a return vector 'c' the same size as 'k'
  c = zeros(Nk,1);
  Xk = fft(x)/Nx;
```

```
  % Copy the coefficients requested
  for i = 1:Nk
    kk = k(i);
    if (kk >= 0)
      c(i) = Xk(kk+1);
    else
      c(i) = Xk(Nx+1+kk);
    end
  end
end
```

Input arguments for the function are x and k. The former is a vector holding samples of the periodic signal $\tilde{x}(t)$ for exactly one period. The latter is a vector with the indices of the EFS coefficients desired. The function returns approximate values of the EFS coefficients with specified indices.

Before testing the function with the script given below, two cautionary issues be brought up:

1. For the quality of approximation to be acceptable, the largest coefficient index requested must be significantly smaller than the number of samples in one period of the signal. The script listed below uses the signal $\tilde{x}(t)$ of Example 3.3 shown in Fig. 3.7. A total of 500 samples are used for describing one period of the signal. The largest EFS coefficient index we request is 5.
2. Line 4 of the script is critical; the set of time instants in vector t starts at $t = 0$, and ends at slightly before $t = T_0 = 3$ s. Specifically, the last time instant at which we evaluate the signal $\tilde{x}(t)$ is $(N-1)T_0/N = 499T_0/500$. The time instant $t = T_0$ belongs to the next period and is not included in the vector t.

```
1 % Script: mexct_3_5a.m
2 N = 500;                 % Number of samples per period
3 T0 = 3;                  % Fundamental period
4 t = [0:N-1]/N*T0;        % Vector of time instants
5 x = (t<=0.5)|(t>=2.5);   % Describe one period of the signal
6 k = [-5:5];              % Vector of desired EFS coefficient indices
7 c = ss_efsapprox(x,k);   % Compute approximate EFS coefficients
8 % Display coefficients
9 c
```

The script prints approximate values of EFS coefficients as shown below. Compare these values to the ones found in Example 3.5, Eqn. (3.74).

```
c =

 -0.0548 + 0.0000i
 -0.0693 - 0.0000i
 -0.0007 - 0.0000i
  0.1375 + 0.0000i
  0.2760 - 0.0000i
  0.3340 + 0.0000i
  0.2760 + 0.0000i
  0.1375 - 0.0000i
 -0.0007 + 0.0000i
 -0.0693 + 0.0000i
 -0.0548 - 0.0000i
```

The script listed below repeats the exercise with approximate EFS coefficients in the range $k = -35, \ldots, 35$. It then computes and graphs a finite-harmonic approximation to the signal using the coefficients obtained. The result is shown in Fig. 3.102.

```
% Script: mexct_3_5b.m
N = 500;                    % Number of samples per period
T0 = 3;                     % Fundamental period
t = [0:N-1]/N*T0;           % Vector of time instants
x = (t<=0.5)|(t>=2.5);      % Describe one period of the signal
k = [-35:35];               % Vector of desired EFS coefficient indices
c = ss_efsapprox(x,k);      % Compute approximate EFS coefficients
% Compute finite-harmonic approximation
t = [-5:0.001:5];           % Create new time vector for graphing
xHat = zeros(size(t));      % Create vector for finite-harmonic approximation
omega0 = 2*pi/T0;           % Fundamental frequency in rad/s
for i = 1:length(k)
  kk = k(i);
  xHat = xHat+c(i)*exp(j*kk*omega0*t);  % Eqn. (3.72)
end
plot(t,real(xHat)); grid;
title('Approximation using m=35 harmonics');
xlabel('Time (sec)');
ylabel('Amplitude');
```

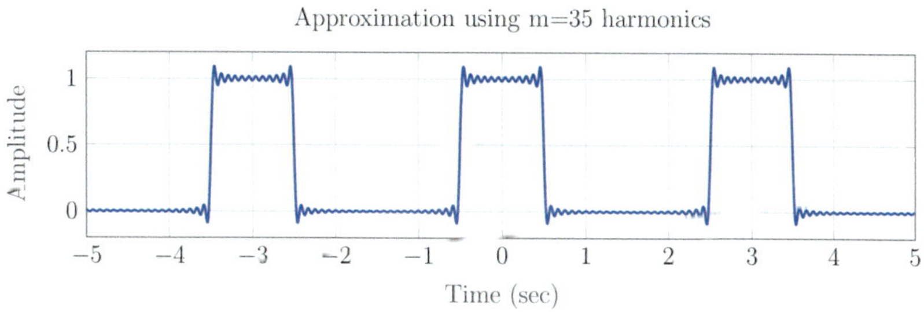

Figure 3.102 – Finite-harmonic approximation for MATLAB Exercise 3.5.

Software resources: mexct_3_5a.m , mexct_3_5b.m , ss_efsapprox.m

MATLAB Exercise 3.6: Developing a function to approximate TFS coefficients

In this exercise we will develop a function `ss_tfsapprox()` to approximate the TFS coefficients of a periodic signal. Instead of writing the function from scratch, an easier method would be to utilize the function `ss_efsapprox()` and then compute the TFS coefficients using Eqns. (3.59) and (3.60). This is the approach we will take. The code for the function is given below.

```
function [a0,a,b] = ss_tfsapprox(x,M)
  k = [-M:M];
  c = ss_efsapprox(x,k);  % Start with EFS coefficients
```

```
  a0 = real(c(M+1));          % Constant term is in the middle of vector 'c'
  a = zeros(M,1);             % Create a vector for 'a' coefficients
  b = zeros(M,1);             % Create a vector for 'b' coefficients
  for i=1:M
    a(i) = real(c(M+1+i)+c(M+1-i));       % Eqn. (3.59)
    b(i) = real(j*(c(M+1+i)-c(M+1-i)));  % Eqn. (3.60)
  end
end
```

In the script mexct_3_6a.m listed below, the function ss_tfsapprox() is used for approximating the TFS coefficients of the signal $\tilde{x}(t)$ of Example 3.1 shown in Fig. 3.5. The caveats raised in MATLAB Exercise 3.5 apply here as well. A total of 1000 time instants is used for describing one period of the signal and the largest harmonic approximated is for index $k = 11$.

```
% Script: mexct_3_6a.m
N = 1000;               % Number of samples per period
T0 = 3;                 % Fundamental period
t = [0:N-1]/N*T0;       % Vector of time instants
x = (t<=1);             % Describe one period of the signal
M = 11;                 % Number of harmonics
[a0,a,b] = ss_tfsapprox(x,M);  % Compute approximate TFS coefficients
% Display coefficients
a0
[a,b]
```

The script prints approximate values of EFS coefficients as shown below. Compare these values to the ones found in Examples 3.1 and 3.2.

```
  a0 =

    0.3340

  ans =

    0.2765    0.4778
   -0.1370    0.2384
    0.0013   -0.0000
    0.0698    0.1197
   -0.0543    0.0952
    0.0013   -0.0000
    0.0402    0.0685
   -0.0336    0.0594
    0.0013   -0.0000
    0.0284    0.0480
   -0.0242    0.0431
```

The script mexct_3_6b.m listed below computes and graphs the finite-harmonic approximation $\tilde{x}^{(10)}(t)$ of the signal $\tilde{x}(t)$. The result is shown in Fig. 3.103. Compare this to part (b) of Fig. 3.6.

```
% Script: mexct_3_6b.m
N = 1000;               % Number of samples per period
T0 = 3;                 % Fundamental period
t = [0:N-1]/N*T0;       % Vector of time instants
```

```
x = (t<=1);             % Describe one period of the signal
M = 10;                 % Number of harmonics
[a0,a,b] = ss_tfsapprox(x,M);  % Compute approximate TFS coefficients
% Compute finite-harmonic approximation
t = [-6:0.01:6];        % Create new time vector for graphing
xHat = a0*ones(size(t));  % Vector for finite-harmonic approximation
omega0 = 2*pi/T0;   % Fundamental frequency in rad/s
for i = 1:M
  xHat = xHat+a(i)*cos(i*omega0*t)+b(i)*sin(i*omega0*t);  % Eqn. (3.47)
end
plot(t,real(xHat)); grid;
title('Approximation using m=10 harmonics');
xlabel('Time (sec)');
ylabel('Amplitude');
```

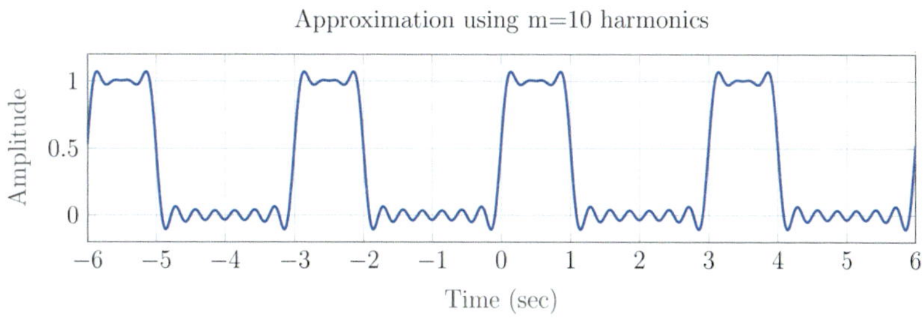

Figure 3.103 – Finite-harmonic approximation for MATLAB Exercise 3.6.

Software resources: mexct_3_6a.m , mexct_3_6b.m , mexct_3_6_Live.mlx , ss_tfsapprox.m

MATLAB Exercise 3.7: Graphing system function for RC circuit

Compute the system function of the RC circuit of Example 3.43 using element values of $R = 1\ \mathrm{M}\Omega$ and $C = 1\ \mu\mathrm{F}$. Graph the system function in the frequency range $-1\ \mathrm{Hz} \le f \le 1\ \mathrm{Hz}$.

Solution: The listing below accomplishes this. Note that line 6 of the code is a direct implementation of Eqn. (3.333) using f and f_c in Hz instead of radian frequencies ω and ω_c.

```
1  % Script: mexct_3_7.m
2  R = 1e6;
3  C = 1e-6;
4  fc = 1/(2*pi*R*C);     % The critical frequency
5  f = [-1:0.01:1];       % Vector of frequency values
6  Hf = 1./(1+j*f/fc);    % Evaluate system function using Eqn. (3.333)
7  % Graph the magnitude and the phase of the system function
8  tiledlayout(2,1);
9  nexttile;
10 plot(f,abs(Hf)); grid;
11 axis([-1,1,-0.2,1.2]);
```

```
ylabel('Magnitude');
xlabel('Frequency (Hz)');
nexttile;
plot(f,angle(Hf)); grid;
ylabel('Phase (rad)');
xlabel('Frequency (Hz)');
```

Software resource: mexct_3_7.m

MATLAB Exercise 3.8: Steady-state response of a system for single-tone input

Consider the CTLTI system of Example 3.44 described by the differential equation

$$\frac{d^2y(t)}{dt^2}+2\frac{dy(t)}{dt}+26\,y(t)=10\,x(t)$$

The system function was found to be

$$H(\omega)=\frac{10}{\left(26-\omega^2\right)+j2\omega}$$

a. Compute and graph the system function in the frequency range $-4\le f\le 4$ Hz.

b. For an input signal $x_1(t)=\cos\left(2\pi f_1 t\right)$ with $f_1=1$ Hz, determine the steady-state output signal $y_1(t)$ and graph it against the input signal. Measure the time delay between the two signals from the graph and compare to the theoretically computed value.

c. Repeat part (b) for the input signal $x_2(t)=\cos\left(2\pi f_2 t\right)$ with $f_2=2$ Hz.

Solution:

a. The script listed below computes and graphs the system function. The result is shown in Fig. 3.104.

```
% Script: mexct_3_8a.m
% Set up an anonymous function for the system function
H = @(omg) 10./((26-omg.*omg)+j*2*omg);
f = [-4:0.01:4];  % Vector of frequencies in Hz
omg = 2*pi*f;     % Vector of frequencies in rad/s
% Graph magnitude and phase of the system function
tiledlayout(2,1);
nexttile;
plot(f,abs(H(omg))); grid
xlabel('f (Hz)');
ylabel('Magnitude');
nexttile;
plot(f,angle(H(omg))); grid;
xlabel('f (Hz)');
ylabel('Phase (rad)');
```

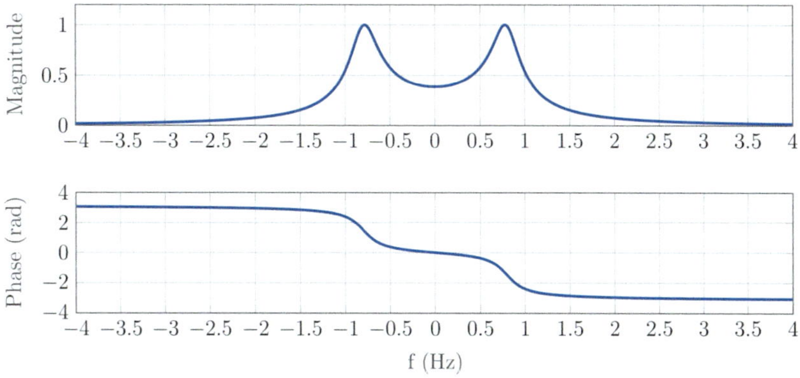

Figure 3.104 – System function for MATLAB Exercise 3.8.

b. In the following script we compute the steady-state output for $f_1 = 1$ Hz and graph it. The resulting graph is in Fig. 3.105.

```
% Script: mexct_3_8b.m
% Set up an anonymous function for the system function
H = @(omg) 10./((26-omg.*omg)+j*2*omg);
t = [0:0.001:4];      % Vector of time instants
omg1 = 2*pi;          % Radian frequency of input signal
x1 = cos(omg1*t);     % Input signal
y1 = abs(H(omg1))*cos(omg1*t+angle(H(omg1)));  % Steady-state output
plot(t,x1,t,y1); grid;
xlabel('Time (sec)');
ylabel('Amplitude');
legend('x_{1}(t)','y_{1}(t)','Location','northeastoutside')
```

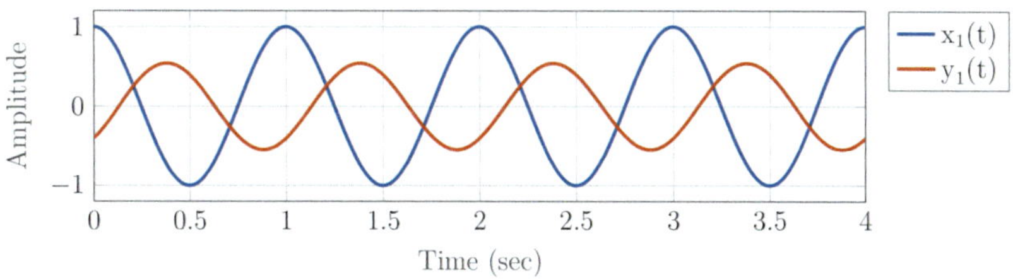

Figure 3.105 – Input and output signals for $f = 1$ Hz.

The phase of the system function at $f_1 = 1$ Hz is $\measuredangle H(f_1) = -2.3912$ radians which corresponds to a time delay of $t_d = 2.3912/(2\pi f_1) = 0.3806$ s. Looking at the graph of input and output signals in Fig. 3.105, we observe that the input signal has a zero crossing at $t_1 = 0.75$ s, and the output signal has a corresponding zero crossing at $t_2 = 1.129$ s, resulting in a time delay value of $t_d = 1.129 - 0.75 = 0.379$ s.

c. The script below computes and graphs the steady-state output for $f_2 = 2$ Hz. The resulting graph is in Fig. 3.106.

```
% Script: mexct_3_8c.m
% Set up an anonymous function for the system function
H = @(omg) 10./((26-omg.*omg)+j*2*omg);
```

```
t = [0:0.001:4];        % Vector of time instants
omg2 = 4*pi;            % Radian frequency of input signal
x2 = cos(omg2*t);       % Input signal
y2 = abs(H(omg2))*cos(omg2*t+angle(H(omg2)));  % Steady-state output
plot(t,x2,t,y2); grid;
xlabel('Time (sec)');
ylabel('Amplitude');
legend('x_{2}(t)','y_{2}(t)','Location','northeastoutside')
```

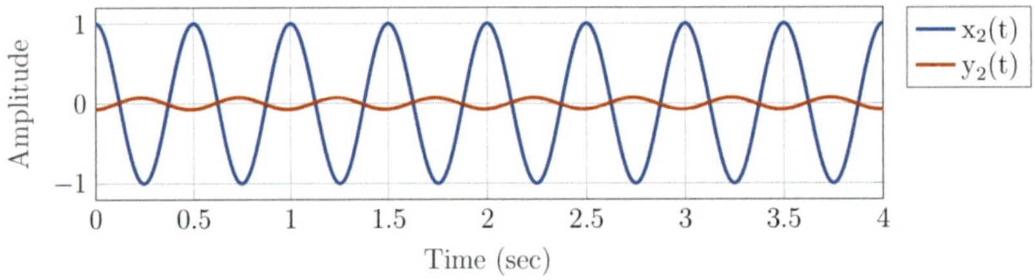

Figure 3.106 – Input and output signals for $f = 2$ Hz.

The phase of the system function at $f_2 = 2$ Hz is $\measuredangle H\left(f_2\right) = -2.9533$ radians, leading to a time delay value of $t_d = 2.9533/\left(2\pi f_2\right) = 0.235$ s. Zero crossings measured from the graph are at $t_1 = 0.375$ s, and $t_2 = 0.611$ s, resulting in a time delay value of $t_d = 0.611 - 0.375 = 0.236$ s.

Software resources: mexct_3_8a.m , mexct_3_8b.m , mexct_3_8c.m

MATLAB Exercise 3.9: Steady-state response of a system to a periodic pulse train

Consider again the CTLTI system of Example 3.44 with system function

$$H(\omega) = \frac{10}{\left(26 - \omega^2\right) + j2\omega}$$

Compute and graph the steady-state response of this system to the periodic pulse train of Example 3.3.

Solution: See the script `mexct_3_9.m` listed below. The first part between lines 2 and 7 creates one period of the signal $\tilde{x}(t)$, and uses the function `ss_efsapprox` to compute approximate EFS coefficients for $k = -35, \ldots, 35$. The next task is to obtain the EFS coefficients for the output signal $\tilde{y}(t)$ as the product of the input signal EFS coefficients and the system function, as described by Eqn. (3.358). This is done in lines 10 and 12. The resulting output signal is shown in Fig. 3.107.

```
1  % Script: mexct_3_9.m
2  N = 500;                    % Number of samples per period
3  T0 = 3;                     % Fundamental period
4  t = [0:N-1]/N*T0;           % Vector of time instants
5  x = (t<=0.5)|(t>=2.5);      % Describe one period of the signal
6  k = [-35:35];               % Vector of desired EFS coefficient indices
7  c = ss_efsapprox(x,k);      % Compute approximate EFS coefficients
8  omega0 = 2*pi/T0;           % Fundamental frequency in rad/s
```

```
% Set up an anonymous function for the system function
H = @(omg) 10./((26-omg.*omg)+j*2*omg);
% Use the system function to modify EFS coefficients
d = c.*H(k*omega0).';    % EFS coefficients for the output signal
% Compute finite-harmonic approximation for the output signal
t = [-5:0.001:5];        % Create new time vector for graphing
yHat = zeros(size(t));  % Create vector for finite-harmonic approximation
for i = 1:length(k)
  kk = k(i);
  yHat = yHat+d(i)*exp(j*kk*omega0*t);  % Eqn. (3.358)
end
plot(t,real(yHat)); grid;
title('Steady-state output signal using m=35 harmonics');
xlabel('Time (sec)');
ylabel('Amplitude');
```

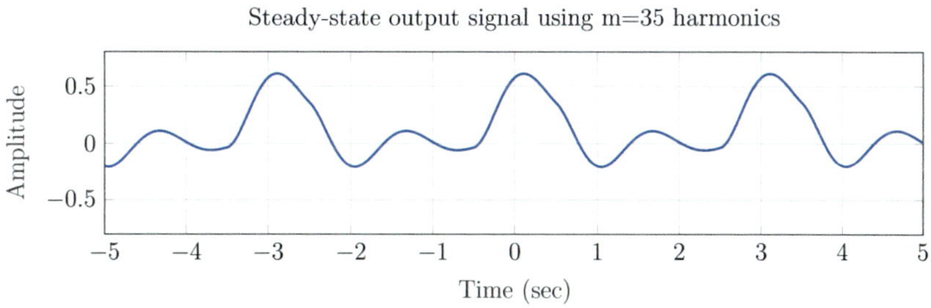

Figure 3.107 – System function for MATLAB Exercise 3.9.

Software resource: mexct_3_9.m

Problems

3.1. Consider the periodic square-wave signal $\tilde{x}(t)$ shown in Fig. 3.1 on Page 202.

a. Set up the cost function for approximating this signal using three terms in the form

$$\tilde{x}^{(3)}(t) = b_1 \sin(\omega_0 t) + b_2 \sin(2\omega_0 t) + b_3 \sin(3\omega_0 t)$$

b. Show that the optimum values of the coefficients are

$$b_1 = \frac{4A}{\pi}, \quad b_2 = 0 \quad \text{and} \quad b_3 = \frac{4A}{3\pi}$$

3.2. Consider the pulse train shown in Fig. P.3.2.

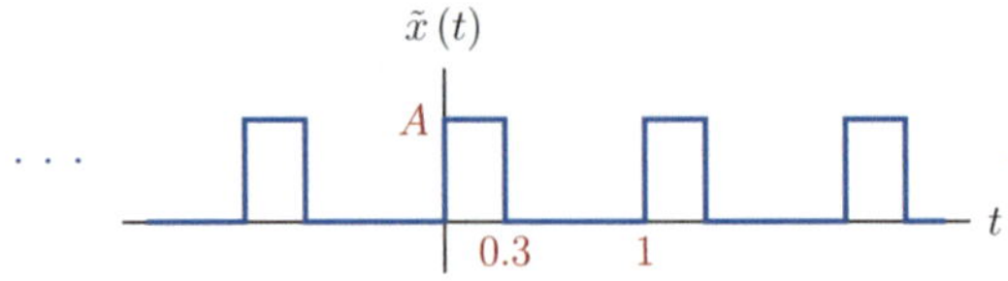

Figure P. 3.2

a. Determine the fundamental period T_0 and the fundamental frequency ω_0 for the signal.

b. Using the technique described in Section 3.2 find an approximation to $\tilde{x}(t)$ in the form

$$\tilde{x}^{(1)}(t) \approx a_0 + a_1 \cos(\omega_0 t) + b_1 \sin(\omega_0 t)$$

Determine the optimum coefficients a_0, a_1 and b_1.

3.3. Consider again the pulse train shown in Fig. P.3.2. Using the technique described in Section 3.2 find an approximation to $\tilde{x}(t)$ in the form

$$\begin{aligned}\tilde{x}^{(2)}(t) = & \; a_0 + a_1 \cos(\omega_0 t) + b_1 \sin(\omega_0 t) \\ & + a_2 \cos(2\omega_0 t) + b_2 \sin(2\omega_0 t)\end{aligned}$$

3.4. Consider again the periodic square-wave signal $\tilde{x}(t)$ shown in Fig. 3.1 on Page 202. The approximation using M trigonometric functions would be in the form

$$\begin{aligned}\tilde{x}^{(M)}(t) = & \; b_1 \sin(\omega_0 t) + b_2 \sin(2\omega_0 t) \\ & + \ldots + b_M \sin(M\omega_0 t) \\ = & \sum_{k=1}^{M} b_k \sin(k\omega_0 t)\end{aligned}$$

Show that the approximation error is equal to $\pm A$ at discontinuities of the signal $\tilde{x}(t)$ independent of the number of terms M, that is,

$$\tilde{\varepsilon}_M\left(n\frac{T_0}{2}\right) = \pm A\,, \quad n\text{: Integer}$$

Hint: It should not be necessary to determine the optimum coefficient values $b_1, b_2, \ldots, b_M$ for this problem.

3.5. Consider the pulse train $\tilde{x}(t)$ shown in Fig. P.3.5.

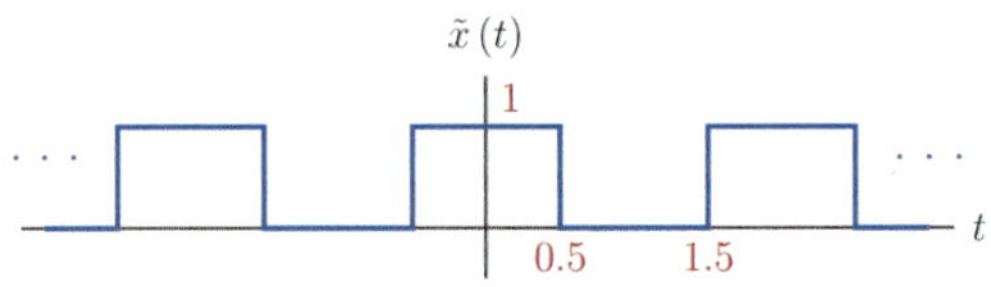

Figure P. 3.5

a. Determine the fundamental period T_0 and the fundamental frequency ω_0 for the signal.

b. Using the approach followed in Section 3.2 determine the coefficients of the approximation

$$\tilde{x}^{(2)}(t) = a_0 + a_1 \cos(\omega_0 t) + a_2 \cos(2\omega_0 t)$$

to the signal $\tilde{x}(t)$ that results in the minimum mean-squared error.

3.6. Refer to the TFS representation of a periodic signal $\tilde{x}(t)$ given by Eqn. (3.28). Using the orthogonality properties of the basis functions show that the coefficients of the sine terms are computed as

$$b_k = \frac{2}{T_0} \int_{t_0}^{t_0+T_0} \tilde{x}(t) \sin(k\omega_0 t)\, dt\,, \quad \text{for} \quad k = 1,\ldots,\infty$$

where T_0 is the fundamental period, $\omega_0 = 2\pi / T_0$ is the fundamental frequency in rad/s and t_0 is an arbitrary time instant.

3.7. Determine the TFS coefficients for the periodic signal shown in Fig. P.3.7.

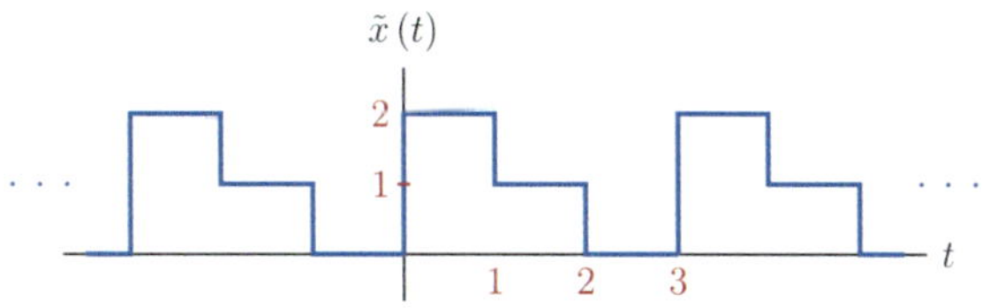

Figure P. 3.7

3.8. Determine the TFS coefficients for the periodic signal shown in Fig. P.3.8. One period of the signal is $\tilde{x}(t) = e^{-2t}$ for $0 < t < 1$ s.

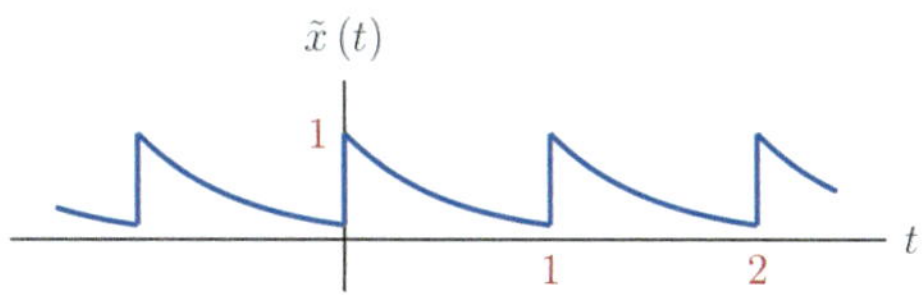

Figure P. 3.8

3.9. Consider the periodic signal

$$\tilde{x}(t) = \cos\left(10\pi t + \frac{\pi}{3}\right) + 2\sin\left(16\pi t + \frac{\pi}{8}\right)$$

a. What is the fundamental frequency f_0 for this signal? What is the fundamental period T_0?

b. Determine the TFS coefficients.

c. Determine the EFS coefficients.

3.10. Consider the periodic signal $\tilde{x}(t)$ shown in Fig. P.3.7.

a. Determine the EFS coefficients from the TFS coefficients obtained in Problem 3.7.

b. Determine the EFS coefficients by direct application of the analysis equation given by Eqn. (3.73).

c. Sketch the line spectrum (magnitude and phase).

3.11. Consider the periodic signal $\tilde{x}(t)$ shown in Fig. P.3.8.

a. Determine the EFS coefficients from the TFS coefficients obtained in Problem 3.8.

b. Determine the EFS coefficients by direct application of the analysis equation given by Eqn. (3.73).

c. Sketch the line spectrum (magnitude and phase).

3.12. Consider the periodic sawtooth signal $\tilde{g}(t)$ shown in Fig. P.3.12.

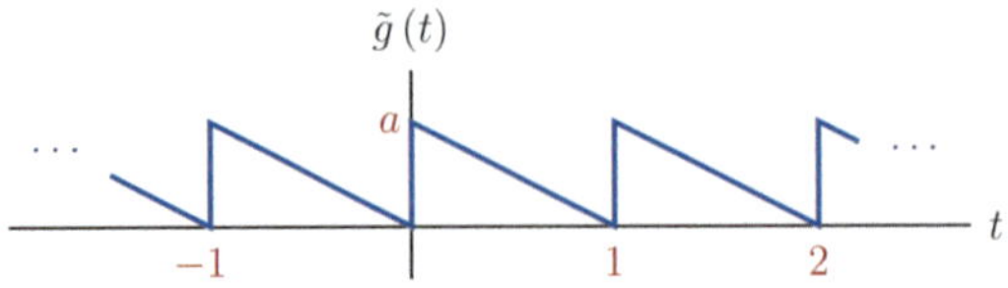

Figure P. 3.12

a. Determine EFS coefficients of this signal using the analysis equation in Eqn. (3.73).

b. Sketch the line spectrum in terms of magnitude and phase.

c. Determine the TFS coefficients of $\tilde{g}(t)$ from the EFS coefficients using the conversion formulas given by Eqns. (3.59) and (3.60).

3.13. Consider the again the periodic sawtooth signal $\tilde{g}(t)$ the EFS coefficients of which were determined in Problem 3.12. The EFS coefficients of a similar sawtooth signal $\tilde{x}(t)$ shown in Fig. 3.13 were determined in Example 3.8 as

$$\text{Re}\{c_k\} = \begin{cases} \dfrac{a}{2}, & k=0 \\ 0, & k \neq 0 \end{cases} \quad \text{and} \quad \text{Im}\{c_k\} = \begin{cases} 0, & k=0 \\ \dfrac{a}{2\pi k}, & k \neq 0 \end{cases}$$

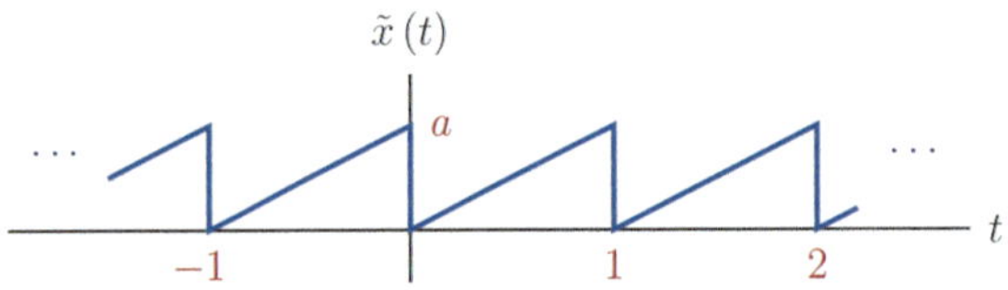

Figure P. 3.13

a. Write the signal $\tilde{g}(t)$ in terms of the signal $\tilde{x}(t)$.

b. Based on the expression found in part (a), explain how the result found in Example 3.8 for the EFS coefficients of $\tilde{x}(t)$ can be modified to obtain the EFS coefficients of $\tilde{g}(t)$. Compare your answer to that found in problem 3.12.

3.14. Let the periodic signal $\tilde{x}(t)$ have the EFS coefficients c_k. A new periodic signal $\tilde{g}(t)$ is defined as a time reversed version of $\tilde{x}(t)$, that is,

$$\tilde{g}(t) = \tilde{x}(-t)$$

a. If $\tilde{g}(t)$ has the EFS coefficients d_k, find the relationship between the two sets of coefficients c_k and d_k.

b. Consider the signal $\tilde{x}(t)$ shown in Fig. 3.19 on Page 227. Its EFS coefficients of were determined in Example 3.8 and given by Eqn. (3.79). Obtain the EFS coefficients of $\tilde{g}(t)$ shown in Fig. P.3.12 from the EFS coefficients of $\tilde{x}(t)$ using the results of part (a).

3.15. The signal $\tilde{x}(t)$ shown in Fig. 3.19 and the signal $\tilde{g}(t)$ shown in Fig. P3.12 add up to a constant:

$$\tilde{x}(t) + \tilde{g}(t) = a$$

Explain how this property can be used for finding the EFS coefficients of $\tilde{g}(t)$ from the EFS coefficients of $\tilde{x}(t)$ that were determined in Example 3.8 and given by Eqn. (3.79).

3.16. Determine the EFS coefficients of the full-wave rectified signal shown in Fig. P.3.16.

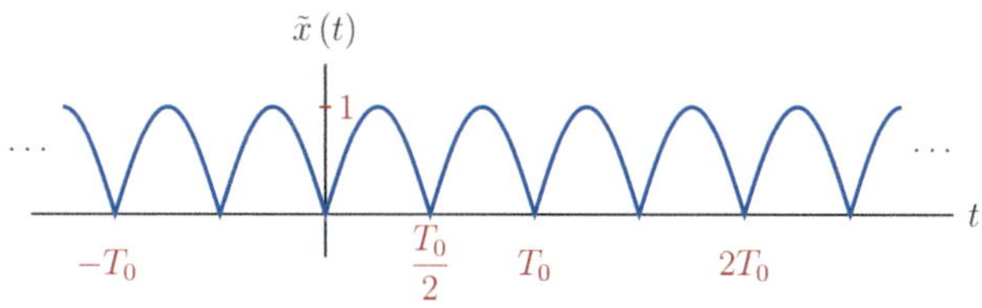

Figure P. 3.16

3.17. Prove the time shifting property of the exponential Fourier series given by Eqn. (3.108). Specifically show that, if the periodic signal $\tilde{x}(t)$ has the EFS coefficients c_k, then the signal $\tilde{x}(t-\tau)$ has the coefficients

$$d_k = c_k\, e^{-jk\omega_0\tau}$$

3.18. Refer to the half-wave rectified sinusoid shown in Fig. 3.23 on Page 230. Its EFS coefficients were determined in Example 3.10. Explain how the EFS coefficients of the full-wave rectified sinusoid in Fig. P.3.16 could be obtained from the results of Example 3.10 through the use of the time shifting property.

3.19. Refer to the half-wave rectified sinusoid shown in Fig. 3.23. Its EFS coefficients were determined in Example 3.10. Using the conversion formulas given by Eqns. (3.87) and (3.88) find the compact Fourier series (CFS) representation of the signal.

3.20. Find the Fourier transform of each of the pulse signals:

a. $x(t) = 3\Pi(t)$

b. $x(t) = 3\Pi(t-0.5)$

c. $x(t) = 2\Pi\left(\dfrac{t}{4}\right)$

d. $x(t) = 2\Pi\left(\dfrac{t-3}{2}\right)$

3.21. Starting with the Fourier transform integral given by Eqn. (3.123) find the Fourier transform of the signal shown in Fig. P.3.21.

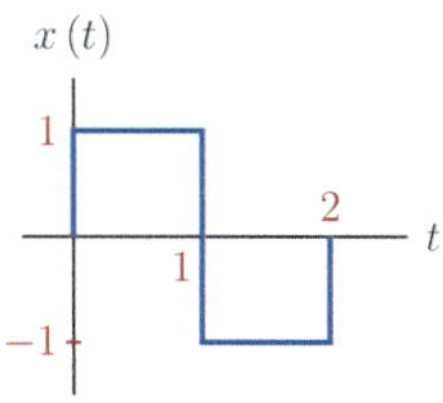

Figure P. 3.21

3.22. Starting with the Fourier transform integral given by Eqn. (3.123) find the Fourier transform of the signal

$$e^{-at}\,u(t) - e^{at}\,u(-t)$$

shown in Fig. P.3.22. Assume $a > 0$.

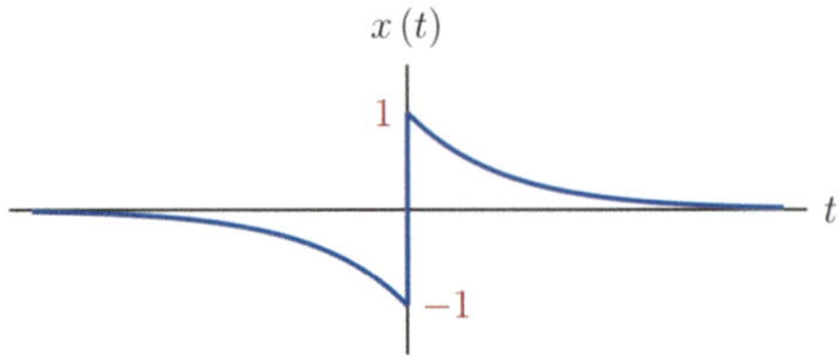

Figure P. 3.22

3.23. Refer to the signal shown in Fig. P.3.21. Find its Fourier transform by starting with the transform of the unit pulse and using linearity and time shifting properties.

3.24. Refer to the signal shown in Fig. P.3.22. Find its Fourier transform by starting with the transform of the right-sided exponential signal and using linearity and time scaling properties.

3.25. The Fourier transform of the triangular pulse with peak amplitude A and two corners at $\pm\tau$ was found in Example 3.17 as

$$A\Lambda\left(\frac{t}{\tau}\right) \overset{\mathcal{F}}{\longleftrightarrow} A\tau\,\text{sinc}^2\left(\frac{\omega\tau}{2\pi}\right)$$

Using this result along with linearity and time shifting properties of the Fourier transform, find the transform of the signal shown in Fig. P.3.25.

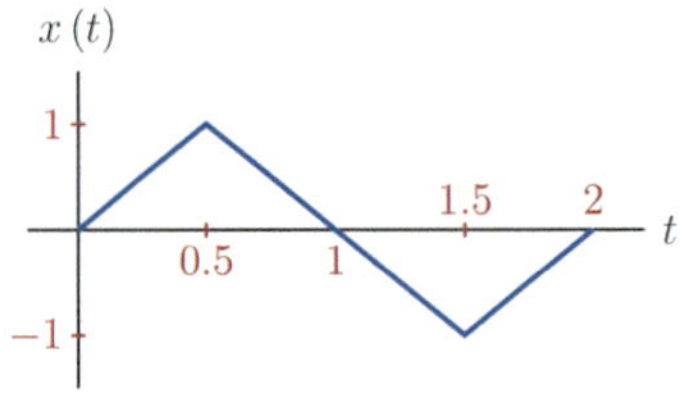

Figure P. 3.25

3.26. The transform pair

$$e^{-a|t|} \overset{\mathcal{F}}{\longleftrightarrow} \frac{2a}{a^2+\omega^2}$$

was obtained in Example 3.16. Using this pair along with the duality property, find the Fourier transform of the signal

$$x(t) = \frac{2}{1+4t^2}$$

3.27. Prove that, for a real-valued signal with odd symmetry, the Fourier transform is purely imaginary.

Hint: Start with the conditions $x^*(t) = x(t)$ and $x(-t) = -x(t)$. Show that $\text{Re}\{X(\omega)\} = 0$.

3.28.

a. Find the Fourier transform $X(\omega)$ of the pulse signal

$$x(t) = \Pi\left(\frac{t-1}{2}\right)$$

b. Express $x(t)$ as the sum of its even and odd components, that is,

$$x(t) = x_e(t) + x_o(t)$$

Determine and sketch each component.

c. Find the Fourier transforms $X_e(\omega)$ and $X_o(\omega)$ of the even and odd components respectively. Show that

$$X_e(\omega) = \text{Re}\left\{X(\omega)\right\} \quad \text{and} \quad X_o(\omega) = j\,\text{Im}\left\{X(\omega)\right\}$$

3.29. Compute and sketch the Fourier transforms of the modulated pulse signals given below:

a. $x(t) = \cos(10\pi t)\,\Pi(t)$

b. $x(t) = \cos(10\pi t)\,\Pi\left(\dfrac{t}{2}\right)$

c. $x(t) = \cos(10\pi t)\,\Pi(2t)$

d. $x(t) = \cos(10\pi t)\,\Pi(4t)$

3.30.

a. Starting with the inverse Fourier transform integral given by Eqn. (3.123) prove the second form of the differentiation-in-time property stated in Eqn. (3.235) for $n = 1$.

b. Show that, if Eqn. (3.235) holds true for $n = k$ where k is any positive integer, then it must also hold true for $n = k + 1$.

3.31.

a. Starting with the Fourier transform integral given by Eqn. (3.124) prove the second form of the differentiation-in-frequency property stated in Eqn. (3.251) for $n = 1$.

b. Show that, if Eqn. (3.251) holds true for $n = k$ where k is any positive integer, then it must also hold true for $n = k + 1$.

3.32. Taking Eqn. (3.235) as a starting point and employing the duality property of the Fourier transform, prove the second form of the differentiation-in-frequency property stated in Eqn. (3.251) for $n = 1$.

3.33. Using the differentiation-in-time property of the Fourier transform, determine the transform of the signal shown in Fig. P.3.33.

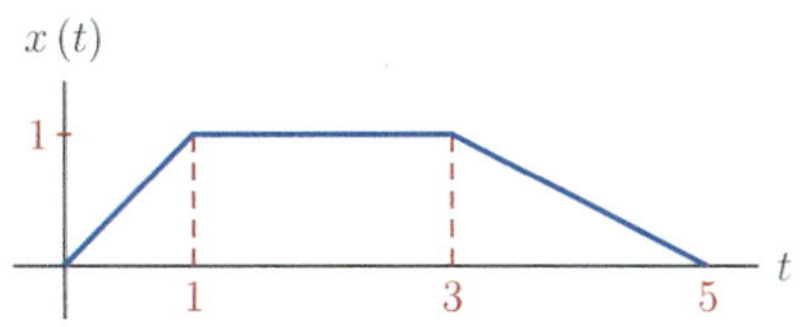

Figure P. 3.33

3.34. The Fourier transform of the trapezoidal signal $x(t)$ shown in Fig. 3.72 on Page 293 was found in Example 3.32 to be

$$X(f) = 2\lambda\,\text{sinc}(f\tau)\,\text{sinc}(2f\lambda)$$

Find the inverse transform of the spectrum shown in Fig. 3.76 on Page 298 by using the result above in conjunction with the duality property.

3.35. Determine the Fourier transform of the signal

$$x(t) = \sin(\pi t)\,\Pi\left(t - \frac{1}{2}\right) = \begin{cases} \sin(\pi t)\,, & 0 \le t \le 1 \\ 0\,, & \text{otherwise} \end{cases}$$

a. using the modulation property of the Fourier transform, and

b. using the multiplication property of the Fourier transform.

3.36. Consider the periodic pulse train shown in Fig. P.3.36.

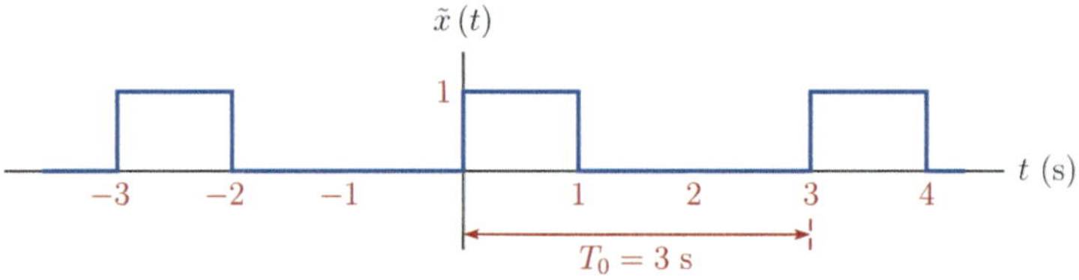

Figure P. 3.36

Recall that the EFS coefficients of this signal were found in Example 3.6. Express the Fourier transform of $\tilde{x}(t)$ using impulse functions.

3.37. Consider the RC circuit shown in Fig. P.3.37.

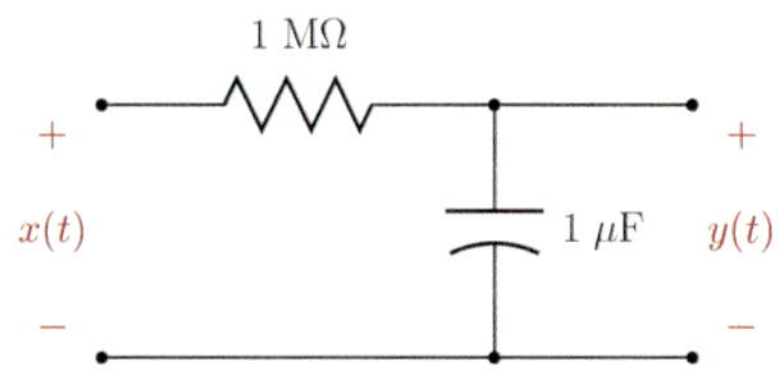

Figure P. 3.37

a. Compute and sketch the system function $H(f)$ for this circuit for the frequency range $-5 \le f \le 5$ Hz.

b. Let the input $x(t)$ to this circuit be the square wave signal shown in Fig. 3.8 with parameter values $A = 1$ and $T_0 = 1$ s. Recall that the TFS coefficients for the square wave signal were found in Example 3.4 and given by Eqn. (3.48). Using the available information, write a general expression for the TFS coefficients of the output signal $y(t)$.

c. Write a general expression for the TFS representation of $y(t)$ in the form of an infinite summation.

3.38. Consider the RC circuit shown in Fig. P.3.38.

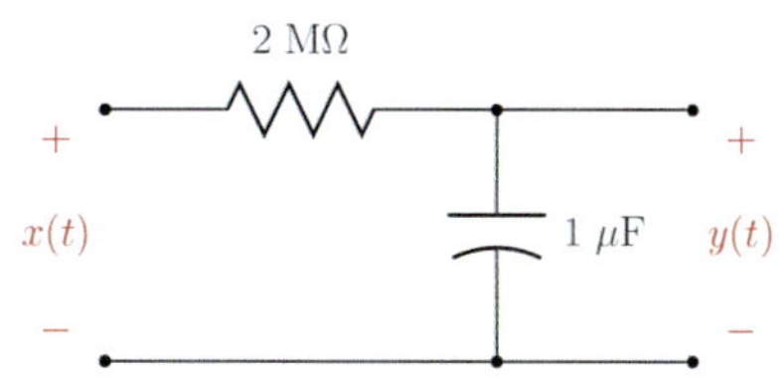

Figure P. 3.38

a. Compute and sketch the system function $H(f)$ for this circuit for the frequency range $-5 \leq f \leq 5$ Hz.

b. Let the input $x(t)$ to this circuit be the periodic sawtooth signal shown in Fig. 3.19 on Page 3.19 with parameter value $a = 1$. Recall that the EFS coefficients for the sawtooth signal were found in Example 3.8. Using the available information, write a general expression for the EFS coefficients of the output signal $y(t)$.

c. Write a general expression for the EFS representation of $y(t)$ in the form of an infinite summation.

3.39. Consider the pulse train with duty cycle d shown in Fig. 3.15 on Page 225. Its EFS coefficients were determined in Example 3.7 to be

$$c_k = d\,\mathrm{sinc}\,(kd)$$

a. Working in the time domain, compute the power of the pulse train as a function of the duty cycle d.

b. Sketch the power spectral density based on the EFS coefficients.

c. Let $d = 0.5$. Suppose this signal is processed through a lowpass system that only retains the first m harmonics and eliminates the others. How many harmonics should be retained if we want to preserve at least 90 percent of the signal power?

d. How many harmonics should be retained to preserve at least 95 percent of the signal power?

e. How many harmonics should be retained to preserve at least 99 percent of the signal power?

3.40. Repeat parts (c)–(e) of Problem 3.39 using $d = 0.2$. Does it take fewer or more harmonics to preserve the same percentage of power when the duty cycle is reduced?

3.41. Use Parseval's theorem to prove that

$$\int_{-\infty}^{\infty} \left|\mathrm{sinc}\,(f)\right|^2 df = 1$$

3.42. Determine and sketch the power spectral density of the following signals:

a. $x(t) = 3\cos(20\pi t)$

b. $x(t) = 2\cos(20\pi t) + 3\cos(30\pi t)$

c. $x(t) = 5\cos(200\pi t) + 5\cos(200\pi t)\cos(30\pi t)$

MATLAB Problems

3.43. Refer to Problem 3.1 in which a periodic square-wave signal was approximated using three terms. Let the parameters of the square-wave be $A = 1$ and $T_0 = 1$.

a. Using MATLAB, compute and graph the three-term approximation superimposed with the original signal, that is, graph $\tilde{x}(t)$ and $\tilde{x}^{(3)}(t)$ on the same coordinate system. Display the signals in the time interval $-1.5 < t < 2.5$ s.

b. Compute the approximation error $\tilde{\varepsilon}^{(3)}(t)$ for the same time interval. Graph $\tilde{x}(t)$ and $\tilde{\varepsilon}^{(3)}(t)$ on the same coordinate system.

3.44. Refer to Problem 3.2 in which a periodic pulse train was approximated using both cosine and sine functions. Using MATLAB, compute and graph the approximation $\tilde{x}^{(1)}(t)$ superimposed with the original signal $\tilde{x}(t)$ and on the same coordinate system. Show the signals in time interval $-1.5 < t < 2.5$ s.

3.45. Compute and graph finite-harmonic approximations to the signal $\tilde{x}(t)$ shown in Fig. P.3.7 using 3, 4, and 5 harmonics. Also graph the approximation error for each case.

3.46. Compute and graph finite-harmonic approximations to the signal $\tilde{x}(t)$ shown in Fig. P.3.8 using 3, 4, and 5 harmonics. Also graph the approximation error for each case.

3.47. Consider the periodic pulse train shown in Fig. 3.15 on Page 225. Its EFS coefficients were determined in Example 3.7 and given in Eqn. (3.78). Write a script to compute and graph the EFS line spectrum for duty cycle values $d = 0.4, 0.6, 0.8$, and 1. Comment on the changes in the spectrum as d is increased.

3.48. Refer to Problem 3.19 in which the CFS representation of a half-wave rectified sinusoidal signal was determined. Write a script to compute and graph finite-harmonic approximations to the signal using $m = 2$, 4, and 6 harmonics.

3.49. Refer to Problem 3.29. Write a script to compute and graph the transform of the modulated pulse

$$x(t) = \cos\left(2\pi f_0 t\right) \Pi\left(\frac{t}{\tau}\right)$$

In your script, define an anonymous function that takes two arguments, namely "`f`" and "`tau`", and returns the spectrum of a pulse with width τ centered around the time origin. Use the result with modulation property of the Fourier transform to compute the transform sought. Use the script to verify the results obtained in each part of Problem 3.29.

3.50. Refer to Problem 3.39.

a. Write a script to compute and graph the finite-harmonic approximation to the signal $\tilde{x}(t)$ using the first m harmonics. Leave m as a parameter that can be adjusted.

b. Using the values of m found in parts (c), (d), and (e) of Problem 3.39 compute and graph the finite-harmonic approximations to the signal. Comment on the correlation between the percentage of power preserved and the quality of the finite-harmonic approximation.

3.51. Refer to Problem 3.37. Develop the MATLAB code to perform the following steps:

a. Create an anonymous function for computing the system function $H(f)$ of the RC circuit. Using this anonymous function, compute and graph the system function for the frequency range $-5 \le f \le 5$ Hz.

b. Let the input $x(t)$ to this circuit be the square wave signal shown in Fig. 3.8 on Page 3.8 with parameter values $A = 1$ and $T_0 = 1$ s. Create a vector to hold the TFS coefficients of the square wave input signal up to the 15th harmonic. Use the general expression for b_k found in Example 3.4.

c. Create a vector to hold the TFS coefficients of the output signal up to the 15th harmonic. These should be obtained from the TFS coefficients of the input signal along with the system function $H(f)$.

d. Using the vector created in part (c), compute an approximation to the output signal $y(t)$. In essence, this would be the approximation $\tilde{y}^{(15)}(t)$. Graph the result obtained for $0 \le t \le 5$ s.

3.52. Refer to Problem 3.38. Develop the MATLAB code to perform the following steps:

a. Create an anonymous function for computing the system function $H(f)$ of the RC circuit. Using this anonymous function, compute and graph the system function for the frequency range $-5 \le f \le 5$ Hz.

b. Let the input $x(t)$ to this circuit be the periodic sawtooth signal shown in Fig. 3.19 on Page 3.19 with parameter value $a = 1$. Create a vector to hold the EFS coefficients of the input signal up to the 15 th harmonic, that is, for $-15 \le k \le 15$. Use the general expression for c_k found in Example 3.8.

c. Create a vector to hold the EFS coefficients of the output signal up to the 15th harmonic. These should be obtained from the EFS coefficients of the input signal along with the system function $H(f)$.

d. Using the vector created in part (c), compute an approximation to the output signal $y(t)$. In essence, this would be the approximation $\tilde{y}^{(15)}(t)$. Graph the result obtained for $0 \le t \le 5$ s.

MATLAB Projects

3.53. Consider the RC circuit shown in Fig. 3.89 on Page 321. Its system function was determined in Example 3.43 to be

$$H(\omega) = \frac{1}{1 + j(\omega/\omega_c)}$$

Assume that the element values of the circuit are adjusted to obtain a critical frequency of $f_c = 80$ Hz or, equivalently, $\omega_c = 160\pi$ rad/s. Let the input signal to this circuit be a periodic pulse train with a period of $T_0 = 20$ milliseconds as shown in Fig. P.3.53.

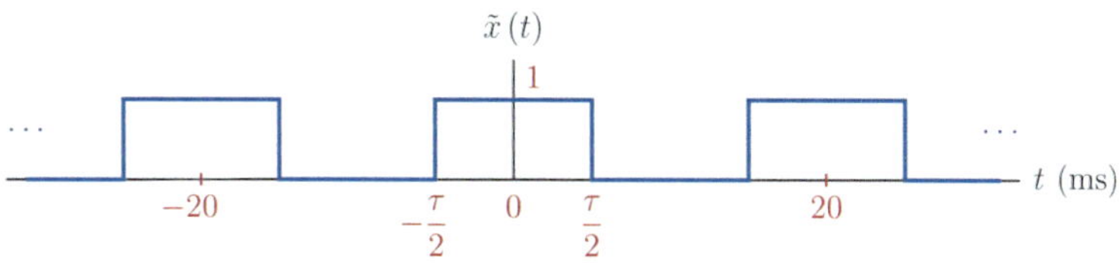

Figure P. 3.53

a. Determine the EFS coefficients of the signal $\tilde{x}(t)$ as a function of the duty cycle d. Write a script to graph the EFS spectrum of the signal for duty cycle values $d = 0.5, 0.3, 0.1$.

b. Let the pulse width be $\tau = 10$ ms corresponding to a duty cycle of $d = 0.5$. The signal $\tilde{x}(t)$ is used as input to the RC circuit. Write a script to compute the EFS coefficients of the output signal $\tilde{y}(t)$. Also graph the EFS spectrum for the output signal.

c. Write a script to construct and graph an approximation to the output signal $\tilde{y}(t)$ using the first 30 harmonics in the EFS spectrum.

d. An alternative to approximating the output of the RC circuit in response to a periodic signal is to determine its response to one isolated period, and then extend the result periodically. Let one period of the input signal be $f(t)$ so that $\tilde{x}(t)$ can be expressed as

$$\tilde{x}(t) = \sum_{r=-\infty}^{\infty} f(t - r\,T_0) = \sum_{r=-\infty}^{\infty} f(t - 0.02\,r)$$

For the input signal we are considering, $f(t)$ would be

$$f(t) = \Pi\left(\frac{t}{\tau}\right)$$

If the response of the RC circuit to one period $f(t)$ is

$$g(t) = \text{Sys}\{f(t)\}$$

then the response to the periodic signal $\tilde{x}(t)$ is

$$\tilde{y}(t) = \sum_{r=-\infty}^{\infty} g(t - r\,T_0) = \sum_{r=-\infty}^{\infty} g(t - 0.02\,r)$$

For $d = 0.5$ manually determine $g(t)$, the response to a single pulse with $\tau = 10$ ms. Write a script to compute the response using this approach and graph it. In your script use an anonymous function to compute the values of $g(t)$. Approximate the summation for $\tilde{y}(t)$ using the terms for $r = -20,\ldots,20$. Compare the result obtained through this approach with the one obtained in part (c). What are the differences and how would you explain them?

e. Repeat parts (b) and (d) for duty cycle values $d = 0.3$ and $d = 0.1$.

CHAPTER 4

Laplace Transform for Continuous-Time Signals and Systems

Chapter Objectives

- Learn the Laplace transform as a more generalized version of the Fourier transform studied in Chapter 3.
- Understand convergence characteristics and the concept of region of convergence. Explore various properties of the Laplace transform.
- Understand the use of the Laplace transform for modeling CTLTI systems. Learn the s-domain system function and its use for solving signal-system interaction problems.
- Learn techniques for obtaining simulation diagrams for CTLTI systems based on the s-domain system function.
- Learn the use of the unilateral Laplace transform for solving differential equations with specified initial conditions.

4.1 Introduction

In Chapter 3 we have focused on representing continuous-time signals in the frequency domain through the use of Fourier series and Fourier transform. We have also adapted frequency

DOI: 10.1201/9781032675251-4

domain analysis techniques for use with linear and time-invariant systems in the form of a system function.

Some limitations exist, however, in the use of Fourier series and Fourier transform for analyzing signals and systems. A signal must be absolute integrable if we are to analyze it using the Fourier transform. Consider, for example, a ramp signal in the form $x(t) = t\,u(t)$ which is not absolute integrable, and therefore has no Fourier transform. Similarly, a system function based on the Fourier transform is only available for systems for which the impulse response is absolute integrable. Consequently, Fourier transform techniques are usable only with stable systems.

Laplace transform, named after the French mathematician Pierre Simon Laplace (1749–1827), overcomes these limitations. It can be seen as an extension, or a generalization, of the Fourier transform. In contrast, the Fourier transform is a special case, or a limited view, of the Laplace transform. A signal that does not have a Fourier transform may still have a Laplace transform for a certain range of the transform variable. Similarly, an unstable system that does not have a system function based on the Fourier transform may still have a system function based on the Laplace transform.

The Fourier transform of a signal, if it exists, can be obtained from its Laplace transform while the reverse is generally not true. In addition to analysis of signals and systems, block diagram and signal flow graph structures for simulating continuous-time systems can be developed using Laplace transform techniques. The unilateral variant of the Laplace transform can be used for solving differential equations subject to specified initial conditions.

We begin with the basic definition of the Laplace transform and its application to some simple signals. The significance of the issue of convergence of the Laplace transform will become apparent throughout this discussion. Section 4.2 is dedicated to the convergence properties of the Laplace transform, and to the concept of region of convergence. We cover the fundamental properties of the Laplace transform in Section 4.3. These properties will prove useful in working with the Laplace transform for analyzing signals and systems, for understanding system characteristics, and for working with signal-system interaction problems. Proofs of significant Laplace transform properties are given not just for the sake of providing proofs, but also to provide further insight and experience on working with transforms. Techniques for computing the inverse Laplace transform are presented in Section 4.4. Use of the Laplace transform for the analysis of linear and time-invariant systems is discussed in Section 4.5 and the s-domain system function concept is introduced. Derivations of block diagram structures for the simulation of continuous-time systems based on the s-domain system function are covered in Section 4.6. In Section 4.7 the unilateral variant of the Laplace transform is introduced, and its usefulness in solving differential equations with specified initial conditions is shown.

Laplace transform

The Laplace transform of a continuous-time signal $x(t)$ is defined as

$$X(s) = \int_{-\infty}^{\infty} x(t)\,e^{-st}\,dt \tag{4.1}$$

where s, the independent variable of the transform, is a complex variable.

Notationally, the relationship between the signal $x(t)$ and the transform $X(s)$ can be expressed in the form

$$X(s) = \mathcal{L}\left\{x(t)\right\} \tag{4.2}$$

or in the form

$$x(t) \overset{\mathcal{L}}{\longleftrightarrow} X(s) \tag{4.3}$$

Since the parameter s is a complex variable, we will represent it graphically as a point in the complex s-plane. It is customary to draw the s-plane with σ as its real axis and ω as its imaginary axis. Based on this convention s is written in Cartesian form as

$$s = \sigma + j\omega \tag{4.4}$$

A specific value $s_1 = \sigma_1 + j\omega_1$ of the variable s can be shown as a point in the s-plane as illustrated in Fig. 4.1 with σ_1 equal to the horizontal displacement of the point from the origin and ω_1 as the vertical displacement from the origin. If the Laplace transform converges at the point $s = s_1$, its value can be computed by evaluating the integral in Eqn. (4.1) at that point.

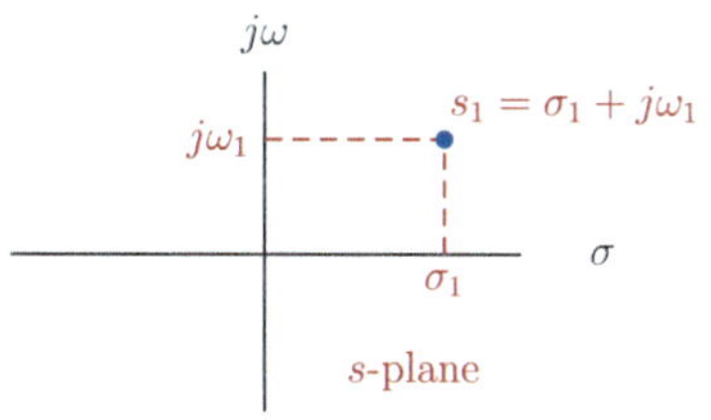

Figure 4.1 – The point $s_1 = \sigma_1 + j\omega_1$ in the complex s-plane.

Using the Cartesian form of the variable s given by Eqn. (4.4), the transform of the signal $x(t)$ becomes

$$X(s)\Big|_{s=\sigma+j\omega} = \int_{-\infty}^{\infty} x(t)\, e^{-(\sigma+j\omega)t}\, dt \tag{4.5}$$

At this point it would be interesting to let the value of σ, the real part of s, be fixed at $\sigma = \sigma_1$. Thus, we have $s = \sigma_1 + j\omega$. If the imaginary part ω is allowed to vary in the range $-\infty < \omega < \infty$, the trajectory of the points s in the complex plane would be a vertical line that passes through the point $s = \sigma_1 + j0$. This is illustrated in Fig. 4.2.

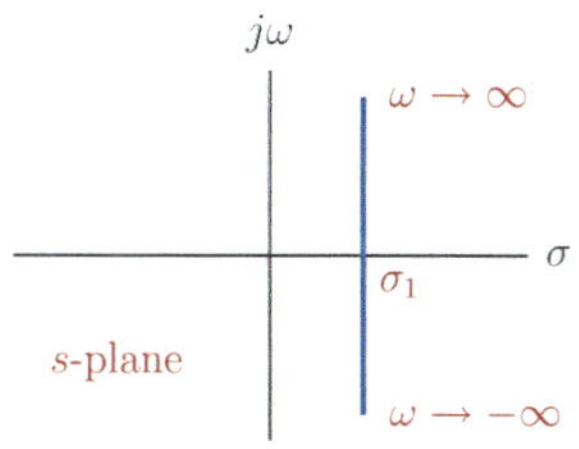

Figure 4.2 – Trajectory of the points $s = \sigma_1 + j\omega$ in the s-plane.

If the transform in Eqn. (4.5) is evaluated at points on the vertical line represented by the trajectory $s=\sigma_1+j\omega$, we get

$$\begin{aligned} X(\sigma_1,\omega) = X(s)\Big|_{s=\sigma_1+j\omega} &= \int_{-\infty}^{\infty} x(t)\,e^{-(\sigma_1+j\omega)t}\,dt \\ &= \int_{-\infty}^{\infty} x(t)\,e^{-\sigma_1 t}e^{-j\omega t}\,dt \\ &= \mathcal{F}\left\{x(t)\,e^{-\sigma_1 t}\right\} \end{aligned} \tag{4.6}$$

Comparison of the result in Eqn. (4.6) with the definition of the Fourier transform in Eqn. (3.122) in Chapter 3 leads to a very important result:

Evaluating $X(s)$ on a vertical line in the s plane:

In the s-plane consider a vertical line passing through the point $s=\sigma_1+j0$ for a fixed value of σ_1. The Laplace transform of a signal $x(t)$ evaluated on this line as the parameter ω is varied from $\omega=-\infty$ to $\omega=\infty$ is the same as the Fourier transform of the modified signal $x(t)\,e^{-\sigma_1 t}$.

$$X(s)\Big|_{s=\sigma_1+j\omega} = \mathcal{F}\left\{x(t)\,e^{-\sigma_1 t}\right\}$$

If we choose $\sigma_1=0$ then the vertical line in Fig. 4.2 coincides with the $j\omega$ axis of the s-plane, and the Laplace transform evaluated on that trajectory becomes

$$X(s)\Big|_{s=0+j\omega} = \int_{-\infty}^{\infty} x(t)\,e^{-j\omega t}\,dt = \mathcal{F}\left\{x(t)\right\} \tag{4.7}$$

This leads us to the next important conclusion:

Evaluating $X(s)$ the $j\omega$ axis of the s plane:

Consider the Laplace transform of a signal $x(t)$ evaluated at all points on the $j\omega$ axis of the s-plane as ω is varied from $\omega=-\infty$ to $\omega=\infty$. The result is the same as the Fourier transform $X(\omega)$ of the signal $x(t)$.

$$X(s)\Big|_{s=j\omega} = \mathcal{F}\left\{x(t)\right\}$$

An easy way to visualize the relationship between the Laplace transform and the Fourier transform is the following: Imagine that the Laplace transform evaluated at every point in the s-plane results in a three-dimensional surface. The transform $X(s)$ is complex-valued in general; therefore, it may be difficult to visualize it as a surface unless we are willing to accept the notion of a complex-valued surface. Alternatively, we may split the complex function $X(s)$ into its magnitude and phase, each defining a separate surface. In any case, if we were to take the surface represented by the Laplace transform and cut it the through the length of the $j\omega$ axis, the profile of the cross-section would be the same as the Fourier transform of the signal. This relationship is illustrated graphically in Fig. 4.3. Consider a particular transform in the form

$$X(s) = \frac{s+0.5}{(s+0.5)^2 + 4\pi^2} \tag{4.8}$$

which is complex-valued. Its magnitude $|X(s)|$ is shown in Fig. 4.3(a) as a three-dimensional surface. Also shown is the set of values of $|X(s)|$ computed on the $j\omega$ axis of the s-plane. In Fig. 4.3(b) the magnitude of the Fourier transform $X(\omega)$ is graphed as a function of the radian frequency ω. Notice how the $j\omega$ axis of the s-plane becomes the frequency axis for the Fourier transform.

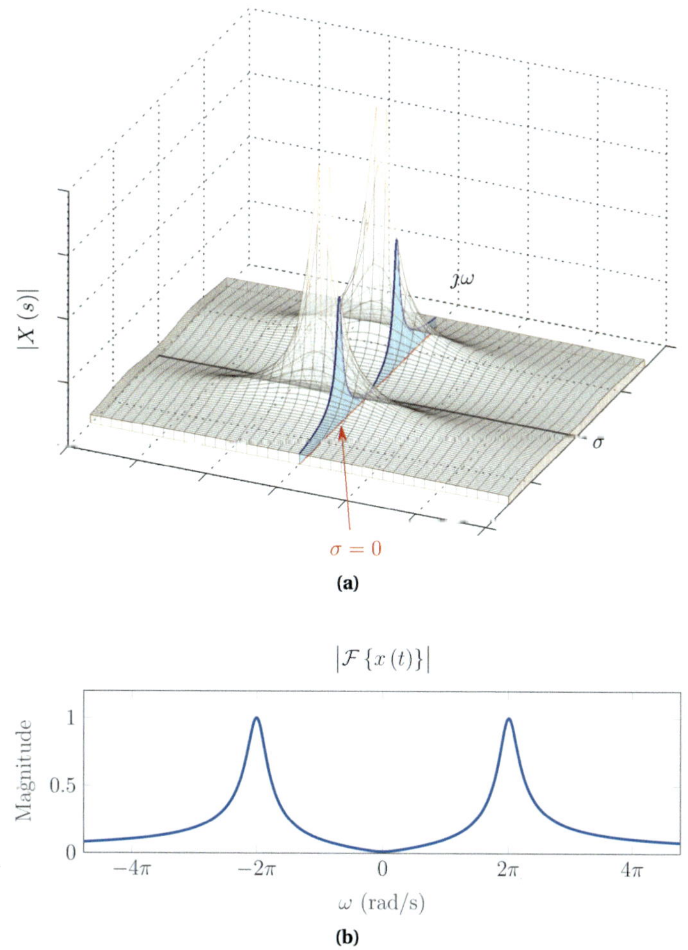

Figure 4.3 – **(a)** The magnitude $|X(s)|$ shown as a surface plot along with the the magnitude computed on the $j\omega$ axis of the s-plane and **(b)** the magnitude of the Fourier transform as a function of ω.

For the same transform under consideration, Fig. 4.4 shows the values on the trajectory $s = 0.5 + j\omega$. Recall that the values of $|X(s)|$ on this trajectory are the same as the magnitude of the Fourier transform of the modified signal $\left[x(t)\, e^{-0.5t}\right]$.

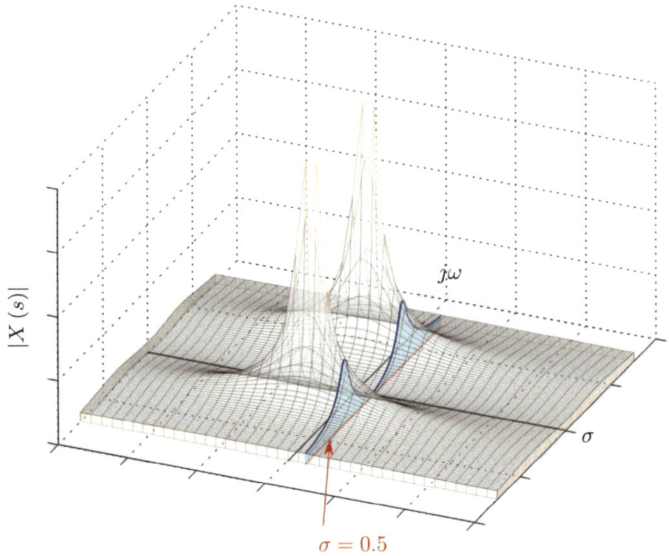

(a)

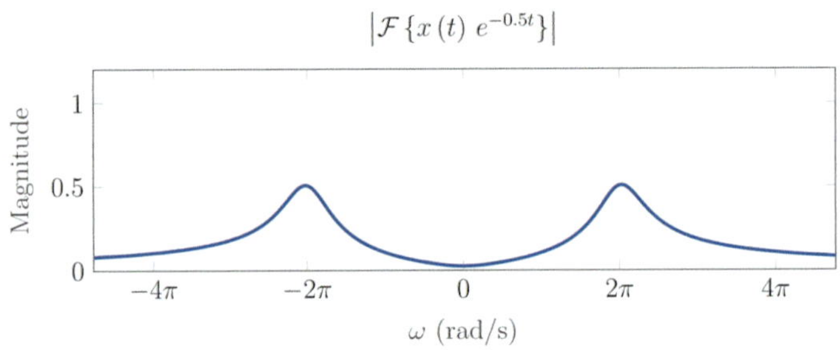

(b)

Figure 4.4 – **(a)** The magnitude $|X(s)|$ shown as a surface plot along with the the magnitude computed on the trajectory $s = 0.5 + j\omega$ on the s-plane and **(b)** the magnitude of the Fourier transform of $x(t)\, e^{-0.5t}$.

Interactive App: Laplace transform in 3-D

The MATLAB app in `appLaplace1.m` is based on the Laplace transform given by Eqn. (4.8). The magnitude $|X(s)|$ of the transform in question is shown in Figs. 4.3(a) and 4.4(a) . This interactive app computes $|X(s)|$ and graphs it as a 3-D surface. It can be rotated freely for viewing from any angle by using the rotation tool in the toolbar. The Laplace transform is also computed at points on the line $s = \sigma_1 + j\omega$ in the s-plane, and the resulting 2-D plot is superimposed with the 3-D surface. The magnitude of the Fourier transform of the modified signal $x(t)\, e^{-\sigma_1 t}$ is graphed separately. The value of σ_1 may be varied through the use of a slider control.

a. Vary σ_1 using the slider control, and observe the changes in the Fourier transform magnitude $\left|\mathcal{F}\left\{x(t)\, e^{-\sigma_1 t}\right\}\right|$. Note that the values $\sigma_1 = 0$ and $\sigma_1 = 0.5$ correspond to Figs. 4.3 and 4.4, respectively.

b. Rotate the 3-D surface using the rotation tool in the toolbar to obtain a birds eye view of the transform, looking down at the s-plane from the top. Pay attention to the locations of the two peaks. Vary the value of σ_1 to bring the vertical line closer to the two peaks, and observe the corresponding changes in the Fourier transform magnitude.

Software resource: `appLaplace1.m`

The Laplace transform as defined by Eqn. (4.1) is referred to as the *bilateral* (two-sided) Laplace transform. A variant of the Laplace transform known as the *unilateral* (one-sided) Laplace transform will be introduced in Section 4.7 as an alternative analysis tool. In this text, when we refer to Laplace transform without the qualifier word “bilateral” or “unilateral”, we will always imply the more general bilateral Laplace transform as defined in Eqn. (4.1).

Software resource: See MATLAB Exercises 4.1 and 4.2.

Example 4.1: Laplace transform of the unit impulse

Find the Laplace transform of the unit impulse signal

$$x(t) = \delta(t)$$

Solution: Applying the definition of the Laplace transform given by Eqn. (4.1) we get

$$X(s) = \mathcal{L}\left\{\delta(t)\right\} = \int_{-\infty}^{\infty} \delta(t)\, e^{-st}\, dt = 1$$

using the sifting property of the unit impulse function. Therefore, the Laplace transform of the unit impulse signal is constant and equal to unity. Since it does not depend on the value of s, it converges at every point in the s-plane with no exceptions.

Example 4.2: Laplace Transform of a time-shifted unit impulse

Find the Laplace transform of the time-shifted unit impulse signal

$$x(t) = \delta(t-\tau)$$

Solution: Application of the Laplace transform definition to the signal at hand and use of the sifting property of the unit impulse function yields

$$X(s) = \int_{-\infty}^{\infty} \delta(t-\tau)\, e^{-st}\, dt = e^{-s\tau} \tag{4.9}$$

Again we have used the sifting property of the unit impulse function. If $s = \sigma + j\omega$, then Eqn. (4.9) becomes

$$X(s)\Big|_{s=\sigma+j\omega} = e^{-\sigma\tau}\, e^{-j\omega\tau}$$

The transform obtained in Eqn. (4.9) converges as long as $\sigma = \text{Re}\left\{s\right\} > -\infty$.

Example 4.3: Laplace transform of the unit step signal

Find the Laplace transform of the unit step signal

$$x(t) = u(t)$$

Solution: Substituting $x(t) = u(t)$ into the Laplace transform definition we obtain

$$X(s) = \int_{-\infty}^{\infty} u(t)\, e^{-st}\, dt$$

Since $u(t) = 1$ for $t > 0$ and $u(t) = 0$ for $t < 0$, the lower limit of the integral can be changed to $t = 0$ and the unit step term can be dropped to yield

$$X(s) = \int_{0}^{\infty} e^{-st}\, dt = -\frac{1}{s}\, e^{-st}\Big|_{0}^{\infty}$$

In order to evaluate the integral for the specified limits we will use the Cartesian form of the complex variable s. Substituting $s = \sigma + j\omega$, we get

$$X(s)\Big|_{s=\sigma+j\omega} = -\frac{1}{\sigma + j\omega}\, e^{-(\sigma+j\omega)t}\Big|_{0}^{\infty}$$

It is obvious that, for the exponential term $e^{-\sigma t}$ to converge as $t \to \infty$, we need $\sigma > 0$:

$$\text{If } \sigma > 0 \quad \Longrightarrow \quad e^{-\sigma t}\Big|_{t\to\infty} = 0$$

The transform becomes

$$X(s) = -\frac{1}{\sigma + j\omega}\left[0 - 1\right] = \frac{1}{\sigma + j\omega} = \frac{1}{s}, \qquad \text{Re}\{s\} > 0 \tag{4.10}$$

The transform expression found in Eqn. (4.10) is valid only for points in the right half of the s-plane. This region is shown shaded in Fig. 4.5. Note that the transform does not converge at points on the $j\omega$ axis. It converges at any point to the right of the $j\omega$ axis regardless of how close to the axis it might be.

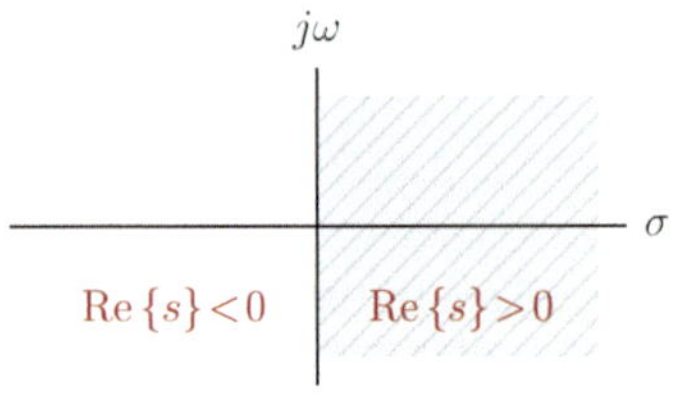

Figure 4.5 – The region in which the Laplace-transform of $x(t) = u(t)$ converges.

Example 4.4: Laplace transform of a time-shifted unit step signal

Find the Laplace transform of the time-shifted unit step signal

$$x(t) = u(t-\tau)$$

Solution: Using $x(t) = u(t-\tau)$ in the Laplace transform definition we get

$$X(s) = \int_{-\infty}^{\infty} u(t-\tau)\, e^{-st}\, dt$$

Since $u(t-\tau) = 1$ for $t > \tau$ and $u(t-\tau) = 0$ for $t < \tau$, the lower limit of the integral can be changed to $t = \tau$ and the unit step function can be dropped without affecting the result.

$$X(s) = \int_{\tau}^{\infty} e^{-st}\, dt = \left. -\frac{1}{s}\, e^{-st} \right|_{\tau}^{\infty}$$

As in Example 4.3, the integral can be evaluated only for $\sigma = \text{Re}\{s\} > 0$ and its value is

$$X(s) = -\frac{1}{s}\left[0 - e^{-s\tau}\right] = \frac{e^{-s\tau}}{s}\,, \qquad \text{Re}\{s\} > 0$$

Convergence conditions for the transform of $u(t-\tau)$ are the same as those found in Example 4.3 for the transform of $u(t)$.

We observe from the last two examples that, when we find the Laplace transform $X(s)$ of a signal, we also need to specify the region in which the transform is valid. The collection of all points in the s-plane for which the Laplace transform converges is called the *region of convergence (ROC)*.

Recall that in Eqn. (4.6) we have represented the Laplace transform of a signal $x(t)$ as equivalent to the Fourier transform of the modified signal $\left[x(t)\, e^{-\sigma t}\right]$. Consequently, the conditions for the convergence of the Laplace transform of $x(t)$ are identical to the conditions for the convergence of the Fourier transform of $\left[x(t)\, e^{-\sigma t}\right]$. Using Dirichlet conditions for the latter, we need the signal $\left[x(t)\, e^{-\sigma t}\right]$ to be absolute integrable for the transform to exist, that is,

$$\int_{-\infty}^{\infty} \left| x(t)\, e^{-\sigma t} \right| dt < \infty \tag{4.11}$$

The convergence condition stated in Eqn. (4.11) highlights the versatility of the Laplace transform over the Fourier transform. The Fourier transform of a signal $x(t)$ exists if the signal is absolute integrable; it does not exist otherwise. Therefore, the existence of the Fourier transform is a binary question; the answer to it is either yes or no. In contrast, if the signal $x(t)$ is not absolute integrable, we may still be able to find values of σ for which the modified signal $\left[x(t)\, e^{-\sigma t}\right]$ is absolute integrable. Therefore, the Laplace transform of the signal $x(t)$ may exist for some values of σ. The question of existence for the Laplace transform is not a binary one; it is a question of which values of σ allow the transform to converge.

The next two examples will further highlight the significance of the region of convergence for the Laplace transform. The fundamental characteristics of the region of convergence will be discussed in detail in Section 4.2.

Example 4.5: Laplace transform of a causal exponential signal

Find the Laplace transform of the signal

$$x(t) = e^{at}\, u(t)$$

where a is any real or complex constant.

Solution: The signal $x(t)$ is causal since $x(t) = 0$ for $t < 0$. Applying the Laplace transform definition to $x(t)$ results in

$$X(s) = \int_{-\infty}^{\infty} e^{at} u(t)\, e^{-st}\, dt$$

Changing the lower limit of the integral to $t = 0$ and dropping the factor $u(t)$ we get

$$X(s) = \int_0^{\infty} e^{at} e^{-st}\, dt = \int_0^{\infty} e^{(a-s)t}\, dt = \frac{1}{a-s}\, e^{(a-s)t}\Big|_0^{\infty}$$

The parameter a may be real or complex-valued. Let us assume that it is in the form $a = a_r + j a_i$ to keep the results general. For the purpose of evaluating the integral above, we will substitute $s = \sigma + j\omega$ and write the transform as

$$\begin{aligned} X(s) &= \frac{1}{a_r + j a_i - \sigma - j\omega}\; e^{(a_r + j a_i - \sigma - j\omega)t}\Big|_0^{\infty} \\ &= \frac{1}{a_r + j a_i - \sigma - j\omega}\; e^{(a_r-\sigma)t}\, e^{j(a_i-\omega)t}\Big|_0^{\infty} \end{aligned} \tag{4.12}$$

For the result of Eqn. (4.12) to converge at the upper limit, we need

$$a_r - \sigma < 0 \quad \Longrightarrow \quad \sigma > a_r$$

or equivalently

$$\mathrm{Re}\{s\} > \mathrm{Re}\{a\} \tag{4.13}$$

which establishes the ROC for the transform. With the condition in Eqn. (4.13) satisfied, the exponential term in Eqn. (4.12) becomes zero as $t \to \infty$, and we obtain

$$X(s) = \frac{1}{a-s}\,[0-1] = \frac{1}{s-a}\,, \qquad \mathrm{Re}\{s\} > \mathrm{Re}\{a\}$$

The ROC is the region to the right of the vertical line that goes through the point $s = a_r + j a_i$ as shown in Fig. 4.6.

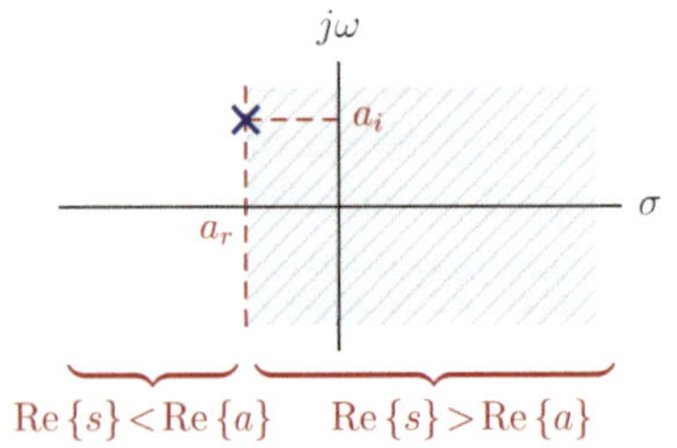

Figure 4.6 – The ROC for the transform $X(s)$ of Example 4.5.

It is interesting to look at the shape of the signal $x(t)$ in conjunction with the ROC.

1. Fig. 4.7 shows the signal for the case where a is real-valued. The signal $x(t)$ is a decaying exponential for $a<0$, and a growing exponential for $a>0$.
2. Real and imaginary parts of the signal $x(t)$ are shown in Figs. 4.8 and 4.9 for the case of parameter a being complex-valued. Fig. 4.8 illustrates the possibility of $\mathrm{Re}\{a\}<0$. Real and imaginary parts of $x(t)$ exhibit oscillatory behavior with exponential damping. In contrast, if $\mathrm{Re}\{a\}>0$, both real and imaginary parts of $x(t)$ grow exponentially as shown in Fig. 4.9.

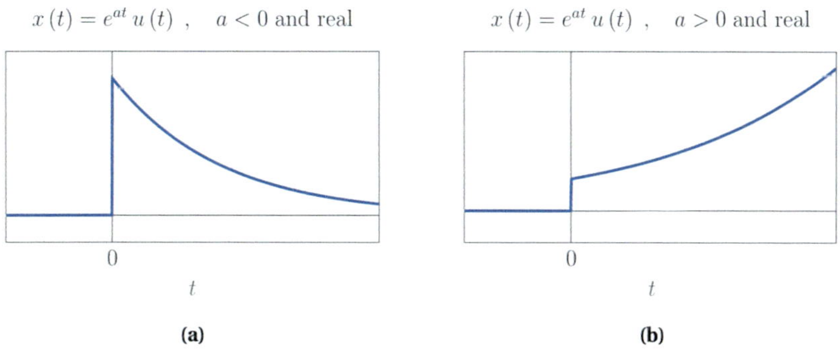

Figure 4.7 – The signal $x(t)$ of Example 4.5 for real-valued parameter a and **(a)** $a<0$ and **(b)** $a>0$.

Since the ROC is to the right of a vertical line, for the case $\mathrm{Re}\{a\}<0$ the $j\omega$ axis is part of the ROC whereas, for the case $\mathrm{Re}\{a\}>0$, it is not. Recall from the previous discussion that the Fourier transform of the signal $x(t)$ is equal to the Laplace transform evaluated on the $j\omega$ axis of the s-plane. Consequently, the Fourier transform of the signal exists only if the region of convergence includes the $j\omega$ axis. We conclude that, for the Fourier transform of the signal $x(t)$ to exist, we need $\mathrm{Re}\{a\}<0$. Signals in Figs. 4.7(a) and 4.8 have Fourier transforms whereas signals in Figs. 4.7(b) and 4.9 do not. This is consistent with the existence conditions discussed in Chapter 3.

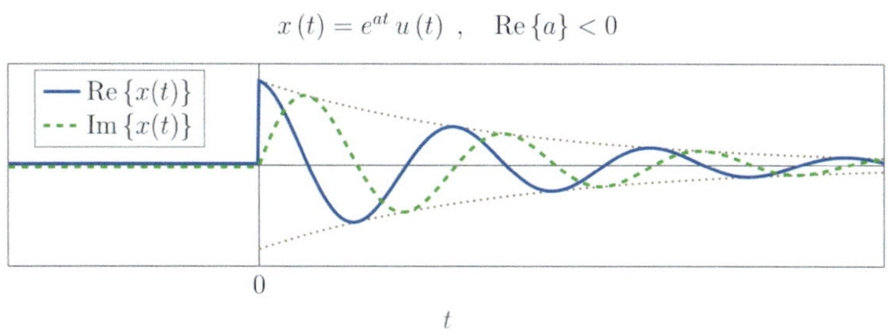

Figure 4.8 – The signal $x(t)$ of Example 4.5 for $\mathrm{Re}\{a\}<0$.

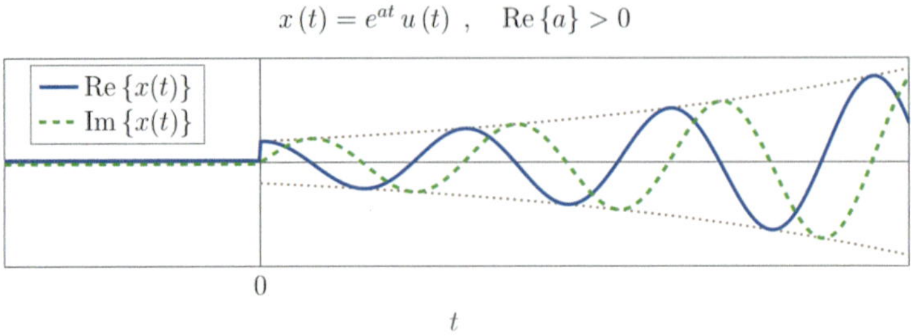

Figure 4.9 – The signal $x(t)$ of Example 4.5 for $\mathrm{Re}\{a\} > 0$.

Software resources: exct_4_5a.m , exct_4_5b.m

Example 4.6: Laplace transform of an anti-causal exponential signal

Find the Laplace transform of the signal

$$x(t) = -e^{at}\,u(-t)$$

where a is any real or complex constant.

Solution: In this case the signal $x(t)$ is anti-causal since it is equal to zero for $t > 0$. Substituting $x(t)$ into the Laplace transform definition we get

$$X(s) = \int_{-\infty}^{\infty} -e^{at}\,u(-t)\,e^{-st}\,dt$$

Since $u(-t) = 1$ for $t < 0$ and $u(-t) = 0$ for $t > 0$, changing the upper limit of the integral to $t = 0$ and dropping the factor $u(-t)$ would have no effect on the result. Therefore

$$X(s) = \int_{-\infty}^{0} -e^{at}e^{-st}\,dt = \int_{-\infty}^{0} -e^{(a-s)t}\,dt = -\frac{1}{a-s}\,e^{(a-s)t}\Big|_{-\infty}^{0}$$

The parameter a may be real or complex valued. We will assume that it is in the general form $a = a_r + j a_i$. For the purpose of evaluating the integral above, let us substitute $s = \sigma + j\omega$ and write the transform as

$$\begin{aligned} X(s) &= -\frac{1}{a_r + j a_i - \sigma - j\omega}\,e^{(a_r + j a_i - \sigma - j\omega)t}\Big|_{-\infty}^{0} \\ &= -\frac{1}{a_r + j a_i - \sigma - j\omega}\,e^{(a_r-\sigma)t}\,e^{j(a_i-\omega)t}\Big|_{-\infty}^{0} \end{aligned} \tag{4.14}$$

In contrast with Example 4.5, the critical end of the integral with respect to convergence is its lower limit. To evaluate the result of Eqn. (4.14) at the lower limit, we need

$$a_r - \sigma > 0 \quad \Longrightarrow \quad \sigma < a_r$$

or equivalently

$$\mathrm{Re}\{s\} < \mathrm{Re}\{a\} \tag{4.15}$$

which establishes the ROC for the transform. With the condition in Eqn. (4.15) satisfied, the exponential term in Eqn. (4.14) becomes zero as $t \to -\infty$, and we obtain

$$X(s) = -\frac{1}{a-s}\left[1-0\right] = \frac{1}{s-a}, \qquad \mathrm{Re}\{s\} < \mathrm{Re}\{a\}$$

The ROC is the region to the left of the vertical line that goes through the point $s = a_r + ja_i$ as shown in Fig. 4.10.

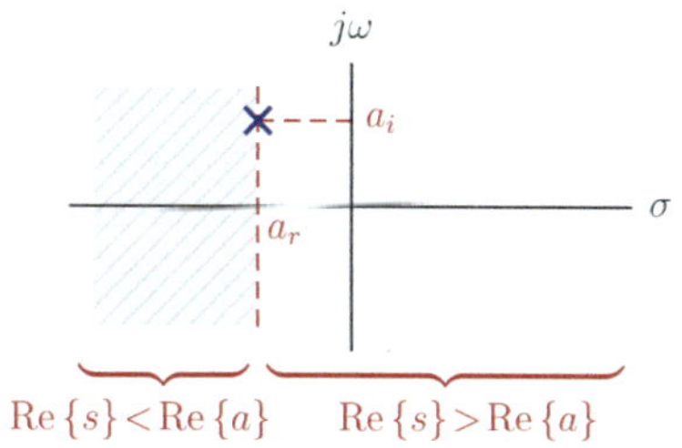

Figure 4.10 – The ROC for the transform $X(s)$ of Example 4.6.

As in Example 4.5 we will look at the shape of the signal $x(t)$ in conjunction with the ROC.

1. Fig. 4.11 shows the signal for the case where a is real-valued. For $a < 0$, the signal grows exponentially as $t \to -\infty$. In contrast, for $a > 0$ the signal decays exponentially.
2. For complex a, real and imaginary parts of $x(t)$ exhibit oscillatory behavior. For $\mathrm{Re}\{a\} < 0$ the signal grows exponentially as shown in Fig. 4.12. In contrast, oscillations are damped exponentially for $\mathrm{Re}\{a\} > 0$. This is shown in Fig. 4.13.

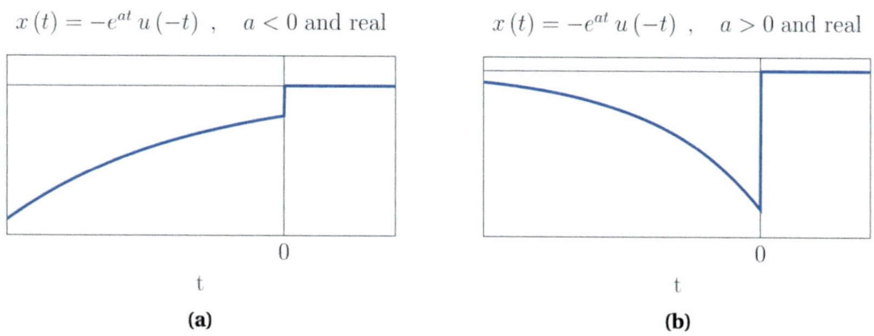

Figure 4.11 – The signal $x(t)$ of Example 4.6 for real-valued parameter a and **(a)** $a < 0$ and **(b)** $a > 0$.

For the case $\mathrm{Re}\{a\} > 0$ the $j\omega$ axis of the s-plane is part of the ROC. For the case $\mathrm{Re}\{a\} < 0$, however, it is not. Thus, the Fourier transform of the signal $x(t)$ exists only when $\mathrm{Re}\{a\} > 0$.

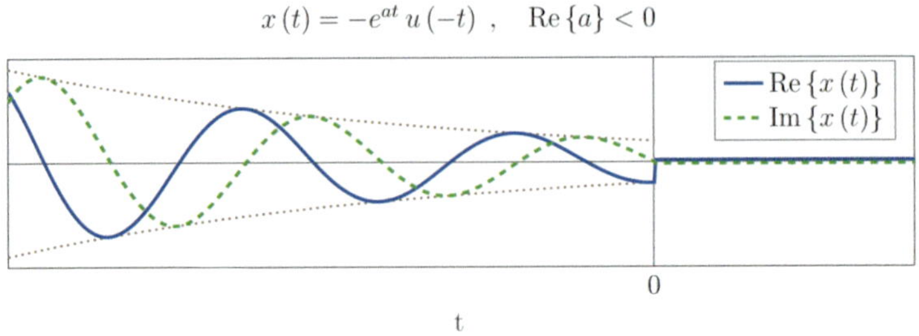

Figure 4.12 – The signal $x(t)$ of Example 4.6 for $\mathrm{Re}\{a\} < 0$.

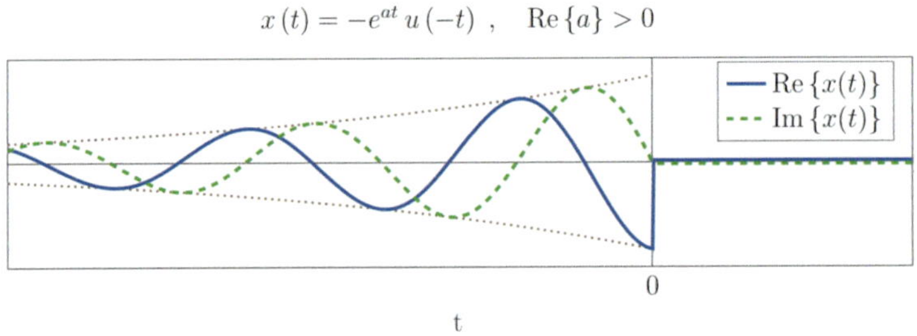

Figure 4.13 – The signal $x(t)$ of Example 4.6 for $\mathrm{Re}\{a\} > 0$.

Software resources: exct_4_6a.m , exct_4_6b.m

Examples 4.5 and 4.6 demonstrate a fundamental concept for the Laplace transform: It is possible for two different signals to have the same transform expression for $X(s)$. In both of the examples above we have found the transform

$$X(s) = \frac{1}{s-a}$$

What separates the two results apart is the ROC. In order for us to uniquely identify which signal among the two led to a particular transform, the ROC must be specified along with the transform. The following two transform pairs will be fundamental in determining to which signal a given transform corresponds:

Significance of the region of convergence:

$$e^{at}\,u(t) \stackrel{\mathcal{L}}{\longleftrightarrow} \frac{1}{s-a}\,, \qquad \text{ROC: } \mathrm{Re}\{s\} > \mathrm{Re}\{a\} \tag{4.16}$$

$$-e^{at}\,u(-t) \stackrel{\mathcal{L}}{\longleftrightarrow} \frac{1}{s-a}\,, \qquad \text{ROC: } \mathrm{Re}\{s\} < \mathrm{Re}\{a\} \tag{4.17}$$

Sometimes we need to solve the inverse problem of finding the signal $x(t)$ with a given Laplace transform $X(s)$. Given a transform

$$X(s) = \frac{1}{s-a}$$

we need to know the ROC to determine if the signal $x(t)$ is the one in Eqn. (4.16) or the one in Eqn. (4.17). This will be very important when we work with inverse Laplace transform in Section 4.4 later in this chapter. In order to avoid ambiguity, we will adopt the convention that the ROC is an integral part of the Laplace transform. It must be specified explicitly, or implied by means of another property of the signal or the system under consideration every time a transform is given.

In each of the Examples 4.3, 4.5, and 4.6 we have obtained the Laplace transform of the specified signal in the form of a rational function of s, that is, a ratio of two polynomials in s. In the general case, a rational transform $X(s)$ is expressed in the form

$$X(s) = K\,\frac{B(s)}{A(s)} \tag{4.18}$$

where the numerator $B(s)$ and the denominator $A(s)$ are polynomials of s. Let the numerator polynomial be written in factored form as

$$B(s) = (s-z_1)\,(s-z_2)\ldots(s-z_M) \tag{4.19}$$

where $z_1, z_2, \ldots, z_M$ are its roots. Similarly, let $p_1, p_2, \ldots, p_N$ be the roots of the denominator polynomial so that

$$A(s) = \left(s-p_1\right)\left(s-p_2\right)\ldots\left(s-p_N\right) \tag{4.20}$$

The transform can be written as

$$X(s) = K\,\frac{(s-z_1)\,(s-z_2)\ldots(s-z_M)}{\left(s-p_1\right)\left(s-p_2\right)\ldots\left(s-p_N\right)} \tag{4.21}$$

In Eqns. (4.20) and (4.21) the parameters M and N are the numerator order and the denominator order, respectively. The larger of M and N is the order of the transform $X(s)$. The roots of the numerator polynomial are referred to as the *zeros* of $X(s)$. In contrast, the roots of the denominator polynomial are the *poles* of $X(s)$. The transform does not converge at a pole; therefore, the ROC cannot contain any poles. In the next section we will see that the poles of the transform also determine the boundaries of the ROC.

Software resource: See MATLAB Exercise 4.4.

Example 4.7: Laplace transform of a pulse signal

Determine the Laplace transform of the pulse signal

$$x(t) = \Pi\left(\frac{t-\tau/2}{\tau}\right)$$

which is shown in Fig. 4.14.

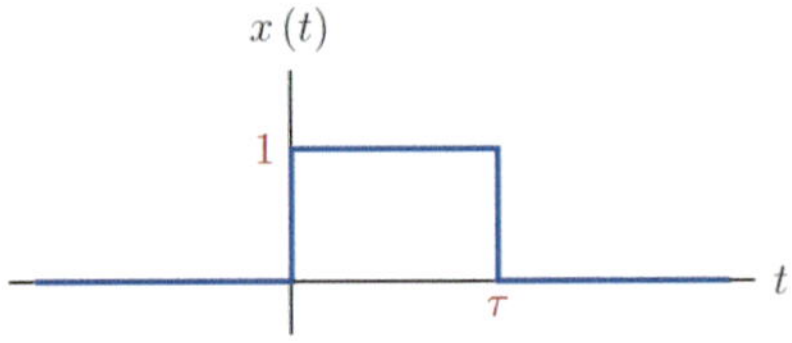

Figure 4.14 – The pulse signal $x(t)$ for Example 4.7.

Solution: The Laplace transform of the signal $x(t)$ is computed as

$$X(s) = \int_0^{\tau} (1)\, e^{-st}\, dt = \left.\frac{e^{-st}}{-s}\right|_0^{\tau} = \frac{1-e^{-s\tau}}{s}$$

At a first glance we may be tempted to think that the transform $X(s)$ might not converge at $s=0$ since the denominator of $X(s)$ becomes equal to zero at $s=0$. We must realize, however, that the numerator of $X(s)$ is also equal to zero at $s=0$. Using L'Hospital's rule on $X(s)$ we obtain its value at $s=0$ as

$$\left. X(s)\right|_{s=0} = \left.\frac{\tau e^{-s\tau}}{1}\right|_{s=0} = \tau$$

confirming the fact that $X(s)$ does indeed converge at $s=0$. As a matter of fact, $X(s)$ converges everywhere on the s-plane with the exception of $s=-\infty \pm j\omega$. In order to observe this, let us write the Taylor series representation of the exponential term $e^{-s\tau}$:

$$e^{-s\tau} = 1 - s\tau + \frac{s^2\tau^2}{2!} - \frac{s^3\tau^3}{3!} + \frac{s^4\tau^4}{4!} - \frac{s^5\tau^5}{5!} + \ldots$$

The numerator of $X(s)$ can be written as

$$1-e^{-s\tau} = s\tau - \frac{s^2\tau^2}{2!} + \frac{s^3\tau^3}{3!} - \frac{s^4\tau^4}{4!} + \frac{s^5\tau^5}{5!} - \ldots$$

and the transform $X(s)$ can be written in series form

$$X(s) = \frac{1-e^{-s\tau}}{s} = \tau - \frac{s\tau^2}{2!} + \frac{s^2\tau^3}{3!} - \frac{s^3\tau^4}{4!} + \frac{s^4\tau^5}{5!} - \ldots$$

Thus, the transform $X(s)$ has an infinite number of zeros and no finite poles. Its ROC is

$$\mathrm{Re}\{s\} > -\infty$$

It will be interesting to determine the zeros of the transform. The roots of the numerator are found by solving the equation

$$1-e^{-s\tau} = 0\,, \quad \Rightarrow e^{-s\tau} = 1 \qquad (4.22)$$

The value $s=0$ is a solution for Eqn. (4.22), however, there must be other solutions as well. Realizing that $e^{j2\pi k}=1$ for all integer values of k, Eqn. (4.22) can be written as

$$e^{-s\tau} = e^{-j2\pi k}$$

which leads to the set of solutions

$$s = \frac{j2\pi k}{\tau}\,, \quad k \text{ integer}$$

The numerator of $X(s)$ has an infinite number of roots, all on the $j\omega$-axis of the s-plane. The roots are uniformly spaced and occur at integer multiples of $j2\pi/\tau$. The denominator polynomial of $X(s)$ is just equal to s and has only one root at $s = 0$. This single root of the denominator cancels the root of the numerator at $s = 0$, so that there is neither a zero nor a pole at the origin of the s-plane. We are left with zeros at

$$z_k = \frac{j2\pi k}{\tau}, \quad k \text{ integer, and } k \neq 0$$

Numerator and denominator roots are shown in Fig. 4.15. The pole-zero diagram for $X(s)$ is shown in Fig. 4.16.

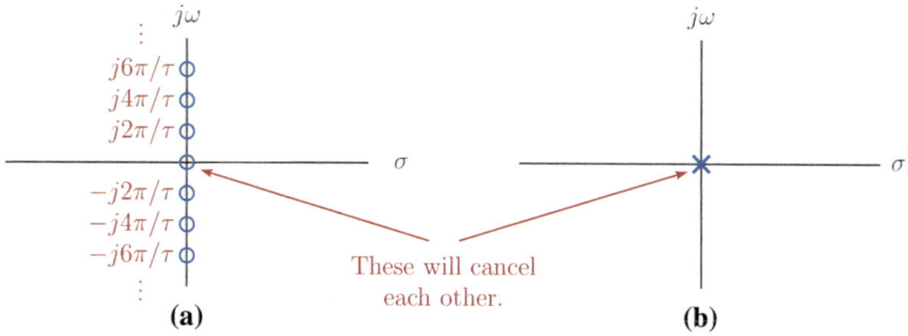

Figure 4.15 – **(a)** Roots of the numerator $\left(1 - e^{-s\tau}\right)$ and **(b)** root of the denominator polynomial s.

To get a sense of what the transform looks like in the s-plane, we will compute and graph the magnitude $|X(s)|$ of the transform as a three-dimensional surface (recall that the transform $X(s)$ is complex-valued, so we can only graph part of it as a surface, that is, we can graph magnitude, phase, real part or imaginary part). Let $s = \sigma + j\omega$, and write the transform as

$$X(s)\Big|_{s=\sigma+j\omega} = \frac{1 - e^{-\sigma\tau} e^{-j\omega\tau}}{\sigma + j\omega}$$

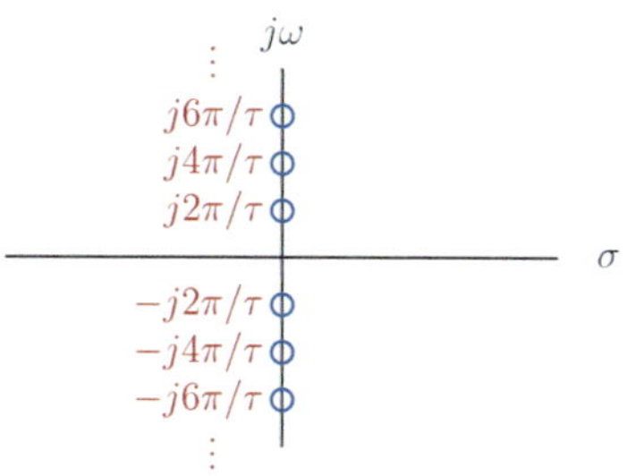

Figure 4.16 – Pole-zero diagram for $X(s)$.

The squared magnitude of a complex function is computed by multiplying it with its own complex conjugate; therefore

$$\begin{aligned}|X(s)|^2 &= X(s)\,X^*(s)\\ &= \left(\frac{1-e^{-\sigma\tau}e^{-j\omega\tau}}{\sigma+j\omega}\right)\left(\frac{1-e^{-\sigma\tau}e^{-j\omega\tau}}{\sigma+j\omega}\right)^*\\ &= \frac{1-2e^{-\sigma\tau}\cos(\omega\tau)+e^{-2\sigma\tau}}{\sigma^2+\omega^2}\end{aligned}$$

and the magnitude of the transform is

$$|X(s)| = \left[\frac{1-2e^{-\sigma\tau}\cos(\omega\tau)+e^{-2\sigma\tau}}{\sigma^2+\omega^2}\right]^{1/2}$$

which is shown in Fig. 4.17(a) as a surface graph.

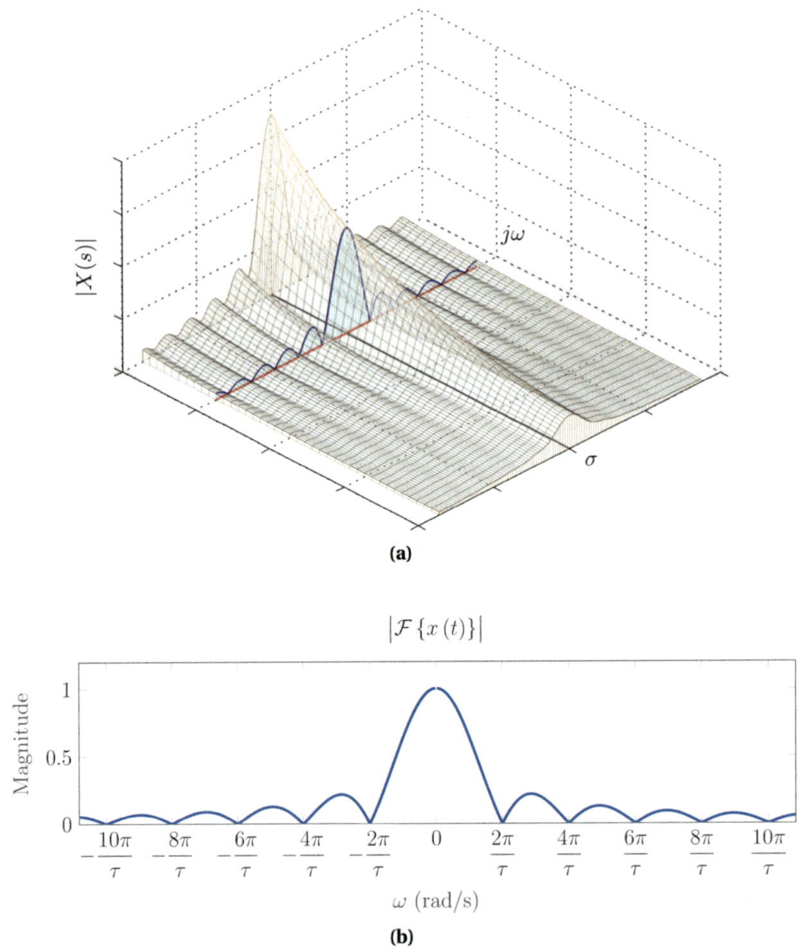

Figure 4.17 – (a) The magnitude $|X(s)|$ for Example 4.7 shown as a surface plot along with the $j\omega$-axis of the s-plane and the magnitude of the transform computed on that axis **(b)** the magnitude of the Fourier transform of $x(t)$ as a function of ω.

Note how the zeros equally spaced on the $j\omega$-axis cause the magnitude surface to dip down at integer multiples of $2\pi/\tau$ radians. In addition to the surface graph, magnitude values computed at points on the $j\omega$-axis of the s-plane are marked on the surface in Fig. 4.17(a). For comparison, the magnitude of the Fourier transform of the signal $x(t)$ computed for the range of angular frequency values $-10\pi/\tau < \omega < 10\pi/\tau$ is shown in Fig. 4.17(b). It should be compared to the values marked on the surface graph that correspond to the magnitude of the Laplace transform evaluated on the $j\omega$-axis.

Software resources: `exct_4_7a.m` , `exct_4_7b.m`

Interactive App: Laplace transform of a pulse signal

The MATLAB app in `appLaplace2.m` is based on the Laplace transform of the pulse signal analyzed in Example 4.7. The magnitude of the transform in question was shown in Fig. 4.17(a). The interactive app computes this magnitude and graphs it as a three-dimensional mesh. It can be rotated freely for viewing from any angle by using the rotation tool in the toolbar. The magnitude of the transform is also evaluated on the vertical line $s = \sigma_1 + j\omega$ in the s-plane and displayed as a two-dimensional graph. This corresponds to the Fourier transform of the modified signal $x(t)\,e^{-\sigma_1 t}$. The value σ_1 and the pulse width τ may be varied through the use of slider controls.

a. Use the slider control to set the pulse width to $\tau = 1$ s.The parameter σ_1 should be set equal to 0 so that the 2-D graph shows $\left|\mathscr{F}\left\{x(t)\right\}\right|$. Pay attention to the dips in the transform. How do they relate to the pole-zero diagram shown in Fig. 4.16?

b. Increase the pulse width to $\tau = 1.5$ s and observe how the dips in the transform change. Again,correlate this behavior with the pole-zero diagram. Repeat with $\tau = 2$ s.

c. Using the appropriate slider control, change the value of the parameter σ_1 to 0.25. Observe the changes in the 3-D graph. Also notice that the 2-D graph now displays $\left|\mathscr{F}\left\{x(t)\,e^{-0.25t}\right\}\right|$ for the Laplace transform evaluated at points $s = 0.25 + j\omega$.

Software resource: `appLaplace2.m`

Example 4.8: Laplace Transform of complex exponential signal

Find the Laplace transform of the signal

$$x(t) = e^{j\omega_0 t}\,u(t)$$

The parameter ω_0 is real and positive, and is in radians.

Solution: Applying the Laplace transform definition directly to the signal $x(t)$ leads to

$$X(s) = \int_{-\infty}^{\infty} e^{j\omega_0 t}\,u(t)\,e^{-st}\,dt$$

Since $x(t)$ is causal, the lower limit of the summation can be set to $t=0$ and the $u(t)$ term can be dropped to yield

$$X(s)=\int_0^{\infty} e^{j\omega_0 t}e^{-st}\,dt=\int_0^{\infty} e^{\left(j\omega_0-s\right)t}\,dt=\left.\frac{e^{\left(j\omega_0-s\right)t}}{j\omega_0-s}\right|_0^{\infty} \tag{4.23}$$

The upper limit is critical in Eqn. (4.23). To be able to evaluate the expression at $s\to\infty$ we need

$$\mathrm{Re}\left\{j\omega_0-s\right\}<0 \quad \Rightarrow \quad \mathrm{Re}\left\{s\right\}>0$$

Provided that this condition is satisfied, the transform in Eqn. (4.23) is computed as

$$X(s)=\frac{1}{j\omega_0-s}\left[0-1\right]=\frac{1}{s-j\omega_0}\,, \qquad \text{ROC: } \mathrm{Re}\left\{s\right\}>0$$

The transform $X(s)$ has one pole at $s=j\omega_0$, and its ROC is the region to the right of the vertical line that passes through its pole. This is illustrated in Fig. 4.18.

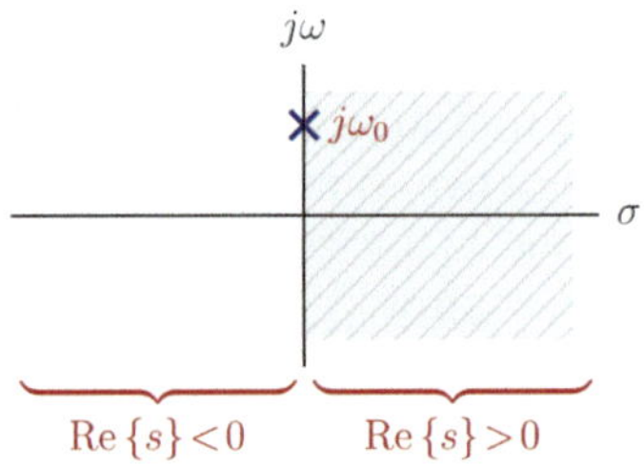

Figure 4.18 – Region of convergence for the transform computed in Example 4.8.

4.2 Characteristics of the Region of Convergence

Through the examples we have worked on so far, we have observed that the Laplace transform of a continuous-time signal always needs to be considered in conjunction with its region of convergence, that is, the region in the s-plane in which the transform converges. In this section we will summarize and justify the fundamental characteristics of the region of convergence.

1. For a finite-duration signal the ROC is the entire s-plane provided that the signal is absolute integrable. The extreme points such as $\mathrm{Re}\{s\}\to\pm\infty$ need to be checked separately.

Let the signal $x(t)$ be equal to zero outside the interval $t_1<t<t_2$. The Laplace transform is

$$X(s)=\int_{t_1}^{t_2} x(t)\,e^{-st}\,dt \tag{4.24}$$

If the signal $x(t)$ is absolute integrable, then the transform given by Eqn. (4.24) converges on the vertical line $s=j\omega$. Since the limits of the integral are finite, it also converges for all other values of s.

2. For a general signal the ROC is in the form of a vertical strip. It is either to the right of a vertical line, to the left of a vertical line, or between two vertical lines as shown in Fig. 4.19.

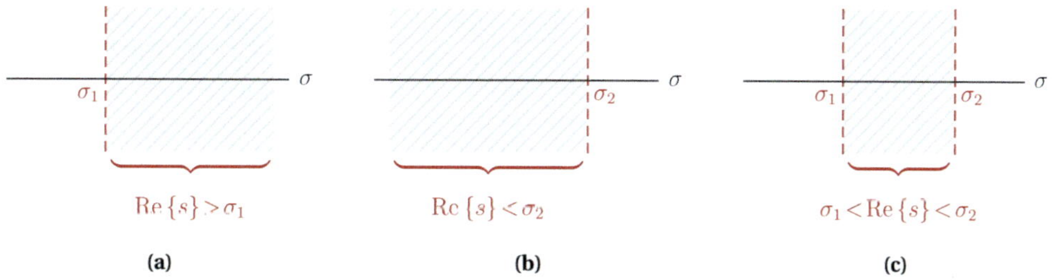

Figure 4.19 – Shape of the region of convergence.

This property is easy to justify. We have established in Eqn. (4.6) of the previous section that, for $s = \sigma + j\omega$, the values of the Laplace transform of the signal $x(t)$ are identical to the values of the Fourier transform of the signal $x(t)\,e^{-\sigma t}$. Therefore, the convergence of the Laplace transform for $s = \sigma + j\omega$ is equivalent to the convergence of the Fourier transform of the signal $x(t)\,e^{-\sigma t}$, and it requires that

$$\int_{-\infty}^{\infty} \left| x(t)\,e^{-\sigma t} \right| dt < \infty \tag{4.25}$$

Thus, the ROC depends only on σ and not on ω, explaining why the region is in the form of a vertical strip. Following are the possibilities for the ROC:

$$\begin{aligned} \mathrm{Re}\{s\} > \sigma_1 &: \quad \text{To the right of a vertical line} \\ \mathrm{Re}\{s\} < \sigma_2 &: \quad \text{To the left of a vertical line} \\ \sigma_1 < \mathrm{Re}\{s\} < \sigma_2 &: \quad \text{Between two vertical lines} \end{aligned}$$

3. For a rational transform $X(s)$ the ROC cannot contain any poles.

By definition, poles of $X(s)$ are values of s that make the transform infinitely large. For rational Laplace transforms, poles are the roots of the denominator polynomial. Since the transform does not converge at a pole, the ROC must naturally exclude all poles of the transform.

4. The ROC for the Laplace transform of a causal signal is the region to the right of a vertical line and is expressed as

$$\mathrm{Re}\{s\} > \sigma_1$$

Since a causal signal equals zero for all $t < 0$, its Laplace transform can be written as

$$X(s) = \int_0^{\infty} x(t)\,e^{-st}\,dt$$

Using $s = \sigma + j\omega$ and remembering that the convergence of the Laplace transform is equivalent to the signal $x(t)\, e^{-\sigma t}$ being absolute integrable leads to the convergence condition

$$\int_0^{\infty} \left| x(t)\, e^{-\sigma t} \right| dt < \infty \tag{4.26}$$

If we can find a value of σ for which Eqn. (4.26) is satisfied, then any larger value of σ will satisfy Eqn. (4.26) as well. All we need to do is find the value $\sigma = \sigma_1$ for the boundary and the ROC will be the region to the right of the vertical line that passes through the point $s = \sigma_1$.

5. The ROC for the Laplace transform of an anti-causal signal is the region to the left of a vertical line and is expressed as

$$\mathrm{Re}\left\{s\right\} < \sigma_2$$

The justification of this property will be similar to that of the previous one. Since an anti-causal signal is equal to zero for all $t > 0$, its Laplace transform can be written as

$$X(s) = \int_{-\infty}^{0} x(t)\, e^{-st}\, dt$$

Using $s = \sigma + j\omega$ the condition for the convergence of the Laplace transform can be expressed through the equivalent condition for the absolute integrability of the signal $x(t)\, e^{-\sigma t}$ as

$$\int_{-\infty}^{0} \left| x(t)\, e^{-\sigma t} \right| dt < \infty \tag{4.27}$$

If we can find a value of σ for which Eqn. (4.27) is satisfied, then any smaller value of σ will satisfy Eqn. (4.27) as well. Once a value $\sigma = \sigma_2$ is found for the boundary, the ROC will be the region to the left of the vertical line that passes through the point $s = \sigma_2$.

6. The ROC for the Laplace transform of a signal that is neither causal nor anti-causal is a strip between two vertical lines and can be expressed as

$$\sigma_1 < \mathrm{Re}\left\{s\right\} < \sigma_2$$

Any signal $x(t)$ can be written as the sum of a causal signal and an anti-causal signal. Let the two signals $x_R(t)$ and $x_L(t)$ be defined in terms of the signal $x(t)$ as

$$x_R(t) = x(t)\, u(t) = \begin{cases} x(t)\,, & t > 0 \\ 0\,, & t < 0 \end{cases}$$

and

$$x_L(t) = x(t)\, u(-t) = \begin{cases} x(t)\,, & t < 0 \\ 0\,, & t > 0 \end{cases}$$

so that $x_R(t)$ is causal, $x_L(t)$ is anti-causal, and

$$x(t) = x_R(t) + x_L(t)$$

Let the Laplace transforms of these two signals be

$$X_R(s) = \mathcal{L}\left\{x_R(t)\right\}, \qquad \text{ROC: } \operatorname{Re}\left\{s\right\} > \sigma_1$$

$$X_L(s) = \mathcal{L}\left\{x_L(t)\right\}, \qquad \text{ROC: } \operatorname{Re}\left\{s\right\} < \sigma_2$$

The Laplace transform of the signal $x(t)$ is

$$X(s) = X_R(s) + X_L(s) \tag{4.28}$$

The ROC for $X(s)$ is at least the overlap of the two regions, that is,

$$\sigma_1 < \operatorname{Re}\left\{s\right\} < \sigma_2 \tag{4.29}$$

provided that $\sigma_2 > \sigma_1$ (otherwise there may be no overlap, and the Laplace transform may not exist at any point in the s-plane).

In some cases the ROC may actually be larger than the overlap in Eqn. (4.29) if the addition of the two transforms in Eqn. (4.28) results in the cancellation of a pole that sets one of the boundaries.

Example 4.9: Laplace transform of two-sided exponential signal

Find the Laplace transform of the signal

$$x(t) = e^{-|t|}$$

Solution: The specified signal exists for all t, and can be written as

$$x(t) = e^{-t}\,u(t) + e^{t}\,u(-t) = \begin{cases} e^{-t}, & t > 0 \\ e^{t}, & t < 0 \end{cases}$$

Thus, $x(t)$ is the sum of a causal signal $x_R(t)$ and an anti-causal signal $x_L(t)$:

$$x(t) = x_R(t) + x_L(t)$$

Two components of $x(t)$ are

$$x_R(t) = e^{-t}\,u(t)$$

$$x_L(t) = e^{t}\,u(-t)$$

When we discuss the properties of the Laplace transform later in Section 4.3 we will show that the Laplace transform is linear. Consequently, the transform of the sum of two signals is equal to the sum of their respective transforms, and $X(s)$ can be written as

$$\begin{aligned} X(s) =& X_R(s) + X_L(s) \\ =& \mathcal{L}\left\{e^{-t}\,u(t)\right\} + \mathcal{L}\left\{e^{t}\,u(-t)\right\} \end{aligned}$$

The individual transforms $X_R(s)$ and $X_L(s)$ that make up $X(s)$ can be determined by adapting the results obtained earlier in Eqns. (4.16) and (4.17) as:

$$X_R(s) = \mathcal{L}\left\{e^{-t}\,u(t)\right\} = \frac{1}{s+1}, \qquad \text{ROC: } \operatorname{Re}\left\{s\right\} > -1$$

$$X_L(s) = \mathcal{L}\left\{e^{t}\,u(-t)\right\} = -\frac{1}{s-1}, \qquad \text{ROC: } \operatorname{Re}\left\{s\right\} < 1$$

The ROC $X_R(s)$ and $X_L(s)$ are shown in Fig. 4.20.

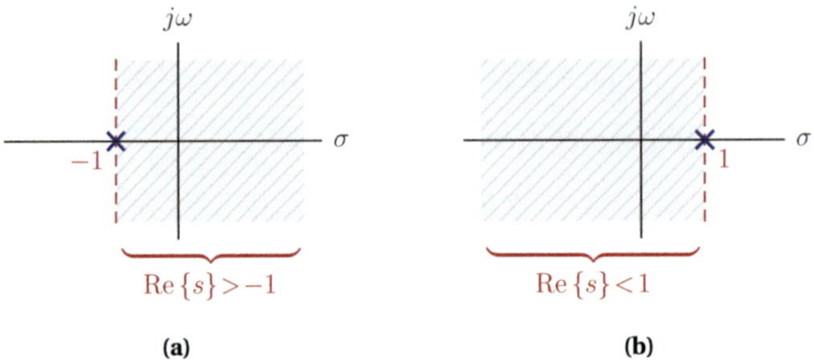

Figure 4.20 – The ROC for **(a)** $X_R(s) = \mathscr{L}\left\{e^{-t}\,u(t)\right\}$ and **(b)** $X_L(s) = \mathscr{L}\left\{e^{t}\,u(-t)\right\}$.

The transform $X(s)$ can be computed as

$$X(s) = \frac{1}{s+1} - \frac{1}{s-1} = \frac{-2}{s^2-1}$$

$X(s)$ has a two poles at $s = \pm 1$. For this transform to converge both $X_R(s)$ and $X_L(s)$ must converge. Therefore, the ROC for $X(s)$ is the overlap of the ROCs of $X_R(s)$ and $X_L(s)$

$$\text{ROC:} \quad -1 < \text{Re}\left\{s\right\} < 1$$

and is shown in Fig. 4.21.

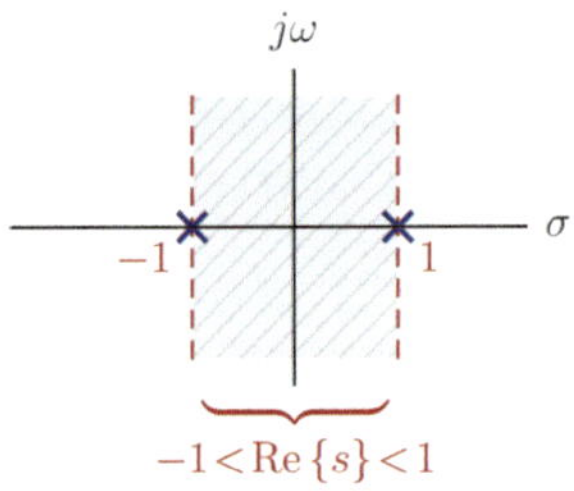

Figure 4.21 – The ROC for the transform $X(s)$.

Software resource: exct_4_9.m

4.3 Properties of the Laplace Transform

The important properties of the Laplace transform will be summarized in this section and the proof of each will be given. The use of these properties greatly simplifies the application of the Laplace transform to problems involving the analysis of continuous-time signals and systems.

4.3.1 Linearity

Laplace transform is linear operator. For any two signals $x_1(t)$ and $x_2(t)$ with their respective transforms

$$x_1(t) \overset{\mathcal{L}}{\longleftrightarrow} X_1(s) \quad \text{and} \quad x_2(t) \overset{\mathcal{L}}{\longleftrightarrow} X_2(s)$$

and any two constants α_1 and α_2, the following relationship holds:

Linearity of the Laplace transform:

$$\alpha_1 x_1(t) + \alpha_2 x_2(t) \overset{\mathcal{L}}{\longleftrightarrow} \alpha_1 X_1(s) + \alpha_2 X_2(s) \tag{4.30}$$

Proof of Eqn. (4.30):

The proof is straightforward using the Laplace transform definition given by Eqn. (4.1). Substituting $x(t) = \alpha_1 x_1(t) + \alpha_2 x_2(t)$ into Eqn. (4.1) we get

$$\mathcal{L}\{\alpha_1 x_1(t) + \alpha_2 x_2(t)\} = \int_{-\infty}^{\infty} \left[\alpha_1 x_1(t) + \alpha_2 x_2(t)\right] e^{-st}\, dt \tag{4.31}$$

The integral in Eqn. (4.31) can be separated into two integrals as

$$\begin{aligned}
\mathcal{L}\{\alpha_1 x_1(t) + \alpha_2 x_2(t)\} &= \int_{-\infty}^{\infty} \alpha_1 x_1(t)\, e^{-st}\, dt + \int_{-\infty}^{\infty} \alpha_2 x_2(t)\, e^{-st}\, dt \\
&= \alpha_1 \int_{-\infty}^{\infty} x_1(t)\, e^{-st}\, dt + \alpha_2 \int_{-\infty}^{\infty} x_2(t)\, e^{-st}\, dt \\
&= \alpha_1\, \mathcal{L}\{x_1(t)\} + \alpha_2\, \mathcal{L}\{x_2(t)\}
\end{aligned} \tag{4.32}$$

The Laplace transform of a weighted sum of two signals is equal to the same weighted sum of their respective transforms $X_1(s)$ and $X_2(s)$. The ROC for the resulting transform is at least the overlap of the two individual transforms, if such an overlap exists. The ROC may be greater than the overlap of the two regions if the addition of the two transforms results in the cancelation of a pole.

The linearity property proven above for two signals can be generalized to any arbitrary number of signals. The Laplace transform of a weighted sum of any number of signals is equal to the same weighted sum of their respective transforms.

Example 4.10: Using the linearity property of the Laplace transform

Determine the Laplace transform of the signal

$$x(t) = 2e^{-t}\, u(t) + 5e^{-2t}\, u(t)$$

Solution: A general expression for the Laplace transform of the causal exponential signal was found in Example 4.5 as

$$\mathcal{L}\{e^{at}\, u(t)\} = \frac{1}{s-a}, \qquad \text{ROC:} \quad \text{Re}\{s\} > \text{Re}\{a\}$$

Applying this result to the exponential terms in $x(t)$ we get

$$\mathcal{L}\left\{e^{-t}\,u(t)\right\} = \frac{1}{s+1}\,, \qquad \text{ROC:} \quad \text{Re}\left\{s\right\} > -1 \tag{4.33}$$

and

$$\mathcal{L}\left\{e^{-2t}\,u(t)\right\} = \frac{1}{s+2}\,, \qquad \text{ROC:} \quad \text{Re}\left\{s\right\} > -2 \tag{4.34}$$

Combining these results and using the linearity property we arrive at the transform sought:

$$X(s) = 2\left(\frac{1}{s+1}\right) + 5\left(\frac{1}{s+2}\right) = \frac{7s+9}{s^2+3s+2}$$

The ROC for $X(s)$ is the overlap of the two regions in Eqns. (4.33) and (4.34), namely

$$\text{Re}\left\{s\right\} > -1$$

Fig. 4.22 illustrates the poles and the ROCs of the individual terms $X_1(s)$ and $X_2(s)$ as well as the ROC of the transform $X(s)$ obtained as the overlap of the two.

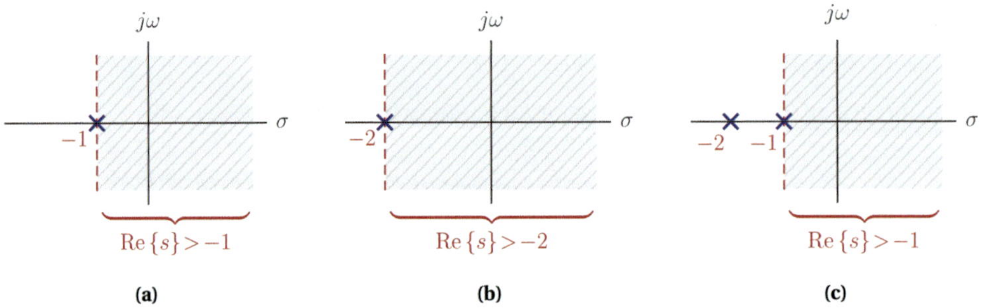

Figure 4.22 – **(a)** The ROC for $\mathcal{L}\left\{e^{-t}\,u(t)\right\}$, **(b)** the ROC for $\mathcal{L}\left\{e^{-2t}\,u(t)\right\}$, and **(c)** the ROC for $X(s)$.

Example 4.11: Laplace transform of a cosine signal

Find the Laplace transform of the signal

$$x(t) = \cos(\omega_0 t)\,u(t)$$

Solution: Using Euler's formula for the $\cos(\omega_0 t)$ term, the signal $x(t)$ can be written as

$$\cos(\omega_0 t)\,u(t) = \frac{1}{2}\,e^{j\omega_0 t}\,u(t) + \frac{1}{2}\,e^{-j\omega_0 t}\,u(t)$$

and its Laplace transform is

$$\mathcal{L}\left\{\cos(\omega_0 t)\,u(t)\right\} = \mathcal{L}\left\{\frac{1}{2}\,e^{j\omega_0 t}\,u(t) + \frac{1}{2}\,e^{-j\omega_0 t}\,u(t)\right\}$$

Using the linearity property of the Laplace transform we can write

$$\begin{aligned}\mathscr{L}\left\{\cos(\omega_0 t)\,u(t)\right\} &= \frac{1}{2}\,\mathscr{L}\left\{e^{j\omega_0 t}\,u(t)\right\} + \frac{1}{2}\,\mathscr{L}\left\{e^{-j\omega_0 t}\,u(t)\right\} \\ &= \frac{1/2}{s-j\omega_0} + \frac{1/2}{s+j\omega_0}\end{aligned}$$

Combining the two terms under a common denominator we have

$$X(s) = \frac{s}{s^2+\omega_0^2} \tag{4.35}$$

The transform $X(s)$ has a zero at $s=0$ and a pair of complex conjugate poles at $s=\pm j\omega_0$. The ROC is the region to the right of the $j\omega$-axis of the s-plane, as illustrated in Fig. 4.23. The $j\omega$-axis itself is not included in the ROC since there are poles on it.

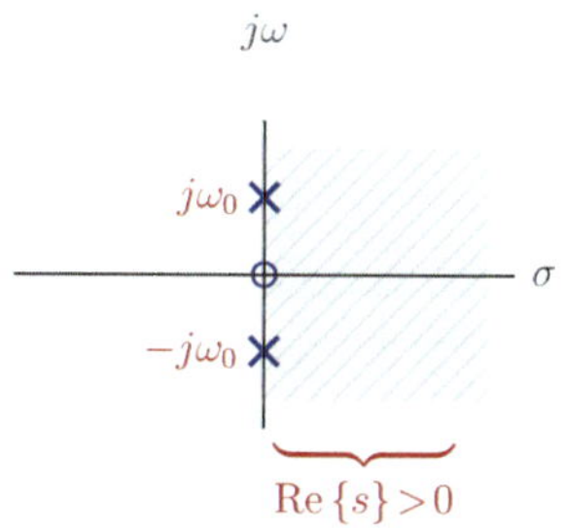

Figure 4.23 – The ROC for the transform in Eqn. (4.35).

Example 4.12: Laplace transform of a sine signal

Find the Laplace transform of the signal

$$x(t) = \sin(\omega_0 t)\,u(t)$$

Solution: This problem is quite similar to that in Example 4.11. Using Euler's formula for the signal $x(t)$ we get

$$\sin(\omega_0 t)\,u(t) = \frac{1}{2j}\,e^{j\omega_0 t}\,u(t) - \frac{1}{2j}\,e^{-j\omega_0 t}\,u(t)$$

and its Laplace transform is

$$\begin{aligned}\mathscr{L}\left\{\sin(\omega_0 t)\,u(t)\right\} &= \mathscr{L}\left\{\frac{1}{2j}\,e^{j\omega_0 t}\,u(t) - \frac{1}{2j}\,e^{-j\omega_0 t}\,u(t)\right\} \\ &= \frac{1}{2j}\,\mathscr{L}\left\{e^{j\omega_0 t}u(t)\right\} - \frac{1}{2j}\,\mathscr{L}\left\{e^{-j\omega_0 t}u(t)\right\} \\ &= \frac{1}{2j}\left(\frac{1}{s-j\omega_0}\right) - \frac{1}{2j}\left(\frac{1}{s+j\omega_0}\right)\end{aligned}$$

Combining the two terms under a common denominator we have

$$X(s) = \frac{\omega_0}{s^2 + \omega_0^2}$$

The transform $X(s)$ has a pair of complex conjugate poles at $s = \pm j\omega_0$. The ROC of the transform is $\text{Re}\{s\} > 0$ as in the previous example.

4.3.2 Time shifting property

Shifting the signal in the time domain is equivalent to multiplying the transform with an exponential factor. This relationship can be useful in certain types of problems where a time- domain signal is expressed in terms of time-shifted basis functions.

Time shifting the signal $x(t)$:

Given the transform pair

$$x(t) \stackrel{\mathscr{L}}{\longleftrightarrow} X(s)$$

the following is also a valid transform pair:

$$x(t-\tau) \stackrel{\mathscr{L}}{\longleftrightarrow} e^{-s\tau}\,X(s) \tag{4.36}$$

Time shifting the signal $x(t)$ in the time domain by τ corresponds to multiplication of the transform $X(s)$ by $e^{-s\tau}$ in the s domain.

Proof of Eqn. (4.36):

The Laplace transform of $x(t-\tau)$ is

$$\mathscr{L}\left\{x(t-\tau)\right\} = \int_{-\infty}^{\infty} x(t-\tau)\,e^{-st}\,dt \tag{4.37}$$

Let us define a new variable $\lambda = t - \tau$, and write the integral in Eqn. (4.37) in terms of this new variable:

$$\begin{aligned}\mathscr{L}\left\{x(t-\tau)\right\} &= \int_{-\infty}^{\infty} x(\lambda)\,e^{-s(\lambda+\tau)}\,d\lambda \\ &= e^{-s\tau}\int_{-\infty}^{\infty} x(\lambda)\,e^{-s\lambda}\,d\lambda \\ &= e^{-s\tau}\,X(s)\end{aligned} \tag{4.38}$$

The ROC for the resulting transform $e^{-s\tau}\,X(s)$ is generally the same as that of $X(s)$.[1]

[1] Care must be taken in certain circumstances. If the time shift makes a causal signal non-causal then the points $s = \infty + j\omega$ would need to be excluded from the ROC. Similarly, if an anti-causal signal loses its anti-causal property as the result of a shift, then the points $s = -\infty + j\omega$ need to be excluded.

Example 4.13: Laplace transform of a pulse signal revisited

The Laplace transform of the pulse signal

$$x(t) = \Pi\left(\frac{t-\tau/2}{\tau}\right)$$

was determined earlier in Example 4.7. Find the same transform through the use of linearity and time-shifting properties.

Solution: The signal $x(t)$ can be expressed as the difference of a unit step signal and a time-shifted unit step signal in the form

$$x(t) = u(t) - u(t-\tau)$$

Using the linearity of the Laplace transform, $X(s)$ can be expressed as

$$\begin{aligned} X(s) &= \mathscr{L}\left\{u(t) - u(t-\tau)\right\} \\ &= \mathscr{L}\left\{u(t)\right\} - \mathscr{L}\left\{u(t-\tau)\right\} \end{aligned}$$

The Laplace transform of the unit step function was found in Example 4.3 to be

$$\mathscr{L}\left\{u(t)\right\} = \frac{1}{s}\,, \qquad \text{ROC:} \quad \text{Re}\left\{s\right\} > 0 \tag{4.39}$$

Using the time-shifting property of the Laplace transform, we find the transform of the time-shifted unit step signal as

$$\mathscr{L}\left\{u(t-\tau)\right\} = \frac{e^{-s\tau}}{s}\,, \qquad \text{ROC:} \quad \text{Re}\left\{s\right\} > 0 \tag{4.40}$$

Subtracting the transform in Eqn. (4.40) from the transform in Eqn. (4.39) yields

$$X(s) = \frac{1-e^{-s\tau}}{s}$$

which matches the earlier result found in Example 4.7. Since $x(t)$ is a finite-duration signal, the ROC is the entire s-plane with the exception of $s = -\infty$. This is one example of the possibility mentioned earlier regarding the boundaries. The ROC for $X(s)$ is larger than the overlap of the two regions given by Eqns. (4.39) and (4.40). The individual transforms $\mathscr{L}\left\{u(t)\right\}$ and $\mathscr{L}\left\{u(t-\tau)\right\}$ each have a pole at $s = 0$, and therefore the individual ROCs are to the right of the $j\omega$-axis. When the two terms are added to construct $X(s)$, however, the pole at $s = 0$ is canceled, resulting in the ROC for $X(s)$ being larger than the overlap of the individual regions.

Example 4.14: Laplace transform of a bipolar pulse signal

Determine the Laplace transform of the bipolar pulse signal shown in Fig. 4.24.

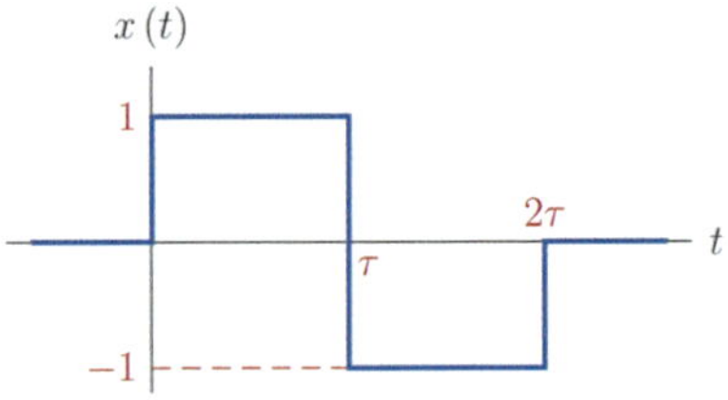

Figure 4.24 – The bipolar pulse signal $x(t)$ for Example 4.14.

Solution: Let $g(t)$ be a unit-amplitude pulse with a width of τ centered around $t=\tau/2$, that is,

$$g(t)=\Pi\left(\frac{t-\tau/2}{\tau}\right)$$

Utilizing the result obtained in Example 4.7, the Laplace transform of $g(t)$ is

$$G(s)=\frac{1-e^{-s\tau}}{s}$$

The bipolar pulse signal $x(t)$ can be expressed in terms of $g(t)$ as

$$x(t)=g(t)-g(t-\tau)$$

Using linearity and time shifting properties of the Laplace transform we have

$$\begin{aligned}X(s)=&G(s)-e^{-s\tau}\,G(s)\\=&\left(1-e^{-s\tau}\right)G(s)\\=&\frac{(1-e^{-s\tau})^2}{s}\end{aligned}$$

The ROC of the transform is

$$\mathrm{Re}\{s\}>-\infty$$

Example 4.15: Laplace transform of a truncated sine function

Determine the Laplace transform of the signal

$$x(t)=\begin{cases}\sin(\pi t)\,, & 0<t<1\\ 0\,, & \text{otherwise}\end{cases}$$

which is graphed in Fig. 4.25.

Solution: We will solve this problem first by direct application of the Laplace transform definition, and then through the use of linearity and shifting properties of the

Laplace transform. Applying the definition of the Laplace transform we get

$$X(s) = \int_0^1 \sin(\pi t)\, e^{-st}\, dt \tag{4.41}$$

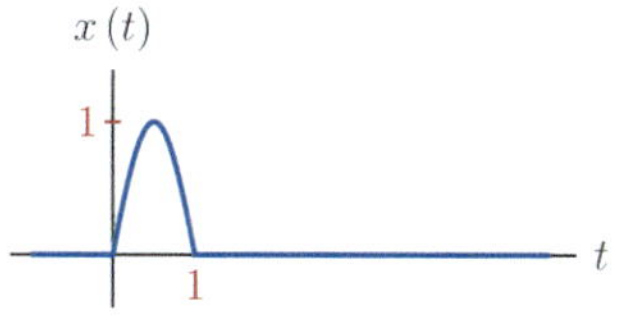

Figure 4.25 – The signal $x(t)$ for Example 4.15.

Writing $\sin(\pi t)$ using Euler's formula leads to

$$\sin(\pi t) = \frac{1}{2j}\, e^{j\pi t} - \frac{1}{2j}\, e^{-j\pi t}$$

and substituting this result into Eqn. (4.41) yields

$$X(s) = \frac{1}{2j} \int_0^1 e^{j\pi t}\, e^{-st}\, dt - \frac{1}{2j} \int_0^1 e^{-j\pi t}\, e^{-st}\, dt$$

which can be evaluated as

$$X(s) = \frac{1}{2j}\, \frac{e^{(j\pi - s)t}}{(j\pi - s)}\bigg|_0^1 + \frac{1}{2j}\, \frac{e^{-(j\pi + s)t}}{(j\pi + s)}\bigg|_0^1 = \frac{\pi\,(1 + e^{-s})}{s^2 + \pi^2} \tag{4.42}$$

In Eqn. (4.42) we have used the fact that $e^{+j\pi} = -1$. Since the signal $x(t)$ is of finite duration, the ROC of the transform is the entire s-plane with just the exception of points where $\mathrm{Re}\{s\} \rightarrow -\infty$.

The solution can be simplified by clever use of the properties of the Laplace transform. The first step is to recognize that the signal $x(t)$ can be expressed as the sum of a causal sinusoidal signal and its time-shifted version as shown in Fig. 4.26.

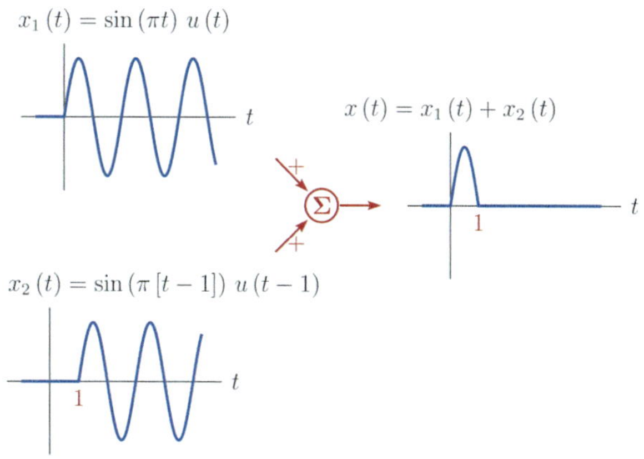

Figure 4.26 – Expressing a truncated sinusoidal signal as the sum of two signals.

Mathematically we have

$$x(t) = \sin(\pi t)\, u(t) + \sin(\pi [t-1])\, u(t-1)$$

Recall that the Laplace transform of a causal sinusoidal signal was found in Example 4.12 as

$$\mathcal{L}\left\{\sin(\omega_0 t)\, u(t)\right\} = \frac{\omega_0}{s^2+\omega_0^2}\,, \qquad \text{ROC:} \quad \text{Re}\left\{s\right\} > 0$$

Using this result with $\omega_0 = \pi$ along with the linearity of the Laplace transform and the time-shifting property, we obtain the transform as

$$\begin{aligned} X(s) =& \mathcal{L}\left\{\sin(\pi t)\, u(t)\right\} + \mathcal{L}\left\{\sin(\pi [t-1])\, u(t-1)\right\} \\ =& \frac{\pi}{s^2+\pi^2} + e^{-s}\,\frac{\pi}{s^2+\pi^2} \\ =& \frac{\pi\left(1+e^{-s}\right)}{s^2+\pi^2} \end{aligned}$$

which matches the result found earlier.

Software resource: exct_4_15.m

4.3.3 Shifting in the s domain

Shifting the Laplace transform in the s domain is equivalent to multiplication of the time-domain signal by an exponential factor. We will see that this property will also be practical to use in solving certain types of problems.

Shifting $X(s)$ in the s-domain:

Given the transform pair

$$x(t) \overset{\mathcal{L}}{\longleftrightarrow} X(s)$$

the following is also a valid transform pair:

$$x(t)\, e^{s_0 t} \overset{\mathcal{L}}{\longleftrightarrow} X(s-s_0) \tag{4.43}$$

Proof of Eqn. (4.43):

Through direct application of the Laplace transform definition given by Eqn. (4.1) we get

$$\begin{aligned} \mathcal{L}\left\{x(t)\, e^{s_0 t}\right\} =& \int_{-\infty}^{\infty} x(t)\, e^{s_0 t}\, e^{-st}\, dt \\ =& \int_{-\infty}^{\infty} x(t)\, e^{-(s-s_0)t}\, dt \\ =& X(s-s_0) \end{aligned}$$

Let the ROC for original transform $X(s)$ be

$$\sigma_1 < \mathrm{Re}\{s\} < \sigma_2$$

For $X(s-s_0)$ to converge, we need

$$\sigma_1 < \mathrm{Re}\{s-s_0\} < \sigma_2$$

Therefore, the ROC for $X(s-s_0)$ is

$$\sigma_1 + \mathrm{Re}\{s_0\} < \mathrm{Re}\{s\} < \sigma_2 + \mathrm{Re}\{s_0\}$$

The ROC for $X(s-s_0)$ is a shifted version of the ROC for $X(s)$, shifted horizontally by an amount equal to the real part of the parameter s_0. This is illustrated in Fig. 4.27.

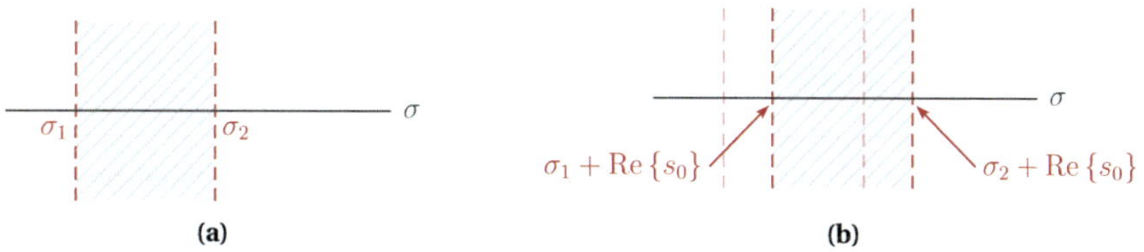

Figure 4.27 – The effect of s-domain shifting on the ROC: **(a)** the ROC for $X(s)$, and **(b)** the ROC for $X(s-s_0)$.

Example 4.16: Laplace transform of exponentially damped sinusoidal signal

Determine the Laplace transform of the signal

$$x(t) = e^{-2t}\cos(3t)\,u(t)$$

Solution: The transform can be found easily by the use of the s-domain shifting property. Let the signal $x_1(t)$ be defined as

$$x_1(t) = \cos(3t)\,u(t)$$

so that

$$x(t) = e^{-2t}\,x_1(t)$$

The Laplace transform of a causal cosine signal was derived in Example 4.11. Using the result found in that example with $\omega_0 = 3$ yields

$$X_1(s) = \mathcal{L}\{\cos(3t)\,u(t)\} = \frac{s}{s^2+9}\,, \qquad \text{ROC:} \quad \mathrm{Re}\{s\} > 0$$

We are now ready to apply the s-domain shifting property.

$$X(s) = X_1(s+2) = \left.\frac{s}{s^2+9}\right|_{s\to s+2} = \frac{s+2}{(s+2)^2+9}$$

The resulting transform has a zero at $s=-2$ and a pair of complex conjugate poles at $s = -2 \pm j3$ as shown in Fig. 4.28. The boundary for the ROC is set by the two poles; therefore the ROC is

$$\text{ROC:} \quad \mathrm{Re}\{s\} > -2$$

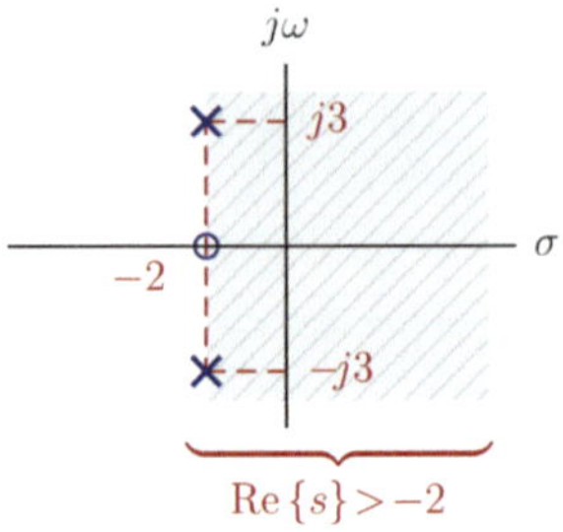

Figure 4.28 – The pole-zero diagram and the ROC for the transform in Example 4.16.

4.3.4 Scaling in time and *s* domains

This property is significant in signal processing. If the time variable is scaled by a factor a, the transform variable is scaled by the same factor, but in the opposite direction.

Scaling in time and *s* domains:

Given the transform pair

$$x(t) \overset{\mathscr{L}}{\longleftrightarrow} X(s)$$

and a real-valued parameter a, the following is also a valid transform pair:

$$x(at) \overset{\mathscr{L}}{\longleftrightarrow} \frac{1}{|a|}\,X\left(\frac{s}{a}\right) \tag{4.44}$$

Proof of Eqn. (4.44):

Once again, we will begin by applying the Laplace transform definition given by Eqn. (4.1) to obtain

$$\mathscr{L}\left\{x(at)\right\} = \int_{-\infty}^{\infty} x(at)\, e^{-st}\, dt \tag{4.45}$$

If $a > 0$, employing a variable change $\lambda = at$ and its consequence $d\lambda = a\,dt$ on the integral of Eqn. (4.45) leads to

$$\begin{aligned}\mathscr{L}\left\{x(at)\right\} &= \int_{-\infty}^{\infty} x(\lambda)\, e^{-s\lambda/a}\, \frac{d\lambda}{a} \\ &= \frac{1}{a}\int_{-\infty}^{\infty} x(\lambda)\, e^{-(s/a)\lambda}\, d\lambda \\ &= \frac{1}{a}\,X\left(\frac{s}{a}\right) \end{aligned} \tag{4.46}$$

On the other hand, if $a<0$, the same variable change leads to

$$\mathcal{L}\left\{x(at)\right\} = \int_{\infty}^{-\infty} x(\lambda)\, e^{-s\lambda/a}\, \frac{d\lambda}{a}$$

$$= -\frac{1}{a}\int_{-\infty}^{\infty} x(\lambda)\, e^{-(s/a)\lambda}\, d\lambda$$

$$= -\frac{1}{a}\, X\left(\frac{s}{a}\right) \tag{4.47}$$

The reason for the negative sign in Eqn. (4.47) is that, for $a<0$, the integration limits for $t\to\pm\infty$ translate to limits $\lambda\to\mp\infty$. In order to account for both possibilities of $a>0$ and $a<0$, the results in Eqns. (4.46) and (4.47) may be combined to yield

$$\mathcal{L}\left\{x(at)\right\} = \frac{1}{|a|}\int_{-\infty}^{\infty} x(\lambda)\, e^{-(s/a)\lambda}\, d\lambda$$

$$= \frac{1}{|a|}\, X\left(\frac{s}{a}\right) \tag{4.48}$$

The use of the absolute value on the scale factor eliminates the need for the negative sign when $a<0$. This completes the proof of the scaling property given by Eqn. (4.44).

The ROC of the result still needs to be related to the ROC for the original transform $X(s)$. Let the latter be

$$\sigma_1 < \mathrm{Re}\left\{s\right\} < \sigma_2$$

For the term $X(s/a)$ in Eqn. (4.48) to converge, we need

$$\sigma_1 < \mathrm{Re}\left\{\frac{s}{a}\right\} < \sigma_2 \tag{4.49}$$

Since the parameter a is real-valued, the ROC in Eqn. (4.49) can be written as

$$\sigma_1 < \frac{\mathrm{Re}\left\{s\right\}}{a} < \sigma_2 \tag{4.50}$$

Depending on the sign of the parameter a, two possibilities need to be considered for the ROC of $\mathcal{L}\left\{x(at)\right\}$:

a. If $a>0$: $a\sigma_1 < \mathrm{Re}\left\{s\right\} < a\sigma_2$

b. If $a<0$: $a\sigma_2 < \mathrm{Re}\left\{s\right\} < a\sigma_1$

Example 4.17: Using the scaling property of the Laplace transform

Use the scaling property to find the Laplace transform of the signal

$$x(t) = e^{2t}\, u(-t)$$

from the knowledge of the transform pair

$$g(t) = e^{-2t}\, u(t) \quad \Longrightarrow \quad G(s) = \frac{1}{s+2}\,, \qquad \text{ROC:} \quad \mathrm{Re}\left\{s\right\} > -2$$

Solution: It is obvious from a comparison of the signals $x(t)$ and $g(t)$ that

$$x(t) = g(-t)$$

so that the scaling property given by Eqn. (4.44) can be used with the scale factor $a = -1$ to yield

$$X(s) = \frac{1}{|-1|}\, G\left(\frac{s}{-1}\right) = G(-s) = \frac{1}{-s+2}$$

and the ROC is

$$\mathrm{Re}\{-s\} > -2 \quad \Rightarrow \quad \mathrm{Re}\{s\} < 2$$

The result found is consistent with Eqn. (4.17).

4.3.5 Differentiation in the time domain

Differentiating the signal in the time domain is equivalent to multiplying its transform by a factor of s. This property will be useful in determining the Laplace transform of each term in a differential equation.

Differentiation in time property of the Laplace transform:

Given the transform pair

$$x(t) \stackrel{\mathcal{L}}{\longleftrightarrow} X(s)$$

the following is also a valid transform pair:

$$\frac{dx(t)}{dt} \stackrel{\mathcal{L}}{\longleftrightarrow} s\,X(s) \tag{4.51}$$

Proof of Eqn. (4.51):

Using the Laplace transform definition in Eqn. (4.1), the transform of $dx(t)/dt$ is

$$\begin{aligned}\mathcal{L}\left\{\frac{dx(t)}{dt}\right\} &= \int_{-\infty}^{\infty} \frac{dx(t)}{dt}\, e^{-st}\, dt \\ &= \int_{-\infty}^{\infty} e^{-st}\, dx(t)\end{aligned} \tag{4.52}$$

Integrating Eqn. (4.52) by parts yields

$$\mathcal{L}\left\{\frac{dx(t)}{dt}\right\} = x(t)\, e^{-st}\Big|_{-\infty}^{\infty} + s\int_{-\infty}^{\infty} x(t)\, e^{-st}\, dt \tag{4.53}$$

The term $x(t)\, e^{-st}$ must evaluate to zero for $t \to \pm\infty$ for the transform $X(s)$ to exist. Therefore, Eqn. (4.53) reduces to

$$\mathcal{L}\left\{\frac{dx(t)}{dt}\right\} = s\int_{-\infty}^{\infty} x(t)\, e^{-st}\, dt = s\,X(s) \tag{4.54}$$

completing the proof.

The ROC for the transform of $dx(t)/dt$ is at least equal to the ROC of the original transform. If the original transform $X(s)$ has a single pole at $s=0$ that sets the boundary of its ROC, then the cancelation of that pole due to multiplication by s causes the ROC of the new transform $s\,X(s)$ to be larger.

Example 4.18: Using the time-domain differentiation property of the Laplace transform

Use the time-domain differentiation property to find the Laplace transform of the signal

$$x(t) = \cos(3t)\,u(t)$$

from the knowledge of the transform pair

$$g(t) = \sin(3t)\,u(t) \quad \Longrightarrow \quad G(s) = \frac{3}{s^2+9}\,, \quad \text{ROC:} \quad \text{Re}\{s\} > 0$$

Solution: Let us differentiate the signal $g(t)$:

$$\frac{dg(t)}{dt} = 3\cos(3t)\,u(t)$$

Based on this result, the signal $x(t)$ can be written as

$$x(t) = \left(\frac{1}{3}\right)\frac{dg(t)}{dt}$$

and its Laplace transform is

$$X(s) = \left(\frac{1}{3}\right) s\,G(s) = \frac{s}{s^2+9}$$

as expected. The ROC of $X(s)$ is

$$\text{Re}\{s\} > 0$$

4.3.6 Differentiation in the s domain

Differentiation in the s domain property of the Laplace transform:

Given the transform pair

$$x(t) \overset{\mathcal{L}}{\longleftrightarrow} X(s)$$

the following is also a valid transform pair:

$$t\,x(t) \overset{\mathcal{L}}{\longleftrightarrow} -\frac{dX(s)}{ds} \tag{4.55}$$

Proof of Eqn. (4.55):

The proof is straightforward by differentiating both sides of the Laplace transform definition given by Eqn. (4.1):

$$\frac{dX(s)}{ds} = \frac{d}{ds}\left[\int_{-\infty}^{\infty} x(t)\, e^{-st}\, dt\right] \tag{4.56}$$

Interchanging the order of integration and differentiation on the right side of Eqn. (4.56) leads to

$$\begin{aligned}\frac{dX(s)}{ds} &= \int_{-\infty}^{\infty} \frac{d}{ds}\left[x(t)\, e^{-st}\right] dt \\ &= -\int_{-\infty}^{\infty} t\,x(t)\, e^{-st}\, dt \\ &= -\mathcal{L}\left\{t\,x(t)\right\} \end{aligned} \tag{4.57}$$

Eqn. (4.55) follows from Eqn. ((4.57).

The ROC for the transform of $t\,x(t)$ is the same as the ROC for the original transform $X(s)$.

Example 4.19: Using the s-domain differentiation property of the Laplace transform

Determine the Laplace transform of the unit ramp signal

$$r(t) = t\,u(t)$$

Solution: The Laplace transform of the unit step function is

$$\mathcal{L}\left\{u(t)\right\} = \frac{1}{s}, \qquad \text{ROC:} \quad \text{Re}\left\{s\right\} > 0$$

Using the s-domain differentiation property, the Laplace transform of the unit ramp signal is

$$R(s) = \mathcal{L}\left\{t\,u(t)\right\} = -\frac{d}{ds}\left[\frac{1}{s}\right] = \frac{1}{s^2} \tag{4.58}$$

and the ROC of the transform is

$$\text{Re}\left\{s\right\} > 0$$

Example 4.20: Transform of a signal using multiple ramp functions

Determine the Laplace transform of the signal $x(t)$ shown in Fig. 4.29.

Solution: The signal $x(t)$ can be expressed in terms of a unit ramp function and its delayed version as

$$\begin{aligned}x(t) &= r(t) - r(t-1) \\ &= t\,u(t) - (t-1)\,u(t-1)\end{aligned}$$

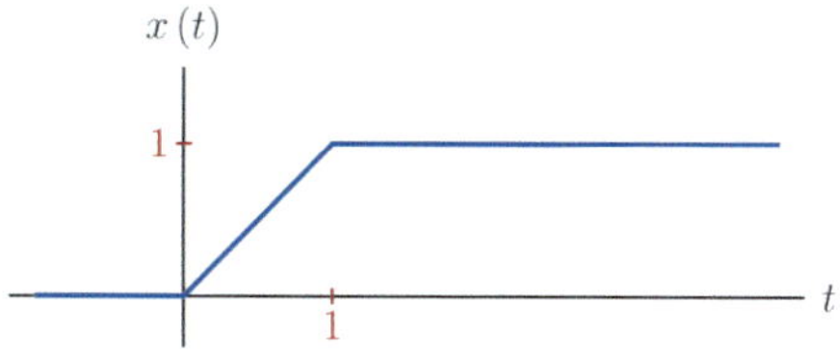

Figure 4.29 – The signal $x(t)$ for Example 4.20.

The transform of the unit ramp function was determined in Example 4.19. Using the result in Eqn. (4.58) along with the time-shifting property we obtain

$$X(s) = \frac{1}{s^2} - e^{-s}\,\frac{1}{s^2} = \frac{1-e^{-s}}{s^2}$$

with the ROC

$$\text{Re}\{s\} > 0$$

Example 4.21: Transform of an exponentially damped ramp function

Determine the Laplace transform of the signal

$$x(t) = t\,e^{-2t}\,u(t)$$

which is shown in Fig. 4.30.

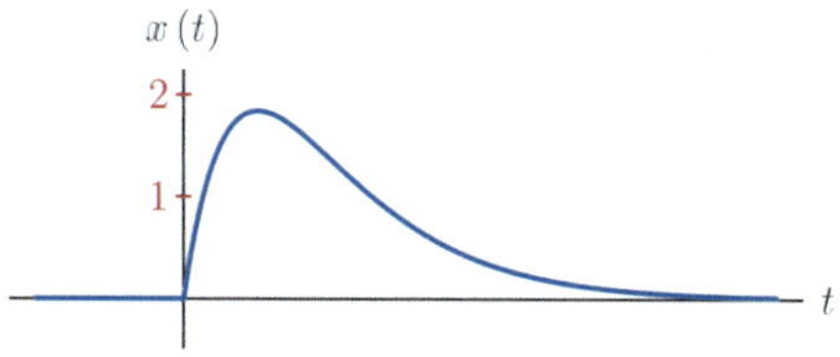

Figure 4.30 – The signal $x(t)$ for Example 4.21.

Solution: The transform of the causal exponential signal $e^{-2t}\,u(t)$ is

$$\mathcal{L}\left\{e^{-2t}\,u(t)\right\} = \frac{1}{s+2}\,, \qquad \text{ROC:} \quad \text{Re}\{s\} > -2$$

Using the s-domain differentiation property

$$X(s) = \mathcal{L}\left\{t\,e^{-2t}\,u(t)\right\} = -\frac{d}{ds}\left[\frac{1}{s+2}\right] = \frac{1}{(s+2)^2}$$

with the ROC

$$\text{Re}\{s\} > -2$$

4.3.7 Convolution property

Convolution property of the Laplace transform will prove to be fundamental in establishing the significance of the s-domain system function concept in Section 4.5.

Convolution property of the Laplace transform:

For any two signals $x_1(t)$ and $x_2(t)$ with their respective transforms

$$x_1(t) \overset{\mathcal{L}}{\longleftrightarrow} X_1(s)$$

and

$$x_2(t) \overset{\mathcal{L}}{\longleftrightarrow} X_2(s)$$

the following transform relationship holds:

$$x_1(t) * x_2(t) \overset{\mathcal{L}}{\longleftrightarrow} X_1(s)\, X_2(s) \tag{4.59}$$

Proof of Eqn. (4.59):

The proof will be carried out by using the convolution result inside the Laplace transform definition. The convolution of two continuous-time signals $x_1(t)$ and $x_2(t)$ is given by

$$x_1(t) * x_2(t) = \int_{-\infty}^{\infty} x_1(\lambda)\, x_2(t-\lambda)\, d\lambda \tag{4.60}$$

Substituting Eqn. (4.60) into the Laplace transform definition leads to

$$\begin{aligned}\mathcal{L}\{x_1(t) * x_2(t)\} &= \mathcal{L}\left\{\int_{-\infty}^{\infty} x_1(\lambda)\, x_2(t-\lambda)\, d\lambda\right\}\\ &= \int_{-\infty}^{\infty}\left[\int_{-\infty}^{\infty} x_1(\lambda)\, x_2(t-\lambda)\, d\lambda\right] e^{-st}\, dt \end{aligned} \tag{4.61}$$

Interchanging the order of the two integrals, Eqn. (4.61) can be written as

$$\begin{aligned}\mathcal{L}\{x_1(t) * x_2(t)\} &= \int_{-\infty}^{\infty}\left[\int_{-\infty}^{\infty} x_1(\lambda)\, x_2(t-\lambda)\, e^{-st}\, dt\right] d\lambda\\ &= \int_{-\infty}^{\infty} x_1(\lambda)\left[\int_{-\infty}^{\infty} x_2(t-\lambda)\, e^{-st}\, dt\right] d\lambda \end{aligned} \tag{4.62}$$

We will focus our attention on the inner integral in Eqn. (4.62). Using the time-shifting property it follows that

$$\int_{-\infty}^{\infty} x_2(t-\lambda)\, e^{-st}\, dt = \mathcal{L}\{x_2(t-\lambda)\} = e^{-s\lambda}\, X_2(s) \tag{4.63}$$

Substituting Eqn. (4.63) into Eqn. (4.62) we obtain

$$\begin{aligned}\mathcal{L}\left\{x_1(t) * x_2(t)\right\} &= \int_{-\infty}^{\infty} x_1(\lambda)\, e^{-s\lambda}\, X_2(s)\, d\lambda \\ &= X_2(s) \int_{-\infty}^{\infty} x_1(\lambda)\, e^{-s\lambda}\, d\lambda \\ &= X_2(s)\, X_1(s) \end{aligned} \tag{4.64}$$

As before, the ROC for the resulting transform is at least the overlap of the ROCs of two individual transforms, if such an overlap exists. It may be greater than the overlap of the two ROCs if the multiplication of $X_1(s)$ and $X_2(s)$ results in the cancelation of a pole that determines the boundary of one of the ROC's.

Convolution property of the Laplace transform is fundamental in its application to CTLTI systems. It provides an alternative to computing the convolution of two signals directly in the time-domain. Instead, the convolution result $x(t) = x_1(t) * x_2(t)$ can be obtained using the procedure outlined below:

Finding convolution result using the Laplace transform:

To compute $x(t) = x_1(t) * x_2(t)$:

1. Find the Laplace transforms of the two signals.

$$X_1(s) = \mathcal{L}\left\{x_1(t)\right\}$$
$$X_2(s) = \mathcal{L}\left\{x_2(t)\right\}$$

2. Multiply the two transforms to obtain $X(s)$.

$$X(s) = X_1(s)\, X_2(s)$$

3. Compute $x(t)$ as the inverse Laplace transform of $X(s)$.

$$x(t) = \mathcal{L}^{-1}\left\{X(s)\right\}$$

Example 4.22: Using the convolution property of the Laplace transform

Consider two signals $x_1(t)$ and $x_2(t)$ given by

$$x_1(t) = e^{-t}\, u(t)$$

and

$$x_2(t) = \delta(t) - e^{-2t}\, u(t)$$

Determine

$$x(t) = x_1(t) * x_2(t)$$

using Laplace transform techniques.

Solution: Individual transforms of the signals $x_1(t)$ and $x_2(t)$ are

$$X_1(s) = \frac{1}{s+1}\,, \qquad \text{ROC: } \text{Re}\{s\} > -1 \tag{4.65}$$

and

$$X_2(s) = 1 - \frac{1}{s+2} = \frac{s+1}{s+2}\,, \qquad \text{ROC: } \text{Re}\{s\} > -2 \tag{4.66}$$

Applying the convolution property, the transform of $x(t)$ is the product of the two transforms:

$$X(s) = X_1(s)\, X_2(s) = \left(\frac{1}{s+1}\right)\left(\frac{s+1}{s+2}\right) = \frac{1}{s+2}$$

The overlap of the two ROCs in Eqns. (4.65) and (4.66) would be $\text{Re}\{s\} > -1$; however, the pole at $s=-1$ is canceled in the process of multiplying the two transforms. Consequently, the ROC for $X(s)$ is

$$\text{Re}\{s\} > -2$$

as shown in Fig. 4.31.

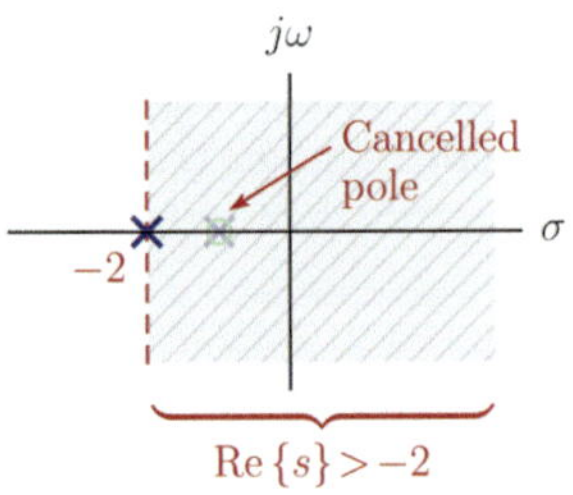

Figure 4.31 – The pole-zero diagram and the ROC for the transform in Example 4.22.

The signal $x(t)$ is the inverse Laplace transform of $X(s)$:

$$x(t) = \mathscr{L}^{-1}\left\{\frac{1}{s+2}\right\} = e^{-2t}\, u(t)$$

4.3.8 Integration property

Integration property of the Laplace transform:

Given the transform pair:

$$x(t) \overset{\mathscr{L}}{\longleftrightarrow} X(s)$$

the following is also a valid transform pair:

$$\int_{-\infty}^{t} x(\lambda)\, d\lambda \overset{\mathscr{L}}{\longleftrightarrow} \frac{1}{s}\, X(s) \tag{4.67}$$

Proof of Eqn. (4.67):

Let a new signal be defined from $x(t)$ as

$$w(t) = \int_{-\infty}^{t} x(t)\, d\lambda \tag{4.68}$$

Using the definition of the Laplace transform in conjunction with the signal $w(t)$ we get

$$\begin{aligned} W(s) &= \int_{-\infty}^{\infty} w(t)\, e^{-st}\, dt \\ &= \int_{-\infty}^{\infty} \left[\int_{-\infty}^{t} x(\lambda)\, d\lambda \right] e^{-st}\, dt \end{aligned}$$

which would be difficult to evaluate directly. Instead, we will make use of other properties of the Laplace transform discussed earlier. Differentiating both sides of Eqn. (4.68) we get

$$\frac{dw(t)}{dt} = x(t) \tag{4.69}$$

Using the time-domain differentiation property of the Laplace transform, Eqn. (4.69) leads to the relationship

$$s\, W(s) = X(s)$$

from which the desired result is obtained:

$$W(s) = \frac{1}{s}\, X(s) \tag{4.70}$$

An alternative method of proving the integration property will also be presented to provide further insight. Let us write the convolution of $x(t)$ with the unit step function:

$$x(t) * u(t) = \int_{-\infty}^{\infty} x(\lambda)\, u(t-\lambda)\, d\lambda \tag{4.71}$$

We know that

$$u(t-\lambda) = \begin{cases} 1, & t-\lambda > 0 \;\Rightarrow\; \lambda < t \\ 0, & t-\lambda < 0 \;\Rightarrow\; \lambda > t \end{cases}$$

Therefore, we can set the upper limit of the integral in Eqn. (4.71) to $\lambda = t$ and drop the unit step function without changing the integration result. Doing so leads to

$$x(t) * u(t) = \int_{-\infty}^{t} x(\lambda)\, d\lambda = w(t) \tag{4.72}$$

Since $w(t)$ is equal to the convolution of $x(t)$ and the unit step function, its Laplace transform must be equal to the product of the respective transforms:

$$W(s) = \mathscr{L}\{u(t)\}\, \mathscr{L}\{x(t)\} = \frac{1}{s}\, X(s) \tag{4.73}$$

The ROC may need to be adjusted. Let the ROC of the original transform $X(s)$ be

$$\sigma_1 < \mathrm{Re}\{s\} < \sigma_2$$

The ROC for the Laplace transform of the unit step function is

$$\text{Re}\{s\} > 0$$

Therefore, the ROC of the transform $W(s)$ must be at least the overlap of these regions. It may be larger than the overlap if $X(s)$ has a zero at $s = 0$ to counter the pole at $s = 0$ introduced by the transform of the unit step function.

Example 4.23: Laplace transform of a triangle signal

Making use of the integration property, determine the Laplace transform of the triangle pulse shown in Fig. 4.32.

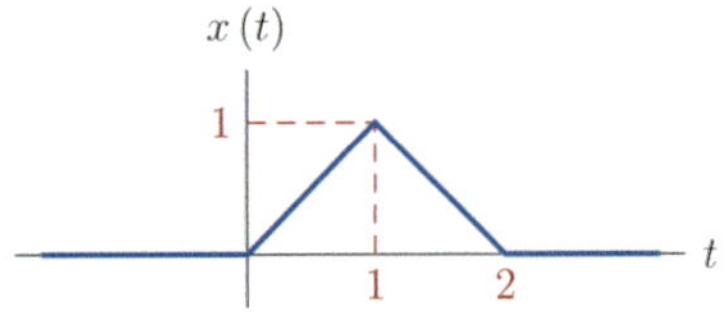

Figure 4.32 – The triangle signal $x(t)$ for Example 4.23.

Solution: Consider the bipolar pulse signal $g(t)$ shown in Fig. 4.33. Clearly, the signal $x(t)$ can be expressed as the running integral of $g(t)$, that is,

$$x(t) = \int_{-\infty}^{t} g(\tau)\, d\tau$$

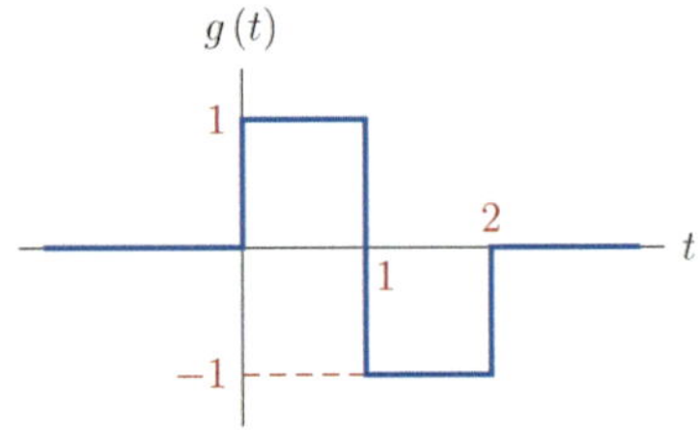

Figure 4.33 – The bipolar pulse signal $g(t)$ for Example 4.23.

Utilizing the result found in Example 4.14, the Laplace transform of $g(t)$ is

$$G(s) = \frac{(1-e^{-s})^2}{s}$$

where we substituted $\tau = 1$. The transform $X(s)$ can now be found through the use of the integration property.

$$X(s) = \frac{1}{s}\, G(s) = \frac{(1-e^{-s})^2}{s^2}$$

The ROC of the transform is

$$\text{Re}\{s\} > -\infty$$

4.4 Inverse Laplace Transform

Consider a transform pair

$$x(t) \overset{\mathscr{L}}{\longleftrightarrow} X(s)$$

If $X(s)$ is given, the signal $x(t)$ can be found using the inverse Laplace transform

$$x(t) = \mathscr{L}^{-1}\left\{X(s)\right\} = \frac{1}{2\pi j}\int_C X(s)\, e^{st}\, ds \tag{4.74}$$

The contour C is a vertical line within the ROC of the transform as shown in Fig. 4.34.

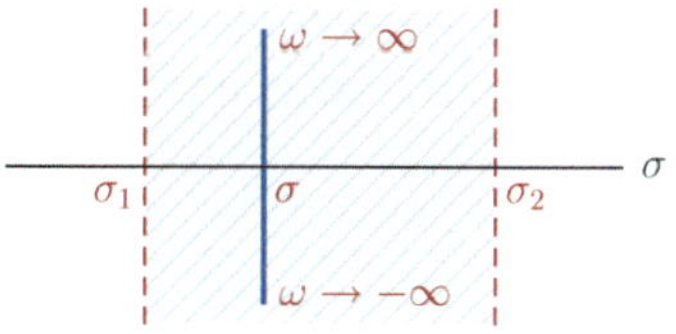

Figure 4.34 – Integration path C in the s-plane.

Even though the contour integral in Eqn. (4.74) will not be used in this text for the actual computation of the inverse Laplace transform, it will be explored a bit further to provide insight.

Inverse Laplace transform through contour integral:

Consider the Laplace transform $X(s)$ evaluated at $s = \sigma + j\omega$, first given by Eqn. (4.5) and repeated below.

$$\text{Eqn. (4.5):} \qquad X(s)\Big|_{s=\sigma+j\omega} = \int_{-\infty}^{\infty} x(t)\, e^{-(\sigma+j\omega)t}\, dt = \int_{-\infty}^{\infty} x(t)\, e^{-\sigma t}\, e^{-j\omega t}\, dt$$

The integral on the right side represents the Fourier transform of the modified signal $x(t)\, e^{-\sigma t}$ provided that it exists. Let us assume that the point $s = \sigma + j\omega$ is in the ROC of $X(s)$ to satisfy the existence condition. The signal $x(t)\, e^{-\sigma t}$ can be found using the inverse Fourier transform:

$$x(t)\, e^{-\sigma t} = \frac{1}{2\pi}\int_{-\infty}^{\infty} X(s)\Big|_{s=\sigma+j\omega} e^{j\omega t}\, d\omega \tag{4.75}$$

Multiplying both sides of Eqn. (4.75) by $e^{\sigma t}$ yields

$$x(t) = \frac{1}{2\pi}\int_{-\infty}^{\infty} X(s)\Big|_{s=\sigma+j\omega} e^{\sigma t}\, e^{j\omega t}\, d\omega = \frac{1}{2\pi}\int_{-\infty}^{\infty} X(s)\Big|_{s=\sigma+j\omega} e^{(\sigma+j\omega)t}\, d\omega \tag{4.76}$$

Substituting

$$s = \sigma + j\omega \qquad \Longrightarrow \qquad ds = j\, d\omega$$

into Eqn. (4.76) we obtain

$$x(t) = \frac{1}{2\pi j} \int_{\sigma - j\infty}^{\sigma + j\infty} X(s)\, e^{st}\, ds \tag{4.77}$$

which is the contour integral given by Eqn. (4.74). The integral needs to be carried out at points on a vertical line in the ROC of the transform.

For a rational function $X(s)$ it is usually easier to compute the inverse Laplace transform through the use of partial fraction expansion (PFE).

4.4.1 Partial fraction expansion with simple poles

Consider a rational transform in the form

$$X(s) = \frac{B(s)}{(s - p_1)(s - p_2)\dots(s - p_N)} \tag{4.78}$$

where the poles $p_1, p_2, \dots, p_N$ are distinct. Furthermore, let the order of the numerator polynomial $B(s)$ be less than N, the order of the denominator polynomial. The transform $X(s)$ can be expanded into partial fractions in the form

$$X(s) = \frac{k_1}{s - p_1} + \frac{k_2}{s - p_2} + \dots + \frac{k_N}{s - p_N} \tag{4.79}$$

The coefficients $k_1, k_2, \dots, k_N$ are called the *residues* of the partial fraction expansion. They can be computed by (see Appendix E)

$$k_i = (s - p_i)\, X(s)\Big|_{s = p_i}, \qquad i = 1, 2, \dots, N \tag{4.80}$$

Once the residues are determined, the inverse transform of each term in the partial fraction expansion is determined using either Eqn. (4.16) or Eqn. (4.17), depending on the placement of each pole relative to the ROC.

$$\mathcal{L}^{-1}\left\{\frac{1}{s - p_i}\right\} = \begin{cases} e^{p_i t}\, u(t), & \text{if the ROC is to the right of } p_i \\ -e^{p_i t}\, u(-t), & \text{if the ROC is to the left of } p_i \end{cases} \tag{4.81}$$

This will be illustrated in the next several examples.

Software resource: See MATLAB Exercise 4.5.

Example 4.24: Inverse Laplace transform using PFE

A causal signal $x(t)$ has the Laplace transform

$$X(s) = \frac{s + 1}{s\,(s + 2)}$$

Determine $x(t)$ using partial fraction expansion.

Solution: Since $x(t)$ is specified to be causal, the ROC of the transform must be to the right of a vertical line. $X(s)$ has poles at $s=-2$ and $s=0$, therefore the ROC is

$$\mathrm{Re}\{s\} > 0$$

Partial fraction expansion of $X(s)$ is in the form

$$X(s) = \frac{k_1}{s} + \frac{k_2}{s+2}$$

Residues are found by the application of residue formulas:

$$k_1 = s\,X(s)\Big|_{s=0} = \frac{s+1}{s+2}\Big|_{s=0} = \frac{1}{2}$$

$$k_2 = (s+2)\,X(s)\Big|_{s=-2} = \frac{s+1}{s}\Big|_{s=-2} = \frac{1}{2}$$

Using the values found, $X(s)$ is

$$X(s) = \frac{1/2}{s} + \frac{1/2}{s+2}$$

Both terms of $X(s)$ correspond to causal terms in the time domain. Use of the transform pair given by Eqn. (4.16) for both terms yields

$$x(t) = \frac{1}{2}\,u(t) + \frac{1}{2}\,e^{-2t}\,u(t)$$

Software resources: exct_4_24a.m , exct_4_24b.m

Example 4.25: Using PFE with complex poles

The Laplace transform of a signal $x(t)$ is

$$X(s) = \frac{s+1}{s\left(s^2+9\right)}$$

with the ROC specified as

$$\mathrm{Re}\{s\} > 0$$

Determine $x(t)$.

Solution: The transform $X(s)$ can be written in factored form as

$$X(s) = \frac{s+1}{s\left(s+j3\right)\left(s-j3\right)}$$

and expanded into partial fractions as

$$X(s) = \frac{k_1}{s} + \frac{k_2}{s+j3} + \frac{k_3}{s-j3}$$

The residues are determined using the residue formulas. The residues associated with complex poles at $\pm j3$ will be complex-valued, however, the method used for computing them is the same.

$$k_1 = s\,X(s)\Big|_{s=0} = \frac{s+1}{(s+j3)\,(s-j3)}\Bigg|_{s=0} = \frac{1}{9}$$

$$k_2 = (s+j3)\,X(s)\Big|_{s=-j3} = \frac{s+1}{s\,(s-j3)}\Bigg|_{s=-j3} = -\frac{1}{18} + j\frac{1}{6}$$

The residue k_3 can be computed using the same method, however, there is a shortcut available. Since the transform $X(s)$ is a rational function with real coefficients, residues of complex conjugate poles must be complex conjugates of each other. Therefore

$$k_3 = k_2^* = -\frac{1}{18} - j\frac{1}{6}$$

Based on the specified ROC, the signal $x(t)$ is causal. Using Eqn. (4.16) for inverting each term of the partial fraction expansion we get

$$\begin{aligned} x(t) =& k_1\,u(t) + k_2\,e^{-j3t}\,u(t) + k_3\,e^{j3t}\,u(t) \\ =& \frac{1}{9}\,u(t) + \left(-\frac{1}{18} + j\frac{1}{6}\right) e^{-j3t}\,u(t) + \left(-\frac{1}{18} - j\frac{1}{6}\right) e^{j3t}\,u(t) \end{aligned}$$

which can be simplified as

$$\begin{aligned} x(t) =& \frac{1}{9}\,u(t) - \frac{1}{18}\left[e^{-j3t} + e^{j3t}\right] u(t) + j\frac{1}{6}\left[e^{-j3t} - e^{j3t}\right] u(t) \\ =& \frac{1}{9}\,u(t) - \frac{1}{9}\cos(3t)\,u(t) + \frac{1}{3}\sin(3t)\,u(t) \end{aligned}$$

An alternative approach would be to combine the two partial fractions with complex poles into a second-order term and write $X(s)$ as

$$X(s) = \frac{1/9}{s} + \frac{-\frac{1}{9}s + 1}{s^2+9}$$

Afterward the transform pairs

$$\cos(3t)\,u(t) \overset{\mathcal{L}}{\longleftrightarrow} \frac{s}{s^2+9} \quad \text{and} \quad \sin(3t)\,u(t) \overset{\mathcal{L}}{\longleftrightarrow} \frac{3}{s^2+9}$$

can be used for arriving at the same result.

Software resources: exct_4_25a.m , exct_4_25b.m

Example 4.26: Using PFE in conjunction with the ROC

The Laplace transform of a signal $x(t)$ is

$$X(s) = \frac{5\,(s-1)}{(s+1)\,(s+2)\,(s-2)\,(s-3)}$$

with the ROC specified as

$$-1 < \mathrm{Re}\{s\} < 2$$

Determine $x(t)$.

Solution: We will first find the partial fraction expansion of $X(s)$. The ROC will be taken into consideration in the next step. Partial fraction expansion of $X(s)$ is in the form

$$X(s) = \frac{k_1}{s+1} + \frac{k_2}{s+2} + \frac{k_3}{s-2} + \frac{k_4}{s-3}$$

The residues are

$$k_1 = (s+1)\,X(s)\Big|_{s=-1} = \left.\frac{5\,(s-1)}{(s+2)\,(s-2)\,(s-3)}\right|_{s=-1} = -0.8333$$

$$k_2 = (s+2)\,X(s)\Big|_{s=-2} = \left.\frac{5\,(s-1)}{(s+1)\,(s-2)\,(s-3)}\right|_{s=-2} = 0.75$$

$$k_3 = (s-2)\,X(s)\Big|_{s=2} = \left.\frac{5\,(s-1)}{(s+1)\,(s+2)\,(s-3)}\right|_{s=2} = -0.4167$$

$$k_4 = (s-3)\,X(s)\Big|_{s=} = \left.\frac{5\,(s-1)}{(s+1)\,(s+2)\,(s-2)}\right|_{s=3} = 0.5$$

The completed form of the partial fraction expansion is

$$X(s) = -\frac{0.8333}{s+1} + \frac{0.75}{s+2} - \frac{0.4167}{s-2} + \frac{0.5}{s-3}$$

Next we need to pay attention to the ROC which is a vertical strip in the s-plane between the poles at $s=-1$ and $s=2$ as shown in Fig. 4.35.

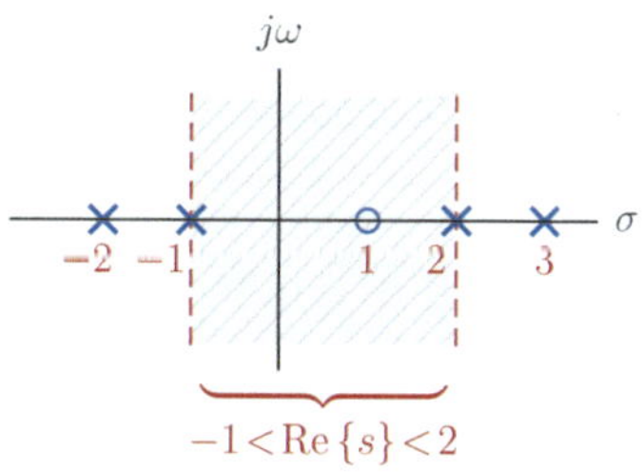

Figure 4.35 – The pole-zero plot and the ROC for $X(s)$ in Example 4.26.

The shape of the ROC indicates a two-sided signal. Accordingly, we will write the transform as the sum of two transforms belonging to causal and anti-causal components of $x(t)$:

$$X(s) = X_R(s) + X_L(s)$$

The poles at $s=-1$ and $s=-2$ are associated with $x_R(t)$, the causal component of $x(t)$. Therefore $X_R(s)$ can be written as

$$X_R(s) = -\frac{0.8333}{s+1} + \frac{0.75}{s+2} = \frac{-0.0833\,(s+11)}{(s+1)\,(s+2)}, \qquad \text{ROC:} \quad \text{Re}\{s\} > -1$$

Conversely the poles at $s=2$ and $s=3$ are associated with $x_L(t)$, the anti-causal component of $x(t)$.

$$X_L(s) = -\frac{0.4167}{s-2} + \frac{0.5}{s-3} = \frac{0.0833\,(s+3)}{(s-2)\,(s-3)}; \qquad \text{ROC:} \quad \text{Re}\{s\} < 2$$

The individual ROCs of the transforms $X_R(s)$ and $X_L(s)$ are shown in Fig. 4.36.

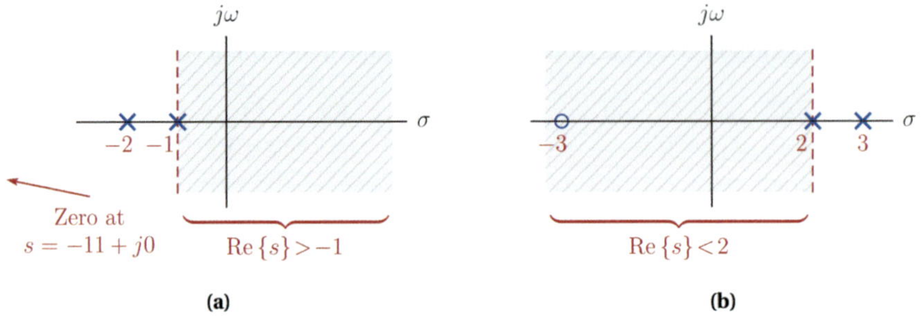

Figure 4.36 – **(a)** ROC for $X_R(s)$ and **(b)** ROC for $X_L(s)$.

The signal $x_R(t)$ is found by using Eqn. (4.16) on the terms of $X_R(s)$:

$$x_R(t) = -0.8333\, e^{-t}\, u(t) + 0.75\, e^{-2t}\, u(t)$$

The signal $x_L(t)$ is found by using Eqn. (4.17) on the terms of $X_L(s)$:

$$x_L(t) = 0.4167\, e^{2t}\, u(-t) - 0.5\, e^{3t}\, u(-t)$$

The signal $x(t)$ is the sum of these two components:

$$\begin{aligned} x(t) =& x_R(t) + x_L(t) \\ =& -0.8333\, e^{-t}\, u(t) + 0.75\, e^{-2t}\, u(t) + 0.4167\, e^{2t}\, u(-t) - 0.5\, e^{3t}\, u(-t) \end{aligned}$$

Software resources: `exct_4_26a.m` , `exct_4_26b.m`

If the order of the numerator polynomial $B(s)$ in Eqn. (4.78) is equal to or greater than the order of the denominator polynomial, special precautions need to be taken before partial fraction expansion can be used. In this case $X(s)$ must be written as

$$X(s) = C(s) + \frac{\bar{B}(s)}{(s-p_1)\,(s-p_2)\ldots(s-p_N)} \tag{4.82}$$

where $C(s)$ is a polynomial of S and the order of the new numerator polynomial $\bar{B}(s)$ is $N-1$.

Example 4.27: Using PFE when numerator order is not less than denominator order

The Laplace transform of a causal signal is

$$X(s) = \frac{s\,(s+1)}{(s+2)\,(s+3)}$$

Determine $x(t)$.

Solution: Since the numerator order is equal to the denominator order, $X(s)$ cannot be expanded into partial fractions directly. Let us first multiply numerator and denominator factors to obtain

$$X(s) = \frac{s^2+s}{s^2+5s+6}$$

which can be written as

$$X(s) = 1 + \frac{-4s-6}{s^2+5s+6}$$

Let $X_1(s)$ be defined as

$$X_1(s) = \frac{-4s-6}{s^2+5s+6} = \frac{-4s-6}{(s+2)\,(s+3)}$$

so that $X(s) = 1 + X_1(s)$. Expanding $X_1(s)$ into partial fractions yields

$$X_1(s) = \frac{2}{s+2} - \frac{6}{s+3}$$

Consequently we have

$$x_1(t) = 2\,e^{-2t}\,u(t) - 6\,e^{-3t}\,u(t)$$

and

$$\begin{aligned} x(t) =& \delta(t) + x_1(t) \\ =& \delta(t) + 2\,e^{-2t}\,u(t) - 6\,e^{-3t}\,u(t) \end{aligned}$$

Software resources: exct_4_27a.m , exct_4_27b.m

4.4.2 Partial fraction expansion with multiple poles

If the transform $X(s)$ has repeated roots, residue calculations become a bit more complicated. Consider a rational transform in the form

$$X(s) = \frac{B(s)}{\left(s-p_1\right)^r \left(s-p_2\right) \dots \left(s-p_N\right)} \tag{4.83}$$

The pole of multiplicity r at $s = p_1$ requires r terms in the partial fraction expansion:

$$X(s) = \frac{k_{1,1}}{s-p_1} + \frac{k_{1,2}}{\left(s-p_1\right)^2} + \dots + \frac{k_{1,r}}{\left(s-p_1\right)^r} + \frac{k_2}{s-p_2} + \dots + \frac{k_N}{s-p_N} \tag{4.84}$$

The residues of single poles at $p_2, \dots, p_N$ are still computed using the residue formulas discussed earlier in the previous section. The residue $k_{1,r}$ is also easy to compute:

$$k_{1,r} = \left(s-p_1\right)^r X(s)\Big|_{s=p_1} \tag{4.85}$$

The residue $k_{1,r-1}$ for the partial fraction with $\left(s-p_1\right)^{r-1}$ in its denominator requires more work, and is computed as

$$k_{1,r-1} = \frac{1}{1!}\,\frac{d}{ds}\left[\left(s-p_1\right)^r X(s)\right]\Big|_{s=p_1} \tag{4.86}$$

The residue $k_{1,r-2}$ is

$$k_{1,r-2} = \frac{1}{2!}\,\frac{d^2}{ds^2}\left[\left(s-p_1\right)^r X(s)\right]\Big|_{s=p_1} \tag{4.87}$$

and so on. See Appendix E for details.

Example 4.28: Multiple-order poles

A causal signal $x(t)$ has the Laplace transform

$$X(s) = \frac{s-1}{(s+1)^3\,(s+2)}$$

Determine $x(t)$ using partial fraction expansion.

Solution: The partial fraction expansion for $X(s)$ is in the form

$$X(s) = \frac{k_{1,1}}{s+1} + \frac{k_{1,2}}{(s+1)^2} + \frac{k_{1,3}}{(s+1)^3} + \frac{k_2}{s+2}$$

The residue k_2 for the single pole at $s=2$ is easily determined using the residue formula:

$$k_2 = (s+2)\,X(s)\Big|_{s=-2} = \frac{s-1}{(s+1)^3}\Big|_{s=-2} = 3$$

The residues of the third-order pole at $s=-1$ are found using Eqns. (4.85) through (4.87).

$$k_{1,3} = (s+1)^3\,X(s)\Big|_{s=-1} = \frac{s-1}{s+2}\Big|_{s=-1} = -2$$

$$k_{1,2} = \frac{d}{ds}\left[(s+1)^3\,X(s)\right]\Big|_{s=-1} = \frac{3}{(s+2)^2}\Big|_{s=-1} = 3$$

$$k_{1,1} = \frac{1}{2}\,\frac{d^2}{ds^2}\left[(s+1)^3\,X(s)\right]\Big|_{s=-1} = \frac{-3}{(s+2)^3}\Big|_{s=-1} = -3$$

and the partial fraction expansion of $X(s)$ is

$$X(s) = -\frac{3}{s+1} + \frac{3}{(s+1)^2} - \frac{2}{(s+1)^3} + \frac{3}{s+2}$$

Using the transform pairs

$$e^{-t}\,u(t) \;\overset{\mathcal{L}}{\longleftrightarrow}\; \frac{1}{s+1}$$

$$t\,e^{-t}\,u(t) \;\overset{\mathcal{L}}{\longleftrightarrow}\; -\frac{d}{ds}\left[\frac{1}{s+1}\right] = \frac{1}{(s+1)^2}$$

and

$$t^2\,e^{-t}\,u(t) \;\overset{\mathcal{L}}{\longleftrightarrow}\; -\frac{d}{ds}\left[\frac{1}{(s+1)^2}\right] = \frac{2}{(s+1)^3}$$

to invert the terms of the partial fraction expansion, we arrive at the solution

$$x(t) = -3\,e^{-t}\,u(t) + 3t\,e^{-t}\,u(t) - t^2\,e^{-t}\,u(t) + 3\,e^{-2t}\,u(t)$$

Software resource: `exct_4_28.m`

Software resource: See MATLAB Exercises 4.5 and 4.10.

4.5 Using the Laplace Transform with CTLTI Systems

We have shown in Chapter 2 that the output signal of a CTLTI system can be computed from its input signal and its impulse response through the use of the convolution integral. If the impulse response of a CTLTI system is $h(t)$, and if the signal $x(t)$ is applied to the system as input, the output signal $y(t)$ is found as

$$y(t) = x(t) * h(t) = \int_{-\infty}^{\infty} x(\lambda)\, h(t-\lambda)\, d\lambda$$

Based on the convolution property of the Laplace transform introduced in Eqn. (4.59) in Section 4.3.7, the transform of the convolution of two signals is equal to the product of their individual transforms. Therefore, the transform of the output signal is

$$Y(s) = X(s)\, H(s) \tag{4.88}$$

Eqn. (4.88) gives us an alternative method of computing the output signal of a CTLTI system instead of direct application of the convolution integral.

Computing the output signal of a CTLTI system:

1. Find the Laplace transforms $X(s)$ and $H(s)$ of the input signal $x(t)$ and the impulse response $h(t)$, respectively.
2. Multiply the two transforms to obtain the transform $Y(s)$.
3. Determine the output signal $y(t)$ from $Y(s)$ by means of the inverse Laplace transform.

Solving Eqn. (4.88) for $H(s)$ we get

$$H(s) = \frac{Y(s)}{X(s)} \tag{4.89}$$

The function $H(s)$ is the *s-domain system function* of the system under consideration. Eqn. (4.89) also provides a method of determining the impulse response of a CTLTI system from the knowledge of its input and output signals.

Determining the impulse response of a CTLTI system:

1. Find the Laplace transforms $X(s)$ and $Y(s)$ of input and output signals $x(t)$ and $y(t)$ respectively.
2. Find the transform $H(s)$ as the ratio of the two transforms as given by Eqn. (4.89).
3. Use the inverse Laplace transform operation on $H(s)$ to find $h(t)$.

We already know that a CTLTI system can be completely and uniquely described by means of its impulse response $h(t)$. Since the system function $H(s)$ is just the Laplace transform of the impulse response $h(t)$, it also represents a complete description of the CTLTI system.

4.5.1 Relating the system function to the differential equation

As discussed in Section 2.4 of Chapter 2, the input-output relationship of a CTLTI system can be modeled by a constant-coefficient linear differential equation given in the standard form

$$\sum_{k=0}^{N} a_k \frac{d^k y(t)}{dt^k} = \sum_{k=0}^{M} b_k \frac{d^k x(t)}{dt^k} \tag{4.90}$$

The order of the CTLTI system is the larger of M and N. Up to this point we have considered three different forms of modeling for a CTLTI system, namely the differential equation, the impulse response and the system function. (A fourth method, state-space modeling, will be introduced in Chapter 5.) It must be possible to obtain any of the three models from the knowledge of any other. In this section we will focus on the problem of determining the system function from the differential equation. If we take the Laplace transform of both sides of Eqn. (4.90) the equality would still be valid:

$$\mathcal{L}\left\{\sum_{k=0}^{N} a_k \frac{d^k y(t)}{dt^k}\right\} = \mathcal{L}\left\{\sum_{k=0}^{M} b_k \frac{d^k x(t)}{dt^k}\right\} \tag{4.91}$$

Laplace transform is linear; therefore, the transform of a summation is equal to the sum of individual terms, allowing us to write Eqn. (4.91) as

$$\sum_{k=0}^{N} \mathcal{L}\left\{a_k \frac{d^k y(t)}{dt^k}\right\} = \sum_{k=0}^{M} \mathcal{L}\left\{b_k \frac{d^k x(t)}{dt^k}\right\} \tag{4.92}$$

and subsequently as

$$\sum_{k=0}^{N} a_k \, \mathcal{L}\left\{\frac{d^k y(t)}{dt^k}\right\} = \sum_{k=0}^{M} b_k \, \mathcal{L}\left\{\frac{d^k x(t)}{dt^k}\right\} \tag{4.93}$$

Using the time-domain differentiation property of the Laplace transform, given by Eqn. (4.51), in Eqn. (4.93) yields

$$\sum_{k=0}^{N} a_k \, s^k \, Y(s) = \sum_{k=0}^{M} b_k \, s^k \, X(s) \tag{4.94}$$

The transforms $X(s)$ and $Y(s)$ do not depend on the summation indices k and can therefore be factored out of the summations in Eqn. (4.94) resulting in

$$Y(s) \sum_{k=0}^{N} a_k \, s^k = X(s) \sum_{k=0}^{M} b_k \, s^k \tag{4.95}$$

The system function can now be obtained from Eqn. (4.95) as

$$H(s) = \frac{Y(s)}{X(s)} = \frac{\sum_{k=0}^{M} b_k s^k}{\sum_{k=0}^{N} a_k s^k} \tag{4.96}$$

Finding the system function from the differential equation:

1. Separate the terms of the differential equation so that $y(t)$ and its derivatives are on the left of the equal sign and $x(t)$ and its derivatives are on the right of the equal sign, as in Eqn. (4.90).
2. Take the Laplace transform of each side of the differential equation, and use the time differentiation property of the Laplace transform as in Eqn. (4.94).
3. Determine the system function as the ratio of $Y(s)$ to $X(s)$ as in Eqn. (4.96).
4. If the impulse response is needed, it can now be determined as the inverse Laplace transform of $H(s)$.

System function, linearity, and initial conditions

Two important observations will be made at this point:

1. The development leading up to the s-domain system function in Eqn. (4.89) has relied heavily on the convolution operation and the convolution property of the Laplace transform. We know from Chapter 2 that the convolution operation is only applicable to problems involving linear and time-invariant systems. Therefore it follows that the system function concept is meaningful only for systems that are both linear and time-invariant. This notion was introduced in earlier discussions involving system functions as well.
2. Furthermore, it was justified in Section 2.4 of Chapter 2 that a constant-coefficient differential equation corresponds to a linear and time-invariant system only if all initial conditions are set equal to zero.

We conclude that, in determining the system function from the differential equation, all initial conditions must be assumed to be zero.

If we need to use Laplace transform techniques to solve a differential equation subject to non-zero initial conditions, that can be done through the use of the unilateral Laplace transform, but not through the use of the system function. The unilateral Laplace transform and its use for solving differential equations will be discussed in Section 4.7.

Example 4.29: Finding the system function from the differential equation

A CTLTI system is defined by means of the differential equation

$$\frac{d^3y(t)}{dt^3}+5\,\frac{d^2y(t)}{dt^2}+17\,\frac{dy(t)}{dt}+13\,y(t)=\frac{d^2x(t)}{dt^2}+x(t)$$

Find the system function $H(s)$ for this system.

Solution: We will assume that all initial conditions are equal to zero and take the Laplace transform of each side of the differential equation to obtain

$$s^3\,Y(s)+5s^2\,Y(s)+17s\,Y(s)+13\,Y(s)=s^2\,X(s)+X(s)$$

The system function can be obtained as

$$H(s) = \frac{Y(s)}{X(s)} = \frac{s^2+1}{s^3+5s^2+17s+13}$$

Characteristic polynomial vs. the denominator of the system function

Another important observation will be made based on the result obtained in Example 4.29: The characteristic equation for the system considered in Example 4.29 is

$$s^3+5s^2+17s+13=0$$

and the solutions of the characteristic equation are the modes of the system as defined in Section 2.5.3 of Chapter 2. When we find the system function $H(s)$ from the differential equation we see that its denominator polynomial is identical to the characteristic polynomial. The roots of the denominator polynomial are the poles of the system function in the s-domain, and consequently, they are identical to the modes of the differential equation of the system.

Recall that in Section 2.5.3 we have reached some conclusions about the relationship between the modes of the differential equation and the natural response of the corresponding system. The same conclusions would apply to the poles of the system function. Specifically, if all poles of the system are real-valued and distinct, then the transient response of the system is in the form

$$y[n] = \sum_{k=1}^{N} c_k e^{p_k t}$$

Complex poles appear in conjugate pairs provided that all denominator coefficients of the system function are real-valued. A pair of complex conjugate poles

$$p_{1a} = \sigma_1 + j\omega_1 \quad \text{and} \quad p_{1b} = \sigma_1 - j\omega_1$$

yields a response of the type

$$y_1(t) = d_1\, e^{\sigma_1 t} \cos(\omega_1 t + \theta_1)$$

Finally, a pole of multiplicity m at $s = p_1$ leads to a response in the form

$$y_1(t) = c_{11}\, e^{p_1 t} + c_{12}\, t\, e^{p_1 t} + \ldots + c_{1m}\, t^{m-1}\, e^{p_1 t} + \text{other terms}$$

regardless of whether p_1 is real or complex-valued. Justifications for these relationships were given in Section 2.5.3 of Chapter 2 through the use of the modes of the differential equation, and will not be repeated here.

Sometimes we need to reverse the problem represented in Example 4.29 and find the differential equation from the knowledge of the system function. The next three examples will demonstrate this.

Software resource: See MATLAB Exercise 4.10.

Example 4.30: Finding the differential equation from the system function

A causal CTLTI system is defined by the system function

$$H(s) = \frac{2s+5}{s^2+5s+6}$$

Find a differential equation for this system.

Solution: The system function is the ratio of the output transform to the input transform, that is,

$$H(s) = \frac{Y(s)}{X(s)}$$

Therefore we can write

$$\left(s^2+5s+6\right) Y(s) = (2s+5)\, X(s)$$

The differential equation follows as

$$\frac{d^2y(t)}{dt^2} + 5\,\frac{dy(t)}{dt} + 6\,y(t) = 2\,\frac{dx(t)}{dt} + 5\,x(t)$$

Example 4.31: Finding the differential equation from input and output signals

The unit step response of a CTLTI system is

$$y(t) = \text{Sys}\left\{u(t)\right\} = \left(2 - 4e^{-t} + 2e^{-2t}\right) u(t)$$

Find a differential equation for this system.

Solution: Since the input signal is a unit step function, that is, $x(t) = u(t)$, its Laplace transform is

$$X(s) = \frac{1}{s}$$

The Laplace transform of the output signal is found as

$$Y(s) = \frac{2}{s} - \frac{4}{s+1} + \frac{2}{s+2} = \frac{4}{s\,(s+1)\,(s+2)}$$

The system function can be obtained as the ratio of the output transform to the input transform:

$$H(s) = \frac{Y(s)}{X(s)} = \frac{4}{(s+1)\,(s+2)} = \frac{4}{s^2+3s+2}$$

The differential equation follows from the system function as

$$\frac{d^2y(t)}{dt^2} + 3\,\frac{dy(t)}{dt} + 2\,y(t) = 4\,x(t)$$

Software resource: exct_4_31.m

Example 4.32: Finding the impulse response from input and output signals

Determine the impulse response of the CTLTI system the unit step response of which was given in Example 4.31.

Solution: The system function was found in Example 4.31 to be

$$H(s) = \frac{4}{s^2 + 3s + 2}$$

which can be expanded into partial fractions to yield

$$H(s) = \frac{4}{s+1} - \frac{4}{s+2}$$

The impulse response is

$$h(t) = \left(4e^{-t} - 4e^{-2t}\right) u(t)$$

It can be easily verified that the unit step response specified in Example 4.31 for this system is indeed the convolution of $h(t)$ found above with the unit step function.

Software resource: `exct_4_32.m`

4.5.2 Response of a CTLTI system to a complex exponential signal

In this section we will consider the response of a CTLTI system to a complex exponential input signal. This will help us gain further insight into the system function concept.

Let a CTLTI system with impulse response $h(t)$ be driven by a complex exponential input signal in the form

$$x(t) = e^{s_0 t}$$

where s_0 represents a point in the s-plane within the ROC of the system function. The output signal can be determined through the use of the convolution integral

$$y(t) = h(t) * x(t) = \int_{-\infty}^{\infty} h(\lambda)\, x(t-\lambda)\, d\lambda \tag{4.97}$$

Substituting $x(t-\lambda) = e^{s_0(t-\lambda)}$ into Eqn. (4.97) and simplifying the resulting integral we obtain

$$\begin{aligned} y(t) = & \int_{-\infty}^{\infty} h(\lambda)\, e^{s_0(t-\lambda)}\, d\lambda \\ = & e^{-s_0 t} \underbrace{\int_{-\infty}^{\infty} h(\lambda)\, e^{s_0 \lambda}\, d\lambda}_{H(s_0)} \end{aligned} \tag{4.98}$$

The integral in Eqn. (4.98) should be recognized as the system function $H(s)$ evaluated at the point $s = s_0$ in the s-plane. Therefore, we reach the following important conclusion:

Response to a complex exponential input signal:

The response of the CTLTI system to the input signal $x(t) = e^{s_0 t}$ is

$$y(t) = \text{Sys}\left\{e^{s_0 t}\right\} = e^{s_0 t}\, H(s_0) \tag{4.99}$$

The CTLTI system responds to the complex exponential signal $x(t) = e^{s_0 t}$ by scaling it with the (generally complex) value of the system function at the point $s = s_0$.

In Eqn. (4.99) the complex exponential input signal is assumed to have been in existence forever; it is not turned on at a specific time instant. Therefore, the response found is the *steady state response* of the system.

Example 4.33: Response to a complex exponential signal

A CTLTI system with the system function

$$H(s) = \frac{s+6}{s^2+5s+6}$$

is driven by the complex exponential input signal

$$x(t) = e^{(-0.4+j4)t}$$

Determine the steady state response of the system.

Solution: The input signal is complex valued and can be written in Cartesian form using Euler's formula:

$$x(t) = e^{-0.4t}\left[\cos(4t) + j\sin(4t)\right]$$

Real and imaginary parts of $x(t)$ are shown in Fig. 4.37.

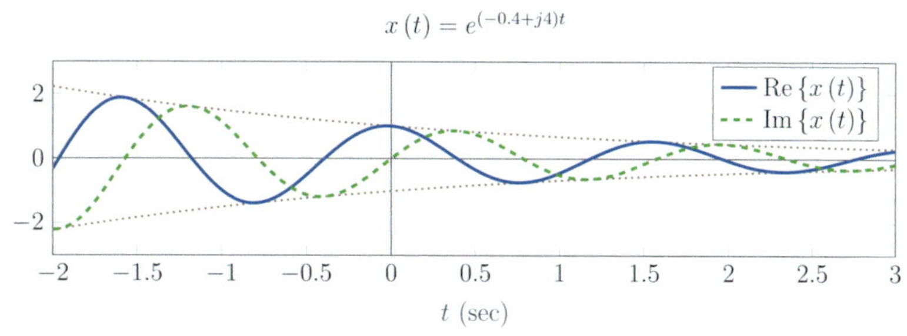

Figure 4.37 – The signal $x(t)$ for Example 4.33.

The value of the system function at $s = s_0 = -0.4 + j4$ is

$$\begin{aligned} H(-0.4+j4) =& \frac{(-0.4+j4)+6}{(-0.4+j4)^2+5(-0.4+j4)+6} \\ =& 0.0021 - j0.3348 \\ =& 0.3348\, e^{-j1.5645} \end{aligned}$$

The output signal is found using Eqn. (4.99) as

$$\begin{aligned} y(t) =& e^{(-0.4+j4)t}\left(0.3348\,e^{-j1.5645}\right) \\ =& 0.3348\,e^{-0.4t}\,e^{j(4t-1.5645)} \end{aligned}$$

or, in Cartesian form as

$$y(t) = 0.3348\,e^{-0.4t}\cos(4t-1.5645) + j\,0.3348\,e^{-0.4t}\sin(4t-1.5645)$$

Real and imaginary parts of the output signal $y(t)$ are shown in Fig. 4.38.

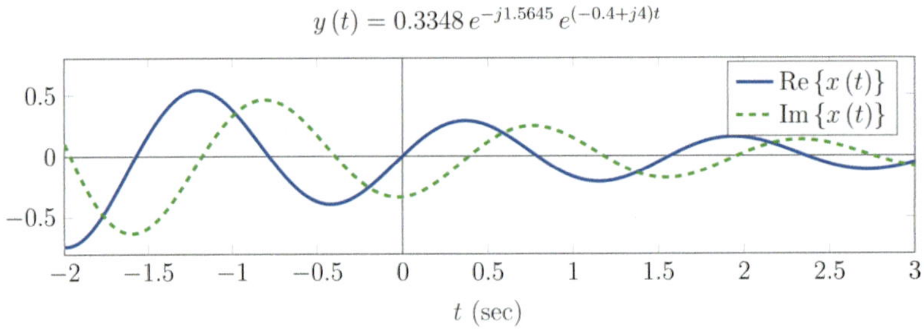

Figure 4.38 – The signal $y(t)$ for Example 4.33.

Software resource: exct_4_33.m

4.5.3 Response of a CTLTI system to an exponentially damped sinusoid

Next we will consider an input signal in the form of an exponentially damped sinusoid

$$x(t) = e^{\sigma_0 t}\cos(\omega_0 t) \tag{4.100}$$

In order to find the response of a CTLTI system to this signal, let us write $x(t)$ using Euler's formula:

$$x(t) = \tfrac{1}{2}\,e^{\sigma_0 t}\,e^{j\omega_0 t} + \tfrac{1}{2}\,e^{\sigma_0 t}\,e^{-j\omega_0 t} \tag{4.101}$$

Let the parameter s_0 be defined as

$$s_0 = \sigma_0 + j\omega_0$$

so that Eqn. (4.101) becomes

$$x(t) = \tfrac{1}{2}\,e^{s_0 t} + \tfrac{1}{2}\,e^{s_0^* t} \tag{4.102}$$

Using the linearity of the system, its response to $x(t)$ can be written as

$$\begin{aligned} y(t) =& \text{Sys}\left\{\tfrac{1}{2}\,e^{s_0 t} + \tfrac{1}{2}\,e^{s_0^* t}\right\} \\ =& \tfrac{1}{2}\,\text{Sys}\left\{e^{s_0 t}\right\} + \tfrac{1}{2}\,\text{Sys}\left\{e^{s_0^* t}\right\} \end{aligned} \tag{4.103}$$

We already know that the response of the system to the term $e^{s_0 t}$ is

$$\text{Sys}\left\{e^{s_0 t}\right\} = e^{s_0 t}\,H(s_0) \tag{4.104}$$

The response to the term $e^{s_0^* t}$ is found similarly:

$$\text{Sys}\left\{e^{s_0^* t}\right\} = e^{s_0^* t}\, H\left(s_0^*\right) \tag{4.105}$$

Using Eqns. (4.104) and (4.105) in Eqn. (4.103) the output signal is

$$y(t) = \tfrac{1}{2}\, e^{s_0 t}\, H(s_0) + \tfrac{1}{2}\, e^{s_0^* t}\, H\left(s_0^*\right) \tag{4.106}$$

It is possible to further simplify the result obtained in Eqn. (4.106). Let the value of the system function evaluated at the point $s = s_0$ be written in polar complex form as

$$H(s_0) = H_0\, e^{j\theta_0} \tag{4.107}$$

where H_0 and θ_0 represent the magnitude and the phase of the system function at the point $s = s_0$ respectively:

$$H_0 = \left|H(s_0)\right| \tag{4.108}$$

and

$$\theta_0 = \measuredangle H(s_0) \tag{4.109}$$

For a real-valued impulse response it can be shown (see Problem 4.27 at the end of this chapter) that the value of the system function at the point $s = s_0^*$ is the complex conjugate of its value at the point $s = s_0$, that is,

$$H\left(s_0^*\right) = \left[H(s_0)\right]^* = H_0\, e^{-j\theta_0} \tag{4.110}$$

Using Eqns. (4.107) and (4.110) in Eqn. (4.106), the output signal $y(t)$ becomes

$$\begin{aligned}
y(t) =& \tfrac{1}{2}\, e^{s_0 t}\, H_0\, e^{j\theta_0} + \tfrac{1}{2}\, e^{s_0^* t}\, H_0\, e^{-j\theta_0} \\
=& \tfrac{1}{2}\left(e^{\sigma_0 t}\, e^{j\omega_0 t}\right) H_0\, e^{j\theta_0} + \tfrac{1}{2}\left(e^{\sigma_0 t}\, e^{-j\omega_0 t}\right) H_0\, e^{-j\theta_0} \\
=& \tfrac{1}{2}\, H_0\, e^{\sigma_0 t}\left[e^{j(\omega_0 t+\theta_0)} + e^{-j(\omega_0 t+\theta_0)}\right] \\
=& H_0\, e^{\sigma_0 t}\cos(\omega_0 t+\theta_0)
\end{aligned} \tag{4.111}$$

The derivation outlined in Eqns. (4.101) through (4.111) can be summarized as shown in Fig. 4.39.

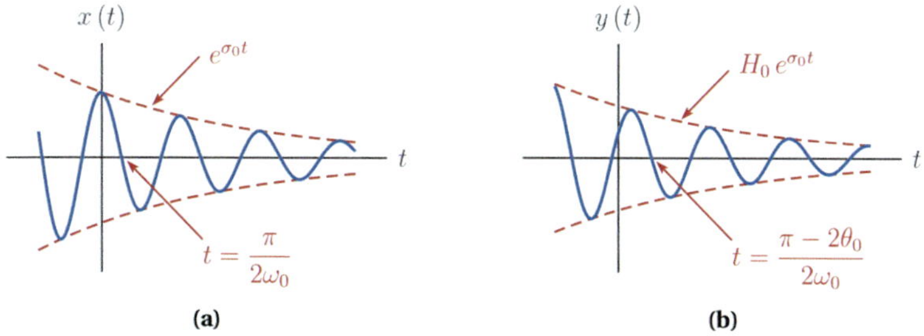

Figure 4.39 – The response of a CTLTI system to an exponentially damped sinusoid: **(a)** input signal and **(b)** output signal.

Comparison of the input signal in Eqn. (4.100) and the output signal Eqn. (4.111) reveals the following:

Response to exponentially damped sinusoid:

1. The amplitude of the signal is multiplied by the magnitude of the system function evaluated at the point $s = s_0 = \sigma_0 + j\omega_0$.
2. The phase of the cosine function is incremented by an amount equal to the phase of the system function evaluated at the point $s = s_0 = \sigma_0 + j\omega_0$.

Example 4.34: Response to an exponentially damped sinusoid

Consider again the system function used in Example 4.33. Determine the response of the system to the input signal

$$x(t) = e^{-0.3t} \cos(2t)$$

Solution: Let s_0 be defined as

$$s_0 = -0.3 + j2$$

The system function evaluated at $s = s_0$ is

$$H(-0.3+j2) = \frac{(-0.3+j2)+6}{(-0.3+j2)^2 + 5(-0.3+j2)+6} = 0.2695 - j0.6297$$

or in polar form

$$H(-0.3+j2) = 0.6849\, e^{-j1.1664}$$

The output signal is found using Eqn. (4.111):

$$y(t) = 0.6849\, e^{-0.3t} \cos(2t - 1.1664)$$

The input and the output signals are shown in Fig. 4.40.

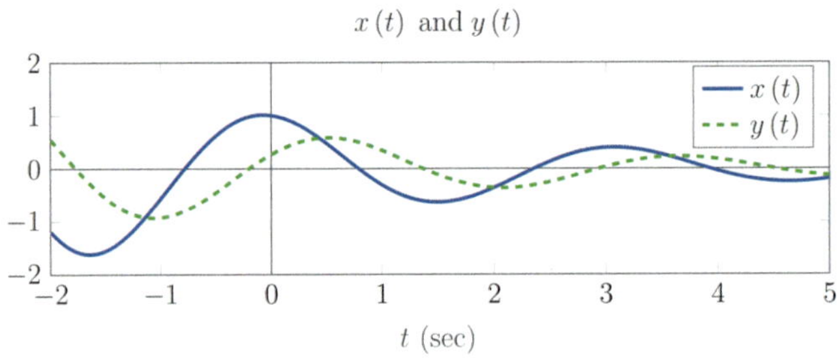

Figure 4.40 – Input and output signals for Example 4.34.

Software resource: exct_4_34.m

4.5.4 Pole-zero plot for a system function

A rational system function $H(s)$ can be expressed in *pole-zero form* as

$$H(s) = K\,\frac{(s-z_1)(s-z_2)\dots(s-z_M)}{(s-p_1)(s-p_2)\dots(s-p_N)} \tag{4.112}$$

The parameters M and N are the numerator order and the denominator order, respectively. The larger of M and N is the order of the system. The roots $z_1,\dots,z_M$ of the numerator polynomial are referred to as the *zeros* of the system function. In contrast, the roots of the denominator polynomial are the *poles* of the system function. A *pole-zero plot* for a system function is obtained by marking the poles and the zeros of the system function on the s-plane. It is customary to use "×" and "o" to mark each pole and each zero, respectively.

Example 4.35: Pole-zero plot for system function

Construct a pole-zero plot for a CTLTI system with system function

$$H(s) = \frac{s^2+1}{s^3+5s^2+17s+13}$$

Solution: The zeros of the system function are the roots of the numerator polynomial, and are found by solving the equation

$$s^2+1=0$$

which yields $z_1 = j$ and $z_2 = -j$. The poles of the system function are the roots of the denominator polynomial or, equivalently, the solutions of the equation

$$s^3+5s^2+17s+13=0$$

The three poles are at $p_1 = -1$, $p_2 = -2+j3$ and $p_3 = -2-j3$. The pole-zero diagram for the system function is shown in Fig. 4.41.

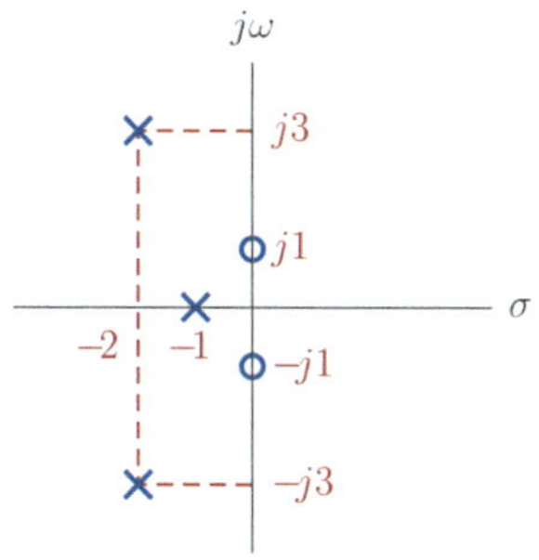

Figure 4.41 – The pole-zero plot for $H(s)$ of Example 4.35.

Software resource: exct_4_35.m

Example 4.36: ROC and the pole-zero plot

Assume that the system function used in Example 4.35 represents a causal system. Indicate the ROC on the pole-zero plot.

Solution: Following is the knowledge we possess:

1. For a causal system, the ROC of the system function must be the area to the right of a vertical line.
2. The ROC must be bound by one or more poles.
3. There may be no poles within the ROC.

Consequently, the ROC of the system function must be

$$\text{ROC:} \qquad \text{Re}\{s\} > -1$$

which is shown in Fig. 4.42.

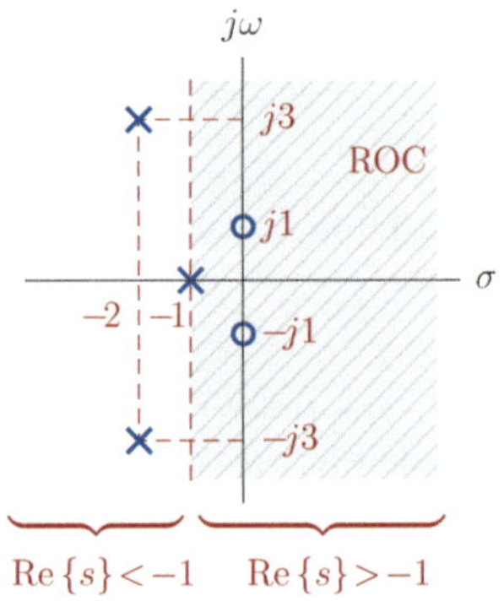

Figure 4.42 – The ROC superimposed on pole-zero plot for Example 4.36.

4.5.5 Graphical interpretation of the pole-zero plot

In this section we will explore the significance of the geometric placement of poles and zeros of the system function. The pole-zero plot, introduced in Section 4.5.4, can be used for understanding the behavior characteristics of a CTLTI system, especially in terms of the magnitude and the phase of the system function. The relationship between the locations of the poles and the zeros of the system function and the frequency-domain behavior of the system is significant.

Let us begin by considering the pole-zero form of the system function given by Eqn. (4.112). Assuming the system is stable, the Fourier transform-based system function $H(\omega)$ exists and can be found by evaluating $H(s)$ for $s = j\omega$:

$$H(\omega) = H(s)\Big|_{s=j\omega} = K\,\frac{(j\omega - z_1)(j\omega - z_2)\dots(j\omega - z_M)}{(j\omega - p_1)(j\omega - p_2)\dots(j\omega - p_N)} \tag{4.113}$$

The numerator of Eqn. (4.113) has M factors each in the form

$$B_i(\omega) = (j\omega - z_i)\,, \qquad i = 1, \dots, M \tag{4.114}$$

Similarly, the denominator of Eqn. (4.113) consists of factors in the form

$$A_i(\omega) = \left(j\omega - p_i\right), \qquad i = 1, \ldots, N \tag{4.115}$$

In a sense, the function $B_i(\omega)$ describes the contribution of the zero at $s = z_i$ to the frequency response of the system. Similarly, the function $A_i(\omega)$ describes the contribution of the pole at $s = p_i$. Using the definitions in Eqns. (4.114) and (4.115), $H(\omega)$ becomes

$$H(\omega) = K\,\frac{B_1(\omega)\,B_2(\omega)\ldots B_M(\omega)}{A_1(\omega)\,A_2(\omega)\ldots A_N(\omega)} \tag{4.116}$$

If we wanted to compute the frequency response of the system at a specific frequency $\omega = \omega_0$, we would get

$$H(\omega_0) = K\,\frac{B_1(\omega_0)\,B_2(\omega_0)\ldots B_M(\omega_0)}{A_1(\omega_0)\,A_2(\omega_0)\ldots A_N(\omega_0)} \tag{4.117}$$

The magnitude of the frequency response at $\omega = \omega_0$ is found by computing the magnitude of each complex function $B_i(\omega_0)$ and $A_i(\omega_0)$, and forming the ratio

$$|H(\omega_0)| = K\,\frac{\left|B_1(\omega_0)\right|.\left|B_2(\omega_0)\right|\ldots\left|B_M(\omega_0)\right|}{\left|A_1(\omega_0)\right|.\left|A_2(\omega_0)\right|\ldots\left|A_N(\omega_0)\right|} \tag{4.118}$$

The phase of the frequency response is computed as the algebraic sum of the phases of numerator and denominator factors:

$$\measuredangle H(\omega_0) = \measuredangle B_1(\omega_0) + \measuredangle B_2(\omega_0) + \ldots + \measuredangle B_M(\omega_0) - \measuredangle A_1(\omega_0) - \measuredangle A_2(\omega_0) - \ldots - \measuredangle A_N(\omega_0) \tag{4.119}$$

We would like to gain a graphical understanding of the computations in Eqns. (4.118) and (4.119). Let us focus on one of the numerator terms, $B_i(\omega_0)$:

$$B_i(\omega_0) = j\omega_0 - z_i \tag{4.120}$$

Clearly, $B(\omega_0)$ is a complex quantity. Its two terms z_i and $j\omega_0$ are shown in Fig. 4.43(a) as points in the s-plane. It is also possible to represent the terms in Eqn. (4.120) with vectors. Treating each term in Eqn. (4.120) as a vector, we obtain the corresponding vector expression

$$\overrightarrow{B_i(\omega_0)} = \overrightarrow{j\omega_0} - \overrightarrow{z_i} \tag{4.121}$$

which can be written in alternative form

$$\overrightarrow{z_i} + \overrightarrow{B_i(\omega_0)} = \overrightarrow{j\omega_0} \tag{4.122}$$

We will draw the vector $\overrightarrow{z_i}$ starting at the origin and ending at the point $s = z_i$. Similarly the vector $\overrightarrow{j\omega_0}$ will be drawn starting at the origin and ending at the point $s = j\omega_0$. These two vectors are shown in Fig. 4.43(b). The relationship expressed by Eqn. (4.122) is illustrated in Fig. 4.43(c).

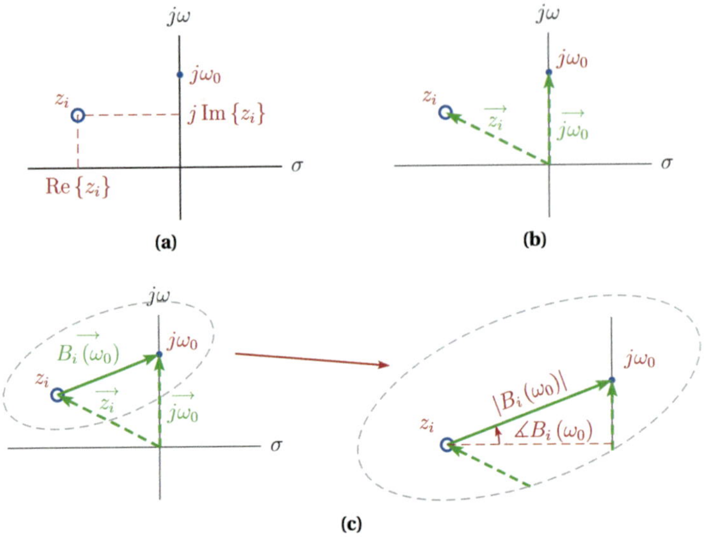

Figure 4.43 – **(a)** The zero $s = z_i$ and the point $s = j\omega_0$ in the complex plane, **(b)** the vectors for z_i and $j\omega_0$ in the complex plane, and **(c)** the relationship expressed by Eqn. (4.122).

The following important conclusions can be drawn from Fig. 4.43(c):

1. The vector $\overrightarrow{B_i(\omega_0)}$ is drawn starting at the zero at $s = z_i$ and ending at the point $s = j\omega_0$ on the $j\omega$-axis.
2. The magnitude $|B_i(\omega_0)|$ is equal to the norm of the vector $\overrightarrow{B_i(\omega_0)}$.
3. The phase $\measuredangle B_i(\omega_0)$ is equal to the angle that the vector $\overrightarrow{B_i(\omega_0)}$ makes with the positive real axis, measured counterclockwise.

If Fig. 4.43(c) is drawn to scale, the norm and the angle of the vector $\overrightarrow{B_i(\omega_0)}$ could simply be measured to determine the contributions of the zero at $s = z_i$ to the magnitude and the phase expressions for $|H(\omega_0)|$ and $\measuredangle H(\omega_0)$ in Eqns. (4.118) and (4.119).

What if we need to determine the contributions of the zero at $s = z_i$ to the magnitude $|H(\omega)|$ and phase $\measuredangle H(\omega)$ not just for a specific frequency ω_0 but for all frequencies ω? Imagine the vector $\overrightarrow{B_i(\omega)}$ to be a piece of rubber band, the tail end of which is permanently attached to the point $s = z_i$. Assume that, as the frequency ω varies, the tip of the rubber band vector moves on the $j\omega$-axis. The length of the rubber band and its angle with the positive real axis vary as the tip is moved. The contributions of the zero at $s = z_i$ to the magnitude and the phase of the frequency response $H(\omega)$ can be graphed by tracking these variations as shown in Fig. 4.44.

Similar analysis can be carried out for determining the contribution of a pole p_i to magnitude and phase characteristics of the system. Fig. 4.45 illustrates the effect of a pole on the frequency response.

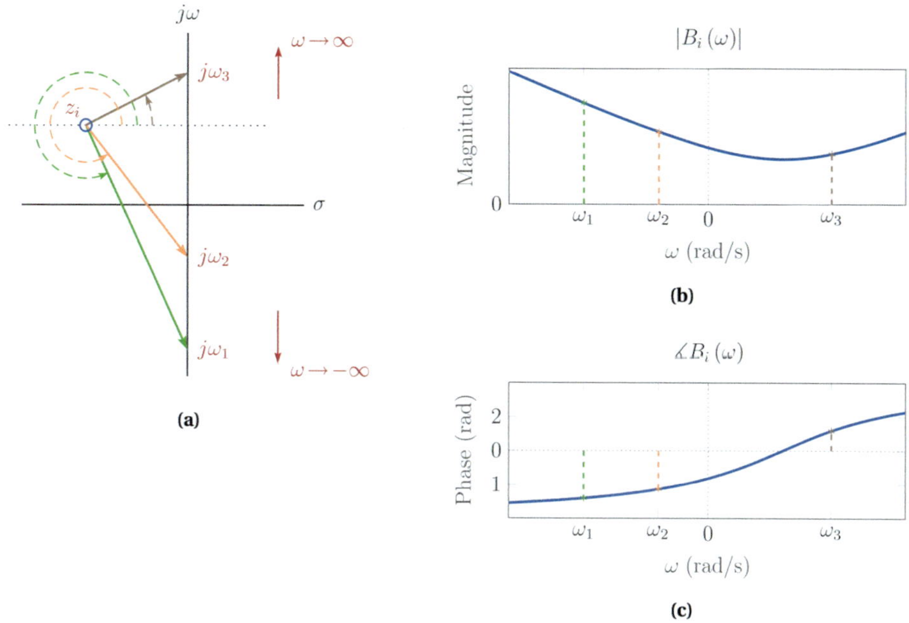

Figure 4.44 – **(a)** Moving the tip of the vector for $B_i(\omega)$ on the $j\omega$-axis using the rubber band analogy, **(b)** contribution of the zero at $s = z_i$ to the magnitude of the frequency response, and **(c)** contribution of the zero at $s = z_i$ to the phase of the frequency response.

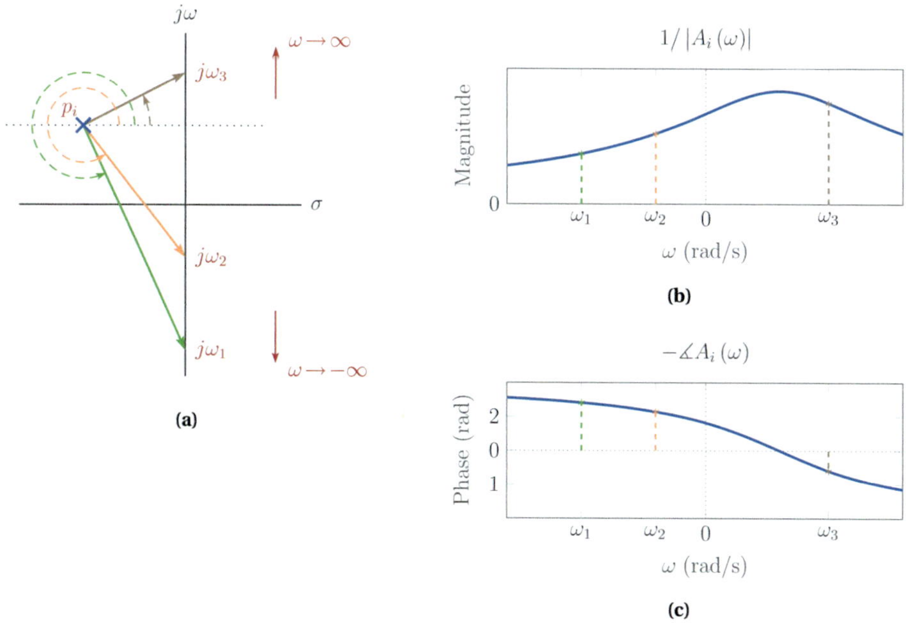

Figure 4.45 – **(a)** Moving the tip of the vector for $A_i(\omega)$ on the $j\omega$-axis using the rubber band analogy, **(b)** contribution of the pole at $s = p_i$ to the magnitude of the frequency response, and **(c)** contribution of the pole at $s = p_i$ to the phase of the frequency response.

For a system with M zeros and N poles, magnitude and phase of the system function at $\omega = \omega_0$ are found by taking into account the contributions of each zero and pole. The next example will illustrate this.

Example 4.37: Frequency response from pole-zero plot

A CTLTI system is described by the system function

$$H(s) = \frac{s^2 + s - 2}{s^2 + 2s + 5}$$

Construct a pole-zero plot and use it to determine the magnitude and the phase of the frequency response of the system at the frequency $\omega_0 = 1.5$ rad/s.

Solution: The system has two zeros and two poles. The zeros of the system function are at

$$z_1 = 1\,, \quad z_2 = -2$$

and the poles are at

$$p_1 = -1 + j2\,, \quad p_2 = -1 - j2$$

Vector forms of the contribution of each zero and pole to the system function at $\omega = \omega_0$ can be written using Eqns. (4.114) and (4.115) as

$$\overrightarrow{B_1(\omega_0)} = \overrightarrow{j\omega_0} - \overrightarrow{z_1} = \overrightarrow{(-1 + j1.5)}$$

$$\overrightarrow{B_2(\omega_0)} = \overrightarrow{j\omega_0} - \overrightarrow{z_2} = \overrightarrow{(2 + j1.5)}$$

$$\overrightarrow{A_1(\omega_0)} = \overrightarrow{j\omega_0} - \overrightarrow{p_1} = \overrightarrow{(1 - j0.5)}$$

$$\overrightarrow{A_2(\omega_0)} = \overrightarrow{j\omega_0} - \overrightarrow{p_2} = \overrightarrow{(1 + j3.5)}$$

The vectors $\overrightarrow{B_1(\omega_0)}$, $\overrightarrow{B_2(\omega_0)}$, $\overrightarrow{A_1(\omega_0)}$, and $\overrightarrow{A_2(\omega_0)}$ are shown in Fig. 4.46.

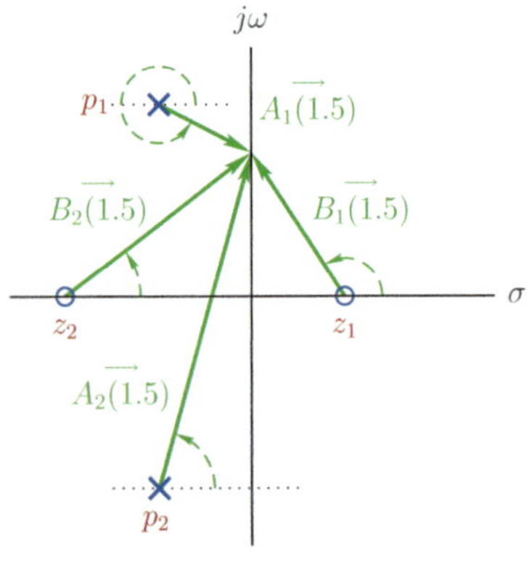

Figure 4.46 – Graphical computation of $H(\omega_0)$ for Example 4.37.

The norm and the phase of each vector are

$$\left|\overrightarrow{B_1(\omega_0)}\right| = 1.8028\,, \qquad \measuredangle\overrightarrow{B_1(\omega_0)} = 2.1588 \text{ rad}$$

$$\left|\overrightarrow{B_2(\omega_0)}\right| = 2.5000\,, \qquad \measuredangle\overrightarrow{B_2(\omega_0)} = 0.6435 \text{ rad}$$

$$\left|\overrightarrow{A_1(\omega_0)}\right| = 1.1180\,, \qquad \measuredangle\overrightarrow{A_1(\omega_0)} = -0.4636 \text{ rad}$$

$$\left|\overrightarrow{A_2(\omega_0)}\right| = 3.6401\,, \qquad \measuredangle\overrightarrow{A_2(\omega_0)} = 1.2925 \text{ rad}$$

Based on the values measured, the magnitude of the frequency response at ω_0 is computed as

$$|H(\omega_0)| = \frac{|B_1(\omega_0)|\,.\,|B_2(\omega_0)|}{|A_1(\omega_0)|\,.\,|A_2(\omega_0)|} = \frac{(1.8028)\,(2.5000)}{(1.1180)\,(3.6401)} = 1.1074$$

The phase of the frequency response is computed as

$$\begin{aligned}\measuredangle H(\omega_0) =& \measuredangle B_1(\omega_0) + \measuredangle B_2(\omega_0) - \measuredangle A_1(\omega_0) - \measuredangle A_2(\omega_0)\\ =& 2.1588 + 0.6435 - (-0.4636) - 1.2925\\ =& 1.9735 \text{ rad}\end{aligned}$$

Complete magnitude and phase characteristics for the system can be obtained by repeating this process for all values of ω that are of interest. Fig. 4.47 shows the magnitude and the phase of the system. The values at $\omega_0 = 1.5$ rad/s are marked on magnitude and phase plots.

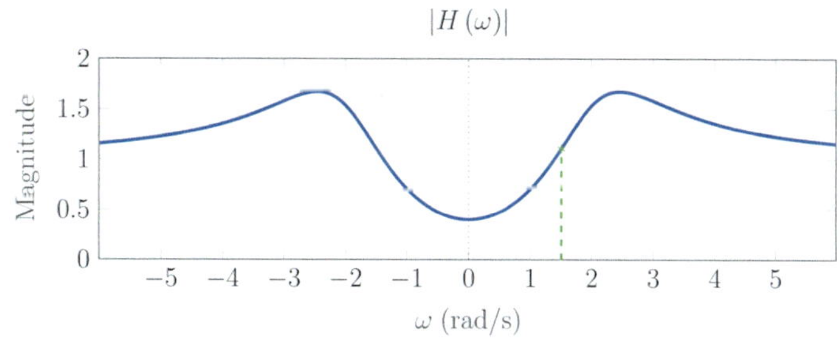

(a)

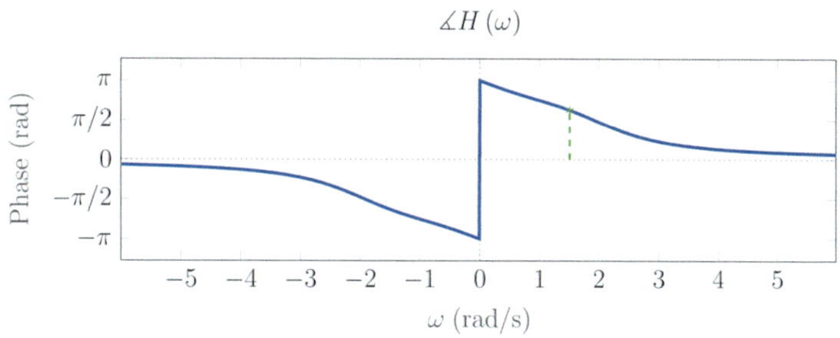

(b)

Figure 4.47 – The frequency response of the system in Example 4.37: **(a)** magnitude and **(b)** phase.

Software resource: exct_4_37.m

Software resource: See MATLAB Exercises 4.6 and 4.7.

Interactive App: Pole-zero explorer

The interactive app in `appPoleZero.m` allows experimentation with the placement of poles and zeros of the system function.

Before using the app, two vectors should be created in MATLAB workspace: one containing the poles of the system and one containing its zeros. In the pole-zero explorer user interface, the "import" button is used for importing these vectors. Pole-zero layout in the s-plane is displayed along with the magnitude and the phase of the system function. The vectors from each zero and each pole to a point on the $j\omega$-axis may optionally be displayed. Individual poles and zeros may be moved, and the effects on magnitude and phase may be observed. Complex conjugate poles and zeros move together to keep the conjugate relationship.

As an example, the system in Example 4.34 may be duplicated by creating and importing the two vectors

$$\texttt{zrs} = [1,\, -2]$$

$$\texttt{pls} = \left[-1+j2,\, -1-j2\right]$$

Software resource: `appPoleZero.m`

4.5.6 System function and causality

Causality in linear and time-invariant systems was discussed in Section 2.8 of Chapter 2. For a CTLTI system to be causal, its impulse response $h(t)$ needs to be equal to zero for $t<0$. Thus, by changing the lower limit of the integral to $t=0$ in the definition of the Laplace transform, the s-domain system function for a causal CTLTI system can be written as

$$H(s) = \int_{-\infty}^{\infty} h(t)\, e^{-st}\, dt = \int_{0}^{\infty} h(t)\, e^{-st}\, dt \tag{4.123}$$

As we have discussed in Section 4.2 of this chapter, the ROC for the system function of a causal system is to the right of a vertical line in the s-plane. As a consequence, the system function must also converge at $\text{Re}\{s\} \to \infty$. Consider a system function in the form

$$H(z) = \frac{B(s)}{A(s)} = \frac{b_M s^M + b_{M-1} s^{M-1} + \ldots + b_1 s + b_0}{a_N s^N + a_{N-1} s^{N-1} + \ldots + a_1 s + a_0}$$

For the system described by $H(s)$ to be causal we need

$$\lim_{s\to\infty} \left[H(s)\right] = \lim_{s\to\infty} \left[\frac{b_M}{a_N}\, s^{M-N}\right] < \infty \tag{4.124}$$

which requires that $M-N \le 0$ and consequently $M \le N$. Thus we arrive at an important conclusion:

Causality condition:

In the s-domain system function of a causal CTLTI system the order of the numerator must not be greater than the order of the denominator.

Note that this condition is necessary for a system to be causal, but it is not sufficient. It is also possible for a non-causal system to have a system function with $M \le N$.

4.5.7 System function and stability

In Section 2.9 of Chapter 2 we have concluded that for a CTLTI system to be stable its impulse response must be absolute integrable, that is,

$$\int_{-\infty}^{\infty} \left| h(t) \right| < \infty$$

Furthermore, we have established in Section 3.3.2 of Chapter 3 that the Fourier transform of a signal exists if the signal is absolute integrable. But the Fourier transform of the impulse response is equal to the s-domain system function evaluated on the $j\omega$-axis of the s-plane, that is,

$$H(\omega) = H(s)\Big|_{s=j\omega}$$

provided that the $j\omega$-axis of the s-plane is within the ROC.

Stability condition:

It follows that, for a CTLTI system to be stable, the ROC of its s-domain system function must include the $j\omega$-axis.

What are the corresponding conditions that must be imposed on the locations of poles and zeros for stability? We will answer this question by taking three-separate cases into account:

1. Causal system:

   The ROC for the system function of a causal system is to the right of a vertical line in the s-plane, and is expressed in the form

   $$\text{Re}\{s\} > \sigma_1$$

   For the ROC to include the $j\omega$-axis we need $\sigma_1 < 0$. Since the ROC cannot have any poles in it, all the poles of the system function must be on or to the left of the vertical line $\sigma = \sigma_1$.

   For a causal system to be stable, the system function must not have any poles on the $j\omega$-axis or in the right half s-plane.

2. Anti-causal system:

   If the system is anti-causal, its impulse response is equal to zero for $t \ge 0$. The ROC for the system function is to the left of a vertical line in the s-plane, and is expressed in the form

   $$\text{Re}\{s\} < \sigma_2$$

For the ROC to include the $j\omega$-axis we need $\sigma_2 > 0$. All the poles of the system function must reside on or to the right of the vertical line $\sigma = \sigma_2$.

> For an anti-causal system to be stable, the system function must not have any poles on the $j\omega$-axis or in the left half s-plane.

3. Neither causal nor anti-causal system:
   In this case the ROC for the system function, if it exists, is the region between two vertical lines at $\sigma = \sigma_1$ and $\sigma = \sigma_2$ and is expressed in the form

$$\sigma_1 < \mathrm{Re}\{s\} < \sigma_2$$

   For stability we need $\sigma_1 < 0$ and $\sigma_2 > 0$. The poles of the system function may be

   a. On or to the left of the vertical line $\sigma = \sigma_1$.
   b. On or to the right of the vertical line $\sigma = \sigma_2$.

Fig. 4.48 illustrates the three possibilities discussed above.

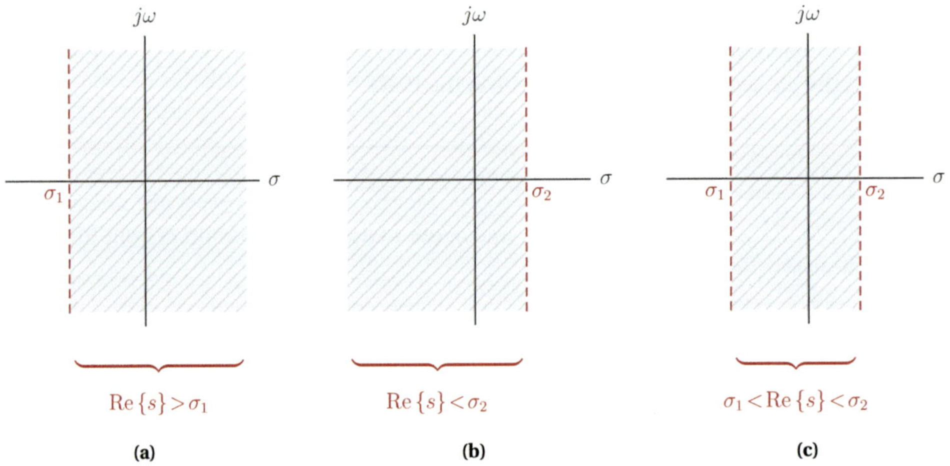

Figure 4.48 – The ROC for a stable system that is **(a)** causal, **(b)** anti-causal, and **(c)** neither causal nor anti-causal.

Example 4.38: Impulse response of a stable system

A stable system is characterized by the system function

$$H(s) = \frac{15\,s\,(s+1)}{(s+3)\,(s-1)\,(s-2)}$$

Determine the ROC of the system function. Afterward find the impulse response of the system.

Solution: The ROC for the system function is not directly stated; however, we are given enough information to deduce it. The three poles of the system function are at $s = -3, 1, 2$.

Since the system is known to be stable, its ROC must include the $j\omega$-axis of the s-plane. The only possible choice is

$$-3 < \text{Re}\{s\} < 1$$

as shown in Fig. 4.49(a). Partial fraction expansion of $H(s)$ is

$$H(s) = \frac{4.5}{s+3} - \frac{7.5}{s-1} + \frac{18}{s-2}$$

Based on the ROC determined above, the first term in the partial fraction expansion corresponds to a causal signal, and the other two terms correspond to anti-causal signals. The impulse response of the system is

$$h(t) = 4.5\, e^{-3t}\, u(t) + 7.5\, e^{t}\, u(-t) - 18\, e^{2t}\, u(-t)$$

which is shown in Fig. 4.49(b). We observe that $h(t)$ tends to zero as t is increased in both directions, consistent with the fact that $h(t)$ must be absolute integrable for a stable system.

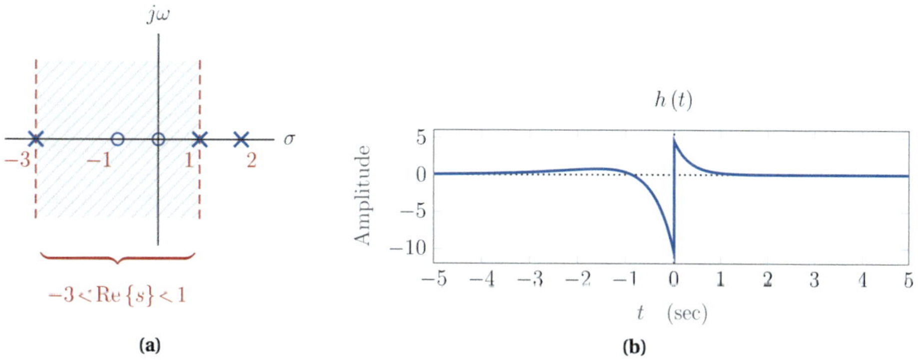

Figure 4.49 – **(a)** The ROC for the system of Example 4.38 and **(b)** the impulse response.

Software resources: exct_4_38a.m , exct_4_38b.m

Example 4.39: Stability of a system described by a differential equation

A causal CTLTI system is characterized by the differential equation

$$\frac{d^2 y(t)}{dt^2} + 2\frac{dy(t)}{dt} + 2\, y(t) = \frac{dx(t)}{dt} + x(t)$$

Determine if the system is stable.

Solution: The ROC for the system function is not directly stated. On the other hand, we are told that the system is causal. This bit of information should allow us to determine the ROC. Taking the Laplace transform of both sides of the differential equation and using the time differentiation property we have

$$s^2\, Y(s) + 2s\, Y(s) + 2\, Y(s) = s\, X(s) + X(s)$$

The system function is found by forming the ratio of the output and the input transforms.

$$H(s) = \frac{Y(s)}{X(s)} = \frac{s+1}{s^2+2s+2}$$

The system has a zero at $s=-1$ and a pair of complex conjugate poles at $s=-1\pm j1$. Since it is causal, the ROC must be to the right of a vertical line going through $\sigma=-1$, that is

$$\text{ROC:} \quad \text{Re}\{s\} > -1$$

Since the ROC includes the $j\omega$-axis of the s-plane, the system is stable. The impulse response can be determined easily by writing the $H(s)$ in the form

$$H(s) = \frac{s+1}{(s+1)^2+1} = \left.\frac{s}{s^2+1}\right|_{s\to s+1}$$

Using s-domain shifting property of the Laplace transform it follows that

$$h(t) = e^{-t}\cos(t)\,u(t)$$

Pole-zero diagram for the system function and the corresponding impulse response are shown in Fig. 4.50.

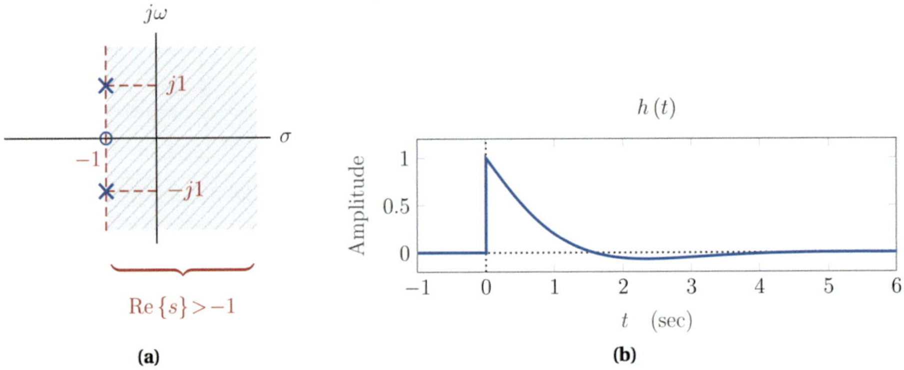

Figure 4.50 – **(a)** Pole-zero diagram and ROC for the system of Example 4.39 and **(b)** impulse response $h(t)$.

Software resources: exct_4_39a.m , exct_4_39b.m

4.5.8 All-pass systems

A system with a constant magnitude characteristic across all frequencies is called an *all-pass system.* Thus, for a system to be considered an all-pass system we need

$$\left|H(\omega)\right| = C \text{ (constant)} \qquad \text{for all } \omega \tag{4.125}$$

Consider a first-order CTLTI system with a pole at $p_1=-\sigma_1+j0$ and a zero at $z_1=\sigma_1+j0$ so that the system function is

$$H(s) = \frac{s-z_1}{s-p_1} = \frac{s-\sigma_1}{s+\sigma_1} \tag{4.126}$$

For the system to be causal and stable the parameter σ_1 needs to be positive so that the resulting pole is in the left half s-plane. The magnitude and the phase of the system function can be expressed using the conventions established in Section 4.5.5. The frequency response $H(\omega)$ is found by substituting $s = j\omega$:

$$H(\omega) = \frac{B_1(\omega)}{A_1(\omega)} = \frac{j\omega - \sigma_1}{j\omega + \sigma_1} \tag{4.127}$$

The magnitude of the frequency response is

$$\left|H(\omega)\right| = \frac{\left|B_1(\omega)\right|}{\left|A_1(\omega)\right|} = \frac{\sqrt{\omega^2 + \sigma_1^2}}{\sqrt{\omega^2 + \sigma_1^2}} = 1 \tag{4.128}$$

The phases of numerator and denominator terms are

$$\measuredangle B_1(\omega) = \pi - \tan^{-1}\left(\frac{\omega}{\sigma_1}\right) \qquad \text{and} \qquad \measuredangle A_1(\omega) = -\tan^{-1}\left(\frac{\omega}{\sigma_1}\right), \tag{4.129}$$

respectively, and the phase of the frequency response is

$$\measuredangle H(\omega) = \measuredangle B_1(\omega) - \measuredangle A_1(\omega) = \pi - 2\tan^{-1}\left(\frac{\omega}{\sigma_1}\right) \tag{4.130}$$

The pole-zero diagram and the vector representation of the system function are shown in Fig. 4.51(a) along with the vectors

$$\overrightarrow{B_1(\omega_0)} = \overrightarrow{j\omega_0} - \overrightarrow{\sigma_1} \qquad \text{and} \qquad \overrightarrow{A_1(\omega_0)} = \overrightarrow{j\omega_0} + \overrightarrow{\sigma_1} \tag{4.131}$$

Fig. 4.51(b) shows the phase characteristics for $\sigma_1 = 1$, $\sigma_1 = 2$, and $\sigma_1 = 3$.

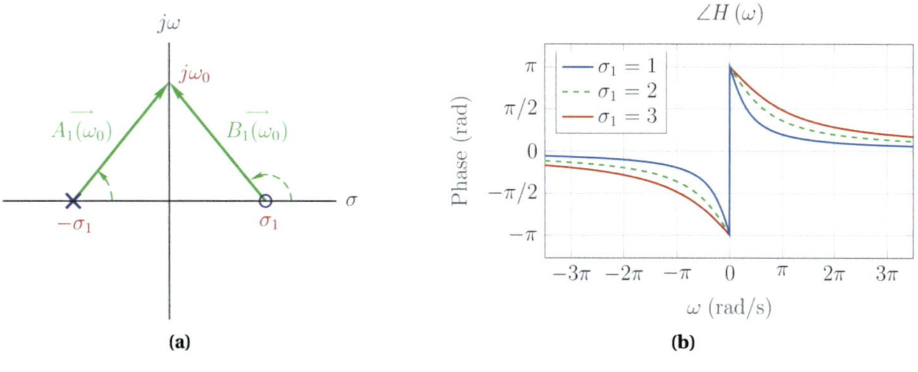

Figure 4.51 – **(a)** Pole-zero diagram for first-order all-pass system in Eqn. (4.126) and **(b)** phase response of the first-order all-pass system.

Since the shape of the phase response can be controlled by the choice of parameter σ_1 while keeping the magnitude response constant, an all-pass system is also referred to as a *phase shifter*. Increased versatility in controlling the phase response can be obtained by choosing the pole and the zero to be complex-valued. Consider a first-order system with a pole at $p_1 = -\sigma_1 + j\omega_1$ and a zero at $z_1 = \sigma_1 + j\omega_1$. Again we choose $\sigma_1 > 0$ to obtain a system that is both causal and stable. The corresponding system function is

$$H(s) = \frac{s - z_1}{s - p_1} = \frac{s - \sigma_1 - j\omega_1}{s + \sigma_1 - j\omega_1} \tag{4.132}$$

The frequency response of the system, found by substituting $s = j\omega$ into the system function, is

$$H(\omega) = \frac{j\omega - z_1}{j\omega - p_1} = \frac{j\omega - \sigma_1 - j\omega_1}{j\omega + \sigma_1 - j\omega_1} \tag{4.133}$$

It is a trivial matter to show that the magnitude of $H(\omega)$ is still equal to unity. Its phase is

$$\measuredangle H(\omega) = \measuredangle\left(j\omega - \sigma_1 - j\omega_1\right) - \measuredangle\left(j\omega + \sigma_1 - j\omega_1\right) = \pi - 2\tan^{-1}\left(\frac{\omega - \omega_1}{\sigma_1}\right) \tag{4.134}$$

The pole-zero diagram and the vector representation of the system function are shown in Fig. 4.52(a) along with the vectors

$$\overrightarrow{B_1}(\omega_0) = \overrightarrow{j\omega_0} - \overrightarrow{z_1} \qquad \text{and} \qquad \overrightarrow{A_1}(\omega_0) = \overrightarrow{j\omega_0} + \overrightarrow{p_1} \tag{4.135}$$

The phase response is shown in Fig. 4.52(b) for $\omega_1 = 1.5$ rad/s and for three different values of σ_1, namely $\sigma_1 = 1$, $\sigma_1 = 2$, and $\sigma_1 = 3$.

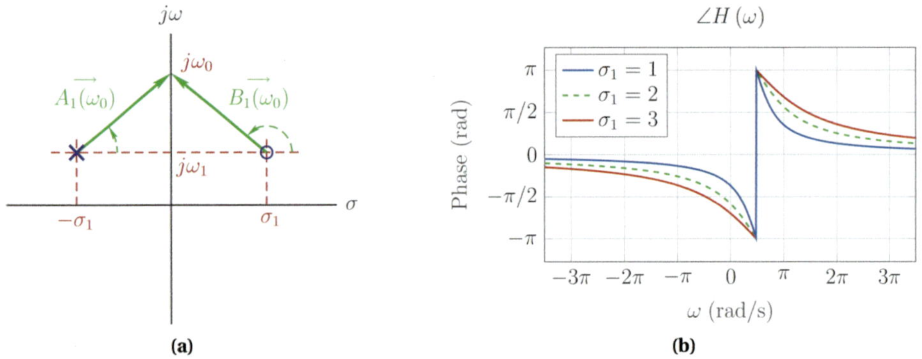

Figure 4.52 – **(a)** Pole-zero diagram for first-order all-pass system in Eqn. (4.132) and **(b)** phase response of the first-order all-pass system.

Naturally, the system function in Eqn. (4.132) has complex coefficients. If an all-pass system with real coefficients is desired, complex poles and zeros must occur in conjugate pairs. A second-order system with zeros at $z_{1,2} = \sigma_1 \pm j\omega_1$ and poles at $p_{1,2} = -\sigma_1 \pm j\omega_1$ has all-pass characteristics. As before, $\sigma_1 > 0$ for a causal and stable system. The system function is

$$H(s) = \frac{\left(s - \sigma_1 - j\omega_1\right)\left(s - \sigma_1 + j\omega_1\right)}{\left(s + \sigma_1 - j\omega_1\right)\left(s + \sigma_1 + j\omega_1\right)} = \frac{(s - \sigma_1)^2 + \omega_1^2}{(s + \sigma_1)^2 + \omega_1^2} \tag{4.136}$$

The pole-zero diagram for a second-order all-pass system is shown in Fig. 4.53(a). The phase response is shown in Fig. 4.53(b) for $\omega_1 = 1.5$ rad/s and for three different values of σ_1.

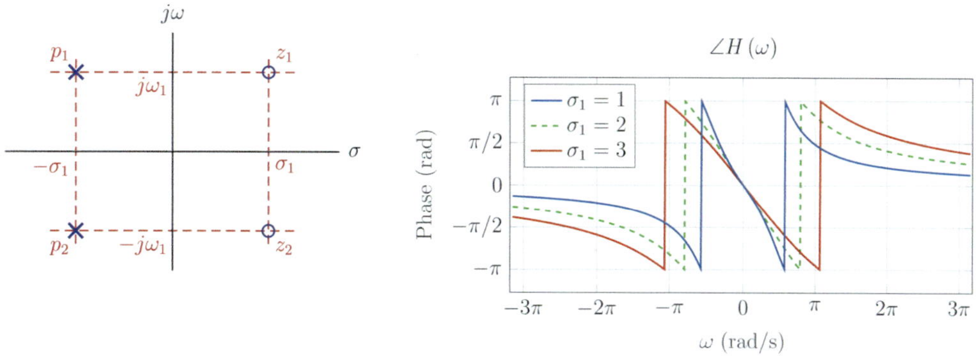

(a) **(b)**

Figure 4.53 – **(a)** Pole-zero diagram for second-order all-pass system in Eqn. (4.136) and **(b)** phase response of the second-order all-pass system.

4.5.9 Inverse systems

The inverse of a system is another system which, when connected in cascade with the original system, forms an identity system. This relationship is depicted in Fig. 4.54.

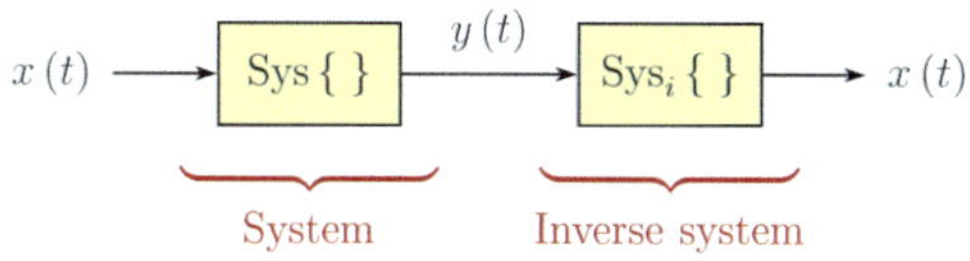

Figure 4.54 – A system and its inverse connected in cascade.

The output signal of the original system is

$$y(t) = \text{Sys}\left\{x(t)\right\} \tag{4.137}$$

We require the inverse system to recover the original input signal $x(t)$ from the output signal $y(t)$, therefore

$$x(t) = \text{Sys}_i\left\{y(t)\right\} \tag{4.138}$$

Combining Eqns. (4.137) and (4.137) yields

$$\text{Sys}_i\left\{\text{Sys}\left\{x(t)\right\}\right\} = x(t) \tag{4.139}$$

Let the original system and its inverse be both CTLTI systems with impulse responses $h(t)$ and $h_i(t)$ respectively as shown in Fig. 4.55. For the output signal of the inverse system to be identical to the input signal of the original system, the impulse response of the cascade combination must be equal to $\delta(t)$, that is,

$$h_{eq}(t) = h(t) * h_i(t) = \delta(t) \tag{4.140}$$

or, using the convolution integral

$$h_{eq}(t) = \int_{-\infty}^{\infty} h(\lambda)\, h_i(t-\lambda)\, d\lambda = \delta(t) \tag{4.141}$$

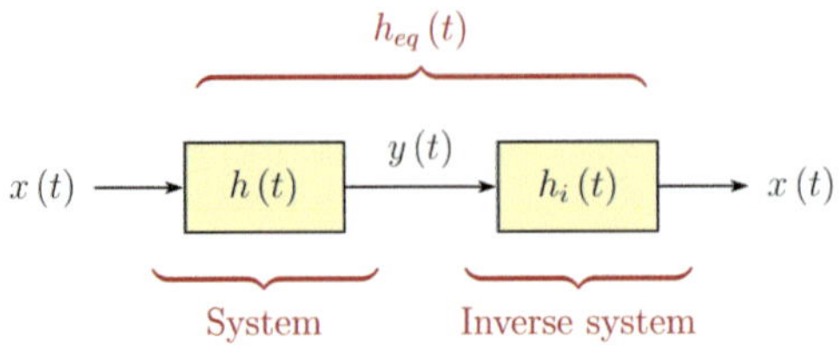

Figure 4.55 – A CTLTI system and its inverse connected in cascade.

The corresponding relationship between the system functions of the original system and the inverse system is found by taking the Laplace transform of Eqn. (4.140):

$$H_{eq}(s) = H(s)\, H_i(s) = 1 \tag{4.142}$$

Consequently, the system function of the inverse system is

$$H_i(s) = \frac{1}{H(s)} \tag{4.143}$$

Two important characteristics of the inverse system are causality and stability. We will first focus on causality. Consider the system $H(s)$ function in the form

$$H(s) = \frac{B(s)}{A(s)} = \frac{b_M s^M + b_{M-1} s^{M-1} + \ldots + b_1 s + b_0}{a_N s^N + a_{N-1} s^{N-1} + \ldots + a_1 s + a_0} \tag{4.144}$$

The system function for the inverse system is

$$H_i(s) = \frac{A(s)}{B(s)} = \frac{a_N s^N + a_{N-1} s^{N-1} + \ldots + a_1 s + a_0}{b_M s^M + b_{M-1} s^{M-1} + \ldots + b_1 s + b_0}$$

If the original system with system function $H(s)$ is causal then $M \le N$ as we have established in Section 4.5.6. By the same token, causality of the inverse system with system function $H_i(s)$ requires $N \le M$. Hence we need $N = M$ if both the original system and its inverse are required to be causal.

To analyze the stability of the inverse system we will find it more convenient to write the system function $H(s)$ in pole zero form. Using $M = N$ we have

$$H(s) = \frac{b_N (s - z_1)(s - z_2)\ldots(s - z_N)}{a_N (s - p_1)(s - p_2)\ldots(s - p_N)} \tag{4.145}$$

If the original system is both causal and stable, all its poles must be in the left half s-plane (see Section 4.5.7), therefore

$$\mathrm{Re}\{p_k\} < 0\,, \qquad k = 1, \ldots, N \tag{4.146}$$

The system function of the inverse system, written in pole-zero form, is

$$H_i(s) = \frac{a_N (s - p_1)(s - p_2)\ldots(s - p_N)}{b_N (s - z_1)(s - z_2)\ldots(s - z_N)} \tag{4.147}$$

For the inverse system to be stable, its poles must also lie in the left half s-plane. The poles of the inverse system are the zeros of the original system. Therefore, for the inverse system to be stable, both zeros and poles of the original system must be in the left half s-plane. In addition to Eqn. (4.146) we also need

$$\mathrm{Re}\{z_k\} < 0\,, \qquad k = 1, \ldots, N \tag{4.148}$$

Minimum-phase system:

A causal CTLTI system that has all of its zeros and poles in the left half s-plane is referred to as a *minimum-phase system.*

A minimum-phase system and its inverse are both causal and stable.

Example 4.40: Inverse of a system described by a differential equation

A causal CTLTI system is described by a differential equation

$$\frac{dy(t)}{dt} + 2\,y(t) = \frac{dx(t)}{dt} + x(t)$$

Determine if a causal and stable inverse can be found for this system. If yes, find a differential equation for the inverse system.

Solution: Taking the Laplace transform of both sides of the differential equation we get

$$(s+2)\;Y(s) = (s+1)\;X(s)$$

The system function for the original system is

$$H(s) = \frac{s+1}{s+2}$$

The system function for the inverse system is found as the reciprocal of $H(s)$:

$$H_i(s) = \frac{1}{H(s)} = \frac{s+2}{s+1}$$

The inverse system is also causal and stable. It leads to the differential equation

$$\frac{dy(t)}{dt} + y(t) = \frac{dx(t)}{dt} + 2\,x(t)$$

4.5.10 Bode plots

Bode plots of the frequency response are used in the analysis and design of feedback control systems. A Bode plot consists of the dB magnitude $20\log_{10}\left|H(\omega)\right|$ and the phase $\measuredangle H(\omega)$, each graphed as a function of $\log_{10}(\omega)$. Because of the use of the logarithm, individual contributions of the zeros and the poles of $H(s)$ to the magnitude of the frequency response are additive rather than multiplicative. Also, since $\log_{10}(\omega)$ is used for the horizontal axis, only positive values of ω are of interest.

Consider again the pole-zero form of the system function $H(s)$ given by Eqn. (4.112). A slightly modified form of it will be more convenient for use in deriving the Bode plot. Let us scale each factor in the numerator and the denominator, and write Eqn. (4.112) as

$$H(s) = K_1\,\frac{(1-s/z_1)\,(1-s/z_2)\ldots(1-s/z_M)}{(1-s/p_1)\,(1-s/p_2)\ldots(1-s/p_N)} \tag{4.149}$$

The new gain factor K_1 is chosen to compensate for all the scale factors used in individual terms. The dB magnitude of $H(\omega)$ is obtained as

$$\begin{aligned}20\log_{10}|H(\omega)| =& 20\log_{10}|K_1| \\ &+20\log_{10}|1-j\omega/z_1|+20\log_{10}|1-j\omega/z_2|+\ldots+20\log_{10}|1-j\omega/z_M| \\ &-20\log_{10}|1-j\omega/p_1|-20\log_{10}|1-j\omega/p_2|-\ldots-20\log_{10}|1-j\omega/p_N|\end{aligned} \tag{4.150}$$

and the phase is

$$\begin{aligned}\measuredangle H(\omega) =& \measuredangle K_1+\measuredangle(1-j\omega/z_1)+\measuredangle(1-j\omega/z_2)+\ldots+\measuredangle(1-j\omega/z_M) \\ &-\measuredangle(1-j\omega/p_1)-\measuredangle(1-j\omega/p_2)-\ldots-\measuredangle(1-j\omega/p_N)\end{aligned} \tag{4.151}$$

To facilitate the analysis of individual contributions from zeros and poles of the system function, let us write $H(s)$ as a cascade combination of $M+N$ subsystems such that

$$H(s) = K_1\, H_1(s)\, H_2(s)\ldots H_M(s)\, H_{M+1}(s)\, H_{M+2}(s)\ldots H_{M+N}(s) \tag{4.152}$$

with

$$H_i(s) = 1-s/z_i\,, \qquad i=1,\ldots,M \tag{4.153}$$

and

$$H_{M+i}(s) = \frac{1}{1-s/p_i}\,, \qquad i=1,\ldots,N \tag{4.154}$$

The Bode plot for magnitude and phase can be constructed by computing the contribution of each term in Eqns. (4.153) and (4.154). We will consider four different cases.

Zero at the origin

Let $H_k(s) = s$. Corresponding dB magnitude and phase are computed as

$$20\log_{10}|H_k(\omega)| = 20\log_{10}|j\omega| = 20\log_{10}(\omega) \tag{4.155}$$

and

$$\measuredangle H_k(\omega) = \measuredangle j\omega = 90° \tag{4.156}$$

The magnitude characteristic, when graphed as a function of $\log_{10}(\omega)$, is a straight line that goes through 0 dB at $\omega = 1$. Its slope is 20 dB per decade.[2] This is illustrated in Fig. 4.56(a).

Pole at the origin

Let $H_k(s) = 1/s$. Corresponding dB magnitude and phase are computed as

$$20\log_{10}|H_k(\omega)| = 20\log_{10}\left|\frac{1}{j\omega}\right| = -20\log_{10}(\omega) \tag{4.157}$$

and

$$\measuredangle H_k(\omega) = \measuredangle\left(\frac{1}{j\omega}\right) = -90° \tag{4.158}$$

The magnitude characteristic, when graphed as a function of $\log_{10}(\omega)$, is a straight line that goes through 0 dB at $\omega = 1$. Its slope is −20 dB per decade. This is illustrated in Fig. 4.56(b).

[2] One decade corresponds to a tenfold change in ω. If $\omega_2 = 10\omega_1$, then $\log_{10}(\omega_2) = \log_{10}(\omega_1)+1$.

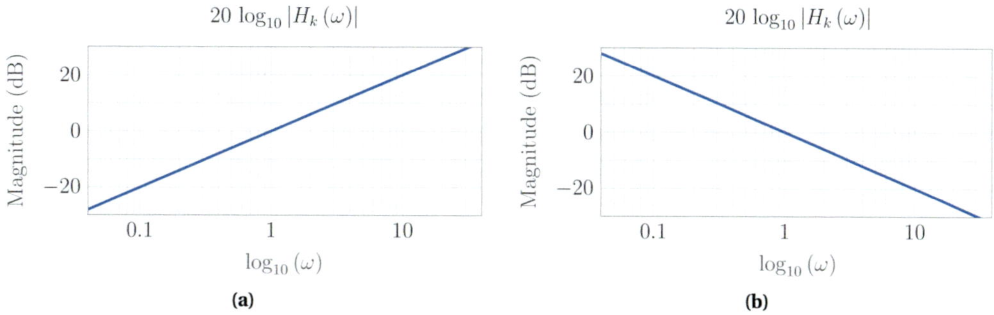

Figure 4.56 **(a)** The dB magnitude for $H_k(s) = s$ and **(b)** dB magnitude for $H_k(s) = 1/s$.

Single real zero

Let z_k be a real-valued zero of the system. Substituting $s = j\omega$ into Eqn. (4.153) we obtain

$$H_k(\omega) = H_k(s)\Big|_{s=j\omega} = 1 - j\omega/z_k \tag{4.159}$$

Corresponding dB magnitude is

$$20\log_{10}\left|H_k(\omega)\right| = 20\log_{10}\sqrt{1+\left(\frac{\omega}{z_k}\right)^2} = 10\log_{10}\left[1+\left(\frac{\omega}{z_k}\right)^2\right] \tag{4.160}$$

and the corresponding phase characteristic is

$$\measuredangle H_k(\omega) = \measuredangle\left(1-\frac{j\omega}{z_k}\right) = \tan^{-1}\left(-\frac{\omega}{z_k}\right) = -\tan^{-1}\left(\frac{\omega}{z_k}\right) \tag{4.161}$$

Keep in mind that we need to graph $20\log_{10}\left|H_k(\omega)\right|$ and $\measuredangle H_k(\omega)$ as functions of $\log_{10}(\omega)$, and we are only interested in positive values of ω.

1. If $\omega << \left|z_k\right|$ we have

$$\left(\frac{\omega}{z_k}\right)^2 << 1 \tag{4.162}$$

The magnitude in Eqn. (4.160) can be approximated as

$$20\log_{10}\left|H_k(\omega)\right| \approx 10\log_{10}(1) = 0 \tag{4.163}$$

The phase angle can be approximated as

$$\measuredangle H_k(\omega) \approx -\tan^{-1}(0) = 0 \tag{4.164}$$

2. At the opposite extreme, for $\omega >> \left|z_k\right|$ we have

$$\left(\frac{\omega}{z_k}\right)^2 >> 1 \tag{4.165}$$

The magnitude in Eqn. (4.160) can be approximated as

$$20\log_{10}\left|H_k(\omega)\right| \approx 10\log_{10}\left(\frac{\omega}{z_k}\right)^2 = 20\log_{10}\left(\frac{\omega}{\left|z_k\right|}\right) \tag{4.166}$$

which can also be written in the form

$$20\log_{10}\left|H_k(\omega)\right| \approx 20\log_{10}(\omega) - 20\log_{10}\left|z_k\right| \quad (4.167)$$

In this case the phase angle depends on the sign of z_k:

$$\measuredangle H_k(\omega) \approx \begin{cases} 90^\circ, & z_k < 0 \\ -90^\circ, & z_k > 0 \end{cases} \quad (4.168)$$

3. At the frequency $\omega = \left|z_k\right|$ the magnitude is

$$20\log_{10}\left|H_k(\omega)\right| = 10\log_{10}(2) \approx 3\text{ dB} \quad (4.169)$$

The phase angle depends on the sign of z_k:

$$\measuredangle H_k(\omega) \approx \begin{cases} 45^\circ, & z_k < 0 \\ -45^\circ, & z_k > 0 \end{cases} \quad (4.170)$$

The conclusions obtained above can be summarized as follows:

For single real zero:

For the term $H_k(s) = 1 - s/z_k$ with real-valued z_k:

1. Magnitude: For $\omega << \left|z_k\right|$ the magnitude characteristic is asymptotic to 0 dB. For $\omega >> \left|z_k\right|$ it becomes asymptotic to a straight line with a slope of 20 dB per decade, which intersects the horizontal axis at $\omega = \left|z_k\right|$. At the frequency $\omega = \left|z_k\right|$ the actual magnitude is approximately equal to 3 dB; therefore, the characteristic passes 3 dB above the intersection of the two asymptotes.
2. Phase: For $\omega << \left|z_k\right|$ the phase is asymptotic to 0 degrees. For $\omega >> \left|z_k\right|$ the phase is 90 degrees for $z_k < 0$ and -90 degrees for $z_k > 0$. At $\omega = z_k$ the phase is 45 degrees for $z_k < 0$ and -45 degrees for $z_k > 0$.

This is illustrated in Figs. 4.57 and 4.58.

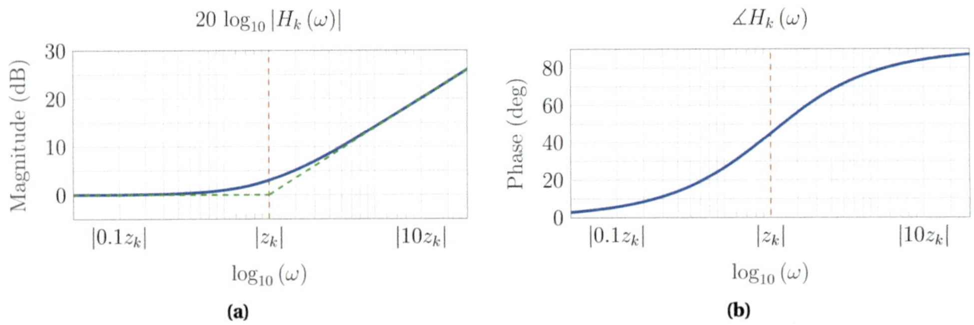

Figure 4.57 – Bode plots for $H_k(s) = (1 - s/z_k)$ with $z_k < 0$: **(a)** magnitude and **(b)** phase.

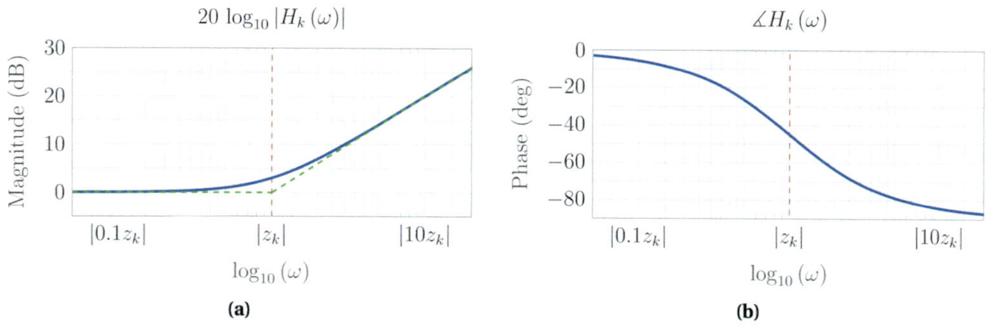

Figure 4.58 – Bode plots for $H_k(s) = \left(1 - s/z_k\right)$ with $z_k > 0$: **(a)** magnitude and **(b)** phase.

Single real pole

Derivation of the Bode plot for a single real pole is similar. Let p_k be a real-valued pole of the system. Substituting $s = j\omega$ into Eqn. (4.154) yields

$$H_k(\omega) = \left.\frac{1}{1 - s/p_k}\right|_{s=j\omega} = \frac{1}{1 - j\omega/p_k} \tag{4.171}$$

Corresponding dB magnitude is

$$20\,\log_{10}\left|H_k(\omega)\right| = 20\,\log_{10}\frac{1}{\sqrt{1+\left(\dfrac{\omega}{p_k}\right)^2}} = -10\,\log_{10}\left[1+\left(\frac{\omega}{p_k}\right)^2\right] \tag{4.172}$$

and the corresponding phase characteristic is

$$\measuredangle H_k(\omega) = \measuredangle\left[\frac{1}{1 - j\omega/p_k}\right] = -\tan^{-1}\left(-\frac{\omega}{p_k}\right) = \tan^{-1}\left(\frac{\omega}{p_k}\right) \tag{4.173}$$

1. If $\omega << \left|p_k\right|$ we have

$$\left(\frac{\omega}{p_k}\right)^2 << 1 \tag{4.174}$$

The magnitude in Eqn. (4.172) can be approximated as

$$20\,\log_{10}\left|H_k(\omega)\right| \approx -10\,\log_{10}(1) = 0 \tag{4.175}$$

The phase angle can be approximated as

$$\measuredangle H_k(\omega) \approx \tan^{-1}(0) = 0 \tag{4.176}$$

2. At the opposite extreme, for $\omega >> \left|p_k\right|$ we have

$$\left(\frac{\omega}{p_k}\right)^2 >> 1 \tag{4.177}$$

The magnitude in Eqn. (4.172) can be approximated as

$$20\,\log_{10}\left|H_k(\omega)\right| \approx -10\,\log_{10}\left(\frac{\omega}{p_k}\right)^2 = -20\,\log_{10}\left(\frac{\omega}{\left|p_k\right|}\right) \tag{4.178}$$

which can also be written in the form

$$20\log_{10}\left|H_k(\omega)\right| \approx -20\log_{10}(\omega) + 20\log_{10}\left|p_k\right| \tag{4.179}$$

In this case the phase angle depends on the sign of p_k:

$$\measuredangle H_k(\omega) \approx \begin{cases} -90^\circ, & p_k < 0 \\ 90^\circ, & p_k > 0 \end{cases} \tag{4.180}$$

3. At the frequency $\omega = \left|p_k\right|$ the magnitude is

$$20\log_{10}\left|H_k(\omega)\right| = -10\log_{10}(2) \approx -3\ \text{dB} \tag{4.181}$$

The phase angle depends on the sign of p_k:

$$\measuredangle H_k(\omega) \approx \begin{cases} -45^\circ, & p_k < 0 \\ 45^\circ, & p_k > 0 \end{cases} \tag{4.182}$$

The conclusions obtained above can be summarized as follows:

For single real pole:

For the term $H_k(s) = 1/\left(1 - s/p_k\right)$ with real-valued p_k:

1. Magnitude: For $\omega << \left|p_k\right|$ the magnitude characteristic is asymptotic to 0 dB. For $\omega >> \left|p_k\right|$ it becomes asymptotic to a straight line with a slope of −20 dB per decade, which intersects the 0-dB axis at $\omega = \left|p_k\right|$. At the frequency $\omega = \left|p_k\right|$ the actual magnitude is approximately equal to −3 dB; therefore, the characteristic passes 3 dB below the intersection of the two asymptotes.
2. Phase: For $\omega << \left|p_k\right|$ the phase is asymptotic to 0 degrees. For $\omega >> \left|p_k\right|$ the phase is −90 degrees for $p_k < 0$ and 90 degrees for $p_k > 0$. At $\omega = p_k$ the phase is −45 degrees for $p_k < 0$ and 45 degrees for $p_k > 0$.

This is illustrated in Figs. 4.59 and 4.60.

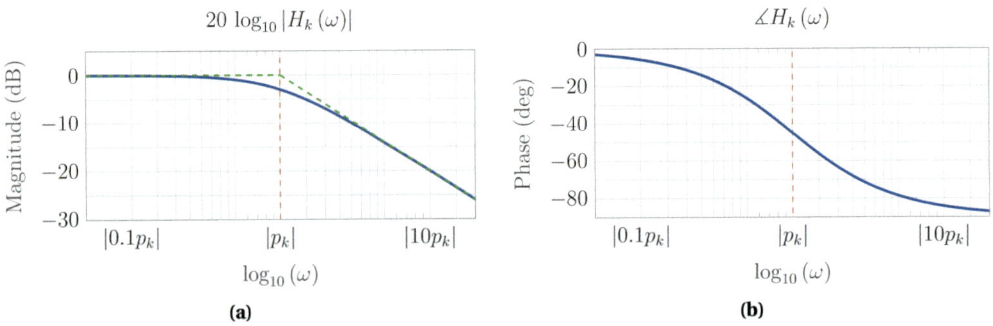

Figure 4.59 – Bode plots for $H_k(s) = 1/\left(1 - s/p_k\right)$ with $p_k < 0$: **(a)** magnitude and **(b)** phase.

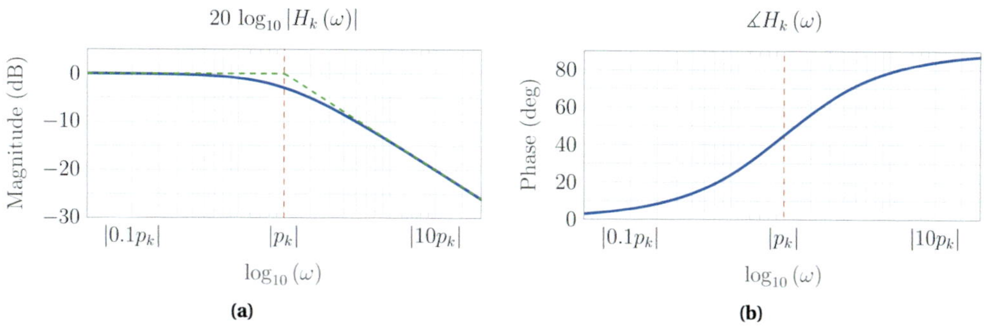

Figure 4.60 – Bode plots for $H_k(s) = 1/(1 - s/p_k)$ with $p_k > 0$: **(a)** magnitude and **(b)** phase.

Example 4.41: Constructing a Bode plot

A CTLTI system is described by the system function

$$H(s) = \frac{s\,(1 + s/300)}{(1 + s/5)\,(1 + s/40)}$$

Plot the Bode magnitude and the phase characteristics of the system.

Solution: Let us express the system function in the form of four subsystems connected in cascade:

$$H(s) = H_1(s)\,H_2(s)\,H_3(s)\,H_4(s)$$

The subsystems are

$$H_1(s) = s\,, \qquad H_2(s) = 1 + s/300\,, \qquad H_3(s) = \frac{1}{1 + s/5}\,, \qquad H_4(s) = \frac{1}{1 + s/40}$$

Magnitude and phase contributions of the four subsystems are shown in Fig. 4.61. Bode plot for the magnitude of the system function is obtained by adding the magnitude contributions in Fig. 4.61(a), (c), (e), and (g). Asymptote lines of the magnitude characteristics can also be added together. Four sections of asymptotes are shown in Fig. 4.62(a) numbered 1 through 4.

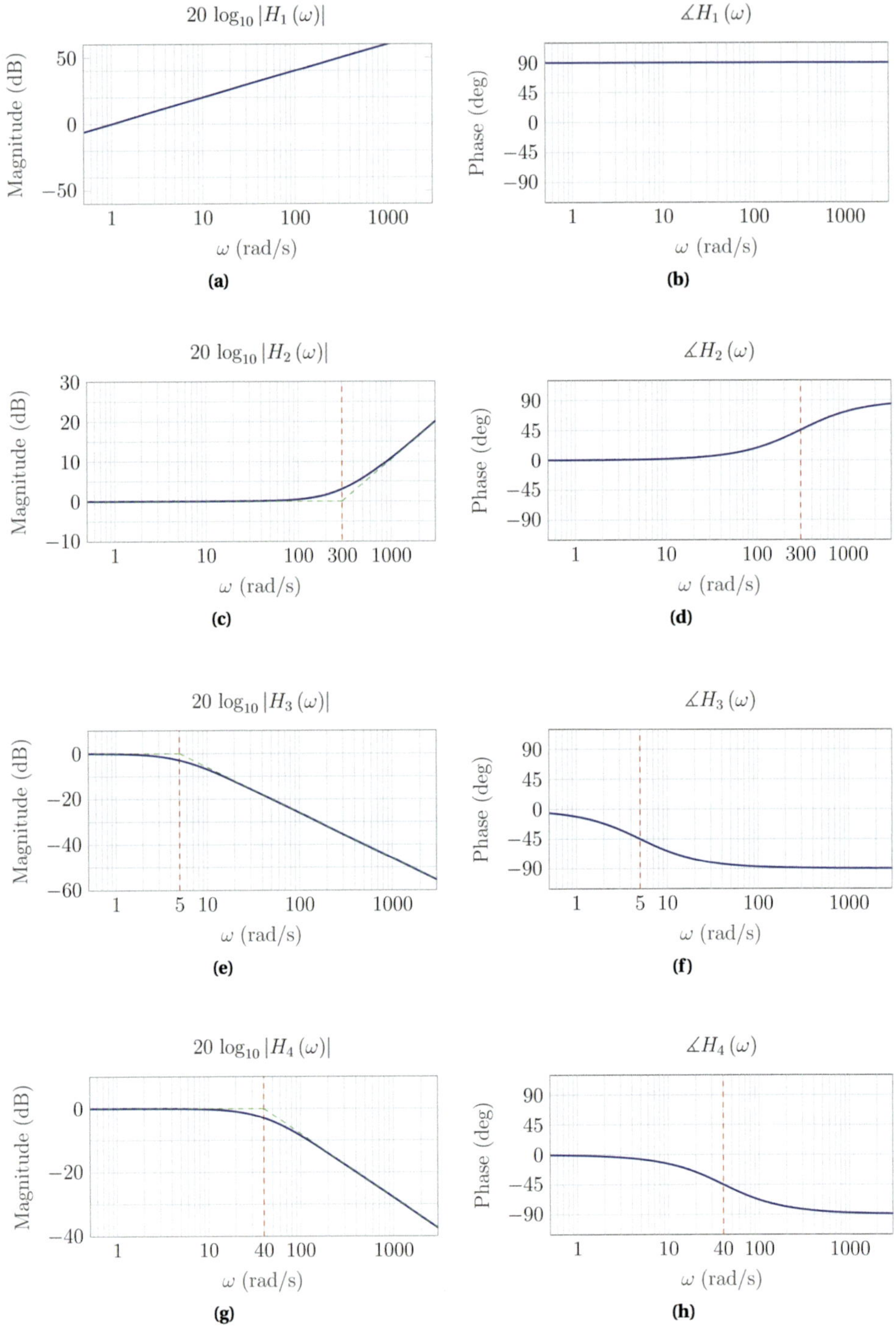

Figure 4.61 – Magnitude and phase contributions of the four subsystems for Example 4.41.

1. Section 1 covers frequencies up to $\omega = 5$ rad/s and has a slope of 20 dB/decade which is the sum of slopes of all four contributing asymptote lines up to that point. At $\omega = 5$ rad/s, its value is 13.98 dB (see Problem 4.43 at the end of this chapter).
2. Section 2 covers frequencies from $\omega = 5$ rad/s to $\omega = 40$ rad/s and has a slope of zero (20 dB per decade from H_1 and -20 dB per decade from H_3.)
3. Section 3 covers frequencies from $\omega = 40$ rad/s to $\omega = 300$ rad/s, and has a slope of -20 dB per decade (20 dB per decade from H_1; -20 dB per decade each from H_3 and H_4). At the endpoint $\omega = 300$ rad/s, its value is -3.52 dB (see Problem 4.43 at the end of this chapter).
4. Section 4 covers frequencies greater than $\omega = 300$ rad/s and has a slope of zero (20 dB per decade each from H_1 and H_2; -20 dB per decade each from H_3 and H_4).

The actual Bode magnitude plot is also shown in Fig. 4.62. Notice how it passes approximately 3 dB above or below each corner point depending on whether it belongs to a zero or a pole.

Bode plot for the phase of the system function is obtained by adding the phase contributions in Fig. 4.61(b), (d), (f), and (h), and is shown in Fig. 4.62(b).

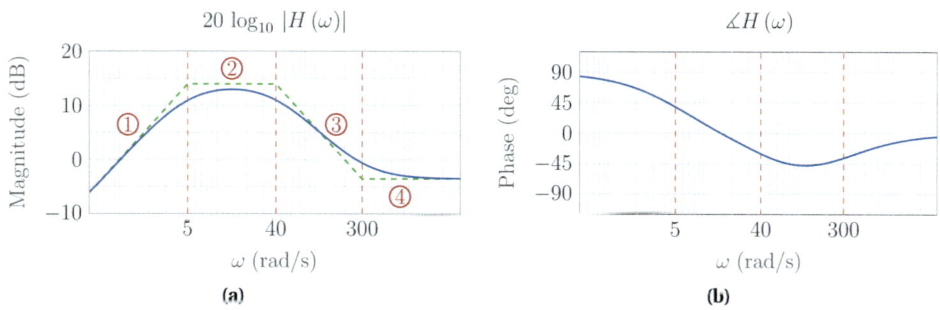

Figure 4.62 – Bode plot for the system of Example 4.41: **(a)** magnitude and **(b)** phase.

Software resources: exct_4_41a.m , exct_4_41b.m

Conjugate pair of poles

Consider a causal and stable second-order system with a pair of complex conjugate poles, that is, $p_2 = p_1^*$. The system function is in the form

$$H(s) = \frac{1}{\left(1 - s/p_1\right)\left(1 - s/p_1^*\right)} \tag{4.183}$$

$H(s)$ can be written in a slightly different form by multiplying both the numerator and the denominator of Eqn. (4.183) with the product $\left(p_1 \, p_1^*\right)$:

$$H(s) = \frac{\left|p_1\right|^2}{\left(s - p_1\right)\left(s - p_1^*\right)} \tag{4.184}$$

Let us put $H(s)$ into the standard form

$$H(s) = \frac{\omega_0^2}{s^2 + 2\zeta\omega_0\, s + \omega_0^2} \tag{4.185}$$

with

$$\omega_0^2 = \left|p_1\right|^2 \quad \text{and} \quad 2\zeta\omega_0 = -2\,\mathrm{Re}\left\{p_1\right\}, \tag{4.186}$$

It follows from Eqn. (4.186) that

$$\zeta = -\frac{\mathrm{Re}\left\{p_1\right\}}{\left|p_1\right|} \tag{4.187}$$

Since the system is causal and stable, $\mathrm{Re}\left\{p_1\right\} < 0$. Consequently, when the poles of the system form a complex conjugate pair, we have $0 < \zeta < 1$.

Analysis of the second-order system

The form of the second-order system function given by Eqn. (4.185) can be used for a system with either two real poles or a complex conjugate pair. Let us write $H(s)$ using two poles p_1 and p_2:

$$H(s) = \frac{1}{\left(1 - s/p_1\right)\left(1 - s/p_2\right)} = \frac{p_1\, p_2}{\left(s - p_1\right)\left(s - p_2\right)} \tag{4.188}$$

Equating $H(s)$ in Eqn. (4.188) with the form given by Eqn. (4.185) we obtain the relationships

$$\omega_0^2 = p_1\, p_2\,, \quad \text{and} \quad 2\zeta\omega_0 = -\,\mathrm{Re}\left\{p_1\right\} - \mathrm{Re}\left\{p_2\right\} \tag{4.189}$$

The parameter $\omega_0 = \sqrt{p_1\, p_2}$ is called the *natural undamped frequency* of the system. The parameter ζ is called the *damping ratio* and is computed as

$$\zeta = \frac{-\,\mathrm{Re}\left\{p_1\right\} - \mathrm{Re}\left\{p_2\right\}}{2\sqrt{p_1\, p_2}} \tag{4.190}$$

The poles of the system function can be related to the parameters ω_0 and ζ as

$$p_{1,2} = -\zeta\omega_0 \pm \omega_0\sqrt{\zeta^2 - 1} \tag{4.191}$$

Since the system is causal and stable, both poles are in the left half of the s-plane. As a result the numerator of Eqn. (4.190) is positive, and therefore the damping ratio ζ must be positive. Locations of the poles depend on the value of ζ. We will observe three distinct possibilities:

Second-order system:

$\zeta > 1$:

The poles p_1 and p_2 are real-valued and distinct. The system is said to be *overdamped.*

$\zeta = 1$:

The expression in square root in Eqn. (4.191) equals zero. The two poles of the system function are both at $p_1 = p_2 = -\zeta\omega_0$. In this case the system is said to be *critically damped.*

$\zeta < 1$:

The expression in square-root in Eqn. (4.191) is negative. The two poles of the system function are a complex conjugate pair:

$$p_{1,2} = -\zeta\omega_0 \pm j\omega_0\sqrt{1-\zeta^2} = -\zeta\omega_0 \pm j\omega_d$$

In this case the system is said to be *underdamped*.

Fig. 4.63 illustrates the placement of the two poles based on the values of the parameters ζ and ω_0.

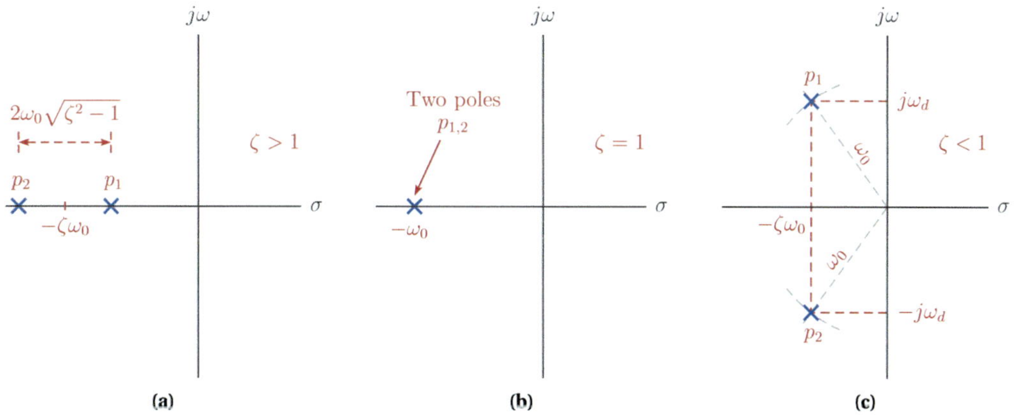

Figure 4.63 – Poles of the second-order system function for **(a)** $\zeta > 1$, overdamped, **(b)** $\zeta = 1$, critically damped, and **(c)** $\zeta < 1$, underdamped.

It is also interesting to look at the movement of the two poles as the parameter ζ is changed. Fig. 4.64 illustrates the trajectories of p_1 and p_2.

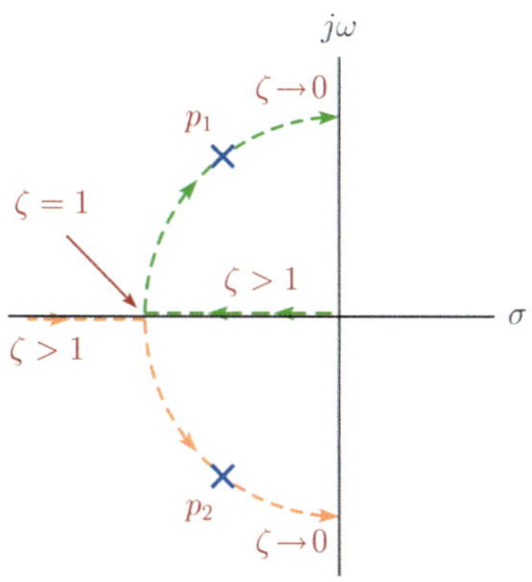

Figure 4.64 – Trajectories of the two poles p_1 and p_2 as ζ is varied.

Let us determine the magnitude and phase responses of the system. Substituting $s = j\omega$ into Eqn. (4.185) yields

$$H(\omega) = \frac{\omega_0^2}{(j\omega)^2 + 2\zeta\omega_0\, j\omega + \omega_0^2} = \frac{1}{1 - \left(\frac{\omega}{\omega_0}\right)^2 + j2\zeta\left(\frac{\omega}{\omega_0}\right)} \tag{4.192}$$

The dB magnitude of the system function is

$$20\,\log_{10}\left|H(\omega)\right| = -10\,\log_{10}\left\{\left[1 - \left(\frac{\omega}{\omega_0}\right)^2\right]^2 + \left[2\zeta\left(\frac{\omega}{\omega_0}\right)\right]^2\right\} \tag{4.193}$$

and the phase characteristic is

$$\measuredangle H(\omega) = -\tan^{-1}\left[\frac{2\zeta\left(\frac{\omega}{\omega_0}\right)}{1 - \left(\frac{\omega}{\omega_0}\right)^2}\right] \tag{4.194}$$

If $\omega << \omega_0$, then we have

$$20\,\log_{10}\left|H(\omega)\right| \approx 0 \quad \text{and} \quad \measuredangle H(\omega) \approx 0.$$

If $\omega >> \omega_0$, the dB magnitude becomes

$$20\,\log_{10}\left|H(\omega)\right| \approx -10\,\log_{10}\left(\frac{\omega}{\omega_0}\right)^4 = -40\,\log_{10}(\omega) + 40\,\log_{10}(\omega)$$

and the phase becomes

$$\measuredangle H(\omega) \approx -\tan^{-1}\left[2\zeta\left(\frac{\omega_0}{\omega}\right)\right] = -180°.$$

It will also be interesting to check the magnitude and the phase at the corner frequency. Substituting $\omega = \omega_0$ into Eqn. (4.192) we obtain

$$H(\omega_0) = \frac{1}{j2\zeta}$$

In this case the phase angle is $\measuredangle H(\omega) = -90$ degrees. Before we compute the dB magnitude at the corner frequency we will define the *quality factor* as

$$Q = \frac{1}{2\zeta} \tag{4.195}$$

so that $H(\omega_0) = -jQ$. The dB magnitude at $\omega = \omega_0$ is

$$20\,\log_{10}\left|H(\omega_0)\right| = 20\,\log_{10} Q$$

The conclusions obtained above can be summarized as follows

Conclusions for the second-order system function of Eqn. (4.185):

1. Magnitude: For $\omega << \omega_0$ the magnitude characteristic is asymptotic to 0 dB. For $\omega >> \omega_0$ it becomes asymptotic to a straight line with a slope of −40 dB per decade, which intersects the 0-dB axis at ω_0. At the corner frequency $\omega = \omega_0$ the actual magnitude is $20 \log_{10} Q = -20 \log_{10} (2\zeta)$.
2. Phase: For $\omega << \omega_0$ the phase is asymptotic to 0 degrees. For $\omega >> \omega_0$ it becomes asymptotic to −180 degrees. At the corner frequency $\omega = \omega_0$ the phase is −90 degrees.

The asymptotic behavior of the dB magnitude characteristic is illustrated in Fig. 4.65 for $\omega_0 = 3$ rad/s and for two different values of ζ, namely $\zeta = 0.1$ and $\zeta = 1.2$.

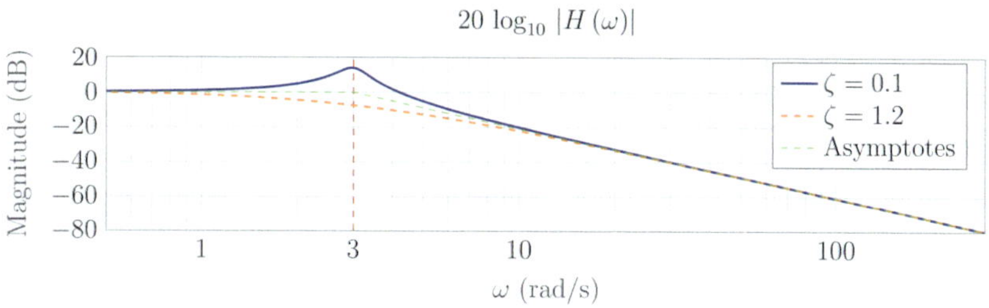

Figure 4.65 – The dB magnitude for the second-order system of Eqn. (4.185) for $\zeta = 0.1$ and $\zeta = 1.2$.

The definitions of overdamped, critically damped and underdamped systems can be related to the new parameter Q as follows:

$$\begin{aligned} \text{Overdamped:} \quad & \zeta > 1 \implies Q < 0.5 \\ \text{Critically damped:} \quad & \zeta = 1 \implies Q = 0.5 \\ \text{Underdamped:} \quad & \zeta < 1 \implies Q > 0.5 \end{aligned}$$

Bode plots for dB magnitude and phase of the second-order system are shown in Fig. 4.66 for several values of Q.

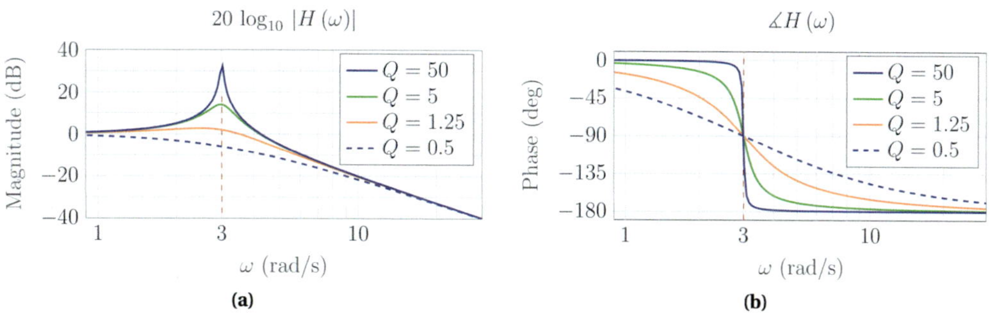

Figure 4.66 – dB Magnitude and phase characteristics of the second-order system of Eqn. (4.185) for $\omega_0 = 3$ rad/s and several values of Q.

The responses of the second-order system to unit impulse and unit step input signals are also of interest. Let us use partial fraction expansion to find the impulse response $h(t)$. For the case of two distinct poles (real or complex) the system function $H(s)$ in partial fraction form is

$$H(s) = \frac{k_1}{s-p_1} + \frac{k2}{s-p_2} \tag{4.196}$$

The residues in Eqn. (4.196) are found as

$$k_1 = \frac{p_1 p_2}{p_1 - p_2} = \frac{\omega_0}{2\sqrt{\zeta^2-1}} \quad \text{and} \quad k_2 = \frac{p_1 p_2}{p_2 - p_1} = -\frac{\omega_0}{2\sqrt{\zeta^2-1}}$$

and the impulse response is

$$\begin{aligned} h(t) =& k_1\, e^{p_1 t} + k_2\, e^{p_1 t} \\ =& \frac{\omega_0}{2\sqrt{\zeta^2-1}}\, e^{-\zeta\omega_0 t} \left[e^{\left(\omega_0\sqrt{\zeta^2-1}\right)t} - e^{-\left(\omega_0\sqrt{\zeta^2-1}\right)t} \right] u(t) \end{aligned} \tag{4.197}$$

If $\zeta < 1$, the result in Eqn. (4.197) becomes

$$h(t) = \frac{\omega_0}{\sqrt{1-\zeta^2}}\, e^{-\zeta\omega_0 t} \sin\left(\omega_0\sqrt{1-\zeta^2}\,t\right) u(t) = \frac{\omega_0}{\sqrt{1-\zeta^2}} \sin(\omega_d t)\, u(t) \tag{4.198}$$

If $\zeta = 1$ then the expression in Eqn. (4.197) or the one in Eqn. (4.198) will not work since $p_1 = p_2$, and the partial fraction expansion in Eqn. (4.196) is not valid in this case. For $\zeta = 1$ we have

$$H(s) = \frac{\omega_0}{(s+\omega_0)^2}$$

Using the s-domain differentiation property of the Laplace transform it can be shown that

$$h(t) = \omega_0^2\, t\, e^{-\omega_0 t}\, u(t) \tag{4.199}$$

The unit step response can be found by convolving the impulse response found with the unit step input signal:

$$\text{Sys}\{u(t)\} = h(t) * u(t) = \int_0^t h(\lambda)\, d\lambda\,, \qquad t \geq 0 \tag{4.200}$$

For $\zeta \neq 1$ the use of Eqn. (4.197) in Eqn. (4.200) yields

$$\begin{aligned} \text{Sys}\{u(t)\} =& 1 + \frac{1}{p_1 - p_2}\left[p_2\, e^{p_1 t} - p_1\, e^{p_2 t}\right] u(t) \\ =& 1 + \frac{e^{-\omega_0\zeta t}}{2\omega_0\sqrt{1-\zeta^2}} \left[\left(-\omega_0\zeta - \omega_0\sqrt{1-\zeta^2}\right) e^{\left(\omega_0\sqrt{1-\zeta^2}\right)t} + \left(\omega_0\zeta - \omega_0\sqrt{1-\zeta^2}\right) e^{\left(-\omega_0\sqrt{1-\zeta^2}\right)t}\right] u(t) \end{aligned} \tag{4.201}$$

If $\zeta = 1$, the unit step response is found by integrating the result Eqn. (4.199):

$$\text{Sys}\{u(t)\} = \left[1 - e^{-\omega_0 t} - \omega_0\, t\, e^{-\omega_0 t}\right] u(t) \tag{4.202}$$

Impulse response of the system is shown in Fig. 4.67(a) for several values of the damping ratio ζ. Fig. 4.67(b) shows the unit step response of the system for the same set of values for ζ.

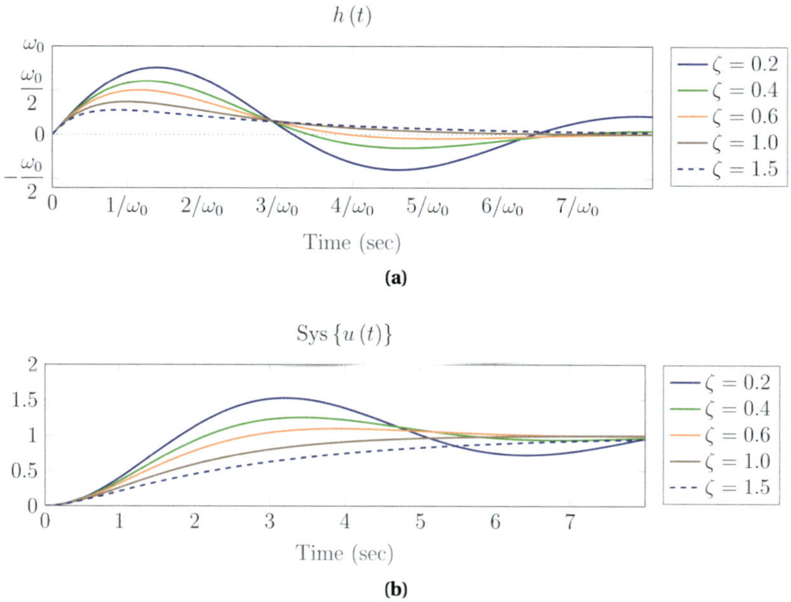

Figure 4.67 – **(a)** Impulse responses of second-order systems with varying damping ratios **(b)** unit step responses of second-order systems with varying damping ratios.

4.6 Simulation Structures for CTLTI Systems

Development of simulation structures for continuous-time systems was discussed briefly in Section 2.6 of Chapter 2 in the context of obtaining a block diagram from a differential equation. Our discussion in this section will parallel that of Section 2.6 and will utilize the s-domain system function as the starting point.

4.6.1 Direct-form implementation

The method of obtaining a block diagram from an s-domain system function will be derived using a third-order system, but its generalization to higher-order system functions is quite straightforward.

Consider a CTLTI system described by a system function $H(s)$.

$$H(s) = \frac{Y(s)}{X(s)} = \frac{b_2 s^2 + b_1 s + b_0}{s^3 + a_2 s^2 + a_1 s + a_0} \tag{4.203}$$

$X(s)$ and $Y(s)$ are the Laplace transforms of the input and the output signals, respectively. Let us use an intermediate function $W(s)$ and express the system function as

$$H(s) = \frac{Y(s)}{W(s)}\,\frac{W(s)}{X(s)} = \frac{b_2 s^{-1} + b_1 s^{-2} + b_0 s^{-3}}{1 + a_2 s^{-1} + a_1 s^{-2} + a_0 s^{-3}} \tag{4.204}$$

where we have also multiplied both the numerator and the denominator of the system function with s^{-3} to ensure that no positive powers of s appear. The relationships described by Eqns. (4.203) and (4.204) are illustrated in Fig. 4.68.

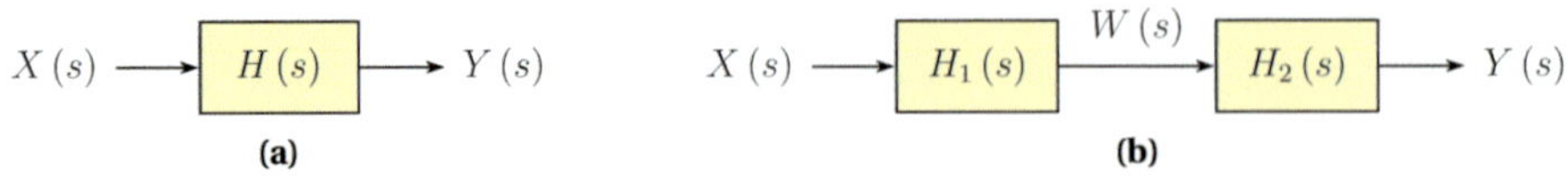

Figure 4.68 – **(a)** CTLTI system with system function $H(s)$ and **(b)** cascade form using an intermediate function $W(s)$.

If we choose to associate $W(s)/X(s)$ with the denominator of Eqn. (4.204), that is,

$$H_1(s) = \frac{W(s)}{X(s)} = \frac{1}{1 + a_2 s^{-1} + a_1 s^{-2} + a_0 s^{-3}} \tag{4.205}$$

then we have

$$H_2(s) = \frac{Y(s)}{W(s)} = b_2 s^{-1} + b_1 s^{-2} + b_0 s^{-3} \tag{4.206}$$

to satisfy Eqn. (4.204). Rearranging the terms in Eqn. (4.205) yields

$$W(s) = X(s) - a_2 s^{-1} W(s) - a_1 s^{-2} W(s) - a_0 s^{-3} W(s) \tag{4.207}$$

The relationship in Eqn. (4.207) can easily be translated to a simulation diagram as shown in Fig. 4.69.

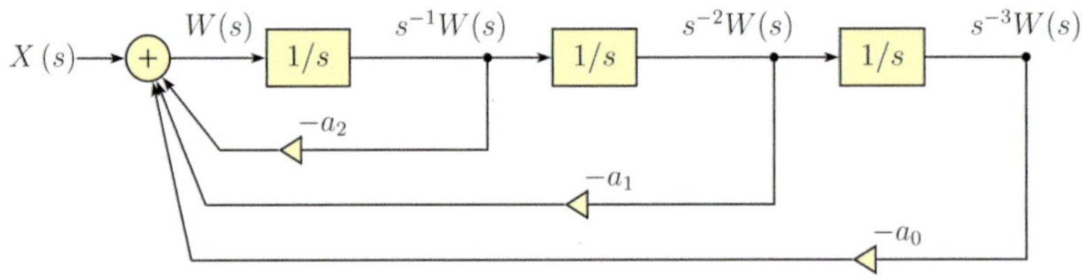

Figure 4.69 – Block diagram implementation of Eqn. (4.207).

Solving Eqn. (4.206) for the output transform $Y(s)$ yields

$$Y(s) = b_2 s^{-1} W(s) + b_1 s^{-2} W(s) + b_0 s^{-3} W(s) \tag{4.208}$$

The terms $s^{-1} W(s)$, $s^{-2} W(s)$, and $s^{-3} W(s)$ are already available in the simulation diagram of Fig. 4.69. Utilizing them to implement the relationship described by Eqn. (4.208), the diagram can be completed as shown in Fig. 4.70.

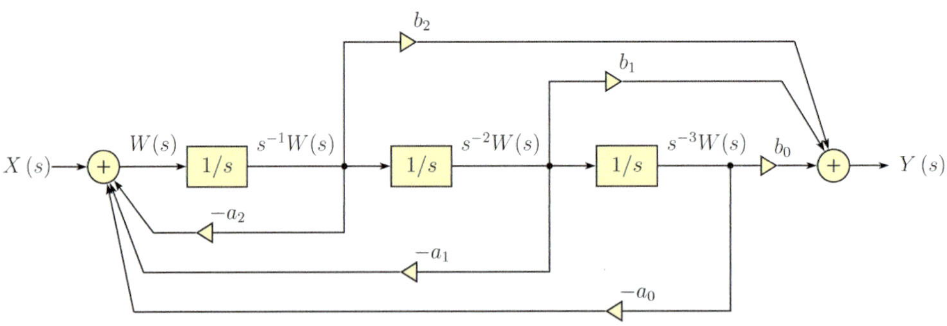

Figure 4.70 – Completed block diagram for simulating the system function $H(s)$ of Eqn. (4.203).

Using the integration property of the Laplace transform stated by Eqn. (4.67), multiplication by s^{-1} in the transform domain corresponds to integration of the signal in the time domain as illustrated in Fig. 4.71. Consequently, the blocks with system function $1/s$ in the diagram of Fig 4.70 represent integrators.

Figure 4.71 – Integrator component **(a)** in the time domain and **(b)** in the transform domain.

It should be noted that the diagram in Fig. 4.70 could easily have been obtained directly from the system function in Eqn. (4.204) by inspection, using the following set of rules:

Drawing a block diagram from the system function:

1. Begin by ordering terms of numerator and denominator polynomials from highest to lowest order of s.
2. Ensure that the leading coefficient in the denominator, that is, the coefficient of the highest order term, is equal to unity. (If it is not equal to unity, simply scale all coefficients to satisfy this rule.)
3. Set gain factors of feed-forward branches equal to the numerator coefficients.
4. Set gain factors of feedback branches equal to the negatives of the denominator coefficients.
5. Be careful to account for any missing powers of s in either polynomial and treat them as terms with their coefficients equal to zero.

Example 4.42: Obtaining a block diagram from system function

A CTLTI system is described through the system function

$$H(s) = \frac{2s^3 - 26s + 24}{s^4 + 7s^3 + 21s^2 + 37s + 30}$$

Draw a block diagram for simulating this system.

Solution: Using the technique outlined above, the block diagram shown in Fig. 4.72 is obtained.

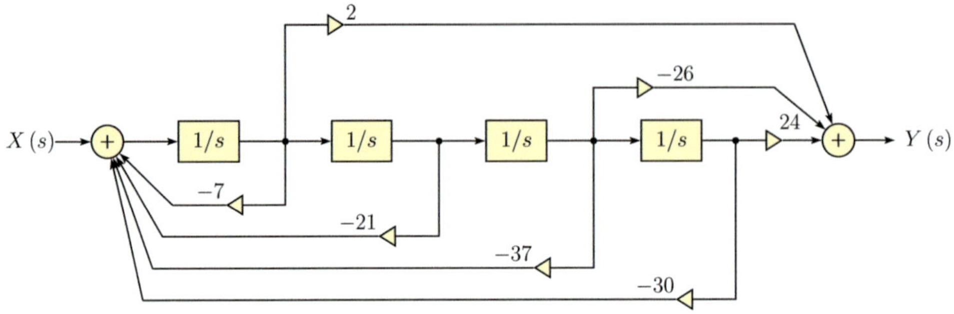

Figure 4.72 – Block diagram for Example 4.42.

4.6.2 Cascade and parallel forms

Instead of simulating a system with the direct-form block diagram discussed in the previous section, it is also possible to express the system function as either the product or the sum of lower order sections, and base the block diagram on cascade or parallel combination of smaller diagrams.

Consider a system function of order M that can be expressed in the form

$$\begin{aligned} H(s) =& H_1(s)\, H_2(s) \ldots H_M(s) \\ =& \frac{W_1(s)}{X(s)} \frac{W_2(s)}{W_1(s)} \cdots \frac{Y(s)}{W_{M-1}(s)} \end{aligned} \tag{4.209}$$

One method of simulating this system would be to build a diagram for each of the subsections $H_i(s)$ using the direct-form approach discussed previously, and then to connect those sections in cascade as shown in Fig. 4.73.

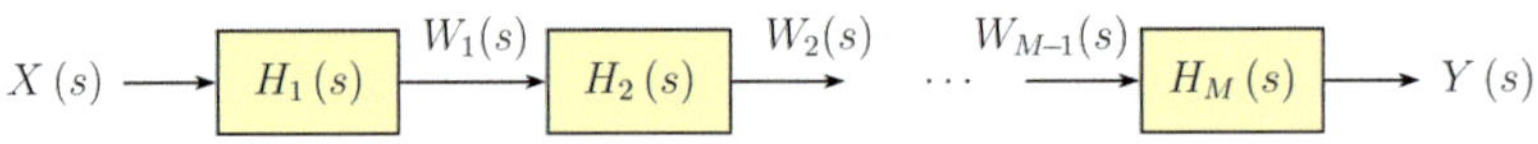

Figure 4.73 – Cascade implementation of $H(s)$.

An easy method of sectioning a system function in the style of Eqn. (4.209) would be to determine the poles and the zeros of the system function, and to use them for factoring numerator and denominator polynomials. Afterward, each section may be constructed by using one of the poles. Each zero is incorporated into one of the sections and some sections may have constant numerators. If some poles and zeros are complex-valued, we may choose to keep conjugate pairs together in second-order sections to avoid the need for complex gain factors in the diagram. The next example will illustrate this process.

Example 4.43: Cascade form block diagram

Develop a cascade form block diagram for simulating the system used in Example 4.42.

Solution: The system function specified in Example 4.42 can be factored into the form

$$H(s) = \frac{2\,(s+4)\,(s-3)\,(s-1)}{\left(s+1-j2\right)\left(s+1+j2\right)(s+3)\,(s+2)}$$

The roots can be found easily using MATLAB (see MATLAB Exercise 4.4). Let us write $H(s)$ as

$$H(s) = H_1(s)\,H_2(s)\,H_3(s)$$

by choosing

$$H_1(s) = \frac{2\,(s+4)}{\left(s+1-j2\right)\left(s+1+j2\right)} = \frac{2s+8}{s^2+2s+5}\,, \qquad H_2(s) = \frac{s-3}{s+3}\,, \qquad H_3(s) = \frac{s-1}{s+2}$$

The cascade form simulation diagram is shown in Fig. 4.74.

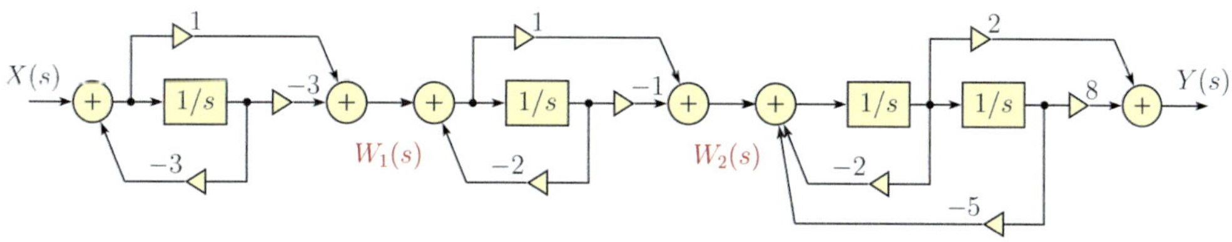

Figure 4.74 – Cascade form block diagram for Example 4.43.

It is also possible to consolidate the neighboring adders although this would cause the intermediate signals $W_1(s)$ and $W_2(s)$ to be lost. The resulting diagram is shown in Fig. 4.75.

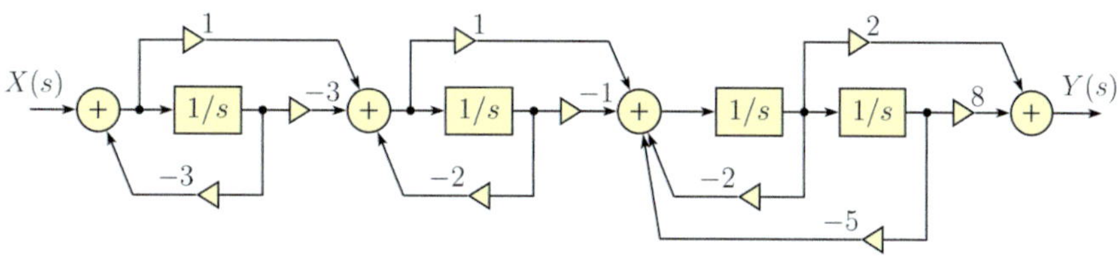

Figure 4.75 – Further simplified cascade form block diagram for Example 4.43.

An alternative to the cascade form simulation diagram is a parallel form diagram which is based on writing the system function as a sum of lower-order functions:

$$\begin{aligned} H(s) &= \bar{H}_1(s) + \bar{H}_2(s) + \ldots + \bar{H}_M(s) \\ &= \frac{\bar{W}_1(s)}{X(s)} + \frac{\bar{W}_2(s)}{X(s)} + \ldots \frac{\bar{W}_M(s)}{X(s)} \end{aligned} \tag{4.210}$$

A simulation diagram can be constructed by implementing each term in Eqn. (4.210) using the direct-form approach, and then connecting the resulting subsystems in a parallel configuration as shown in Fig. 4.76.

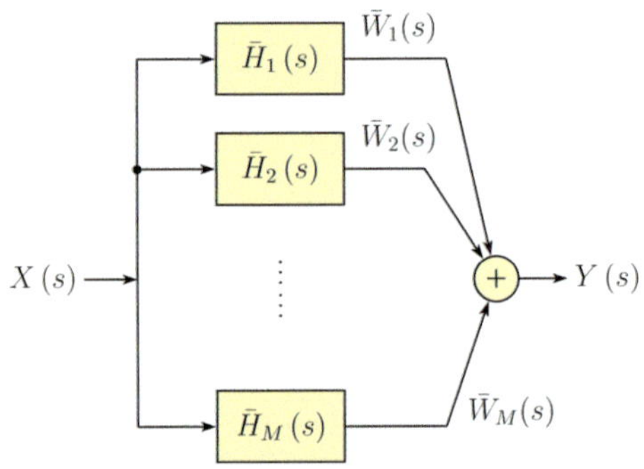

Figure 4.76 – Parallel implementation of $H(s)$.

A rational system function $H(s)$ can be sectioned in the form of Eqn. (4.210) using partial fraction expansion. If some poles and zeros are complex-valued, we may choose to keep conjugate pairs together in second-order sections to avoid the need for complex gain factors in the diagram. This process will be illustrated in the next example.

Example 4.44: Parallel form block diagram

Develop a parallel form block diagram for simulating the system used in Example 4.42.

Solution: The system function specified in Example 4.42 can be expanded into partial fractions as

$$H(s) = \frac{-2+j3}{s+1-j2} + \frac{-2-j3}{s+1+j2} + \frac{12}{s+2} + \frac{-6}{s+3}$$

Since the first two terms have complex poles and complex conjugate residues, we will combine them back into a second-order section to avoid the need for complex gain factors in the diagram. This results in

$$H(s) = \bar{H}_1(s) + \bar{H}_2(s) + \bar{H}_3(s)$$

with

$$\bar{H}_1(s) = \frac{-2+j3}{s+1-j2} + \frac{-2-j3}{s+1+j2} = \frac{-4s-16}{s^2+2s+5}, \qquad \bar{H}_2(s) = \frac{12}{s+2}, \qquad \bar{H}_3(s) = \frac{-6}{s+3}$$

The parallel form simulation diagram is shown in Fig. 4.77.

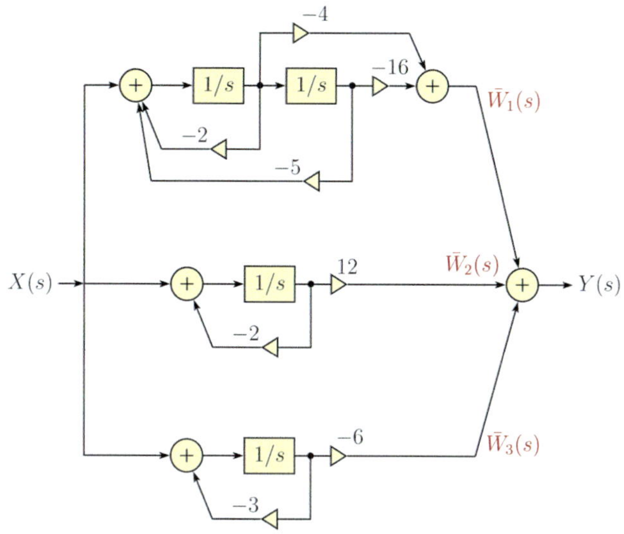

Figure 4.77 – Parallel form block diagram for Example 4.44.

4.7 Unilateral Laplace Transform

It was mentioned in earlier discussion that the Laplace transform as defined by Eqn. (4.1) is sometimes referred to as the *bilateral Laplace transform*. In this section we will discuss an alternative version of the Laplace transform, known as the *unilateral Laplace transform*.

Unilateral Laplace transform:

$$X_u(s) = \mathscr{L}_u\{x(t)\} = \int_{0^+}^{\infty} x(t)\, e^{-st}\, dt \qquad (4.211)$$

We use the subscript "u" to distinguish the unilateral Laplace transform from its bilateral counterpart. Comparing the definition in Eqn. (4.211) with the definition of the bilateral transform given by Eqn. (4.1) it is clear that the only difference is the lower limit of the integral. In fact, if the signal $x(t)$ is causal, that is, if $x(t) = 0$ for $t < 0$, then both definitions of the Laplace transform produce the same result.

$$X_u(s) = X(s) \quad \text{if} \quad x(t): \text{causal} \qquad (4.212)$$

In many engineering applications, we work with causal signals and may not need to pay attention to which Laplace transform definition we use. On the other hand, for a non-causal signal, the two definitions produce different results. Consider a signal $x(t)$ that is non-causal. The unilateral

Laplace transform of $x(t)$ is

$$\mathcal{L}_u\{x(t)\} = \int_{0^+}^{\infty} x(t)\, e^{-st}\, dt$$

$$= \int_{-\infty}^{\infty} x(t)\, u(t)\, e^{-st}\, dt = \mathcal{L}\{x(t)\, u(t)\} \tag{4.213}$$

Thus, the unilateral Laplace transform of a signal $x(t)$ is the same as the bilateral Laplace transform of the signal $[x(t)\, u(t)]$.

Because of the way $X_u(s)$ is defined, its region of convergence is always to the right of a vertical line in the s-plane and does not have to be explicitly stated. The ambiguity that we have observed with the bilateral Laplace transform in Eqns. (4.16) and (4.17) does not exist with the unilateral transform.

Most of the properties discussed in Section 4.3 for the bilateral Laplace transform apply to the unilateral Laplace transform as well. A few of the properties need to be modified and a few new ones need to be introduced. These will be discussed briefly.

4.7.1 Time shifting property

The use of the time shifting property with the unilateral Laplace transform requires special care. Recall the time shifting property derived in Eqn. (4.36) and repeated here:

$$\text{Eqn. (4.36):} \qquad \mathcal{L}\{x(t-\tau)\} = e^{-s\tau}\, \mathcal{L}\{x(t)\}$$

Using Eqn. (4.213), the unilateral Laplace transform of $x(t-\tau)$ is

$$\mathcal{L}_u\{x(t-\tau)\} = \mathcal{L}\{x(t-\tau)\, u(t)\} \tag{4.214}$$

Correspondingly, the time shifting property will work for the unilateral Laplace transform only if the shift by τ does not cause any signal components to move from the negative time territory to positive time territory or vice versa. Mathematically we have the following:

Time shifting for the unilateral Laplace transform:

Given the transform pair

$$x(t) \stackrel{\mathcal{L}_u}{\longleftrightarrow} X_u(s),$$

it can be shown that

$$x(t-\tau) \stackrel{\mathcal{L}_u}{\longleftrightarrow} e^{-s\tau}\, X_u(s) \tag{4.215}$$

provided that

$$x(t-\tau)\, u(t) = x(t-\tau)\, u(t-\tau)\,, \quad \text{all } t.$$

Fig. 4.78 depicts a signal $x(t)$ and three of its shifted versions. It can be shown that the time shifting property in Eqn. (4.215) holds for $x_1(t)$ and $x_2(t)$, but not for $x_3(t)$.

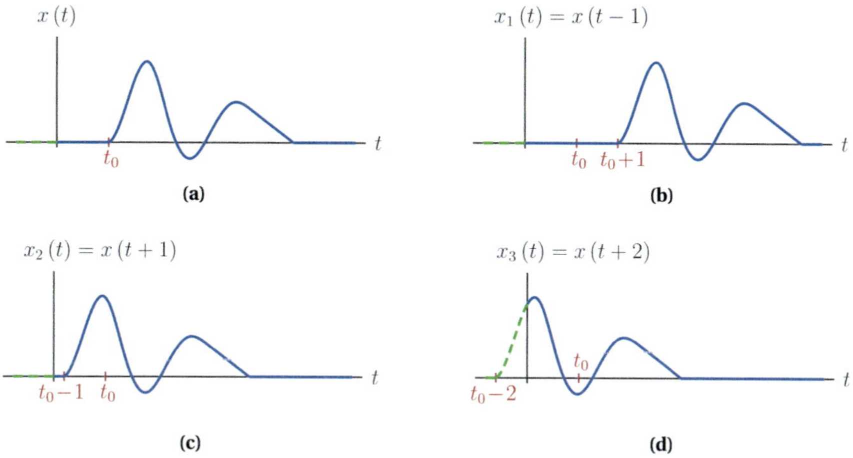

Figure 4.78 – **(a)** The signal $x(t)$, **(b)** the signal $x_1(t) = x(t-1)$, **(c)** the signal $x_2(t) = x(t+1)$, and **(d)** the signal $x_3(t) = x(t+2)$.

In contrast, the time shifting property does not hold for any amount of shift of the signal $x(t)$ in Fig. 4.79 due to signal components that cross from left of the vertical axis to the right or vice versa.

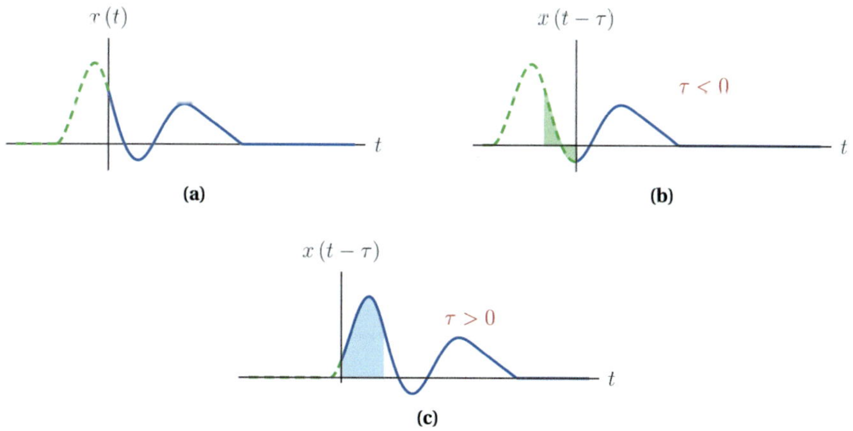

Figure 4.79 – **(a)** The signal $x(t)$, **(b)** the signal $x_1(t) = x(t-1)$, and **(c)** the signal $x_2(t) = x(t+1)$.

4.7.2 Differentiation in time

Differentiation in time property works a bit differently for the unilateral Laplace transform compared to its bilateral counterpart.

Differentiation in time for the unilateral Laplace transform:

Given the transform pair

$$x(t) \overset{\mathcal{L}_u}{\longleftrightarrow} X_u(s),$$

the following is also a valid transform pair:

$$\frac{dx(t)}{dt} \overset{\mathcal{L}_u}{\longleftrightarrow} s\,X_u(s) - x(0^+) \tag{4.216}$$

This property will be very important in the use of unilateral Laplace transform for solving differential equations with initial conditions. The proof is similar to the proof of the corresponding property for the bilateral Laplace transform and will be given here.

Proof of Eqn. (4.216):

Using the Laplace transform definition in Eqn. (4.1), the transform of $dx(t)/dt$ is

$$\begin{aligned}\mathcal{L}_u\left\{\frac{dx(t)}{dt}\right\} &= \int_{0^+}^{\infty} \frac{dx(t)}{dt}\, e^{-st}\, dt \\ &= \int_{0^+}^{\infty} e^{-st}\, d\left[x(t)\right]\end{aligned} \tag{4.217}$$

Integrating Eqn. (4.217) by parts yields

$$\mathcal{L}_u\left\{\frac{dx(t)}{dt}\right\} = x(t)\,e^{-st}\Big|_{0^+}^{\infty} + s\int_{0^+}^{\infty} x(t)\,e^{-st}\,dt \tag{4.218}$$

The term $x(t)\,e^{-st}$ must evaluate to zero for $t\to\infty$ for the transform $X_u(s)$ to exist. Therefore, Eqn. (4.218) reduces to

$$\mathcal{L}_u\left\{\frac{dx(t)}{dt}\right\} = -x(0^+) + s\int_{0^+}^{\infty} x(t)\,e^{-st}\,dt = -x(0^+) + s\,X_u(s) \tag{4.219}$$

to complete the proof.

The unilateral Laplace transforms of higher-order derivatives can be found through repeated use of Eqn. (4.216). For example,

$$\begin{aligned}\mathcal{L}_u\left\{\frac{dx^2(t)}{dt^2}\right\} &= \mathcal{L}_u\left\{\frac{d}{dt}\left(\frac{dx(t)}{dt}\right)\right\} \\ &= s\,\mathcal{L}_u\left\{\frac{dx(t)}{st}\right\} - \frac{dx(t)}{dt}\bigg|_{t=0^+}\end{aligned} \tag{4.220}$$

Using Eqn. (4.216) in Eqn. (4.220) yields

$$\begin{aligned}\mathcal{L}_u\left\{\frac{dx^2(t)}{dt^2}\right\} &= s\left[s\,X_u(s) - x(0^+)\right] - \frac{dx(t)}{dt}\bigg|_{t=0^+} \\ &= s^2\,X_u(s) - s\,x(0^+) - \frac{dx(t)}{dt}\bigg|_{t=0^+}\end{aligned} \tag{4.221}$$

Similarly it can be shown that

$$\mathscr{L}_u\left\{\frac{dx^3(t)}{dt^3}\right\} = s^3\,X_u(s) - s^2\,x\left(0^+\right) - s\,\left.\frac{dx(t)}{dt}\right|_{t=0^+} - \left.\frac{d^2x(t)}{dt^2}\right|_{t=0^+} \tag{4.222}$$

and for the general case

$$\mathscr{L}_u\left\{\frac{dx^n(t)}{dt^n}\right\} = s^n\,X_u(s) - s^{n-1}\,x\left(0^+\right) - s^{n-2}\,\left.\frac{dx(t)}{dt}\right|_{t=0^+} - \ldots - \left.\frac{d^{n-1}x(t)}{dt^{n-1}}\right|_{t=0^+} \tag{4.223}$$

As mentioned before, the primary utility of the unilateral Laplace transform is in solving differential equations with specified initial conditions. The next couple of examples will illustrate this.

Example 4.45: Using Laplace transform to solve a differential equation

Consider the circuit shown in Fig. 4.80(a) driven by the pulse signal shown in Fig. 4.80(b). Determine the output signal $y(t)$ for $t>0$ subject to the initial condition $y\left(0^+\right) = -2$ V.

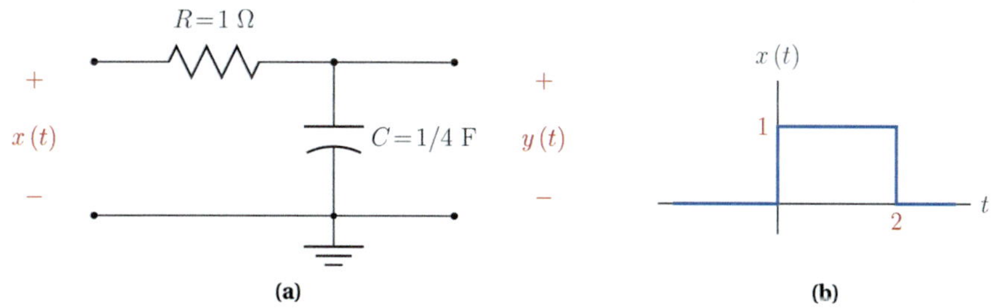

Figure 4.80 – **(a)** The RC circuit for Example 4.45 and **(b)** its input signal $x(t)$.

Solution: The system can be modeled with the following differential equation (see Example 2.9 in Chapter 2 for derivation):

$$\frac{dy(t)}{dt} + 4\,y(t) = 4\,x(t)$$

Taking the unilateral Laplace transform of each side of the differential equation we obtain

$$s\,Y_u(s) - y\left(0^+\right) + 4\,Y_u(s) = 4\,X_u(s) \tag{4.224}$$

The input signal $x(t)$ can be written as

$$x(t) = u(t) - u(t-2)$$

and its unilateral Laplace transform is

$$X_u(s) = \left(1 - e^{-2s}\right)\frac{1}{s} \tag{4.225}$$

Using Eqn. (4.225) in Eqn. (4.224) and substituting the initial value $y\left(0^+\right)$ leads to

$$Y_u(s) = \frac{-2}{s+4} + \left(1 - e^{-2s}\right)\frac{4}{s\,(s+4)}$$

which can be written in the form

$$Y_u(s) = \frac{-2s+4}{s(s+4)} - e^{-2s}\,\frac{4}{s(s+4)}$$

Let

$$Y_{1u}(s) = \frac{-2s+4}{s(s+4)} \quad \text{and} \quad Y_{2u}(s) = \frac{4}{s(s+4)}$$

so that

$$Y_u(s) = Y_{1u}(s) - e^{-2s}\,Y_{2u}(s)$$

and consequently

$$y(t) = y_1(t) - y_2(t-2)$$

Using partial fraction expansion, we can write

$$Y_{1u}(s) = \frac{1}{s} - \frac{3}{s+4} \quad \Longrightarrow \quad y_1(t) = \left(1 - 3\,e^{-4t}\right)u(t)$$

$$Y_{2u}(s) = \frac{1}{s} - \frac{1}{s+4} \quad \Longrightarrow \quad y_2(t) = \left(1 - e^{-4t}\right)u(t)$$

The output signal is

$$y(t) = \left(1 - 3\,e^{-4t}\right)u(t) - \left(1 - e^{-4(t-2)}\right)u(t-2)$$

and is shown in Fig. 4.81.

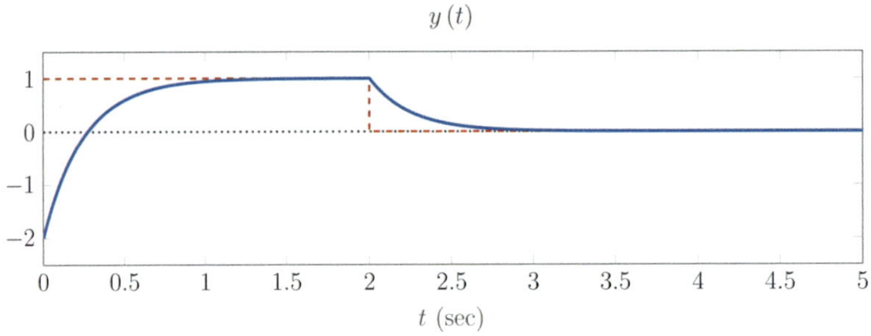

Figure 4.81 – The output signal $y(t)$ for Example 4.45.

Software resource: See MATLAB Exercise 4.11.

Example 4.46: Solving second-order differential equation using Laplace transform

Find the solution of the second-order homogeneous differential equation

$$\frac{d^2y(t)}{dt^2}+9\,y(t)=0$$

subject to initial conditions

$$y(0)=1\,,\quad \text{and}\quad \left.\frac{dy(t)}{dt}\right|_{t=0^+}=1$$

Solution: Taking the unilateral Laplace transform of the differential equation leads to

$$s^2\,Y_u(s)-s\,y\left(0^+\right)-\left.\frac{dy(t)}{dt}\right|_{t=0^+}+9\,Y_u(s)=0$$

Substituting initial conditions we obtain

$$s^2\,Y_u(s)-s-1+9\,Y_u(s)=0$$

which can be solved for $Y_u(s)$ to yield

$$Y_u(s)=\frac{s+1}{s^2+9}$$

The transform $Y_u(s)$ can be written in partial fraction form as

$$Y_u(s)=\frac{k_1}{s+j3}+\frac{k_2}{s-j3}$$

with residues

$$k_1=\frac{1}{2}+j\,\frac{1}{6}\quad\text{and}\quad k_2=\frac{1}{2}-j\,\frac{1}{6}$$

Therefore the solution is

$$\begin{aligned}y(t)&=\left(\frac{1}{2}+j\,\frac{1}{6}\right)e^{-j3t}+\left(\frac{1}{2}-j\,\frac{1}{6}\right)e^{j3t}\\&=\cos(3t)+\frac{1}{3}\,\sin(3t)\,,\qquad t\ge 0\end{aligned}$$

The solution $y(t)$ is shown in Fig. 4.82. It can easily be verified that $y(t)$ satisfies the differential equation and the specified initial conditions.

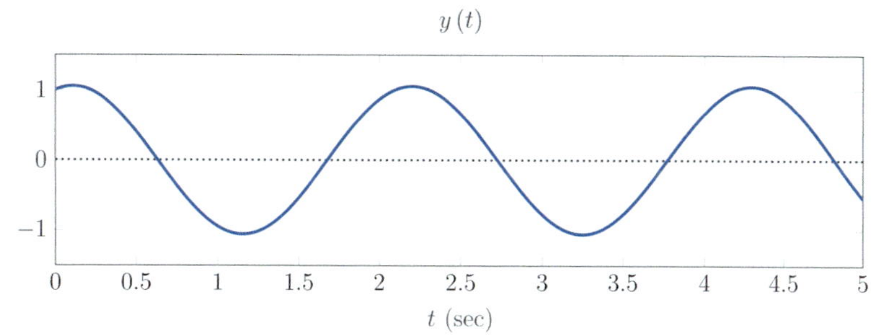

Figure 4.82 – The solution $y(t)$ for Example 4.46.

4.7.3 Initial and final value theorems

Initial value theorem:

Given the transform pair

$$x(t) \stackrel{\mathcal{L}_u}{\longleftrightarrow} X_u(s)\,,$$

we have

$$\lim_{t\to 0^+} x(t) = \lim_{s\to\infty} \left[s\,X_u(s) \right] \tag{4.226}$$

provided that the limit exists.

The initial value theorem does not apply to rational transforms in which the order of the numerator is equal to or greater than that of the denominator, since the limit in Eqn. (4.226) would not exist in that case.

Final value theorem:

Given the transform pair

$$x(t) \stackrel{\mathcal{L}_u}{\longleftrightarrow} X_u(s)$$

we have

$$\lim_{t\to\infty} x(t) = \lim_{s\to 0} \left[s\,X_u(s) \right] \tag{4.227}$$

The final value theorem does not apply to a transform $X_u(t)$ if the corresponding signal $x(t)$ is unbounded, or if it has undamped oscillations. An unbounded signal indicates the presence of poles in the right half s-plane. Undamped oscillations are associated with poles on the $j\omega$ axis of the s-plane other than at the origin $s=0$. Consequently, the final value theorem should not be used with transforms that have poles on the $j\omega$ axis or in the right half s-plane. A single pole at the origin $s=0$ is permissible.

Example 4.47: Application of initial and final value theorems

Consider the transform

$$X_u(s) = \frac{s}{s^2+4}$$

Can initial and final values be determined using initial and final value theorems?

Solution: Since the numerator order is lower than the denominator order, the initial value theorem can be used.

$$x(0^+) = \lim_{s\to\infty} \left[s\left(\frac{s}{s^2+4}\right) \right] = \lim_{s\to\infty} \left[\frac{s^2}{s^2} \right] = 1$$

The final value theorem does not apply since $X_u(s)$ has poles on the $j\omega$ axis at $s=\pm j2$. We may also recognize that $X_u(s)$ is the Laplace transform of the signal

$$x(t) = \cos(2t)\,u(t)$$

which does not have a final value due to the undamped oscillating term.

Example 4.48: Further exploring initial and final value theorems

Determine initial and final values of the signal with the Laplace transform

$$X_u(s) = \frac{2}{s\left(s^2+4s+13\right)}$$

Solution: The numerator order is lower than the denominator order; therefore, the initial value theorem can be used.

$$x\left(0^+\right) = \lim_{s\to\infty}\left[\frac{2}{s^2+4s+13}\right] = 0$$

The transform $X_u(s)$ has a pair of complex conjugate poles at $s = -2 \pm j3$ and a single pole at $s = 0$. Complex conjugate poles are in the left half s-plane. The single pole at the origin does not cause any problems in the use of the final value theorem since it is canceled by the factor s in Eqn. (4.227). The final value of $x(t)$ is

$$\lim_{t\to\infty}\left[x(t)\right] = \lim_{s\to 0}\left[\frac{2}{s^2+4s+13}\right] = \frac{2}{13}$$

Summary of Key Points

- The Laplace transform of a continuous-time signal $x(t)$ is defined as

  Eqn. (4.1): $$X(s) = \int_{-\infty}^{\infty} x(t)\, e^{-st}\, dt$$

  where s, the independent variable of the transform, is a complex variable.

- The Laplace transform of a signal $x(t)$ evaluated for $s = \sigma_1 + j\omega$ as the parameter ω is varied from $\omega = -\infty$ to $\omega = \infty$ is the same as the Fourier transform of the modified signal $x(t)\, e^{-\sigma_1 t}$.

  $$X(s)\Big|_{s=\sigma_1+j\omega} = \mathcal{F}\left\{x(t)\, e^{-\sigma_1 t}\right\}$$

- The Laplace transform of a signal $x(t)$ evaluated for $s = j\omega$ as the parameter ω is varied from $\omega = -\infty$ to $\omega = \infty$ is the same as the Fourier transform of the signal $x(t)$.

  $$X(s)\Big|_{s=j\omega} = \mathcal{F}\left\{x(t)\right\}$$

- The region of convergence is the region in the s plane where the Laplace transform converges. For the transform to exist, we need the signal $\left[x(t)\, e^{-\sigma t}\right]$ to be absolute integrable, that is,

  Eqn. (4.11): $$\int_{-\infty}^{\infty} \left|x(t)\, e^{-\sigma t}\right| dt < \infty$$

☞ The following are two fundamental transform pairs:

Eqn. (4.16): $e^{at}\,u(t) \stackrel{\mathcal{L}}{\longleftrightarrow} \dfrac{1}{s-a}\,,$ ROC: $\text{Re}\{s\} > \text{Re}\{a\}$

Eqn. (4.17): $-e^{at}\,u(-t) \stackrel{\mathcal{L}}{\longleftrightarrow} \dfrac{1}{s-a}\,,$ ROC: $\text{Re}\{s\} < \text{Re}\{a\}$

☞ Characteristics of the region of convergence:

1. For a finite-duration signal the ROC is the entire s-plane provided that the signal is absolute integrable. The extreme points such as $\text{Re}\{s\} \to \pm\infty$ need to be checked separately.
2. For a general signal the ROC is in the form of a vertical strip. It is either to the right of a vertical line, to the left of a vertical line, or between two vertical lines.
3. For a rational transform $X(s)$ the ROC cannot contain any poles.
4. The ROC for the Laplace transform of a causal signal is the region to the right of a vertical line expressed as $\text{Re}\{s\} > \sigma_1$.
5. The ROC for the Laplace transform of an anti-causal signal is the region to the left of a vertical line expressed as $\text{Re}\{s\} < \sigma_2$.
6. The ROC for the Laplace transform of a signal that is neither causal nor anti-causal is a strip between two vertical lines expressed as $\sigma_1 < \text{Re}\{s\} < \sigma_2$.

☞ Given the transform pair

$$x(t) \stackrel{\mathcal{L}}{\longleftrightarrow} X(s)\,,$$

the following are also valid transform pairs:

Time shifting:

Eqn. (4.36): $$x(t-\tau) \stackrel{\mathcal{L}}{\longleftrightarrow} e^{-s\tau}\,X(s)$$

Shifting in the s domain:

Eqn. (4.43): $$x(t)\,e^{s_0 t} \stackrel{\mathcal{L}}{\longleftrightarrow} X(s-s_0)$$

Scaling in time and s domains

Eqn. (4.44): $$x(at) \stackrel{\mathcal{L}}{\longleftrightarrow} \frac{1}{|a|}\,X\left(\frac{s}{a}\right)$$

Differentiation in time:

Eqn. (4.51): $$\frac{dx(t)}{dt} \stackrel{\mathcal{L}}{\longleftrightarrow} s\,X(s)$$

Differentiation in the s domain:

Eqn. (4.55): $$t\,x(t) \stackrel{\mathcal{L}}{\longleftrightarrow} -\frac{dX(s)}{ds}$$

Integration:

Eqn. (4.67): $$\int_{-\infty}^{t} x(\lambda)\,d\lambda \stackrel{\mathcal{L}}{\longleftrightarrow} \frac{1}{s}\,X(s)$$

☞ Given two transform pairs

$$x_1(t) \overset{\mathcal{L}}{\longleftrightarrow} X_1(s) \quad \text{and} \quad x_2(t) \overset{\mathcal{L}}{\longleftrightarrow} X_2(s)$$

the following are also valid transform pairs:

Linearity:

Eqn. (4.30): $$\alpha_1 x_1(t) + \alpha_2 x_2(t) \overset{\mathcal{L}}{\longleftrightarrow} \alpha_1 X_1(s) + \alpha_2 X_2(s) \qquad \text{for any } \alpha_1, \alpha_2$$

Convolution:

Eqn. (4.59): $$x_1(t) * x_2(t) \overset{\mathcal{L}}{\longleftrightarrow} X_1(s)\, X_2(s)$$

☞ One method of finding the signal $x(t)$ from the transform $X(s)$ is to use the contour integral

Eqn. (4.74): $$x(t) = \mathcal{L}^{-1}\{X(s)\} = \frac{1}{2\pi j}\int_C X(s)\, e^{st}\, ds$$

where the contour C is a vertical line within the ROC of the transform.

☞ A more practical method of finding the signal $x(t)$ from a rational transform

Eqn. (4.78): $$X(s) = \frac{B(s)}{(s-p_1)(s-p_2)\dots(s-p_N)}$$

is to use partial fraction expansion in the form

Eqn. (4.79): $$X(s) = \frac{k_1}{s-p_1} + \frac{k_2}{s-p_2} + \dots + \frac{k_N}{s-p_N}$$

The residues $k_1, k_2, \dots, k_N$ for single poles can be computed as

Eqn. (4.80): $$k_i = (s-p_i)\, X(s)\Big|_{s=p_i}, \qquad i = 1, 2, \dots, N$$

☞ Given the input signal $x(t)$ and the impulse response $h(t)$, the output signal of a CTLTI system can be determined using the following steps:

1. Find the Laplace transforms $X(s)$ and $H(s)$ of the input signal $x(t)$ and the impulse response $h(t)$ respectively.
2. Multiply the two transforms to obtain the transform $Y(s)$.
3. Determine the output signal $y(t)$ from $Y(s)$ by means of the inverse Laplace transform.

☞ Given the input signal $x(t)$ and the output signal $y(t)$, the impulse response of a CTLTI system can be determined using the following steps:

1. Find the Laplace transforms $X(s)$ and $Y(s)$ of input and output signals $x(t)$ and $y(t)$, respectively.
2. Find the transform $H(s)$ as the ratio of the two transforms, that is, $H(s) = Y(s)/X(s)$.
3. Use the inverse Laplace transform operation on $H(s)$ to find $h(t)$.

☞ The system function for a CTLTI system can be determined from its differential equation using the following steps:

1. Separate the terms of the differential equation so that $y(t)$ and its derivatives are on the left of the equal sign, and $x(t)$ and its derivatives are on the right of the equal sign.
2. Take the Laplace transform of each side of the differential equation and use the time differentiation property of the Laplace transform.
3. Determine the system function as the ratio of $Y(s)$ to $X(s)$.

In determining the system function from the differential equation, all initial conditions must be assumed zero.

☞ The response of a CTLTI system to the input signal $x(t) = e^{s_0 t}$ is

Eqn. (4.99): $$y(t) = \text{Sys}\left\{e^{s_0 t}\right\} = e^{s_0 t}\, H(s_0)$$

☞ The response of a CTLTI system to the exponentially damped sinusoidal signal $x(t) = e^{\sigma_0 t} \cos(\omega_0 t)$ is

Eqn. (4.111): $$H_0\, e^{\sigma_0 t} \cos(\omega_0 t + \theta_0)$$

where

$$H_0 = \left|H\left(\sigma_0 + j\omega_0\right)\right|, \quad \text{and} \quad \theta_0 = \measuredangle H\left(\sigma_0 + j\omega_0\right)$$

☞ In the s-domain system function of a causal CTLTI system the order of the numerator must not be greater than the order of the denominator. This condition is necessary for a system to be causal, but it is not sufficient.

☞ For a CTLTI system to be stable, the ROC of its s-domain system function must include the $j\omega$-axis.

1. For a causal system to be stable, its system function must not have any poles on the $j\omega$-axis or in the right half s-plane.
2. For an anti-causal system to be stable, its system function must not have any poles on the $j\omega$-axis or in the left half s-plane.

☞ A system with a constant magnitude characteristic across all frequencies is called an all-pass system. A first-order CTLTI system with a pole at $p_1 = -\sigma_1 + j0$ and a zero at $z_1 = \sigma_1 + j0$ has an all-pass magnitude characteristic. Higher-order all-pass systems can be constructed by cascading first-order all-pass sections.

☞ The inverse of a system is another system which, when connected in cascade with the original system, forms an identity system.

☞ A causal CTLTI system that has all of its zeros and poles in the left half s-plane is referred to as a minimum-phase system. A minimum-phase system and its inverse are both causal and stable.

☞ Bode plots allow the dB magnitude and the phase characteristics of the system function to be obtained from the knowledge of its poles and zeros. The use of a logarithmic scale for the magnitude characteristic allows contributions from individual cascaded sections to be added rather than multiplied.

☞ Unilateral Laplace transform of a signal $x(t)$ is defined as

Eqn. (4.211): $$X_u(s) = \mathcal{L}_u\{x(t)\} = \int_{0^+}^{\infty} x(t)\, e^{-st}\, dt$$

If the signal $x(t)$ is causal, then the unilateral Laplace transform is identical to the bilateral Laplace transform.

☞ The unilateral Laplace transform of a time-shifted signal is

Eqn. (4.211): $$x(t-\tau) \overset{\mathcal{L}_u}{\longleftrightarrow} e^{-s\tau}\, X_u(s)$$

provided that

$$x(t-\tau)\, u(t) = x(t-\tau)\, u(t-\tau)\,, \quad \text{all } t$$

☞ The unilateral Laplace transform of the time derivative of a signal is

Eqn. (4.216): $$\frac{dx(t)}{dt} \overset{\mathcal{L}_u}{\longleftrightarrow} s\,X_u(s) - x\left(0^+\right)$$

☞ Given the transform pair

$$x(t) \overset{\mathcal{L}_u}{\longleftrightarrow} X_u(s)$$

the initial value of the signal can be expressed as

Eqn. (4.226): $$\lim_{t\to 0^+} x(t) = \lim_{s\to\infty} \left[\, s\,X_u(s)\,\right]$$

provided that the limit exists. The final value of the signal is

Eqn. (4.227): $$\lim_{t\to\infty} x(t) = \lim_{s\to 0} \left[\, s\,X_u(s)\,\right]$$

The final value theorem does not apply to a transform $X_u(t)$ if the corresponding signal $x(t)$ is unbounded, or if it has undamped oscillations.

Further Reading

[1] J. Bak and D.J. Newman. *Complex analysis.* Undergraduate Texts in Mathematics. Springer, 2010.

[2] R.J. Beerends. *Fourier and Laplace Transforms.* Cambridge University Press, 2003.

[3] Frank M Callier and Charles A Desoer. *Linear system theory.* Springer Science & Business Media, 2012.

[4] Alan M Cohen. *Numerical Methods for Laplace Transform Inversion.* Vol. 5. Springer Science & Business Media, 2007.

[5] Gustav Doetsch. *Introduction to the Theory and Application of the Laplace Transformation.* Springer Science & Business Media, 2012.

[6] Phil Dyke. *An Introduction to Laplace Transforms and Fourier Series.* Springer Science & Business Media, 2014.

[7] Vladimir Eiderman. *An Introduction to Complex Analysis and the Laplace Transform.* CRC Press, 2021.

[8] Daniel Fleisch. *A Student's Guide to Laplace Transforms.* Cambridge University Press, 2022.

[9] W.R.L. Page. *Complex Variables and the Laplace Transform for Engineers.* Dover Books on Electrical Engineering Series. Dover Publications, 1980.

[10] J.L. Schiff. *The Laplace Transform: Theory and Applications.* Springer Undergraduate Texts in Mathematics and Technology. Springer, 1999.

[11] J.L. Taylor. *Complex Variables.* Pure and Applied Undergraduate Texts. American Mathematical Society, 2011.

MATLAB Exercises with Solutions

MATLAB Exercise 4.1: Three-dimensional plot of Laplace transform

In Fig. 4.3 the magnitude of the transform

$$\text{Eqn. (4.8):} \qquad X(s) = \frac{s+0.5}{(s+0.5)^2+4\pi^2}$$

was graphed as a three-dimensional surface. In this exercise we will reproduce that figure using MATLAB and display a cutout of the Laplace transform surface on the line $s = j\omega$. The first step is to produce a set of complex values of s on a rectangular grid in the s-plane.

```
>> [sr,si] = meshgrid([-6:0.3:6],[-15:0.5:15]);
>> s = sr+j*si;
```

The next step is to compute the magnitude of the transform at each point on the grid.

```
>> Xs = @(s) (s+0.5)./((s+0.5).^2+4*pi*pi);
>> XsMag = abs(Xs(s));
>> XsMag = XsMag.*(XsMag<=2)+2.*(XsMag>2);
```

The last line above facilitates clipping of $|X(s)|$ values greater than 2. Recall that the denominator of $X(s)$ becomes zero for $s = -1/2 \mp j2\pi$, and clipping is necessary for graphing purposes. A three-dimensional mesh plot of $|X(s)|$ can now be generated with the following lines:

```
>> meshz(sr,si,XsMag);
>> axis([-6,6,-15,15]);
```

The script `mexct_4_1a.m` listed below produces a mesh plot complete with axis labels and color specifications.

```
% Script: mexct_4_1a.m
[sr,si] = meshgrid([-6:0.15:6],[-15:0.25:15]);
s = sr+j*si;
Xs = @(s) (s+0.5)./((s+0.5).^2+4*pi*pi);  % Eqn. (4.8)
XsMag = abs(Xs(s));
XsMag = XsMag.*(XsMag<=2)+2.*(XsMag>2);
```

```
7  shading interp;                            % Interpolated shading
8  colormap copper;                           % Specify the color map used
9  m1 = meshz(sr,si,XsMag);                   % Draw mesh plot
10 axis([-6,6,-15,15]);
11 % Adjust transparency of surface lines
12 set(m1,'EdgeAlpha',0.4','FaceAlpha',0.4);
13 % Specify x,y,z axis labels
14 xlabel('\sigma');
15 ylabel('j\omega');
16 zlabel('|X(s)|');
17 % Specify viewing angles
18 view(gca,[23.5,38]);
```

In line 9 of the script, the handle returned by the function `meshz()` is assigned to the variable `m1` so that it can be used in line 12 for adjusting the transparency of the surface. Alternatively, a contour plot of $\left|X(s)\right|$ can be produced by slightly modifying the code. The script `mexct_4_1b.m` listed below gives a bird's eye view of the magnitude of the transform by plotting points that have the same magnitude value as contours.

```
% Script: mexct_4_1b.m
[sr,si] = meshgrid([-6:0.15:6],[-15:0.25:15]);
s = sr+j*si;
Xs = @(s) (s+0.5)./((s+0.5).^2+4*pi*pi);  % Eqn. (4.8)
XsMag = abs(Xs(s));
XsMag = XsMag.*(XsMag<=2)+2.*(XsMag>2);
colormap copper;                           % Specify the color map used
values = [[0:0.04:0.2],[0.3:0.1:2]];      % z value for each contour
m2 = contour(sr,si,XsMag,values); grid;
axis([-6,6,-15,15]);
% Specify x,y axis labels.
xlabel('\sigma');
ylabel('j\omega');
```

We know from the discussion in Section 4.1 that the Fourier transform $X(\omega)$ is equal to the Laplace transform evaluated on the $j\omega$ axis of the s-plane, that is,

$$\text{Eqn. (4.7):}\qquad X(s)\Big|_{s=0+j\omega} = \mathcal{F}\left\{x(t)\right\}$$

Applying this relationship to the magnitudes of the two transforms involved, $\left|\mathcal{F}\left\{x(t)\right\}\right|$ is obtained by evaluating $|X(s)|$ on the $j\omega$ axis. The script `mexct_4_1c.m` listed below demonstrates this.

```
% Script: mexct_4_1c.m
[sr,si] = meshgrid([-6:0.15:6],[-15:0.25:15]);
s = sr+j*si;
Xs = @(s) (s+0.5)./((s+0.5).^2+4*pi*pi);  % Eqn. (4.8)
XsMag = abs(Xs(s));
XsMag = XsMag.*(XsMag<=2)+2.*(XsMag>2);
% Define the trajectory s=j*omega
omega = [-15:0.01:15];
tr = j*omega;
shading interp;                            % Interpolated shading
colormap copper;                           % Specify the color map used
```

```
12 % Produce a mesh plot and hold it
13 m1 = meshz(sr,si,XsMag);
14 hold on;
15 % Superimpose a plot of X(s) magnitude values evaluated on the
16 % imaginary axis of the s-plane in blue
17 m2 = plot3(real(tr),imag(tr),abs(Xs(tr)),'b-','LineWidth',1);
18 % Superimpose a plot of the imaginary axis of the s-plane in red
19 m3 = plot3(real(tr),imag(tr),zeros(size(tr)),'r-','LineWidth',1);
20 hold off;
21 axis([-6,6,-15,15]);
22 % Adjust transparency of surface lines
23 set(m1,'EdgeAlpha',0.4','FaceAlpha',0.4);
24 % Specify x,y,z axis labels.
25 xlabel('\sigma');
26 ylabel('j\omega');
27 zlabel('|X(s)|');
28 % Specify viewing angles.
29 view(gca,[23.5,38]);
```

Lines 8 and 9 create a vector of s values on the $j\omega$ axis. Line 17 graphs the values of $|X(s)|$ along the $j\omega$ axis using the `plot3()` function. It is also possible, with a few changes in the code, to cut the Laplace transform surface along the $j\omega$ axis and display the profile of the cutout. The modified script `mexct_4_1d.m` to accomplish this is listed below. Notice how the magnitude values in the right half of the s-plane are suppressed in line 14 of the script.

```
1 % Script: mexct_4_1d.m
2 [sr,si] = meshgrid([-6:0.15:6],[-15:0.25:15]);
3 s = sr+j*si;
4 Xs = @(s) (s+0.5)./((s+0.5).^2+4*pi*pi);  % Eqn. (4.8)
5 XsMag = abs(Xs(s));
6 XsMag = XsMag.*(XsMag<=2)+2.*(XsMag>2);
7 % Define the trajectory s=j*omega
8 omega = [-15:0.01:15];
9 tr = j*omega;
10 % Produce a mesh plot and hold it
11 shading interp;
12 colormap copper;
13 % Set the surface equal to zero in the right half s-plane
14 XsMag = XsMag.*(sr<=0);
15 m1 = mesh(sr,si,XsMag);
16 hold on;
17 % Superimpose a plot of X(s) magnitude values evaluated on the
18 % imaginary axis of the s-plane in blue
19 plot3(real(tr),imag(tr),abs(Xs(tr)),'b-','LineWidth',1);
20 plot3(real(tr),imag(tr),zeros(size(tr)),'r-','LineWidth',1);
21 % Stem plot on the trajectory for a painted profile look
22 stem3(real(tr([1:25:3000])),imag(tr([1:25:3000])),..
23     abs(Xs(tr([1:25:3000]))),'g-','Marker','none');
24 hold off;
25 axis([-6,6,-15,15]);
26 % Adjust transparency of surface lines
27 set(m1,'EdgeAlpha',0.4','FaceAlpha',0.4);
28 % Specify x,y,z axis labels
29 xlabel('\sigma');
```

```
ylabel('j\omega');
zlabel('|X(s)|');
% Specify viewing angles
view(gca,[23.5,38]);
```

Software resources: mexct_4_1a.m, mexct_4_1b.m, mexct_4_1c.m, mexct_4_1d.m

MATLAB Exercise 4.2: Computing the Fourier transform from the Laplace transform

The Fourier transform of a signal is equal to its Laplace transform evaluated on the $j\omega$-axis of the s-plane.

$$\mathcal{F}\left\{x(t)\right\} = X(s)\Big|_{s=j\omega}$$

Consider the Laplace transform

$$X(s) = \frac{s+0.5}{(s+0.5)^2+4\pi^2}$$

The first method of computing and graphing the Fourier transform of the signal is to use an anonymous function for $X(s)$ and evaluate it on the $j\omega$-axis. The magnitude $\left|X(\omega)\right|$ is graphed using the following statements:

```
>> Xs = @(s) (s+0.5)./((s+0.5).^2+4*pi*pi);
>> omg = [-15:0.05:15];
>> Xomg = Xs(j*omg);
>> plot(omg,abs(Xomg)); grid;
```

If the phase $\measuredangle X(\omega)$ is needed, it can be graphed using

```
>> plot(omg,angle(Xomg)); grid;
```

In the above development care must be taken to ensure that $X(s)$ is computable at every point on the $j\omega$ axis of the s-plane, that is, there are no poles on the $j\omega$ axis. MATLAB Exercise 4.3 will illustrate what may happen when $X(s)$ has poles on the $j\omega$ axis.

An alternative method of computing the Fourier transform from the Laplace transform is to use the built-in MATLAB function `freqs()`. We will begin by writing $X(s)$ in rational form with numerator and denominator polynomials ordered in descending powers of s.

$$X(s) = \frac{s+0.5}{s^2+s+39.7284}$$

Vectors `num` and `den` to hold numerator and denominator coefficients should be entered as

```
>> num = [1,0.5];
>> den = [1,1,39.7284];
```

Afterward the magnitude and the phase of the Fourier transform may be computed and graphed with the statements

```
>> omg = [-15:0.05:15];
>> Xomg = freqs(num,den,omg);
>> plot(omg,abs(Xomg)); grid;
>> plot(omg,angle(Xomg)); grid;
```

Software resources: mexct_4_2a.m , mexct_4_2b.m

MATLAB Exercise 4.3: Handling poles on the imaginary axis

Consider the Laplace transform

$$X(s)=\frac{s+1}{s^2+4}$$

Clearly $X(s)$ has poles on the $j\omega$ axis of the s-plane at $s=\mp j2$, and is not computable at those two points. Let us try the same approach we have used in MATLAB Exercise 4.2:

```
>> Xs = @(s) (s+1)./(s.^2+4);
>> omg = [-4:0.04:4];
>> Xomg = Xs(j*omg);
>> plot(omg,abs(Xomg)); grid;
```

We have deliberately included the troublesome values of ω in the vector `omg`. Listing the contents of `omg` should reveal that elements with indices `51` and `151` hold the radian frequencies $\omega=-2$ and $\omega=2$, respectively. The resulting MATLAB graph is shown in Fig. 4.83.

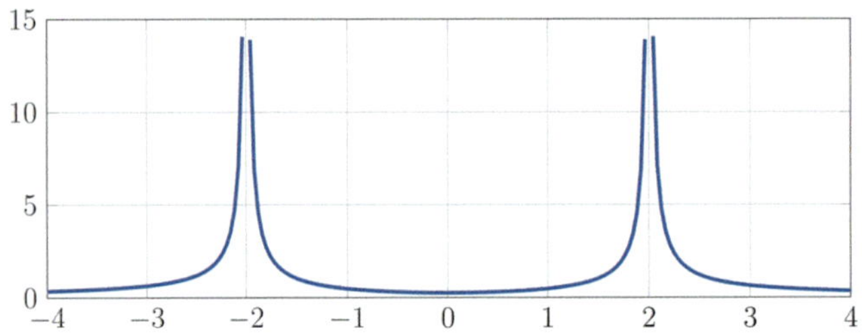

Figure 4.83 – Graph produced by the code in MATLAB Exercise 4.3.

Notice the gaps in the graph at the two frequencies coresponding to the pole locations. MATLAB correctly indicates asymptotic behavior at $\omega=\pm 2$. Checking elements of the vector `Xomg` at indices `51` and `151` confirms this:

```
>> Xomg(51)

ans =
      Inf -    Infi

>> Xomg(151)

ans =
      Inf +    Infi
```

In this case MATLAB handles the singularities gracefully, but we still need to exercise care. If we had picked frequencies not exactly at singularities but very close to them, we would have ended up with very large magnitude values affecting the scaling of the graph and making it difficult to see the details of the transform at other frequencies. For example, try the following set of statements:

```
>> Xs = @(s) (s+1)./(s.^2+4);
>> omg = [-4:0.005:4];
>> Xomg = Xs(j*omg);
>> plot(omg,abs(Xomg)); grid;
```

The result is shown in Fig. 4.84.

Observe how the scaling of the graph changes compared to Fig. 4.83.

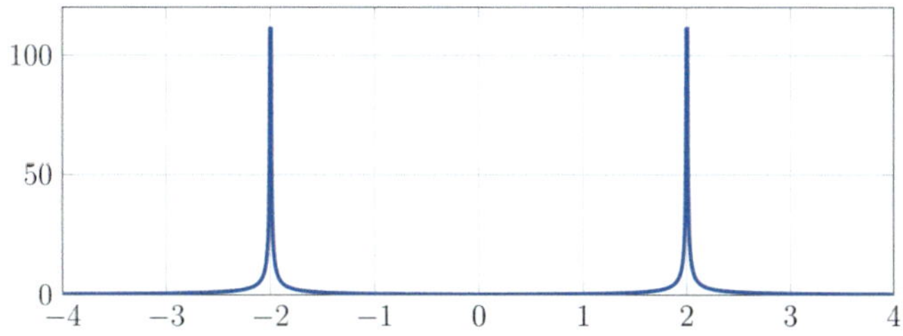

Figure 4.84 – Effect of computing the transform at points very close to a singularity.

Software resources: mexct_4_3a.m , mexct_4_3b.m

MATLAB Exercise 4.4: Graphing poles and zeros

Consider a CTLTI system described by the system function

$$X(s) = \frac{s^3 - s^2 - 4s + 4}{s^4 + 7s^3 + 21s^2 + 37s + 30}$$

Poles and zeros of the system function can be graphed on the s-plane by entering coefficients of numerator and denominator polynomials as vectors and then computing the roots.

```
>> num = [1,-1,-4,4];
>> den = [1,7,21,37,30];
>> z = roots(num);
>> p = roots(den);
```

To produce the graph we need

```
>> plot(real(z),imag(z),'o',real(p),imag(p),'x');
```

which uses "o" for a zero and "x" for a pole. The graph produced does not display the real and imaginary axes in the s-plane. A more complete pole-zero plot may be generated with the following lines:

```
>> plot(real(z),imag(z),'o',real(p),imag(p),'x',..
     [-4,4],[0,0],'k:',[0,0],[-3,3],'k:');
>> axis([-4,4,-3,3]);
>> xlabel('\sigma');
>> ylabel('j\omega');
```

The result is shown in Fig. 4.85.

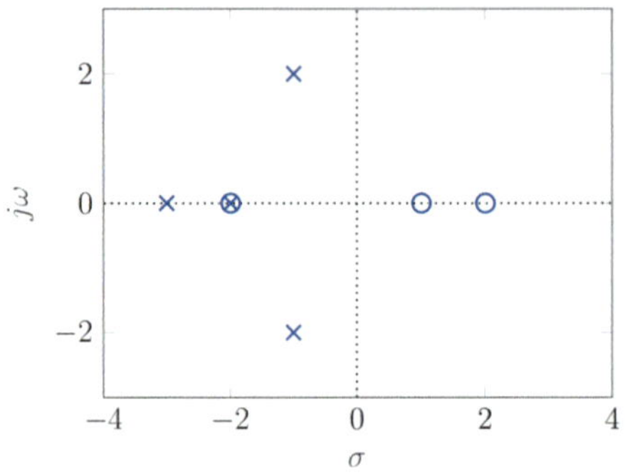

Figure 4.85 – The pole-zero plot obtained in MATLAB Exercise 4.4.

Software resource: mexct_4_4.m

MATLAB Exercise 4.5: Residue calculations

Consider the transform

$$X_1(s) = \frac{s\,(s-2)}{s^4 + 9\,s^3 + 30\,s^2 + 42\,s + 20}$$

In expanding $X(s)$ to partial fractions, MATLAB function `residue()` can be used. Numerator and denominator polynomials are specified by means of two vectors that list the coefficients of these polynomials in the order of descending powers of s.

```
>> num = conv([1,0],[1,-2]);
>> den = [1,9,30,42,20];
>> [res,poles,qt] = residue(num,den)

res =
    1.7000 - 1.9000i
    1.7000 + 1.9000i
   -4.0000
    0.6000

poles =
   -3.0000 + 1.0000i
   -3.0000 - 1.0000i
   -2.0000
   -1.0000

qt =
    []
```

Notice how the convolution function `conv()` is used for multiplying the numerator factors s and $(s-2)$ expressed through the vectors $[1,0]$ and $[1,-2]$ respectively. Returned vectors `res` and `poles` hold the residues and the poles of the partial fraction expansion. The *quotient* vector `qt` is empty since the numerator order is less than the denominator order in this case.

The order of residues and poles in these two vectors is important. We have

$$\begin{aligned} p_1 &= -3+j1\,, & k_1 &= 1.7-j1.9 \\ p_2 &= -3-j1\,, & k_2 &= 1.7+j1.9 \\ p_3 &= -2\,, & k_3 &= -4 \\ p_4 &= -1\,, & k_4 &= 0.6 \end{aligned}$$

The partial fraction expansion we seek is in the form

$$X_1(s) = \frac{(1.7-j1.9)}{s+3-j1} + \frac{(1.7+j1.9)}{s+3+j1} - \frac{4}{s+2} + \frac{0.6}{s+1}$$

Consider another transform

$$X_2(s) = \frac{3\,s^2+2\,s+5}{s^2+4\,s+20}$$

the residues for which can be computed through

```
>> num = [3,2,5];
>> den = [1,4,20];
>> [res,poles,qt] = residue(num,den)

res =
  -5.0000 + 4.3750i
  -5.0000 - 4.3750i

poles =
  -2.0000 + 4.0000i
  -2.0000 - 4.0000i

qt =
  3
```

The vector qt has a single element and corresponds to the quotient polynomial $Q(s) = 3$. The partial fraction expansion for $X_2(s)$ is

$$X_2(s) = 3 + \frac{(-5+j4.375)}{s+2-j4} + \frac{(-5-j4.375)}{s+2+j4}$$

Software resources: mexct_4_5a.m , mexct_4_5b.m

MATLAB Exercise 4.6: Computing frequency response from pole-zero layout

The problem of obtaining the frequency response of a CTLTI system from the placement of its poles and zeros in the s-domain was discussed in Section 4.5.5. In this exercise we will use MATLAB to compute the frequency response for a system characterized by the system function

$$H(s) = \frac{s-1}{s+2}$$

The system has one zero at $s = 1$ and one pole at $s = -2$. Suppose we need the frequency response of this system at the frequency $\omega = 3$ rad/s, which is equal to the system function evaluated for $s = j3$:

$$H(3) = H(s)\Big|_{s=j3} = \frac{j3-1}{j3+2}$$

Using vector notation

$$\overrightarrow{H(3)} = \frac{\overrightarrow{B}}{\overrightarrow{A}} = \frac{\overrightarrow{(j3-1)}}{\overrightarrow{(j3+2)}}$$

where we have defined the vectors $\overrightarrow{A}$ and $\overrightarrow{B}$ for notational convenience. The script `mexct_4_6a.m` listed below computes the vectors $\overrightarrow{A}$ and $\overrightarrow{B}$ and uses them for computing the magnitude and the phase of the system function at $\omega = 3$ rad/s:

```
1  % Script: mexct_4_6a.m
2  omega = 3;
3  s = j*omega;
4  B = s-1;                 % Numerator
5  A = s+2;                 % Denominator
6  mag = abs(B)/abs(A);     % Magnitude of the transform at 3 rad/s
7  phs = angle(B)-angle(A); % Magnitude of the transform at 3 rad/s
```

This script computes the frequency response of the system at one specific frequency. It would be more interesting if we could use the same idea to compute the frequency response of the system at a large number of frequencies so that its magnitude and phase can be graphed as functions of ω. Let us change the variable `omega` into a vector by editing line 1 of the script:

```
1  omega = [-5:0.01:5];
```

This change causes the variable `s` to become a complex vector with 1001 elements. Also, in line 6, the standard division operator "`/`" needs to be changed to the element-by-element division operator "`./`" to read

```
5  mag = abs(B)./abs(A);
```

The script `mexct_4_6b.m` is listed below with these modifications and the addition of graphing statements:

```
1  % Script: mexct_4_6b.m
2  omega = [-5:0.01:5];     % Vector of radian frequencies
3  s = j*omega;             % Vector of s values on the imaginary axis
4  B = s-1;                 % Numerator
5  A = s+2;                 % Denominator
6  mag = abs(B)./abs(A);    % Magnitude of the transform
7  phs = angle(B)-angle(A); % Phase of the transform
8  % Graph the results
9  tiledlayout(2,1);
10 nexttile;
11 plot(omega,mag); grid;
12 xlabel('\omega (rad/s)');
13 ylabel('Magnitude');
14 nexttile;
```

```
plot(omega,phs); grid;
xlabel('\omega (rad/s)');
ylabel('Phase (rad)');
```

The graph produced by the script is shown in Fig. 4.86.

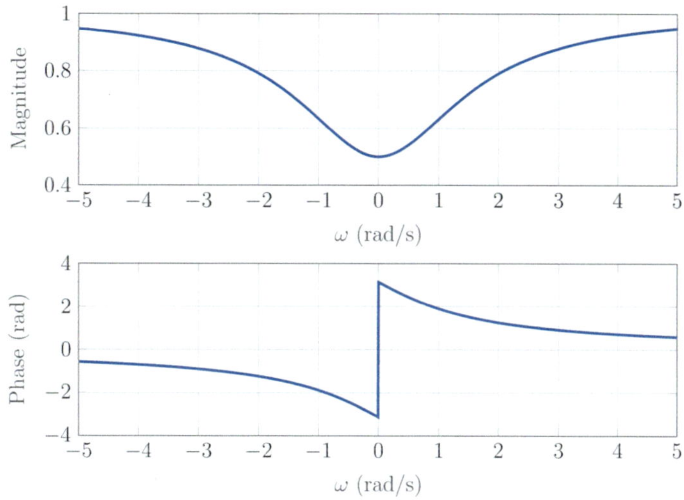

Figure 4.86 – Frequency response for the system in MATLAB Exercise 4.6.

Software resources: mexct_4_6a.m , mexct_4_6b.m

MATLAB Exercise 4.7: Frequency response from pole-zero layout revisited

In MATLAB Exercise 4.6 we have explored a method of computing the frequency response of a CTLTI system based on the graphical interpretation of the pole-zero layout of the system function, discussed in Section 4.5.5. The idea can be generalized into the development of a MATLAB function `ss_freqs()` for computing the frequency response.

We have seen in MATLAB Exercise 4.2 that the built-in function `freqs()` can be used for computing the frequency response of a CTLTI system from the knowledge of numerator and denominator polynomial coefficients. Developing our own version is still worthwhile, however, since it is instructive of how the concepts developed in Section 4.5.5 can be used in practice.

```
function [mag,phs] = ss_freqs(zrs,pls,gain,omega)
  nz = length(zrs);         % Number of zeros.
  np = length(pls);         % Number of poles.
  nomg = length(omega);     % Number of frequency points.
  s = j*omega;              % Get points on the imaginary axis.
  mag = ones(1,nomg);
  phs = zeros(1,nomg);
  if (nz > 0)               % Compute contributions from zeros
```

```
9       for n = 1:nz
10        mag = mag.*abs(s-zrs(n));
11        phs = phs+angle(s-zrs(n));
12      end
13    end
14    if (np > 0)              % Compute contributions from poles
15      for n = 1:np
16        mag = mag./abs(s-pls(n));
17        phs = phs-angle(s-pls(n));
18      end
19    end
20    mag = mag*gain;          % Account for the constant gain factor
21    phs = wrapToPi(phs);     % Wrap phase angles to stay in interval -pi,pi
22  end
```

It is useful to refer to Eqns. (4.118) and (4.119) to understand how the code is developed. In lines 6 and 7 we start with a unit magnitude response and a zero phase response for all frequencies. In the code block from line 8 to line 13 the variables `mag` and `phs` are modified using the contributions from each numerator factor $(j\omega - z_k)$. Similarly, in the code block from line 14 to line 19, contributions from the denominator factors $(j\omega - p_k)$ are used for modifying the variables for magnitude and phase. Line 21 of the function causes all phase angles to be contained in the interval $(-\pi, \pi)$.

The script `mexct_4_7.m` listed below may be used for testing `ss_freqs()` with the system function

$$H(s) = \frac{s + 0.5}{s^2 + s + 39.7284}$$

and graphing the resulting frequency response. The resulting graph is shown in Fig. 4.87.

```
% Script: mexct_4_7.m
num = [1,0.5];
den = [1,1,39.7284];
zrs = roots(num);       % Compute zeros.
pls = roots(den);       % Compute poles.
omg = [-15:0.05:15];    % Vector of frequencies.
[mag,phs] = ss_freqs(zrs,pls,1,omg);
tiledlayout(2,1);
nexttile;
plot(omg,mag); grid;
xlabel('\omega (rad/s)');
ylabel('Magnitude');
nexttile;
plot(omg,phs); grid;
xlabel('\omega (rad/s)');
ylabel('Phase (rad)');
```

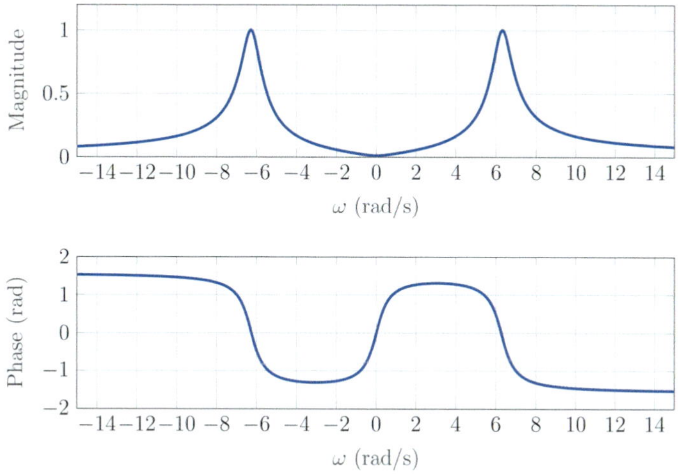

Figure 4.87 – Frequency response for the system in MATLAB Exercise 4.7.

Software resources: `mexct_4_7.m`, `ss_freqs.m`

MATLAB Exercise 4.8: Dynamic system models in MATLAB

MATLAB has some functions for creating and working with *dynamic system models* that represent linear and time-invariant systems. Consider, for example, a CTLTI system described by the system function

$$H(s) = \frac{s^2+1}{s^3+5\,s^2+17\,s+13}$$

An object named `sys1` representing this system can be created in MATLAB with the following lines:

```
>> num = [1,0,1];
>> den = [1,5,17,13];
>> sys1 = tf(num,den)

Transfer function:
        s^2 + 1
-----------------------
s^3 + 5 s^2 + 17 s + 13
```

The object `sys1` created with a call to the function `tf()` may be used with other functions that accept system models. For example, a pole-zero plot can be generated through

```
>> pzmap(sys1)
```

The resulting graph is shown in Fig. 4.88.

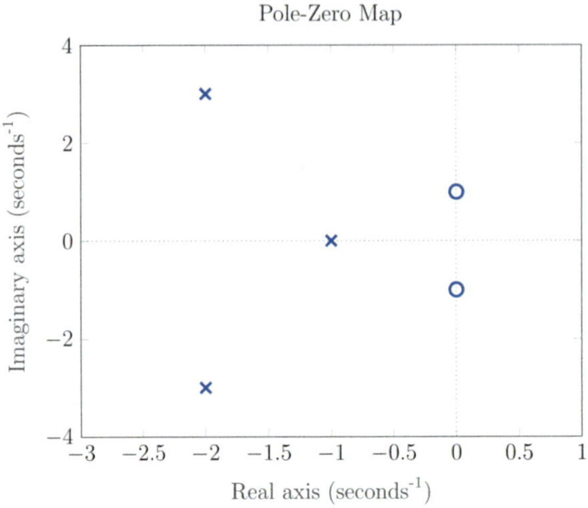

Figure 4.88 – Pole-zero map for the dynamic system model `sys1`.

The impulse response of the system can be graphed using the following code, and produces the graph shown in Fig. 4.89:

```
>> t = [0:0.01:5];
>> h = impulse(sys1,t);
>> plot(t,h); grid;
```

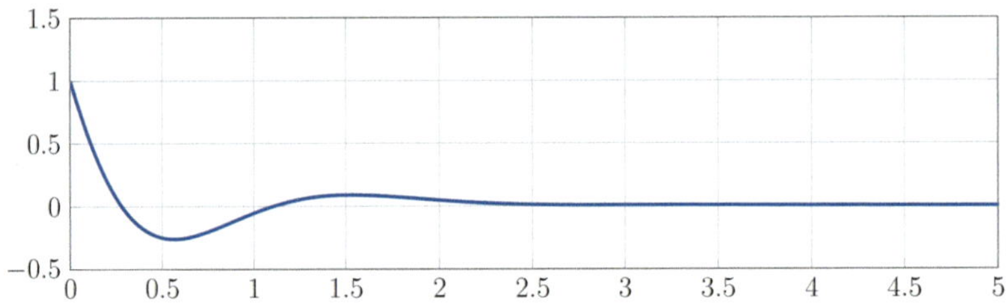

Figure 4.89 – Impulse response for the dynamic system model `sys1`.

Similarly, the unit step response can be graphed by typing the following lines

```
>> t = [0:0.01:5];
>> y = step(sys1,t);
>> plot(t,y); grid;
```

The resulting graph is shown in Fig. 4.90. A system can be specified in a variety of ways. An alternative to using the numerator and the denominator coefficients is to use the zeros and the poles of the system through the function `zpk()`. For example, consider the system function

$$H(s) = \frac{10\,s\,(s-1)}{(s+1)\,(s+2)\,(s+3)}$$

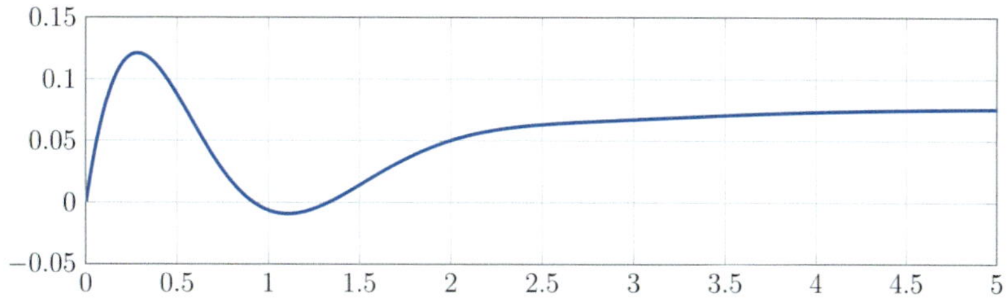

Figure 4.90 – Unit step response for the dynamic system model sys1.

A dynamic model sys2 for this system function can be created as follows:

```
>> zrs = [0,1];
>> pls = [-1,-2,-3];
>> sys2 = zpk(zrs,pls,10)
```

Let the input signal to this system be

$$x(t) = \cos(4\pi t)\; u(t)$$

The response of the system may be computed and graphed as shown in Fig. 4.91 with the following code:

```
>> t = [0:0.01:3];
>> x = cos(2*pi*2*t);
>> [y,t] = lsim(sys2,x,t);
>> plot(t,x,t,y);
```

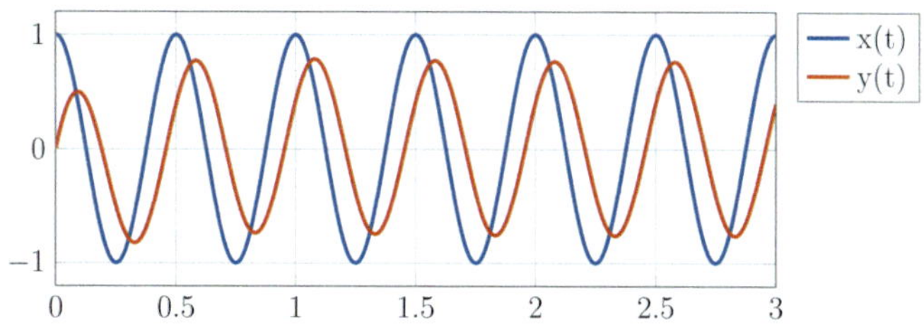

Figure 4.91 – Response for the dynamic system model sys2 to input signal $x(t) = \cos(4\pi t)\; u(t)$.

Software resources: mexct_4_8a.m , mexct_4_8b.m , mexct_4_8c.m

MATLAB Exercise 4.9: Bode diagram for a CTLTI system

In this exercise we will construct Bode diagram showing magnitude and phase of the system with the system function

$$H(s) = \frac{10\, s\, (s-1)}{(s+1)\, (s+2)\, (s+3)}$$

The easiest method of accomplishing this is to use the dynamic system models we have explored in MATLAB Exercise 4.8. The system function $H(s)$ has zeros at $s = 0,\ 1$ and poles at $s = -1,\ -2,\ -3$. Using these we will create a system model `sys`, and use it with the function `bode()`.

```
>> zrs = [0,1];
>> pls = [-1,-2,-3];
>> sys = zpk(zrs,pls,10);
>> bode(sys); grid;
```

The resulting magnitude and phase plots are shown in Fig. 4.92.

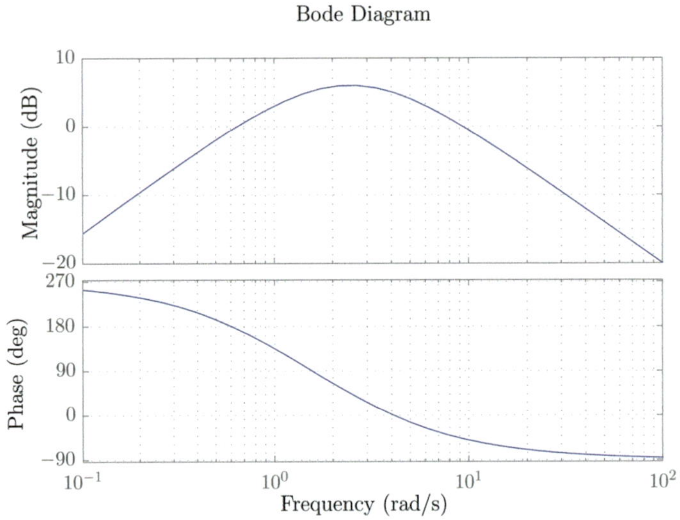

Figure 4.92 – Bode diagram produced by the code in MATLAB Exercise 4.9.

Software resource: `mexct_4_9.m`

MATLAB Exercise 4.10: Symbolic calculations for Laplace transform

It is also possible to do symbolic processing with MATLAB. As a first step we need to specify that `s` and `t` are symbolic variables rather than numeric variables. This is accomplished with the statement

```
>> syms s t
```

Notice that there is no comma, but rather just space, separating the two symbolic variables. The next step is to define the signal $x(t)$ whose Laplace transform we seek. For example, the signal

$$x(t) = e^{-t} \cos(2t)\, u(t)$$

is defined with the statement

```
>> xt = exp(-t)*cos(2*t);
```

The resulting variable `xt` is also a symbolic variable since it utilizes the variable `t` defined with the previous statement. The Laplace transform can now be computed using the function `laplace()`.

```
>> Xs = laplace(xt)

Xs =
(s + 1)/((s + 1)^2 + 4)
```

MATLAB returns the answer as a symbolic expression. To display the answer in a way closer to its natural form, use

```
>> pretty(Xs)

    s + 1
 ------------
          2
 (s + 1)  + 4
```

which corresponds to

$$X(s) = \frac{s+1}{(s+1)^2+4}$$

Combining the statements listed above, the transform can be computed in a compact form as

```
>> syms s t;
>> Xs = laplace(exp(-t)*cos(2*t));
```

Some observations are in order:

1. In using the function `laplace()` with just one argument we assume that the independent variable of the signal is named `t`, and the independent variable of the transform is named `s`. If, for some reason, the names `v` and `w` need to be used in place of `t` and `s`, respectively, then an alternative syntax of the function with three arguments can be employed:

   ```
   >> Xs = laplace(xt,v,w)
   ```

2. The function `laplace()` computes the Laplace transform integral starting with the lower limit $t = 0$. In other words, the transform computed is the *unilateral Laplace transform* discussed in Section 4.7.

Inverse Laplace transform can be computed symbolically using the function `ilaplace()`. Consider the problem of finding the inverse Laplace transform of

$$X(s) = \frac{s+1}{s(s+2)}$$

which was solved in Example 4.24 through the use of partial fractions. The following set of statements produce the solution using symbolic processing in MATLAB:

```
>> syms s t
>> Xs = (s+1)/(s*(s+2));
>> xt = ilaplace(Xs)

xt =
1/(2*exp(2*t)) + 1/2

>>  pretty(xt)

      1          1
  ---------- + -
  2 exp(2 t)   2
```

The result displayed corresponds to

$$x(t) = \frac{1}{2}e^{-2t}\,u(t) + \frac{1}{2}\,u(t)$$

The symbolic result obtained for the signal $x(t)$ can be graphed in the time interval $0 \le t \le 5$ using the function `fplot()` as follows:

```
>> fplot(xt,[0,5]); grid;
>> axis([0,5,0,1.2]);
```

The graph produced is shown in Fig. 4.93.

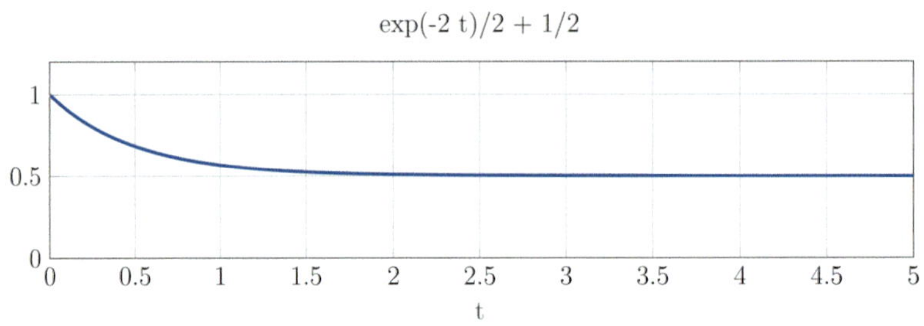

Figure 4.93 – The graph produced using function `fplot()`.

Software resources: mexct_4_10a.m , mexct_4_10b.m

MATLAB Exercise 4.11: Solving a differential equation using Laplace transform

Two examples of using the unilateral Laplace transform for solving differential equations with specified initial conditions were given in Section 4.7. In this exercise we will explore the use of symbolic processing capabilities of MATLAB for solving similar problems. The functions `laplace()` and `ilaplace()` were explored in MATLAB Exercise 4.10. They compute the forward and the inverse Laplace transform with the assumption that the time-domain signal involved is causal. In effect they implement the unilateral variant of the Laplace transform.

Consider the RC circuit problem solved in Example 4.45. The governing differential equation is

$$\frac{dy(t)}{dt} + 4\,y(t) = 4\,x(t)$$

with the initial value of the solution specified as $y(0^+) = -2$ V. The input signal is a unit amplitude pulse with a duration of 2 seconds.

$$x(t) = u(t) - u(t-2)$$

The following script can be used for solving the problem and graphing the result:

```
1 % Script: mexct_4_11.m
2 syms s t Ys
3 xt = heaviside(t)-heaviside(t-2);  % x(t)=u(t)-u(t-2)
4 Xs = laplace(xt);                  % Laplace transform of x(t)
5 Y1 = s*Ys+2;                       % Laplace transform of dy/dt
6 Ys = solve(Y1+4*Ys-4*Xs,Ys);       % Solve for Y(s)
7 yt = ilaplace(Ys)                  % Inv. Laplace transform of Y(s)
8 fplot(yt,[0,5]); grid;
9 axis([0,5,-2.5,1.5]);
```

In line 2 we declare three symbolic variables. The variable `Ys` corresponds to the yet unknown transform $Y_u(s)$. In line 3 the input signal is specified in symbolic form. Note that we make use of the symbolic function `heaviside()` to represent the unit step function. The symbolic result `Y1` represents the Laplace transform

$$Y_1(s) = \mathscr{L}\left\{\frac{dy(t)}{dt}\right\} = s\,Y_u(s) - y(0^+) = s\,Y_u(s) + 2$$

Line 6 of the script uses the function `solve()` to solve the equation

$$Y_1(s) + 4\,Y_u(s) - 4\,X(s) = 0$$

for $Y_u(s)$. Line 7 prints the symbolic result

```
yt =
4*heaviside(t - 2)*(exp(8 - 4*t)/4 - 1/4) - 3*exp(-4*t) + 1
```

which can be written as

$$4\,u(t-2)\left(e^{(8-4t)/4} - 1/4\right) - 3\,e^{-4t} + 1$$

Keeping in mind that this is the solution for $t \geq 0$, we have the same expression that was found earlier in Example 4.45. The graph produced by the function `fplot()` on line 8 is shown in Fig. 4.94, and it matches Fig. 4.81 as well.

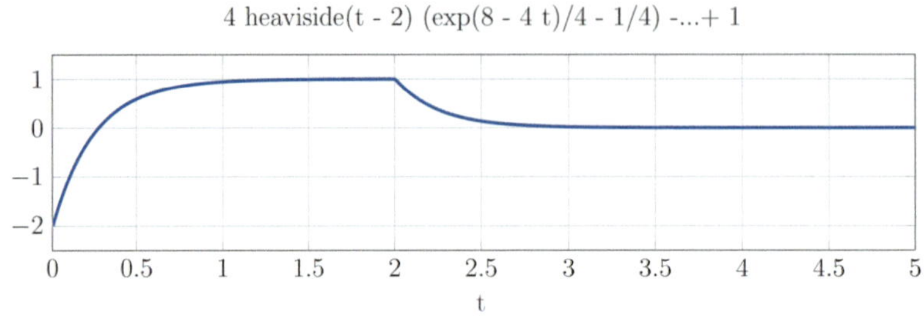

Figure 4.94 – Graph of $y(t)$ produced through symbolic calculations in MATLAB Exercise 4.11.

Software resource: mexct_4_11.m

Problems

4.1. Using the Laplace transform definition given by Eqn. (4.1) determine the transform of each signal listed below. For each transform construct a pole-zero diagram and specify the ROC.

a. $x(t) = e^{-2t}\, u(t)$

b. $x(t) = e^{-2t}\, u(t-1)$

c. $x(t) = e^{2t}\, u(-t)$

d. $x(t) = e^{2t}\, u(-t+1)$

e. $x(t) = \begin{cases} 1, & 0 < t < 1 \\ -1, & 1 < t < 2 \\ 0, & t < 0 \text{ or } t > 2 \end{cases}$

f. $x(t) = \begin{cases} 1, & 0 < t < 1 \\ -1, & t > 1 \\ 0, & t < 0 \end{cases}$

4.2. Using the Laplace transform definition given by Eqn. (4.1) determine the transform of the signal

$$x(t) = \sum_{n=0}^{\infty} e^{-anT}\, \delta(t - nT)$$

where $a > 0$ and $T > 0$. Put the transform into a closed form using the appropriate formula. Afterward construct a pole-zero plot and indicate the ROC.

4.3. Pole-zero diagrams for four transforms are shown in Fig. P.4.3. For each, determine the ROC if it is known that the Fourier transform of $x(t)$ exists.

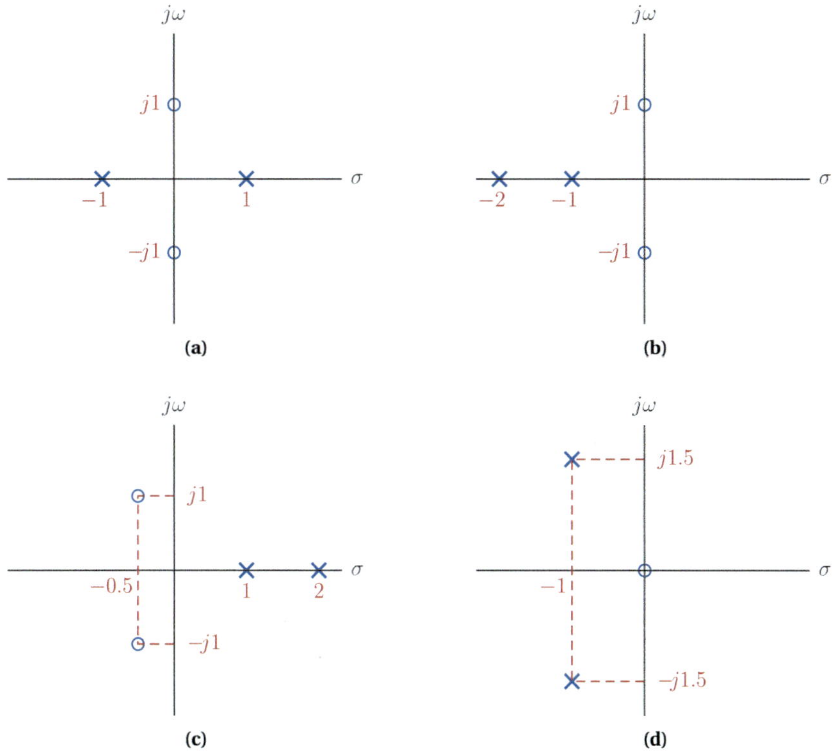

Figure P. 4.3

4.4. Consider again the four transforms with pole zero diagrams shown in Fig. P.4.3. Determine the ROC for each of the transforms if $x(t)$ is known to be causal in each case.

4.5. Repeat Problem 4.3 if $x(t)$ is known to be anti-causal for each case.

4.6. The Laplace transform of a signal $x(t)$ is given by

$$X(s) = \frac{s-1}{s^2-s-2}\,, \qquad \text{ROC:} \quad -1 < \text{Re}\{s\} < 2$$

Which of the following Fourier transforms can be obtained from $X(s)$ without actually determining the signal $x(t)$? In each case, either determine the indicated Fourier transform or explain why it cannot be determined.

a. $\mathscr{F}\{x(t)\}$

b. $\mathscr{F}\{x(t)\,e^{t}\}$

c. $\mathscr{F}\{x(t)\,e^{-3t}\}$

4.7. The Laplace transform of a signal $x(t)$ is given by

$$X(s) = \frac{1}{s^2+2s+5}\,, \qquad \text{ROC:} \quad \text{Re}\{s\} > -1$$

Which of the following Fourier transforms can be obtained from $X(s)$ without actually determining the signal $x(t)$? In each case, either determine the indicated Fourier transform or explain why it cannot be determined.

a. $\mathcal{F}\left\{x(t)\right\}$

b. $\mathcal{F}\left\{x(t)\,e^{-t}\right\}$

c. $\mathcal{F}\left\{x(t)\,e^{t}\right\}$

d. $\mathcal{F}\left\{x(t)\,e^{3t}\right\}$

4.8. Construct a pole-zero diagram and specify the ROC for each of the transforms given below. Also, determine the Fourier transform $X(\omega)$ if it exists.

a. $X(s)=\dfrac{s-2}{s^2+3s+2}\,,\quad x(t)$ is causal

b. $X(s)=\dfrac{s}{s^2-1}\,,\quad x(t)$ is causal

c. $X(s)=\dfrac{s+1}{s^2-4s+3}\,,\quad x(t)$ is anti-causal

d. $X(s)=\dfrac{s^2-s}{s^2-s-6}\,,\quad x(t)$ is anti-causal

e. $X(s)=\dfrac{s^2-s}{s^2+8s+15}\,,\quad \mathcal{F}\left\{x(t)\,e^{4t}\right\}$ exists

4.9. Using the linearity property of the Laplace transform, determine $X(s)$ for each of the signals listed below. Also indicate the ROC in each case.

a. $x(t)=3\,e^{-t}\,u(t)-5\,e^{-3t}\,u(t)$

b. $x(t)=3\,e^{-t}\,u(t)+2\,e^{3t}\,u(-t)$

c. $x(t)=\delta(t)+2\,e^{-t}\,u(t)$

d. $x(t)=\left(1-e^{-t}\right)\,u(t)$

e. $x(t)=\cos(2t)\,u(t)+2\sin(3t)\,u(t)$

f. $x(t)=e^{-2t}\cos(3t)\,u(t)$

4.10. Using time shifting and linearity properties of the Laplace transform as needed, determine $X(s)$ for each of the signals listed below. Also indicate the ROC in each case.

a. $x(t)=e^{-2(t-1)}\,u(t-1)$

b. $x(t)=e^{-2t}\,u(t-1)$

c. $x(t)=e^{2(t+1)}\,u(-t-1)$

d. $x(t)=e^{2t}\,u(-t-1)$

e. $x(t)=e^{2t}\,u(-t+1)$

4.11. Determine the Laplace transform of the signal

$$x(t)=\cos(3t+\pi/6)\,u(t)$$

using two different methods as specified below. Show that the same result can be obtained through each.

a. Use a trigonometric identity to express $x(t)$ as the sum of a cosine and a sine term. Afterward use the entries for $\cos(3t)$ and $\sin(3t)$ from the Laplace transform table.

b. Use Euler's formula to express $x(t)$ as a sum of complex exponentials. Afterward use the entry for e^{at} from the Laplace transform table.

4.12. Determine the Laplace transform of each signal listed below. Specify the ROC in each case. Sketch each signal and justify the ROC based on the characteristics of each signal.

a. $x(t) = u(t) - u(t-1)$

b. $x(t) = u(t) - 2u(t-1) + u(t-2)$

c. $x(t) = u(t) - 2u(t-1)$

4.13. Using the linearity and time shifting properties, determine the Laplace transform of each signal shown graphically in Fig. P.4.13. Specify the ROC in each case.

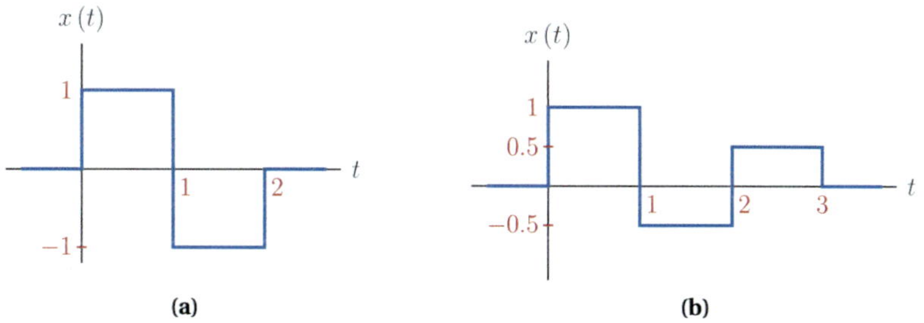

Figure P. 4.13

4.14. Express each of the signals shown in Fig. P.4.14 in terms of the unit ramp function and its scaled and/or time shifted versions. Afterward, determine the Laplace transform of each signal using linearity and time shifting properties, Specify the ROC in each case.

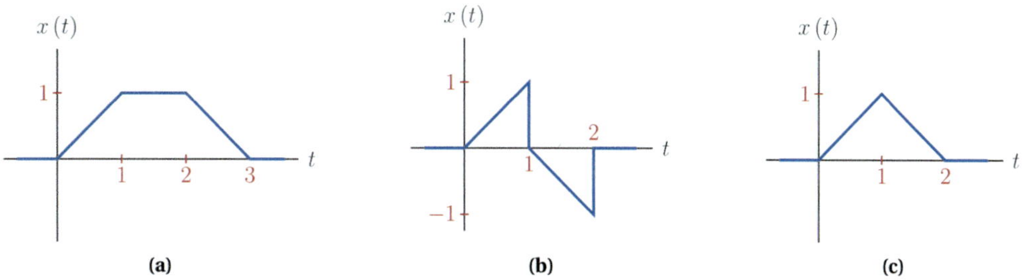

Figure P. 4.14

4.15. The signal $x(t)$ is in the form of one cycle of a sinusoidal waveform as shown in Fig. P.4.15.

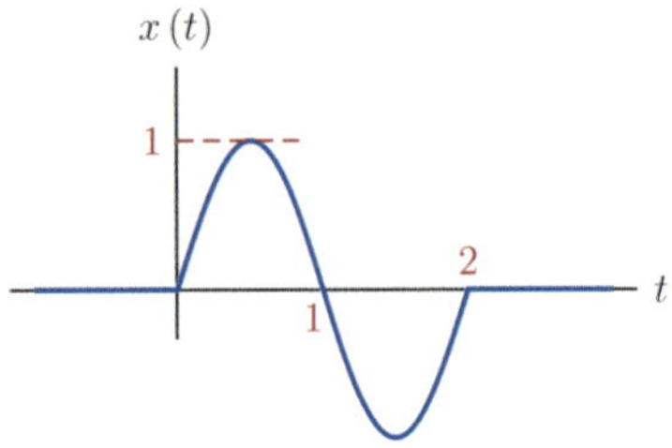

Figure P. 4.15

Determine its Laplace transform using two different methods:

a. Apply Laplace transform definition directly. Hint: Express the sinusoidal signal using Euler's formula.

b. Use the transform for $x_1(t) = \sin(\omega_0 t)\, u(t)$ and then apply linearity and time shifting properties.

4.16. The transforms given below correspond to causal signals. Using the table of Laplace transforms, determine the inverse Laplace transform of each. Hint: For each case find a similar entry in the table and adjust its parameters to obtain the desired result.

a. $X(s) = \dfrac{1}{(s+2)^2}$

b. $X(s) = \dfrac{s+3}{(s+3)^2+2}$

c. $X(s) = \dfrac{1}{(s+1)^2+6}$

4.17. Determine the inverse Laplace transform of each function listed below using partial fraction expansion.

a. $X(s) = \dfrac{1}{s^2+3s+2}, \quad \text{ROC: } \operatorname{Re}\{s\} > -1$

b. $X(s) = \dfrac{1}{s^2+3s+2}, \quad \text{ROC: } \operatorname{Re}\{s\} < -2$

c. $X(s) = \dfrac{1}{s^2+3s+2}, \quad \text{ROC: } -2 < \operatorname{Re}\{s\} < -1$

d. $X(s) = \dfrac{(s-1)(s-2)}{(s+1)(s+2)(s+3)}, \quad x(t)\text{: causal}$

e. $X(s) = \dfrac{(s+1)(s-2)}{(s-1)(s+2)(s+3)}, \quad \mathcal{F}\{x(t)\} \text{ exists}$

4.18. Use partial fraction expansion to find the inverse Laplace transform of each function listed below.

a. $X(s) = \dfrac{2s+3}{s+2}, \quad \text{ROC: } \operatorname{Re}\{s\} > -2$

b. $X(s) = \dfrac{s(s+1)}{(s+2)(s+3)}, \quad \text{ROC: } \operatorname{Re}\{s\} < -3$

c. $X(s) = \dfrac{s+5}{s^2+4}\,, \qquad \text{ROC: } \operatorname{Re}\{s\} > 0$

d. $X(s) = \dfrac{s+6}{(s+1)^2+4}\,, \qquad \text{ROC: } \operatorname{Re}\{s\} > 0$

e. $X(s) = \dfrac{s\,(s-1)}{(s+1)^2\,(s+2)}\,, \qquad \text{ROC: } \operatorname{Re}\{s\} > -1$

4.19. Given the transform

$$X(s) = \frac{s\,(s-3)}{(s-1)\,(s+1)\,(s+2))}$$

a. Construct a pole-zero diagram for $X(s)$.

b. List all possibilities for the ROC.

c. For each choice of the ROC listed in part (b) determine the inverse transform $x(t)$ and indicate whether it is square integrable or not.

4.20. For each Laplace transform given below, find all possible signals that may have led to that transform. (Consider all ROC choices.)

a. $X(s) = \dfrac{s+1}{(s+2)\,(s+3)}$

b. $X(s) = \dfrac{s+1}{(s+2)^2\,(s+3)}$

c. $X(s) = \dfrac{s\,(s-1)}{(s+3)^2}$

d. $X(s) = \dfrac{s}{(s+2)^2+9}$

4.21. Use partial fraction expansion and the time shifting property to find the inverse Laplace transforms of functions listed below. Assume each transform corresponds to a causal signal.

a. $X(s) = \dfrac{1-e^{-s}}{s+1}$

b. $X(s) = \dfrac{(1-e^{-s})\,s}{s+1}$

c. $X(s) = \dfrac{1-e^{-s}}{s\,(s+1)}$

d. $X(s) = \dfrac{1}{s}\left(1-e^{-s}+e^{-2s}-e^{-3s}\right)$

4.22. Find a system function for each CTLTI system described below by means of a differential equation.

a. $3\,\dfrac{dy(t)}{dt} + 2\,y(t) = 7\,x(t)$

b. $\dfrac{d^2y(t)}{dt^2} + 4\,\dfrac{dy(t)}{dt} + 3\,y(t) = \dfrac{dx(t)}{dt} - x(t)$

c. $\dfrac{d^2y(t)}{dt^2} + 4\,y(t) = \dfrac{d^2x(t)}{dt^2} + \dfrac{dx(t)}{dt} + 3\,x(t)$

4.23. Find a differential equation for each CTLTI system described below by means of a system function.

a. $H(s) = \dfrac{1}{s+4}$

b. $H(s) = \dfrac{s}{s+4}$

c. $H(s) = \dfrac{s+1}{s^2+5s+6}$

d. $H(s) = \dfrac{s(s-1)}{s^3+5s^2+8s+6}$

e. $H(s) = \dfrac{(s-1)(s+2)}{(s+1)\left(s^2+6s+13\right)}$

4.24. Let $y_u(t)$ represent the response of a CTLTI system to a unit step input signal $x(t) = u(t)$. Determine the system function for each system whose unit step response given below. Afterward find a differential equation for each system.

a. $y_u(t) = e^{-t}\,u(t)$

b. $y_u(t) = \left(1-e^{-t}\right)u(t)$

c. $y_u(t) = \left(e^{-t}-e^{-2t}\right)u(t)$

d. $y_u(t) = \left(1-e^{-t}+2e^{-2t}\right)u(t)$

e. $y_u(t) = \left[1-0.3\,e^{-t}\cos(2t)\right]u(t)$

4.25. Let $y_r(t)$ represent the response of a CTLTI system to a unit ramp input signal $x(t) = t\,u(t)$. Determine the system function for each system whose unit ramp response is given below. Afterward find a differential equation for each system.

a. $y_r(t) = \left(1-e^{-t}\right)u(t)$

b. $y_r(t) = \left(e^{-t}+2e^{-2t}-4e^{-3t}\right)u(t)$

c. $y_r(t) = \left[1-0.3\,e^{-t}\cos(2t)\right]u(t)$

4.26. A CTLTI system is described by means of the system function

$$H(s) = \frac{s+3}{s^2+3s+2}$$

Determine the response of the system to the following signals:

a. $x(t) = e^{-0.5t}$

b. $x(t) = e^{(-0.5+j2)t}$

c. $x(t) = e^{j3t}$

d. $x(t) = e^{-j3t} + e^{j3t}$

e. $x(t) = e^{j2t} + e^{j3t}$

4.27. Show that, for a CTLTI system with a real-valued impulse response, the value of the system function at the point $s = s_0^*$ is the complex conjugate of its value at the point $s = s_0$. Given

$$H(s_0) = H_0\,e^{j\Theta_0}$$

where

$$H_0 = \left|H(s_0)\right|$$

and

$$\Theta_0 = \measuredangle H(s_0)$$

prove that

$$H\left(s_0^*\right) = \left[H(s_0)\right]^* = H_0\, e^{-j\Theta_0}$$

Hint: Use the Laplace transform definition with the impulse response $h(t)$ and evaluate the result at $s = s_0^*$. Manipulate the resulting expression to obtain the desired proof.

4.28. A CTLTI system is described by means of the system function

$$H(s) = \frac{s\,(s+1)}{s^2+4s+13}$$

Determine the response of the system to the following signals:

a. $x(t) = \cos(2t)$

b. $x(t) = e^{-0.5t}\cos(2t)$

c. $x(t) = e^{-t}\sin(2t)$

4.29. A CTLTI system has the impulse response

$$h(t) = e^{-t}\,u(t)$$

Using Laplace transform techniques, determine the response of the system to each input signal listed below. Identify transient and steady-state components of the output signal in each case.

a. $x(t) = \cos(2t)\,u(t)$

b. $x(t) = \sin(3t)\,u(t)$

4.30. Determine the impulse response of each system specified below by means of a system function.

a. $H(s) = \dfrac{s+1}{s+3}$

b. $H(s) = \dfrac{s^2+1}{s^2+3s+2}$

c. $H(s) = \dfrac{s^2-1}{(s+2)\left(s^2+2s+2\right)}$

d. $H(s) = \dfrac{s-3}{s^2+2s+1}$

4.31. Consider again the system used in Problem 4.29. Using the Laplace transform, determine the response of the system to the signal

$$x(t) = \cos(2t)$$

which is similar to the input signal in part (a) of Problem 4.29 except it exists for all t instead of being turned on at $t = 0$. Compare the output signal to that obtained in Problem 4.29(a) and comment on the result.

4.32. A CTLTI system has the impulse response

$$h(t) = e^{-t}\left[u(t) - u(t-2)\right]$$

Using Laplace transform techniques, determine the response of the system to the input signal $x(t) = e^{-0.5t}\cos(2t)\,u(t)$.

4.33. Construct a pole-zero diagram for each of the system functions given below.

a. $H(s) = \dfrac{s+1}{(s+2)(s+3)}$

b. $H(s) = \dfrac{s^2+1}{s^2+5s+6}$

c. $H(s) = \dfrac{s^2-1}{(s+2)\left(s^2+4s+13\right)}$

4.34. Pole-zero diagrams for four system functions are shown in Fig. P.4.34. For each, determine the system function $H(s)$. Set $|H(0)| = 1$ for each system.

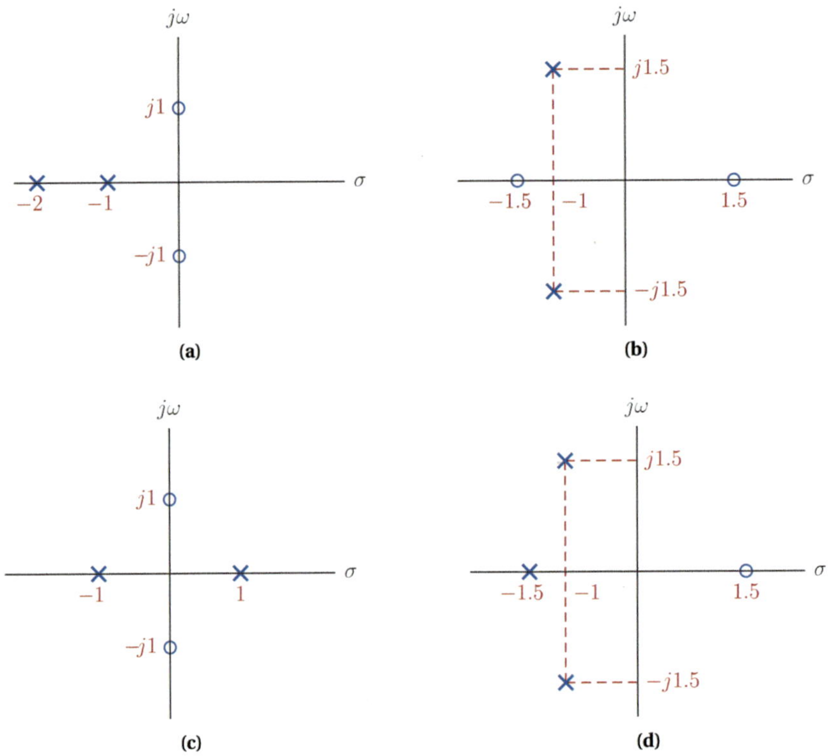

Figure P. 4.34

4.35. For each system function given below, sketch the magnitude and phase characteristics using the graphical method outlined in Section 4.5.5.

a. $H(s) = \dfrac{s-1}{s+2}$

b. $H(s) = \dfrac{s^2+1}{s^2+5s+6}$

c. $H(s) = \dfrac{s^2-1}{s^2+2s+5}$

4.36. The system functions given below correspond to causal systems. For each system, construct a pole-zero plot and determine whether the system is stable or not.

a. $H(s) = \dfrac{s+1}{s^3+7s^2+16s+12}$

b. $H(s) = \dfrac{s^2-1}{s^2+s-6}$

c. $H(s) = \dfrac{s^2+1}{s^2+2s+10}$

d. $H(s) = \dfrac{s^2+1}{s^2-2s+10}$

4.37. Determine which of the system functions given below could represent a system that is both causal and stable.

a. $H(s) = \dfrac{s-1}{s^2+3s+2}$

b. $H(s) = \dfrac{s(s+1)}{s^2+2s+5}$

c. $H(s) = \dfrac{s^2+1}{s\left(s^2+5s+6\right)}$

d. $H(s) = \dfrac{s+3}{s^2-6s+10}$

4.38. A causal system is described by the differential equation

$$\frac{d^2y(t)}{dt^2} = -2\frac{dy(t)}{dt} - a\,y(t) + x(t)$$

a. Determine the system function $H(s)$.

b. Determine which values of a lead to a stable system.

4.39. Consider a first-order all-pass system with system function

$$H(s) = \frac{s-2}{s+2}$$

a. Compute and sketch the phase response of this system.

b. Determine the response $y(t)$ of the system to the input signal $x(t) = e^{-t}\,u(t)$.

c. Find analytical expressions for the magnitude responses $\left|X(\omega)\right|$ and $\left|Y(\omega)\right|$ of the input and the output signals.

d. Sketch each magnitude spectrum.

4.40. Repeat Problem 4.39 using the second-order all-pass system with the system function

$$H(s) = \frac{s^2-2s+5}{s^2+2s+5}$$

4.41. A CTLTI system has the system function

$$H(s) = \frac{(s+1)(s-2)}{(s+3)(s+4)}$$

a. Does the system have an inverse that is both causal and stable?

b. Express the system function in the form of the product of two system functions

$$H(s) = H_1(s)\, H_2(s)$$

such that $H_1(s)$ corresponds to a minimum-phase system and $H_2(s)$ corresponds to a first-order all-pass system.

c. Find the inverse system $H_1^{-1}(s)$. Is it causal? Is it stable?

d. Find the overall system function of the cascade combination of $H(s)$ and $H_1^{-1}(s)$.

4.42. For each CTLTI system defined by means of a differential equation below, determine whether a causal and stable inverse exists. If the answer is yes, find a differential equation for the inverse system.

a. $$\frac{dy(t)}{dt} + 5\,y(t) = 2\,\frac{dx(t)}{dt} + 3\,x(t)$$

b. $$\frac{d^2y(t)}{dt^2} + 7\,\frac{dy(t)}{dt} + 12\,y(t) = \frac{dx(t)}{dt} + x(t)$$

c. $$\frac{d^2y(t)}{dt^2} + 3\,y(t) = \frac{dx(t)}{dt} + 5\,x(t)$$

d. $$\frac{dy(t)}{dt} - 3\,y(t) = \frac{d^2x(t)}{dt^2} + 2\,\frac{dx(t)}{dt} + x(t)$$

4.43. Refer to the Bode magnitude plot in Example 4.41.

a. Show that the value of the asymptote for the Bode magnitude plot is 13.98 dB at $\omega = 5$.

b. Also show that, at $\omega = 300$ rad/s, the corner of the asymptote is at −3.52 dB.

4.44. Develop a direct-form block diagram for each system specified below by means of a system function. Assume that each system is causal and initially relaxed.

a. $$H(s) = \frac{s+3}{s+2}$$

b. $$H(s) = \frac{s^2+1}{s^3+9\,s^2+26\,s+24}$$

c. $$H(s) = \frac{s+5}{2\,s^3+4\,s+7}$$

4.45. Develop a cascade-form block diagram for each system specified below by means of a system function. Assume that each system is causal and initially relaxed. Use first- and second-order cascade sections and ensure that all coefficients are real.

a. $$H(s) = \frac{s^2+3\,s+2}{s^3+5\,s^2+8\,s+6}$$

b. $$H(s) = \frac{s^2+1}{s^3+9\,s^2+26\,s+24}$$

c. $H(s) = \dfrac{s^2 - 3s + 2}{s^2 + 3s + 2}$

4.46. Develop a parallel-form block diagram for each system specified in Problem 4.45. Assume that each system is causal and initially relaxed. Use first- and second-order parallel sections and ensure that all coefficients are real.

4.47. Let $X_u(s)$ be the unilateral Laplace transform of the signal $x(t)$, that is

$$x(t) \overset{\mathcal{L}_u}{\longleftrightarrow} X_u(s)$$

For each of the signals listed below, write the unilateral Laplace transform $G_u(s)$ in terms of $X_u(s)$. In each case, indicate the condition for obtaining the transform directly from $X_u(s)$ without having to find $x(t)$ first.

a. $g(t) = x(t-1)$

b. $g(t) = x(t+2)$

c. $g(t) = x(2t)$

d. $g(t) = e^{-2t}\, x(t)$

e. $g(t) = t\, x(t)$

4.48. For the RLC circuit shown in Fig. P.4.48, the initial values are

$$y(0^+) = 2\,\text{V}, \quad \text{and} \quad i(0^+) = 0.5\,\text{A}$$

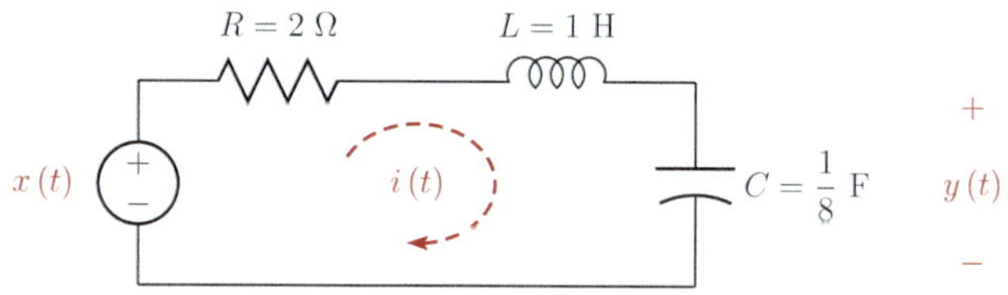

Figure P. 4.48

a. Find a differential equation between the input $x(t)$ and the output $y(t)$.

b. Obtain the system function $H(s)$ from the differential equation found in part (a).

c. Using unilateral Laplace transform determine the output signal if $x(t) = u(t)$.

d. Determine the output signal if $x(t) = e^{-2t}\, u(t)$.

4.49. For the circuit shown in Fig. P4.49, the initial values are

$$y(0^+) = 2\,\text{V}, \quad \text{and} \quad v_0(0^+) = -3\,\text{V}$$

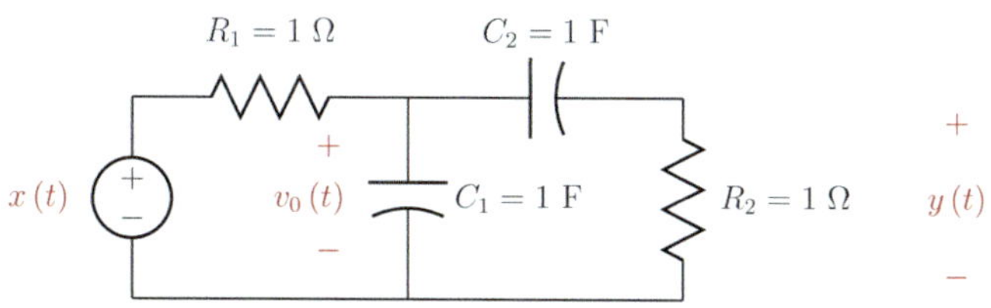

Figure P. 4.49

a. Find a differential equation between the input $x(t)$ and the output $y(t)$.

b. Obtain the system function $H(s)$ from the differential equation found in part (a).

c. Using unilateral Laplace transform determine the output signal if $x(t) = u(t)$.

d. Determine the output signal if $x(t) = e^{-t}\, u(t)$.

MATLAB Problems

4.50. Consider the transform

$$X(s) = \frac{s\,(s+2)}{(s+1)^2+9}$$

a. Determine the poles and zeros of $X(s)$. Manually construct a pole-zero diagram.

b. Write a MATLAB script to evaluate the magnitude of $X(s)$ at a grid of complex points in the s-plane. Use the function `meshgrid()` to generate the grid of complex points within the ranges $-5 < \sigma < 5$ and $-5 < \omega < 5$ with increments of 0.1 in each direction.

c. Use the function `mesh()` to produce a three-dimensional mesh plot of $|X(s)|$.

d. Evaluate the Laplace transform for $s = j\omega$ and use the function `plot3()` to plot it over the three-dimensional mesh plot.

4.51. Refer to Problem 4.8. Using MATLAB, construct a pole-zero plot for each transform listed.

4.52. Refer to Problem 4.8. For each signal that has a Fourier transform, write a script to compute $X(\omega)$ from the Laplace transform $X(s)$ given, and to graph it in the frequency range $-10 \le \omega \le 10$ rad/s.

4.53. Each transform given below corresponds to a causal signal $x(t)$. Use symbolic mathematics capabilities of MATLAB to find the inverse Laplace transform of each. Refer to MATLAB Exercise 4.10 for an example of the solution technique.

a. $X(s) = \dfrac{1}{s^2+3s+2}$

b. $X(s) = \dfrac{(s-1)\,(s-2)}{(s+1)\,(s+2)\,(s+3)}$

c. $X(s) = \dfrac{s\,(s-1)}{(s+1)^2\,(s+2)}$

4.54. A CTLTI system is described by means of the system function

$$H(s) = \frac{s\,(s+1)}{s^2+4s+13}$$

Compute and graph the response of the system to the signal

$$x(t) = e^{-0.2t}\, u(t)$$

using the following steps:

a. Define a system object "`sys`" for $H(s)$.

b. Create a vector with samples of the input signal.

c. Use the function `lsim()` to compute the output signal.

4.55. Using the function `residue()` obtain a partial fraction expansion for the transform $X(s)$ given by

$$X(s) = \frac{s^4 - 9s^3 + 30s^2 - 42s + 20}{s^5 + 12s^4 + 59s^3 + 152s^2 + 200s + 96}$$

Use the results from MATLAB to write $X(s)$ in partial fraction format.

4.56. Using system objects and the function `bode()`, construct Bode magnitude and phase plots for the system functions given below:

a. $H(s) = \dfrac{s+1}{s^2 + 5s + 6}$

b. $H(s) = \dfrac{s(s-1)}{s^3 + 5s^2 + 8s + 6}$

c. $H(s) = \dfrac{(s-1)(s+2)}{(s+1)\left(s^2 + 6s + 13\right)}$

4.57. Consider the second-order system function

$$H(s) = \frac{\omega_0^2}{s^2 + 2\zeta\omega_0 s + \omega_0^2}$$

Let $\omega_0 = 5$ rad/s. Using system objects and the function `bode()`, construct Bode magnitude and phase plots for the system for $\zeta = 0.01,\ 0.1,\ 0.5,\ 1,\ 2$.

4.58. Refer to Problem 4.41.

a. Using system objects of MATLAB, construct objects to represent $H(s)$, $H_1(s)$, and $H_1^{-1}(s)$.

b. Compute and graph the frequency responses (magnitude and phase) for $H(s)$, $H_1(s)$, $H_1^{-1}(s)$, and $H(s)\,H_1^{-1}(s)$.

c. Compute and graph the unit step responses of $H(s)$ and $H(s)\,H_1^{-1}(s)$. Comment on the results.

4.59. Refer to the RLC circuit in Problem 4.48. Using symbolic mathematics capabilities of MATLAB, write a script to compute and graph the response of the circuit for the input signal and the initial conditions specified in the problem statement. Use MATLAB Exercise 4.11 for an example of the solution technique.

4.60. Refer to the RLC circuit in Problem 4.49. Using symbolic mathematics capabilities of MATLAB, write a script to compute and graph the response of the circuit for the input signal and the initial conditions specified in the problem statement. Use MATLAB Exercise 4.11 for an example of the solution technique.

MATLAB Projects

4.61. Consider the second-order system analyzed in Section 4.5.10. Its system function is

$$H(s) = \frac{\omega_0^2}{s^2 + 2\zeta\omega_0 s + \omega_0^2}$$

Let $y(t)$ be the unit step response of this system.

a. Determine the Laplace transform of the unit step response $y(t)$. Afterward use the final value theorem to verify that the final value of the unit step response is unity.

b. Develop a MATLAB script to compute the unit step response of this system. Construct a dynamic system model for specified values of parameters ω_0 and ζ. Use the function `step()` to compute the unit step response for $0 \le t \le 30$ s with increments of $T = 0.01$ s.

c. One performance measure of the responsiveness of a system is the *rise time* which is defined as the time it takes for the unit step response to progress from 10 percent to 90 percent of its final value. Investigate the dependency of the rise time (t_r) on parameter ζ. Set $\omega_0 = 1$ rad/s and use the script from part (b) repeatedly with varying values of ζ. Measure the rise time from the graph and/or the data vector for each ζ value. Roughly sketch the relationship between t_r and ζ.

4.62. Refer to Problem 4.61. Another performance measure for a system is the *settling time* which is defined as the time it takes for the unit step response to settle within a certain percentage of its final value. Let t_s represent the 5 percent settling time, that is, the time it takes for the response to settle in the interval $0.95 < y(t) < 1.05$ and stay. Using the MATLAB script developed in problem 4.61 investigate the dependency of the settling time on parameter ζ. Use $\omega_0 = 1$ rad/s. Roughly sketch the relationship between t_s and ζ.

CHAPTER 5

State-Space Analysis of Continuous-Time Systems

Chapter Objectives

- Understand the concepts of *state variables* and *state-space representation* of systems.
- Learn the advantages of state-space modeling over other methods of representing continuous-time systems.
- Learn to derive state-space models for CTLTI systems from the knowledge of other system descriptions such as a circuit diagram, a system function, a differential equation or a block diagram. Understand how state-space models can be developed for multiple-input multiple-output systems.
- Learn how to obtain a system function or a differential equation from the state-space model of a system. Explore similarity transformations for obtaining alternative state-space models for a given system.
- Develop techniques for solving state-space models for determining the state variables and the output signal of a system. Learn homogeneous and forced solutions. Understand the significance of the state transition matrix. Learn its fundamental properties and how it is used in solving the state equation.
- Learn how to find the appropriate similarity transformation for obtaining a diagonal state matrix that leads to a set of uncoupled first-order equations for each state variable.

DOI: 10.1201/9781032675251-5

5.1 Introduction

In previous chapters of this text we have considered three different methods for describing a CTLTI system:

1. A linear constant-coefficient differential equation involving the input and the output variables of the system in the time domain;
2. A system function $H(s)$ that describes the frequency-domain behavior characteristics of the system;
3. An impulse response $h(t)$ that describes how the system responds to a unit impulse function $\delta(t)$.

We know that these three descriptions are equivalent. For a CTLTI system, the system function $H(s)$ can be obtained from the differential equation using Laplace transform techniques, specifically using the linearity and the differentiation-in-time properties of the Laplace transform. Afterward, the impulse response $h(t)$ can be determined from the system function using inverse Laplace transform techniques such as partial fraction expansion. By reversing the steps described above, it is also possible to obtain the system function from the impulse response, or to obtain the differential equation from the system function. Thus, the three definition forms listed above for a linear and time-invariant system are interchangeable since we are able to go back and forth between them freely.

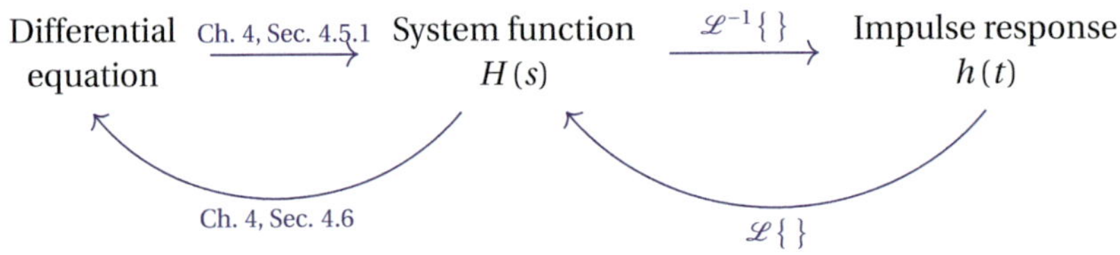

For signal-system interaction problems, that is, for problems involving the actions of a system on an input signal to produce an output signal, the solution of a general N-th order differential equation may be difficult or impossible to carry out analytically for anything other than fairly simple systems and input signals. Sometimes we can avoid the need to solve the differential equation, and attempt to solve the problem using frequency-domain techniques instead. This involves first translating the problem into the frequency domain, solving it in terms of transform-based descriptions of the signals and the systems involved, and finally translating the results back to the time domain. For systems that have nonzero initial conditions, this approach requires the use of the unilateral variant of the Laplace transform, If the system under consideration is nonlinear or time-varying or both, then the system function based approach cannot be used at all, since system functions are only available for CTLTI systems.

The system descriptions discussed above each represent the system in terms of its input-output relationship. Essentially, the system is treated like a box with input and output ports, and user interaction with the system takes place through those ports only. No attention is paid to what happens internally. Imagine a scenario where an input signal in the time interval $0 \le t \le t_0$ drives the system and we determine the output signal for the same duration. Right after that, a second input signal is applied to the system, this time in the interval $t_0 \le t \le t_1$, and we would like to compute the output signal in the same time interval. How would we approach this problem? Clearly, we cannot treat the second part of the problem as a fresh start, since the system has been brought to a certain internal *state* by the first input signal. The solution of the second part of the problem must take this into account.

In this chapter we will introduce yet another method for describing a system, namely *state-space representation* based on a set of variables called the *state variables* of the system. Consider a continuous-time system that is originally described by means of a N-th order differential equation. A state-space model for such a system can be obtained by replacing the N-th order differential equation with N first-order differential equations that involve as many state variables and their derivatives. In general, the N first-order differential equations are coupled with each other, and must be solved together as a set rather than as individual equations.

As it should be evident from the foregoing discussion, the state-space model of a system is a time-domain model rather than a frequency-domain one since it is derived from a differential equation. State-space modeling provides a number of advantages over other modeling techniques:

1. State-space modeling provides a standard method of formulating the time-domain input-output relationship of a system independent of the order of the system.
2. State-space model of a system describes the internal structure of the system, in contrast with the other description forms which simply describe the relationship between input and output signals.
3. Systems with multiple inputs and/or outputs can be handled within the standard formulation we will develop in the rest of this chapter without the need to resort to special techniques that may be dependent on the topology of the system or on the number of input or output signals.
4. State-space models for CTLTI systems are naturally suitable for use in computer simulation of these systems.
5. State-space modeling also lends itself to the representation of nonlinear and/or time-varying systems, a task that is not achievable through modeling with the use of the system function or the impulse response. A N-th order differential equation can be used for representing nonlinear and/or time-varying signals, however, computer solutions for these types of systems still necessitate the use of state-space modeling techniques.

In Section 5.2 we discuss general characteristics of state-space models. Section 5.3 focuses on the formulation of state space models for CTLTI systems. The discussion starts with the problem of obtaining a state-space model for a CTLTI system described by means of a circuit diagram, a differential equation or a system function. The state-space model of a system is not unique; a method for obtaining alternative state-space models for the same system are discussed in Section 5.3.4. Extension to multiple-input multiple-output systems is the subject of Section 5.3.5. Methods for solving state equations are discussed in Section 5.4. The concept of state transition matrix is introduced; homogeneous and forced solutions of the system are formulated. Section 5.5 covers the method for obtaining a system function from the state-space description. The problem of diagonalizing the state matrix is examined in Section 5.6.

5.2 State-Space Modeling of Continuous-Time Systems

Consider a single-input single-output continuous-time system described by a differential equation of order N. Let $r(t)$ and $y(t)$ be the input and the output signals of the system, respectively. Note that in this chapter we will change our notation a bit, and use $r(t)$ for an input signal instead of the usual $x(t)$. This change is needed in order to reserve the use of $x_i(t)$ for state variables as it is customary. The system under consideration can be thought of as a *black box* with input signal $r(t)$, output signal $y(t)$, and N internal variables $x_i(t)$ for $i = 1, 2, \ldots, N$ as shown in Fig. 5.1.

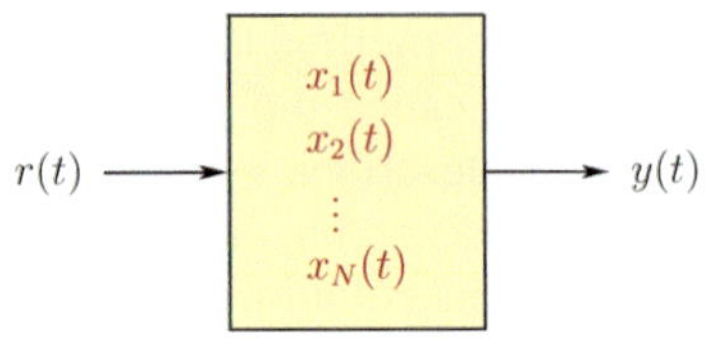

Figure 5.1 – State-space representation of a system.

These internal variables are the *state variables* or, more compactly, the *states* of the system under consideration. The system can be described through a set of first-order differential equations written for each state variable:

$$\begin{aligned}
\frac{dx_1(t)}{dt} &= f_1\left[x_1(t),\ldots,x_N(t),r(t)\right] \\
\frac{dx_2(t)}{dt} &= f_2\left[x_1(t),\ldots,x_N(t),r(t)\right] \\
&\vdots \\
\frac{dx_N(t)}{dt} &= f_N\left[x_1(t),\ldots,x_N(t),r(t)\right]
\end{aligned} \tag{5.1}$$

The first derivative of each state variable appears on the left side of each equation, and is set equal to some function $f_i[\ldots]$ of all state variables and the input $r(t)$. The equations given in Eqn. (5.1) are referred to as the *state equations* of the system. They describe how each state variable changes in the time domain in response to

1. the input signal $r(t)$ and
2. the current values of all state variables.

In a sense, observing the time-domain behavior of the state variables provides us with information about the internal operation of the system. Once the N state equations are solved for the N state variables, the output signal can be found through the use of the *output equation*

$$y(t) = g\left[x_1(t),\ldots,x_N(t),r(t)\right] \tag{5.2}$$

Eqns. (5.1) and (5.2) together form a *state-space model* for the system.

Up to this point in our discussion we focused on a system with a single input signal $r(t)$ and a single output signal $y(t)$. State-space models can be extended to systems with multiple input and/or output signals in a straightforward manner. Consider a system with K input signals $\left\{r_i(t);\ i=1,\ldots,K\right\}$ and M output signals $\left\{y_j(t);\ i=1,\ldots,M\right\}$ as shown in Fig. 5.2.

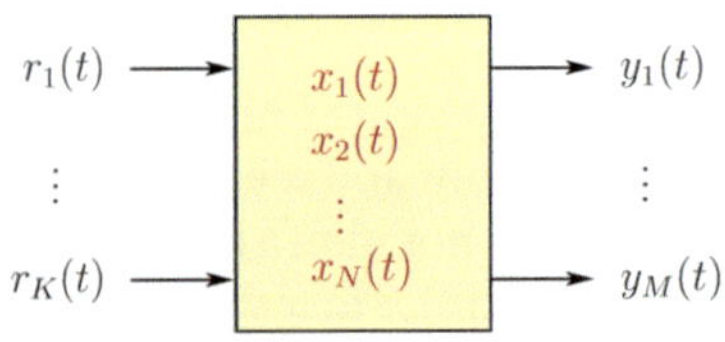

Figure 5.2 – State-space representation of a multiple-input multiple-output system.

The state equations given by Eqn. (5.1) can be modified as follows to represent this system through a state-space model:

$$\begin{aligned}
\frac{dx_1(t)}{dt} &= f_1\left[x_1(t),\ldots,x_N(t),r_1(t),\ldots,r_K(t)\right]\\
\frac{dx_2(t)}{dt} &= f_2\left[x_1(t),\ldots,x_N(t),r_1(t),\ldots,r_K(t)\right]\\
&\vdots\\
\frac{dx_N(t)}{dt} &= f_N\left[x_1(t),\ldots,x_N(t),r_1(t),\ldots,r_K(t)\right]
\end{aligned} \tag{5.3}$$

Each right-side function $f_i[\ldots]$ becomes a function of all input signals as well as the state variables. In addition to this change, the output equation given by Eqn. (5.2) needs to be replaced with a set of output equations for computing all output signals:

$$\begin{aligned}
y_1(t) &= g_1\left[x_1(t),\ldots,x_N(t),r_1(t),\ldots,r_K(t)\right]\\
y_2(t) &= g_2\left[x_1(t),\ldots,x_N(t),r_1(t),\ldots,r_K(t)\right]\\
&\vdots\\
y_M(t) &= g_M\left[x_1(t),\ldots,x_N(t),r_1(t),\ldots,r_K(t)\right]
\end{aligned} \tag{5.4}$$

Thus, the functions $g_1[\ldots], g_2[\ldots], \ldots, g_M[\ldots]$ allow the output signals to be computed from the state variables and the input signals.

In most signal-system interaction problems our main focus is on the relationship between the input and the output signals of the system. Often we are faced with one of two types of questions:

1. What is the output signal of a system in response to a specified input signal?
2. What system produces a specified output signal for a specified input signal?

The first question represents an analysis perspective and the second question represents a design perspective. Even though both questions are posed with a single-input single-output system in mind, they can easily be adapted to multiple-input multiple-output systems. In answering either question, the role of state variables is an intermediate one. We will see in later parts of this chapter that a given system can have multiple state-space models that are equivalent in terms of the input-output relationships they define. They would differ from each other, however, in terms of how they model the internal structure of the system.

In some state-space models, the state variables $x_1(t),\ldots,x_N(t)$ are chosen to correspond to actual physical signals within the system. In other models they may be purely mathematical variables that do not correspond to any physical signals. If we are in a situation where we need to understand the internal workings of a system, we may want to choose state variables that correspond to actual signals within the system. If, on the other hand, our focus is solely on the input-output relationship of the system, the role of state variables is intermediate. In such a case we would choose the state variables that lead to the simplest possible structure in the state equations given by Eqn. 5.1.

5.3 State-Space Models for CTLTI Systems

The formulation given in Eqns. (5.1) through (5.4) was not based on any assumptions about the system under consideration. Therefore, it will apply to systems that do not necessarily have to be linear or time-invariant. If the continuous-time system is also linear and time-invariant (CTLTI), the right-side functions $f_i[\ldots]$, $i = 1,2,\ldots,N$ and $g_i[\ldots]$, $i = 1,2,\ldots,M$ are linear functions of the state variables and the input signals. For a single-input single-output CTLTI system, the state-space representation is in the form

$$\begin{aligned}
\frac{dx_1(t)}{dt} &= a_{11}\,x_1(t) + a_{12}\,x_2(t) + \ldots + a_{1N}\,x_N(t) + b_1\,r(t) \\
\frac{dx_2(t)}{dt} &= a_{21}\,x_1(t) + a_{22}\,x_2(t) + \ldots + a_{2N}\,x_N(t) + b_2\,r(t) \\
&\vdots \\
\frac{dx_N(t)}{dt} &= a_{N1}\,x_1(t) + a_{N2}\,x_2(t) + \ldots + a_{NN}\,x_N(t) + b_N\,r(t)
\end{aligned} \tag{5.5}$$

and the output equation is in the form

$$y(t) = c_1\,x_1(t) + c_2\,x_2(t) + \ldots, c_N\,x_N(t) + d\,r(t) \tag{5.6}$$

Coefficients a_{ij}, b_i, c_j, and d used in the state equations and the output equation are all constants independent of the time variable. Eqns. (5.5) and (5.6) can be expressed in matrix notation as follows:

State-space representation of a single-input single-output CTLTI system:

$$\text{State equation:} \quad \dot{\mathbf{x}}(t) = \mathbf{A}\mathbf{x}(t) + \mathbf{B}\,r(t) \tag{5.7}$$

$$\text{Output equation:} \quad y(t) = \mathbf{C}\mathbf{x}(t) + d\,r(t) \tag{5.8}$$

The coefficient matrices used in Eqns. (5.7) and (5.8) are

$$\mathbf{A} = \begin{bmatrix} a_{11} & a_{12} & \ldots & a_{1N} \\ a_{21} & a_{22} & \ldots & a_{2N} \\ \vdots & & & \\ a_{N1} & a_{N2} & \ldots & a_{NN} \end{bmatrix}, \quad \mathbf{B} = \begin{bmatrix} b_1 \\ b_2 \\ \vdots \\ b_N \end{bmatrix}, \quad \mathbf{C} = \begin{bmatrix} c_1 \\ c_2 \\ \vdots \\ c_N \end{bmatrix}^T \tag{5.9}$$

The vector $\mathbf{x}(t)$ is the *state vector* that consists of the N state variables:

$$\mathbf{x}(t) = \begin{bmatrix} x_1(t) \\ x_2(t) \\ \vdots \\ x_N(t) \end{bmatrix} \tag{5.10}$$

The vector $\dot{\mathbf{x}}(t)$ contains the derivatives of the N state variables with respect to time:

$$\dot{\mathbf{x}}(t) = \begin{bmatrix} dx_1(t)/dt \\ dx_2(t)/dt \\ \vdots \\ dx_N(t)/dt \end{bmatrix} \tag{5.11}$$

Thus, a single-input single-output CTLTI system can be uniquely described by specifying the matrix **A**, the vectors **B** and **C**, and the scalar d. If the initial value $\mathbf{x}(0)$ of the state vector is known, the state equation given by Eqn. (5.7) can be solved for $\mathbf{x}(t)$. Once $\mathbf{x}(t)$ is determined, the output signal can be found through the use of Eqn. (5.8).

Some authors use a variant of the state equation in place of Eqn. (5.7), and express $\mathbf{x}(t)$ as follows:

$$\dot{\mathbf{x}}(t) = \mathbf{A}\mathbf{x}(t) + \mathbf{B}\,r(t) + \mathbf{E}\frac{dr(t)}{dt} \tag{5.12}$$

It can be shown (see Problem 5.2 at the end of this chapter) that this alternative form of the state equation can be converted into the standard form used in Eqn. (5.7) by the use of a linear transformation of the state vector, and therefore does not warrant separate analysis.

The derivation presented above can be easily adopted to systems with multiple inputs and/or outputs. This will be discussed in Section 5.3.5.

5.3.1 Obtaining a state-space model from physical description

In this section we present two examples of obtaining a state-space model for a system starting with the physical relationships that exist between the signals within the system.

Example 5.1: State-space model for RLC circuit

Find a state-space model for the RLC circuit shown in Fig. 5.3 by choosing the inductor current $i_L(t)$ and the capacitor voltage $v_C(t)$ as the state variables.

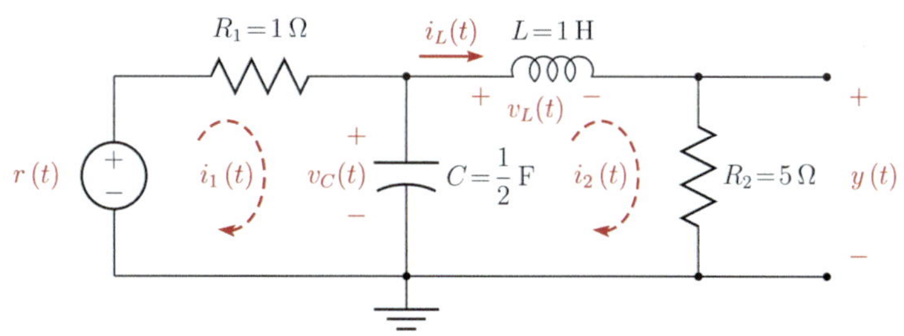

Figure 5.3 – The RLC circuit for Example 5.1.

Solution: We will begin by writing KVL for the two loops:

$$-r(t) + R_1\,i_1(t) + v_C(t) = 0 \tag{5.13}$$

$$-v_C(t) + v_L(t) + R_2\,i_2(t) = 0 \tag{5.14}$$

The voltage of an inductor is proportional to the derivative of its current:

$$v_L(t) = L\frac{di_2(t)}{dt} = L\frac{di_L(t)}{dt} \tag{5.15}$$

The current of a capacitor is proportional to the derivative of its voltage, therefore we have

$$i_1(t) - i_2(t) = C\frac{dv_c(t)}{dt}$$

which can be solved for $i_1(t)$ to yield

$$i_1(t) = C\,\frac{dv_c(t)}{dt} + i_2(t) = C\,\frac{dv_c(t)}{dt} + i_L(t) \tag{5.16}$$

Substituting Eqns. (5.15) and (5.16) into Eqns. (5.13) and (5.14) and rearranging terms we obtain

$$\begin{aligned}\frac{dv_C(t)}{dt} &= -\frac{1}{R_1 C}\,v_C(t) - \frac{1}{C}\,i_L(t) + \frac{1}{R_1 C}\,r(t)\\ \frac{di_L(t)}{dt} &= \frac{1}{L}\,v_C(t) - \frac{R_2}{L}\,i_L(t)\end{aligned}$$

The output signal is

$$y(t) = R_2\,i_L(t)$$

Let the two state variables $x_1(t)$ and $x_2(t)$ be defined as

$$x_1(t) = v_C(t) \quad \text{and} \quad x_2(t) = i_L(t)$$

Using the state variables, the state equations are

$$\frac{dx_1(t)}{dt} = -\frac{1}{R_1 C}\,x_1(t) - \frac{1}{C}\,x_2(t) + \frac{1}{R_1 C}\,r(t) \tag{5.17}$$

$$\frac{dx_2(t)}{dt} = \frac{1}{L}\,x_1(t) - \frac{R_2}{L}\,x_2(t) \tag{5.18}$$

and the output equation is

$$y(t) = R_2\,x_2(t) \tag{5.19}$$

The state-space model in Eqns. (5.17) through (5.19) can be expressed in matrix form as

$$\dot{\mathbf{x}}(t) = \begin{bmatrix} -1/R_1 C & -1/C \\ 1/L & -R_2/L \end{bmatrix} \mathbf{x}(t) + \begin{bmatrix} 1/R_1 C \\ 0 \end{bmatrix} r(t)$$

and

$$y(t) = \begin{bmatrix} 0 & R_2 \end{bmatrix} \mathbf{x}(t) + (0)\,r(t)$$

With the substitution of the numerical values for circuit elements we get

$$\dot{\mathbf{x}}(t) = \begin{bmatrix} -2 & -2 \\ 1 & -5 \end{bmatrix} \mathbf{x}(t) + \begin{bmatrix} 1 \\ 0 \end{bmatrix} r(t)$$

and

$$y(t) = \begin{bmatrix} 0 & 5 \end{bmatrix} \mathbf{x}(t) + (0)\,r(t)$$

Example 5.2: State-space model for series RLC circuit

Find a state-space model for the series RLC circuit shown in Fig. 5.4.

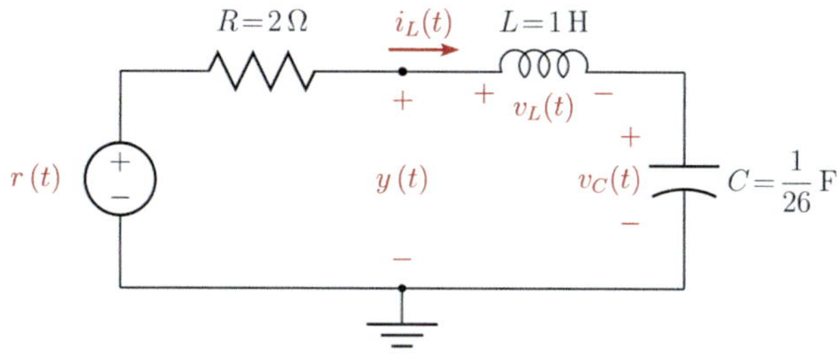

Figure 5.4 – The series RLC network for Example 5.2.

Solution: Applying KVL to the circuit, we obtain

$$r(t) = R\,i_L(t) + v_L(t) + v_C(t) \tag{5.20}$$

Using the fundamental properties of the capacitor and the inductor we have

$$v_L(t) = L\,\frac{di_L(t)}{dt} \quad \text{and} \quad i_L(t) = C\,\frac{dv_C(t)}{dt}$$

Using these relationships with Eqn. (5.20) we get the two state equations:

$$\frac{di_L(t)}{dt} = -\frac{R}{L}\,i_L(t) - \frac{1}{L}\,v_C(t) + \frac{1}{L}\,r(t)$$

and

$$\frac{dv_C(t)}{dt} = \frac{1}{C}\,i_L(t)$$

The output voltage $y(t)$ is obtained as

$$y(t) = -R\,i_L(t) + r(t)$$

We will choose the inductor current and the capacitor voltage as the state variables of the system, so we have $x_1(t) = i_L(t)$ and $x_2(t) = v_C(t)$. The state equations are

$$\begin{aligned}
\frac{dx_1(t)}{dt} &= -\frac{R}{L}\,x_1(t) - \frac{1}{L}\,x_2(t) + \frac{1}{L}\,r(t) \\
\frac{dx_2(t)}{dt} &= \frac{1}{C}\,x_1(t)
\end{aligned}$$

and the output equation is

$$y(t) = -R\,x_1(t) + r(t)$$

The state equations and the output equation can be written in matrix form as

$$\begin{bmatrix} dx_1(t)/dt \\ dx_2(t)/dt \end{bmatrix} = \begin{bmatrix} -R/L & -1/L \\ 1/C & 0 \end{bmatrix} \begin{bmatrix} x_1(t) \\ x_2(t) \end{bmatrix} + \begin{bmatrix} 1/L \\ 0 \end{bmatrix} r(t)$$

and

$$y(t) = \begin{bmatrix} -R & 0 \end{bmatrix} \begin{bmatrix} x_1(t) \\ x_2(t) \end{bmatrix} + (1)\, r(t)$$

Substituting the numerical values, we get

$$\dot{\mathbf{x}}(t) = \begin{bmatrix} -2 & -1 \\ 26 & 0 \end{bmatrix} \mathbf{x}(t) + \begin{bmatrix} 1 \\ 0 \end{bmatrix} r(t)$$

and

$$y(t) = \begin{bmatrix} -2 & 0 \end{bmatrix} \mathbf{x}(t) + (1)\, r(t)$$

5.3.2 Obtaining a state-space model from differential equation

Converting a Nth-order differential equation to a state-space model requires finding N first-order differential equations that represent the same system, and an output equation that links the output signal to the state variables. The solution is not unique. In this section we will present just one method of accomplishing this task. Consider a CTLTI system described by a differential equation in the form

$$\frac{d^N y(t)}{dt^N} + a_{N-1} \frac{d^{N-1} y(t)}{dt^{N-1}} + \ldots + a_1 \frac{dy(t)}{dt} + a_0\, y(t) = b_0\, r(t) \tag{5.21}$$

We will rewrite the differential equation so that the highest order derivative is computed from the remaining terms.

$$\frac{d^N y(t)}{dt^N} = -a_{N-1} \frac{d^{N-1} y(t)}{dt^{N-1}} - \ldots - a_1 \frac{dy(t)}{dt} - a_0\, y(t) + b_0\, r(t) \tag{5.22}$$

Let us choose the state variables as follows:

$$\begin{aligned}
x_1(t) &= y(t) \\
x_2(t) &= \frac{dy(t)}{dt} & \Longrightarrow \quad \frac{dx_1(t)}{dt} &= x_2(t) \\
x_3(t) &= \frac{d^2 y(t)}{dt^2} & \Longrightarrow \quad \frac{dx_2(t)}{dt} &= x_3(t) \\
&\vdots \\
x_N(t) &= \frac{d^{N-1} y(t)}{dt^{N-1}} & \Longrightarrow \quad \frac{dx_{N-1}(t)}{dt} &= x_N(t)
\end{aligned}$$

The resulting state equations are

$$\begin{aligned}
\frac{dx_1(t)}{dt} &= x_2(t)\\
\frac{dx_2(t)}{dt} &= x_3(t)\\
&\vdots\\
\frac{dx_{N-1}(t)}{dt} &= x_N(t)\\
\frac{dx_N(t)}{dt} &= -a_0\,x_1(t) - a_1\,x_2(t) - \ldots - a_{N-1}\,x_N(t) + b_0\,r(t)
\end{aligned} \tag{5.23}$$

The last state equation is obtained by using the state variables $x_1(t)$ through $x_N(t)$ defined above in the differential equation given by Eqn. (5.22), and writing $dx_N(t)/dt$ in terms of the state variables $x_1(t)$ through $x_N(t)$ and the input signal $r(t)$. The output equation is easily found from the assignment of the first state variable:

$$y(t) = x_1(t) \tag{5.24}$$

In matrix form, the state-space model is

$$\begin{aligned}
\dot{\mathbf{x}}(t) &= \mathbf{A}\mathbf{x}(t) + \mathbf{B}\,r(t)\\
y(t) &= \mathbf{C}\mathbf{x}(t) + d\,r(t)
\end{aligned}$$

with

$$\mathbf{A} = \begin{bmatrix} 0 & 1 & 0 & 0 & \ldots & 0\\ 0 & 0 & 1 & 0 & \ldots & 0\\ 0 & 0 & 0 & 1 & \ldots & 0\\ \vdots & \vdots & \vdots & \vdots & \ldots & \vdots\\ 0 & 0 & 0 & 0 & \ldots & 1\\ -a_0 & -a_1 & -a_2 & -a_3 & \ldots & -a_{N-1} \end{bmatrix}, \quad \mathbf{B} = \begin{bmatrix} 0\\ 0\\ 0\\ \vdots\\ 0\\ b_0 \end{bmatrix} \tag{5.25}$$

for the state equation, and

$$\mathbf{C} = \begin{bmatrix} 1 & 0 & 0 & \ldots & 0 & 0 \end{bmatrix}, \quad d = 0 \tag{5.26}$$

for the output equation. This form of the state-space model is called the *phase-variable canonical form.* It has special significance in the study of control systems. Pay attention to the structure of the coefficient matrix **A**, repeated below in partitioned form:

$$\mathbf{A} = \left[\begin{array}{c|ccccc} 0 & 1 & 0 & 0 & \ldots & 0\\ 0 & 0 & 1 & 0 & \ldots & 0\\ 0 & 0 & 0 & 1 & \ldots & 0\\ \vdots & \vdots & \vdots & \vdots & \ldots & \vdots\\ 0 & 0 & 0 & 0 & \ldots & 1\\ \hline -a_0 & -a_1 & -a_2 & -a_3 & \ldots & -a_{N-1} \end{array}\right] \tag{5.27}$$

The top left partition is a $(N-1)\times 1$ vector of zeros; the top right partition is an identity matrix of order $(N-1)$; the bottom row is a vector of differential equation coefficients in reverse order and with a sign change. This is handy for obtaining the state-space model by directly inspecting the differential equation. It is important to remember, however, that the coefficient of the highest order derivative in Eqn. (5.21) must be equal to unity for this to work.

Example 5.3: State-space model from differential equation

Find a state-space model for a CTLTI system described by the differential equation

$$\frac{d^3y(t)}{dt^3}+5\,\frac{d^2y(t)}{dt^2}+11\,\frac{dy(t)}{dt}+15\,y(t)=3\,r(t)$$

Solution: Expressing the highest order derivative as a function of the other terms leads to

$$\frac{d^3y(t)}{dt^3}=-5\,\frac{d^2y(t)}{dt^2}-11\,\frac{dy(t)}{dt}-15\,y(t)+3\,r(t) \tag{5.28}$$

State variables can be defined as

$$\begin{aligned} x_1(t) &= y(t) \\ x_2(t) &= \frac{dy(t)}{dt} &\quad\Longrightarrow\quad x_2(t) &= \frac{dx_1(t)}{dt} \\ x_3(t) &= \frac{d^2y(t)}{dt^2} &\quad\Longrightarrow\quad x_3(t) &= \frac{dx_2(t)}{dt} \end{aligned}$$

Recognizing that

$$\frac{d^3y(t)}{dt^3}=\frac{dx_3(t)}{dt}$$

the differential equation in Eqn. (5.28) can be written as

$$\frac{dx_3(t)}{dt}=-5\,x_3(t)-11\,x_2(t)-15\,x_1(t)+3\,r(t)$$

In matrix form, the state-space model is

$$\dot{\mathbf{x}}(t)=\begin{bmatrix} 0 & 1 & 0 \\ 0 & 0 & 1 \\ -15 & -11 & -5 \end{bmatrix}\mathbf{x}(t)+\begin{bmatrix} 0 \\ 0 \\ 3 \end{bmatrix} r(t)$$

$$y(t)=\begin{bmatrix} 1 & 0 & 0 \end{bmatrix}\mathbf{x}(t)$$

Example 5.4: State-space model from differential equation – revisited

Find a state-space model for a CTLTI system described by the differential equation

$$\frac{d^3y(t)}{dt^3}+5\,\frac{d^2y(t)}{dt^2}+11\,\frac{dy(t)}{dt}+15\,y(t)=3\,r(t)+7\,\frac{dr(t)}{dt}$$

Solution: This differential equation is similar to that of Example 5.3 except for the derivative of the input signal that appears on the right side of the equal sign. While it is possible to use the alternative form of state equation suggested by Eqn. (5.12) for this system, we will choose to stay with the standard form of Eqn. (5.7). Let us first rewrite the differential equation in a rearranged form.

$$\frac{d^3y(t)}{dt^3} - 7\frac{dr(t)}{dt} = -5\frac{d^2y(t)}{dt^2} - 11\frac{dy(t)}{dt} - 15\,y(t) + 3\,r(t) \tag{5.29}$$

State variables will be defined as

$$x_1(t) = y(t)$$

$$x_2(t) = \frac{dy(t)}{dt} \quad \Longrightarrow \quad x_2(t) = \frac{dx_1(t)}{dt}$$

$$x_3(t) = \frac{d^2y(t)}{dt^2} - 7\,r(t) \quad \Longrightarrow \quad x_3(t) = \frac{dx_2(t)}{dt} - 7\,r(t)$$

We have defined the last state variable $x_3(t)$ differently than the way we did in Example 5.3 so that the left side of Eqn. (5.29) is equal to the derivative of $x_3(t)$. Using these definitions, Eqn. (5.29) becomes

$$\frac{dx_3(t)}{dt} = -5\left[x_3(t) + 7\,r(t)\right] - 11\,x_2(t) - 15\,x_1(t) + 3\,r(t)$$

In matrix form, the state-space model is

$$\dot{\mathbf{x}}(t) = \begin{bmatrix} 0 & 1 & 0 \\ 0 & 0 & 1 \\ -15 & -11 & -5 \end{bmatrix} \mathbf{x}(t) + \begin{bmatrix} 0 \\ 7 \\ 32 \end{bmatrix} r(t)$$

$$y(t) = \begin{bmatrix} 1 & 0 & 0 \end{bmatrix} \mathbf{x}(t)$$

When we compare Examples 5.3 and 5.4 we see that coefficient matrix **A** is the same in both of them, but the vector **B** is different. The inspection method for deriving the state-space model directly from the differential equation would not be as straightforward to use when we have derivatives of the input signal appearing on the right side. We will, however, generalize the relationship between the coefficients of the differential equation and the state-space model when we consider the derivation of the latter from the system function in the next section.

5.3.3 Obtaining a state-space model from system function

If the CTLTI system under consideration has a system function that is simple in the sense that it has no multiple poles and all of its poles are real-valued, then a state-space model can be easily found using partial fraction expansion. This idea will be explored in the next example.

Example 5.5: State-space model from s-domain system function

It can be shown that the RLC circuit that was used in Example 5.1 and shown in Fig. 5.3 has the system function

$$G(s) = \frac{Y(s)}{R(s)} = \frac{10}{(s+3)\,(s+4)}$$

Find an alternative, but equivalent, state-space model for the system based on this system function.

Solution: Expanding the system function into partial fractions and then solving for the transform $Y(s)$ we get

$$Y(s) = \frac{10}{s+3}\,R(s) - \frac{10}{s+4}\,R(s)$$

Let us define $Z_1(s)$ and $Z_2(s)$ as

$$Z_1(s) = \frac{10}{s+3}\,R(s) \tag{5.30}$$

and

$$Z_2(s) = \frac{-10}{s+4}\,R(s) \tag{5.31}$$

so that $Y(s) = Z_1(s) + Z_2(s)$. Rearranging the terms in Eqns. (5.30) and (5.31) we get

$$s\,Z_1(s) = -3\,Z_1(s) + 10\,R(s) \tag{5.32}$$

and similarly

$$s\,Z_2(s) = -4\,Z_2(s) - 10\,R(s) \tag{5.33}$$

Taking the inverse Laplace transform of Eqns. (5.32) and (5.33) leads to the two first-order differential equations

$$\frac{dz_1(t)}{dt} = -3z_1(t) + 10\,r(t) \tag{5.34}$$

$$\frac{dz_2(t)}{dt} = -4z_2(t) - 10\,r(t) \tag{5.35}$$

Using $z_1(t)$ and $z_2(t)$ as the new state variables, the state-space model can be written as

$$\begin{bmatrix} dz_1(t)/dt \\ dz_2(t)/dt \end{bmatrix} = \begin{bmatrix} -3 & 0 \\ 0 & -4 \end{bmatrix} \begin{bmatrix} z_1(t) \\ z_2(t) \end{bmatrix} + \begin{bmatrix} 10 \\ -10 \end{bmatrix} r(t)$$

and

$$y(t) = \begin{bmatrix} 1 & 1 \end{bmatrix} \begin{bmatrix} z_1(t) \\ z_2(t) \end{bmatrix}$$

Example 5.5 was made simple by the fact that the partial fraction expansion of the system function contains only first-order terms with real coefficients. An interesting consequence of the simple technique utilized is that the state equations obtained in Eqns. (5.34) and (5.35) are uncoupled, that is, each state equation involves only one of the state variables. This corresponds to a diagonal coefficient matrix **A**. Consequently, we have the advantage of being able to solve the state equations independently of each other.

If the system function does not have the properties that allowed the simple solution of Example 5.5, finding a state-space model for a system function involves the use of a block diagram as an intermediate step. The development of this idea will be presented using a third-order system function as an example; generalization to higher-order system functions will be readily apparent. Consider the system function

$$G(s) = \frac{Y(s)}{X(s)} = \frac{b_3 s^3 + b_2 s^2 + b_1 s + b_0}{s^3 + a_2 s^2 + a_1 s + a_0} \tag{5.36}$$

Using the techniques discussed in Section 4.6.1 of Chapter 4, the simulation diagram shown in Fig. 5.5 can be obtained for $G(s)$.

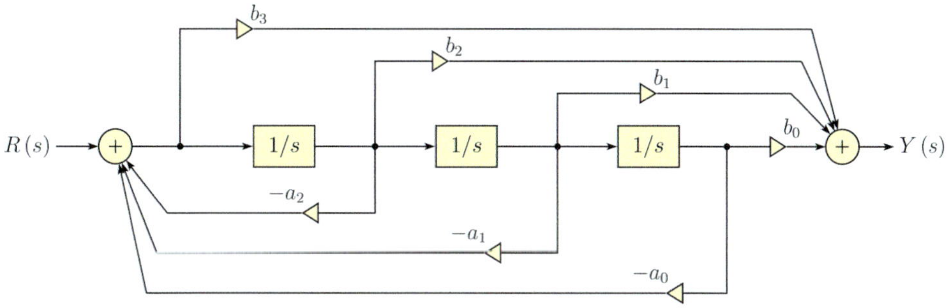

Figure 5.5 – Simulation diagram for $G(s)$ of Eqn. (5.36).

We know from the integration property of the Laplace transform (see Section 4.3.8 of Chapter 4) that multiplying a transform by $1/s$ is equivalent to integrating the corresponding signal in the time domain. Therefore, a time-domain equivalent of the block diagram in Fig. 5.5 can be obtained by replacing $1/s$ components with integrators and using time-domain versions of input and output quantities. The resulting diagram is shown in Figure 5.6.

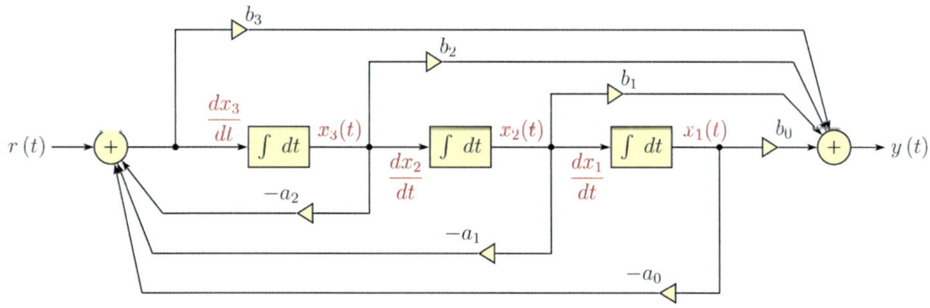

Figure 5.6 – Selecting states from the simulation diagram for $G(s)$.

Obtaining a state-space model from a block diagram:

A state-space model can be obtained from the diagram of Fig. 5.6 as follows:

1. Designate the output signal of each integrator as a state variable. These are marked in Fig. 5.6 as $x_1(t)$, $x_2(t)$, and $x_3(t)$.
2. The input of each integrator becomes the derivative of the corresponding state variable. These are marked in Fig. 5.6 as dx_1/dt, dx_2/dt, and dx_3/dt.
3. Express the signal at the input of each integrator as a function of state variables and the input signal $r(t)$ as dictated by the diagram.

Following state equations are obtained by inspecting the diagram in Fig. 5.6:

$$\frac{dx_1(t)}{dt} = x_2(t) \tag{5.37}$$

$$\frac{dx_2(t)}{dt} = x_3(t) \tag{5.38}$$

$$\frac{dx_3(t)}{dt} = -\,a_0\,x_1(t) - a_1\,x_2(t) - a_0\,x_3(t) + r(t) \tag{5.39}$$

The output signal can be written as

$$y(t) = b_0\,x_1(t) + b_1\,x_2(t) + b_2\,x_3(t) + b_3\,\frac{dx_3(t)}{dt} \tag{5.40}$$

We do not like to see the derivative of a state variable on the right side of the output equation. Substituting Eqn. (5.39 into Eqn. (5.40) yields the proper form of the output equation

$$y(t) = (b_0 - b_3\,a_0)\,x_1(t) + (b_1 - b_3\,a_1)\,x_2(t) + (b_2 - b_3\,a_2)\,x_3(t) + b_3\,r(t) \tag{5.41}$$

Expressed in matrix form, the state-space model is

$$\dot{\mathbf{x}}(t) = \begin{bmatrix} 0 & 1 & 0 \\ 0 & 0 & 1 \\ -a_0 & -a_1 & -a_2 \end{bmatrix} \mathbf{x}(t) + \begin{bmatrix} 0 \\ 0 \\ 1 \end{bmatrix} r(t) \tag{5.42}$$

$$y(t) = \begin{bmatrix} (b_0 - b_3\,a_0) & (b_1 - b_3\,a_1) & (b_2 - b_3\,a_2) \end{bmatrix} \mathbf{x}(t) + b_3\,r(t) \tag{5.43}$$

The results can easily be extended to a N-th order system function in the form

$$G(s) = \frac{Y(s)}{X(s)} = \frac{b_N s^N + b_{N-1} s^{N-1} + \ldots + b_1 s + b_0}{s^N + a_{N-1} s^{N-1} + \ldots + a_1 s + a_0} \tag{5.44}$$

to produce the state-space model

$$\dot{\mathbf{x}}(t) = \begin{bmatrix} 0 & 1 & 0 & 0 & \ldots & 0 \\ 0 & 0 & 1 & 0 & \ldots & 0 \\ 0 & 0 & 0 & 1 & \ldots & 0 \\ \vdots & \vdots & \vdots & \vdots & \ldots & \vdots \\ 0 & 0 & 0 & 0 & \ldots & 1 \\ -a_0 & -a_1 & -a_2 & -a_3 & \ldots & -a_{N-1} \end{bmatrix} \mathbf{x}(t) + \begin{bmatrix} 0 \\ 0 \\ 0 \\ \vdots \\ 0 \\ 1 \end{bmatrix} r(t) \tag{5.45}$$

$$y(t) = \begin{bmatrix} (b_0 - b_N\,a_0) & (b_1 - b_N\,a_1) & \ldots & (b_{N-2} - b_N\,a_{N-2}) & (b_{N-1} - b_N\,a_{N-1}) \end{bmatrix} \mathbf{x}(t) + b_N\,r(t) \tag{5.46}$$

It is interesting to see that the structure of the coefficient matrix **A** is the same as in Eqn. (5.27). Similarly, the vector **B** is identical to that found in Section 5.3.2.

Software resource: See MATLAB Exercise 5.1.

Example 5.6: State-space model from block diagram

Find a state-space model for a CTLTI system described through the system function

$$G(s) = \frac{2s^3 - 26s + 24}{s^4 + 7s^3 + 21s^2 + 37s + 30}$$

Solution: A block diagram for this system was constructed in Example 4.42 of Chapter 4 and was shown in Fig. 4.72. That diagram is shown again in Fig. 5.7 with notation adapted to time domain and with the designations of the four state variables marked.

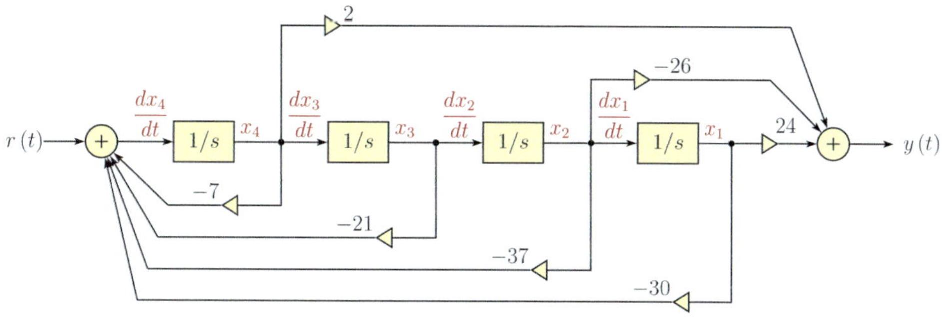

Figure 5.7 – Block diagram for Example 5.6.

The state equations are

$$\frac{dx_1(t)}{dt} = x_2(t)$$

$$\frac{dx_2(t)}{dt} = x_3(t)$$

$$\frac{dx_3(t)}{dt} = x_4(t)$$

$$\frac{dx_4(t)}{dt} = -30\,x_1(t) - 37\,x_2(t) - 21\,x_3(t) - 7\,x_4(t) + r(t)$$

and the output equation is

$$y(t) = 24\,x_1(t) - 26\,x_2(t) + 2\,x_4(t)$$

In matrix form we get

$$\dot{\mathbf{x}}(t) = \begin{bmatrix} 0 & 1 & 0 & 0 \\ 0 & 0 & 1 & 0 \\ 0 & 0 & 0 & 1 \\ -30 & -37 & -21 & -7 \end{bmatrix} \mathbf{x}(t) + \begin{bmatrix} 0 \\ 0 \\ 0 \\ 1 \end{bmatrix} r(t)$$

and

$$y(t) = \begin{bmatrix} 24 & -26 & 0 & 2 \end{bmatrix} \mathbf{x}(t)$$

Example 5.7: State-space model from block diagram – revisited

Find a state-space model for a CTLTI system with the system function

$$G(s) = \frac{2s^3 - 26s + 24}{s^4 + 7s^3 + 21s^2 + 37s + 30}$$

using the cascade form block diagram.

Solution: A cascade form block diagram for this system was constructed in Example 4.43 of Chapter 4 and was shown in Fig. 4.75. That diagram is shown again in Fig. 5.8 with notation adapted to time domain and with the designations of the four state variables marked.

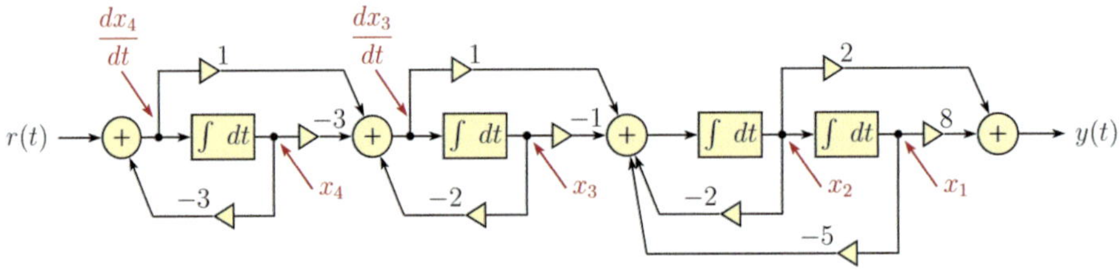

Figure 5.8 – Block diagram for Example 5.7.

Writing the input signal of each integrator in terms of the other signals in the diagram we obtain

$$\frac{dx_1(t)}{dt} = x_2(t) \tag{5.47}$$

$$\frac{dx_2(t)}{dt} = -5\,x_1(t) - 2\,x_2(t) - x_3(t) + \frac{dx_3(t)}{dt} \tag{5.48}$$

$$\frac{dx_3(t)}{dt} = -2\,x_3(t) - 3\,x_4(t) + \frac{dx_4(t)}{dt} \tag{5.49}$$

$$\frac{dx_4(t)}{dt} = -3\,x_4(t) + r(t) \tag{5.50}$$

and the output signal is

$$y(t) = 8\,x_1(t) + 2\,x_2(t)$$

State equations should not contain any derivatives on the right side of the equal sign. Using Eqn. (5.50) in Eqn. (5.49) yields

$$\frac{dx_3(t)}{dt} = -2\,x_3(t) - 6\,x_4(t) + r(t) \tag{5.51}$$

Similarly, using Eqn. (5.51) in Eqn. (5.48) we get

$$\frac{dx_2(t)}{dt} = -5\,x_1(t) - 2\,x_2(t) - 3\,x_3(t) - 6\,x_4(t) + r(t) \tag{5.52}$$

The state-space model can now be put in matrix form using Eqns. (5.47), (5.50), (5.51), and (5.52). The result is

$$\dot{\mathbf{x}}(t) = \begin{bmatrix} 0 & 1 & 0 & 0 \\ -5 & -2 & -3 & -6 \\ 0 & 0 & -2 & -6 \\ 0 & 0 & 0 & -3 \end{bmatrix} \mathbf{x}(t) + \begin{bmatrix} 0 \\ 1 \\ 1 \\ 1 \end{bmatrix} r(t)$$

and

$$y(t) = \begin{bmatrix} 8 & 2 & 0 & 0 \end{bmatrix} \mathbf{x}(t)$$

5.3.4 Alternative state-space models

In Examples 5.1 and 5.2 we have defined the state variables of each system as the capacitor voltage and the inductor current. These choices were mainly motivated by the fact that the derivatives of these two quantities readily appear in KVL equations for either circuit.

Other state variables could have been defined for these systems as well, leading to alternative state-space models. Consider a CTLTI system defined through Eqns. (5.7) and (5.8). Let $\mathbf{z}(t)$ be a length-N column vector of variables $z_1(t), \ldots, z_N(t)$, related to the original state vector $\mathbf{x}(t)$ of the CTLTI system through

$$\mathbf{x}(t) = \mathbf{P}\mathbf{z}(t) \tag{5.53}$$

where $\mathbf{P}$ is a constant N by N transformation matrix. The only requirement on the matrix $\mathbf{P}$ is that it be non-singular (invertible). Substituting Eqn. (5.53) into Eqn. (5.7) leads to

$$\mathbf{P}\dot{\mathbf{z}}(t) = \mathbf{A}\mathbf{P}\mathbf{z}(t) + \mathbf{B}\,r(t) \tag{5.54}$$

Multiplication of both sides of Eqn. (5.54) with the inverse of $\mathbf{P}$ from the left yields

$$\dot{\mathbf{z}}(t) = \mathbf{P}^{-1}\mathbf{A}\mathbf{P}\mathbf{z}(t) + \mathbf{P}^{-1}\mathbf{B}\,r(t) \tag{5.55}$$

Finally, using Eqn. (5.53) with Eqn. (5.8) we get

$$y(t) = \mathbf{C}\mathbf{P}\mathbf{z}(t) + d\,r(t) \tag{5.56}$$

Eqns. (5.55) and (5.56) provide us with an alternative state-space model for the same CTLTI system, using a new state vector $\mathbf{z}(t)$. They can be written in the standard form of Eqns. (5.7) and (5.8) as

$$\dot{\mathbf{z}}(t) = \tilde{\mathbf{A}}\mathbf{z}(t) + \tilde{\mathbf{B}}\,r(t) \tag{5.57}$$

$$y(t) = \tilde{\mathbf{C}}\mathbf{z}(t) + d\,r(t) \tag{5.58}$$

with newly defined coefficient matrices

$$\tilde{\mathbf{A}} = \mathbf{P}^{-1}\mathbf{A}\mathbf{P} \tag{5.59}$$

$$\tilde{\mathbf{B}} = \mathbf{P}^{-1}\mathbf{B} \tag{5.60}$$

and

$$\tilde{\mathbf{C}} = \mathbf{C}\mathbf{P} \tag{5.61}$$

The transformation relationship $\tilde{\mathbf{A}} = \mathbf{P}^{-1}\mathbf{A}\mathbf{P}$ is known in control system theory as a *similarity transformation.* It can be used for obtaining alternative state-space models of a system. Each choice of the transformation matrix $\mathbf{P}$ leads to a different, but equivalent, state-space model. In each of these alternative models the input-output relationship remains the same although the state variables are different. Since any non-singular matrix $\mathbf{P}$ can be used in a similarity transformation, and infinite number of different state-space models can be found for a given system. In the next several examples we will explore this concept further.

Software resource: See MATLAB Exercise 5.2.

Example 5.8: Alternative state-space model for series RLC circuit

Refer to the system in Example 5.2 again. Since the system is of second order, the input-output relationship can be expressed through a second-order differential equation between $y(t)$ and $r(t)$. Find such a differential equation from the state-space model found in Example 5.2. Afterward obtain an alternative state-space model for the same system directly from the second-order differential equation.

Solution: The state-space model found in Example 5.2 was

$$\dot{\mathbf{x}}(t) = \begin{bmatrix} -2 & -1 \\ 26 & 0 \end{bmatrix} \mathbf{x}(t) + \begin{bmatrix} 1 \\ 0 \end{bmatrix} r(t)$$

and

$$y(t) = \begin{bmatrix} -2 & 0 \end{bmatrix} \mathbf{x}(t) + r(t)$$

If we write the state equations and the output equation in open form, we obtain

$$\frac{dx_1(t)}{dt} = -2\,x_1(t) - x_2(t) + r(t) \tag{5.62}$$

$$\frac{dx_2(t)}{dt} = 26\,x_1(t) \tag{5.63}$$

$$y(t) = -2\,x_1(t) + r(t) \tag{5.64}$$

Let us differentiate Eqn. (5.64) to obtain

$$\frac{dy(t)}{dt} = -2\,\frac{dx_1(t)}{dt} + \frac{dr(t)}{dt} \tag{5.65}$$

Substituting Eqn. (5.62) into Eqn. (5.65)

$$\frac{dy(t)}{dt} = 4\,x_1(t) + 2\,x_2(t) - 2\,r(t) + \frac{dr(t)}{dt} \tag{5.66}$$

Differentiating Eqn. (5.66) one more time and using the result with Eqns. (5.62) and (5.63) leads to

$$\frac{d^2y(t)}{dt^2} = 44\,x_1(t) - 4\,x_2(t) + 4\,r(t) - 2\,\frac{dr(t)}{dt} + \frac{d^2r(t)}{dt^2} \tag{5.67}$$

Finally, eliminating the state variables $x_1(t)$ and $x_2(t)$ between Eqns. (5.64), (5.66), and (5.67) yields the second-order differential equation

$$\frac{d^2y(t)}{dt^2} = -2\,\frac{dy(t)}{dt} - 26\,y(t) + \frac{d^2r(t)}{dt^2} + 26\,r(t) \tag{5.68}$$

A state-space model in phase-variable canonical form can be obtained from the differential equation in Eqn. (5.68) using the technique discussed in Section 5.3.3. The result is

$$\dot{\mathbf{z}}(t) = \begin{bmatrix} 0 & 1 \\ -26 & -2 \end{bmatrix} \mathbf{z}(t) + \begin{bmatrix} 0 \\ 1 \end{bmatrix} r(t)$$

$$y(t) = \begin{bmatrix} 0 & -2 \end{bmatrix} \mathbf{z}(t) + r(t)$$

Software resource: exct_5_8.m

Example 5.9: State variable transformation

Show that the alternative state-space model found in Example 5.8 could have been obtained from the state-space model of Example 5.2 through the transformation

$$\mathbf{x}(t) = \mathbf{P}\mathbf{z}(t) = \begin{bmatrix} 0 & 1 \\ 26 & 0 \end{bmatrix} \mathbf{z}(t)$$

Solution: The original state-space model found in Example 5.1 has the coefficient matrices

$$\mathbf{A} = \begin{bmatrix} -2 & -1 \\ 26 & 0 \end{bmatrix}, \quad \mathbf{B} = \begin{bmatrix} 1 \\ 0 \end{bmatrix}, \quad \mathbf{C} = \begin{bmatrix} -2 & 0 \end{bmatrix}, \quad d = 1$$

The inverse of the transformation matrix **P** specified above is

$$\mathbf{P}^{-1} = \begin{bmatrix} 0 & 1/26 \\ 1 & 0 \end{bmatrix}$$

Applying the similarity transformation given byEqn. (5.59) to the coefficient matrix **A** yields

$$\tilde{\mathbf{A}} = \mathbf{P}^{-1}\mathbf{A}\mathbf{P} = \begin{bmatrix} 0 & 1/26 \\ 1 & 0 \end{bmatrix} \begin{bmatrix} -2 & -1 \\ 26 & 0 \end{bmatrix} \begin{bmatrix} 0 & 1 \\ 26 & 0 \end{bmatrix} = \begin{bmatrix} 0 & 1 \\ -26 & -2 \end{bmatrix}$$

which is identical to the matrix $\tilde{\mathbf{A}}$ found in Example 5.8. Vectors $\tilde{\mathbf{B}}$ and $\tilde{\mathbf{C}}$ can also be found using Eqns. (5.60) and (5.61) as

$$\tilde{\mathbf{B}} = \mathbf{P}^{-1}\mathbf{B} = \begin{bmatrix} 0 & 1/26 \\ 1 & 0 \end{bmatrix} \begin{bmatrix} 1 \\ 0 \end{bmatrix} = \begin{bmatrix} 0 \\ 1 \end{bmatrix}$$

and

$$\tilde{\mathbf{C}} = \mathbf{C}\mathbf{P} = \begin{bmatrix} -2 & 0 \end{bmatrix} \begin{bmatrix} 0 & 1 \\ 26 & 0 \end{bmatrix} = \begin{bmatrix} 0 & -2 \end{bmatrix}$$

Vectors $\tilde{\mathbf{B}}$ and $\tilde{\mathbf{C}}$ also agree with those found in Example 5.8.

Software resource: exct_5_9.m

5.3.5 CTLTI systems with multiple inputs and/or outputs

A CTLTI system with multiple input and/or output signals can also be represented with a state-space model in the standard form of Eqns. (5.7) and (5.8). Consider a CTLTI system with K input signals $\{\, r_i(t);\ i = 1,\ldots,K \,\}$ and M output signals $\{\, y_j(t);\ i = 1,\ldots,M \,\}$. Let $\mathbf{r}(t)$ be the $K \times 1$

vector of input signals:

$$\mathbf{r}(t)=\begin{bmatrix} r_1(t) \\ r_2(t) \\ \vdots \\ r_K(t) \end{bmatrix} \tag{5.69}$$

Similarly, $\mathbf{y}(t)$ is the $M \times 1$ vector of output signals.

$$\mathbf{y}(t)=\begin{bmatrix} y_1(t) \\ y_2(t) \\ \vdots \\ y_M(t) \end{bmatrix} \tag{5.70}$$

With these new vectors, the state equation is in the form

$$\dot{\mathbf{x}}(t)=\mathbf{A}\mathbf{x}(t)+\mathbf{B}\mathbf{r}(t) \tag{5.71}$$

The matrix $\mathbf{A}$ is still the $N \times N$ coefficient matrix and the vector $\mathbf{x}(t)$ is still the $N \times 1$ state vector as before. Dimensions of the matrix $\mathbf{B}$ will need to be changed, however, to make it compatible with the $K \times 1$ vector $\mathbf{r}(t)$. Therefore, $\mathbf{B}$ is now a $N \times K$ matrix.

The output equation is in the form

$$\mathbf{y}(t)=\mathbf{C}\mathbf{x}(t)+\mathbf{D}\mathbf{r}(t) \tag{5.72}$$

Dimensions of $\mathbf{C}$ need to be updated as well. Since $\mathbf{y}(t)$ is a $M \times 1$ vector, $\mathbf{C}$ is a $M \times N$ matrix. Additionally, the scalar d in the output equation in Eqn. (5.8) is replaced with a $M \times K$ matrix $\mathbf{D}$.

Example 5.10: Multiple-input multiple-output system

Consider a CTLTI system with two inputs and two outputs that is described by a pair of differential equations:

$$\frac{d^2y_1(t)}{dt^2}+3\frac{dy_1(t)}{dt}+2\,y_1(t)+y_2(t)=3\,r_1(t)+r_2(t)$$

$$\frac{dy_2(t)}{dt}+3\,y_2(t)+4\,y_1(t)=r_1(t)+4\frac{dr_1(t)}{dt}-2\,r_2(t)$$

Find a state-space model for this system.

Solution: We will begin by rearranging the terms of the two differential equations to obtain

$$\frac{d^2y_1(t)}{dt^2}=-3\frac{dy_1(t)}{dt}-2\,y_1(t)-y_2(t)+3\,r_1(t)+r_2(t) \tag{5.73}$$

$$\frac{dy_2(t)}{dt}-4\frac{dr_1(t)}{dt}=-3\,y_2(t)-4\,y_1(t)+r_1(t)-2\,r_2(t) \tag{5.74}$$

Let us define the state variables as follows:

$$x_1(t)=y_1(t)\,,\qquad x_2(t)=\frac{dy_1(t)}{dt}\,,\qquad x_3(t)=y_2(t)-4\,r_1(t) \tag{5.75}$$

It naturally follows that

$$\frac{dx_1(t)}{dt}=x_2(t) \tag{5.76}$$

Using the definitions of the state variables in Eqns. (5.75) and (5.76) we obtain

$$\begin{aligned}\frac{dx_2(t)}{dt} &= -3\,x_2(t) - 2\,x_1(t) - \left[x_3(t) + 4\,r_1(t)\right] + 3\,r_1(t) + r_2(t)\\ &= -2\,x_1(t) - 3\,x_2(t) - x_3(t) - r_1(t) + r_2(t)\end{aligned} \tag{5.77}$$

and

$$\begin{aligned}\frac{dx_3(t)}{dt} &= -3\left[x_3(t) + 4\,r_1(t)\right] - 4\,x_1(t) + r_1(t) - 2\,r_2(t)\\ &= -4\,x_1(t) - 3\,x_3(t) - 11\,r_1(t) - 2\,r_2(t)\end{aligned} \tag{5.78}$$

Eqns. (5.76), (5.77) and (5.78) can be put in matrix form as

$$\dot{\mathbf{x}}(t) = \begin{bmatrix} 0 & 1 & 0 \\ -2 & -3 & -1 \\ -4 & 0 & -3 \end{bmatrix} \mathbf{x}(t) + \begin{bmatrix} 0 & 0 \\ -1 & 1 \\ -11 & -2 \end{bmatrix} \mathbf{r}(t)$$

and

$$\mathbf{y}(t) = \begin{bmatrix} 1 & 0 & 0 \\ 0 & 0 & 1 \end{bmatrix} \mathbf{x}(t) + \begin{bmatrix} 0 & 0 \\ 0 & 4 \end{bmatrix} \mathbf{r}(t)$$

5.4 Solution of State-Space Model

Before deriving the general solution for the state-space model of a continuous-time system, it will be useful to consider a special case in which the coefficient matrix **A** is upper-triangular, that is, all of its elements below the main diagonal are equal to zero. Example 5.11 will demonstrate that a step by step solution can be obtained in this case.

Example 5.11: Solving state equations with upper-triangular coefficient matrix

A system is described by the state-space model

$$\dot{\mathbf{x}}(t) = \begin{bmatrix} -3 & 1 & 0 \\ 0 & -2 & 5 \\ 0 & 0 & -1 \end{bmatrix} \mathbf{x}(t) + \begin{bmatrix} 0 \\ 2 \\ 1 \end{bmatrix} r(t)$$

and

$$y(t) = \begin{bmatrix} 1 & 0 & 0 \end{bmatrix} \mathbf{x}(t)$$

The initial value of the state vector is

$$\mathbf{x}(0) = \begin{bmatrix} 0 \\ 3 \\ 2 \end{bmatrix}$$

Determine the output signal in response to a unit step input signal.

Solution: The differential equation for the state variable $x_3(t)$ does not include either of the other two state variables, therefore it is convenient to start with it.

$$\frac{dx_3(t)}{dt} = -x_3(t) + r(t) = -x_3(t) + u(t)\,, \qquad x_3(0) = 2$$

Let us apply the solution technique developed in Section 2.5.1 of Chapter 2. Using Eqn. (2.49) with $\alpha = 1$ and $t_0 = 0$, the solution for $x_3(t)$ is

$$x_3(t) = e^{-t}\,x_3(0) + \int_0^t e^{-(t-\tau)}\,u(\tau)\,d\tau\,, \qquad t \ge 0$$

Using the initial value $x_3(0) = 2$ and realizing that $u(\tau) = 1$ within the integration interval we obtain

$$x_3(t) = 2\,e^{-t} + \int_0^t e^{-(t-\tau)}\,d\tau = \left(1 + e^{-t}\right)u(t)$$

Having determined $x_3(t)$, the differential equation for $x_2(t)$ is

$$\begin{aligned}\frac{dx_2(t)}{dt} &= -2\,x_2(t) + 5\,x_3(t) + 2\,r(t)\\ &= -2\,x_2(t) + 5\left(1 + e^{-t}\right)u(t) + 2\,u(t)\\ &= -2\,x_2(t) + \left(7 + 5\,e^{-t}\right)u(t)\,, \qquad x_2(0) = 3\end{aligned}$$

Repeating the procedure outlined above, $x_2(t)$ is determined as

$$\begin{aligned}x_2(t) =& 3\,e^{-2t} + \int_0^t e^{-2(t-\tau)}\left(7 + 5e^{-\tau}\right)u(\tau)\,d\tau\\ =& \left(\frac{7}{2} + 5\,e^{-t} - \frac{11}{2}\,e^{-2t}\right)u(t)\end{aligned}$$

Finally, the differential equation for $x_1(t)$ is

$$\begin{aligned}\frac{dx_1(t)}{dt} &= -3\,x_1(t) + x_2(t)\\ &= -3\,x_1(t) + \left(\frac{7}{2} + 5\,e^{-t} - \frac{11}{2}\,e^{-2t}\right)u(t)\,, \qquad x_1(0) = 0\end{aligned}$$

which can be solved to yield

$$x_1(t) = \left(\frac{7}{6} + \frac{5}{2}\,e^{-t} - \frac{11}{2}\,e^{-2t} + \frac{11}{6}\,e^{-3t}\right)u(t)$$

The output of the system is found using the output equation:

$$y(t) = x_1(t) = \left(\frac{7}{6} + \frac{5}{2}\,e^{-t} - \frac{11}{2}\,e^{-2t} + \frac{11}{6}\,e^{-3t}\right)u(t)$$

The solution of Example 5.11 was made simple by the fact that the coefficient matrix is upper-triangular. Starting with the last state equation and working our way up we were able to resolve the dependencies between state equations one at a time. Next, we will develop a general solution method that does not rely on an upper-triangular coefficient matrix. The solution will be obtained in two steps: First, the solution of the homogeneous state equation will be found. Next, a particular solution will be found based on the input signal $r(t)$. The complete solution of the state equation is obtained as the sum of the homogeneous solution and the particular solution. The derivation of the solution procedure should be compared to the development in Section 2.5.2 of Chapter 2 for an ordinary differential equation.

5.4.1 Homogeneous solution

Consider the homogeneous state equation obtained by setting $r(t) = 0$:

$$\dot{\mathbf{x}}(t) = \mathbf{A}\mathbf{x}(t) \tag{5.79}$$

Each of the state variables $x_i(t)$ in the state vector $\mathbf{x}(t)$ can be expressed by a Taylor series in the form

$$x_i(t) = x_i(0) + \left.\frac{dx_i(t)}{dt}\right|_{t=0} t + \frac{1}{2!}\left.\frac{d^2x_i(t)}{dt^2}\right|_{t=0} t^2 + \ldots + \frac{1}{n!}\left.\frac{d^n x_i(t)}{dt^n}\right|_{t=0} t^n + \ldots \tag{5.80}$$

Alternatively, Eqn. (5.80) can be generalized to construct a vector form of Taylor series for expressing the state vector $\mathbf{x}(t)$ as

$$\mathbf{x}(t) = \mathbf{x}(0) + \left.\frac{d\mathbf{x}(t)}{dt}\right|_{t=0} t + \frac{1}{2!}\left.\frac{d^2\mathbf{x}(t)}{dt^2}\right|_{t=0} t^2 + \ldots + \frac{1}{n!}\left.\frac{d^n\mathbf{x}(t)}{dt^n}\right|_{t=0} t^n + \ldots \tag{5.81}$$

Each vector derivative in Eqn. (5.81) corresponds to a vector of corresponding derivatives of all state variables:

$$\left.\frac{d^n\mathbf{x}(t)}{dt^n}\right|_{t=0} = \begin{bmatrix} d^n x_1(t)/dt^n \\ d^n x_2(t)/dt^n \\ \vdots \\ d^n x_N(t)/dt^n \end{bmatrix}_{t=0} \tag{5.82}$$

Using the homogeneous state equation, the first derivative evaluated at $t = 0$ is

$$\left.\frac{d\mathbf{x}(t)}{dt}\right|_{t=0} = \left.\mathbf{A}\mathbf{x}(t)\right|_{t=0} = \mathbf{A}\mathbf{x}(0) \tag{5.83}$$

The second derivative of the state vector evaluated at time $t = 0$ is obtained by differentiating both sides of Eqn. (5.79) and using it in conjunction with Eqn. (5.83):

$$\left.\frac{d^2\mathbf{x}(t)}{dt^2}\right|_{t=0} = \mathbf{A}\left.\frac{d\mathbf{x}(t)}{dt}\right|_{t=0} = \mathbf{A}\left(\mathbf{A}\mathbf{x}(0)\right) = \mathbf{A}^2\mathbf{x}(0) \tag{5.84}$$

Continuing in this fashion, the n-th derivative of the state vector evaluated at $t = 0$ is

$$\left.\frac{d^n\mathbf{x}(t)}{dt^n}\right|_{t=0} = \mathbf{A}^n\mathbf{x}(0) \tag{5.85}$$

Using Eqn. (5.85), the Taylor series expansion given by Eqn. (5.81) for the state vector becomes

$$\mathbf{x}(t) = \left[\mathbf{I} + \mathbf{A}\,t + \frac{1}{2!}\mathbf{A}^2 t^2 + \ldots + \frac{1}{n!}\mathbf{A}^n t^n + \ldots\right]\mathbf{x}(0) \tag{5.86}$$

In Eqn. (5.86) the infinite series in square brackets is the matrix exponential function $e^{\mathbf{A}t}$, that is,

$$e^{\mathbf{A}t} = \mathbf{I} + \mathbf{A}\,t + \frac{1}{2!}\mathbf{A}^2 t^2 + \ldots + \frac{1}{n!}\mathbf{A}^n t^n + \ldots \tag{5.87}$$

Therefore, the homogeneous solution of the state equation is

$$\mathbf{x}(t) = e^{\mathbf{A}t}\,\mathbf{x}(0) \tag{5.88}$$

As in the case of scalar differential equations, the homogeneous solution represents the behavior of the state variables of the system due to the initial conditions alone. In the absence of an external input signal $r(t)$, the matrix $e^{\mathbf{A}t}$ transforms the state vector at time $t = 0$ to the state vector at time t. Consequently it is referred to as the *state transition matrix*. A method for computing the state transition matrix of a state-space model will be presented in Section 5.4.3.

State transition matrix:

$$\boldsymbol{\phi}(t) = e^{\mathbf{A}t} \tag{5.89}$$

5.4.2 General solution

The general solution technique for the state equation will parallel the development of the general solution for a first-order differential equation in Section 2.5.1 of Chapter 2. Consider the non-homogeneous state equation

$$\dot{\mathbf{x}}(t) = \mathbf{A}\mathbf{x}(t) + \mathbf{B}\,r(t) \tag{5.90}$$

with the initial state of the system specified at time $t = t_0$.

Derivation of the general solution of state equation:

We will begin with an assumed solution in the form

$$\mathbf{x}(t) = e^{\mathbf{A}t}\,\mathbf{s}(t) \tag{5.91}$$

where $\mathbf{s}(t)$ is a yet undetermined column vector of time functions $s_1(t), \ldots, s_N(t)$. Differentiation of both sides of Eqn. (5.91) with respect to time results in

$$\dot{\mathbf{x}}(t) = \mathbf{A}\,e^{\mathbf{A}t}\,\mathbf{s}(t) + e^{\mathbf{A}t}\,\dot{\mathbf{s}}(t) \tag{5.92}$$

Comparing Eqns. (5.92) and (5.90) leads us to the conclusion

$$\mathbf{B}\,r(t) = e^{\mathbf{A}t}\,\dot{\mathbf{s}}(t) \tag{5.93}$$

Solving Eqn. (5.93) for $\dot{\mathbf{s}}(t)$ yields

$$\dot{\mathbf{s}}(t) = e^{-\mathbf{A}t}\,\mathbf{B}\,r(t) \tag{5.94}$$

Integrating both sides of Eqn. (5.94) over time starting at $t = t_0$ we obtain

$$\mathbf{s}(t) = \mathbf{s}(t_0) + \int_{t_0}^{t} e^{-\mathbf{A}\tau}\,\mathbf{B}\,r(\tau)\,d\tau \tag{5.95}$$

Substituting Eqn. (5.95) into Eqn. (5.91) results in

$$\mathbf{x}(t) = e^{\mathbf{A}t}\,\mathbf{s}(t_0) + e^{\mathbf{A}t}\int_{t_0}^{t} e^{-\mathbf{A}\tau}\,\mathbf{B}\,r(\tau)\,d\tau \tag{5.96}$$

The term $\mathbf{s}(t_0)$ is obtained from Eqn. (5.91) as

$$\mathbf{s}(t_0) = e^{-\mathbf{A}\,t_0}\,\mathbf{x}(t_0) \tag{5.97}$$

and Eqn. (5.96) becomes

$$\mathbf{x}(t) = e^{\mathbf{A}(t-t_0)}\,\mathbf{x}(t_0) + \int_{t_0}^{t} e^{\mathbf{A}(t-\tau)}\,\mathbf{B}\,r(\tau)\,d\tau \tag{5.98}$$

General solution of the state equation:

The state equation

$$\dot{\mathbf{x}}(t) = \mathbf{A}\mathbf{x}(t) + \mathbf{B}\,r(t)\,, \qquad \mathbf{x}(t_0): \text{specified} \tag{5.99}$$

is solved as

$$\mathbf{x}(t) = e^{\mathbf{A}(t-t_0)}\,\mathbf{x}(t_0) + \int_{t_0}^{t} e^{\mathbf{A}(t-\tau)}\,\mathbf{B}\,r(\tau)\,d\tau \tag{5.100}$$

Often the initial state of a system is specified at time instant $t_0 = 0$. In that case the general solution becomes

$$\mathbf{x}(t) = e^{\mathbf{A}\,t}\,\mathbf{x}(0) + \int_{0}^{t} e^{\mathbf{A}(t-\tau)}\,\mathbf{B}\,r(\tau)\,d\tau \tag{5.101}$$

5.4.3 Computation of the state transition matrix

In the previous section it became apparent that the state transition matrix plays an important role in solving the state equations. The definition of the state transition matrix $\boldsymbol{\phi}(t)$ is based on the infinite series

$$\boldsymbol{\phi}(t) = e^{\mathbf{A}\,t} = \mathbf{I} + \mathbf{A}\,t + \frac{1}{2!}\mathbf{A}^2\,t^2 + \ldots + \frac{1}{n!}\mathbf{A}^n\,t^n + \ldots$$

While the series is convergent, and can be used for approximating $\phi(t)$ for specific values of t, we still need a method for determining the analytical form of $\boldsymbol{\phi}(t)$. Such a method will be presented in this section. We will begin with the homogeneous state equation given by Eqn. (5.79) and take the unilateral Laplace transform of each side of it to obtain

$$s\mathbf{X}(s) - \mathbf{x}(0) = \mathbf{A}\mathbf{X}(s) \tag{5.102}$$

The vector $\mathbf{X}(s)$ is a column vector of Laplace transforms of state variables, that is,

$$\mathbf{X}(s) = \begin{bmatrix} X_1(s) \\ X_2(s) \\ \vdots \\ X_N(s) \end{bmatrix}$$

Rearranging terms of Eqn. (5.102) we get

$$\left[s\mathbf{I} - \mathbf{A}\right]\mathbf{X}(s) = \mathbf{x}(0) \tag{5.103}$$

The matrix **I** in Eqn. (5.103) represents the identity matrix or order N. Its inclusion is necessary to keep the two terms in square brackets dimensionally compatible, since we could not subtract a matrix from a scalar. Solving Eqn. (5.103) for $\mathbf{X}(s)$ results in

$$\mathbf{X}(s) = \left[s\mathbf{I}-\mathbf{A}\right]^{-1}\mathbf{x}(0) \tag{5.104}$$

The state vector can be found by taking inverse Laplace transform of Eqn. (5.104) as

$$\mathbf{x}(t) = \mathscr{L}^{-1}\left\{\left[s\mathbf{I}-\mathbf{A}\right]^{-1}\right\}\mathbf{x}(0) \tag{5.105}$$

The state transition matrix $\boldsymbol{\phi}(t)$ is found by comparing Eqns. (5.105) and (5.88):

$$\boldsymbol{\phi}(t) = \mathscr{L}^{-1}\left\{\left[s\mathbf{I}-\mathbf{A}\right]^{-1}\right\} \tag{5.106}$$

The matrix

$$\boldsymbol{\Phi}(s) = \left[s\mathbf{I}-\mathbf{A}\right]^{-1} \tag{5.107}$$

is called the *resolvent matrix* and is the Laplace transform of the state transition matrix.

Computation of the state transition matrix:

1. Form the matrix $\left[s\mathbf{I}-\mathbf{A}\right]$.
2. Find the resolvent matrix $\boldsymbol{\Phi}(s)$ by inverting the matrix $\left[s\mathbf{I}-\mathbf{A}\right]$.
3. Find the state transition matrix $\boldsymbol{\phi}(t)$ by computing the inverse Laplace transform of the resolvent matrix $\boldsymbol{\Phi}(s)$.

Example 5.12: Finding the state transition matrix

Determine the state transition matrix $\boldsymbol{\phi}(t)$ for a system with the coefficient matrix

$$\mathbf{A} = \begin{bmatrix} -1 & -1 & 0 \\ 1 & 0 & -1 \\ 5 & 7 & -6 \end{bmatrix}$$

Solution: The first step is to construct the matrix $[s\mathbf{I}-\mathbf{A}]$:

$$\begin{aligned}[s\mathbf{I}-\mathbf{A}] &= s\begin{bmatrix} 1 & 0 & 0 \\ 0 & 1 & 0 \\ 0 & 0 & 1 \end{bmatrix} - \begin{bmatrix} -1 & -1 & 0 \\ 1 & 0 & -1 \\ 5 & 7 & -6 \end{bmatrix} \\ &= \begin{bmatrix} s+1 & 1 & 0 \\ -1 & s & 1 \\ -5 & -7 & s+6 \end{bmatrix}\end{aligned}$$

In the next step we find the resolvent matrix $\mathbf{\Phi}(s)$:

$$\mathbf{\Phi}(s) = [s\mathbf{I} - \mathbf{A}]^{-1} = \frac{1}{\det[s\mathbf{I} - \mathbf{A}]}\,\text{adj}\left[s\mathbf{I} - \mathbf{A}\right]$$

$$= \begin{bmatrix} \dfrac{s^2+6s+7}{(s+1)(s+2)(s+4)} & \dfrac{-s-6}{(s+1)(s+2)(s+4)} & \dfrac{1}{(s+1)(s+2)(s+4)} \\ \dfrac{1}{(s+2)(s+4)} & \dfrac{s+6}{(s+2)(s+4)} & \dfrac{-1}{(s+2)(s+4)} \\ \dfrac{5s+7}{(s+1)(s+2)(s+4)} & \dfrac{7s+2}{(s+1)(s+2)(s+4)} & \dfrac{s^2+s+1}{(s+1)(s+2)(s+4)} \end{bmatrix}$$

The state transition matrix $\boldsymbol{\phi}(t)$ is found by taking the inverse Laplace term of each element of the resolvent matrix using partial fraction expansion.

$$\boldsymbol{\phi}(t) = \begin{bmatrix} \left(\frac{2}{3}e^{-t} + \frac{1}{2}e^{-2t} - \frac{1}{6}e^{-4t}\right) & \left(-\frac{5}{3}e^{-t} + 2e^{-2t} - \frac{1}{3}e^{-4t}\right) & \left(\frac{1}{3}e^{-t} - \frac{1}{2}e^{-2t} + \frac{1}{6}e^{-4t}\right) \\ \left(\frac{1}{2}e^{-2t} - \frac{1}{2}e^{-4t}\right) & \left(2e^{-2t} - e^{-4t}\right) & \left(-\frac{1}{2}e^{-2t} + \frac{1}{2}e^{-4t}\right) \\ \left(\frac{2}{3}e^{-t} + \frac{3}{2}e^{-2t} - \frac{13}{6}e^{-4t}\right) & \left(-\frac{5}{3}e^{-t} + 6e^{-2t} - \frac{13}{3}e^{-4t}\right) & \left(\frac{1}{3}e^{-t} - \frac{3}{2}e^{-2t} + \frac{13}{6}e^{-4t}\right) \end{bmatrix}$$

Example 5.13: Solving state equations for RLC circuit

A state-space model for the RLC circuit in Fig. 5.3 was found in Example 5.1 as

$$\dot{\mathbf{x}}(t) = \begin{bmatrix} -2 & -2 \\ 1 & -5 \end{bmatrix}\mathbf{x}(t) + \begin{bmatrix} 1 \\ 0 \end{bmatrix} r(t)$$

$$y(t) = \begin{bmatrix} 0 & 5 \end{bmatrix}\mathbf{x}(t)$$

The initial value of the state vector is

$$\mathbf{x}(0) = \begin{bmatrix} 3 \\ -2 \end{bmatrix}$$

Compute the two state variables $x_1(t)$ and $x_2(t)$ as well as the output signal $y(t)$ in response to a unit step input signal.

Solution: We will begin by constructing the matrix $[s\mathbf{I} - \mathbf{A}]$:

$$\left[s\mathbf{I} - \mathbf{A}\right] = s\begin{bmatrix} 1 & 0 \\ 0 & 1 \end{bmatrix} - \begin{bmatrix} -2 & -2 \\ 1 & -5 \end{bmatrix}$$

$$= \begin{bmatrix} s+2 & 2 \\ -1 & s+5 \end{bmatrix}$$

The resolvent matrix $\mathbf{\Phi}(s)$ is the inverse of $[s\mathbf{I} - \mathbf{A}]$:

$$\mathbf{\Phi}(s) - [s\mathbf{I} - \mathbf{A}]^{-1} = \begin{bmatrix} \dfrac{s+5}{(s+3)(s+4)} & \dfrac{-2}{(s+3)(s+4)} \\ \dfrac{1}{(s+3)(s+4)} & \dfrac{s+2}{(s+3)(s+4)} \end{bmatrix}$$

State transition matrix $\boldsymbol{\phi}(t)$ is found by taking the inverse Laplace transform of each element of the resolvent matrix through partial fraction expansion:

$$\boldsymbol{\phi}(t) = \mathscr{L}^{-1}\{\boldsymbol{\Phi}(s)\} = \begin{bmatrix} \left(2e^{-3t} - e^{-4t}\right) & \left(-2e^{-3t} + 2e^{-4t}\right) \\ \left(e^{-3t} - e^{-4t}\right) & \left(-e^{-3t} + 2e^{-4t}\right) \end{bmatrix} \tag{5.108}$$

The state vector $\mathbf{x}(t)$ can now be computed through the use of Eqn. (5.101). The first term in the solution is the homogeneous solution which is

$$\boldsymbol{\phi}(t)\,\mathbf{x}(0) = \begin{bmatrix} \left(2e^{-3t} - e^{-4t}\right) & \left(-2e^{-3t} + 2e^{-4t}\right) \\ \left(e^{-3t} - e^{-4t}\right) & \left(-e^{-3t} + 2e^{-4t}\right) \end{bmatrix} \begin{bmatrix} 3 \\ -2 \end{bmatrix} = \begin{bmatrix} 10e^{-3t} - 7e^{-4t} \\ 5e^{-3t} - 7e^{-4t} \end{bmatrix}$$

In preparation for the computation of the second term in Eqn. (5.101), which is the particular solution, we first need to compute

$$\boldsymbol{\phi}(t)\,\mathbf{B} = \begin{bmatrix} \left(2e^{-3t} - e^{-4t}\right) & \left(-2e^{-3t} + 2e^{-4t}\right) \\ \left(e^{-3t} - e^{-4t}\right) & \left(-e^{-3t} + 2e^{-4t}\right) \end{bmatrix} \begin{bmatrix} 1 \\ 0 \end{bmatrix} = \begin{bmatrix} 2e^{-3t} - e^{-4t} \\ e^{-3t} - e^{-4t} \end{bmatrix}$$

The particular solution is

$$\int_0^t \boldsymbol{\phi}(t-\tau)\,\mathbf{B}\,r(\tau)\,d\tau = \int_0^t \begin{bmatrix} 2e^{-3(t-\tau)} - e^{-4(t-\tau)} \\ e^{-3(t-\tau)} - e^{-4(t-\tau)} \end{bmatrix} d\tau = \begin{bmatrix} \frac{5}{12} - \frac{2}{3}e^{-3t} + \frac{1}{4}e^{-4t} \\ \frac{1}{12} - \frac{1}{3}e^{-3t} + \frac{1}{4}e^{-4t} \end{bmatrix}$$

The state vector is obtained by adding the homogeneous and particular solutions:

$$\mathbf{x}(t) = \begin{bmatrix} 10e^{-3t} - 7e^{-4t} \\ 5e^{-3t} - 7e^{-4t} \end{bmatrix} + \begin{bmatrix} \frac{5}{12} - \frac{2}{3}e^{-3t} + \frac{1}{4}e^{-4t} \\ \frac{1}{12} - \frac{1}{3}e^{-3t} + \frac{1}{4}e^{-4t} \end{bmatrix} = \begin{bmatrix} \frac{5}{12} + \frac{28}{3}e^{-3t} - \frac{27}{4}e^{-4t} \\ \frac{1}{12} + \frac{14}{3}e^{-3t} - \frac{27}{4}e^{-4t} \end{bmatrix}$$

and the output signal is obtained through the use of the output equation:

$$y(t) = \begin{bmatrix} 0 & 5 \end{bmatrix} \begin{bmatrix} \frac{5}{12} + \frac{28}{3}e^{-3t} - \frac{27}{4}e^{-4t} \\ \frac{1}{12} + \frac{14}{3}e^{-3t} - \frac{27}{4}e^{-4t} \end{bmatrix} = \frac{5}{12} + \frac{70}{3}e^{-3t} - \frac{135}{4}e^{-4t}$$

$$t \geq 0$$

Software resource: `exct_5_13.m`

Software resource: See MATLAB Exercises 5.3, 5.4 and 5.5.

5.4.4 Properties of the state transition matrix

The state transition matrix defined by Eqn. (5.89) controls the natural response of the system with no external input signals applied. It is the solution that is only dependent on the initial state of the system. If the state of the system is known at time $t = t_0$, it can be determined at any later time instant t through

$$\mathbf{x}(t) = \boldsymbol{\phi}(t - t_0)\,\mathbf{x}(t_0) \tag{5.109}$$

Some of the important properties of the state transition matrix are listed below.

1. $\boldsymbol{\phi}(0) = \mathbf{I}$

   This property simply follows from Eqn. (5.87) by substituting $t = 0$.

   $$\boldsymbol{\phi}(0) = e^{\mathbf{A}(0)} = I$$

2. $\boldsymbol{\phi}^{-1}(t) = \boldsymbol{\phi}(-t)$

   The proof is straightforward. Using the definition of the state transition matrix given by Eqn. (5.87) we have

   $$\boldsymbol{\phi}(t) = e^{\mathbf{A}t} \quad \longrightarrow \quad \boldsymbol{\phi}(-t) = e^{-\mathbf{A}t} \tag{5.110}$$

   which leads to

   $$e^{\mathbf{A}t}\,e^{-\mathbf{A}t} = \mathbf{I} \quad \longrightarrow \quad \boldsymbol{\phi}(t)\,\boldsymbol{\phi}(-t) = \mathbf{I}$$

3. $\boldsymbol{\phi}(t_2 - t_0) = \boldsymbol{\phi}(t_2 - t_1)\,\boldsymbol{\phi}(t_1 - t_0)$

   The proof of this property is also fairly simple. Using Eqn. (5.109) with $t - t_1$ yields

   $$\mathbf{x}(t_1) = \boldsymbol{\phi}(t_1 - t_0)\,\mathbf{x}(t_0) \tag{5.111}$$

   Using Eqn. (5.109) with $t = t_2$ and $t_0 = t_1$, we get

   $$\mathbf{x}(t_2) = \boldsymbol{\phi}(t_2 - t_1)\,\mathbf{x}(t_1) \tag{5.112}$$

   Substituting Eqn. (5.111) into Eqn. (5.112) leads to

   $$\begin{aligned}\mathbf{x}(t_2) &= \boldsymbol{\phi}(t_2 - t_1)\,\boldsymbol{\phi}(t_1 - t_0)\,\mathbf{x}(t_0)\\ &= \boldsymbol{\phi}(t_2 - t_0)\,\mathbf{x}(t_0)\end{aligned} \tag{5.113}$$

   completing the proof.

5.5 Obtaining System Function from State-Space Model

The problem of obtaining a state-space model of a CTLTI system from its s-domain system function was considered in Section 5.3.3. In this section we will discuss the inverse problem: Given a state-space model for a CTLTI system, determine the system function $G(s)$.

Initially we will focus on a single-input single-output system. Let the system be defined by a state-space model in the standard form of Eqns. (5.7) and (5.8). Taking the Laplace transform of both sides of the state equation yields

$$s\mathbf{X}(s) = \mathbf{A}\mathbf{X}(s) + \mathbf{B}\,R(s) \tag{5.114}$$

$\mathbf{X}(s)$ is a column vector that holds the Laplace transform of each state variable:

$$\mathbf{X}(s) = \begin{bmatrix} X_1(s) \\ X_2(s) \\ \vdots \\ X_N(s) \end{bmatrix}$$

In taking the Laplace transform of the state equation we have assumed zero initial conditions, that is,

$$\mathbf{x}(0) = \mathbf{0} \tag{5.115}$$

since the system is CTLTI. Recall that the system function concept is only valid for a linear and time-invariant system. Rearranging terms of Eqn. (5.114) yields

$$(s\mathbf{I} - \mathbf{A})\,\mathbf{X}(s) = \mathbf{B}\,R(s) \tag{5.116}$$

The $N \times N$ identity matrix $\mathbf{I}$ is necessary to keep the two terms in parentheses compatible. Solving Eqn. (5.117) for $\mathbf{X}(s)$ we obtain

$$\mathbf{X}(s) = (s\mathbf{I} - \mathbf{A})^{-1}\,\mathbf{B}\,R(s) \tag{5.117}$$

Next we will take the Laplace transform of the output equation and then use Eqn. (5.117) in it to obtain

$$\begin{aligned} Y(s) &= \mathbf{C}\mathbf{X}(s) + d\,R(s) \\ &= \mathbf{C}\,(s\mathbf{I} - \mathbf{A})^{-1}\,\mathbf{B}\,R(s) + d\,R(s) \\ &= \left[\mathbf{C}\,(s\mathbf{I} - \mathbf{A})^{-1}\,\mathbf{B} + d\right]\,R(s) \end{aligned} \tag{5.118}$$

The system function is found from Eqn. (5.118) as

$$G(s) = \frac{Y(s)}{R(s)} = \mathbf{C}\,(s\mathbf{I} - \mathbf{A})^{-1}\,\mathbf{B} + d \tag{5.119}$$

We recognize the term $(s\mathbf{I} - \mathbf{A})^{-1}$ in Eqn. (5.119) as the resolvent matrix $\mathbf{\Phi}(s)$. Therefore

$$G(s) = \mathbf{C}\mathbf{\Phi}(s)\,\mathbf{B} + d \tag{5.120}$$

Example 5.14: Obtaining the system function

Determine the system function for a CTLTI system described by the state-space model

$$\dot{\mathbf{x}}(t) = \begin{bmatrix} -2 & -2 \\ 1 & -5 \end{bmatrix} \mathbf{x}(t) + \begin{bmatrix} 1 \\ 0 \end{bmatrix} r(t)$$

$$y(t) = \begin{bmatrix} 0 & 5 \end{bmatrix} \mathbf{x}(t)$$

Solution: For the specified coefficient matrix $\mathbf{A}$, the resolvent matrix was determined in Example 5.13 as

$$\mathbf{\Phi}(s) = \left[s\mathbf{I} - \mathbf{A}\right]^{-1} = \begin{bmatrix} \dfrac{s+5}{(s+3)(s+4)} & \dfrac{-2}{(s+3)(s+4)} \\ \dfrac{1}{(s+3)(s+4)} & \dfrac{s+2}{(s+3)(s+4)} \end{bmatrix}$$

Using Eqn. (5.120) the system function is

$$\begin{aligned} G(s) &= \mathbf{C}\,\boldsymbol{\Phi}(s)\,\mathbf{B}+d \\ &= \begin{bmatrix} 0 & 5 \end{bmatrix} \begin{bmatrix} \dfrac{s+5}{(s+3)(s+4)} & \dfrac{-2}{(s+3)(s+4)} \\ \dfrac{1}{(s+3)(s+4)} & \dfrac{s+2}{(s+3)(s+4)} \end{bmatrix} \begin{bmatrix} 1 \\ 0 \end{bmatrix} \\ &= \frac{5}{(s+3)(s+4)} \end{aligned}$$

Software resource: See MATLAB Exercises 5.6.

The results obtained for a single-input single-output system can easily be adapted to multiple input multiple-output systems. Consider a CTLTI system with K input signals $\{r_i(t);\ i=1,\ldots,K\}$ and M output signals $\{y_j(t);\ i=1,\ldots,M\}$. The matrix $\mathbf{B}$ is $(N\times K)$, and the matrix $\mathbf{C}$ is $(M\times N)$. The scalar d turns into a matrix $\mathbf{D}$ that is $(M\times K)$. As a result, Eqn. (5.120) turns into

$$\mathbf{G}(s) = \mathbf{C}\,\boldsymbol{\Phi}(s)\,\mathbf{B}+\mathbf{D} \tag{5.121}$$

$\mathbf{G}(s)$ is a $(M\times K)$ matrix of system functions from each input to each output.

$$\begin{bmatrix} Y_1(s) \\ Y_2(s) \\ \vdots \\ Y_M(s) \end{bmatrix} = \begin{bmatrix} G_{11}(s) & G_{12}(s) & \ldots & G_{1K}(s) \\ G_{21}(s) & G_{22}(s) & \ldots & G_{2K}(s) \\ \vdots & & & \\ G_{M1}(s) & G_{M2}(s) & \ldots & G_{MK}(s) \end{bmatrix} \begin{bmatrix} R_1(s) \\ R_2(s) \\ \vdots \\ R_K(s) \end{bmatrix} \tag{5.122}$$

Example 5.15: System function for multiple-input multiple-output system

A 2-input 2-output system is characterized by the state-space model

$$\dot{\mathbf{x}}(t) = \begin{bmatrix} 0 & 1 & 0 \\ -2 & -3 & -1 \\ -4 & 0 & -3 \end{bmatrix} \mathbf{x}(t) + \begin{bmatrix} 0 & 0 \\ -1 & 1 \\ -11 & -2 \end{bmatrix} \mathbf{r}(t)$$

$$\mathbf{y}(t) = \begin{bmatrix} 1 & 0 & 0 \\ 0 & 0 & 1 \end{bmatrix} \mathbf{x}(t) + \begin{bmatrix} 0 & 0 \\ 0 & 4 \end{bmatrix} \mathbf{r}(t)$$

Determine the system function matrix $\mathbf{G}(s)$.

Solution: We will begin by constructing the matrix $[s\mathbf{I}-\mathbf{A}]$.

$$\begin{aligned} [s\mathbf{I}-\mathbf{A}] &= s \begin{bmatrix} 1 & 0 & 0 \\ 0 & 1 & 0 \\ 0 & 0 & 1 \end{bmatrix} - \begin{bmatrix} 0 & 1 & 0 \\ -2 & -3 & -1 \\ -4 & 0 & -3 \end{bmatrix} \\ &= \begin{bmatrix} s & -1 & 0 \\ 2 & s+3 & 1 \\ 4 & 0 & s+3 \end{bmatrix} \end{aligned}$$

The state transition matrix is found by inverting $[s\mathbf{I}-\mathbf{A}]$.

$$\mathbf{\Phi}(s) = [s\mathbf{I}-\mathbf{A}]^{-1} = \frac{1}{s^3+6s^2+11s+2}\begin{bmatrix} (s+3)^2 & (s+3) & -1 \\ -2(s+1) & s(s+3) & -s \\ -4(s+3) & -4 & (s+1)(s+2) \end{bmatrix}$$

The system function matrix is

$$\mathbf{G}(s) = \mathbf{C}\mathbf{\Phi}(s)\,\mathbf{B}+\mathbf{D}$$

$$= \frac{1}{s^3+6s^2+11s+2}\begin{bmatrix} -(s-8) & (s+5) \\ -\left(11s^2+33s+18\right) & 2s\left(2s^2+11s+19\right) \end{bmatrix}$$

Software resource: exct_5_15.m

5.6 Diagonalization of the Matrix A

In Section 5.3.4 we have discussed alternative state-space models for a system that can be obtained through the use of a similarity transformation. For a system described by means of state variables $\mathbf{x}(t)$, a new set of state variables $\mathbf{z}(t)$ can be obtained through the transformation

$$\mathbf{x}(t) = \mathbf{P}\mathbf{z}(t)$$

where **P** could be any non-singular, that is, invertible, square matrix. Since an infinite number of such transformation matrices can be found, an infinite number of alternative state-space models exist for a given system. Coefficient matrices **A**, **B**, and **C** are transformed to $\tilde{\mathbf{A}}$, $\tilde{\mathbf{B}}$, and $\tilde{\mathbf{C}}$ as described by Eqns. (5.59) through (5.61).

$$\text{Eqns. (5.59) through (5.61):} \qquad \tilde{\mathbf{A}} = \mathbf{P}^{-1}\mathbf{A}\mathbf{P}\,, \quad \tilde{\mathbf{B}} = \mathbf{P}^{-1}\mathbf{B}\,, \quad \tilde{\mathbf{C}} = \mathbf{C}\mathbf{P}$$

An interesting question that can be raised is this: Could a transformation matrix **P** can be found such that the resulting coefficient matrix $\tilde{\mathbf{A}}$ is diagonal? The motivation behind the desire to have a diagonal coefficient matrix is that it would cause the corresponding first-order differential equations to be uncoupled from each other. Since elements on the diagonal would be the only nonzero elements of the matrix $\tilde{\mathbf{A}}$, we could write

$$\frac{dx_i(t)}{dt} = \tilde{a}_{ii}\,x_i(t) + \tilde{b}_i\,r(t)\,, \qquad i = 1,\ldots,N \tag{5.123}$$

Each equation can be solved independently. The state transition matrix $\boldsymbol{\phi}(t)$ would also be diagonal.

5.6.1 Characteristic equation of the system

Consider a CTLTI system the behavior of which is governed by the differential equation given by Eqn. (5.21). The characteristic equation of the system is found by setting the denominator polynomial of the system function equal to zero:

$$s^N + a_{N-1}\,s^{N-1} + \ldots + a_1\,s + a_0 = 0$$

The system function can be determined from the state-space model using Eqn. (5.119) which is repeated here for convenience:

$$\text{Eqn. (5.119):} G(s) = \frac{Y(s)}{R(s)} = \mathbf{C}\,(s\mathbf{I}-\mathbf{A})^{-1}\,\mathbf{B}+d$$

We know from matrix algebra that the inverse of a matrix is found as its adjoint matrix divided by its determinant. Applying this to Eqn. (5.119) we get

$$(s\mathbf{I}-\mathbf{A})^{-1} = \frac{\text{adj}\,(s\mathbf{I}-\mathbf{A})}{\left|s\mathbf{I}-\mathbf{A}\right|} \tag{5.124}$$

which leads to the system function

$$G(s) = \frac{\mathbf{C}\left[\text{adj}\,(s\mathbf{I}-\mathbf{A})\right]B}{\left|s\mathbf{I}-\mathbf{A}\right|}+d \tag{5.125}$$

Setting the denominator of the system function in Eqn. (5.125) equal to zero provides us with another expression for the characteristic equation:

$$\left|s\mathbf{I}-\mathbf{A}\right| = \mathbf{0} \tag{5.126}$$

An important conclusion is that the roots of Eqn. (5.126) must be the poles of the system function. Furthermore, if $\mathbf{A}$ is transformed into $\tilde{\mathbf{A}}$ through a similarity transformation, Eqn. (5.126) must still be the characteristic equation since the system did not change.

$$\left|s\mathbf{I}-\mathbf{P}^{-1}\mathbf{A}\mathbf{P}\right| = \mathbf{0} \tag{5.127}$$

5.6.2 Eigenvalues and eigenvectors

Eigenvalues λ_i and eigenvectors $\mathbf{v_i}$ of a square matrix $\mathbf{A}$ are defined through the relationship

$$\mathbf{A}\,\mathbf{v_i} = \lambda_i\,\mathbf{v_i} \tag{5.128}$$

An eigenvector $\mathbf{v_i}$ is a vector which, when multiplied by the matrix $\mathbf{A}$ from the left, produces a result that is a scaled version of itself with no change in direction. The eigenvalue λ_i is the scale factor. Rearranging terms of Eqn. (5.128) leads to

$$\left(\lambda_i\mathbf{I}-\mathbf{A}\right)\mathbf{v_i} = \mathbf{0} \tag{5.129}$$

which can be solved as

$$\left|\lambda_i\mathbf{I}-\mathbf{A}\right| = \mathbf{0} \tag{5.130}$$

This is identical to Eqn. (5.126). Therefore, the eigenvalues of the matrix $\mathbf{A}$ are equal to the poles of the system function $G(s)$. It can be shown that, if the eigenvalues $\lambda_1, \lambda_2, \ldots, \lambda_N$ of the matrix $\mathbf{A}$ are distinct, then the transformation matrix constructed using the eigenvectors of $\mathbf{A}$ in the form

$$\mathbf{P} = \begin{bmatrix} \mathbf{v_1} & \mathbf{v_2} & \ldots & \mathbf{v_N} \end{bmatrix} \tag{5.131}$$

would transform $\mathbf{A}$ into a diagonal state matrix in the form

$$\mathbf{\Lambda} = \mathbf{P}^{-1}\mathbf{A}\mathbf{P} = \begin{bmatrix} \lambda_1 & 0 & \ldots & 0 \\ 0 & \lambda_2 & \ldots & 0 \\ \ldots & \ldots & \ldots & \ldots \\ 0 & 0 & \ldots & \lambda_N \end{bmatrix} \tag{5.132}$$

with eigenvalues on the diagonal.

Example 5.16: Obtaining a state-space model with a diagonal A matrix

Consider again the state-space model found for the RLC circuit in Example 5.1:

$$\dot{\mathbf{x}}(t) = \begin{bmatrix} -2 & -2 \\ 1 & -5 \end{bmatrix} \mathbf{x}(t) + \begin{bmatrix} 1 \\ 0 \end{bmatrix} r(t)$$

$$y(t) = \begin{bmatrix} 0 & 5 \end{bmatrix} \mathbf{x}(t) + (0)\, r(t)$$

Find a similarity transformation matrix **P** that would convert this state-space model to one with a diagonal coefficient matrix. Determine the resulting state-space model.

Solution: The first step is to form the matrix $(\lambda\mathbf{I} - \mathbf{A})$.

$$\lambda\mathbf{I} - \mathbf{A} = \lambda \begin{bmatrix} 1 & 0 \\ 0 & 1 \end{bmatrix} - \begin{bmatrix} -2 & -2 \\ 1 & -5 \end{bmatrix} = \begin{bmatrix} \lambda+2 & 2 \\ -1 & \lambda+5 \end{bmatrix}$$

Eigenvalues are found by setting the determinant of the matrix $(\lambda\mathbf{I} - \mathbf{A})$ equal to zero:

$$|\lambda\mathbf{I} - \mathbf{A}| = 0 \quad \longrightarrow \quad (\lambda+2)\,(\lambda+5) - (2)\,(-1) = 0$$

$$(\lambda+3)\,(\lambda+4) = 0$$

The eigenvalues of **A** are therefore $\lambda_1 = -3$ and $\lambda_2 = -4$. Let the corresponding eigenvectors be

$$\mathbf{v_1} = \begin{bmatrix} v_{11} \\ v_{21} \end{bmatrix} \quad \text{and} \quad \mathbf{v_2} = \begin{bmatrix} v_{12} \\ v_{22} \end{bmatrix}$$

Using Eqn. (5.129) for λ_1 and $\mathbf{v_1}$, we get

$$(-3\mathbf{I} - \mathbf{A}) \begin{bmatrix} v_{11} \\ v_{21} \end{bmatrix} = \mathbf{0} \quad \longrightarrow \quad \begin{bmatrix} -1 & 2 \\ -1 & 2 \end{bmatrix} \begin{bmatrix} v_{11} \\ v_{21} \end{bmatrix} = \mathbf{0} \quad \longrightarrow \quad v_{11} = 2v_{21}$$

If we choose $v_{21} = 1$, then $v_{11} = 2$, and the first eigenvector is

$$\mathbf{v_1} = \begin{bmatrix} 2 \\ 1 \end{bmatrix}$$

Similarly, for λ_2 and $\mathbf{v_2}$ we have

$$(-4\mathbf{I} - \mathbf{A}) \begin{bmatrix} v_{12} \\ v_{22} \end{bmatrix} = \mathbf{0} \quad \longrightarrow \quad \begin{bmatrix} -2 & 2 \\ -1 & 1 \end{bmatrix} \begin{bmatrix} v_{12} \\ v_{22} \end{bmatrix} = \mathbf{0} \quad \longrightarrow \quad v_{12} = v_{22}$$

If we choose $v_{22} = 1$, then $v_{12} = 1$, and the second eigenvector is

$$\mathbf{v_2} = \begin{bmatrix} 1 \\ 1 \end{bmatrix}$$

The transformation matrix **P** can now be formed as

$$\mathbf{P} = \begin{bmatrix} \mathbf{v_1} & \mathbf{v_2} \end{bmatrix} = \begin{bmatrix} 2 & 1 \\ 1 & 1 \end{bmatrix}$$

The matrix $\mathbf{\Lambda}$ is

$$\mathbf{\Lambda} = \mathbf{P}^{-1}\mathbf{A}\mathbf{P} = \begin{bmatrix} 1 & -1 \\ -1 & 2 \end{bmatrix} \begin{bmatrix} -2 & -2 \\ 1 & -5 \end{bmatrix} \begin{bmatrix} 2 & 1 \\ 1 & 1 \end{bmatrix} = \begin{bmatrix} -3 & 0 \\ 0 & -4 \end{bmatrix}$$

Vectors $\tilde{\mathbf{B}}$ and $\tilde{\mathbf{C}}$ can be found as

$$\tilde{\mathbf{B}} = \mathbf{P}^{-1}\mathbf{B} = \begin{bmatrix} 1 & -1 \\ -1 & 2 \end{bmatrix} \begin{bmatrix} 1 \\ 0 \end{bmatrix} = \begin{bmatrix} 1 \\ -1 \end{bmatrix}$$

and

$$\tilde{\mathbf{C}} = \mathbf{C}\mathbf{P} = \begin{bmatrix} 0 & 5 \end{bmatrix} \begin{bmatrix} 2 & 1 \\ 1 & 1 \end{bmatrix} = \begin{bmatrix} 5 & 5 \end{bmatrix}$$

Using these results, the new state-space model for the system is

$$\dot{\mathbf{z}}(t) = \begin{bmatrix} -3 & 0 \\ 0 & -4 \end{bmatrix} \mathbf{z}(t) + \begin{bmatrix} 1 \\ -1 \end{bmatrix} r(t)$$

and

$$y(t) = \begin{bmatrix} 5 & 5 \end{bmatrix} \mathbf{z}(t) + (0)\, r(t)$$

Software resource: exct_5_16.m

Summary of Key Points

- State-space model of a system describes the internal structure of the system, in contrast with the other description forms which simply describe the relationship between input and output signals. It can also be used for representing systems with multiple input and output signals.

- State-space representation of a single-input single-output CTLTI system is in the form

  Eqn. 5.7: State equation: $\dot{\mathbf{x}}(t) = \mathbf{A}\mathbf{x}(t) + \mathbf{B}\, r(t)$

  Eqn. 5.8: Output equation: $y(t) = \mathbf{C}\mathbf{x}(t) + d\, r(t)$

- Phase variable canonical form of the state-space model for a CTLTI system can be obtained from its differential equation using the technique outlined in Section 5.3.2.

- A state-space model for a CTLTI system can also be obtained from a block diagram by designating the output of each integrator as a state variable, and expressing the input signal of each integrator as a linear combination of state variables and system inputs.

- Alternative state-space descriptions of a system can be obtained by transforming the state variables through the transformation

  Eqn. 5.53: $\mathbf{x}(t) = \mathbf{P}\mathbf{z}(t)$

  where $\mathbf{P}$ is a non-singular N by N transformation matrix. New coefficients matrices are

  $$\tilde{\mathbf{A}} = \mathbf{P}^{-1}\mathbf{A}\mathbf{P}\,, \qquad \tilde{\mathbf{B}} = \mathbf{P}^{-1}\mathbf{B}\,, \qquad \tilde{\mathbf{C}} = \mathbf{C}\mathbf{P}$$

- The solution of the homogeneous state equation is

  Eqn. 5.88: $\mathbf{x}(t) = \boldsymbol{\phi}(t)\,\mathbf{x}(0)$

where $\boldsymbol{\phi}(t)$ is the state transition matrix defined as

Eqn. 5.89: $$\boldsymbol{\phi}(t) = e^{\mathbf{A}t} = \mathbf{I} + \mathbf{A}\,t + \frac{1}{2!}\mathbf{A}^2\,t^2 + \ldots + \frac{1}{n!}\mathbf{A}^n\,t^n + \ldots$$

☞ The general solution of the state equation with input signal $r(t)$ and initial state vector $\mathbf{x}(t_0)$ is found as

Eqn. 5.100: $$\mathbf{x}(t) = e^{\mathbf{A}(t-t_0)}\,\mathbf{x}(t_0) + \int_{t_0}^{t} e^{\mathbf{A}(t-\tau)}\,\mathbf{B}\,r(\tau)\,d\tau$$

☞ The state transition matrix can be computed as the inverse Laplace transform of the resolvent matrix:

Eqn. 5.106: $$\boldsymbol{\phi}(t) = \mathcal{L}^{-1}\left\{\left[s\mathbf{I} - \mathbf{A}\right]^{-1}\right\}$$

☞ The system function can be obtained from the state-space description of a CTLTI system as

Eqn. 5.119: $$G(s) = \frac{Y(s)}{R(s)} = \mathbf{C}\,(s\mathbf{I} - \mathbf{A})^{-1}\,\mathbf{B} + d$$

☞ A coefficient matrix **A** with distinct eigenvalues $\lambda_1, \lambda_2, \ldots, \lambda_N$ can be transformed into a diagonal matrix $\boldsymbol{\Lambda}$ using a similarity transformation as

Eqn. 5.132: $$\boldsymbol{\Lambda} = \mathbf{P}^{-1}\mathbf{A}\mathbf{P} = \begin{bmatrix} \lambda_1 & 0 & \ldots & 0 \\ 0 & \lambda_2 & \ldots & 0 \\ \ldots & \ldots & \ldots & \ldots \\ 0 & 0 & \ldots & \lambda_N \end{bmatrix}$$

provided that the transformation matrix **P** is in the form

Eqn. 5.131: $$\mathbf{P} = \begin{bmatrix} \mathbf{v_1} & \mathbf{v_2} & \ldots & \mathbf{v_N} \end{bmatrix}$$

where $\mathbf{v_i}$ are the eigenvectors of **A**.

Further Reading

[1] R.C. Dorf and R.H. Bishop. *Modern Control Systems.* Prentice Hall, 2011.

[2] B. Friedland. *Control System Design: An Introduction to State-Space Methods.* Dover Publications, 2012.

[3] Benjamin C Kuo. *Automatic Control Systems.* Prentice Hall PTR, 1987.

[4] K. Ogata. *Modern Control Engineering.* Instrumentation and Controls Series. Prentice Hall, 2010.

[5] L.A. Zadeh and C.A. Desoer. *Linear System Theory: The State Space Approach.* Dover Civil and Mechanical Engineering Series. Dover Publications, 2008.

MATLAB Exercises with Solutions

MATLAB Exercise 5.1: Obtaining state-space model from system function

MATLAB function `tf2ss()` facilitates conversion of a rational system function in the s domain to a state-space model. Consider a DTLTI system with the system function

$$H(s) = \frac{s^2+s-2}{s^2+6s+5}$$

A state-space model for the system is obtained using the following:

```
>> num = [1,1,-2];
>> den = [1,6,5];
>> [A,B,C,d] = tf2ss(num,den)

A =
    -6    -5
     1     0

B =
     1
     0

C =
    -5    -7

d =
     1
```

The results correspond to the state space model

$$\dot{\mathbf{x}}(t) = \begin{bmatrix} -6 & -5 \\ 1 & 0 \end{bmatrix} \mathbf{x}(t) + \begin{bmatrix} 1 \\ 0 \end{bmatrix} r(t)$$

$$y(t) = \begin{bmatrix} -5 & -7 \end{bmatrix} \mathbf{x}(t) + r(t)$$

This state-space model is different than what would be obtained using the technique developed in Section 5.3.3, Eqns. (5.45) and (5.46), but it is equivalent. As an exercise we will develop our own version of the function `tf2ss()` to produce the result obtained in Section 5.3.3, and name it `ss_tf2ss()`. The listing for the function `ss_tf2ss()` is given below:

```
function [A,B,C,d] = ss_tf2ss(num,den)
  nden = length(den);   % Size of denominator vector
  N = nden - 1;         % Order of the system
  % If 'num' has fewer elements than 'den', put zeros in front
  while (length(num)<nden)
    num = [0,num];
  end;
  % Divide all coefficients by leading denominator coefficient
  num = num/den(1);
  den = den/den(1);
  num = fliplr(num);    % Reorder numerator coefficients
```

```
12  den = fliplr(den);    % Reorder denominator coefficients
13  % Construct the state-space model
14  A = [zeros(N-1,1),eye(N-1);-den(1:N)];  % Eqn. (5.45)
15  B = [zeros(N-1,1);1];                   % Eqn. (5.45)
16  C = num(1:N)-num(N+1)*den(1:N);         % Eqn. (5.46)
17  d = num(N+1);                           % Eqn. (5.46)
18 end
```

After line 12 of the code is executed, the vectors num and den are structured as follows:

$$\text{num} = \left[a_0, a_1, \ldots, a_{N-1}, 1\right]$$

$$\text{den} = \left[b_0, b_1, \ldots, b_{N-1}, b_N\right]$$

The construction of state-space model in lines 13 through 16 follows the development in Section 5.3.3 Eqns. (5.45) and (5.46). Using the function `ss_tf2ss()` another state-space model for the same system can be obtained:

```
>> num = [1,1,-2];
>> den = [1,6,5];
>> [A,B,C,d] = ss_tf2ss(num,den)

A =
     0     1
    -5    -6

B =
     0
     1

C =
    -7    -5

d =
     1
```

The results correspond to the state space model

$$\dot{\mathbf{z}}(t) = \begin{bmatrix} 0 & 1 \\ -5 & -6 \end{bmatrix} \mathbf{z}(t) + \begin{bmatrix} 0 \\ 1 \end{bmatrix} r(t)$$

$$y(t) = \begin{bmatrix} -7 & -5 \end{bmatrix} \mathbf{z}(t) + r(t)$$

It can easily be shown (see Problem 5.8 at the end of this chapter) that this state-space model is related to the one found using MATLAB function `tf2ss()` by the transformation

$$\mathbf{x}(t) = \begin{bmatrix} 0 & 1 \\ 1 & 0 \end{bmatrix} \mathbf{z}(t)$$

Software resources: `mexct_5_1a.m` , `mexct_5_1b.m` , `ss_tf2ss.m`

MATLAB Exercise 5.2: Diagonalizing the state matrix

The state-space description of a system is not unique. It was discussed in Section 5.3.4 that, for a given state-space model that uses the state vector $\mathbf{x}(t)$, alternative models can be obtained through similarity transformations. In Section 5.6 a method was presented for finding a similarity transformation that results in a diagonal matrix $\mathbf{A}$. Let the state-space model of a system be

$$\dot{\mathbf{x}}(t) = \begin{bmatrix} 6.5 & 2 & 1.5 \\ -10.5 & -2 & -1.5 \\ 5.5 & 2 & 2.5 \end{bmatrix} \mathbf{x}(t) + \begin{bmatrix} 1 \\ -2 \\ 0 \end{bmatrix} r(t)$$

$$y(t) = \begin{bmatrix} -1 & -1 & -2 \end{bmatrix} \mathbf{x}(t)$$

Find an alternative model

$$\dot{\mathbf{z}}(t) = \tilde{\mathbf{A}}\mathbf{z}(t) + \tilde{\mathbf{B}}\, r(t)$$

$$y(t) = \tilde{\mathbf{C}}\mathbf{z}(t) + d\, r(t)$$

where the matrix $\tilde{\mathbf{A}}$ is diagonal.

Solution: We will begin by entering the state-space model into MATLAB and determining eigenvalues and eigenvectors of the matrix $\mathbf{A}$. Built-in function `eig()` can be used for computing eigenvalues and eigenvectors.

```
>> A = [6.5,2,1.5;-10.5,-2,-1.5;5.5,2,2.5];
>> B = [1;,-2;0];
>> C = [-1,-1,-2];
>> [eigVec,eigVal] = eig(A)

eigVec =
   -0.4082   -0.1690   -0.3015
    0.8165    0.8452    0.9045
   -0.4082   -0.5071   -0.3015

eigVal =
    4.0000         0         0
         0    1.0000         0
         0         0    2.0000

>> P = eigVec;
>> A_tilde = inv(P)*A*P   % Eqn. (5.59)

A_tilde =
    4.0000   -0.0000   -0.0000
    0.0000    1.0000    0.0000
    0.0000    0.0000    2.0000

>> B_tilde = inv(P)*B   % Eqn. (5.60)

B_tilde =
   -0.0000
    2.9580
```

```
  -4.9749

>> C_tilde = C*P    % Eqn. (5.61)

C_tilde =

   0.4082    0.3381   -0.0000
```

Note that the function `eig()` returns two matrices. The first one, `eigVec`, is a matrix in which each column is an eigenvector. Recall from Section 5.6 that the diagionalizing transformation matrix **P** constructed in Eqn. (5.131) has the same exact format. The second returned matrix `eigVal` has is a diagonal matrix that has the eigenvectors on the main diagonal. It has exactly the same format as the matrix **Λ** in Eqn. (5.132). We could have saved a few lines of code by simply using

```
>> [P,A_tilde] = eig(A);
```

The results lead to a new state-space model as follows:

$$\dot{\mathbf{z}}(t) = \begin{bmatrix} 4 & 0 & 0 \\ 0 & 1 & 0 \\ 0 & 0 & 2 \end{bmatrix} \mathbf{z}(t) + \begin{bmatrix} 0 \\ 2.9580 \\ -4.9749 \end{bmatrix} r(t)$$

$$y(t) = \begin{bmatrix} 0.4082 & 0.3381 & 0 \end{bmatrix} \mathbf{z}(t)$$

Software resource: `mexct_5_2.m`

MATLAB Exercise 5.3: Computation of the state transition matrix

In this exercise we will explore various approaches to the problem of computing the state transition matrix through the use of MATLAB. The definition of the state transition matrix was given by Eqn. (5.89) which is repeated here:

$$\text{Eqn. (5.89):} \qquad \boldsymbol{\phi}(t) = e^{\mathbf{A}t}$$

An infinite series form of the state transition matrix was also given, and is repeated below:

$$\text{Eqn. (5.87):} \qquad e^{\mathbf{A}t} = \mathbf{I} + \mathbf{A}\,t + \frac{1}{2!}\mathbf{A}^2\,t^2 + \ldots + \frac{1}{n!}\mathbf{A}^n\,t^n + \ldots$$

Consider the coefficient matrix **A** that was used in Example 5.12:

$$\mathbf{A} = \begin{bmatrix} -1 & -1 & 0 \\ 1 & 0 & -1 \\ 5 & 7 & -6 \end{bmatrix}$$

It is possible to compute the complex exponential function $e^{\mathbf{A}t}$ at one specific time instant using the MATLAB function `expm()`. The following segment of code computes $\boldsymbol{\phi}(0.2)$, the value of the state transition matrix at time $t = 0.2$:

```
>> A = [-1,-1,0; 1,0,-1; 5,7,-6];
>> expm(A*0.2)

ans =
   0.8061   -0.1737    0.0126
   0.1105    0.8913   -0.1105
   0.5778    0.7103    0.2410
```

It is important to remember to use the function `expm()` and not the function `exp()` for this purpose. The latter function computes the exponential function of each element of the matrix which is not the same as computing $e^{\mathbf{A}t}$.

It is also possible to approximate the state transition matrix by using the first few terms of the infinite series for it. The infinite series can be written in nested form as

$$\phi(t) = e^{\mathbf{A}t} = \mathbf{I} + \mathbf{A}t\left[\mathbf{I} + \frac{1}{2}\mathbf{A}t\left[\mathbf{I} + \frac{1}{3}\mathbf{A}t\left[\mathbf{I} + \frac{1}{4}\mathbf{A}t\left[\mathbf{I} + \ldots\ldots\ldots\right]\right]\right]\right]$$

The following code segment computes the first five terms of the series for $\boldsymbol{\phi}(0.2)$:

```
>> id = eye(3);
>> At = A*0.2;
>> id+At*(id+At/2*(id+At/3*(id+At/4)))

ans =
   0.8057   -0.1743    0.0130
   0.1093    0.8891   -0.1093
   0.5727    0.7003    0.2461
```

Even though we have used only the first five terms, the result obtained is quite close to that obtained through the use of the function `expm()`. The accuracy of the approximation can be improved by including more terms.

Both code segments listed above compute the state transition matrix at one specific time instant. If elements of the state transition matrix are needed as functions of time over a specified interval, the function `expm()` can be used within a loop. The script listed below computes and graphs $\phi_{12}(t)$, the row-1 column-2 element of the state transition matrix, in the time interval $0 \le t \le 4$ sec.

```
% Script: mexct_5_3.m
A = [-1,-1,0; 1,0,-1; 5,7,-6];  % Construct matrix 'A'
t = [0:0.01:4];                 % Vector of time instants
phi12 =[];                      % Placeholder for row-1 column-2 element
% Compute the STM for each element in vector 't'
for i=1:length(t)
  time = t(i);                  % Pick a time instant
  stm = expm(A*time);           % Compute the STM at time instant t(i)
  phi12 = [phi12,stm(1,2)];     % Append to vector 'phi12'
end;
% Graph element (1,2) of the STM
plot(t,phi12); grid;
title('Element 1,2 of the state transition matrix');
xlabel('t (sec)');
```

The graph produced is shown in Fig. 5.9.

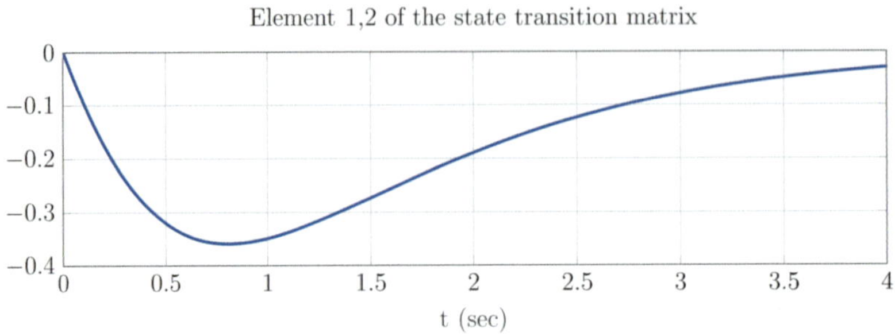

Figure 5.9 – Graph obtained in MATLAB Exercise 5.3.

This result should be compared to the analytical expression obtained in Example 5.12 for the element $\phi_{12}(t)$:

$$\phi_{12}(t) = -\tfrac{5}{3}\,e^{-t} + 2\,e^{-2t} - \tfrac{1}{3}\,e^{-4t}$$

The following code segment can be used for computing and graphing the analytical form of $\phi_{12}(t)$ for comparison:

```
>> phi12a = -5/3*exp(-t)+2*exp(-2*t)-1/3*exp(-4*t);
>> plot(t,phi12a);
```

Software resource: `mexct_5_3.m`

MATLAB Exercise 5.4: Solving the homogeneous state equation

Consider the homogeneous state equation

$$\dot{\mathbf{x}}(t) = \begin{bmatrix} 0 & 5 \\ -2 & -2 \end{bmatrix} \mathbf{x}(t)$$

Let the initial value of the state vector be

$$\mathbf{x}(0) = \begin{bmatrix} 1 \\ -2 \end{bmatrix}$$

Using MATLAB, compute and graph the state transition matrix $\boldsymbol{\phi}(t)$ in the time interval $0 \le t \le 4$ s. Afterward, compute and graph the state vector $\mathbf{x}(t)$ in the same time interval.

Solution: The solution of the homogeneous state equation is computed as

$$\mathbf{x}(t) = \boldsymbol{\phi}(t)\,\mathbf{x}(0)\,, \qquad t \ge 0$$

The state transition matrix is the form

$$\boldsymbol{\phi}(t) = \begin{bmatrix} \phi_{11}(t) & \phi_{12}(t) \\ \phi_{21}(t) & \phi_{22}(t) \end{bmatrix}$$

Using the state transition matrix with the initial value of the state vector we obtain

$$\mathbf{x}(t) = \begin{bmatrix} \phi_{11}(t) & \phi_{12}(t) \\ \phi_{21}(t) & \phi_{22}(t) \end{bmatrix} \begin{bmatrix} 1 \\ -2 \end{bmatrix} = \begin{bmatrix} \phi_{11}(t) - 2\phi_{12}(t) \\ \phi_{21}(t) - 2\phi_{22}(t) \end{bmatrix}$$

The function `expm()` was first used in MATLAB Exercise 5.3 for computing the state transition matrix at for specific value of t. In the script listed below, it is used in a loop structure for computing the state transition matrix at a set of time instants. The elements of the state transition matrix are computed and graphed. The result is shown in Fig. 5.10.

```
% Script: mexct_5_4a.m
A = [0,5;-2,-2];   % State matrix
t = [0:0.01:4];    % Vector of time instants
% Create placeholders for 'phi11', 'phi12', 'phi21' and 'phi22'
phi11 =[];
phi12 =[];
phi21 =[];
phi22 =[];
% Compute the STM for each element of vector 't'
for i=1:length(t)
  time = t(i);                 % Pick a time instant
  stm = expm(A*time);          % Compute the STM at time instant t(i)
  phi11 = [phi11,stm(1,1)];    % Append element (1,1) to vector 'phi11'
  phi12 = [phi12,stm(1,2)];    % Append element (1,2) to vector 'phi12'
  phi21 = [phi21,stm(2,1)];    % Append element (2,1) to vector 'phi21'
  phi22 = [phi22,stm(2,2)];    % Append element (2,2) to vector 'phi22'
end;
% Graph elements of the STM
tiledlayout(2,2);
nexttile;
plot(t,phi11); grid;
title('\phi_{11}(t)');
nexttile;
plot(t,phi12); grid;
title('\phi_{12}(t)');
nexttile;
plot(t,phi21); grid;
title('\phi_{21}(t)');
nexttile;
plot(t,phi22); grid;
title('\phi_{22}(t)');
```

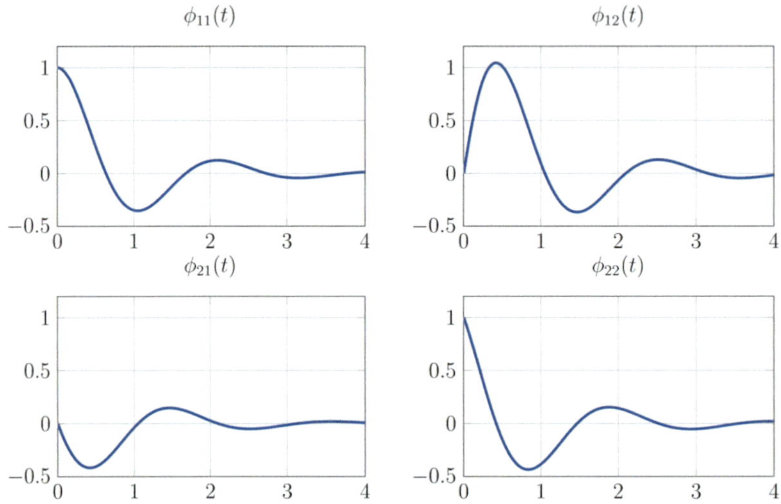

Figure 5.10 – Elements of state transition matrix for MATLAB Exercise 5.4.

The script listed below computes and graphs the elements of the state vector. It relies on the previous script to be run first, so that vectors `t`, `phi11`, `phi12`, `phi21`, and `phi22` are already in MATLAB workspace. The graph produced by the script `mexct_5_4b.m` is shown in Fig. 5.11.

```
% Script: mexct_5_4b.m
% Compute elements of the state vector
x1 = phi11-2*phi12;
x2 = phi21-2*phi22;
% Graph elements of the state vector
tiledlayout(2,1)
nexttile;
plot(t,x1); grid;
title('x_{1}(t)');
nexttile;
plot(t,x2); grid;
title('x_{2}(t)');
```

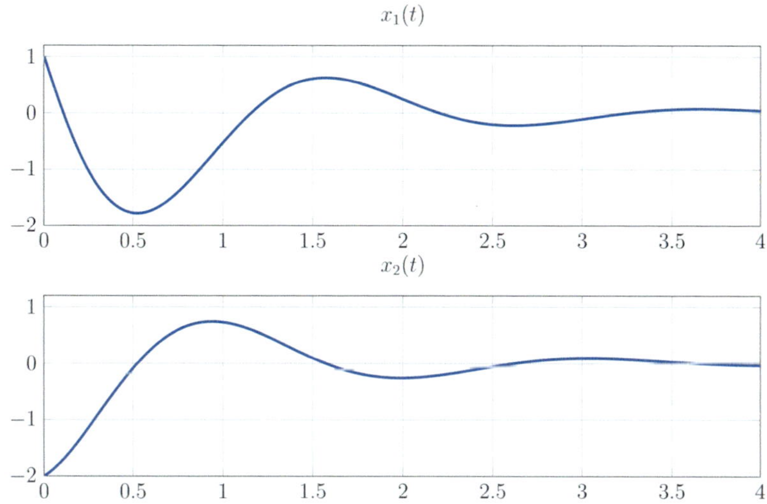

Figure 5.11 – Elements of state vector for MATLAB Exercise 5.4.

Software resources: mexct_5_4a.m , mexct_5_4b.m

MATLAB Exercise 5.5: Symbolic computation of the state transition matrix

It is also possible to compute the state transition matrix using symbolic mathematics capabilities of MATLAB. This is especially useful for checking solutions to homework problems. Consider the system for which a state-space model was given in Example 5.13. The coefficient matrix is

$$\mathbf{A} = \begin{bmatrix} -2 & -2 \\ 1 & -5 \end{bmatrix}$$

The state transition matrix $\boldsymbol{\phi}(t)$ can be computed as a symbolic expression using the following lines of code:

```
>> A = [-2,-2; 1,-5];
>> t = sym('t');    % Define symbolic variable.
>> expm(A*t)        % Compute the state transition matrix.

ans =
[ 2/exp(3*t) - 1/exp(4*t), 2/exp(4*t) - 2/exp(3*t)]
[ 1/exp(3*t) - 1/exp(4*t), 2/exp(4*t) - 1/exp(3*t)]
```

The symbolic answer produced should match the result found in Example 5.13, Eqn. (5.108). An alternative approach would be to

1. obtain the resolvent matrix $\boldsymbol{\Phi}(s)$ and
2. to obtain the state transition matrix as the inverse Laplace transform of the resolvent matrix.

Mathematically we have

$$\text{Eqn. (5.107):} \qquad \boldsymbol{\Phi}(s) = [s\mathbf{I} - \mathbf{A}]^{-1}$$

$$\text{Eqn. (5.106):} \qquad \boldsymbol{\phi}(t) = \mathscr{L}^{-1}\left\{[s\mathbf{I} - \mathbf{A}]^{-1}\right\}$$

The following lines of code implement this approach using the symbolic mathematics capabilities of MATLAB:

```
>> s = sym('s');          % Define symbolic variable.
>> tmp = s*eye(2)-A;
>> rsm = inv(tmp)         % Resolvent matrix.

rsm =
[ (s + 5)/(s^2 + 7*s + 12),       -2/(s^2 + 7*s + 12)]
[        1/(s^2 + 7*s + 12), (s + 2)/(s^2 + 7*s + 12)]

>> stm = ilaplace(rsm)  % State transition matrix.

stm =
[ 2/exp(3*t) - 1/exp(4*t), 2/exp(4*t) - 2/exp(3*t)]
[ 1/exp(3*t) - 1/exp(4*t), 2/exp(4*t) - 1/exp(3*t)]
```

Software resources: mexct_5_5a.m , mexct_5_5b.m

MATLAB Exercise 5.6: Obtaining the system function from state-space model

In Example 5.14 we have obtained the system function of a second-order CTLTI system from its state-space model. Verify the solution using symbolic mathematics capabilities of MATLAB.

Solution: Symbolic computation of the state transition matrix was discussed in MATLAB Exercise 5.5. Extending the technique used in that exercise, the system function can be found with the following statements:

```
>> A = [-2,-2;1,-5];
>> B = [1;0];
>> C = [0,5];
>> d = [0];
>> s = sym('s');
>> tmp = s*eye(2)-A

tmp =
[ s + 2,     2]
[    -1, s + 5]

>> rsm = inv(tmp)   % Resolvent matrix, Eqn. (5.107)

rsm =
[ (s + 5)/(s^2 + 7*s + 12),       -2/(s^2 + 7*s + 12)]
[        1/(s^2 + 7*s + 12), (s + 2)/(s^2 + 7*s + 12)]

>> G = C*rsm*B+d    % Eqn. (5.119)
```

```
G =
5/(s^2 + 7*s + 12)
```

The system function obtained matches the result in Example 5.14.

Software resource: `mexct_5_6.m`

Problems

5.1. A continuous-time system is described by the following set of differential equations.

$$\frac{dx_1(t)}{dt} = 2\,x_1(t) + 2\,x_2(t) - x_3(t)(t) + r_2(t)$$

$$\frac{dx_2(t)}{dt} = -3\,x_1(t) + 2\,x_3(t) - r_2(t)$$

$$\frac{dx_3(t)}{dt} = x_1(t) + x_2(t) - 2\,x_3(t) + r_1(t) + r_2(t)$$

$$y_1(t) = x_1(t) + x_3(t) + r_1(t)$$

$$y_2(t) = x_2(t) + x_3(t) + r_1(t)$$

Write the state-space model of the system in matrix form.

5.2. Consider the alternative form of the matrix state equation given by Eqn. (5.12):

$$\dot{\mathbf{x}}(t) = \mathbf{A}\mathbf{x}(t) + \mathbf{B}\,r(t) + \mathbf{E}\frac{dr(t)}{dt}$$

a. Show that this alternative form can be converted to the standard form given by Eqn. (5.7) by defining a new state vector

$$\mathbf{z} = \mathbf{x} - \mathbf{F}\,r(t)$$

Express the vector **F** in terms of **A**, **B** and **E**.

b. The state equation for a CTLTI system is

$$\dot{\mathbf{x}}(t) = \begin{bmatrix} -1 & 3 \\ -2 & -2 \end{bmatrix}\mathbf{x}(t) + \begin{bmatrix} 0 \\ 1 \end{bmatrix} r(t) + \begin{bmatrix} 2 \\ 1 \end{bmatrix}\frac{dr(t)}{dt}$$

Convert the state equation into the standard form by defining a new state vector $\mathbf{z}(t)$.

5.3. Find a state-space model for the circuit shown in Fig. P.5.3. Choose the capacitor voltages $v_{c1}(t)$ and $v_{c2}(t)$ as the state variables $x_1(t)$ and $x_2(t)$, respectively.

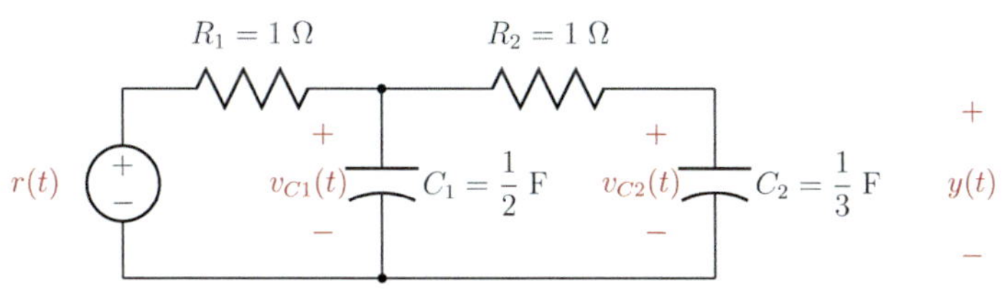

Figure P. 5.3

5.4. Find a state-space model for the RLC circuit shown in Fig. P.5.4. Choose the capacitor voltage $v_c(t)$ and the inductor current $i_L(t)$ as the state variables $x_1(t)$ and $x_2(t)$, respectively.

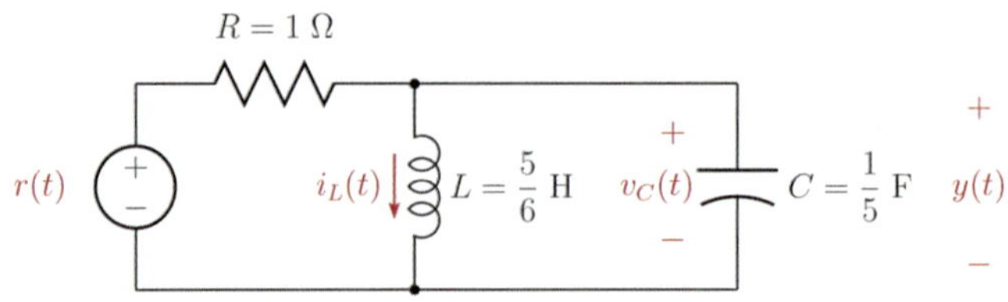

Figure P. 5.4

5.5. Using the approach taken in Examples 5.3 and 5.4 find a state-space model for each CTLTI system described by a differential equation below:

a. $\dfrac{d^2y(t)}{dt^2} + 3\dfrac{dy(t)}{dt} + 2\,y(t) = 2\,r(t)$

b. $\dfrac{d^2y(t)}{dt^2} + 4\dfrac{dy(t)}{dt} + 3\,y(t) = 3\,r(t)$

c. $\dfrac{d^2y(t)}{dt^2} - y(t) = r(t)$

d. $\dfrac{d^3y(t)}{dt^3} + 6\dfrac{d^2y(t)}{dt^2} + 11\dfrac{dy(t)}{dt} + 6\,y(t) = r(t)$

e. $\dfrac{d^3y(t)}{dt^3} + 6\dfrac{d^2y(t)}{dt^2} + 11\dfrac{dy(t)}{dt} + 6\,y(t) = r(t) + 2\dfrac{dr(t)}{dt}$

5.6. Find a state-space model for each CTLTI system described below by means of a system function. Use the partial fraction expansion of the system function so that the resulting state-space model has a diagonal coefficient matrix **A**.

a. $G(s) = \dfrac{2}{(s+1)\,(s+2)}$

b. $G(s) = \dfrac{10\,s+1}{(s+2)\,(s+4)}$

c. $G(s) = \dfrac{(s-1)\,(s+4)}{(s+1)\,(s+2)\,(s+3)}$

5.7. In Example 5.4 a state space model was obtained using an ad-hoc method for the CTLTI system defined by the differential equation

$$\frac{d^3y(t)}{dt^3} + 5\frac{d^2y(t)}{dt^2} + 11\frac{dy(t)}{dt} + 15\,y(t) = 3\,r(t) + 7\frac{dr(t)}{dt}$$

Obtain a same state space model for the system using a more formal approach by

a. first finding a system function $G(s)$ for the system, and

b. deriving a state space model from the system function.

5.8. Refer to MATLAB Exercise 5.1 in which two different state-space models were obtained for the same system. The first model was

$$\dot{\mathbf{x}}(t) = \begin{bmatrix} -6 & -5 \\ 1 & 0 \end{bmatrix} \mathbf{x}(t) + \begin{bmatrix} 1 \\ 0 \end{bmatrix} r(t)$$

$$y(t) = \begin{bmatrix} -5 & -7 \end{bmatrix} \mathbf{x}(t) + r(t)$$

and the second model was

$$\dot{\mathbf{z}}(t) = \begin{bmatrix} 0 & 1 \\ -5 & -6 \end{bmatrix} \mathbf{z}(t) + \begin{bmatrix} 0 \\ 1 \end{bmatrix} r(t)$$

$$y(t) = \begin{bmatrix} -7 & -5 \end{bmatrix} \mathbf{z}(t) + r(t)$$

Show that the second state-space model can be obtained from the first by means of the similarity transformation

$$\mathbf{x}(t) = \begin{bmatrix} 0 & 1 \\ 1 & 0 \end{bmatrix} \mathbf{z}(t)$$

5.9. Refer to Example 5.1. The state-space model obtained for the RLC circuit in Fig. 5.3 was

$$\dot{\mathbf{x}}(t) = \begin{bmatrix} -2 & -2 \\ 1 & -5 \end{bmatrix} \mathbf{x}(t) + \begin{bmatrix} 1 \\ 0 \end{bmatrix} r(t)$$

$$y(t) = \begin{bmatrix} 0 & 5 \end{bmatrix} \mathbf{x}(t) + (0)\, r(t)$$

a. Write the state equations in open form. Express $dy(t)/dt$ and $d^2y(t)/dt^2$ in terms of the state variables $x_1(t)$ and $x_2(t)$ and the input signal $r(t)$.

b. Find a second-order differential equation between $y(t)$ and $r(t)$ by eliminating the state variables between the equations for $y(t)$, $dy(t)/dt$, and $d^2y(t)/dt^2$.

c. Using the second-order differential equation obtained in part (b), find an alternative state-space model in phase-variable canonical form.

5.10. Refer to Problem 5.9. Find a similarity transformation

$$\mathbf{x}(t) = \mathbf{P}\mathbf{z}(t)$$

that would transform the original state-space model to the alternative model in phase-variable canonical form that was found.

5.11. Refer again to Example 5.1 in which a state-space model for the RLC circuit in Fig. 5.3 was found to be

$$\dot{\mathbf{x}}(t) = \begin{bmatrix} -2 & -2 \\ 1 & -5 \end{bmatrix} \mathbf{x}(t) + \begin{bmatrix} 1 \\ 0 \end{bmatrix} r(t)$$

$$y(t) = \begin{bmatrix} 0 & 5 \end{bmatrix} \mathbf{x}(t) + (0)\, r(t)$$

In this problem we will explore an alternative approach for finding a differential equation between the input and the output signals of the system.

a. Write the state equations in open form. Take the Laplace transform of each equation (two state equations and one output equation) by making use of linearity and time differentiation properties of the Laplace transform.

b. Eliminate $X_1(s)$ and $X_2(s)$ to find an equation between $Y(s)$ and $R(s)$. Determine the system function $G(s)$.

c. Find a differential equation for the system from the system function found in part (b).

5.12. A CTLTI system with two inputs and two outputs is described by the following pair of differential equations:

$$\frac{dy_1(t)}{dt} = -2\,y_1(t) - y_2(t) + r_1(t) + r_2(t)$$

$$\frac{dy_2(t)}{dt} = -3\,y_2(t) + y_1(t) + \frac{dr_1(t)}{dt}$$

Following the approach used in Example 5.10 find a state-space model for this system.

5.13. A CTLTI system with two inputs and two outputs is described by the state-space model

$$\dot{\mathbf{x}}(t) = \begin{bmatrix} -1 & 2 \\ -3 & -6 \end{bmatrix} \mathbf{x}(t) + \begin{bmatrix} 1 & 2 \\ 0 & -1 \end{bmatrix} \mathbf{r}(t)$$

$$\mathbf{y}(t) = \begin{bmatrix} -1 & -3 \\ 2 & 0 \end{bmatrix} \mathbf{x}(t) + \begin{bmatrix} 0 & 1 \\ 4 & -1 \end{bmatrix} \mathbf{r}(t)$$

a. Using the appropriate similarity transformation, find an equivalent state-space model for this system with a diagonal state matrix.

b. Using the state-space model found, express output transforms $Y_1(s)$ and $Y_2(s)$ in terms of input transforms $R_1(s)$ and $R_2(s)$.

c. Using the results found in part (b), write the input-output relationships of the system in the s-domain as

$$\begin{bmatrix} Y_1(s) \\ Y_2(s) \end{bmatrix} = \begin{bmatrix} G_{11}(s) & G_{12}(s) \\ G_{21}(s) & G_{22}(s) \end{bmatrix} \begin{bmatrix} R_1(s) \\ R_2(s) \end{bmatrix}$$

5.14. Refer to Example 5.11 in which the state variables and the output signal of a third-order system were determined in response to unit step input signal and subject to specified initial conditions. Verify that the solutions found for each of the state variables and the output signal satisfy the state-space model and the initial conditions.

5.15. A continuous-time system is described by the state-space model

$$\dot{\mathbf{x}}(t) = \begin{bmatrix} -3 & 2 \\ 0 & -1 \end{bmatrix} \mathbf{x}(t) + \begin{bmatrix} 0 \\ 1 \end{bmatrix} r(t)$$

$$y(t) = \begin{bmatrix} 1 & 1 \end{bmatrix} \mathbf{x}(t) + r(t)$$

The initial value of the state vector is

$$\mathbf{x}(0) = \begin{bmatrix} -2 \\ 3 \end{bmatrix}$$

Taking advantage of the upper-triangular property of the state matrix determine the output signal in response to a unit step input signal. Follow the approach taken in Example 5.11.

5.16. Refer to the system in Problem 5.15.

a. Using the appropriate similarity transformation, obtain an equivalent state-space model for the system in which the state matrix is diagonal.

b. Use the alternative state-space model with the diagonal state matrix to determine the output signal in response to a unit step input signal.

5.17. Consider the homogeneous state-space model

$$\dot{\mathbf{x}}(t)=\begin{bmatrix}-3 & 2\\ 0 & -1\end{bmatrix}\mathbf{x}(t)$$

a. Determine the state transition matrix $\boldsymbol{\phi}(t)$.

b. Using the initial state vector

$$\mathbf{x}(0)=\begin{bmatrix}-2\\ 3\end{bmatrix}$$

and the state transition matrix found in part (a), find the $\mathbf{x}(2)$, the value of the state vector at time instant $t=2$.

c. Find $\mathbf{x}(5)$.

5.18. The general solution of the continuous-time state-space model can also be obtained using Laplace transform techniques. Let $\mathbf{X}(s)$ be the vector that holds the Laplace transform of each state variable.

a. Take the unilateral Laplace transform of both sides of the state equation given by Eqn. (5.7) taking $\mathbf{x}(0)$, the initial value of the state vector, into account.

b. Solve for $\mathbf{X}(s)$ in terms of the Laplace transform $R(s)$ of the input signal.

c. Take the Laplace transform of the output equation given by Eqn. (5.8) and substitute the solution found for $\mathbf{X}(s)$ to obtain the relationship between $Y(s)$ and $R(s)$.

d. Write the time-domain relationship between the output signal $y(t)$ and the input signal $r(t)$ remembering that

$$\boldsymbol{\phi}(t)=e^{\mathbf{A}t}=\mathscr{L}^{-1}\left\{\left[s\mathbf{I}-\mathbf{A}\right]^{-1}\right\}$$

and that the product of two Laplace transforms corresponds to convolution of the corresponding signals in the time domain.

5.19. Refer to Example 5.13. Verify that the solutions found for the state variables and the output signal satisfy the state-space model and the specified initial conditions.

5.20. A state-space model for the RLC circuit of Fig. 5.4 was found in Example 5.2 as

$$\dot{\mathbf{x}}(t)=\begin{bmatrix}-2 & -1\\ 26 & 0\end{bmatrix}\mathbf{x}(t)+\begin{bmatrix}1\\ 0\end{bmatrix}r(t)$$

$$y(t)=\begin{bmatrix}-2 & 0\end{bmatrix}\mathbf{x}(t)+(1)\,r(t)$$

Let the initial state of the system be

$$\mathbf{x}(0)=\begin{bmatrix}2\\ 5\end{bmatrix}$$

a. Find the state transition matrix $\boldsymbol{\phi}(t)$.

b. The input to the system is a unit step signal, that is, $r(t)=u(t)$. Using the solution technique outlined in Eqn. (5.100), find an expression for the state vector $\mathbf{x}(t)$. Afterward find an expression for the output signal $y(t)$.

5.21. Refer to the system in Problem 5.20. Determine the state vector and the output signal again, this time using Laplace transform techniques. Specifically, take the unilateral Laplace transform of the state equation to find the relationship between $\mathbf{X}(s)$ and $R(s)$. Take the transform of the output equation to find a relationship between $Y(s)$ and $R(s)$. Determine $\mathbf{x}(t)$ and $y(t)$ using inverse Laplace transform. Compare with the results found in Problem 5.20.

5.22. A state-space model for the RLC circuit of Fig. 5.4 was found in Example 5.2 as

$$\dot{\mathbf{x}}(t) = \begin{bmatrix} -2 & -1 \\ 26 & 0 \end{bmatrix} \mathbf{x}(t) + \begin{bmatrix} 1 \\ 0 \end{bmatrix} r(t)$$

$$y(t) = \begin{bmatrix} -2 & 0 \end{bmatrix} \mathbf{x}(t) + (1)\, r(t)$$

Find the system function $G(s) = Y(s)/R(s)$.

5.23. A CTLTI system is described by the state-space model

$$\dot{\mathbf{x}}(t) = \begin{bmatrix} -4 & 3 \\ -1 & 0 \end{bmatrix} \mathbf{x}(t) + \begin{bmatrix} 2 \\ 1 \end{bmatrix} r(t)$$

$$y(t) = \begin{bmatrix} 3 & 1 \end{bmatrix} \mathbf{x}(t) + (2)\, r(t)$$

a. Find a transformation matrix in the form

$$\mathbf{P} = \begin{bmatrix} 1 & a \\ b & 1 \end{bmatrix}$$

such that the transformation

$$\mathbf{x}(t) = \mathbf{P}\mathbf{z}(t)$$

converts the state-space model into one with a diagonal coefficient matrix.

b. Write the state-space model in terms of the new state vector $\mathbf{z}(t)$.

MATLAB Problems

5.24. Consider the CTLTI system with system function

$$G(s) = \frac{2s^3 - 26s + 24}{s^4 + 7s^3 + 21s^2 + 37s + 30}$$

A state-space model for this system was obtained in Exercise 5.7 from its cascade-form block diagram. An alternative model can be found using a parallel-form block diagram instead.

a. Write a script to expand $G(s)$ into partial fractions. Use the function `residue()` to find poles of $G(s)$ along with corresponding residues. Real-valued poles should lead to first-order parallel sections. Complex-conjugate pole pair should be combined into a second-order section.

b. Draw the parallel-form block diagram. Select the state variables, and write the differential equations for the corresponding state-space model.

c. Express the differential equations found in part (b) in matrix form.

5.25. A CTLTI system is described by the state-space model

$$\dot{\mathbf{x}}(t) = \begin{bmatrix} 49 & 25 & -9 \\ -96 & -49 & 17 \\ 48 & 24 & -9 \end{bmatrix} \mathbf{x}(t) + \begin{bmatrix} -1 \\ 2 \\ -1 \end{bmatrix} r(t)$$

$$y(t) = \begin{bmatrix} 5 & 3 & 1 \end{bmatrix} \mathbf{x}(t)$$

Write a MATLAB script to perform the following steps:

a. Using the built-in function `ss2tf()`, determine the system function $G(s)$.

b. Find an alternative state-space model for the same system in phase-variable canonical form. Hint: After finding the system function, use the function `ss_tf2ss()` developed in MATLAB Exercise 5.1.

5.26. Consider again the CTLTI system with state-space description given in Problem 5.25. Write a script to determine its system function using symbolic mathematics capabilities of MATLAB. Compare the result to the system function found in part (a) of Problem 5.25.

5.27. Consider the system described by the state-space model

$$\dot{\mathbf{x}}(t) = \begin{bmatrix} -3 & 1 & 0 \\ 0 & -2 & 5 \\ 0 & 0 & -1 \end{bmatrix} \mathbf{x}(t) + \begin{bmatrix} 0 \\ 2 \\ 1 \end{bmatrix} r(t)$$

$$y(t) = \begin{bmatrix} 1 & 0 & 0 \end{bmatrix} \mathbf{x}(t)$$

The initial value of the state vector is

$$\mathbf{x}(0) = \begin{bmatrix} 0 \\ 3 \\ 2 \end{bmatrix}$$

The response of this system to a unit step input signal was determined in Example 5.11 to be

$$y(t) = \left(\frac{7}{6} + \frac{5}{2} e^{-t} - \frac{11}{2} e^{-2t} + \frac{11}{6} e^{-3t} \right) u(t)$$

An approximate solution for the state vector can be found by writing Eqn. (5.100) for $t = t_0 + T$ where T is a sufficiently small step size.

$$\begin{aligned}\hat{\mathbf{x}}(t_0 + T) &= e^{\mathbf{A}T} \hat{\mathbf{x}}(t_0) + \frac{1}{2} \left[e^{\mathbf{A}(0)} \mathbf{B} r(t_0 + T) + e^{\mathbf{A}T} \mathbf{B} r(t_0) \right] T \\ &= \boldsymbol{\phi}(T) \hat{\mathbf{x}}(t_0) + \frac{1}{2} \left[\mathbf{B} r(t_0 + T) + \boldsymbol{\phi}(T) \mathbf{B} r(t_0) \right] T\end{aligned}$$

where we have used the trapezoidal approximation to the integral. An approximation for the output signal can then be obtained as

$$\hat{y}(t_0 + T) = \mathbf{C} \hat{\mathbf{x}}(t_0 + T) + d\, r(t_0 + T)$$

Write a script to perform the following steps.

a. Select a step size T, and compute $\boldsymbol{\phi}(T)$, the value of the state transition matrix at $t = T$, using the function `expm()`.

b. Set a loop and iterate through the approximation, computing $\hat{\mathbf{x}}(T)$ from the knowledge of $\hat{\mathbf{x}}(0)$, then computing $\hat{\mathbf{x}}(2T)$ from the knowledge of $\hat{\mathbf{x}}(T)$, and so on. Advance the state vector from $t = 0$ to $t = 4$ s. Compute the output $\hat{y}(t)$ for the same time interval.

c. Graph the approximate output signal $\hat{y}(t)$ against the actual output signal $y(t)$ found analytically in Example 5.11. Do this for $T = 0.1$ and $T = 0.01$. Comment on the results.

5.28. A CTLTI system is described by the state-space model

$$\dot{\mathbf{x}}(t) = \begin{bmatrix} 49 & 25 & -9 \\ -96 & -49 & 17 \\ 48 & 24 & -9 \end{bmatrix} \mathbf{x}(t) + \begin{bmatrix} -1 \\ 2 \\ -1 \end{bmatrix} r(t)$$

$$y(t) = \begin{bmatrix} 5 & 3 & 1 \end{bmatrix} \mathbf{x}(t)$$

Write a MATLAB script to find an alternative state-space model for the system in which the matrix **A** is diagonal.

CHAPTER 6

Analysis and Design of Continuous-Time Filters

Chapter Objectives

- Become familiar with frequency selective analog filters.
- Learn the concept of distortion and conditions for distortionless transmission.
- Learn ideal filter behavior. Starting with the concept of distortionless transmission, develop mathematical relationships for ideal frequency selective filters.
- Learn design specifications for realizable filters.
- Explore methods for designing realizable analog filters using Butterworth, Chebyshev, and elliptic approximation formulas.
- Learn transformation methods for converting an analog lowpass filter to a highpass, bandpass, or bandstop filter.

6.1 Introduction

In many signal processing applications the need arises to change the strength, or the relative significance, of various frequency components in a given signal. Sometimes we may need to eliminate certain frequency components in a signal; at other times we may need to boost the strength of a range of frequencies over others. The act of changing relative amplitudes of frequency components of a signal is referred to as *filtering*, and the system that facilitates this is referred to as a *filter*.

DOI: 10.1201/9781032675251-6

In Chapters 3 and 4 we have developed techniques for transform-domain analysis of CTLTI systems. Specifically, the relationship between the input signal $x(t)$ and the output signal $y(t)$ of a CTLTI system is expressed in the transform domain as

$$Y(\omega) = H(\omega)\, X(\omega) \tag{6.1}$$

in terms of the Fourier transforms of the signals involved, or as

$$Y(s) = H(s)\, X(s) \tag{6.2}$$

using the Laplace transforms of the signals. In either case, the system function $H(\omega)$ or $H(s)$ serves as a multiplier that shapes the input spectrum to create the resulting output spectrum.

In a general sense, any CTLTI system can be seen as a filter. Consider, for example, a communication channel the purpose of which is to transmit an information bearing signal from one point to another. Assuming that the particular channel we are considering can be reasonably modeled as a CTLTI system, we have the model shown in Fig. 6.1.

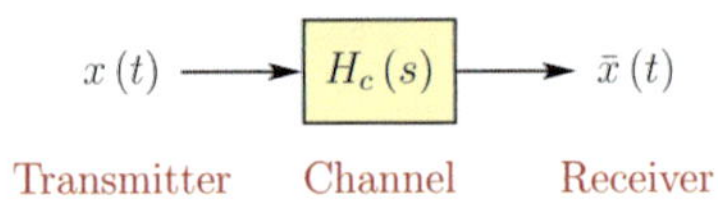

Figure 6.1 – CTLTI model for a communication channel.

Ideally we would like the signal $\bar{x}(t)$ at the output of the channel to be identical to the signal $x(t)$ that is transmitted. This is generally not the case, however, and the signal $\bar{x}(t)$ ends up being a *distorted* version of $x(t)$. Therefore in this case the channel acts as a filter the action of which is an undesired one. In order to compensate for the distortion effect of the channel, a second system may be connected in cascade with it as shown in Fig. 6.2. This second system is referred to as an *equalizer*. Its purpose is to eliminate or at least reduce the distortion effect caused by the channel so that the output signal $y(t)$ is a better representation of the transmitted signal $x(t)$.

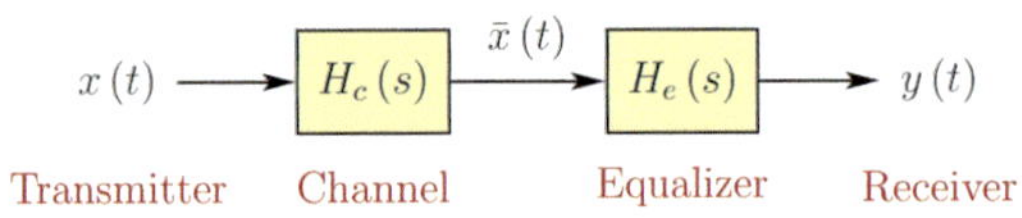

Figure 6.2 – Cascade connection of the channel with an equalizer.

For the arrangement in Fig. 6.2 we can write the following relationships:

$$\left.\begin{aligned} \bar{X}(s) &= H_c(s)\, X(s) \\ Y(s) &= H_e(s)\, \bar{X}(s) \end{aligned}\right\} \quad \Longrightarrow \quad Y(s) = H_c(s)\, H_e(s)\, X(s) \tag{6.3}$$

If we want the output signal $y(t)$ to be identical to the original input signal $x(t)$, it follows from Eqn. 6.3 that the cascade combination of the channel and the equalizer must form an identity system, that is,

$$H_c(s)\, H_e(s) = 1 \tag{6.4}$$

This requires the equalizer system function to be the inverse of the channel system function, that is

$$H_e(s) = \frac{1}{H_c(s)} \tag{6.5}$$

6.2 Distortionless Transmission

As discussed in the previous section, a system driven by an input signal $x(t)$ produces an output signal $y(t)$ that is, in general, different in shape than the input signal. Thus the system modifies the signal as it transmits it from its input to its output. This modification may be a desired effect of the system or an undesired side-effect. Regardless, it is referred to as *distortion.*

Consider again a CTLTI system with impulse response $h(t)$ driven by an input signal $x(t)$. The input and the output signals are related to each other through the convolution relationship

$$y(t) = h(t) * x(t) = \int_{-\infty}^{\infty} h(\lambda)\, x(t-\lambda)\, d\lambda \tag{6.6}$$

The corresponding relationship between the frequency spectra of input and output signals was given by Eqns. 6.1 and 6.2. The type of distortion caused by a CTLTI system is *linear distortion* and it exhibits itself in two distinct forms, namely *amplitude distortion* and *phase distortion.* Let us write the magnitude spectrum of the output signal from Eqn. (6.1):

$$\left|Y(\omega)\right| = \left|H(\omega)\right|\,\left|X(\omega)\right| \tag{6.7}$$

Considering that the input signal $x(t)$ is a specific mixture of sinusoidal components at various frequencies, Eqn. (6.7) describes how the system modifies the amplitude of each of these sinusoidal components. The scale factor applied to a signal component at a particular frequency may be different than the scale factor applied to a component at another frequency. This non-uniform scaling of sinusoidal components of the input signal leads to *amplitude distortion.*

The phase spectrum of the output signal is found from Eqn. (6.1) as

$$\measuredangle Y(\omega) = \measuredangle H(\omega) + \measuredangle X(\omega) \tag{6.8}$$

The system adds an offset to the phase of each sinusoidal component of the input signal. Phase-shifting a sinusoidal component of the input signal causes a corresponding time-shift (delay). If signal components are delayed by differing amounts, their relative alignment in time is disturbed, causing the output signal to be different than the input signal. This effect is known as *phase distortion.*

What type of a system function would be needed in order to obtain a *distortionless* version of the input signal at the output? As a starting point for discussion let us assume that, what we mean by distortionless transmission is the output signal being identical to the input signal, that is,

$$y(t) = x(t) \tag{6.9}$$

This clearly requires that we have

$$Y(\omega) = X(\omega) \tag{6.10}$$

which in turn requires

$$H(\omega) = 1\,, \quad \text{all } \omega \tag{6.11}$$

and consequently

$$h(t) = \delta(t) \tag{6.12}$$

Thus, we need the impulse response of the system to be an impulse. While Eqns. (6.11) and (6.12) represent a mathematically satisfying result, the corresponding system is physically unrealizable. No physical system is capable of responding to an input signal instantaneously, without a time delay between the input and the output. In order to come up with a realizable result, we will relax our definition of distortionless transmission and accept an output signal in the form

$$y(t) = K\,x(t - t_d) \tag{6.13}$$

where K is a constant scale factor and t_d is a constant time-delay. Thus, $y(t)$ in Eqn. (6.13) is a scaled and delayed version of the input signal $x(t)$. Taking the Fourier transform of both sides of Eqn. (6.13) yields

$$Y(\omega) = K\,e^{-j\omega t_d}\,X(\omega) \tag{6.14}$$

and the system function needed for distortionless transmission is

$$H(\omega) = K\,e^{-j\omega t_d} \tag{6.15}$$

Magnitude and phase characteristics of the system function are

$$\left|H(\omega)\right| = K \tag{6.16}$$

and

$$\Theta(\omega) = \measuredangle H(\omega) = -\omega\,t_d \tag{6.17}$$

A distortion-free system has a magnitude characteristic that is constant, and a phase characteristic that is a linear function of frequency as shown in Fig. 6.3.

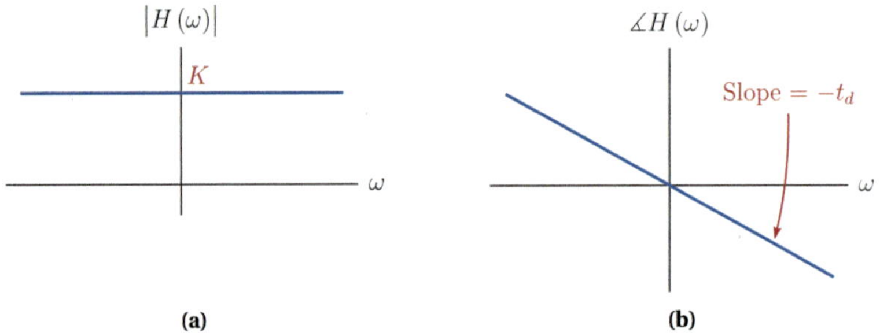

Figure 6.3 – Frequency-domain characteristics for a distortionless system: **(a)** Magnitude and **(b)** phase.

Example 6.1: Distortion caused by the RC circuit

Consider again the simple RC circuit in Fig. 6.4 that was used in several examples in Chapters 2, 3, and 4.

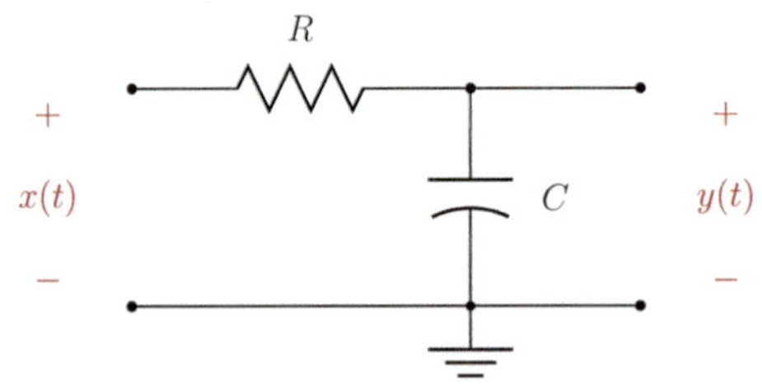

Figure 6.4 – RC circuit for Example 6.1.

In some cases the RC circuit can be used as a simple model for a wired electrical connection between two points. Such a model would provide a reasonably good approximation to the characteristics of the wired connection provided that the distance between the two points being connected is not too long. (For long distance connections the resistance and the capacitance would have to be modeled as distributed parameters rather than being concentrated at one point each.)

The system function for the RC circuit is

$$H(\omega) = \frac{1}{1 + j\omega RC}$$

Clearly this is not a distortionless system as defined by Eqn. (6.15), however, under certain circumstances it can be considered near distortion-free. The magnitude of the system function is

$$\left|H(\omega)\right| = \frac{1}{\sqrt{1 + (\omega/\omega_c)^2}} \tag{6.18}$$

with the 3-dB cutoff frequency defined as

$$\omega_c \triangleq \frac{1}{RC} \tag{6.19}$$

This magnitude characteristic is shown in Fig. 6.5(a). The phase of the system function is

$$\Theta(\omega) = \measuredangle H(\omega) = -\tan^{-1}\left(\frac{\omega}{\omega_c}\right) \tag{6.20}$$

and the corresponding time delay is

$$t_d(\omega) = -\frac{\Theta(\omega)}{\omega} = \frac{1}{\omega}\tan^{-1}\left(\frac{\omega}{\omega_c}\right) \tag{6.21}$$

The phase characteristic is shown in Fig. 6.5(b).

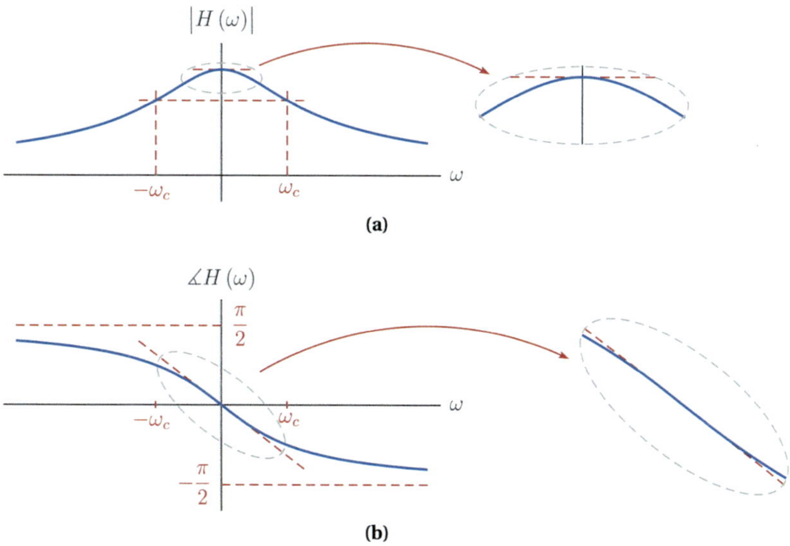

Figure 6.5 – Magnitude and phase characteristics of the RC circuit of Example 6.1 at low frequencies.

The magnitude characteristic of the system seems to be relatively flat for frequencies in the vicinity of $\omega = 0$. Similarly, the phase characteristic is close to a straight line, and the time-delay is relatively constant for low frequencies. Thus, if we limit the use of the system to input signals that contain only the frequencies $\omega << \omega_c$, we could consider the system as near distortion-free.

What is the frequency range (in terms of ω_c) in which one can consider the RC circuit approximately distortionless? Of course, the answer depends on the amount of variation in magnitude and/or time-delay that we consider negligible. Since both the magnitude and the time-delay characteristics exhibit their peak values at $\omega = 0$, and both are symmetric around that point, we could simply look for a frequency at which the characteristic drops down by a certain percentage. Let ω_0 be the frequency at which the magnitude of the system function drops below the peak value by p percent. Using Eqn. (6.18) and realizing that the maximum magnitude is $H(0) = 1$, we can write

$$\frac{1}{\sqrt{1+(\omega_0/\omega_c)^2}} \le \frac{100-p}{100} \tag{6.22}$$

Solving this inequality for ω_0 yields

$$\omega_0 \le \omega_c \sqrt{\left(\frac{100}{100-p}\right)^2 - 1} \tag{6.23}$$

For the frequencies $|\omega| \le 0.1425\,\omega_c$, the variation in the magnitude of the system function will be 1 percent or less. If we are willing to tolerate up to 5 percent magnitude variation, then the usable frequency range is $|\omega| \le 0.3287\,\omega_c$.

Next consider the phase characteristic given by Eqn. (6.20). Differentiating it with respect to frequency, we obtain

$$\frac{d\Theta}{d\omega} = -\frac{1}{\omega_c}\left[\frac{1}{1+(\omega/\omega_c)^2}\right] \tag{6.24}$$

and the slope of the phase characteristic at frequency $\omega = 0$ is

$$\left.\frac{d\Theta}{d\omega}\right|_{\omega=0} = -\frac{1}{\omega_c} \tag{6.25}$$

The line that is tangent to the phase characteristic at the frequency $\omega = 0$ is also shown in Fig. 6.5(b) for comparison. The phase remains fairly close to the tangent line for low frequencies. Similar analysis can be performed for the time-delay characteristic of the system, however, because of the form of Eqn. (6.21), arriving at an analytical result is tedious. Instead of trying to duplicate the methodology used in Eqns. (6.22) and (6.23), let us just take the frequency ranges obtained for specified magnitude variations above and determine the corresponding variation in time-delay. It can easily be shown that

$$t_d(0) = 0.1592/f_c$$

$$t_d\left(0.1425 f_c\right) = 0.1581/f_c$$

$$t_d\left(0.3287 f_c\right) = 0.1538/f_c$$

If we accept 1 percent variation in magnitude, the corresponding variation in time delay is 0.67 percent. For a 5 percent magnitude variation, the time-delay varies by 3.38 percent.

Software resources: exct_6_1.m , exct_6_1_Live.mlx

Interactive App: Distortion caused by the RC circuit

The interactive app in `appDistort.m` is modeled after Example 6.1 and Fig. 6.5. Magnitude, phase, and time-delay characteristics of the RC circuit system function are graphed. Circuit parameters may be varied using slider controls. Allowed percentage deviation of the magnitude response from its peak value may also be specified. Based on this percent tolerance, the range of frequencies within which the distortion is considered negligible is indicated with dashed lines on magnitude, phase, and time-delay characteristics. Zoom-in button allows a closer look at the critical values on each graph.

a. Begin with $R = 1$ MΩ and $C = 1$ ţF. Let the tolerance limit be 1 percent. Observe the frequency range in which the magnitude stays within 1 percent of the dc value. Also observe deviations from ideal behavior for the phase and the time-delay, and compare them to theoretical values.

b. Gradually reduce the value of R down to 0.5 MΩ and observe the changes in the frequency range.

c. While keeping $R = 0.5$ MΩ increase the tolerance limit to 2, 3, 4, and 5 percent, and observe the changes in the frequency interval.

Software resource: `appDistort.m`

6.3 Ideal Filters

The conditions for a CTLTI system to be distortionless were identified in Eqns. (6.16) and (6.17) of the previous section and were graphically illustrated in Fig. 6.3. We observe that, in order to be truly distortionless, a system must have infinite bandwidth. In other words, it must be able to respond to signals containing an infinite range of frequencies. No physically realizable system is capable of doing this.

Furthermore, signals encountered in engineering applications have limited bandwidth, that is, they contain a limited range of frequencies. Consider such a signal $x(t)$. A CTLTI system that satisfies Eqns. (6.16) and (6.17) not necessarily at every frequency, but at all frequencies contained in the signal $x(t)$ would still facilitate distortionless transmission of this signal. Ideal frequency-selective filters are systems that allow distortionless transmission of signal components in a particular frequency range while eliminating all other signal components.

One type of ideal filter is an *ideal lowpass filter* that is characterized by the system function

$$\begin{aligned} H_{LP}(\omega) &= \Pi\left(\frac{\omega}{2\omega_0}\right) e^{-j\omega t_d} \\ &= \begin{cases} e^{-j\omega t_d}\,, & |\omega| \le \omega_0 \\ 0\,, & |\omega| > \omega_0 \end{cases} \end{aligned} \tag{6.26}$$

Any signal that contains only the frequencies in the range $|\omega| \le \omega_0$ is transmitted through this system without any distortion of its magnitude or phase characteristics. The magnitude and the phase of the system function $H_{LP}(\omega)$ of an ideal lowpass filter are graphed in Fig. 6.6.

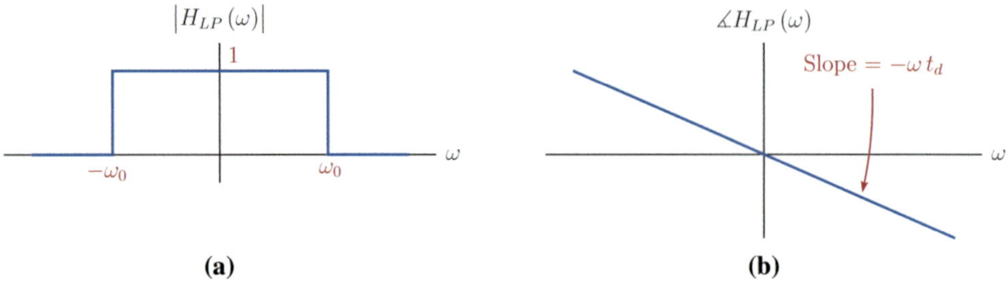

Figure 6.6 – The frequency spectrum of the ideal lowpass filter: **(a)** magnitude and **(b)** phase.

The magnitude of the system function is constant for frequencies in the range $-\omega_0 \le \omega \le \omega_0$. The phase of the system function is a straight line that passes through the origin, with the slope $-\omega\, t_d$. Signals that do not contain any frequencies outside the range $|\omega| \le \omega_0$ would be transmitted through this system without experiencing any distortion. The frequency ω_0 in Eqn. (6.26) is the *cutoff frequency* of the ideal lowpass filter. The parameter t_d indicates the time delay caused by the filter.

Some signals are bandpass in nature, that is, they contain frequencies in a bandpass frequency range $\omega_1 \le |\omega| < \omega_2$. Examples of these signals are encountered often in the study of communication systems. The system function of an *ideal bandpass filter* that would allow distortionless transmission of the appropriate bandpass signal is given by

$$H_{BP}(\omega) = \begin{cases} e^{-j\omega\, t_d}\,, & \omega_1 \le |\omega| \le \omega_2 \\ 0\,, & \text{otherwise} \end{cases} \tag{6.27}$$

and is illustrated in Fig. 6.7.

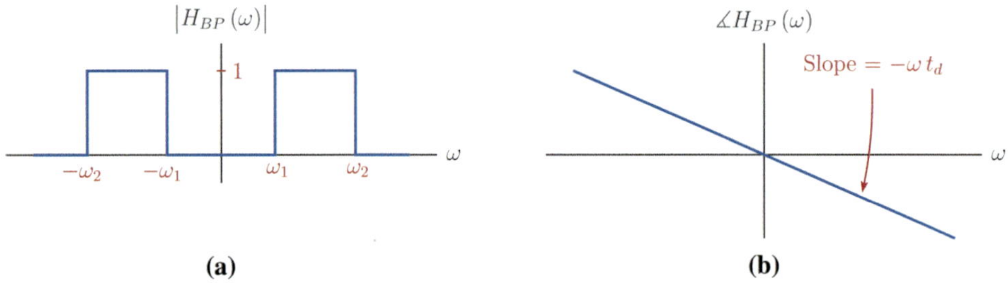

Figure 6.7 – The frequency spectrum of the ideal bandpass filter: **(a)** magnitude and **(b)** phase.

Similarly, ideal highpass and bandstop filters can also be defined. An *ideal highpass filter* transmits, without distortion, signals that contain only frequencies greater than a minimum value, that is, $|\omega| \ge \omega_0$. An *ideal bandstop filter,* on the other hand, facilitates distortionless transmission of signals that contain only those frequencies that are outside a specified range. Frequencies that satisfy either $|\omega| \le \omega_1$ or $|\omega| \ge \omega_2$ are transmitted; all other frequencies are suppressed. Figs. 6.8 and 6.9 illustrate the system functions of ideal highpass and bandstop filters, respectively.

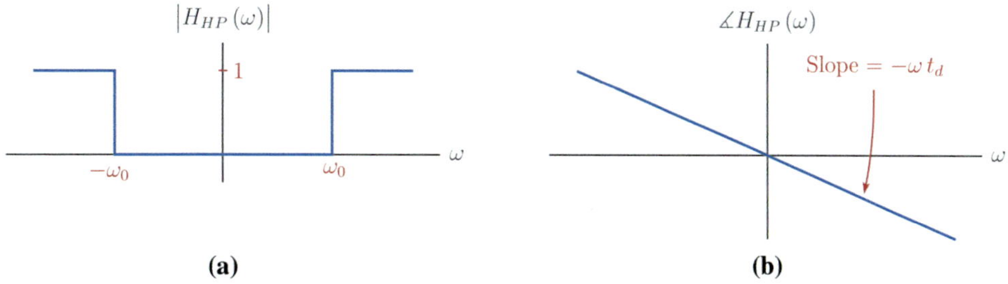

Figure 6.8 – The frequency spectrum of the ideal highpass filter: **(a)** magnitude and **(b)** phase.

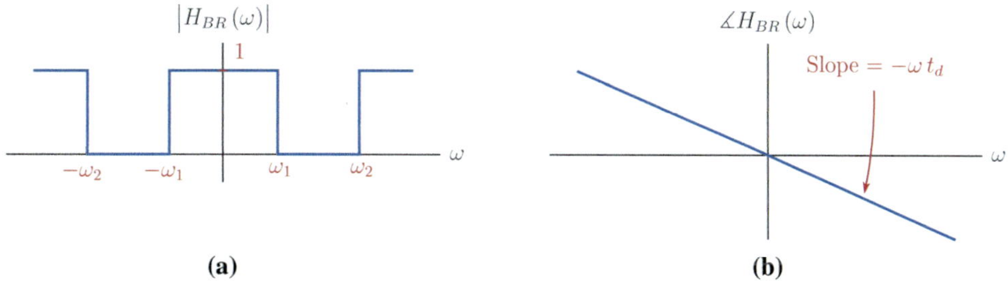

Figure 6.9 – The frequency spectrum of the ideal bandstop filter: **(a)** magnitude and **(b)** phase.

The ideal filters discussed above are not practically realizable. In order to see why this is the case, let us determine the impulse response of an ideal filter. For the ideal lowpass filter of Eqn. (6.26), the impulse response can be easily computed from the system function through inverse Fourier transform. Using the inverse Fourier transform relationship given by Eqn. (3.121) in Chapter 3 on the ideal lowpass filter system function in Eqn. (6.26) we obtain

$$h_{LP}(t) = \mathscr{F}^{-1}\left\{H_{LP}(\omega)\right\} = \frac{\omega_0}{\pi}\,\text{sinc}\left(\frac{\omega_0}{\pi}(t-t_d)\right) \tag{6.28}$$

For convenience, let us express $h_{LP}(t)$ using f_0 instead of ω_0:

$$h_{LP}(t) = 2f_0\,\text{sinc}\left(2f_0\,(t-t_d)\right) \tag{6.29}$$

The resulting impulse response is shown in Fig. 6.10.

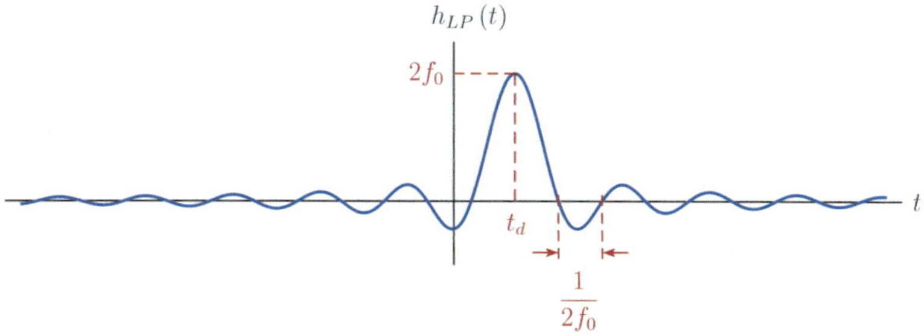

Figure 6.10 – Impulse response of the ideal lowpass filter.

A few observations are in order regarding the impulse response of the ideal lowpass filter:

1. The impulse response $h_{LP}(t)$ given by Eqn. (6.29) is in the form of a sinc function with a peak amplitude of $2f_0$. This peak occurs at the time instant $t = t_d$.
2. The zero-crossings of the impulse response are uniformly spaced. Consecutive zero-crossings are $1/(2f_0)$ apart from each other.
3. The impulse response exists for all time instants including $t < 0$. Thus, the ideal lowpass filter is a non-causal system, and is therefore physically unrealizable.
4. Furthermore, no amount of additional time delay would make the impulse response in Eqn. (6.29) causal.

It is also interesting and instructive to derive the impulse response of an ideal bandpass filter the spectrum of which is shown in Fig. 6.7. We could compute the impulse response as the inverse Fourier transform of the system function in Eqn. (6.27), using the approach that was employed in finding the impulse response of the ideal lowpass filter. Instead we will take a simpler approach and express the ideal bandpass filter spectrum in terms of frequency shifts applied to the ideal lowpass filter spectrum. In order to accomplish this, we will initially consider an ideal lowpass filter with zero-phase and no time delay, i.e., $t_d = 0$. A comparison of Figs. 6.6 and 6.7 reveals that the ideal bandpass filter spectrum with zero-phase can be written as

$$\begin{aligned} H_{BP}(\omega) =& H_{LP}(\omega - \omega_b) + H_{LP}(\omega + \omega_b) \\ =& \Pi\left(\frac{\omega - \omega_b}{2\omega_0}\right) + \Pi\left(\frac{\omega + \omega_b}{2\omega_0}\right) \end{aligned} \tag{6.30}$$

with parameter ω_b set to the midpoint frequency of the passband

$$\omega_b = \frac{\omega_2 + \omega_1}{2} \tag{6.31}$$

and the cutoff frequency of the lowpass filter chosen as

$$\omega_0 = \frac{\omega_2 - \omega_1}{2} \tag{6.32}$$

The construction of the bandpass filter spectrum from the lowpass filter spectrum is illustrated in Fig. 6.11. Using the frequency-shifting theorem of the Fourier transform, we are now ready to write the impulse response of the ideal bandpass filter as

$$\begin{aligned} h_{BP}(t) =& h_{LP}(t)\, e^{j2\pi f_b t} + h_{LP}(t)\, e^{-j2\pi f_b t} \\ =& 2\, h_{LP}(t) \cos\left(2\pi f_b t\right) \\ =& 4 f_0 \operatorname{sinc}\left(2 f_0 t\right) \cos\left(2\pi f_b t\right) \end{aligned} \tag{6.33}$$

For convenience we have used frequencies in Hz rather than the radian frequencies in Eqn. (6.33). This impulse response is for an ideal bandpass filter with no time delay. It may be generalized by adding a linear phase term and the corresponding constant time delay to obtain

$$h_{BP}(t) = 4 f_0 \operatorname{sinc}\left[2 f_0 (t - t_d)\right] \cos\left[2\pi f_b (t - t_d)\right] \tag{6.34}$$

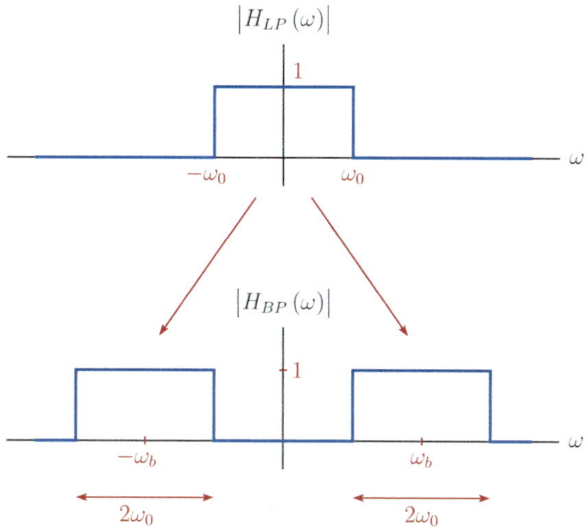

Figure 6.11 – Obtaining the spectrum of the ideal bandpass filter from the spectrum of the ideal lowpass filter.

Finally, by substituting Eqns. (6.31) and (6.32) into Eqn. (6.34) we get

$$h_{BP}(t) = 2\left(f_2 - f_1\right)\,\text{sinc}\left[\left(f_2 - f_1\right)(t - t_d)\right]\,\cos\left[\pi\left(f_2 + f_1\right)(t - t_d)\right] \tag{6.35}$$

which is shown graphically in Fig. 6.12.

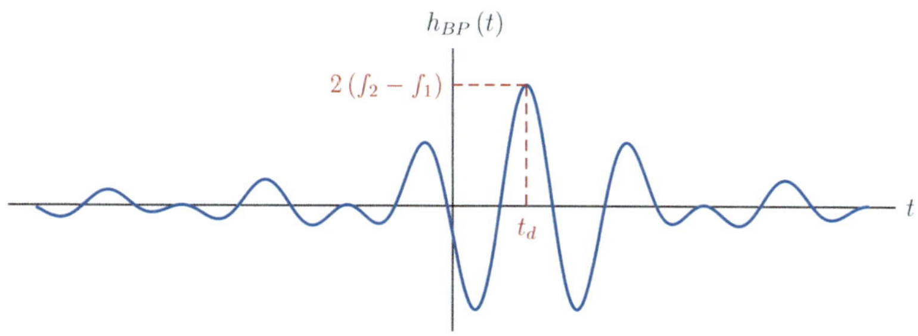

Figure 6.12 – Impulse response of the ideal bandpass filter.

Interactive App: Ideal lowpass filter

The interactive app in `appIdealLP.m` illustrates the impulse response of the ideal lowpass filter as defined in Eqn. (6.29), and shown in Fig. 6.10. The magnitude and the phase of the filter transfer function $H_{LP}(f)$ are displayed along with the impulse response $h_{LP}(t)$. The cutoff frequency f_0 and the time delay t_d may be varied using slider controls.

a. Set the filter cutoff frequency to $f_0 = 20$ Hz and the time delay to $t_d = 0$. Examine the impulse response, and pay special attention to the spacing of zero crossings. How does the spacing relate to the cutoff frequency?

b. With the time delay set to $t_d = 0$ vary the cutoff frequency f_0 and observe its effects on the impulse response. Make a note of the spacing of zero crossings for $f_0 = 25$ Hz and $f_0 = 50$ Hz.

c. With the cutoff frequency fixed at $f_0 = 50$ Hz vary the time delay and observe its effects on the impulse response as well as the phase of the system function.

Software resource: `appIdealLP.m`

Interactive App: Ideal bandpass filter

The interactive app in `appIdealBP.m` illustrates the impulse response of the ideal bandpass filter the system function of which was described by Eqn. (6.35) and Fig. 6.12. The magnitude and the phase of the filter transfer function $H_{BP}(f)$ are displayed along with the impulse response $h_{BP}(t)$. Slider controls allow adjustments to the passband center frequency f_b and the passband half-width f_0 as defined by Eqns. (6.31) and (6.32).

a. With the time delay set to $t_d = 0$ and passband half-width set to $f_0 = 30$ Hz vary the passband center frequency f_b and observe its effects on the impulse response.

b. With $t_d = 0$ and passband center frequency set to $f_b = 100$ Hz vary the passband half-width f_0 and observe its effects on the impulse response.

c. With $f_b = 100$ Hz and $f_0 = 30$ Hz vary the time delay and observe its effects on the impulse response as well as the phase of the system function.

Software resource: `appIdealBP.m`

6.4 Specifications for Realizable Filters

In a typical filter design problem, our focus is on finding a system function the characteristics of which meet a set of user specifications. We have seen in Section 6.3 that filters with ideal characteristics are not realizable since they are non-causal. We seek realizable approximations to ideal filter behavior. Causality and stability are implicit requirements that don't even need to be mentioned in the specifications. Furthermore, the requirements on the frequency spectrum of the filter must be relaxed and some tolerances must be allowed in order to obtain realizable filters.

The specifications for the desired filter are usually given in terms of a set of tolerance limits for the magnitude response $|H(s)|$. Specification diagrams for the four frequency-selective filter types, namely lowpass, highpass, bandpass, and bandstop filters, are shown in Fig. 6.13. Consider, for example, the lowpass filter specifications illustrated by Fig. 6.13(a). The frequency interval $0 < \omega < \omega_1$ is referred to as the *passband* of the filter. Within the passband the magnitude of the filter spectrum is required to remain in the range $1 - \Delta_1 \le |G(\omega)| \le 1$. For frequencies greater than ω_2, that is, in the *stopband* of the filter, the magnitude is required to stay below Δ_2. The parameters Δ_1 and Δ_2 are called the *passband tolerance* and the *stopband tolerance*, respectively.

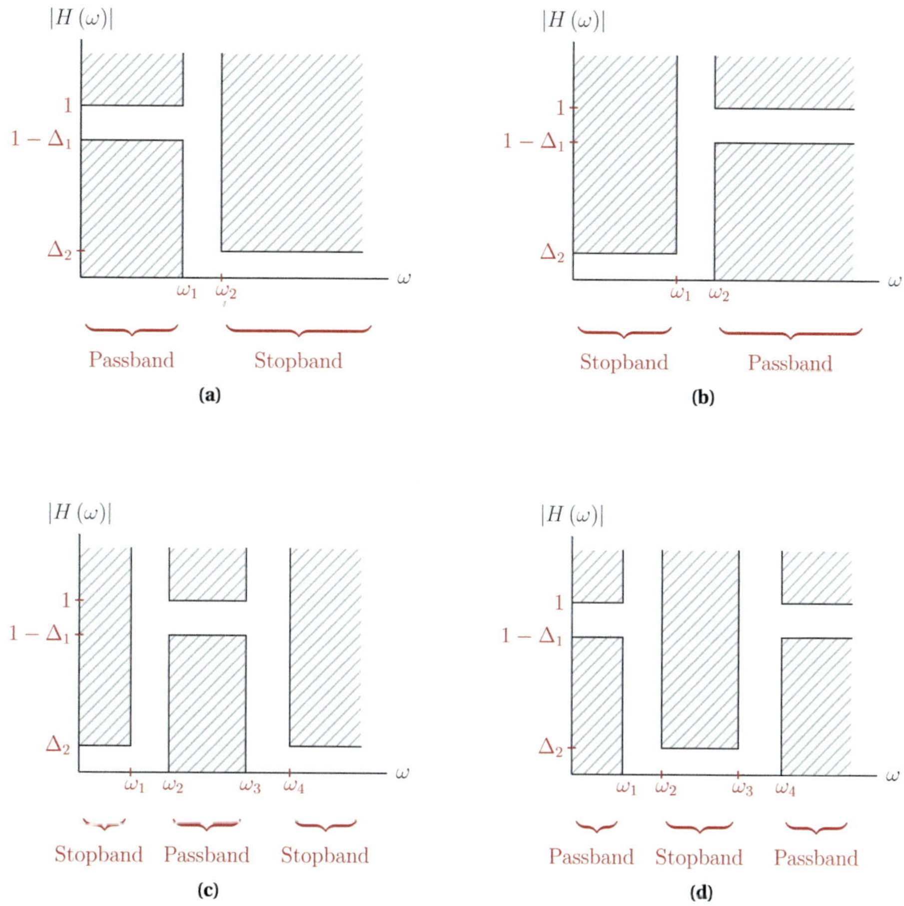

Figure 6.13 – Specification diagrams for frequency selective filters: **(a)** lowpass, **(b)** highpass, **(c)** bandpass, and **(d)** bandstop.

The frequency range $\omega_1 < \omega < \omega_2$ between the passband and the stopband is referred to as the *transition band* where there are no constraints on the magnitude behavior of the filter. The inclusion of a transition band along with the tolerance limits on passband and stopband behavior should allow us to come up with practically realizable designs. It should be noted that the impulse response $h(t)$ is required to be real-valued. Recall from the discussion in Section 3.3.5 of Chapter 3 that if the impulse response is real then the system function is conjugate symmetric, and therefore the magnitude of the system function is an even function of ω. This is the reason why the specification diagrams in Fig. 6.13 display only the positive frequencies.

Specifications for a highpass filter, shown in Fig. 6.13(b) mirror those of the lowpass filter. Bandpass and bandstop filter specifications, shown in Fig. 6.13(c),(d) have two transition bands each.

Tolerance limits for a desired filter can also be given on a decibel (dB) scale as shown in Fig. 6.14. The maximum value of the magnitude response is set as 0 dB. The maximum allowed dB *passband ripple* R_p and the minimum required dB *stopband attenuation* A_s are related to the parameters Δ_1 and Δ_2 as

$$R_p = 20\log_{10}\left(\frac{1}{1-\Delta_1}\right) \tag{6.36}$$

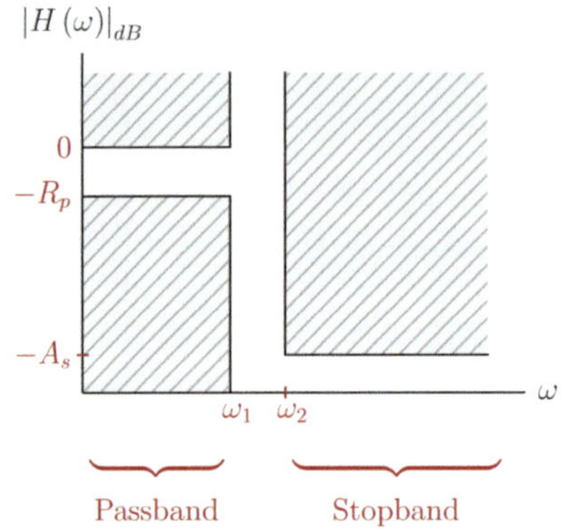

Figure 6.14 – dB Tolerance specifications for analog lowpass filter.

and

$$A_s = 20\log_{10}\left(\frac{1}{\Delta_2}\right) \tag{6.37}$$

Given the specifications of the desired filter in the form of one of the specification diagrams in Fig. 6.13, the design problem can be stated as follows:

Analog filter design problem:

Find the system function $H(s)$ of a filter the magnitude response of which remains in the allowed area of the specification diagram.

6.5 Design of Analog Lowpass Filters

In general, the analog filter design problem begins with the design of a lowpass prototype filter regardless of the type of filter that is ultimately needed. If a highpass, bandpass, or bandstop filter is desired, it is obtained subsequently by means of a *frequency transformation* to be applied to the lowpass prototype filter. This approach is motivated by the availability of well established methods for the design of analog lowpass filters.

A number of approximation formulas exist in the literature for specifying the squared-magnitude of a realizable lowpass filter spectrum. Some of the better known formulas will be mentioned here.

Butterworth lowpass filters are characterized by the squared-magnitude function

$$|H(\omega)|^2 = \frac{1}{1+(\omega/\omega_c)^{2N}} \tag{6.38}$$

where the parameters N and ω_c are the filter order and the cutoff frequency, respectively.

As an alternative to the Butterworth approximation formula, the *Chebyshev type-I* approximation formula for the squared-magnitude function of a lowpass filter is

$$|H(\omega)|^2 = \frac{1}{1+\varepsilon^2 C_N^2(\omega/\omega_1)} \tag{6.39}$$

where ε is a positive constant. The frequency ω_1 is the passband edge frequency. The function $C_N(v)$ in the denominator represents the *Chebyshev polynomial* of order N.

Chebyshev type-II approximation formula for the squared-magnitude function is similar to the Chebyshev type-I formula.

$$\left|H(\omega)\right|^2 = \frac{\varepsilon^2\, C_N^2(\omega_2/\omega)}{1+\varepsilon^2\, C_N^2(\omega_2/\omega)} \tag{6.40}$$

The parameter ω_2 is the *stopband edge frequency.*

Finally, an *elliptic* lowpass filter is characterized by the squared-magnitude function

$$\left|H(\omega)\right|^2 = \frac{1}{1+\varepsilon^2\, \psi_N^2(\omega/\omega_1)} \tag{6.41}$$

The parameter ε is a positive constant. The function $\psi_N(v)$ is called a *Chebyshev rational function* and is defined in terms of *Jacobi elliptic functions.*

In the rest of this section we will provide a brief introduction to analog filter design methods. A more detailed treatment of analog filter design and implementation can be found in [4], [8], [5], [3].

6.5.1 Butterworth lowpass filters

Consider again the Butterworth squared-magnitude function given by Eqn. (6.38). At the frequency $\omega=\omega_c$ the magnitude is equal to $1/\sqrt{2}$. Since this approximately corresponds to the -3 dB point on a logarithmic scale, the parameter ω_c is also referred to as the *3-dB cutoff frequency* of the Butterworth filter. Furthermore, it can easily be shown that the first $2N-1$ derivatives of the magnitude spectrum $|H(\omega)|$ are equal to zero at $\omega=0$, that is,

$$\left.\frac{d^n}{d\omega^n}\left\{\left|H(\omega)\right|\right\}\right|_{\omega=0} = 0\,, \qquad n=0,\ldots,2N-1 \tag{6.42}$$

The magnitude characteristic $|H(\omega)|$ for a Butterworth filter is said to be *maximally flat.* Magnitude spectra for Butterworth filters of various orders are shown in Fig. 6.15.

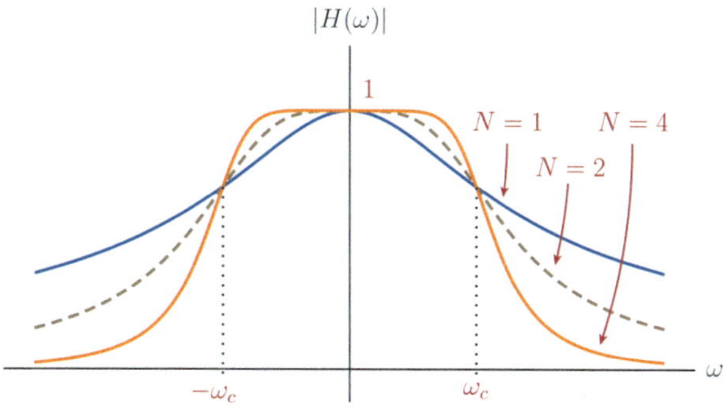

Figure 6.15 – Magnitude spectra for Butterworth lowpass filters or order $N=1$, $N=2$, and $N=4$.

We know that the squared-magnitude function can be expressed as the product of the system function and its complex conjugate, that is

$$\left|H(\omega)\right|^2 = H(\omega)\, H^*(\omega) \tag{6.43}$$

In addition, since the impulse response $h(t)$ of the filter is real-valued, its system function exhibits conjugate symmetry (see Section 3.3.5 of Chapter 3):

$$H^*(\omega) = H(-\omega) \tag{6.44}$$

Combining Eqns. (6.43) and (6.44) yields

$$\left|H(\omega)\right|^2 = H(\omega)\, H(-\omega) \tag{6.45}$$

Using the s-domain system function $H(s)$, the problem of designing a Butterworth filter can be stated as follows: Given the values of the two filter parameters ω_c and N, find the s-domain system function $H(s)$ for which the squared-magnitude of the function $H(\omega)$ matches the right side of Eqn. (6.38).

For a system function $H(s)$ with real coefficients, it can be shown that

$$\left|H(\omega)\right|^2 = H(s)\, H(-s)\Big|_{s=j\omega} \tag{6.46}$$

The procedure in Eqn. (6.46) needs to be reversed to find the system function $H(s)$. Substituting s for $j\omega$, or equivalently, $-s^2$ for ω^2 in Eqn. (6.38) we obtain

$$H(s)\, H(-s) = \frac{1}{1+\left(-s^2/\omega_c^2\right)^N} \tag{6.47}$$

If $H(s)$ has a pole in the left half s-plane at $p_1 = \sigma_1 + j\omega_1$, then $H(-s)$ has a corresponding pole at $\bar{p}_1 = -\sigma_1 - j\omega_1$. Thus the poles of $H(s)$ are mirror images of the poles of $H(-s)$ with respect to the origin. In order to extract $H(s)$ from the product $H(s)\,H(-s)$, we need to find the poles of this product, and separate the two sets of poles that belong to $H(s)$ and $H(-s)$ respectively. The cases of even and odd filter orders N need to be treated separately.

Case 1: N is odd

The poles of the product $H(s)\,H(-s)$ are the values of s that satisfy the equation

$$s^{2N} - \omega_c^{2N} = 0 \tag{6.48}$$

which can be written in the alternative form

$$s^{2N} = \omega_c^{2N}\, e^{j2\pi k} \tag{6.49}$$

where we have used the fact that $e^{j2\pi k} = 1$ for all integer k. Poles of $H(s)\,H(-s)$ are

$$p_k = \omega_c\, e^{j\pi k/N}\,, \qquad k = 0,\ldots,2N-1 \tag{6.50}$$

A few observations are in order:

1. All the poles of the product $H(s)\,H(-s)$ are located on a circle with radius equal to ω_c.
2. Furthermore, the poles are equally spaced on the circle and the angle between any two adjacent poles is π/N radians.
3. Since $H(s)$ and $H(-s)$ have only real coefficients, all complex poles appear in conjugate pairs.

Fig. 6.16 depicts the situation for $\omega_c = 2$ rad/s, and $N = 5$.

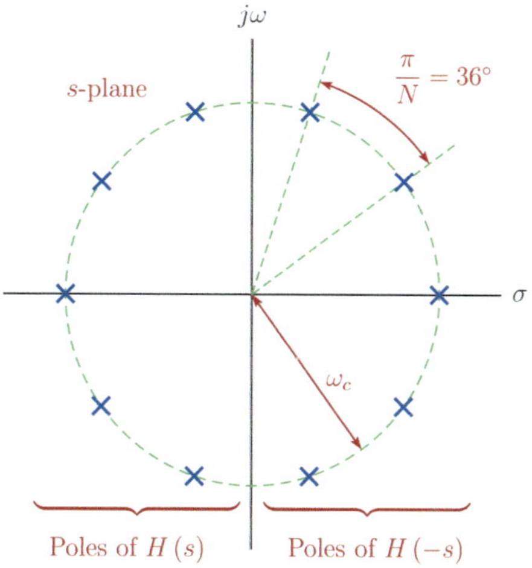

Figure 6.16 – Poles of the product $H(s)\,H(-s)$ for Butterworth lowpass filter of order $N = 5$ with 3-dB cutoff frequency $\omega_c = 2$ rad/s.

We are interested in obtaining a filter that is both causal and stable. It is therefore necessary to use the poles in the left half of the s-plane for $H(s)$. The remaining poles, the ones in the right half of the s-plane, must be associated with $H(-s)$. The system function $H(s)$ for the Butterworth lowpass filter is constructed in the form

$$H(s) = \frac{\omega_c^N}{\prod_k \left(s - p_k\right)}, \qquad \text{for all } k \text{ that satisfy } \quad \frac{\pi}{2} < \frac{k\pi}{N} < \frac{3\pi}{2} \tag{6.51}$$

Case 2: N is even

The poles of the product $H(s)\,H(-s)$ are the solutions of the equation

$$s^{2N} + \omega_c^{2N} = 0 \tag{6.52}$$

which can be written as

$$s^{2N} = \omega_c^{2N}\, e^{j\pi(2k+1)} \tag{6.53}$$

In this case, the general solution for the poles is

$$p_k = \omega_c\, e^{j\pi(2k+1)/2N}, \qquad \text{for} \quad k = 0,\ldots,2N-1 \tag{6.54}$$

For even filter orders, poles of the magnitude-squared transfer function are still equally spaced on a circle. Compared to odd filter orders, the only difference in this case is that no poles appear on the real axis. The system function $H(s)$ can be constructed in the same way as it was done for odd values of N.

Poles of $H(s)\,H(-s)$ for the Butterworth lowpass filter:

$$p_k = \begin{cases} \omega_c\, e^{jk\pi/N}, & k = 0,\ldots,2N-1 \quad \text{if } N \text{ is odd} \\ \omega_c\, e^{j(2k+1)\pi/2N}, & k = 0,\ldots,2N-1 \quad \text{if } N \text{ is even} \end{cases}$$

Interactive App: Magnitude response of Butterworth lowpass filter

The interactive app in `appButter.m` illustrates the magnitude characteristic of a Butterworth lowpass filter. Recall that the squared magnitude of the system function for a Butterworth lowpass filter was given by Eqn. (6.38). Desired 3-dB cutoff frequency f_c and filter order N may be specified. Observe the following:

a. The edges of the magnitude characteristic become sharper as the filter order is increased. Set the 3-dB cutoff frequency at $f_c = 40$ Hz. Start with filter order set to $N = 1$ and gradually increment it to $N = 2, 3, \ldots, 10$ while observing the magnitude characteristic, and paying attention to the sharpness of the transition region.

b. Set filter order to $N = 4$. Vary the cutoff frequency f_c from 20 Hz to 60 Hz while observing the magnitude characteristic.

c. Notice how at the frequency $f = f_c$ the magnitude is equal to $1/\sqrt{2}$ times its peak value independent of the cutoff frequency or the filter order.

Software resource: `appButter.m`

Example 6.2: Butterworth filter design

Determine the system function $H(s)$ of a third-order Butterworth lowpass filter with a 3-dB cutoff frequency of $\omega_c = 20\pi$ rad/s.

Solution: Using the Butterworth approximation formula in Eqn. (6.38), the squared-magnitude function of the desired filter can be written as

$$\left|H(\omega)\right|^2 = \frac{1}{1 + (\omega/20\pi)^6}$$

Substituting $\omega^2 \rightarrow -s^2$ in $H|(\omega)|^2$ we obtain

$$H(s)\,H(-s) = \frac{1}{1 + \left(-s^2/400\pi^2\right)^3}$$

or equivalently

$$H(s)\,H(-s) = \frac{(20\pi)^6}{(20\pi)^6 - s^6}$$

The poles of the function $H(s)\,H(-s)$ are those values of s that satisfy

$$s^6 = (20\pi)^6 \tag{6.55}$$

Solutions of Eqn. (6.55) are

$$p_k = (20\pi)\, e^{j2\pi k/6}\,, \qquad k = 0,\ldots,5$$

as shown in Fig. 6.17.

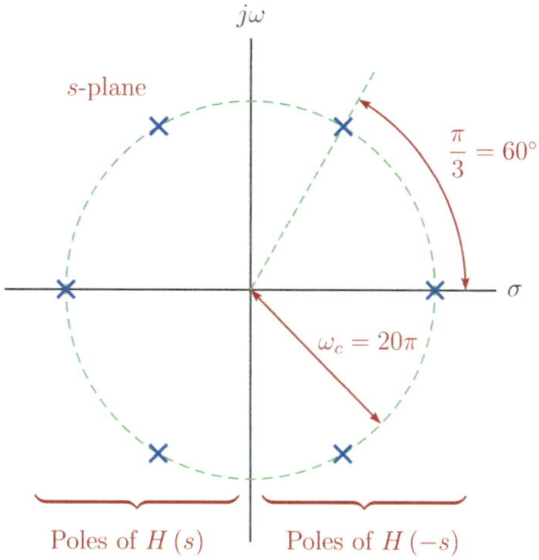

Figure 6.17 – Poles of $H(s)\,H(-s)$ for the third-order Butterworth lowpass filter of Example 6.2.

We are particularly interested in the poles that lie in the left half of the s-plane since those are the poles that are associated with $H(s)$. By inspection of Fig. 6.17 it is clear that the poles of $H(s)$ are

$$p_2 = (20\pi)\, e^{j2\pi/3}$$

$$p_3 = (20\pi)\, e^{j\pi} = -20\pi$$

$$p_4 = (20\pi)\, e^{j4\pi/3} = (20\pi)\, e^{-j2\pi/3}$$

We are now ready to construct the transfer function by using these poles:

$$\begin{aligned} H(s) &= \frac{(20\pi)^3}{\left(s-p_2\right)\left(s-p_3\right)\left(s-p_4\right)} \\ &= \frac{(20\pi)^3}{(s+20\pi)\left(s^2+20\pi\, s+400\pi^2\right)} \end{aligned} \tag{6.56}$$

In order to compute the magnitude and the phase characteristics of the system we need to substitute $s = j\omega$ into Eqn. (6.56) to obtain

$$H(\omega) = \frac{(20\pi)^3}{\left(j\omega+20\pi\right)\left(-\omega^2+j\,20\pi\omega+400\pi^2\right)}$$

The magnitude of the system function is

$$\left|H(\omega)\right| = \frac{(20\pi)^3}{\sqrt{\omega^2+(20\pi)^2}\,\sqrt{\left(-\omega^2+400\pi^2\right)^2+(20\pi\omega)^2}}$$

and the phase of the system function is

$$\measuredangle H(\omega) = -\tan^{-1}\left(\frac{\omega}{20\pi}\right) - \tan^{-1}\left(\frac{20\pi\omega}{-\omega^2+400\pi^2}\right)$$

Magnitude and phase responses of the designed filter are shown in Fig. 6.18.

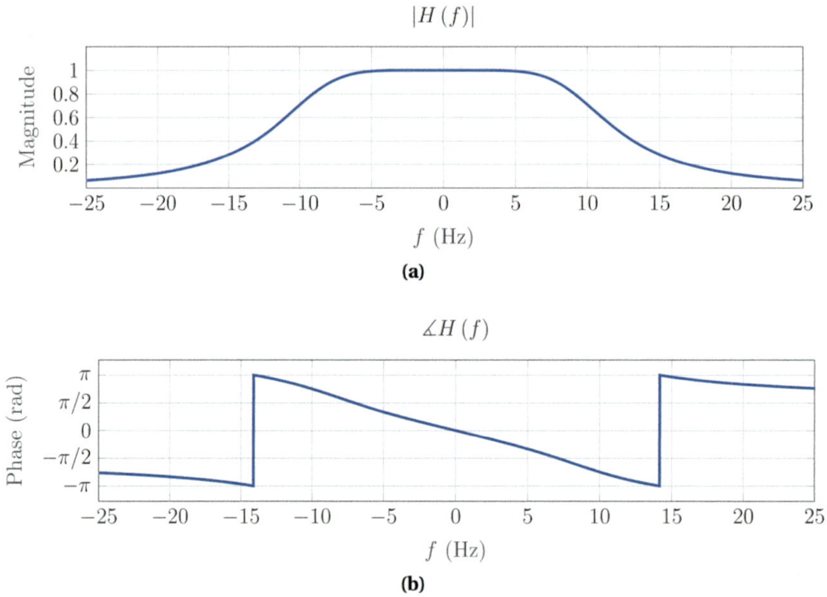

Figure 6.18 – The frequency spectrum of the third-order Butterworth lowpass filter designed in Example 6.2: **(a)** magnitude and **(b)** phase.

Software resource: `exct_6_2.m`

Example 6.3: Responses of ideal vs. realizable filters

To get a sense of how the behavior of an ideal filter may be approximated in practice, we will compare the output signals of an ideal filter and a realizable filter in response to the periodic pulse train. The pulse train $x(t)$ that is used as the input signal is shown in Fig. 6.19. This signal is used as input to two filters:

1. An ideal lowpass filter with a cutoff frequency of $f_0 = 10$ Hz and no time delay ($t_d = 0$), and
2. A third-order Butterworth lowpass filter with 3-dB cutoff frequency $f_c = 10$ Hz. (Note that this is the same filter that was designed in Example 6.2.)

A general expression for the exponential Fourier series coefficients of the pulse train was found in Example 3.7 of Chapter 3 and is

$$c_k = Ad\,\mathrm{sinc}(kd) \tag{6.57}$$

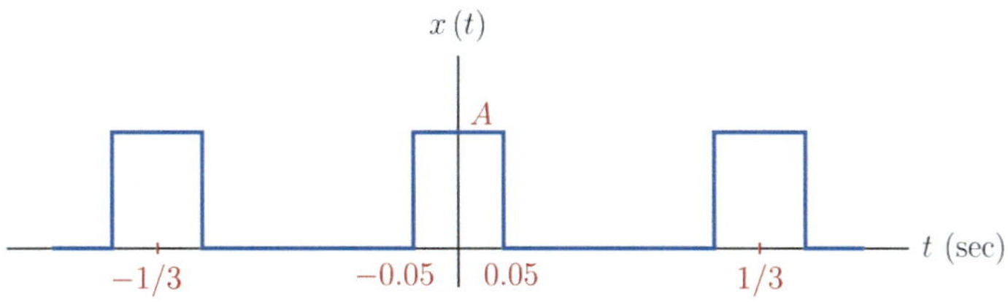

Figure 6.19 – The pulse train used in Example 6.3.

The parameter A corresponds to the amplitude of each pulse. The parameter d is the duty cycle of the pulse train. It is also important to realize that the expression in Eqn. (6.57) is valid only for the case where one of the pulses is centered around the origin, making the resulting pulse train is an even function of time. The signal $x(t)$ shown in Fig. 6.19 clearly satisfies this condition. For our specific signal $x(t)$ we have $A = 1$ and $d = 0.3$. Thus, the exponential Fourier series coefficients for $x(t)$ can be expressed as

$$c_k = 0.3\,\text{sinc}(0.3k)$$

We know from the defining equation of exponential Fourier series that, once the coefficients $\{c_k,\quad k = -\infty,\ldots,\infty\}$ are determined, the input signal $x(t)$ can be expressed as

$$x(t) = \sum_{k=-\infty}^{\infty} c_k\, e^{j2\pi k f_0 t}$$

In this case the fundamental frequency of the signal $x(t)$ is $f_0 = 1/0.3333 = 3$ Hz. If this signal is processed through the ideal lowpass filter with a cutoff frequency of $f_0 = 10Hz$, only the first three harmonics on either side of the frequency axis fall into the passband of the filter, and therefore are retained. All higher harmonics are eliminated from the output signal. Thus, the output of the ideal lowpass filter is

$$y_{\text{ideal}}(t) = \sum_{k=-3}^{3} c_k\, e^{j2\pi k f_0 t}$$

Using the numerical values of the exponential Fourier series coefficients, the output of the ideal lowpass filter is

$$\begin{aligned} y_{\text{ideal}}(t) &= 0.3 + 0.2575\left(e^{j2\pi f_0 t} + e^{-j2\pi f_0 t}\right) + \\ &\quad 0.1514\left(e^{j4\pi f_0 t} + e^{-j4\pi f_0 t}\right) + 0.0328\left(e^{j6\pi f_0 t} + e^{-j6\pi f_0 t}\right) \end{aligned}$$

Using Eulers formula, the expression for $y_{ideal}(t)$ can be simplified to

$$\begin{aligned} y_{\text{ideal}}(t) &= 0.3 + 0.5150\,cos\left(2\pi f_0 t\right) + 0.3027\,cos\left(4\pi f_0 t\right) + \\ &\quad 0.0656\,cos\left(6\pi f_0 t\right) \end{aligned}$$

In order to compute the output of the third-order Butterworth filter driven by the same signal, we need to consider the effect of the system function on Fourier series coefficients. Recall that in Example 3.46 of Chapter 3 we found the exponential Fourier series coefficients of a periodic output signal to be

$$\tilde{c}_k = c_k\, H\left(k f_0\right)$$

$H(f)$ needs to be obtained from the s-domain system function found in Eqn. (6.56) through the substitution $s \to j2\pi f$. Thus we have

$$H(f) = \frac{(20\pi)^3}{(j2\pi f + 20\pi)(-4\pi^2 f^2 + j40\pi^2 f + 400\pi^2)}$$
$$= \frac{1000}{(jf + 10)(-f^2 + j10f + 100)}$$

Values of $H(f)$ at can be evaluated at integer multiples of the fundamental frequency $f_0 = 3$ Hz as

$$H(kf_0) = \frac{1000}{(jkf_0 + 10)(-k^2 f_0^2 + j10kf_0 + 100)}$$

The output signal is

$$y_{\text{bw}}(t) = \sum_{k=-\infty}^{\infty} c_k\, H(kf_0)\, e^{j2\pi k f_0 t} \tag{6.58}$$

A comparison of the output signals for the ideal filter and the third-order Butterworth filter is given in Fig. 6.20.

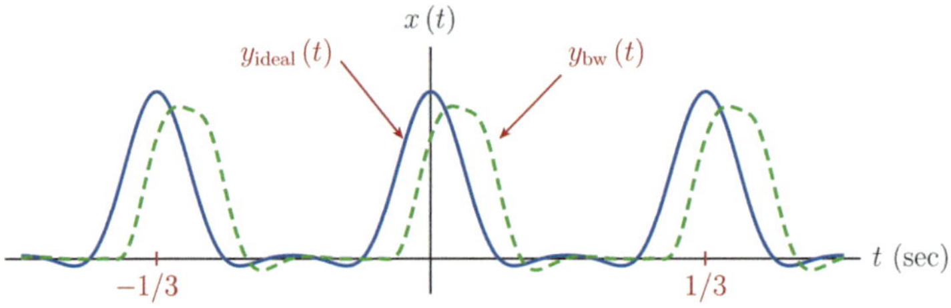

Figure 6.20 – The responses of the two filters of Example 6.3 to the periodic pulse train of Fig. 6.19.

Even though the summation in Eqn. (6.58) contains an infinite number of terms, only the middle 21 terms $\{c_{-10}, \ldots, c_{10}\}$ were used in graphing the output of the Butterworth filter.

Software resources: exct_6_3.m , exct_6_3_Live.mlx

Software resource: See MATLAB Exercise 6.1.

Obtaining N and ω_c for the Butterworth lowpass filter

The desired behavior of a lowpass filter to be designed is usually specified in terms of the two critical frequencies ω_1 and ω_2 as well as the dB tolerance values R_p and A_s as shown in Fig. 6.14. The corresponding parameter values for ω_c and N need to be determined from the provided set of specifications so that the filter can be designed. If $R_p = 3$ dB, then we have $\omega_c = \omega_1$. Otherwise the requirements at the two critical frequencies can be expressed as inequalities.

At $\omega = \omega_1$ we require

$$-10\log_{10}\left[1+\left(\frac{\omega_1}{\omega_c}\right)^{2N}\right] \ge -R_p \tag{6.59}$$

Similarly at $\omega = \omega_2$ we require

$$-10\log_{10}\left[1+\left(\frac{\omega_2}{\omega_c}\right)^{2N}\right] \le -A_s \tag{6.60}$$

Rearranging Eqns. (6.59) and (6.60) yields

$$\left(\frac{\omega_1}{\omega_c}\right)^{2N} \le 10^{R_p/10} - 1 \tag{6.61}$$

and

$$\left(\frac{\omega_2}{\omega_c}\right)^{2N} \ge 10^{A_s/10} - 1 \tag{6.62}$$

Dividing both sides of Eqn. (6.61) by the corresponding terms of Eqn. (6.62) we obtain

$$\left(\frac{\omega_1}{\omega_2}\right)^{2N} \le \frac{10^{R_p/10} - 1}{10^{A_s/10} - 1} \tag{6.63}$$

which can be solved for the filter order N that is needed:

$$N \ge \frac{\log_{10}\sqrt{\left(10^{A_s/10} - 1\right) / \left(10^{R_p/10} - 1\right)}}{\log_{10}\left(\omega_2/\omega_1\right)} \tag{6.64}$$

The right side of the inequality in Eqn. (6.64) may or may not be an integer. If it is not, then the value of N needs to be rounded up. The excess tolerance created by rounding N up to the next integer can be used in either in the passband or the stopband.

In summary, the parameters of the Butterworth lowpass filter are computed as follows:

Given dB tolerances R_p and A_s and critical frequencies ω_1 and ω_2:

$$N = \frac{\log_{10}\sqrt{\left(10^{A_s/10} - 1\right) / \left(10^{R_p/10} - 1\right)}}{\log_{10}\left(\omega_2/\omega_1\right)} \quad \Longrightarrow \quad \text{Round up to next integer}$$

After determining N, compute ω_c from one of the following:

$$\left(\frac{\omega_1}{\omega_c}\right)^{2N} = 10^{R_p/10} - 1 \qquad \text{or} \qquad \left(\frac{\omega_2}{\omega_c}\right)^{2N} = 10^{A_s/10} - 1$$

6.5.2 Chebyshev lowpass filters

The squared-magnitude function for a Chebyshev lowpass filter is given by

$$\left|H(\omega)\right|^2 = \frac{1}{1+\varepsilon^2\, C_N^2(\omega/\omega_1)} \tag{6.65}$$

Parameter ε is a positive constant, N is the order of the filter, and ω_1 is the passband edge frequency in rad/s (see Fig. 6.13a). The term $C_N(\nu)$ represents the *Chebyshev polynomial* of order N. Chebyshev filters of this type are sometimes called *Chebyshev type-I* filters. A variant of the squared-magnitude response in Eqn. (6.65) will be introduced in the next section as the *inverse Chebyshev filter* or the *Chebyshev type-II filter.*

Once the values of the parameters ε, N, and ω_1 are specified, the design procedure proceeds as follows:

1. Find $C_N(\nu)$, the Chebyshev polynomial of order N.
2. Substitute $\nu = \omega/\omega_1$ in the polynomial $C_N(\nu)$.
3. Form the Chebyshev type I squared-magnitude function as specified in Eqn. (6.65).
4. Find the poles of the system function $H(s)$ from the squared-magnitude function. As in the case of the Butterworth lowpass filter, this involves substituting $j\omega = s$ into the squared-magnitude function, finding the roots of the denominator polynomial, and separating those poles into two sets, one to be associated with $H(s)$ and the other to be associated with $H(-s)$.

Before we carry out the steps outlined above, it will be useful to review the properties of Chebyshev polynomials.

Chebyshev polynomials

The Chebyshev polynomial of order N is defined as

$$C_N(\nu) = \cos\left(N\cos^{-1}(\nu)\right) \tag{6.66}$$

A better approach for understanding the definition in Eqn. (6.66) would be to split it into two equations as

$$\nu = \cos(\theta) \tag{6.67}$$

and

$$C_N(\nu) = \cos(N\theta) \tag{6.68}$$

The Chebyshev polynomial of a specified order N can be obtained by

1. Using trigonometric identities to express $\cos(N\theta)$ as a function of $\cos(\theta)$ and
2. Replacing each $\cos(\theta)$ term with ν.

The first two polynomials are easy to obtain from this definition:

$$C_0(\nu) = 1$$

$$C_1(\nu) = \nu$$

To obtain $C_2(v)$ let us write $\cos(2\theta)$ in terms of $\cos(\theta)$ using a trigonometric identity:

$$\cos(2\theta) = 2\cos^2(\theta) - 1$$

Thus, the second-order Chebyshev polynomial is

$$C_2(v) = 2v^2 - 1$$

Higher-order Chebyshev polynomials may be obtained by continuing in this fashion. As the order increases, however, the procedure outlined above becomes increasingly tedious. Fortunately, it is possible to derive a recursive formula to facilitate the derivation of higher-order Chebyshev polynomials.

Let us write $\cos\big((N+1)\theta\big)$ and $\cos\big((N-1)\theta\big)$ using trigonometric identities:

$$\cos\big((N+1)\theta\big) = \cos(\theta)\cos(N\theta) - \sin(\theta)\sin(N\theta)$$

$$\cos\big((N-1)\theta\big) = \cos(\theta)\cos(N\theta) + \sin(\theta)\sin(N\theta)$$

Adding the two equations results in

$$\cos\big((N+1)\theta\big) + \cos\big((N-1)\theta\big) = 2\cos(\theta)\cos(N\theta)$$

which can be rearranged to produce the recursive formula we seek:

$$\cos\big((N+1)\theta\big) = 2\cos(\theta)\cos(N\theta) - \cos\big((N-1)\theta\big)$$

Thus, the recursive formula for obtaining Chebyshev polynomials is

$$C_{N+1}(v) = 2v\,C_N(v) - C_{N-1}(v) \qquad (6.69)$$

The recursive formula in Eqn. (6.69) allows any order Chebyshev polynomial to be found provided that the polynomials for the previous two orders are known. With the knowledge of $C_0(v) = 1$ and $C_1(v) = v$, the polynomial $C_2(v)$ can be found as

$$\begin{aligned} C_2(v) &= 2v\,C_1(v) - C_0(v) \\ &= 2v^2 - 1 \end{aligned}$$

Similarly $C_3(v)$ is found as

$$\begin{aligned} C_3(v) &= 2v\,C_2(v) - C_1(v) \\ &= 2v\left(2v^2 - 1\right) - v \\ &= 4v^3 - 3v \end{aligned}$$

The behavior of the first few orders of Chebyshev polynomials are illustrated in Fig. 6.21.

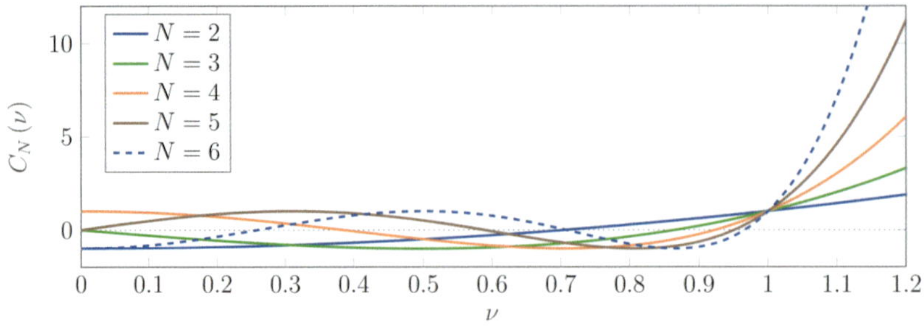

Figure 6.21 – Some Chebyshev polynomials.

A few interesting characteristics of Chebyshev polynomials can be readily observed:

1. All Chebyshev polynomials pass through the point $(1, 1)$, that is, $C_N(1) = 1$ for all N.
2. For all Chebyshev polynomials, if $|v| \leq 1$, then $|C_N(v)| \leq 1$.
3. For $|v| > 1$, all Chebyshev polynomials grow monotonically without bound.
4. At $v = 0$ the behavior of Chebyshev polynomials depend on the order N.

$$C_N(0) = \begin{cases} 0, & N \text{ odd} \\ \pm 1, & N \text{ even} \end{cases}$$

Fig. 6.22 shows the squared Chebyshev polynomials for the first few orders N.

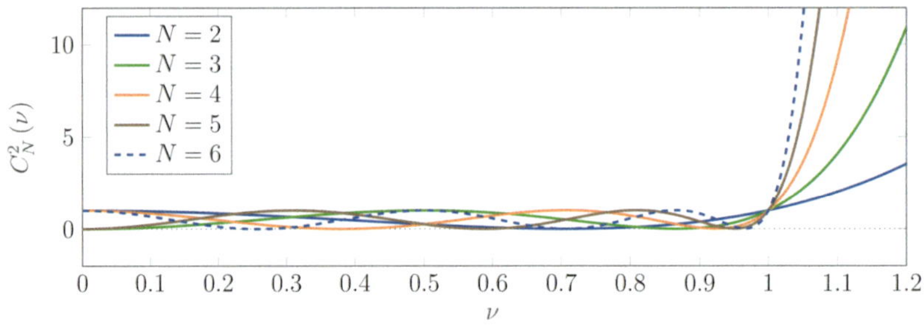

Figure 6.22 – Squared Chebyshev polynomials.

Squared Chebyshev polynomials $C_N^2(v)$ oscillate in the amplitude range $[0, 1]$ for $|v| \leq 1$ and $N \geq 2$. For $|v| > 1$ they grow monotonically without bound.

Software resource: See MATLAB Exercise 6.2.

The squared magnitude function given by Eqn. (6.65) is shown in Fig. 6.23 for several values of N using $\varepsilon = 0.4$ and $\omega_1 = 1$ rad/s.

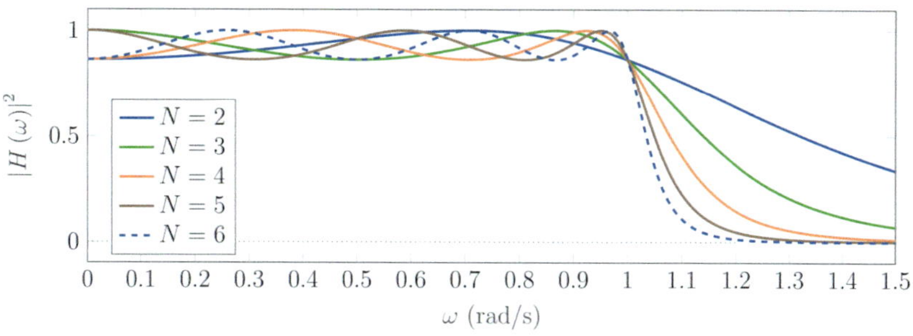

Figure 6.23 – Squared magnitude for Chebyshev filters with $\omega_1 = 1$ rad/s.

Poles for the Chebyshev lowpass filter

In constructing the system function for a Chebyshev lowpass filter we will use an approach similar to that taken with a Butterworth lowpass filter in Section 6.5.1. First we will determine the poles of the product $H(s)\,H(-s)$. Recall that the product $H(s)\,H(-s)$ is obtained by starting with the squared magnitude function $|H(\omega)|^2$ and replacing $j\omega$ with s:

$$H(s)\,H(-s) = \frac{1}{1+\varepsilon^2\,C_N^2\left(\dfrac{s}{j\omega_1}\right)} \tag{6.70}$$

The poles p_k of $H(s)\,H(-s)$ given by Eqn. (6.70) are the solutions of the equation

$$1+\varepsilon^2\,C_N^2\left(\frac{p_k}{j\omega_1}\right) = 0 \tag{6.71}$$

for $k = 0,\ldots,2N-1$.

Let us define

$$v_k = \frac{p_k}{j\omega_1} \tag{6.72}$$

so that

$$1+\varepsilon^2\,C_N^2(v_k) = 0 \tag{6.73}$$

Using the definition of the Chebyshev polynomial given by Eqn. (6.68) we can rewrite Eqn. (6.73) as

$$1+\varepsilon^2\cos^2(N\theta_k) = 0 \tag{6.74}$$

where $v_k = \cos(\theta_k)$. Solving Eqn. (6.74) for $\cos(\theta_k)$ yields

$$\cos(N\theta_k) = \pm\frac{j}{\varepsilon} \tag{6.75}$$

Let $\theta_k = \alpha_k + j\beta_k$ with α_k and β_k both as real parameters. Eqn. (6.75) becomes

$$\cos\left(N\alpha_k + jN\beta_k\right) = \pm\frac{j}{\varepsilon} \tag{6.76}$$

which, using the appropriate trigonometric identity, can be written as

$$\cos(N\alpha_k)\cos\left(jN\beta_k\right) - \sin(N\alpha_k)\sin\left(jN\beta_k\right) = \pm\frac{j}{\varepsilon} \tag{6.77}$$

Recognizing that $\cos\left(jN\beta_k\right)=\cosh\left(N\beta_k\right)$ and $\sin\left(jN\beta_k\right)=j\,\sinh\left(N\beta_k\right)$, we obtain

$$\cos\left(N\alpha_k\right)\cosh\left(N\beta_k\right)-j\,\sin\left(N\alpha_k\right)\sinh\left(N\beta_k\right)=\pm\frac{j}{\varepsilon} \tag{6.78}$$

Equating real and imaginary parts of both sides of Eqn. (6.78) yields

$$\cos\left(N\alpha_k\right)\cosh\left(N\beta_k\right)=0 \tag{6.79}$$

and

$$\sin\left(N\alpha_k\right)\sinh\left(N\beta_k\right)=\pm\frac{1}{\varepsilon} \tag{6.80}$$

To satisfy Eqn. (6.79) the cosine term must be set equal to zero, leading to

$$\cos\left(N\alpha_k\right)=0 \quad \Longrightarrow \quad \alpha_k=\frac{(2k+1)\,\pi}{2N}\,, \quad k=0,\ldots,2N-1 \tag{6.81}$$

Using this value of α_k in Eqn. (6.80) results in

$$\sin\left(N\alpha_k\right)=\pm 1 \tag{6.82}$$

and

$$\beta_k=\frac{\sinh^{-1}(1/\varepsilon)}{N} \tag{6.83}$$

The poles of the product $H(s)\,H(-s)$ can now be determined. Using Eqn. (6.72)

$$p_k=j\omega_1\,v_k=j\omega_1\,\cos\left(\theta_k\right)\,, \quad k=0,\ldots,2N-1 \tag{6.84}$$

Using the values of α_k and β_k found, the poles of $H(s)\,H(-s)$ are

$$p_k=j\omega_1\left[\cos\left(\alpha_k\right)\cosh\left(\beta_k\right)-j\,\sin\left(\alpha_k\right)\sinh\left(\beta_k\right)\right] \tag{6.85}$$

for $k=0,\ldots,2N-1$. It can be shown that the poles given by Eqn. (6.85) are on an elliptical trajectory. The ones in the left half s-plane must be associated with $H(s)$ in order to obtain a causal and stable filter.

Poles of $H(s)\,H(-s)$ for the Chebyshev type-I lowpass filter:

$$\text{Using:} \quad \alpha_k=\frac{(2k+1)\,\pi}{2N} \quad \text{and} \quad \beta_k=\frac{\sinh^{-1}(1/\varepsilon)}{N}$$

$$p_k=j\omega_1\left[\cos\left(\alpha_k\right)\cosh\left(\beta_k\right)-j\,\sin\left(\alpha_k\right)\sinh\left(\beta_k\right)\right]\,, \quad k=0,\ldots,2N-1$$

Interactive App: Magnitude response of Chebyshev type-I lowpass filter

The interactive app in `appCheb1.m` illustrates the magnitude characteristic of a Chebyshev type-I lowpass filter. Recall that the squared magnitude of the system function for a Chebyshev type-I lowpass filter was given by Eqn. (6.39). The passband edge frequency f_1, the parameter ε, and the filter order N may be specified as desired. Observe the following:

a. Set the passband edge frequency to $f_1 = 30$ Hz. Initially set the filter order to $N = 1$ and gradually increase it to $N = 2, 3, \ldots, 10$ while watching the magnitude characteristic. Observe how the edges of the magnitude characteristic become sharper as the filter order is increased.

b. At the frequency $f = f_1$ the magnitude is equal to

$$H(f_1) = \frac{1}{\sqrt{1+\varepsilon^2}}$$

Set $f_1 = 30$ Hz and $N = 4$. Vary the parameter ε from $\varepsilon = 0.1$ to $\varepsilon = 0.5$ gradually, and observe the changes in the magnitude characteristic.

Software resource: `appCheb1.m`

Example 6.4: Design of a Chebyshev lowpass filter

Design a third-order Chebyshev type-I analog lowpass filter with a passband edge frequency of $\omega_1 = 1$ rad/s and $\varepsilon = 0.4$. Afterward compute and graph the magnitude response of the designed filter.

Solution: Using Eqn. (6.81) the parameter α_k is found as

$$\alpha_k = \frac{(2k+1)\,\pi}{6}\,, \qquad k = 0, \ldots, 5$$

and the parameter β_k is obtained from Eqn. (6.83) as

$$\beta_k = \frac{\sinh^{-1}(1/0.4)}{3} = 0.5491$$

Poles of $H(s)\,H(-s)$ are found through the use of Eqn. (6.85):

$$p_k = j\left[\cos\left(\frac{(2k+1)\,\pi}{6}\right)\cosh(0.5491) - j\,\sin\left(\frac{(2k+1)\,\pi}{6}\right)\sinh(0.5491)\right], \qquad k = 0, \ldots, 5 \tag{6.86}$$

Evaluating Eqn. (6.86) for $k = 0, \ldots, 5$ yields the following poles:

$$\begin{aligned} p_0 &= 0.2885 + j\,1 \\ p_1 &= 0.5771 \\ p_2 &= 0.2885 - j\,1 \\ p_3 &= -0.2885 - j\,1 \\ p_4 &= -0.5771 \\ p_5 &= -0.2885 + j\,1 \end{aligned}$$

The first three poles, namely p_0, p_1, and p_2 are in the right half s-plane; therefore they belong to $H(-s)$. The system function $H(s)$ should be constructed using the remaining three poles.

$$H(s) = \frac{0.6250}{\left(s+0.2885+j\,1\right)\left(s+0.5771\right)\left(s+0.2885-j\,1\right)}$$
$$= \frac{0.6250}{s^3+1.1542\,s^2+1.4161\,s+0.6250}$$

The numerator of $H(s)$ was adjusted to achieve $|H(0)| = 0$. The magnitude and the phase of $H(s)$ are shown in Fig. 6.24.

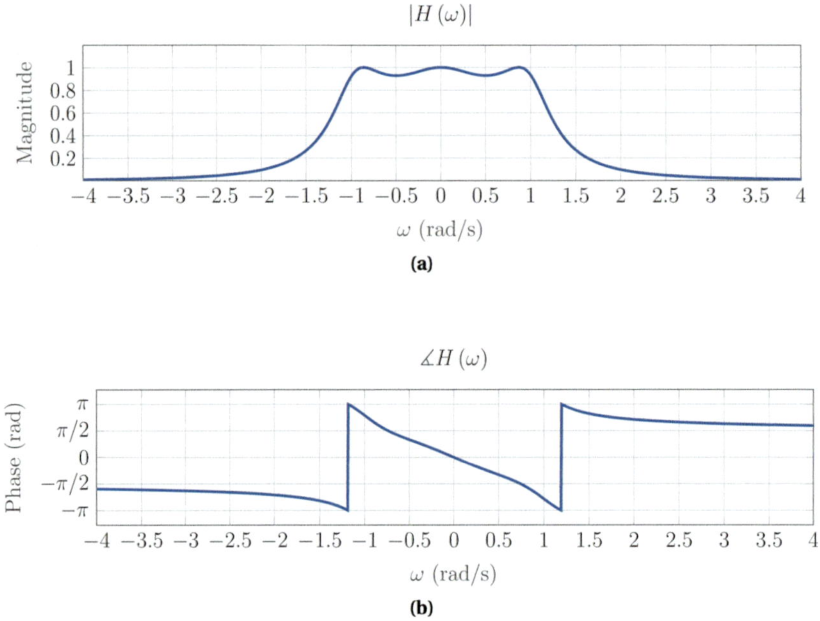

Figure 6.24 – **(a)** The magnitude and **(b)** the phase of the third-order Chebyshev lowpass filter designed in Example 6.4.

Software resources: `exct_6_4.m` , `exct_6_4_Live.mlx`

Software resource: See MATLAB Exercise 6.3.

Obtaining N and ε for the Chebyshev lowpass filter

In some design problems it is desired to find the lowest order filter that satisfies a set of design specifications. The desired behavior of the lowpass filter is typically specified in terms of the two critical frequencies ω_1 and ω_2 as well the dB tolerance values R_p and A_s for the passband and the stopband, respectively. Refer to Fig. 6.14.

At the frequency $\omega = \omega_1$ we need

$$20\log_{10}\left[\frac{|H(\omega_1)|}{|H(\omega)|_{\max}}\right] = -R_p \tag{6.87}$$

Recall that $C_N(1) = 1$ for all Chebyshev polynomials. Therefore Eqn. (6.87) becomes

$$10\log_{10}\left(\frac{1}{1+\varepsilon^2}\right) = -R_p \tag{6.88}$$

which can be solved for ε to yield

$$\varepsilon = \sqrt{10^{R_p/10}-1} \tag{6.89}$$

In the next step we need to determine the filter order N. At the stopband edge frequency we have the inequality

$$20\log_{10}|H(\omega_2)| \le -A_s \tag{6.90}$$

In order to simplify the notation let us define $\omega_0 = \omega_2/\omega_1$. Eqn. (6.90) can be expressed as

$$10\log_{10}\left(\frac{1}{1+\varepsilon^2\,C_N^2(\omega_0)}\right) \le -A_s \tag{6.91}$$

In the worst case Eqn. (6.91) can be taken with equality and solved for $C_N(\omega_0)$ to yield

$$C_N(\omega_0) = \sqrt{\frac{10^{A_s/10}-1}{10^{R_p/10}-1}} \tag{6.92}$$

in which we have used the value of ε found in Eqn. (6.89). In order to solve Eqn. (6.92) for N we will define some intermediate variables. Let

$$F = \sqrt{\frac{10^{A_s/10}-1}{10^{R_p/10}-1}} \tag{6.93}$$

and

$$\cos(\theta_0) = \omega_0 \tag{6.94}$$

The equation that needs to be solved for N is

$$\cos(N\theta_0) = F \tag{6.95}$$

Since θ_0 is complex (remember that $\omega_0 > 1$), we will express it in cartesian complex form as $\theta_0 = \alpha_0 + j\,\beta_0$, and write (6.95) as

$$\cos(N\alpha_0)\cosh(N\beta_0) - j\,\sin(N\alpha_0)\sinh(N\beta_0) = F \tag{6.96}$$

Since F is real, the left side of Eqn. (6.96) must be real as well. Therefore we must have

$$\alpha_0 = 0 \quad\Longrightarrow\quad \theta_0 = j\,\beta_0 \tag{6.97}$$

and

$$\cosh(N\beta_0) = F \tag{6.98}$$

Using Eqn. (6.94) we can write

$$\beta_0 = -j\,\theta_0 = \cosh^{-1}(\omega_0) \tag{6.99}$$

Solving Eqns. (6.98) and (6.99) for the filter order N we obtain

$$N = \frac{\cosh^{-1}(F)}{\cosh^{-1}(\omega_0)} \tag{6.100}$$

The value of N obtained through Eqn. (6.100) may not necessarily be an integer, and must be rounded up to the next integer. The excess tolerance obtained by rounding N up causes the stopband inequality given by Eqn. (6.91) to improve. If it is desired to use the excess tolerance for improving the passband response instead, then ε may be recalculated from Eqn. (6.91) using the rounded-up value of N.

In summary, the parameters of a Chebyshev type-I lowpass filter are computed as follows:

Given dB tolerances R_p and A_s and critical frequencies ω_1 and ω_2:

$$\omega_0 = \frac{\omega_2}{\omega_1}, \quad \text{and} \quad F = \sqrt{\frac{10^{A_s/10}-1}{10^{R_p/10}-1}}$$

$$N = \frac{\cosh^{-1}(F)}{\cosh^{-1}(\omega_0)} \quad \Longrightarrow \quad \text{Round up to next integer}$$

Compute ε from one of the following:

$$10\log_{10}\left(\frac{1}{1+\varepsilon^2}\right) = -R_p \quad \text{or} \quad 10\log_{10}\left(\frac{1}{1+\varepsilon^2\,C_N^2(\omega_0)}\right) = -A_s$$

Software resource: See MATLAB Exercise 6.4.

6.5.3 Inverse Chebyshev lowpass filters

The squared-magnitude function for an *inverse Chebyshev* lowpass filter, also referred to as a *Chebyshev type-II* lowpass filter, is

$$\left|H(\omega)\right|^2 = \frac{\varepsilon^2\,C_N^2(\omega_2/\omega)}{1+\varepsilon^2\,C_N^2(\omega_2/\omega)} \tag{6.101}$$

As in the case of Chebyshev type-I filters, ε is a positive constant and $C_N(v)$ represents the Chebyshev polynomial of order N. The parameter ω_2 is the stopband edge frequency (see Fig. 6.13a). Magnitude responses of several Chebyshev type-II filters are shown in Fig. 6.25. The magnitude response of the Chebyshev type-II filter is smooth in the passband and has equiripple behavior in the stopband.

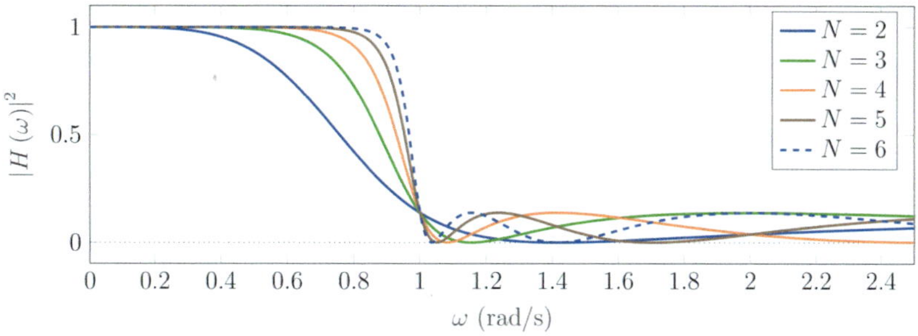

Figure 6.25 – Squared magnitude for inverse Chebyshev filters with $\omega_2 = 1$ rad/s.

Poles and zeros of the inverse Chebyshev filter

The denominator of the squared magnitude function $|H(\omega)|^2$ in Eqn. (6.101) is very similar to that of the type-I Chebyshev filter squared magnitude response given by Eqn. (6.65). Consequently, most of the results obtained in the previous section through the derivation of the poles of Chebyshev type-I lowpass filter will be usable. For an inverse Chebyshev filter the poles p_k of the product $H(s)\,H(-s)$ are the solutions of the equation

$$1+\varepsilon^2\, C_N^2\left(\frac{j\omega_2}{p_k}\right)=0 \qquad (6.102)$$

for $k = 0,\ldots,2N-1$.

Let

$$v_k = \frac{j\omega_2}{p_k} \qquad (6.103)$$

so that Eqn. (6.102) becomes

$$1+\varepsilon^2\, C_N^2\,(v_k)=0 \qquad (6.104)$$

This result is identical to that obtained in Eqn. (6.73) for computing the poles of the Chebyshev type-I lowpass filter, therefore, the solution obtained for p_k may be used here as well. We have

$$v_k = \cos(\theta_k) \qquad (6.105)$$

with

$$\alpha_k = \mathrm{Re}\{\theta_k\} = \frac{(2k+1)\,\pi}{2N} \qquad (6.106)$$

and

$$\beta_k = \mathrm{Im}\{\theta_k\} = \frac{\sinh^{-1}(1/\varepsilon)}{N} \qquad (6.107)$$

The poles p_k are determined from α_k and β_k as

$$\begin{aligned} p_k &= \frac{j\omega_2}{v_k} = \frac{j\omega_2}{\cos\left(\alpha_k + j\beta_k\right)} \\ &= \frac{j\omega_2}{\cos(\alpha_k)\cosh\left(\beta_k\right) - j\,\sin(\alpha_k)\sinh\left(\beta_k\right)} \end{aligned} \qquad (6.108)$$

As discussed before, the poles in the left half s-plane are associated with $H(s)$ in order to obtain a causal and stable filter. The zeros of $H(s)\,H(-s)$ are found by solving

$$C_N^2\left(\frac{j\omega_2}{z_k}\right)=0 \tag{6.109}$$

It can be shown that the zeros of the inverse Chebyshev lowpass filter are

$$z_k=\frac{\pm j\omega_2}{\cos\left(\dfrac{(2k-1)\,\pi}{2N}\right)}\,,\qquad k=1,\ldots,K \tag{6.110}$$

where the upper limit K is computed as

$$K=\begin{cases}(N-1)/2\,, & N \text{ odd}\\ N/2\,, & N \text{ even}\end{cases} \tag{6.111}$$

A summary of the results is given below:

Poles and zeros for the Chebyshev type-II lowpass filter:

Poles of $H(s)\,H(-s)$:

$$\text{Using:}\qquad \alpha_k=\frac{(2k+1)\,\pi}{2N}\qquad \text{and}\qquad \beta_k=\frac{\sinh^{-1}(1/\varepsilon)}{N}$$

$$p_k=\frac{j\omega_2}{\cos(\alpha_k)\cosh(\beta_k)-j\sin(\alpha_k)\sinh(\beta_k)}\,,\qquad k=0,\ldots,2N-1$$

Zeros of $H(s)$:

$$z_k=\frac{\pm j\omega_2}{\cos\left(\dfrac{(2k-1)\,\pi}{2N}\right)}\,,\qquad k=1,\ldots,K$$

where

$$K=\begin{cases}(N-1)/2\,, & N \text{ odd}\\ N/2\,, & N \text{ even}\end{cases}$$

Interactive App: Magnitude response of Chebyshev type-II lowpass filter

The interactive app in `appCheb2.m` illustrates the magnitude characteristic of a Chebyshev type-II lowpass filter. Recall that the squared magnitude of the system function for a Chebyshev type-II lowpass filter was given by Eqn. (6.40). The stopband edge frequency f_2, the parameter ϵ and the filter order N may be specified as desired. Observe the following:

a. Set the stopband edge frequency to $f_2=30$ Hz. Initially set the filter order to $N=1$ and gradually increase it to $N=2,3,\ldots,10$ while watching the magnitude characteristic.

Observe how the edges of the magnitude characteristic become sharper as the filter order is increased.

b. At the frequency $f = f_2$ the magnitude is equal to

$$H(f_2) = \frac{\varepsilon^2}{\sqrt{1+\varepsilon^2}}$$

Set $f_2 = 30$ Hz and $N = 4$. Vary the parameter ε from $\varepsilon = 0.1$ to $\varepsilon = 0.5$ gradually, and observe the changes in the magnitude characteristic.

Software resource: `appCheb2.m`

Software resource: See MATLAB Exercise 6.5.

Example 6.5: Design of an inverse Chebyshev lowpass filter

Design a third-order Chebyshev type-II analog lowpass filter with $\omega_2 = 3$ rad/s and $\varepsilon = 0.2$. Afterward compute and graph the magnitude response of the designed filter.

Solution: Using Eqn. (6.106) the parameter α_k is found as

$$\alpha_k = \frac{(2k+1)\,\pi}{6}\,, \qquad k = 0,\ldots,5$$

and the parameter β_k is obtained from Eqn. (6.107) as

$$\beta_k = \frac{\sinh^{-1}(1/0.2)}{3} = 0.7708$$

Poles of $H(s)\,H(-s)$ are found through the use of Eqn. (6.108):

$$p_k = \frac{j3}{\cos\left(\dfrac{(2k+1)\,\pi}{6}\right)\cosh(0.7708) - j\,\sin\left(\dfrac{(2k+1)\,\pi}{6}\right)\sinh(0.7708)}\,, \qquad k = 0,\ldots,5 \tag{6.112}$$

Evaluating Eqn. (6.112) for $k = 0,\ldots,5$ yields the following values for the poles of $H(s)\,H(-s)$:

$$\begin{aligned}
p_0 &= -0.8659 + j2.3165\\
p_1 &= -3.5316\\
p_2 &= -0.8659 - j2.3165\\
p_3 &= 0.8659 - j2.3165\\
p_4 &= 3.5317\\
p_5 &= 0.8659 + j2.3165
\end{aligned}$$

The stable poles in the left half s-plane are p_0, p_1, and p_2, therefore, those must be associated with $H(s)$. Zeros of the system function are found using Eqn. (6.110) as

$$z_k = \frac{\pm j3}{\cos(\pi/6)} = \pm j3.4641$$

The system function can now be constructed as

$$H(s) = \frac{A\left(s - j3.4641\right)\left(s + j3.4641\right)}{\left(s + 0.8659 - j2.3165\right)(s + 3.5317)\left(s + 0.8659 + j2.3165\right)}$$

$$= \frac{A\left(s^2 + 12\right)}{s^3 + 5.2635\,s^2 + 12.232\,s + 21.6}$$

where A is a constant gain factor that can be adjusted to set the peak value of the magnitude response equal to unity.

$$\left|H(0)\right| = \frac{12\,A}{21.6} = 1 \qquad \Longrightarrow \qquad A = 1.8$$

The completed system function is

$$H(s) = \frac{1.8\left(s^2 + 12\right)}{s^3 + 5.2635\,s^2 + 12.232\,s + 21.6}$$

Magnitude and phase responses are shown in Fig. 6.26.

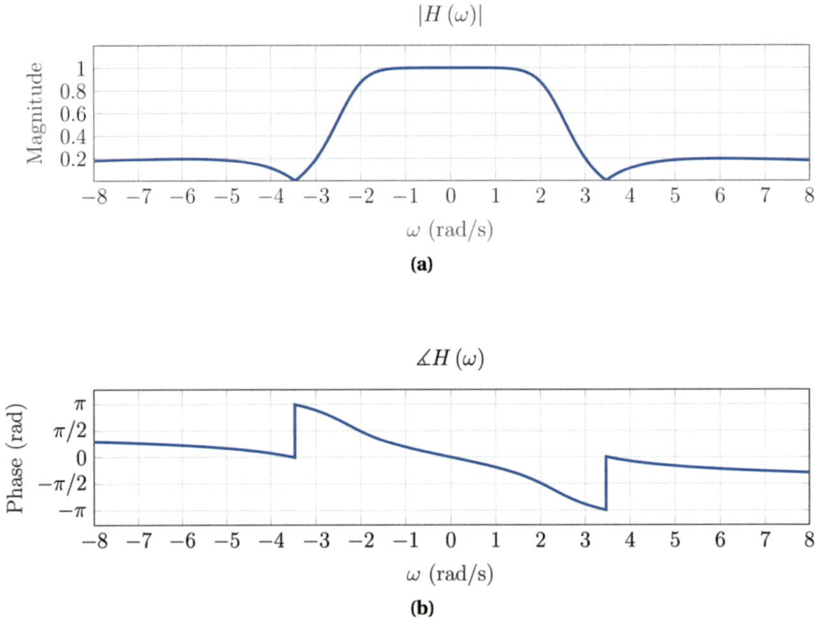

Figure 6.26 – **(a)** The magnitude and **(b)** the phase of the third-order Chebyshev type-II lowpass filter designed in Example 6.5.

Software resources: exct_6_5.m , exct_6_5_Live.mlx

Obtaining N and ε for the inverse Chebyshev lowpass filter

The inverse Chebyshev lowpass filter is uniquely described by the choice of parameters ω_2, ε, and N. If the desired filter is described by the specification diagram of Fig. 6.14 then we have the set of parameters R_p, A_s, ω_1, and ω_2 from which ε and N must be determined.

The first step is to choose the parameter ε so that the magnitude response of the filter fills the allowed area completely in the stopband.

Evaluating the squared magnitude response given by Eqn. (6.101) at the stopband edge frequency $\omega = \omega_2$ and remembering that $C_N(1) = 1$ for all Chebyshev polynomials we obtain

$$10\log_{10}\left(\frac{\varepsilon^2}{1+\varepsilon^2}\right) = -A_s \tag{6.113}$$

which can be solved for ε to yield

$$\varepsilon = \frac{1}{\sqrt{10^{A_s/10}-1}} \tag{6.114}$$

Next the filter order N needs to be determined. At the passband edge frequency we have the following inequality:

$$20\log_{10}\left|H(j\omega_1)\right| \ge -R_p \tag{6.115}$$

Let us define $\omega_0 = \omega_2/\omega_1$ to simplify the notation. Eqn. (6.115) can be written as

$$10\log_{10}\left[\frac{\varepsilon^2 C_N^2(\omega_0)}{1+\varepsilon^2 C_N^2(\omega_0)}\right] \ge -R_p \tag{6.116}$$

In the worst case Eqn. (6.116) can be taken with equality and solved for $C_N(\omega_0)$ to yield

$$C_N(\omega_0) = \sqrt{\frac{10^{A_s/10}-1}{10^{R_p/10}-1}} \tag{6.117}$$

after the substitution of the result found in Eqn. (6.114). Incidentally this is identical to Eqn. (6.92), therefore, its solution must be identical as well. Using the definition of parameter F as given by Eqn. (6.93) the minimum filter order that satisfies the specifications is found as xx

$$N = \frac{\cosh^{-1}(F)}{\cosh^{-1}(\omega_0)} \tag{6.118}$$

If Eqn. (6.118) does not yield an integer, it must be rounded up to the next integer. The excess tolerance created by the rounding up of N can be used in either frequency band by solving Eqn. (6.113) or Eqn. (6.116) for ε.

In summary, the parameters of a Chebyshev type-II lowpass filter are computed as follows:

Given dB tolerances R_p and A_s and critical frequencies ω_1 and ω_2:

$$\omega_0 = \frac{\omega_2}{\omega_1}, \quad \text{and} \quad F = \sqrt{\frac{10^{A_s/10}-1}{10^{R_p/10}-1}}$$

$$N = \frac{\cosh^{-1}(F)}{\cosh^{-1}(\omega_0)} \quad \Longrightarrow \quad \text{Round up to next integer}$$

Compute ε from one of the following:

$$\varepsilon = \frac{1}{\sqrt{10^{A_s/10}-1}} \quad \text{or} \quad \varepsilon = \frac{1}{\sqrt{C_N^2(\omega_0)\left(10^{R_p/10}-1\right)}}$$

6.5.4 Elliptic lowpass filters

An alternative to Butterworth and Chebyshev lowpass filter formulas discussed in previous sections is the elliptic lowpass filter described through the squared magnitude function

$$\left|H(\omega)\right|^2 = \frac{1}{1+\epsilon^2\,\psi_N^2(\omega/\omega_1)} \tag{6.119}$$

where ω_1 is the passband edge frequency and ε is a positive constant. The function $\psi_N^2(x)$ is called the *Chebyshev rational function* and is defined in terms of *Jacobi elliptic functions.* A thorough treatment of Jacobi elliptic functions is well beyond the scope of this text, and will not be given here. Details of elliptic functions, related formulas, and derivations can be found in references [5], [3].

Typical elliptic lowpass filter magnitude response is illustrated in Fig. 6.27. As seen in the figure, neither the passband nor the stopband response is flat. Elliptic filters exhibit equiripple behavior in both the passband and the stopband. What is gained in return is sharper transition behavior compared to Butterworth and Chebyshev filters of the same order. Alternatively, a given set of specifications can often be satisfied with a lower order elliptic filter compared to other filter types.

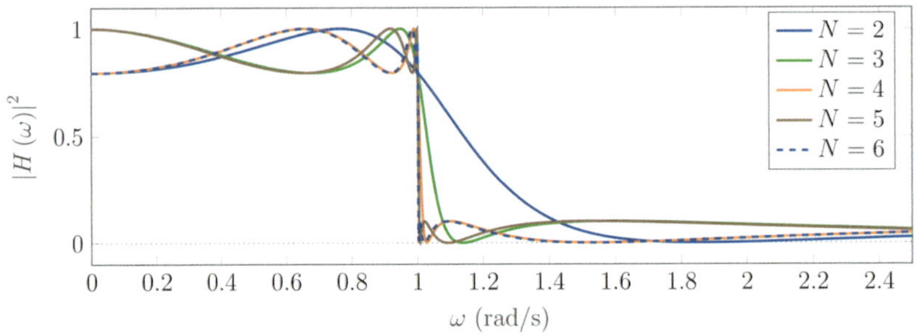

Figure 6.27 – Squared magnitude for elliptic filters with $\omega_1 = 1$ rad/s.

Software resource: See MATLAB Exercise 6.6.

6.6 Analog Filter Transformations

In preceding sections we have discussed the design of analog lowpass filters. Design formulas for Butterworth, Chebyshev and elliptic squared-magnitude functions were presented only for lowpass filters. If a different filter type such as a highpass, bandpass, or bandstop filter is needed, it is obtained from a lowpass filter, referred to as a *lowpass prototype*, by means of an *analog filter transformation*.

Let $G(s)$ be the system function of an analog lowpass filter and let $H(\lambda)$ represent the new filter to be obtained from it. For the new filter we use λ as the Laplace transform variable. $H(\lambda)$ is obtained from $G(s)$ through a transformation such that

$$H(\lambda) = G(s)\Big|_{s=f(\lambda)} \tag{6.120}$$

The function $s = f(\lambda)$ is the transformation that converts the lowpass filter into the type of filter desired. In the following sections we will discuss specific functions $f(\lambda)$ for obtaining highpass, bandpass, and bandstop filters from lowpass prototypes.

6.6.1 Lowpass to highpass transformation

Consider the specification diagrams shown in Fig. 6.28 for a lowpass filter with magnitude response $|G(\omega)|$ and a highpass filter with magnitude response $|H(\omega)|$.

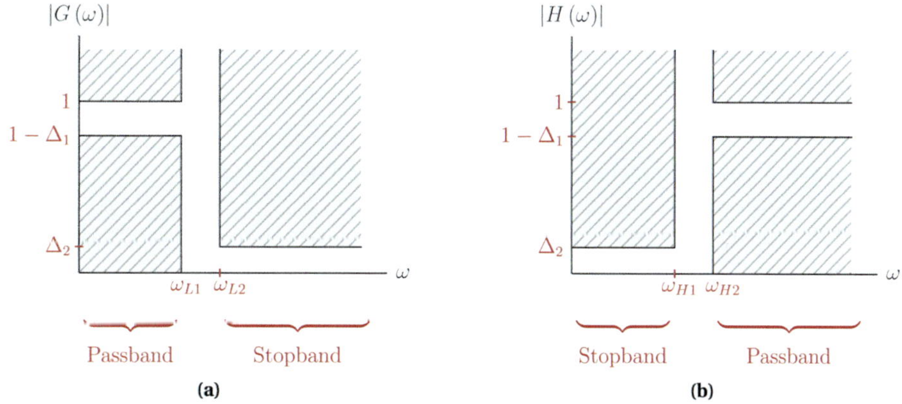

Figure 6.28 – Specification diagrams for **(a)** lowpass filter and **(b)** highpass filter.

It is desired to obtain the highpass filter system function $H(\lambda)$ from the lowpass filter system function $G(s)$. The transformation

$$\frac{s}{\omega_{L1}} = \frac{\omega_{H2}}{\lambda} \qquad \Longrightarrow \qquad s = \frac{\omega_{L1}\,\omega_{H2}}{\lambda} = \frac{\omega_0^2}{\lambda} \tag{6.121}$$

can be used for this purpose. The magnitudes of the two filters are identical at their respective passband edge frequencies, that is,

$$\left|H(\omega_{H2})\right| = \left|G(\omega_{L1})\right| \tag{6.122}$$

The stopband edges of the two filters are related by

$$\omega_{L2}\,\omega_{H1} = \omega_{L1}\,\omega_{H2} \tag{6.123}$$

so that we have

$$\left|H(\omega_{H1})\right| = \left|G(\omega_{L2})\right| \tag{6.124}$$

Example 6.6: Design of a Chebyshev highpass filter

Recall that a third-order Chebyshev type-I analog lowpass filter was designed in Example 6.4 with a cutoff frequency of $\omega_1 = 1$ rad/s and $\varepsilon = 0.4$. The system function of the filter was found to be

$$G(s) = \frac{0.6250}{s^3 + 1.1542\,s^2 + 1.4161\,s + 0.6250}$$

Convert this filter to a highpass filter with a passband edge frequency of $\omega_{H2} = 3$ rad/s. Afterward compute and graph magnitude and phase responses of the highpass filter obtained.

Solution: The critical frequency of the lowpass filter is $\omega_{L1} = 1$ rad/s, therefore

$$\omega_0^2 = \omega_{L1}\,\omega_{H2} = (1)\,(3) = 3$$

We will use the transformation

$$s = \frac{3}{\lambda}$$

which leads to the system function

$$H(\lambda) = \frac{0.6250}{\left(\dfrac{3}{\lambda}\right)^3 + 1.1542\left(\dfrac{3}{\lambda}\right)^2 + 1.4161\left(\dfrac{3}{\lambda}\right) + 0.6250}$$

$$= \frac{0.6250\,\lambda^3}{0.6250\,\lambda^3 + 4.2482\,\lambda^2 + 10.3875\,\lambda + 27}$$

Magnitude and phase characteristics of $H(\lambda)$ are shown in Fig. 6.29.

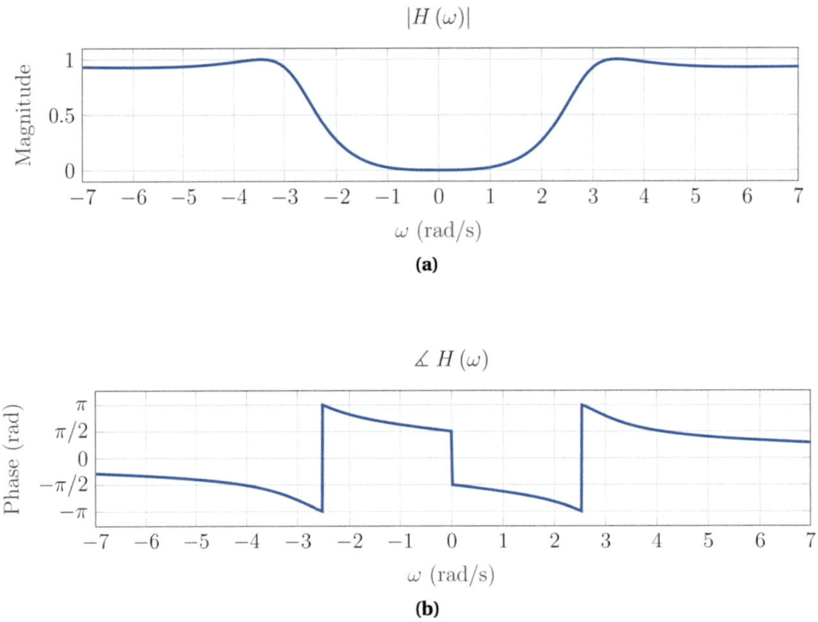

Figure 6.29 – The magnitude and the phase of the third-order Chebyshev highpass filter designed in Example 6.6.

Software resource: exct_6_6.m

Software resource: See MATLAB Exercise 6.7.

6.6.2 Lowpass to bandpass transformation

Specification diagrams in Fig. 6.30 show a lowpass filter with magnitude response $|G(\omega)|$ and a bandpass filter with magnitude response $|H(\omega)|$.

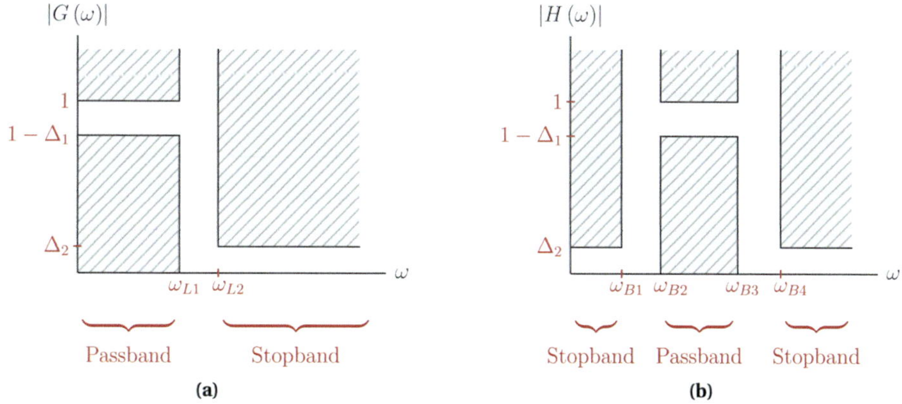

Figure 6.30 – Specification diagrams for **(a)** lowpass filter and **(b)** bandpass filter.

It is desired to obtain the bandpass filter system function $H(\lambda)$ from the lowpass filter system function $G(s)$. The transformation

$$s = \frac{\lambda^2 + \omega_0^2}{B\,\lambda} \tag{6.125}$$

can be used for this purpose. We require

$$\left|H(\omega_{B2})\right| = \left|G(-\omega_{L1})\right| \tag{6.126}$$

and

$$\left|H(\omega_{B3})\right| = \left|G(\omega_{L1})\right| \tag{6.127}$$

It can be shown that (see Problem 6.11) these requirements are met if the parameters ω_0 and B are chosen as

$$\omega_0 = \sqrt{\omega_{B2}\,\omega_{B3}} \qquad \text{and} \qquad B = \frac{\omega_{B3} - \omega_{B2}}{\omega_{L1}} \tag{6.128}$$

The parameter ω_0 is the *geometric mean* of the passband edge frequencies of the bandpass filter. The parameter B is the ratio of the bandwidth of the bandpass filter to the bandwidth of the lowpass filter.

Example 6.7: Design of a Chebyshev type-I bandpass filter

Consider again the third-order Chebyshev type-I analog lowpass filter designed in Example 6.4 with a cutoff frequency of $\omega_1 = 1$ rad/s and $\varepsilon = 0.4$. The system function of the filter was

$$G(s) = \frac{0.6250}{s^3 + 1.1542\,s^2 + 1.4161\,s + 0.6250}$$

Convert this lowpass filter to a bandpass filter with critical frequencies of $\omega_{B2} = 2$ rad/s and $\omega_{B3} = 3$ rad/s. Afterward compute and graph magnitude and phase responses of the bandpass filter obtained.

Solution: We will use the transformation given by Eqn. (6.125). Using Eqn. (6.128) we obtain the parameters of the transformation as

$$\omega_0^2 = (2)\,(3) = 6 \qquad \text{and} \qquad B = \frac{3-2}{1} = 1$$

leading to the transformation

$$s = \frac{\lambda^2 + 6}{\lambda} \tag{6.129}$$

Substituting Eqn. (6.129) into the lowpass prototype system function $G(s)$ we obtain

$$H(\lambda) = \frac{0.6250}{\left(\dfrac{\lambda^2+6}{\lambda}\right)^3 + 1.1542\left(\dfrac{\lambda^2+6}{\lambda}\right)^2 + 1.4161\left(\dfrac{\lambda^2+6}{\lambda}\right) + 0.6250}$$

which can be simplified to

$$H(\lambda) = \frac{0.6250\,\lambda^3}{\lambda^6 + 1.1542\,\lambda^5 + 19.4161\,\lambda^4 + 14.4754\,\lambda^3 + 116.4963\,\lambda^2 + 41.5502\,\lambda + 216}$$

Magnitude and phase of $H(\lambda)$ are shown in Fig. 6.31.

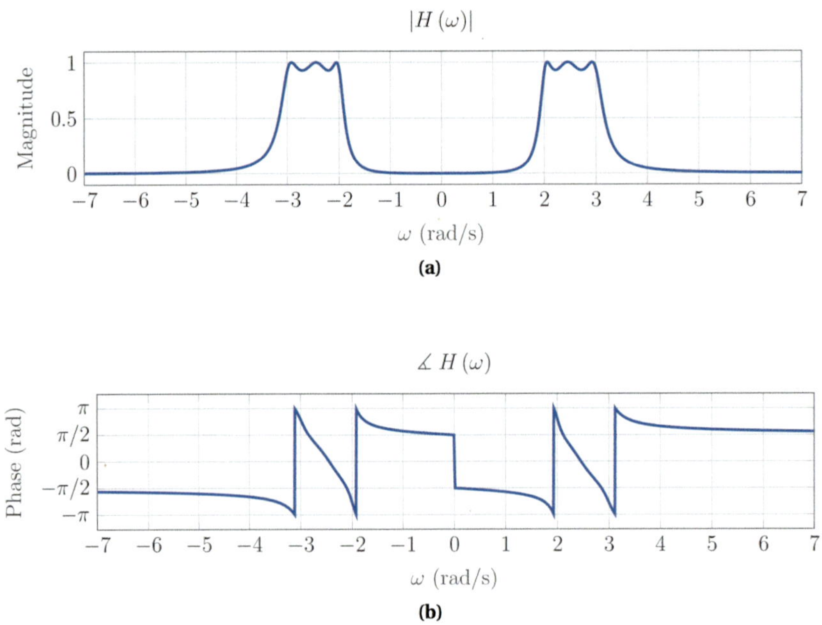

Figure 6.31 – The magnitude and the phase of the third-order Chebyshev bandpass filter designed in Example 6.7.

Software resource: `exct_6_7.m`

6.6.3 Lowpass to bandstop transformation

Specification diagrams in Fig. 6.32 show a lowpass filter with magnitude response $|G(\omega)|$ and a bandstop filter with magnitude response $|H(\omega)|$.

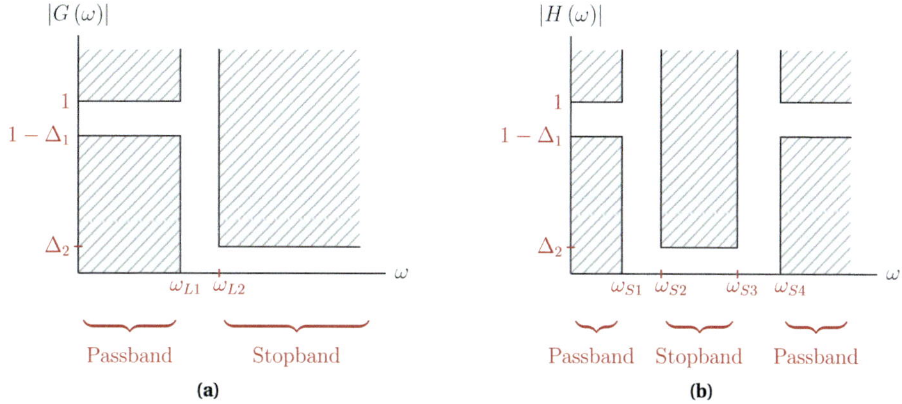

Figure 6.32 – Specification diagrams for **(a)** lowpass filter and **(b)** bandstop filter.

In order to obtain the bandstop filter system function $H(\lambda)$ from the lowpass filter system function $G(s)$, the transformation to be used is in the form

$$s = \frac{B\lambda}{\lambda^2+\omega_0^2} \tag{6.130}$$

For a lowpass filter to be converted into a bandstop filter we require

$$\left|H(\omega_{S4})\right| = \left|G(-\omega_{L1})\right| \tag{6.131}$$

and

$$\left|H(\omega_{S1})\right| = \left|G(\omega_{L1})\right| \tag{6.132}$$

It can be shown that (see Problem 6.12) these requirements are met if the parameters ω_0 and B are chosen as

$$\omega_0 = \sqrt{\omega_{S1}\,\omega_{S4}} \qquad \text{and} \qquad B = (\omega_{S4}-\omega_{S1})\,\omega_{L1} \tag{6.133}$$

Example 6.8: Design of a Chebyshev type-I bandstop filter

Once again we will work with the third-order Chebyshev type-I analog lowpass filter which was designed in Example 6.4 with a cutoff frequency of $\omega_{L1} = 1$ rad/s and $\varepsilon = 0.4$. The system function of the filter was obtained as

$$G(s) = \frac{0.6250}{s^3+1.1542\,s^2+1.4161\,s+0.6250}$$

Convert this lowpass filter to a bandstop filter with stopband edge frequencies of $\omega_{S1} = 2$ rad/s and $\omega_{S4} = 4$ rad/s. Afterward compute and graph magnitude and phase responses of the bandstop filter obtained.

Solution: We will use the transformation given by Eqn. (6.130). Using Eqn. (6.133) we obtain the parameters of the transformation as

$$\omega_0^2 = \omega_{S1}\,\omega_{S4} = (2)\,(4) = 8 \qquad \text{and} \qquad B = (\omega_{s4} - \omega_{S1})\,\omega_{L1} = (4-2)\,(1) = 2$$

leading to the transformation

$$s = \frac{2\,\lambda}{\lambda^2 + 8} \tag{6.134}$$

Substituting Eqn. (6.134) into the lowpass prototype system function $G(s)$ we obtain

$$H(\lambda) = \frac{0.6250}{\left(\dfrac{2\,\lambda}{\lambda^2+8}\right)^3 + 1.1542\left(\dfrac{2\,\lambda}{\lambda^2+8}\right)^2 + 1.4161\left(\dfrac{2\,\lambda}{\lambda^2+8}\right) + 0.6250}$$

which can be simplified to

$$H(\lambda) = \frac{\lambda^6 + 24\,\lambda^4 + 192\,\lambda^2 + 512}{\lambda^6 + 4.5314\,\lambda^5 + 31.3867\,\lambda^4 + 85.3021\,\lambda^3 + 251.0936\,\lambda^2 + 290.0083\,\lambda + 512}$$

Magnitude and phase of $H(\lambda)$ are shown in Fig. 6.33.

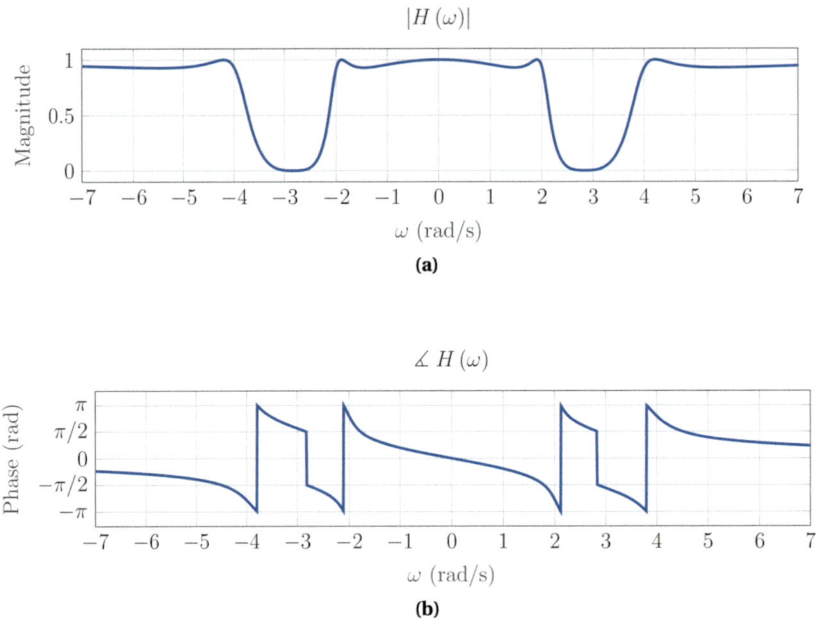

Figure 6.33 – The magnitude and the phase of the third-order Chebyshev bandstop filter designed in Example 6.8.

Software resource: exct_6_8.m

Software resource: See MATLAB Exercise 6.8.

Summary of Key Points

☞ The input and output signals $x(t)$ and $y(t)$ of a CTLTI system with impulse response $h(t)$ are related to each other through the convolution relationship

Eqn. (6.6): $$y(t) = h(t) * x(t) = \int_{-\infty}^{\infty} h(\lambda)\, x(t-\lambda)\, d\lambda$$

☞ The type of distortion caused by a CTLTI system is *linear distortion* and it exhibits itself in two distinct forms, namely *amplitude distortion* and *phase distortion.* The magnitude spectrum is

Eqn. (6.7): $$|Y(\omega)| = |H(\omega)|\, |X(\omega)|$$

and the phase spectrum is

Eqn. (6.8): $$\measuredangle Y(\omega) = \measuredangle H(\omega) + \measuredangle X(\omega)$$

☞ For distortionless transmission we need

Eqn. (6.13): $$y(t) = K\, x(t-t_d)$$

which requires the system function to be in the form

Eqn. (6.15): $$H(\omega) = K\, e^{-j\omega t_d}$$

This corresponds to a constant magnitude and a linear phase characteristic.

☞ Ideal lowpass filter that is characterized by the system function

Eqn. (6.26): $$H_{LP}(\omega) = \Pi\left(\frac{\omega}{2\omega_0}\right) e^{-j\omega t_d} = \begin{cases} e^{-j\omega t_d}\,, & |\omega| \le \omega_0 \\ 0\,, & |\omega| > \omega_0 \end{cases}$$

☞ An ideal bandpass filter has the system function

Eqn. (6.27): $$H_{BP}(\omega) = \begin{cases} e^{-j\omega t_d}\,, & \omega_1 \le |\omega| \le \omega_2 \\ 0\,, & \text{otherwise} \end{cases}$$

☞ Ideal highpass and bandstop filters are defined similarly:

$$H_{HP}(\omega) = \begin{cases} e^{-j\omega t_d}\,, & |\omega| \ge \omega_0 \\ 0\,, & |\omega| < \omega_0 \end{cases}\,,\quad H_{BS}(\omega) = \begin{cases} e^{-j\omega t_d}\,, & |\omega| \le \omega_1 \text{ or } |\omega| \ge \omega_2 \\ 0\,, & \text{otherwise} \end{cases}$$

☞ Specifications for realizable filters are typically given in terms of critical frequencies (two for lowpass and highpass filters, four for bandpass and bandstop filters), and tolerance limits for the magnitude characteristics. These are shown graphically using specification diagrams as shown in Fig. 6.13.

☞ Butterworth lowpass filters are characterized by the squared-magnitude function

Eqn. (6.38): $$|H(\omega)|^2 = \frac{1}{1+(\omega/\omega_c)^{2N}}$$

where the parameters N and ω_c are the filter order and the cutoff frequency, respectively. Butterworth filter magnitude response exhibits smooth behavior in both the passband and the stopband.

☞ Chebyshev type-I lowpass filters have the squared-magnitude function

Eqn. (6.39): $$\left|H(\omega)\right|^2 = \frac{1}{1+\varepsilon^2\, C_N^2\,(\omega/\omega_1)}$$

where ε is a positive constant. The frequency ω_1 is the passband edge frequency. The function $C_N(v)$ in the denominator represents the *Chebyshev polynomial* of order N. Chebyshev type-I filter magnitude response has equiripple behavior in the passband, and is smooth in the stopband.

☞ Chebyshev type-II lowpass filters have the squared-magnitude function

Eqn. (6.40): $$\left|H(\omega)\right|^2 = \frac{\varepsilon^2\, C_N^2\,(\omega_2/\omega)}{1+\varepsilon^2\, C_N^2\,(\omega_2/\omega)}$$

The parameter ω_2 is the *stopband edge frequency*. Chebyshev type-II filter magnitude response is smooth in the passband and has equiripple behavior in the stopband.

☞ Elliptic lowpass filters have the squared-magnitude function

Eqn. (6.41): $$\left|H(\omega)\right|^2 = \frac{1}{1+\varepsilon^2\, \psi_N^2\,(\omega/\omega_1)}$$

The parameter ε is a positive constant. The function $\psi_N(v)$ is called a *Chebyshev rational function*, and is defined in terms of *Jacobi elliptic functions*. Elliptic filter magnitude response shows equiripple behavior in both the passband and the stopband.

Further Reading

[1] Amal Banerjee. *Automated Electronic Filter Design*. Springer, 2017.

[2] Hercules G Dimopoulos. *Analog Electronic Filters: Theory, Design and Synthesis*. Springer Science & Business Media, 2011.

[3] Larry D Paarmann. *Design and Analysis of Analog Filters: A Signal Processing Perspective*. Vol. 617. Springer Science & Business Media, 2005.

[4] R. Schaumann and M.E. Van Valkenburg. *Design of Analog Filters*. Oxford University Press, 2010.

[5] Kendall L Su. *Analog Filters*. Springer Science & Business Media, 2012.

[6] F. Taylor. *Digital Filters: Principles and Applications with MATLAB*. IEEE Series on Digital & Mobile Communication. Wiley, 2011.

[7] L.D. Thede. *Practical Analog and Digital Filter Design*. Artech House, 2005.

[8] L. Wanhammar. *Analog Filters Using MATLAB*. Springer London, Limited, 2009.

[9] Arthur B Williams and Fred J Taylor. *Electronic Filter Design Handbook*. McGraw-Hill Education, 2006.

[10] Steve Winder. *Analog and Digital Filter Design*. Elsevier, 2002.

MATLAB Exercises with Solutions

MATLAB Exercise 6.1: Butterworth analog lowpass filter design

MATLAB function `butter()` may be used for designing analog Butterworth filters. For example, a fourth-order Butterworth lowpass filter with a 3-dB cutoff frequency of $\omega_c = 3$ rad/s is designed as follows:

```
>> [num,den] = butter(4,3,'s')

num =
         0         0         0         0   81.0000

den =
    1.0000    7.8394   30.7279   70.5544   81.0000
```

The last argument `'s'` is needed for specifying that an *analog* filter is being designed since the function `butter()` can be used for designing both analog and discrete-time Butterworth filters. Without the argument `'s'`, a discrete-time filter is assumed. This is also the case for functions that design Chebyshev and elliptic filters as we will see in later exercises.

The vectors `num` and `den` returned by the function contain coefficients of numerator and denominator polynomials. They correspond to the s-domain system function

$$H(s) = \frac{81}{s^4 + 7.8394\,s^3 + 30.7279\,s^2 + 70.5544\,s + 81}$$

for the designed filter. Magnitude and phase characteristics of the designed filter may now be graphed with the following statements:

```
>> omg = [-10:0.01:10];
>> H = freqs(num,den,omg);
>> plot(omg,abs(H));
>> plot(omg,angle(H));
```

Software resource: `mexct_6_1.m`

MATLAB Exercise 6.2: Exploring Chebyshev polynomials

In this exercise we will develop and test a function for computing the coefficients of the Chebyshev polynomial of a specified order N. The Nth-order Chebyshev polynomial $C_N(\nu)$ can be computed from the knowledge of the two lower order Chebyshev polynomials using the recursive formula given by Eqn. (6.69) and repeated below in a slightly modified form for convenience.

$$\text{Eqn. (6.69):} \qquad C_N(\nu) = 2\nu\, C_{N-1}(\nu) \quad C_{N-2}(\nu)\,, \qquad N \ge 2$$

The function `ss_chebpol()` listed below.

```
1  function [c] = ss_chebpol(N)
2    if (N==0)
3      c = [1];          % Chebyshev polynomial of order 0
4    elseif (N==1)
5      c = [1,0];        % Chebyshev polynomial of order 1
6    else
7      cNm2 = [1];       % Chebyshev polynomial of order N-2
8      cNm1 = [1,0];     % Chebyshev polynomial of order N-1
9      for i = 2:N
10       c = 2*conv([1,0],cNm1)-[0,0,cNm2];  % Eqn. (6.69)
11       cNm2 = cNm1; % Update polynomial of order N-2
12       cNm1 = c;    % Update polynomial of order N-1
13     end
14   end
15 end
```

The loop between lines 9 and 13 implements the recursive formula. The script `mexct_6_2.m` listed below uses the function `ss_chebpol()` along with the function `polyval()` to evaluate and graph the Chebyshev polynomials for $N = 2, \ldots, 6$.

```
% Script: mexct_6_2.m
nu = [0:0.005:1.1];
c2 = ss_chebpol(2);    % Polynomial coefficients
p2 = polyval(c2,nu);   % Evaluate polynomial
% Combine two lines into one for computing polynomial coefficients
% and evaluating the polynomial.
p3 = polyval(ss_chebpol(3),nu);
p4 = polyval(ss_chebpol(4),nu);
p5 = polyval(ss_chebpol(5),nu);
p6 = polyval(ss_chebpol(6),nu);
plot(nu,p2,nu,p3,nu,p4,nu,p5,nu,p6); grid;
axis([0,1.1,-2,8]);
legend('N=2','N=3','N=4','N=5','N=6','Location','NorthWest');
title('Chebyshev polynomials for N=2,3,4,5,6');
xlabel('\nu');
ylabel('C_N(\nu)');
```

The graph produced is shown in Fig. 6.34.

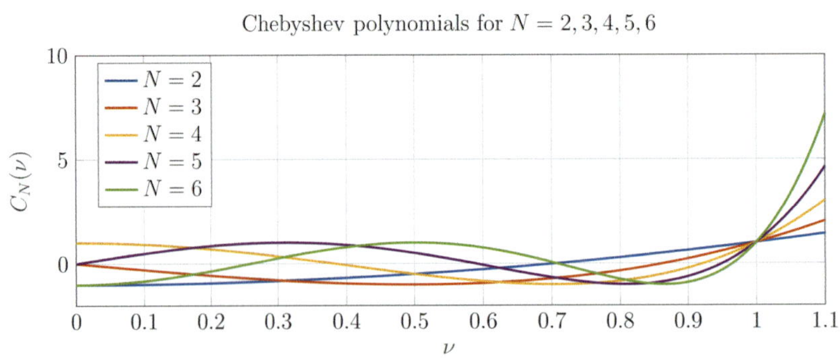

Figure 6.34 – MATLAB graph of Chebyshev polynomials for $N = 2, \ldots, 6$.

Software resources: mexct_6_2.m , ss_chebpol.m

MATLAB Exercise 6.3: Chebyshev type-I analog lowpass filter design

In Example 6.4 a Chebyshev type-I lowpass filter with $N = 3$ and $\varepsilon = 0.4$ was designed. In this exercise we will verify the results using function cheby1(). The second argument for the function cheby1() is the dB passband tolerance R_p which is computed as

$$R_p = 10\,\log_{10}\left(1+\varepsilon^2\right)$$

The script mexct_6_3.m listed below will obtain the desired filter and graph its magnitude and phase characteristics.

```
% Script: mexct_6_3.m
N = 3;                                  % Filter order
omega1 = 1;                             % Passband edge frequency
epsilon = 0.4;
Rp = 10*log10(1+epsilon^2);             % Passband ripple in dB
[num,den] = cheby1(N,Rp,omega1,'s');    % Design the filter
tf(num,den)                             % Display system function
omg = [-4:0.01:4];                      % Vector of radian frequencies
H = freqs(num,den,omg);                 % Evaluate system function
tiledlayout(2,1);
nexttile;
plot(omg,abs(H)); grid;                 % Graph the magnitude response
title('|H(\omega)|');
xlabel('\omega (rad/s)');
ylabel('Magnitude');
nexttile;
plot(omg,angle(H)); grid;               % Graph the phase response
title('\angle H(\omega)');
xlabel('\omega (rad/s)');
ylabel('Phase (rad)');
```

Software resource: mexct_6_3.m

MATLAB Exercise 6.4: Chebyshev type-I analog lowpass filter design revisited

Design a Chebyshev type-I lowpass filter with the set of specifications

$$\omega_1 = 3\ \text{rad/s}\,, \qquad \omega_2 = 4\ \text{rad/s}\,, \qquad R_p = 1\ \text{dB}\,, \qquad A_s = 30\ \text{dB}$$

Solution: The design can be carried out using MATLAB functions cheb1ord() and cheby1() as follows:

```
% Script: mexct_6_4.m
omg1 = 3;                               % Passband edge frequency
```

```
omg2 = 4;                            % Stopband edge frequency
Rp = 1;                              % Passband ripple in dB
As = 30;                             % Stopband attenuation in dB
N = cheb1ord(omg1,omg2,Rp,As,'s');   % Compute necessary filter order
[num,den] = cheby1(N,Rp,omg1,'s');   % Design the filter
tf(num,den)                          % Display system function
omg = [-7:0.01:7];                   % Vector of radian frequencies
H = freqs(num,den,omg);              % Evaluate system function
tiledlayout(2,1);
nexttile;
plot(omg,20*log10(abs(H))); grid;    % Graph the dB magnitude response
axis([-7,7,-60,10]);
title('|H(\omega)|');
xlabel('\omega (rad/s)');
ylabel('Magnitude (dB)');
nexttile;
plot(omg,angle(H)); grid;            % Graph the phase response
axis([-7,7,-4,4]);
title('\angle H(\omega)');
xlabel('\omega (rad/s)');
ylabel('Phase (rad)');
```

The results correspond to a seventh-order system function

$$H(s) = \frac{67.1555}{s^7 + 2.7694\,s^6 + 19.5847\,s^5 + 38.5774\,s^4 + 119.961\,s^3 + 133.315\,s^2 + 155.767\,s + 67.1555}$$

Note that the magnitude of the system function is graphed on a dB scale using the logarithm of $|H(\omega)|$, and is shown in Fig. 6.35.

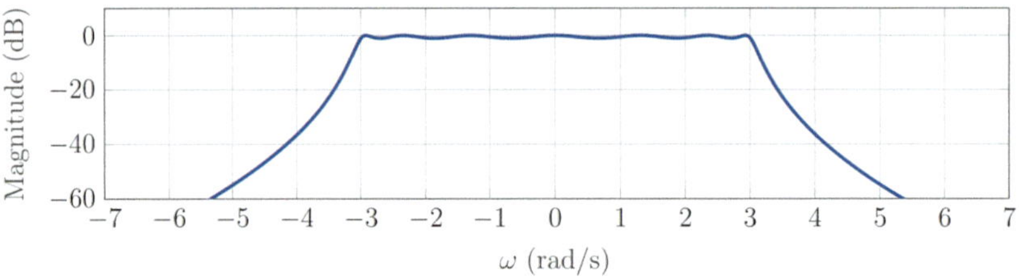

Figure 6.35 – Magnitude response of the filter designed in MATLAB Exercise 6.4.

Software resource: mexct_6_4.m

MATLAB Exercise 6.5: Chebyshev type-II analog lowpass filter design

Design a Chebyshev type-II lowpass filter with the set of specifications

$$\omega_1 = 3\ \text{rad/s}\,, \qquad \omega_2 = 4\ \text{rad/s}\,, \qquad R_p = 1\ \text{dB}\,, \qquad A_s = 30\ \text{dB}$$

Solution: The design can be carried out using MATLAB functions `cheb2ord()` and `cheby2()` as follows:

```
% Script: mexct_6_5.m
omega1 = 3;                              % Passband edge frequency
omega2 = 4;                              % Stopband edge frequency
Rp = 1;                                  % Passband ripple in dB
As = 30;                                 % Stopband attenuation in dB
N = cheb2ord(omega1,omega2,Rp,As,'s');   % Compute necessary filter order
[num,den] = cheby2(N,As,omega2,'s');     % Design the filter
tf(num,den)                              % Display system function
omg = [-7:0.01:7];                       % Vector of radian frequencies
H = freqs(num,den,omg);                  % Evaluate system function
tiledlayout(2,1);
nexttile;
plot(omg,20*log10(abs(H))); grid;        % Graph the dB magnitude
axis([-7,7,-60,10]);
title('|H(\omega)|');
xlabel('\omega (rad/s)');
ylabel('Magnitude (dB)');
nexttile;
plot(omg,angle(H)); grid;                % Graph the phase
axis([-7,7,-4,4]);
title('\angle H(\omega)');
xlabel('\omega (rad/s)');
ylabel('Phase (rad)');
```

The results correspond to a seventh order system function

$$H(s) = \frac{0.89\,s^6 + 113.39\,s^4 + 3628.6\,s^2 + 33176}{s^7 + 18.09\,s^6 + 163.21\,s^5 + 959.29\,s^4 + 3933\,s^3 + 11877\,s^2 + 23394\,s + 33176}$$

The dB magnitude of the system function is shown in Fig. 6.36.

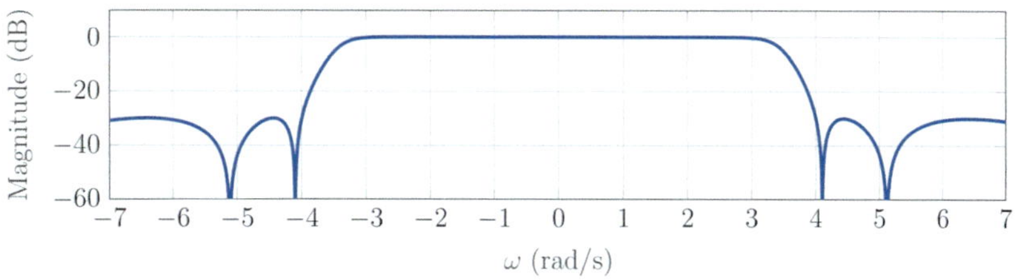

Figure 6.36 – Magnitude response of the filter designed in MATLAB Exercise 6.5.

Software resource: mexct_6_5.m

MATLAB Exercise 6.6: Elliptic analog lowpass filter design

Design an elliptic lowpass filter with the set of specifications

$$\omega_1 = 2 \text{ rad/s}, \qquad \omega_2 = 2.5 \text{ rad/s}, \qquad R_p = 3 \text{ dB}, \qquad A_s = 30 \text{ dB}$$

Solution: We will use MATLAB functions `ellipord()` and `ellip()` for this design problem.

```
% Script: mexct_6_6.m
omg1 = 2;                                  % Passband edge frequency
omg2 = 2.5;                                % Stopband edge frequency
Rp = 3;                                    % Passband ripple in dB
As = 30;                                   % Stopband attenuation in dB
[N,omg1] = ellipord(omg1,omg2,Rp,As,'s');  % Compute necessary filter order
[num,den] = ellip(N,Rp,As,omg1,'s');       % Design the filter
tf(num,den)                                % Display system function
omg = [-7:0.01:7];                         % Vector of radian frequencies
H = freqs(num,den,omg);                    % Evaluate system function
tiledlayout(2,1);
nexttile;
plot(omg,20*log10(abs(H))); grid;          % Graph the dB magnitude
axis([-7,7,-60,10]);
title('|H(\omega)|');
xlabel('\omega (rad/s)');
ylabel('Magnitude (dB)');
nexttile;
plot(omg,angle(H)); grid;                  % Graph the phase
axis([-7,7,-4,4]);
title('\angle H(\omega)');
xlabel('\omega (rad/s)');
ylabel('Phase (rad)');
```

The results correspond to a fourth-order system function

$$H(s) = \frac{0.0316\,s^4 + 0.8385\,s^2 + 3.7362}{s^4 + 1.1348\,s^3 + 5.3763\,s^2 + 3.9221\,s + 5.2775}$$

The dB magnitude of the system function is shown in Fig. 6.37.

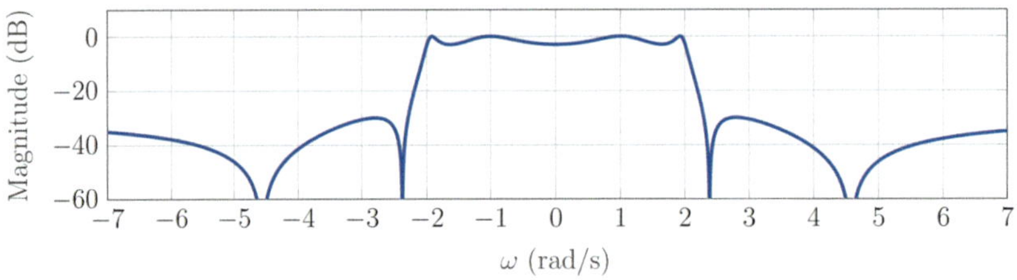

Figure 6.37 – Magnitude response of the filter designed in MATLAB Exercise 6.6.

Software resource: mexct_6_6.m

MATLAB Exercise 6.7: Lowpass to highpass filter transformation

Recall that a Chebyshev type-I analog lowpass filter was designed in MATLAB Exercise 6.3. Using it as the starting point, obtain a highpass filter with a passband edge at $\omega_{H2} = 3$ rad/s.

Solution: Before we solve this problem, it is worthwhile to take a quick look at three MATLAB functions we will utilize for analog frequency transformations. Each of the three functions provide multiple syntax options. We will only focus on the basic syntax of each that corresponds to our formulation of frequency transformations discussed in Section 6.6. MATLAB help information, at least in version 2023a, is somewhat vague about the arguments used with these functions, so we will attempt to clarify usage in the table below.

`lp2hp()`	Lowpass to highpass transformation	`[numT,denT] = lp2hp(num,den,Wo)` Argument `Wo` corresponds to ω_0^2 in Eqn. (6.121).
`lp2bp()`	Lowpass to bandpass transformation	`[numT,denT] = lp2bp(num,den,Wo,Bw)` Arguments `Wo` and `Bw` correspond to ω_0 and B respectively in Eqn. (6.128).
`lp2bs()`	Lowpass to bandstop transformation	`[numT,denT] = lp2bs(num,den,Wo,Bw)` Arguments `Wo` and `Bw` correspond to ω_0 and B respectively in Eqn. (6.133).

Important thing to remember is that the argument `Wo` has a different meaning in `lp2hp()` compared to the other two functions. Also, what is referred to as the *center frequency* in MATLAB help information is actually the *geometric center* as defined in Eqns. (6.128) and (6.133).

In this particular problem we will use the function `lp2hp()`. The filter of MATLAB Exercise 6.3 can be redesigned and then converted to a highpass filter with the following script:

```
% Script: mexct_6_7.m
N = 3;                                  % Filter order
omgL1 = 1;                              % Passband edge frequency
epsilon = 0.4;
Rp = 10*log10(1+epsilon^2);             % Passband ripple in dB
[num,den] = cheby1(N,Rp,omgL1,'s');     % Design the lowpass filter
omgH2 = 3;                              % Passband edge freq. for hp filter
[numhp,denhp] = lp2hp(num,den,omgH2);
tf(numhp,denhp)                         % Display system function
omg = [-7:0.01:7];                      % Vector of radian frequencies
H = freqs(numhp,denhp,omg);             % Evaluate system function
plot(omg,abs(H)); grid;                 % Graph the magnitude response
axis([-7,7,-0.1,1.1]);
title('|H(\omega)|');
xlabel('\omega (rad/s)');
ylabel('Magnitude');
```

Software resource: mexct_6_7.m

MATLAB Exercise 6.8: Transformations from lowpass to bandpass and bandstop filters

Using the Chebyshev filter designed in MATLAB Exercise 6.3 as a basis, design the following two filters:

a. A bandpass filter with the critical frequencies

$$\omega_{B2} = 1.5 \text{ rad/s}, \qquad \omega_{B3} = 2.5 \text{ rad/s}$$

b. A bandstop filter with the critical frequencies

$$\omega_{S1} = 1 \text{ rad/s}, \qquad \omega_{S4} = 3.5 \text{ rad/s}$$

Solution:

a. The script `mexct_6_8a.m` listed below repeats the lowpass filter design of MATLAB Exercise 6.3 and then converts the result to a bandpass filter through the use of the function `lp2bp()`. The magnitude response of the resulting bandpass filter is graphed, as shown in Fig. 6.38.

```
% Script: mexct_6_8a.m
N = 3;                                 % Filter order
omgL1 = 1;                             % Passband edge frequency
epsilon = 0.4;
Rp = 10*log10(1+epsilon^2);            % Passband ripple in dB
[num,den] = cheby1(N,Rp,omgL1,'s');    % Design the lowpass filter
omgB2 = 1.5;                           % Passband edge freq. for bp filter
omgB3 = 2.5;                           % Passband edge freq. for bp filter
omg0 = sqrt(omgB2*omgB3);              % Geometric ctr. freq. for trans.
B = (omgB3-omgB2)/omgL1;               % Parameter for transformation
[numbp,denbp] = lp2bp(num,den,omg0,B);
tf(numbp,denbp)                        % Display system function
omg = [-10:0.01:10];                   % Vector of radian frequencies
H = freqs(numbp,denbp,omg);            % Evaluate system function
plot(omg,abs(H)); grid;                % Graph the magnitude response
axis([-10,10,-0.1,1.1]);
title('|H(\omega)|');
xlabel('\omega (rad/s)');
ylabel('Magnitude');
```

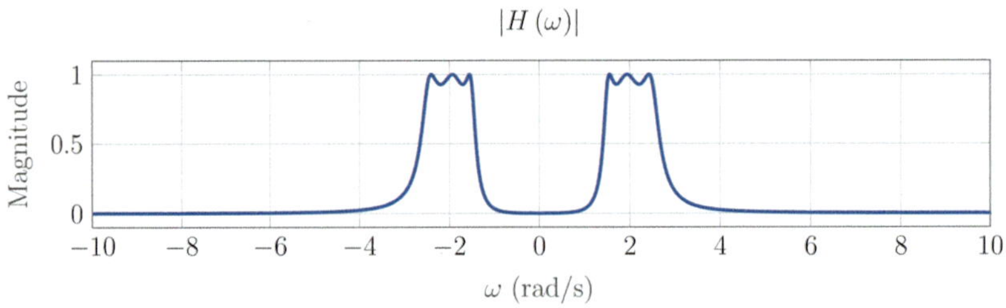

Figure 6.38 – Magnitude response of the bandpass filter designed in MATLAB Exercise 6.8.

b. See the script `mexct_6_8b.m` listed below for the design and subsequent conversion of the lowpass filter of MATLAB Exercise 6.3. The final bandstop filter is obtained through the use of the function `lp2bs()`. Its magnitude response is shown in Fig. 6.39.

```
% Script: mexct_6_8b.m
N = 3;                                 % Filter order
omgL1 = 1;                             % Passband edge frequency
epsilon = 0.4;
Rp = 10*log10(1+epsilon^2);            % Passband ripple in dB
[num,den] = cheby1(N,Rp,omgL1,'s');    % Design the lowpass filter
```

```
omgS1 = 1;                            % Stopband edge freq. for bs filter
omgS4 = 3.5;                          % Stopband edge freq. for bs filter
omg0 = sqrt(omgS1*omgS4);             % Geometric ctr. freq. for trans.
B = (omgS4-omgS1)*omgL1;              % Parameter for transformation
[numbs,denbs] = lp2bs(num,den,omg0,B);
tf(numbs,denbs)                       % Display system function
omg = [-10:0.01:10];                  % Vector of radian frequencies
H = freqs(numbs,denbs,omg);           % Evaluate system function
plot(omg,abs(H)); grid;               % Graph the magnitude response
axis([-10,10,-0.1,1.1]);
title('|H(\omega)|');
xlabel('\omega (rad/s)');
ylabel('Magnitude');
```

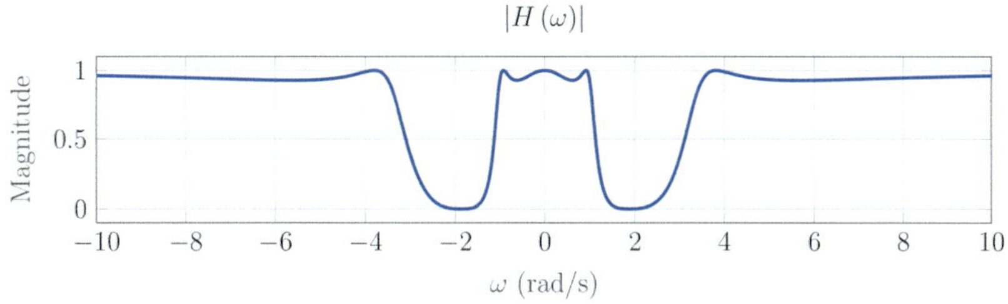

Figure 6.39 – Magnitude response of the bandstop filter designed in MATLAB Exercise 6.8.

Software resources: mexct_6_8a.m , mexct_6_8b.m

MATLAB Exercise 6.9: Bandpass filter design with various approximation formulas

An analog bandpass filter is to be designed such that its dB magnitude response satisfies the following conditions:

$$-3 \le |H(f)|_{dB} \le 0 \quad \text{for} \quad 2 \le f \le 3\,\text{Hz}$$

$$|H(f)|_{dB} \le -25.0 \quad \text{for} \quad f \le 1.5\,\text{Hz or } f \ge 3.5\,\text{Hz}$$

Using each of the four approximation formulas discussed in Section (6.5) design the minimum order filter to satisfy the requirements. Graph dB magnitude responses of all four designed filters on the same coordinate system.

Solution: Four critical frequencies for the filters in question are

$$f_1 = 1.5 \text{ Hz}, \qquad f_2 = 2 \text{ Hz}, \qquad f_3 = 3 \text{ Hz}, \qquad f_4 = 3.5 \text{ Hz},$$

and the dB tolerance values are

$$R_p - 3 \text{ dB}, \qquad A_s = 25 \text{ dB}$$

The script `mexct_6_9.m` listed below illustrates the design process.

```
% Script: mexct_6_9.m
Rp = 3;
As = 25;
fp = [2,3];          % Passband edge frequencies in Hz
fs = [1.5,3.5];      % Stopband edge frequencies in Hz
Wp = 2*pi*fp;        % Passband edge frequencies in rad/s
Ws = 2*pi*fs;        % Stopband edge frequencies in rad/s
f = [0:0.005:7];     % Set vector of frequencies for graphing
omg = 2*pi*f;
% Butterworth filter
[N,Wn] = buttord(Wp,Ws,Rp,As,'s');    % Compute Butterworth parameters
[num1,den1] = butter(N,Wn,'s');       % Design Butterworth filter
H1 = tf(num1,den1)
H1dB = dBMagnitude(num1,den1,omg);
% Chebyshev type-I filter
[N,Wp] = cheb1ord(Wp,Ws,Rp,As,'s');  % Compute Chebyshev type-I parameters
[num2,den2] = cheby1(N,Rp,Wp,'s');   % Design Chebyshev type-I filter
H2 = tf(num2,den2)
H2dB = dBMagnitude(num2,den2,omg);
% Chebyshev type-II filter
[N,Ws] = cheb2ord(Wp,Ws,Rp,As,'s');  % Compute Chebyshev type-II parameters
[num3,den3] = cheby2(N,As,Ws,'s');   % Design Chebyshev type-II filter
H3 = tf(num3,den3)
H3dB = dBMagnitude(num3,den3,omg);
% Elliptic filter
[N,Wp] = ellipord(Wp,Ws,Rp,As,'s');  % Compute elliptic filter parameters
[num4,den4] = ellip(N,Rp,As,Wp,'s'); % Design elliptic filter
H4 = tf(num4,den4)
H4dB = dBMagnitude(num4,den4,omg);
% Graph the results
p = plot(f,H1dB,f,H2dB,f,H3dB,f,H4dB); grid;
axis([0,7,-40,5]);
xlabel('Frequency (Hz)');
ylabel('Magnitude (dB)');
legend('Butterworth','Chebyshev type-I','Chebyshev type-II','Elliptic',..
    'Location','NorthEast');

function HdB = dBMagnitude(num,den,omg)
  % Function to compute dB magnitude from numerator and denominator
  H = freqs(num,den,omg);
  HdB = 20*log10(abs(H));
end
```

It is evident from the results that the requirements can be satisfied by a tenth-order Butterworth filter, an eighth-order Chebyshev filter (type-I or II), or a sixth-order elliptic filter. The dB magnitude response of each of these filters is shown Fig. 6.40. In addition, the figure includes shaded areas to indicate the design specifications. This last part is not included in the script listed above, but was just added to the figure.

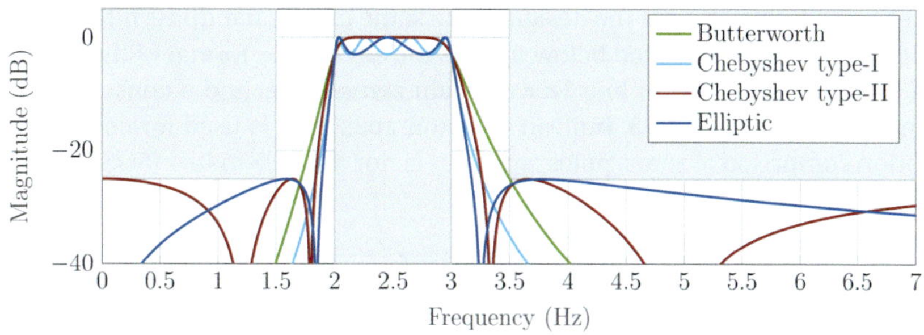

Figure 6.40 – Magnitude responses of the filters designed in MATLAB Exercise 6.9.

Software resources: mexct_6_9.m , mexct_6_9_Live.mlx

MATLAB Exercise 6.10: Filter design using second-order sections

In Matlab Exercise 6.9 we have designed bandpass filters using MATLAB functions that return numerator and denominator polynomial coefficients as vectors. This works well for low-order lowpass filters. For other filter types or for higher filter orders, however, the use of numerator and denominator polynomials may lead to numerical instability problems in analysis and implementation due to very large coefficient values. Let's take, for example, the sixth-order elliptic filter that was one of the designs in MATLAB Exercise 6.9. Type in the following lines *after running the script of the previous exercise.*

```
>> format long
>> num4

num4 =
1.0e+04 *
Columns 1 through 4
0   0.000109315913539  -0.000000000000000   0.060348503096405
Columns 5 through 7
-0.000000000000000   6.133457533460496   0.000000000000000

>>  den4

den4 =
1.0e+07 *
Columns 1 through 4
0.000000100000000   0.000000368505692   0.000074954143587   0.000183137330812
Columns 5 through 7
0.017754425890145   0.020675983386591   1.329024421198502
```

Looking at MATLAB's responses, we see that numerator coefficients are on the order of 10^4, and denominator coefficients are on the order of 10^7. There is a huge gap between the smallest and the largest coefficients of each polynomial, making accurate computation difficult. Butterworth and Chebyshev filters designed in the same script are of even higher order and have larger coefficient values.

In this exercise we will repeat the design of the same elliptic bandpass filter to yield second-order sections. The script listed below utilizes the alternative format of the built-in function `ellip()` for this purpose. In line 12 we obtain zeros, poles and a constant gain factor for the bandpass filter. In line 13, built-in function `zp2sos()` is used for converting this filter description comprised of zeros, poles, and gain factor to a SOS matrix (SOS stands for *second-order sections*).

```
1  % Script: mexct_6_10a.m
2  Rp = 3;
3  As = 25;
4  fp = [2,3];           % Passband edge frequencies in Hz
5  fs = [1.5,3.5];       % Stopband edge frequencies in Hz
6  Wp = 2*pi*fp;         % Passband edge frequencies in rad/s
7  Ws = 2*pi*fs;         % Stopband edge frequencies in rad/s
8  f = [0:0.005:7];      % Set vector of frequencies for graphing
9  omg = 2*pi*f;
10 % Elliptic filter
11 [N,Wp] = ellipord(Wp,Ws,Rp,As,'s');  % Compute elliptic filter parameters
12 [zeros,poles,gain] = ellip(N,Rp,As,Wp,'s');  % Design elliptic filter
13 [sos,gain] = zp2sos(zeros,poles,gain);       % Convert to SOS format
14 H4dB = dBMagnitudeFromSOS(sos,gain,omg);     % Compute dB magnitude
15 % Graph the results
16 plot(f,H4dB); grid;
17 axis([0,7,-40,5]);
18 xlabel('Frequency (Hz)');
19 ylabel('Magnitude (dB)');
20
21 function HdB = dBMagnitudeFromSOS(sos,gain,omg)
22   % Function to compute dB magnitude from SOS matrix.
23   [nRows,nCols] = size(sos);                % Get the size of SOS matrix
24   HdB = 20*log10(gain)*ones(size(omg));     % Start with gain factor
25   for k = 1:nRows
26     num = sos(k,1:3);  % Numerator polynomial of section k
27     den = sos(k,4:6);  % Denominator polynomial of section k
28     HdB = HdB + dBMagnitude(num,den,omg);  % Add the contribution
29   end
30 end
31
32 function HdB = dBMagnitude(num,den,omg)
33   % Function to compute dB magnitude from numerator and denominator
34   H = freqs(num,den,omg);
35   HdB = 20*log10(abs(H));
36 end
```

Local function `dBMagnitudeFromSOS()` is used for computing the dB magnitude response of the designed filter which is shown in Fig. 6.41. It matches the elliptic filter dB magnitude from Fig. 6.40.

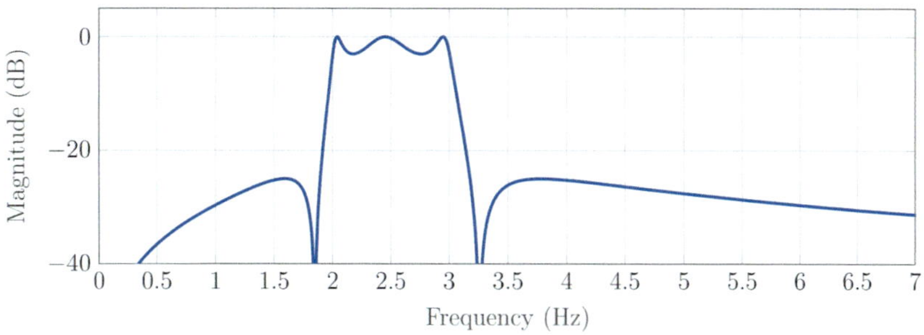

Figure 6.41 – Magnitude responses of the filters designed in MATLAB Exercise 6.10.

It is important to understand the structure of the SOS matrix. After running the script above, type in the following:

```
>> zeros,poles,gain

zeros =
   0.0000 +20.4388i
   0.0000 -20.4388i
   0.0000 +11.5892i
   0.0000 -11.5892i
  -0.0000 + 0.0000i

poles =
  -0.3855 +18.6478i
  -0.3855  18.6478i
  -1.1946 +15.3442i
  -1.1946 -15.3442i
  -0.2625 +12.6969i
  -0.2625 -12.6969i

gain =
   1.0932
```

followed by

```
>> sos

sos =
        0    1.0000    0.0000    1.0000    0.7710  347.8904
   1.0000   -0.0000  417.7454    1.0000    2.3891  236.8705
   1.0000   -0.0000  134.3106    1.0000    0.5250  161.2796
```

The matrix sos has 3 rows and 6 columns. Each row corresponds to a second-order section. Within each row, the first three elements are the numerator coefficients, and the next three elements are the denominator coefficients. If we wanted to print out the system function of each cascade section, we could use the script listed below:

```
% Script: mexct_6_10b.m
Rp = 3;
As = 25;
fp = [2,3];          % Passband edge frequencies in Hz
```

```
fs = [1.5,3.5];        % Stopband edge frequencies in Hz
Wp = 2*pi*fp;          % Passband edge frequencies in rad/s
Ws = 2*pi*fs;          % Stopband edge frequencies in rad/s
% Elliptic filter
[N,Wp] = ellipord(Wp,Ws,Rp,As,'s');  % Compute elliptic filter parameters
[zeros,poles,gain] = ellip(N,Rp,As,Wp,'s');  % Design elliptic filter
[sos,gain] = zp2sos(zeros,poles,gain);       % Convert to SOS format
[nRows,nCols] = size(sos);
for k = 1:nRows
  num = sos(k,1:3);         % Numerator polynomial of section k
  den = sos(k,4:6);         % Denominator polynomial of section k
  section = tf(num,den)     % Display system function for section k
end
```

Software resources: mexct_6_10a.m , mexct_6_10b.m , mexct_6_10_Live.mlx

Problems

6.1. Consider the RC circuit used in Example 6.1 and shown in Fig. 6.4. Let the parameter values be $R = 1$ kΩ and $C = 1$ ţF.

a. Find the frequency range in Hz in which the magnitude of the system function exhibits 1 percent or less deviation from its value at $\omega = 0$.

b. For the frequency range found in part (a) determine how much the phase characteristic deviates from a straight line. Determine the endpoints of the phase characteristic in the range of interest, and compare to the endpoints of a straight line with the appropriate slope.

c. Find the variation in the time delay over the frequency range established in part (a).

d. Repeat parts (a) through (c) if 2 percent variation in magnitude is allowed.

6.2. Consider a 2nd-order lowpass filter with the system function

$$H(s) = \frac{1}{s^2 + \sqrt{2}\,s + 1}$$

a. Find the frequency range in Hz in which the magnitude of the system function exhibits 1 percent or less deviation from its value at $\omega = 0$.

b. For the frequency range found in part (a) determine how much the phase characteristic deviates from a straight line. Determine the endpoints of the phase characteristic in the range of interest, and compare to the endpoints of a straight line with the appropriate slope.

c. Find the variation in the time delay over the frequency range established in part (a).

d. Repeat parts (a) through (c) if 2 percent variation in magnitude is allowed.

6.3. The magnitude spectrum of an ideal lowpass filter is shown in Fig. P.6.3. The time delay is to be $t_d = 0.1$ s for all frequencies.

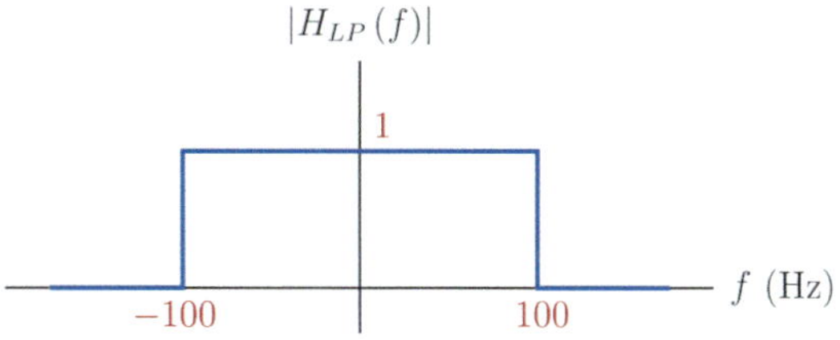

Figure P. 6.3

a. Write a mathematical expression for the system function $H(f)$.

b. Determine the impulse response $h(t)$.

c. Sketch the impulse response found in part (b).

6.4. The magnitude spectrum of an ideal bandpass filter is shown in Fig. P.6.4. The time delay is to be $t_d = 0.3$ s for all frequencies.

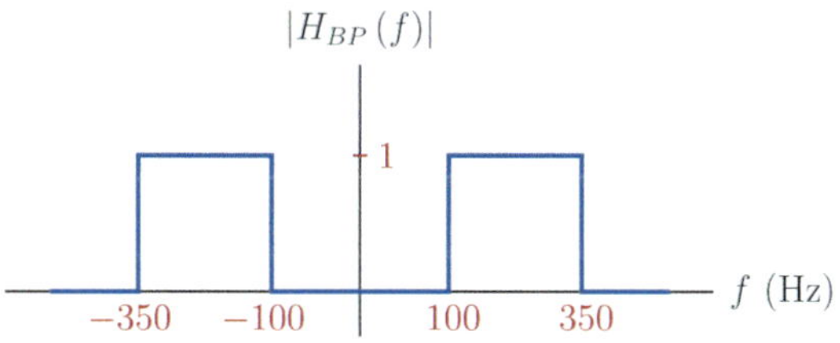

Figure P. 6.4

a. Write a mathematical expression for the system function $H(f)$.

b. Determine the impulse response $h(t)$.

c. Sketch the impulse response found in part (b).

6.5. Design a fourth-order Butterworth analog lowpass filter with 3-dB cutoff frequency of $\omega_c = 2$ rad/s. Find its poles and construct the system function $H(s)$.

6.6. A Butterworth analog lowpass filter is to be designed with the following specifications:

$$\omega_1 = 3 \text{ rad/s} \qquad \omega_2 = 4.5 \text{ rad/s} \qquad R_p = 1 \text{ dB} \qquad A_s = 30 \text{ dB}$$

Using the method presented in Section 6.5.1 determine the filter order N and the 3-dB cutoff frequency ω_c. Do not actually design the filter.

6.7. Consider the problem of determining Butterworth analog lowpass filter parameters from design specifications. First N is determined from Eqn. (6.64). In general it is not an integer, and must be rounded up to the nearest integer. This rounding up causes the tolerance specifications to be exceeded in the designed filter. If Eqn. (6.61) is used for computing ω_c, then specifications are met exactly at the passband edge $\omega = \omega_1$ and exceeded at the stopband edge $\omega = \omega_2$. If Eqn. (6.62) is used for computing ω_c, then specifications are met exactly at the stopband edge $\omega = \omega_2$ and exceeded at the passband edge $\omega = \omega_1$.

An alternative would be to divide the excess tolerance equally (in a dB sense) between the passband and the stopband. Derive the general solution for ω_c to accomplish this.

6.8. Design a fourth-order Chebyshev type-I analog filter with a passband edge frequency of $\omega_1 = 2$ rad/s and $\varepsilon = 0.3$. Find the poles of the filter and then construct the system function $H(s)$.

6.9. Design a fourth-order Chebyshev type-II analog filter with a stopband edge frequency of $\omega_2 = 2$ rad/s and $\varepsilon = 0.3$. Find the poles of the filter and then construct the system function $H(s)$.

6.10. An analog lowpass filter is needed with the following specifications:

$$\omega_1 = 2 \text{ rad/s} \qquad \omega_1 = 3.5 \text{ rad/s} \qquad R_p\omega_1 = 1 \text{ dB} \qquad A_s\omega_1 = 20 \text{ dB}$$

a. Determine N and ω_c for a Butterworth filter that meets the requirements. Find the system function for the corresponding filter.

b. Determine N and ε for a Chebyshev type-I filter that meets the requirements. Find the system function for the corresponding filter.

c. Determine N and ε for a Chebyshev type-II filter that meets the requirements. Find the system function for the corresponding filter.

6.11. Consider the lowpass to bandpass filter transformation given by Eqn. (6.125). Show that the parameters ω_0 and B must be chosen as given by Eqn. (6.128) in order to satisfy the conditions given by Eqns. (6.126) and (6.127). Hint: Let $s = j\omega$ and $\lambda = jp$. Write the relationship between the radian frequencies ω and p and sketch ω as a function of p. Impose the conditions of Eqns. (6.126) and (6.127) and solve for the unknown parameters.

6.12. Consider the lowpass to bandstop filter transformation given by Eqn. (6.130). Show that the parameters ω_0 and B must be chosen as given by Eqn. (6.133) in order to satisfy the conditions given by Eqns. (6.131) and (6.132). Hint: Let $s = j\omega$ and $\lambda = jp$. Write the relationship between the radian frequencies ω and p and sketch ω as a function of p. Impose the conditions of Eqns. (6.131) and (6.132) and solve for the unknown parameters.

6.13. Consider a lowpass filter with the system function

$$G(s) = \frac{2}{s+2}$$

This filter is to be converted to a highpass filter with system function $H(\lambda)$ by means of the transformation $s = \omega_0/\lambda$.

a. Determine ω_0 so that the frequency 2 rad/s for the lowpass filter corresponds to the frequency 5 rad/s for the highpass filter.

b. Using the value of ω_0 found in part (a), determine the frequency for the highpass filter that corresponds to the frequency 6 rad/s for the lowpass filter.

c. Determine the system function $H(\lambda)$ for the highpass filter. Evaluate its magnitude at the frequency $\omega = 5$ rad/s and verify that it is equal to the magnitude of the lowpass filter at $\omega = 2$ rad/s.

6.14. A Butterworth analog highpass filter is to be designed with the following specifications:

$$\omega_{H1} = 2 \text{ rad/s} \qquad \omega_{H2} = 5 \text{ rad/s} \qquad R_p = 1 \text{ dB} \qquad A_s = 30 \text{ dB}$$

a. Find the corresponding specifications ω_{L1}, ω_{L2}, R_p, and A_s for an appropriate lowpass prototype filter.

b. Determine the parameters N and ω_c for the lowpass prototype filter.

c. Design the lowpass prototype filter. Find its poles and construct its system function $G(s)$.

d. Convert the lowpass prototype filter to a highpass filter using the appropriate transformation. Determine the system function $H(\lambda)$.

e. Evaluate the dB magnitude of the highpass filter at the frequencies ω_{H1} and ω_{H2}. Verify that the designed filter meets the requirements.

MATLAB Problems

6.15. Consider again the RC circuit of Problem 6.1. Develop a MATLAB script to graph magnitude, phase, and time delay characteristics of the circuit in a specified range of frequency $-f_0 < f < f_0$.

- **a.** Using the script developed, graph the characteristics for the frequency range found in part (a) of Problem 6.1.
- **b.** Repeat for the frequency range found in part (d) of Problem 6.1.

6.16. Consider the second-order lowpass filter of Problem 6.2. Develop a MATLAB script to graph magnitude, phase, and time delay characteristics of the system in a specified range of frequency $-f_0 < f < f_0$.

- **a.** Using the script developed, graph the characteristics for the frequency range found in part (a) of Problem 6.2.
- **b.** Repeat for the frequency range found in part (d) of Problem 6.2.

6.17.

- **a.** Develop a MATLAB function `ss_ilp()` to compute the impulse response of an ideal lowpass filter with the system function

$$H(\omega) = \begin{cases} 1\, e^{-j\omega t_d}\,, & |\omega| \le \omega_0 \\ 0\,, & \text{otherwise} \end{cases}$$

  The syntax of your function should be

```
h = ss_ilp(omg0,td,t)
```

  where `omg0` and `td` are parameters of the ideal lowpass filter, and `t` is the vector of time instants at which the impulse response if to be evaluated.

- **b.** Write a script to test the function `ss_ilp()` for several sets of parameter values.

6.18.

- **a.** Develop a MATLAB function `ss_ibp()` to compute the impulse response of an ideal bandpass filter with the system function

$$H(\omega) = \begin{cases} 1\, e^{-j\omega t_d}\,, & \omega_1 \le |\omega| \le \omega_2 \\ 0\,, & \text{otherwise} \end{cases}$$

  The syntax of your function should be

```
h = ss_ibp(omg1,omg2,td,t)
```

  where `omg1`, `omg2`, and `td` are parameters of the ideal lowpass filter, and `t` is the vector of time instants at which the impulse response if to be evaluated. The function `ss_ibp()` should internally utilize the function `ss_ilp()` developed in Problem 6.17.

- **b.** Write a script to test the function `ss_ibp()` for several sets of parameter values.

6.19. Refer to the Butterworth analog lowpass filter design considered in Problem 6.5. Design the same filter using MATLAB. Graph the magnitude and the phase characteristics of the designed filter.

6.20. Refer to Problem 6.6. Using the values of N and ω_c found in Problem 6.6 and the MATLAB function `butter()` design the filter. Compute and graph the dB magnitude response of the filter and verify that it satisfies the design specifications.

6.21.

- **a.** Design the filter specified in Problem 6.8 using MATLAB.
- **b.** Compute and graph the dB magnitude response and verify that it meets the requirements of the design.
- **c.** Graph the pole-zero diagram of the system function.

6.22.

- **a.** Design the filter specified in Problem 6.9 using MATLAB.
- **b.** Compute and graph the dB magnitude response and verify that it meets the requirements of the design.
- **c.** Graph the pole-zero diagram of the system function.

6.23. Refer to Problem 6.10.

- **a.** Repeat each design (Butterworth, Chebyshev type I and Chebyshev type-II) using appropriate MATLAB functions. Compare to the results of hand calculations obtained in Problem 6.10.
- **b.** Compute and graph the magnitude response of each design.
- **c.** Compute and graph the dB magnitude responses of all three designs on the same coordinate system.

6.24. Repeat the Butterworth analog highpass design in Problem 6.14 using MATLAB functions `buttord()`, `butter()` and `lp2hp()`. Afterward compute and graph the dB magnitude of the highpass filter and check the dB tolerance values at the two critical frequencies.

CHAPTER 7

COMMUNICATIONS APPLICATION: AMPLITUDE MODULATION

Chapter Objectives

- Learn the significance of and the need for modulation in analog and digital communication systems.
- Understand amplitude modulation as one method of embedding an information bearing signal into a high frequency sinusoidal carrier.
- Explore time- and frequency-domain characteristics of amplitude modulated signals. Understand variants of amplitude modulation (DSB and SSB).
- Study techniques for obtaining modulated signals as well as for extracting information bearing signals from modulated signals.

7.1 Introduction

The main purpose of any communication system is to facilitate the transfer of an information bearing signal from one point in space to another, with an acceptable level of quality.

Consider, for example, two people communicating through their cell phones. The sound of the person who acts as information source is detected by the microphone on his or her phone, converted to an electrical signal for processing within the phone, and eventually converted to an electromagnetic signal which is transmitted from the antenna of the phone. This electromagnetic signal is received by the antennas within the cell phone network, is transmitted from cell

DOI: 10.1201/9781032675251-7

to cell as needed, possibly travels through various types of connections such as fiber optics and satellite links, and finally reaches the antenna of the phone that is designated as its destination. Within the receiving phone, the electromagnetic signal is converted back to an electrical signal, and is eventually converted to sound.

Thus, any communication system has a transmitter and a receiver. The collection of all systems between the transmitter and the receiver is referred to as the *communication channel*. In general, a channel has two undesirable effects on the signals that are transmitted through it:

1. It distorts them and
2. It contaminates them with noise.

As a result, the electrical signal reproduced in the receiving phone and the sound reproduced through its speaker are only approximations of the original signals transmitted. In the design of communication systems we take steps to ensure that the signals at the receiving end are reasonably good representations of the signals transmitted.

In this chapter we will present an introduction to modulation which is one of the steps taken in the design of a communication system to ensure reliable transmission of signals. In doing so, we will utilize the analysis tools and transforms developed in previous chapters for linear systems.

Modulation is the process of embedding an information bearing signal, referred to as the *message signal*, in another signal known as the *carrier*. The carrier by itself does not contain any information, yet it may have other features that make it desirable. Some examples of these desirable features are: 1) suitability for transmission through a particular medium, 2) better performance in the presence of noise and interference, 3) suitability to more efficient utilization of resources, and so on.

Fig. 7.1 illustrates the general structure of a very simple analog communication system.

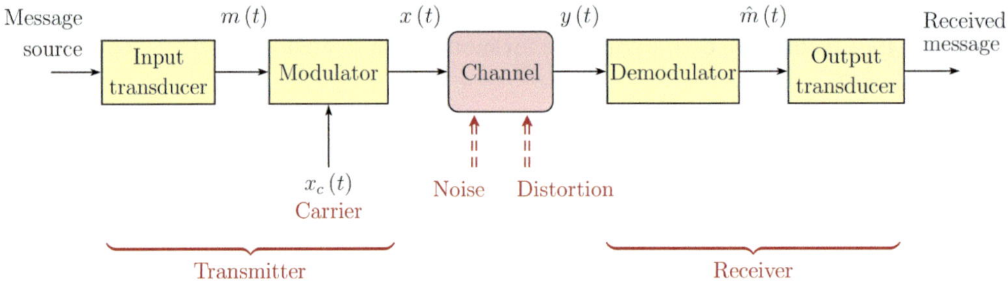

Figure 7.1 – Structure of a simple analog communication system.

The overall operation of the communication system can be summarized as follows:

1. The role of the input transducer is to convert the source message from its native format to the form of an electrical signal $m(t)$, referred to as the *message signal*. Some examples of input transducers are microphone, video camera, pressure gauge, temperature sensor, etc.
2. The modulator embeds the message signal $m(t)$ into the carrier signal $x_c(t)$ to create the modulated carrier $x(t)$ which is then transmitted through the channel.
3. The output signal $y(t)$ of the channel is a distorted version of $x(t)$ which is also contaminated with noise. In simple terms we have

$$y(t) = \text{Sys}\{x(t)\} + w(t)$$

where Sys{...} represents the effect of the channel on the modulated signal, and $w(t)$ represents the additive channel noise[1].

4. At the receiver side of the system the demodulator reverses the action of the modulator, and extracts an approximate version $\hat{m}(t)$ of the original message signal $m(t)$ from the received modulated signal $y(t)$. The demodulator may or may not need a local copy of the carrier signal $x_c(t)$.
5. Finally, the output transducer converts the electrical signal $\hat{m}(t)$ to its native format. Examples of output transducers are speaker, video display, etc.

7.2 The Need for Modulation

We need to address the question of why a modulator and a demodulator are needed in a communication system. Using a modulated carrier for transmitting a message has a number of benefits over transmitting the message signal $m(t)$ directly through the channel without modulation:

1. Compatibility of the signal with the channel:

   Modulation gives us the ability to produce a signal that is compatible with the characteristics of the channel that will transmit it.

   Most message signals we encounter in our daily lives involve relatively low frequencies, referred to as the *baseband* frequencies. For example, voice signals used in telephone communications occupy the frequency range from about 300 Hz to about 3600 Hz. High-fidelity music signals are typically in the frequency range from about 20 Hz to about 20 kHz. Similarly, baseband video signals such as those encountered on a TV monitor occupy roughly the frequency range from 0 to about 4.2 MHz. Signals such as these can be transmitted using wire links or optical fiber connections, but they are not suitable for over-the-air transmission due to their large wavelengths. The antennas required for over-the-air transmission of these signals in their original form would have to be unreasonably long. Modulating a high frequency carrier with these message signals makes over-the-air broadcast possible while keeping required antenna lengths at practical levels.
2. Reduced susceptibility to noise and interference:

   Modulation of a carrier with the message signal results in a shift of the range of signal frequencies. This frequency shift can be used in a way that allows us to position the frequency spectrum of the signal where noise and interference conditions are more favorable compared to the conditions at baseband frequencies.

   Additionally, some modulation techniques allow performance improvements in terms of noise susceptibility provided that increased bandwidth is available.
3. Multiplexing:

   Modulation allows for more efficient use of available resources by multiplexing different signals in the same medium.

   In sinusoidal modulation, the frequency shift that results from modulation can also be used for mixing multiple messages in one medium, with each message signal modulating a carrier at a different frequency. These modulated carrier signals can be combined and transmitted together. They can later be separated from each other provided that the

[1] This is actually an oversimplification. In an actual communication channel, the signal is not necessarily distorted first and contaminated by noise afterward; distortion and contamination take place simultaneously in a distributed fashion, resulting in a much more difficult model for analysis.

frequency ranges occupied by different carriers do not overlap with each other. This effect is known as *frequency division multiplexing (FDM)*. On radio and television, FDM forms the basis of simultaneous broadcast of multiple stations in the same air space.

In digital pulse modulation, the gaps between the successive pulses of a carrier can be used for accommodating messages from multiple sources, a concept known as *time division multiplexing (TDM)*. Multiple digital information sources can utilize the same channel simultaneously, by taking turns in sending data.

7.3 Types of Modulation

The type of modulation used in a communication system depends on

1. The type of carrier used and
2. The particular parameter of the carrier that is utilized for embedding the message signal into it.

We will focus on two commonly used types of carrier signals, namely a sinusoidal carrier and a pulse train. Consider the sinusoidal carrier signal:

$$x_c(t) = A_c \cos\left(2\pi f_c t + \phi_c\right) \tag{7.1}$$

The carrier described by Eqn. (7.1) has three parameters, namely the amplitude A_c, the frequency f_c, and the phase ϕ_c as shown in Fig. 7.2.

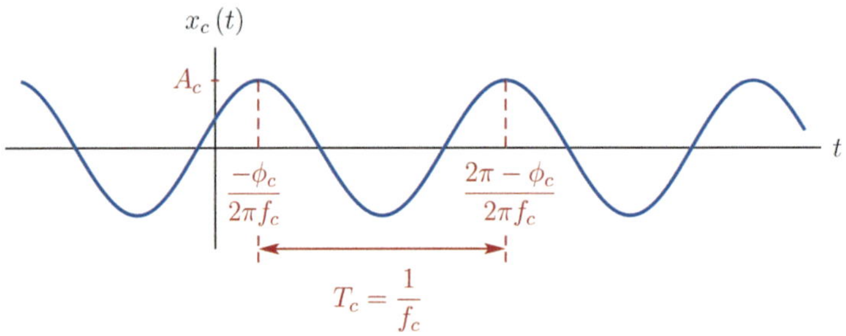

Figure 7.2 – Sinusoidal carrier signal.

If the values of these three parameters are fixed, then the carrier clearly has no information value since its future behavior is completely predictable. Such a carrier signal is referred to as a *blank carrier* or *unmodulated carrier*. If, on the other hand, we allow one of these parameters to vary in time, then the resulting signal will be a more interesting one, and certainly not as predictable. Furthermore, if the time variations of the parameter in question are somehow related to an information bearing message signal $m(t)$, then these variations can be tracked at the receiving end of the communication system, and the message signal $m(t)$ can be extracted from the variations of carrier parameters. This is the essence of sinusoidal modulation. The modulated carrier can be used as a vehicle for transmitting the message signal $m(t)$ from one point to another. Different modulation schemes can be devised depending on which of the three parameters is varied in time:

1. In *amplitude modulation*, the amplitude A_c is varied as a function of the message signal $m(t)$.
2. In *angle modulation*, the angle $2\pi f_c t + \phi_c$ of the sinusoidal carrier varied as a function of the message signal $m(t)$. This can be achieved by varying either the phase or the frequency, and leads to two variants of angle modulation known as *phase modulation* and *frequency modulation* respectively.

An alternative to the sinusoidal carrier described in Eqn. (7.1) is a pulse train in the form

$$x_c(t) = \sum_{k=-\infty}^{\infty} A_c \Pi\left(\frac{t - kT_0}{\tau}\right) \tag{7.2}$$

which is shown in Fig. 7.3.

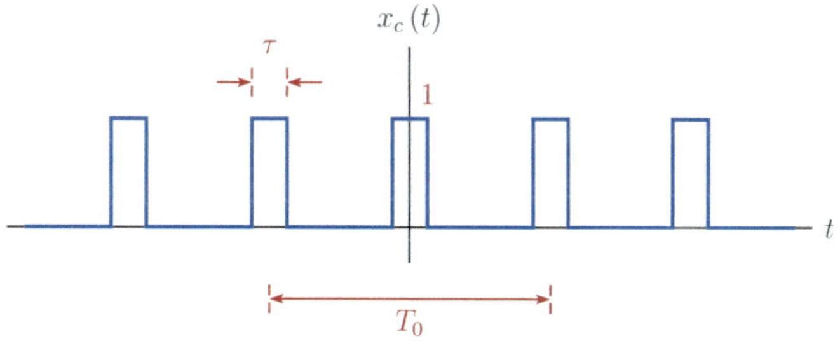

Figure 7.3 – Pulse train as a carrier signal.

If the pulse train is used for analog modulation, it presents three opportunities for a message signal to be embedded in it.

1. The amplitudes of pulses can be varied as a function of the message signal, resulting in *pulse-amplitude modulation (PAM)*.
2. The duration of each pulse can be varied as a function of the message signal. This leads to *pulse-width modulation (PWM)*.
3. Finally, the position of each pulse within its allocated period can be varied. The resulting modulation scheme is known as *pulse-position modulation (PPM)*.

The pulse train can also be used as the basis of digital modulation. The first step is to modulate the amplitude of the pulse train with the message signal using pulse-amplitude modulation. Afterward the amplitude of each pulse is rounded up or down to be associated with one of the discrete amplitude levels from a predetermined set, an operation known as *quantization*. Finally, each quantized amplitude is *encoded* so that it can be represented using a set of binary digits (bits). The combination of pulse-amplitude modulation, quantization, and encoding is referred to as *pulse-code modulation (PCM)*.

7.4 Amplitude Modulation

Consider the unmodulated sinusoidal carrier given by Eqn. (7.1). Its amplitude A_c is constant. One method of incorporating the message signal $m(t)$ into the carrier is to allow the amplitude of the carrier vary in a way that is related to $m(t)$. Let us replace the constant amplitude term A_c with $[A_c + m(t)]$ to obtain

$$\begin{aligned} x_{AM}(t) &= \left[A_c + m(t)\right] \cos\left(2\pi f_c t\right) \\ &= A_c \cos\left(2\pi f_c t\right) + m(t) \cos\left(2\pi f_c t\right) \end{aligned} \tag{7.3}$$

The result is an *amplitude-modulated (AM)* carrier. To set the terminology straight we will list the roles of the signals involved in Eqn. (7.3) as follows:

$m(t)$:	Modulating signal
$x_c(t)$:	Unmodulated (blank) carrier
$x_{AM}(t)$:	Modulated carrier

The parameter f_c is referred to as the *carrier frequency.* Note that the phase term ϕ_c in Eqn. (7.1) is not important for this discussion, and we will simplify the formulation of amplitude modulation without losing any generality by assuming that $\phi_c = 0$. One example of an AM signal is depicted in Fig. 7.4.

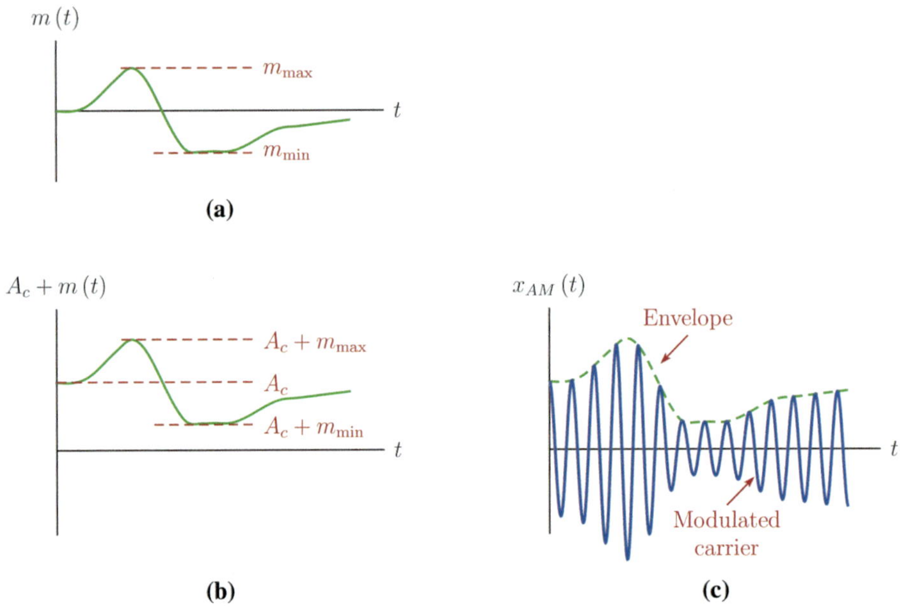

Figure 7.4 – Waveforms involved in amplitude modulation: **(a)** The message signal, **(b)** the message signal with offset A_c, and **(c)** the amplitude-modulated carrier and its envelope.

In this particular example we have assumed that the amplitudes of the message signal are constrained to remain in the interval $m_{\min} \le m(t) \le m_{\max}$. The new amplitude function $[A_c + m(t)]$ will remain in the range

$$A_c + m_{\min} \le \left[A_c + m(t)\right] \le A_c + m_{\max}$$

Furthermore, the relationship between A_c and $m(t)$ is chosen such that $[A_c + m(t)] \ge 0$ for all t. As a result, the envelope of the modulated carrier, that is, the waveform obtained by connecting the positive peaks of $x_{AM}(t)$, matches $[A_c + m(t)]$. This will be important as we consider ways of recovering the message signal $m(t)$ from the AM signal $x_{AM}(t)$.

In contrast, consider the example in Figure 7.5 where $[A_c + m(t)]$ is allowed to be negative for some values of t. In this case the envelope of the AM waveform does not equal $[A_c + m(t)]$, but rather its absolute value $|A_c + m(t)|$.

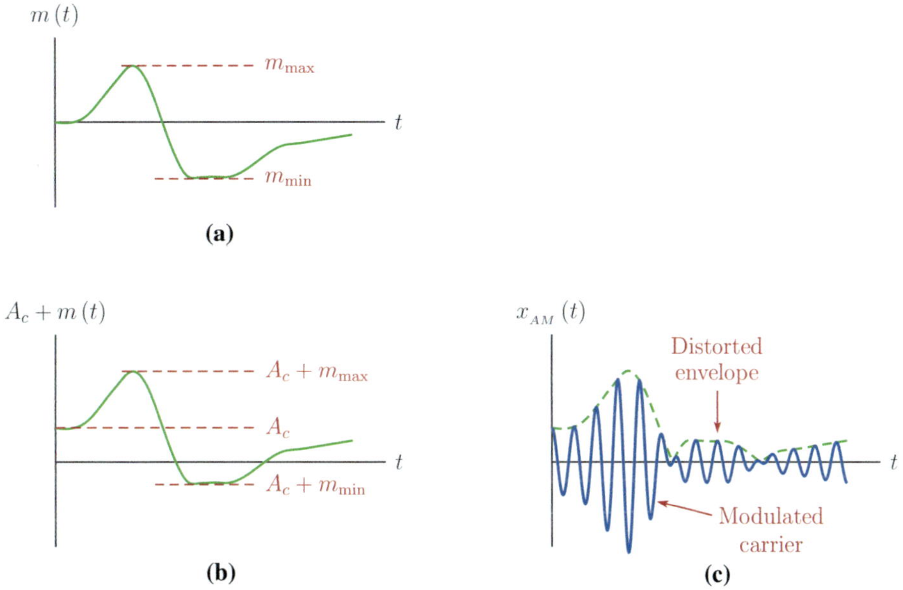

Figure 7.5 – Waveforms involved in amplitude modulation when $[A_c + m(t)] \geq 0$ is not satisfied for all t: **(a)** The message signal, **(b)** the message signal with offset A_c, and **(c)** the amplitude-modulated carrier and its distorted envelope.

Combining the two cases considered above we obtain a general expression for the envelope of the AM signal as

$$x_e(t) = \left|A_c + m(t)\right| \tag{7.4}$$

If $[A_c + m(t)] \geq 0$ for all t, then the envelope is a replica of the message signal $m(t)$ subject to a constant offset. Otherwise, the envelope is a distorted version of the message signal. In order to preserve the shape of the message signal in the envelope of the AM signal, we need

$$A_c + m_{\text{min}} \geq 0 \tag{7.5}$$

Let us define the modulation index as

$$\mu = \frac{|m_{\text{min}}|}{A_c} \tag{7.6}$$

which can be expressed either as a fraction or as a percentage. For the envelope to display the correct shape of the message signal, we need $0 \leq \mu \leq 1$. An amplitude modulated carrier with a modulation index $\mu < 1$ is said to be *under-modulated*. A modulation index of $\mu = 1$ leads to the *critically-modulated* case. Both of these will allow a simple system known as an *envelope detector* to be used in recovering the message signal $m(t)$ from the AM signal $x_{AM}(t)$. On the other hand, if $\mu > 1$, the carrier is said to be *over-modulated*. In this case the envelope is a distorted version of the message signal, and the envelope detector is no longer a viable option for the recovery of $m(t)$. A more complicated synchronous detector will be needed to extract the message.

Interactive App: Effect of modulation index on envelope

The interactive app in `ampMod1.m` uses the signals in Figs. 7.4 and 7.5. The message signal and the blank carrier are shown on the left side of the graphical user interface. The main graph on the right side displays the modulated carrier computed as in Eqn. (7.3), as well as its envelope given by Eqn. (7.4). A slider control is provided for allowing the modulation index (refer to Eqn. (7.6)) to be varied in the range $0.4 \le \mu \le 1.8$. Alternatively, the desired value of the modulation index may be typed into the edit field (the same range restrictions still apply). Note that, in the implementation of this app, the peak amplitudes m_{min} and m_{max} of the message signal are fixed. Change of modulation index μ is facilitated by changing the parameter A_c as dictated by Eqn. (7.6).

a. Set $\mu = 0.5$. Observe the modulated signal and compare it to Fig. 7.4.

b. Gradually increase the modulation index to about $\mu = 0.9$. Observe the changes in the AM signal and its envelope as μ is increased.

c. Keep increasing μ up to about $\mu = 1.2$. Pay attention to the envelope crossover that begins just as μ exceeds 1. Compare the modulated signal to that shown in Fig. 7.5.

Software resource: `ampMod1.m`

Example 7.1: Tone modulation

Sometimes it is insightful to analyze a modulation technique by using a message signal that has a simple analytical description. One particular message signal that can be used in this way is a single tone. Let the modulating signal be

$$m(t) = A_m \cos(2\pi f_m t)$$

Substituting $m(t)$ in Eqn. (7.3) the corresponding AM waveform is

$$x_{AM}(t) = \left[A_c + A_m \cos(2\pi f_m t)\right] \cos(2\pi f_c t) \tag{7.7}$$

which can also be written in the form

$$x_{AM}(t) = A_c \left[1 + \mu \cos(2\pi f_m t)\right] \cos(2\pi f_c t) \tag{7.8}$$

with the modulation index defined as $\mu = A_m / A_c$. The envelope of the modulated carrier is

$$x_e(t) = \left|A_c \left[1 + \mu \cos(2\pi f_m t)\right]\right| \tag{7.9}$$

The tone-modulated carrier is shown in Fig. 7.6 for various values of the modulation index. Notice how the envelope is distorted when $\mu > 1$.

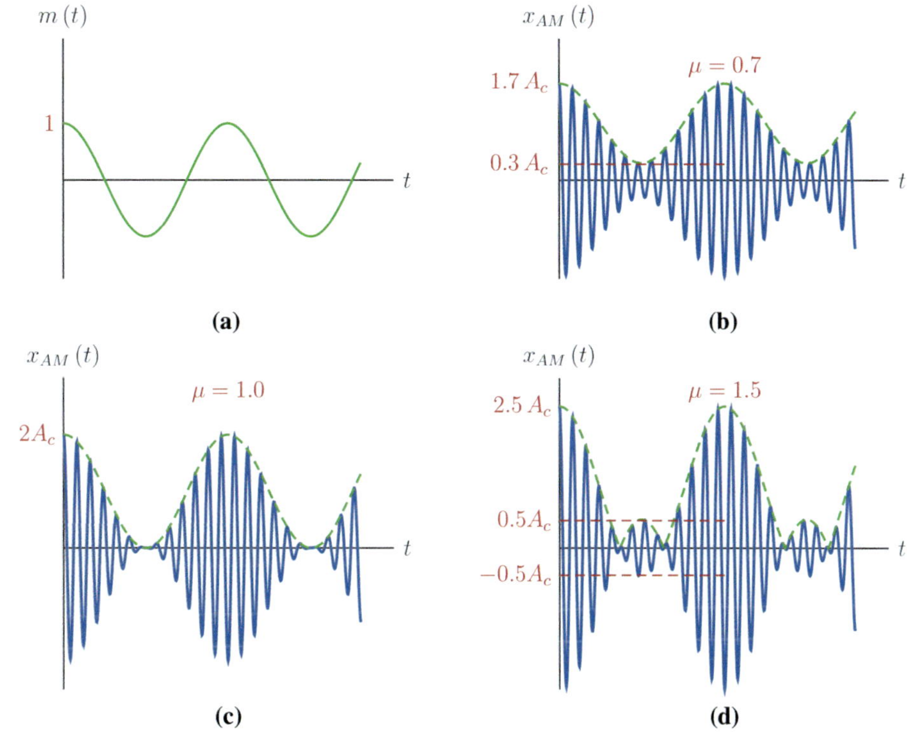

Figure 7.6 – Amplitude modulation using a single-tone message signal in Example 7.1: **(a)** The message signal, **(b)** under-modulated AM signal with $\mu = 0.7$, **(c)** critically-modulated AM signal with $\mu = 1$, and **(d)** over-modulated AM signal with $\mu = 1.5$.

Software resource: `exct_7_1.m`

Interactive App: Amplitude modulation of a carrier with a single tone

The interactive app in `ampMod2.m` illustrates amplitude modulation of a carrier by a single tone as discussed in Example 7.1. Three sliders are provided for varying the message frequency f_m, the carrier frequency f_c, and the modulation index μ. Sliders allow f_m to be changed from 0.5 kHz to 3 kHz, and f_c to be changed from 10 kHz to 25 kHz. Modulation index may be varied between 0.4 and 1.8. It is also possible to vary parameters by typing their values into corresponding edit fields.

a. Set parameter values as $f_m = 1.5$ Hz, $f_c = 10$ Hz, and $\mu = 0.7$. Observe the modulated signal $x_{AM}(t)$ and its envelope $x_e(t)$.

b. Gradually increase the carrier frequency to $f_c = 25$ Hz. Observe the effect of the change on the AM signal and its envelope.

c. With the carrier frequency set at $f_c = 25$ Hz, gradually increase the modulation index toward $\mu = 1.5$. Pay attention to the envelope crossover that begins just as μ exceeds 1. Compare the modulated signal to those shown in Fig. 7.6.

d. Also pay attention to the phase reversal of the carrier at the points of envelope crossover.

Software resource: `ampMod2.m`

Software resource: See MATLAB Exercise 7.1.

7.4.1 Frequency spectrum of the AM signal

Consider again the expression for an amplitude modulated sinusoidal carrier given by Eqn. (7.3), repeated here for convenience:

$$\text{Eqn. (7.3):} \qquad x_{AM}(t) = A_c \cos\left(2\pi f_c t\right) + m(t)\cos\left(2\pi f_c t\right)$$

Taking the Fourier transforms of both sides, we obtain the frequency spectrum of the AM signal as

$$\begin{aligned} X_{AM}(f) =& \frac{1}{2}A_c\,\delta\left(f+f_c\right)+\frac{1}{2}A_c\,\delta\left(f-f_c\right)+ \\ & \frac{1}{2}M\left(f+f_c\right)+\frac{1}{2}M\left(f-f_c\right) \end{aligned} \tag{7.10}$$

where we made use of the modulation theorem of the Fourier transform discussed in Section 3.3.5 of Chapter 3. The first two terms in Eqn. (7.10) are due to the carrier frequency. The term $\delta(f+f_c)$ represents an impulse function shifted to the frequency $f=-f_c$. Similarly, $\delta(f-f_c)$ is an impulse function shifted to the frequency $f=f_c$. The last two terms in Eqn. (7.10) contain shifted versions of the message spectrum. The amount of frequency shift is controlled by the carrier frequency. The term $M\left(f+f_c\right)$ is a left-shifted version of the message spectrum, and $M\left(f-f_c\right)$ a right-shifted one. This is depicted in Fig. 7.7. Note that the shape of the message spectrum $M\left(f\right)$ in Fig. 7.7 is chosen arbitrarily for illustration purposes only; different message signals will have different looking spectra.

For positive frequencies, the part of the AM spectrum that occupies frequencies greater than the carrier frequency f_c is referred to as the *upper sideband.* Conversely, the part of the spectrum that occupies frequencies less than f_c is referred to as the *lower sideband.* For the negative side of the frequency axis, these definitions are mirrored. Another interesting point to observe is that, for positive frequencies, lower and upper sidebands occupy the frequency range from f_c-W to f_c+W. Thus, if the spectrum of the message signal has a one-sided bandwidth of W Hz, then the spectrum of the amplitude-modulated signal has a one-sided bandwidth of $B_T=2W$ Hz, that is, twice the message bandwidth. B_T is called the *transmission bandwidth* of the AM signal.

This illuminates one of the shortcomings of amplitude modulation: It is wasteful of bandwidth. A message signal which occupies a bandwidth of W Hz in its baseband form requires twice that bandwidth when it is embedded into the amplitude of a carrier. We will see in Section 7.6 that *single sideband modulation* provides a solution to the problem of wasting bandwidth. The relationship between the carrier frequency f_c and the message bandwidth W is also important. To avoid overlaps between spectral components when the message spectrum is shifted right and left, we need $f_c > W$, that is, we need the carrier frequency to be greater than the highest frequency in the message spectrum.

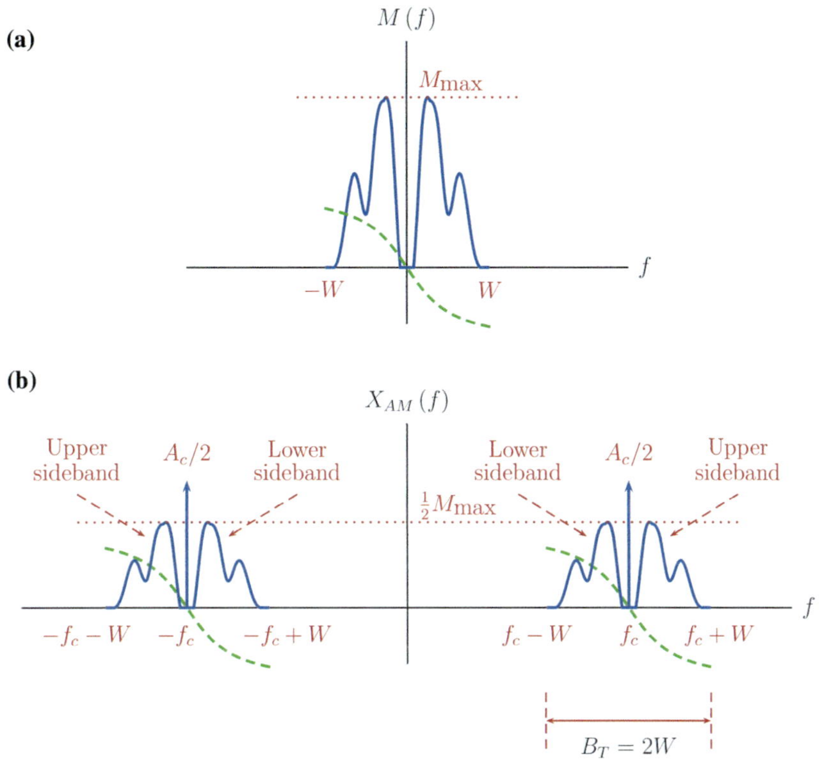

Figure 7.7 – **(a)** Frequency spectrum $M(f)$ of the message signal; magnitude (solid) and phase (dashed), **(b)** frequency spectrum $X_{AM}(f)$ of the amplitude-modulated signal; magnitude (solid) and phase (dashed).

Going one step further, if the envelope of the AM signal is required to be a good-quality representation of the message signal, then we need $f_c >> W$, that is, we need the carrier frequency to be significantly greater than the message bandwidth. In practical applications, f_c is often several magnitudes greater than W.

Interactive App: AM frequency spectrum

The interactive app in `ampMod3.m` illustrates the frequency spectrum of the AM signal, and is based on Fig. 7.7. The sample message spectrum picked for this app has a bandwidth of 1 kHz and a peak magnitude of $\max\{|M(f)|\} = 1$. Amplitude parameter for the carrier is set as $A_c = 2$. As a result, the modulation index is fixed. A slider control allows the carrier frequency to be varied in the range $2\,\text{kHz} \le f_c \le 9\,\text{kHz}$.

a. Set the carrier frequency at $f_c = 2$ kHz. Compare the AM spectrum to that shown in Fig. 7.7.

b. Gradually increase the carrier frequency, and observe the changes in the AM spectrum. Pay attention to the shifting of the spectrum when f_c is changed.

Software resource: `ampMod3.m`

Example 7.2: Frequency spectrum of the tone-modulated carrier

Compute the frequency spectrum of the tone-modulated AM carrier of Example 7.1 in terms of the carrier frequency f_c, the message frequency f_m, the modulation index μ, and the amplitude A_c.

Solution: Recall that the tone-modulated signal is in the form

$$x_{AM}(t) = A_c\left[1+\mu\cos(2\pi f_m t)\right]\cos\left(2\pi f_c t\right)$$

which can also be written as

$$x_{AM}(t) = A_c\cos\left(2\pi f_c t\right) + \mu A_c\cos\left(2\pi f_m t\right)\cos\left(2\pi f_c t\right) \tag{7.11}$$

Applying the appropriate trigonometric identity

$$\cos(a)\cos(b) = \frac{1}{2}\cos(a+b) + \frac{1}{2}\cos(a-b)$$

to the second term of Eqn. (7.11) we obtain

$$\begin{aligned} x_{AM}(t) =& A_c\cos\left(2\pi f_c t\right) + \frac{\mu A_c}{2}\cos\left(2\pi[f_c+f_m]\,t\right) \\ &+ \frac{\mu A_c}{2}\cos\left(2\pi[f_c-f_m]\,t\right) \end{aligned} \tag{7.12}$$

Eqn. (7.12) contains three sinusoidal terms at frequencies f_c, f_c+f_m, and f_c-f_m, respectively. Individual Fourier transforms of the three terms are

$$\begin{aligned} A_c\cos\left(2\pi f_c t\right) &\overset{\mathcal{F}}{\longleftrightarrow} \frac{A_c}{2}\,\delta\left(f+f_c\right) + \frac{A_c}{2}\,\delta\left(f-f_c\right) \\ \frac{\mu A_c}{2}\cos\left(2\pi\left[f_c+f_m\right]t\right) &\overset{\mathcal{F}}{\longleftrightarrow} \frac{\mu A_c}{4}\,\delta\left(f+\left[f_c+f_m\right]\right) + \frac{\mu A_c}{4}\,\delta\left(f-\left[f_c+f_m\right]\right) \\ \frac{\mu A_c}{2}\cos\left(2\pi\left[f_c-f_m\right]t\right) &\overset{\mathcal{F}}{\longleftrightarrow} \frac{\mu A_c}{4}\,\delta\left(f+\left[f_c-f_m\right]\right) + \frac{\mu A_c}{4}\,\delta\left(f-\left[f_c-f_m\right]\right) \end{aligned}$$

Thus, the spectrum of the tone-modulated AM signal is obtained as

$$\begin{aligned} X_{AM}(f) =& \frac{A_c}{2}\,\delta\left(f+f_c\right) + \frac{A_c}{2}\,\delta\left(f-f_c\right) \\ &+ \frac{\mu A_c}{4}\,\delta\left(f+\left[f_c+f_m\right]\right) + \frac{\mu A_c}{4}\,\delta\left(f-\left[f_c+f_m\right]\right) \\ &+ \frac{\mu A_c}{4}\,\delta\left(f+\left[f_c-f_m\right]\right) + \frac{\mu A_c}{4}\,\delta\left(f-\left[f_c-f_m\right]\right) \end{aligned} \tag{7.13}$$

Frequency spectra for the single-tone message signal and the amplitude-modulated signal are shown in Figure 7.8. The first two terms in Eqn. (7.13) are due to the carrier alone. The remaining four terms are the sidebands due to the message signal, and they are around positive and negative carrier frequencies with displacements of $\pm f_m$ in either direction.

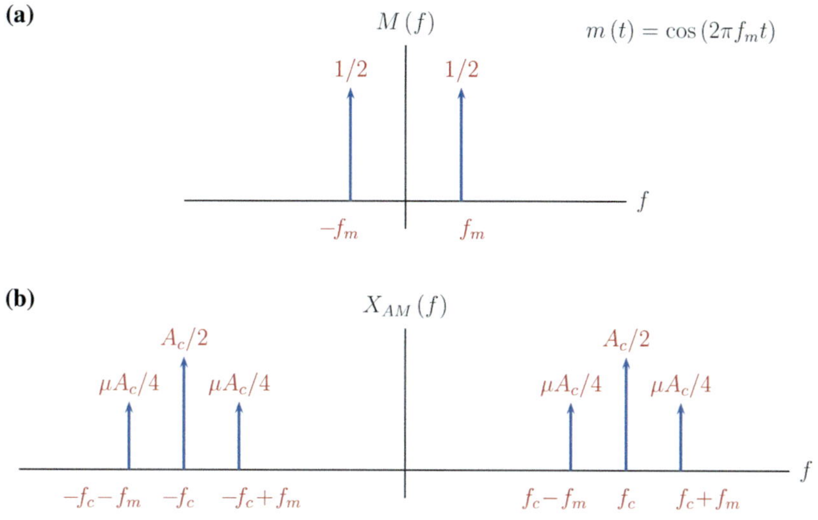

Figure 7.8 – **(a)** Frequency spectrum $M(f)$ of the single-tone message signal and **(b)** frequency spectrum of the tone-modulated AM signal of Example 7.2.

Interactive App: AM spectrum for a single-tone message signal

The interactive app in `ampMod4.m` illustrates the effects of various parameters on the frequency spectrum of the tone-modulated AM signal and is based on Fig. 7.8 and Eqn. (7.13) of Example 7.2. Slider controls can be used for varying the message frequency f_m, the carrier frequency f_c, and the modulation index μ. Sliders allow f_m to be changed from 0.5 kHz to 3 kHz, and f_c to be changed from 10 to 25 kHz. Modulation index may be varied between 0.4 and 1.8. The amplitude is fixed at $A_c = 1$.

a. Start with parameters set at $f_m = 1.5$ kHz, $f_c = 10$ kHz, and $\mu = 0.6$. Gradually increase the carrier frequency f_c, and observe the frequency shifting of the spectrum .

b. Reset the carrier frequency back to $f_c = 10$ kHz. Gradually increase the message frequency f_m, and observe how the sidebands move away from the carrier in both directions.

c. Reset the message frequency back to $f_m = 1.5$ kHz. Gradually increase the modulation index μ, and observe the envelope crossover that occurs in the AM signal as $\mu = 1$ threshold is passed.

Software resource: `ampMod4.m`

Example 7.3: AM spectrum for a multi-tone modulating signal

Determine the frequency spectrum of the AM signal produced by modulating a carrier at frequency $f_c = 1$ kHz with the two-tone signal

$$m(t) = 3\cos\left(2\pi(100)\,t - \pi/6\right) + 2\sin\left(2\pi(200)\,t\right)$$

Use the modulation index $\mu = 0.7$.

Solution: First we need to determine the message spectrum. In order to get the phase relationship between the two terms correctly, the sine function in $m(t)$ needs to be converted to a cosine function. Using the trigonometric identity

$$\sin(a) = \cos(a - \pi/2)$$

the signal $m(t)$ can be written as

$$m(t) = 3\cos\left(2\pi(100)t - \pi/6\right) + 2\cos\left(2\pi(200)t - \pi/2\right) \tag{7.14}$$

Using Euler's formula, we get

$$\begin{aligned} m(t) =& \frac{3}{2}\, e^{j\left(2\pi(100)t - \pi/6\right)} + \frac{3}{2}\, e^{-j\left(2\pi(100)t-\pi/6\right)} \\ &+ e^{j\left(2\pi(200)t-\pi/2\right)} + e^{-j\left(2\pi(200)t-\pi/2\right)} \end{aligned}$$

which yields the spectrum shown in Fig. 7.9.

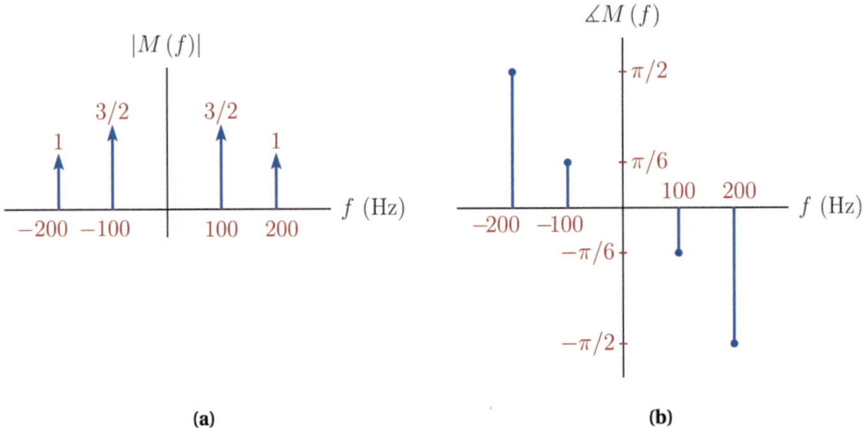

Figure 7.9 – Frequency spectrum $M(f)$ of the two-tone message signal of Example 7.3: **(a)** magnitude and **(b)** phase.

The next step is to determine the carrier amplitude A_c that will yield the desired modulation index value of $\mu = 0.7$. This, in turn, requires the minimum value of the message signal which is found from Eqn. (7.14) to be

$$m_{\text{min}} = -3.2536$$

and occurs at time instant $t = 4.28$ ms as shown in Fig. 7.10.

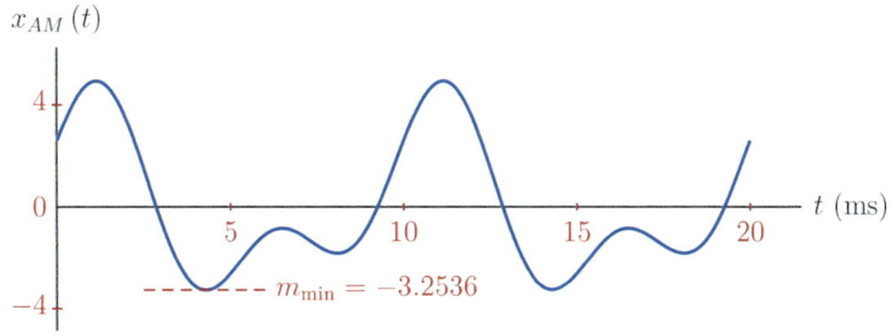

Figure 7.10 – The two-tone message signal of Example 7.3.

Using Eqn. (7.6) the carrier parameter A_c is found as

$$A_c = \frac{|m_{\text{min}}|}{\mu} = \frac{|-3.2536|}{0.7} = 4.648$$

Thus, the AM signal is

$$\begin{aligned} x_{AM}(t) &= \left[A_c + m(t)\right] \cos\left(2\pi f_c t\right) \\ &= \left[4.648 + 3\cos\left(2\pi(100)t - \pi/6\right) + 2\cos\left(2\pi(200)t - \pi/2\right)\right] \cos\left(2\pi(1000)t\right) \end{aligned}$$

and is shown in Fig. 7.11(a). The frequency spectrum of the AM signal can be easily found using Eqn. (7.10) along with the message spectrum in Fig. 7.9. The magnitude and the phase spectra of the AM signal are shown in Fig. 7.11(b),(c).

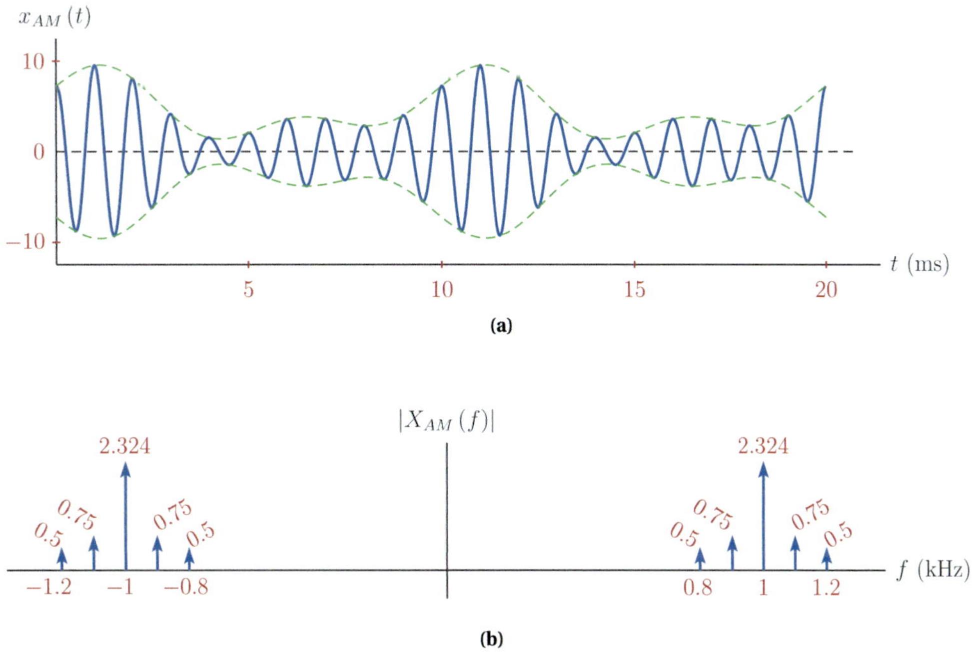

Figure 7.11 – **(a)** the AM signal for Example 7.3, **(b)** the magnitude of the frequency spectrum, and **(c)** the phase of the frequency spectrum.

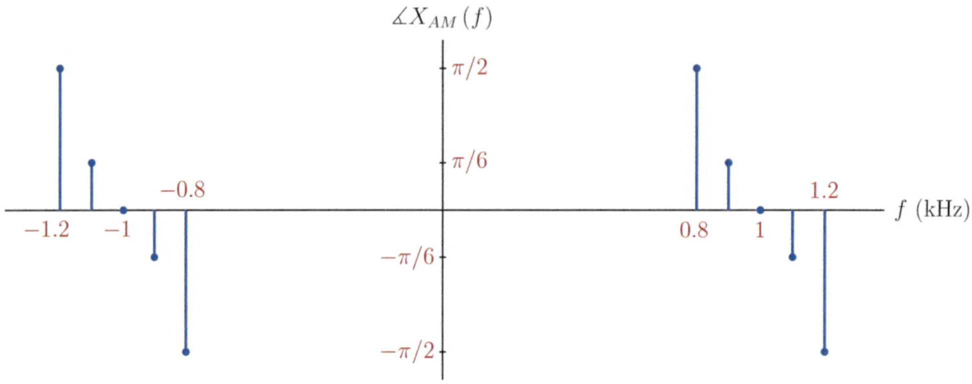

(c)

Figure 7.11 – (Continued)

Software resource: `exct_7_3.m`

Software resource: See MATLAB Exercise 7.2.

Example 7.4: Phasor analysis of the tone-modulated AM signal

The mathematical expression for a carrier with frequency f_c that is amplitude-modulated by a single tone with frequency f_m was derived in Eqn. (7.8) as

$$\text{Eqn. (7.8):} \qquad x_{AM}(t) = A_c\left[1 + \mu\cos\left(2\pi f_m t\right)\right]\cos\left(2\pi f_c t\right)$$

which, using trigonometric identities, can be written as

$$x_{AM}(t) = A_c\cos\left(2\pi f_c t\right) + \frac{\mu A_c}{2}\cos\left(2\pi\left(f_c + f_m\right)\right) + \frac{\mu A_c}{2}\cos\left(2\pi\left(f_c - f_m\right)\right) \tag{7.15}$$

The first term in Eqn. (7.15) is the carrier term. The remaining two terms are the upper sideband and lower sideband terms respectively as we have seen in Fig. 7.8. Let

$$x_{\text{car}}(t) \triangleq A_c\cos\left(2\pi f_c t\right) \tag{7.16}$$

$$x_{\text{usb}}(t) \triangleq \frac{\mu A_c}{2}\cos\left(2\pi\left(f_c + f_m\right)t\right) \tag{7.17}$$

and

$$x_{\text{lsb}}(t) \triangleq \frac{\mu A_c}{2}\cos\left(2\pi\left(f_c - f_m\right)t\right) \tag{7.18}$$

so that the amplitude modulated signal can be written as

$$x_{AM}(t) = x_{\text{car}}(t) + x_{\text{usb}}(t) + x_{\text{lsb}}(t) \tag{7.19}$$

In Section 1.10 of Chapter 1 we have discussed means of representing sinusoidal signals with phasors. The techniques developed in that section can be applied to the three terms of the AM signal in Eqns. (7.16) through (7.18) to obtain

$$x_{\text{car}}(t) = \text{Re}\left\{A_c\, e^{j2\pi f_c t}\right\} \tag{7.20}$$

$$x_{\text{usb}}(t) = \text{Re}\left\{\frac{\mu A_c}{2}\, e^{j2\pi(f_c+f_m)t}\right\} \tag{7.21}$$

$$x_{\text{lsb}}(t) = \text{Re}\left\{\frac{\mu A_c}{2}\, e^{-j2\pi(f_c-f_m)t}\right\} \tag{7.22}$$

If we wanted to apply the phasor notation of Section 1.10 in a strict sense, we would need to express Eqns. (7.20), (7.21) and (7.22) as

$$\boldsymbol{X}_{\text{car}} = A_c \measuredangle 0^\circ \quad \Longrightarrow \quad x_{\text{car}}(t) = \text{Re}\left\{\boldsymbol{X}_{\text{car}}\, e^{j2\pi f_c t}\right\} \tag{7.23}$$

$$\boldsymbol{X}_{\text{usb}} = \frac{\mu A_c}{2} \measuredangle 0^\circ \quad \Longrightarrow \quad x_{\text{usb}}(t) = \text{Re}\left\{\boldsymbol{X}_{\text{usb}}\, e^{j2\pi(f_c+f_m)t}\right\} \tag{7.24}$$

$$\boldsymbol{X}_{\text{lsb}} = \frac{\mu A_c}{2} \measuredangle 0^\circ \quad \Longrightarrow \quad x_{\text{lsb}}(t) = \text{Re}\left\{\boldsymbol{X}_{\text{lsb}}\, e^{j2\pi(f_c-f_m)t}\right\} \tag{7.25}$$

The phasor $\boldsymbol{X}_{\text{car}} = A_c \measuredangle 0^\circ$ rotates counterclockwise at the rate of f_c Hz, that is, f_c revolutions per second. The phasors $\boldsymbol{X}_{\text{usb}} = (\mu A_c/2) \measuredangle 0^\circ$ and $\boldsymbol{X}_{\text{lsb}} = (\mu A_c/2) \measuredangle 0^\circ$ rotate at the rates of $(f_c + f_m)$ and $(f_c + f_m)$ Hz, respectively. It is impractical to work with three phasors when each rotates at a different speed than the others. Instead, we will take a slightly unconventional approach and rewrite Eqns. (7.24) and (7.25) as

$$x_{\text{usb}}(t) = \text{Re}\left\{\boldsymbol{X}_{\text{usb}}\, e^{j2\pi f_m t}\, e^{j2\pi f_c t}\right\} = \text{Re}\left\{\tilde{\boldsymbol{X}}_{\text{usb}}\, e^{j2\pi f_c t}\right\} \tag{7.26}$$

and

$$x_{\text{lsb}}(t) = \text{Re}\left\{\boldsymbol{X}_{\text{lsb}}\, e^{-j2\pi f_m t}\, e^{j2\pi f_c t}\right\} = \text{Re}\left\{\tilde{\boldsymbol{X}}_{\text{lsb}}\, e^{j2\pi f_c t}\right\}$$

The modified phasors

$$\tilde{\boldsymbol{X}}_{\text{usb}} = (\mu A_c/2) \measuredangle 2\pi f_m t \tag{7.27}$$

and

$$\tilde{\boldsymbol{X}}_{\text{lsb}} = (\mu A_c/2) \measuredangle -2\pi f_m t \tag{7.28}$$

have the same base rotation rate of f_c Hz as the phasor $\boldsymbol{X}_{\text{car}}$. This allows us to add the three phasors together as vectors. In addition to the base rotation at the rate of f_c Hz, the modified phasors $\tilde{\boldsymbol{X}}_{\text{usb}}$ and $\tilde{\boldsymbol{X}}_{\text{lsb}}$ each have a secondary rotation at the rate of $\pm f_m$ Hz relative to the carrier phasor $\boldsymbol{X}_{\text{car}}$. This is illustrated in Fig. 7.12 for a modulation index of $\mu < 1$. The phasor $\boldsymbol{X}_{\text{car}}$ is drawn starting at the origin. The phasors $\tilde{\boldsymbol{X}}_{\text{usb}}$ and $\tilde{\boldsymbol{X}}_{\text{lsb}}$ are shown attached to the tip of the phasor $\boldsymbol{X}_{\text{car}}$, and they rotate around the tip of $\boldsymbol{X}_{\text{car}}$ with speeds of $\pm f_m$. The entire system of three phasors rotates around the origin at a speed of f_c. The AM signal can be written as

$$x_{AM}(t) = \text{Re}\left\{\left(\boldsymbol{X}_{\text{car}} + \tilde{\boldsymbol{X}}_{\text{usb}} + \tilde{\boldsymbol{X}}_{\text{lsb}}\right) e^{j2\pi f_c t}\right\} \tag{7.29}$$

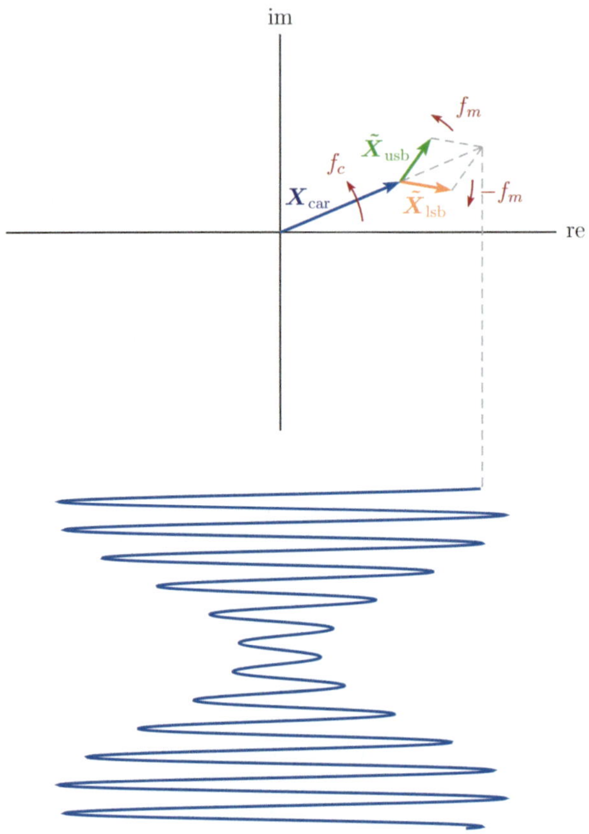

Figure 7.12 – Phasor representation of the tone-modulated AM signal.

One interesting use of the phasor representation derived above is in finding the envelope of the AM signal. In the time domain, we know that the envelope of the AM signal is obtained by connecting the positive peaks of the high-frequency carrier. The equivalent of this action in phasor domain would be to freeze the base rotation (around the origin) of the system of three phasors at a 0 degree angle, and allow only the slower rotations of the phasors $\tilde{\boldsymbol{X}}_{usb}$ and $\tilde{\boldsymbol{X}}_{lsb}$ at the rates $\pm f_m$ Hz around the tip of $\boldsymbol{X}_{car}$. The use of this idea for determining the envelope is shown in Fig. 7.13. The resulting envelope is

$$\begin{aligned} x_e(t) =& \mathrm{Re}\left\{A_c + \frac{\mu A_c}{2}\, e^{j2\pi f_m t} + \frac{\mu A_c}{2}\, e^{-j2\pi f_m t}\right\} \\ =& A_c + \mu A_c \cos\left(2\pi f_m t\right) \end{aligned}$$

which is consistent with the expression found earlier for $\mu < 1$.

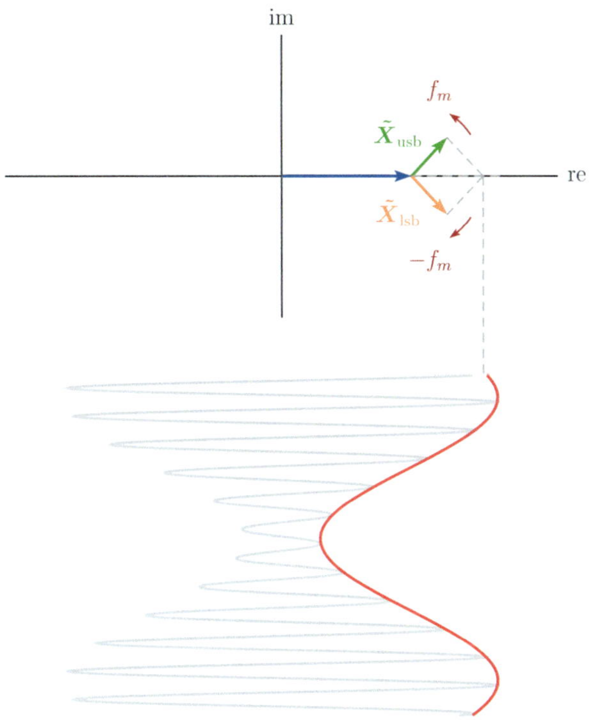

Figure 7.13 – Using phasor representation to determine the envelope of the AM signal.

Interactive App: Phasor representation of tone-modulated AM signal

The interactive app in `ampMod5.m` illustrates the concepts discussed in Example 7.4. The system of three phasors $\boldsymbol{X}_{\text{car}}$, $\tilde{\boldsymbol{X}}_{\text{usb}}$, and $\tilde{\boldsymbol{X}}_{\text{lsb}}$ that make up the phasor description of the AM signal is shown in the upper right part of the graphical user interface. The carrier frequency f_c, the single-tone message frequency f_m, and the modulation index μ of the AM signal may each be controlled using the appropriate slider control. The time variable may be advanced using a slider control, and the rotation of the system of phasors may be observed. Pay particular attention to

a. The base rotation around the origin, and the slower rotation of the two sideband phasors $\tilde{\boldsymbol{X}}_{\text{usb}}$ and $\tilde{\boldsymbol{X}}_{\text{lsb}}$ around the tip of the carrier phasor $\boldsymbol{X}_{\text{car}}$, and

b. How the real part of the vector sum of three phasors corresponds to the amplitude of the AM signal $x_{AM}(t)$ shown in the graph at the bottom.

Software resource: `ampMod5.m`

Interactive App: Using phasors to find the envelope of tone-modulated AM signal

The interactive app in `ampMod6.m` shows how the system of three phasors discussed in Example 7.4 relates to the envelope of the tone-modulated AM signal. In order to see just the envelope without the carrier, the base rotation of the system of phasors is stopped, that is, the carrier phasor $\boldsymbol{X}_{\text{car}}$ lays on the horizontal axis without rotating. The sideband phasors $\tilde{\boldsymbol{X}}_{\text{usb}}$ and $\tilde{\boldsymbol{X}}_{\text{lsb}}$ rotate with the frequency f_m. In this case, the real part of the vector sum of three phasors corresponds to the envelope of the AM signal.

Software resource: `ampMod6.m`

7.4.2 Power balance and modulation efficiency

In the AM signal of Eqn. (7.3), the first term $A_c \cos(2\pi f_c t)$, referred to as the *carrier component*, is independent of the message signal. The sole purpose of transmitting the carrier component along with the message-dependent second term is to ensure that the envelope of the resulting signal resembles the message signal. From a power perspective, the transmission of the carrier component represents a waste of power. In contrast, the second term in Eqn. (7.3) includes the message signal $m(t)$, and thus represents the useful component of the AM signal. It would be interesting to find what percentage of the power in the transmitted AM signal is useful, and what percentage is wasted. The average power of the AM signal is

$$P_x = \langle x_{AM}^2(t) \rangle = \Big\langle \left[A_c + m(t)\right]^2 \cos^2\left(2\pi f_c t\right) \Big\rangle \tag{7.30}$$

where the $\langle \ldots \rangle$ operator indicates time-averaging as we have discussed in Section 1.8 of Chapter 1. Using the appropriate trigonometric identity[2], Eqn. (7.30) can be written as

$$P_x = \left\langle \frac{1}{2} \left[A_c + m(t)\right]^2 \right\rangle + \left\langle \frac{1}{2} \left[A_c + m(t)\right]^2 \cos\left(4\pi f_c t\right) \right\rangle \tag{7.31}$$

Assuming that the carrier frequency is significantly greater than the bandwidth of the message signal, the second term in Eqn. (7.31) averages to zero, and we have

$$\begin{aligned} P_x &= \left\langle \frac{1}{2} \left[A_c + m(t)\right]^2 \right\rangle \\ &= \left\langle \frac{1}{2} A_c^2 + A_c\, m(t) + \frac{1}{2} m^2(t) \right\rangle \\ &= \frac{1}{2} A_c^2 + A_c \langle m(t) \rangle + \frac{1}{2} \langle m^2(t) \rangle \end{aligned} \tag{7.32}$$

For most message signals $\langle m(t) \rangle = 0$, and Eqn. (7.32) reduces to

$$P_x = \frac{1}{2} A_c^2 + \frac{1}{2} \langle m^2(t) \rangle = P_c + P_m \tag{7.33}$$

The first term in Eqn. (7.33), $P_c = \frac{1}{2} A_c^2$, is the power in the carrier component, that is, the wasted power. In contrast, the second term $P_m = \frac{1}{2} \langle m^2(t) \rangle$ is the power in the sidebands, or the useful

[2] $\cos^2(a) = \frac{1}{2}\left[1 + \cos(2a)\right]$

power. We will define the modulation efficiency as the percentage of the useful power within the total power of the AM signal:

$$\eta = \frac{\text{Power in sidebands}}{\text{Total power}} = \frac{P_m}{P_c + P_m} = \frac{\left\langle m^2(t) \right\rangle}{A_c^2 + \left\langle m^2(t) \right\rangle} \tag{7.34}$$

As expected, modulation efficiency depends on the significance of A_c^2 next to $\langle m^2(t) \rangle$. For practical message signals, lower and upper bounds of the allowable amplitude range are often symmetric, that is, $|m_{\min}| = |m_{\max}|$. For the case $\mu \le 1$, this would lead to

$$\left| m_{\min} \right|, \left| m_{\max} \right| \le A_c \tag{7.35}$$

As a result, the numerator term is bound by

$$\langle m^2(t) \rangle \le A_c^2 \tag{7.36}$$

with the equality achieved only when the message signal is in the form of a square-wave type signal with amplitudes of $\pm A_c$ only. Thus, the maximum theoretical efficiency that can be attained with amplitude modulation (when the envelope is preserved) is $\eta = 0.5$, or 50 percent. Actual efficiency values obtained in practice will be significantly lower.

In summary, amplitude modulation is wasteful of power. This is the second major shortcoming of amplitude modulation besides the issue of inefficient bandwidth utilization discussed in the previous section.

Example 7.5: Efficiency in tone modulation

Determine the modulation efficiency for the tone-modulated AM signal of Example 7.1.

Solution: Recall that the modulating signal was given by

$$m(t) = \mu A_c \cos\left(2\pi f_m t\right)$$

Thus, the average normalized power of the message signal is

$$\left\langle m^2(t) \right\rangle = \left\langle \mu^2 A_c^2 \cos^2\left(2\pi f_m t\right) \right\rangle = \frac{\mu^2 A_c^2}{2}$$

Using this result in Eqn. (7.34) we find the modulation efficiency of a tone-modulated AM signal to be

$$\eta = \frac{\mu^2}{2 + \mu^2}$$

In the case of tone-modulation, the maximum efficiency that can be achieved while preserving the signal envelope is to be $\eta = 0.333$, or 33.3 percent for the critically-modulated case $\mu = 1$. For smaller values of μ the efficiency will be even lower.

Example 7.6: Multi-tone modulating signal revisited

Determine the power efficiency of the AM signal of Example 7.3.

Solution: Recall that the AM signal obtained in Example 7.3 was

$$x_{AM}(t) = \left[4.648 + 3\cos\left(2\pi(100)t - \pi/6\right) + 2\cos\left(2\pi(200)t - \pi/2\right)\right]\cos\left(2\pi(1000)t\right)$$

The average normalized power in the message signal is

$$\begin{aligned}\left\langle m^2(t)\right\rangle &= \left\langle 9\cos^2\left(2\pi(100)t - \pi/6\right)\right\rangle + \left\langle 4\sin^2\left(2\pi(200)t\right)\right\rangle \\ &= \frac{9}{2} + \frac{4}{2} = 6.5\end{aligned}$$

The average normalized power of the AM signal including the carrier is

$$\begin{aligned}\left\langle x_{AM}^2(t)\right\rangle &= \left\langle A_c^2\cos^2\left(2\pi(1000)t\right)\right\rangle + \left\langle m^2(t)\right\rangle \\ &= \frac{A_c^2}{2} + \left\langle m^2(t)\right\rangle \\ &= \frac{(4.648)^2}{2} + 6.5 = 17.3\end{aligned}$$

Power efficiency is found as the ratio of the two results:

$$\eta = \frac{6.5}{17.3} = 0.3756 = 37.6 \text{ percent}$$

7.4.3 Generation of AM signals

Consider again the basic form of the AM signal given by Eqn. (7.3). Theoretically it is possible to produce this signal by using a *product modulator* as shown in Figure 7.14.

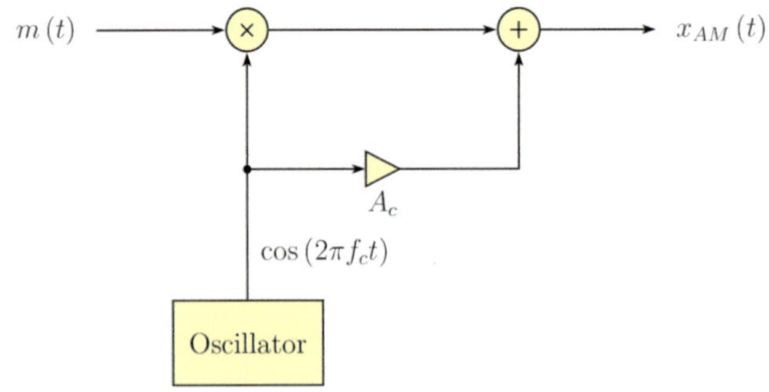

Figure 7.14 – Product modulator block diagram for generating AM signal.

The system uses an analog multiplier, an analog adder, and an amplifier. An oscillator generates the sinusoidal carrier signal $\cos\left(2\pi f_c t\right)$. The message signal and the carrier signal are first multiplied to obtain the second term of Eqn. (7.3). Afterward, a properly scaled version of the carrier signal is added into the mix to complete the AM signal. The adder and the amplifier in Figure 7.14 can be implemented easily, however, multiplication of two analog signals in time is an operation than is only practical at relatively low power levels. Radio broadcast stations produce AM signals at power levels typically in the 10 kW to 50 kW range. Implementation of analog multipliers is impractical at such power levels, hence we need to look into alternative means of generating high-power AM signals. We will consider two of these alternatives, namely a switching modulator and a square-law modulator.

Switching modulator

Consider the arrangement shown in Fig. 7.15 where the carrier signal $x_c(t) = B_c \cos(2\pi f_c t)$ and the message signal $m(t)$ are added together, and their sum is applied to a diode rectifier that consists of a diode and a resistor.

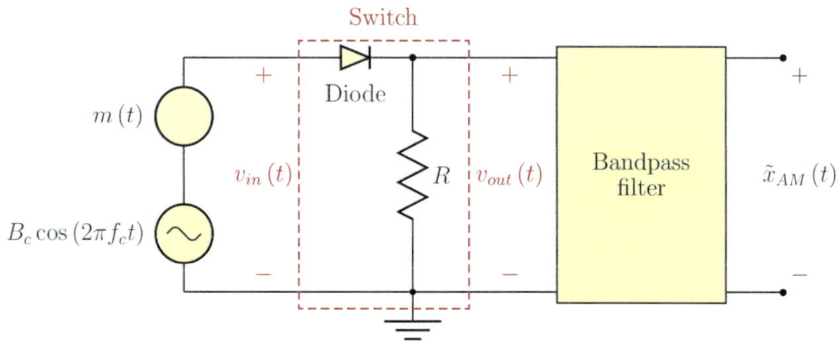

Figure 7.15 – Switching modulator circuit.

Ideally, the role of the diode rectifier would be to replace negative signal amplitudes with zero while keeping positive amplitudes unchanged. It is assumed that the amplitude of the sum signal is large enough for the diode to be modeled as an ideal switch. Furthermore, we will ensure that the amplitude of the carrier term within the sum is significantly greater than the message term, that is, $B_c >> m(t)$, so that the switching action of the diode and resistor combination is controlled largely by the carrier term alone. With these assumptions, the combination of the diode and the load resistor can be modeled as an ideal switch with the characteristic as shown in Fig. 7.16

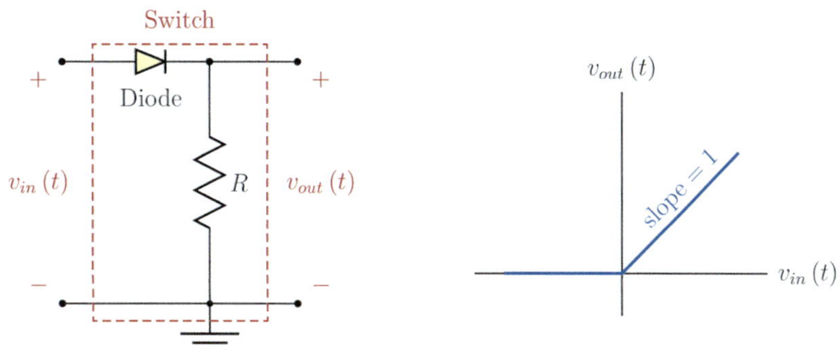

Figure 7.16 – Diode-based switch and its input-output characteristic.

The input to the switching circuit with the diode and the load resistor is

$$v_{in}(t) = B_c \cos\left(2\pi f_c t\right) + m(t) \tag{7.37}$$

and the output voltage of the switching circuit is

$$v_{out}(t) = \begin{cases} v_{in}(t), & \text{if } x_c(t) \ge 0 \\ 0, & \text{otherwise} \end{cases} \tag{7.38}$$

The assumption $B_c >> m(t)$ ensures that the switching action in Eqn. (7.38) is controlled by $x_c(t)$, and the effect of the message signal $m(t)$ on switching is negligibly small. Mathematically, the switching relationship in Eqn. (7.38) is equivalent to multiplying $v_i(t)$ with a pulse train:

$$v_{out}(t) = v_{in}(t)\, p(t) \tag{7.39}$$

This relationship is depicted in Fig. 7.17.

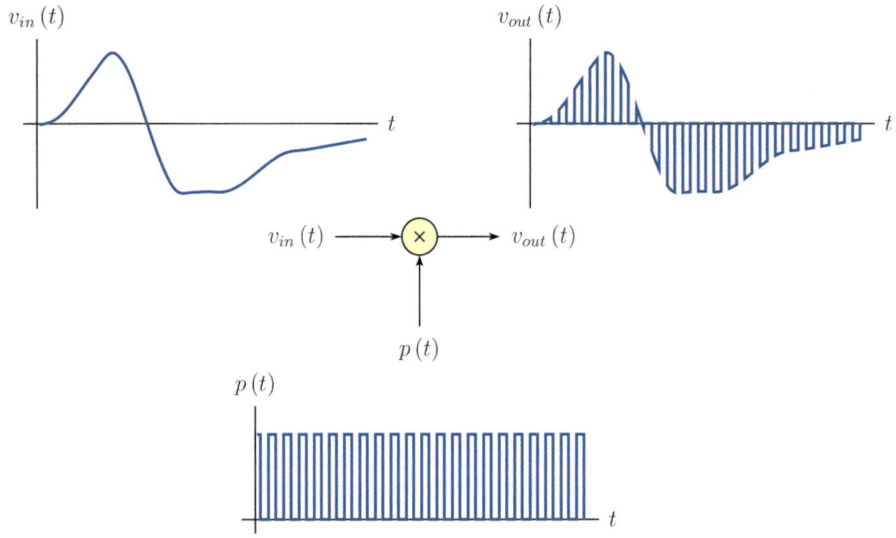

Figure 7.17 – Modeling the switching circuit through multiplication by a pulse train.

The pulse train $p(t)$ is periodic with the same period $T_c = 1/f_c$ as the carrier, and it has a duty cycle of $d = 0.5$. (The assumption $B_c >> m(t)$ that we made earlier was needed for this purpose.) An alternative yet equivalent perspective is given in Fig. 7.18 where the relationship in Eqn. (7.39) is implemented in the form of an electronic switch controlled by the pulse train $p(t)$.

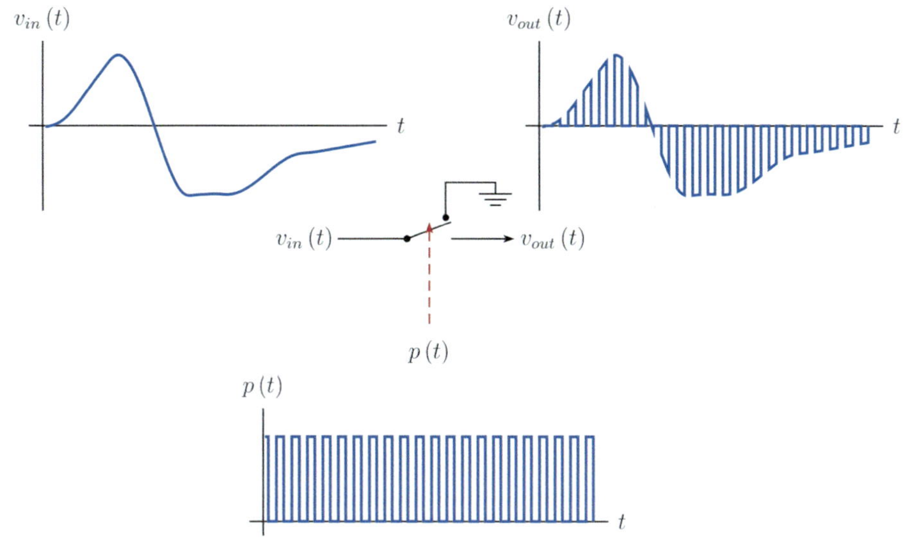

Figure 7.18 – Switching modulator circuit using a switch controlled by a pulse-train.

For the duration of each pulse, that is, when $p(t) = 1$, the switch is in the top position, connecting the signal $v_{in}(t)$ to the input of the bandpass filter. Between pulses, the switch is in the lower position, grounding the input of the bandpass filter. The trigonometric Fourier series (TFS) representation of the pulse train $p(t)$ in Eqn. (7.39) is (see Section 3.2.2 of Chapter 3)

$$p(t) = a_0 + \sum_{k=1}^{\infty} a_k \cos(2\pi k f_c t) \tag{7.40}$$

Note that the fundamental frequency in Eqn. (7.40) is the same as the carrier frequency f_c. The coefficients of the Fourier series expansion are found as

$$a_0 = \frac{1}{2} \tag{7.41}$$

and

$$\begin{aligned} a_k &= \frac{2}{\pi n} \sin\left(\frac{\pi n}{2}\right), \quad k = 1, 2, \ldots \\ &= \left\{\frac{2}{\pi},\ 0,\ -\frac{2}{3\pi},\ 0,\ \frac{2}{5\pi}, \ldots\right\}, \quad k = 1, 2, \ldots \end{aligned} \tag{7.42}$$

Using Eqns. (7.40) through (7.42) along with Eqn. (7.38), we can write Eqn. (7.39) as

$$\begin{aligned} v_{out}(t) = &\left[B_c \cos(2\pi f_c t) + m(t)\right] \\ &\left[\frac{1}{2} + \frac{2}{\pi}\cos(2\pi f_c t) - \frac{2}{3\pi}\cos(6\pi f_c t) + \frac{2}{5\pi}\cos(10\pi f_c t) - \ldots\right] \end{aligned} \tag{7.43}$$

or as

$$\begin{aligned} v_{out}(t) = &\frac{B_c}{\pi} + \frac{1}{2} m(t) + \frac{B_c}{2}\cos(2\pi f_c t) + \frac{2}{\pi} m(t)\cos(2\pi f_c t) + \\ &\{\text{high frequency terms}\} \end{aligned} \tag{7.44}$$

The purpose of the bandpass filter in Fig. 7.15 becomes apparent from Eqn. (7.44). The first two terms are at low frequencies: B_c/π is a constant and $\frac{1}{2}\,m(t)$ has frequencies extending up to W Hz. The next two terms correspond to an AM signal, and they are the terms we need to keep so that we have

$$\tilde{x}_{AM}(t) = \frac{B_c}{2}\cos\left(2\pi f_c t\right) + \frac{2}{\pi}\,m(t)\cos\left(2\pi f_c t\right) \tag{7.45}$$

The remaining terms in Eqn. (7.43) are at higher frequencies. The bandpass filter is needed to eliminate the undesired low-frequency and high-frequency terms. It needs to be designed such that the content in the frequency range

$$f_c - W \le f \le f_c + W$$

is kept, and the frequencies outside that range are eliminated. Also, in order to avoid any overlap between the spectrum of the AM signal and the spectra of low-frequency terms, we need $f_c > 2W$. This is not a very restricting requirement since, in most practical cases, $f_c >> W$. In Eqn. (7.45), the parameter B_c can be adjusted to obtain the desired modulation index. Comparing Eqn. (7.45) with Eqn. (7.8) it is obvious that $A_c = B_c/2$, and the modulation index can be determined by

$$\mu = \frac{(2/\pi)\max\left\{|m(t)|\right\}}{B_c/2} = \frac{4\max\left\{|m(t)|\right\}}{\pi B_c} \tag{7.46}$$

Interactive App: Switching modulator simulation

The interactive app in `ampMod7.m` implements a simulation of the switching modulator discussed above. Implementation is based on Eqns. (7.37) through (7.46), using a single-tone modulating signal $m(t) = \cos\left(2\pi f_m t\right)$. The parameters B_c, f_c, and f_m are adjustable through the use of slider controls in the interface. The undesired terms of Eqn. (7.44) are filtered out using a fifth order Butterworth bandpass filter. The passband edge frequencies f_1 and f_2 of this filter (refer to MATLAB Example 7.3) are controlled indirectly by means of a center frequency f_b and a passband width f_v so that the lower edge of the passband is at

$$f_1 = f_b - \frac{1}{2}f_v$$

and its higher edge is at

$$f_2 = f_b + \frac{1}{2}f_v$$

It should be understood that, by "edge frequency" for a Butterworth filter, we are referring to a frequency where the magnitude of the system function decreases from its peak value by 3 dB.

The modulation index that results from this process is given by Eqn. (7.46). For the single-tone modulating signal used in this case $\max\{|m(t)|\} = 1$, and Eqn. (7.46) reduces to

$$\mu = \frac{4}{\pi B_c}$$

which is displayed as the parameter B_c is varied. The signals $v_{in}(t)$, $v_{out}(t)$, and $\tilde{x}_{AM}(t)$ are computed and graphed. Additionally, the ideal envelope

$$x_e(t) = \frac{B_c}{2}\left[1 + \mu\, m(t)\right]$$

is displayed superimposed with $\tilde{x}_{AM}(t)$ for comparison. Vary the simulation parameters and observe the following:

a. The relationship between the modulation index μ and the carrier amplitude B_c,
b. The summing of the carrier and the message signals to obtain the signal $v_{in}(t)$,
c. The relationship between the signals $v_{in}(t)$ and $v_{out}(t)$,
d. The time lag between the ideal envelope and the synthesized AM signal that is due to the Butterworth bandpass filter,
e. The effect of the edge frequencies of the Butterworth filter on the quality of the synthesized signal and on the time lag,
f. The relationship between f_c, f_m, and the bandpass filter edge frequencies as dictated by Eqn. (7.44).

Software resource: `ampMod7.m`

Software resource: See MATLAB Exercises 7.3 and 7.4.

Square-law modulator

The basis of a square-law modulator is a nonlinear device with a strong second-order nonlinearity. Consider an ideal second-order nonlinear device with the following input-output characteristic:

$$v_{out}(t) = a\, v_{in}(t) + b\, v_{in}^2(t) \tag{7.47}$$

In practice, a diode or a transistor operating in small-signal mode would produce a characteristic that is fairly close to Eqn. (7.47). As in the case of the switching modulator, let the input voltage to this nonlinear device be the sum of the carrier and the message signal:

$$v_{in}(t) = B_c \cos(2\pi f_c t) + m(t) \tag{7.48}$$

Using Eqn. (7.48) in Eqn. (7.47), the output voltage of the nonlinear device is found to be

$$v_{out}(t) = a\left[B_c \cos(2\pi f_c t) + m(t)\right] + b\left[B_c \cos(2\pi f_c t) + m(t)\right]^2 \tag{7.49}$$

which can be written as

$$\begin{aligned} v_{out}(t) =& a B_c \cos(2\pi f_c t) + a\, m(t) + b B_c^2 \cos^2(2\pi f_c t) + \\ & b\, m^2(t) + 2b B_c\, m(t) \cos(2\pi f_c t) \end{aligned} \tag{7.50}$$

In Eqn. (7.50) we recognize the first and the last terms as the ones that we need for an amplitude modulated carrier. The three terms in between are the terms that we will need to eliminate. Let us rearrange Eqn. (7.50) by combining the useful terms together to obtain

$$\begin{aligned} v_{out}(t) =& a B_c \left[1 + \frac{2b}{a}\, m(t)\right] \cos(2\pi f_c t) \\ & + a\, m(t) + b B_c^2 \cos^2(2\pi f_c t) + b\, m^2(t) + 2b B_c\, m(t) \cos(2\pi f_c t) \end{aligned} \tag{7.51}$$

It is obvious that the first term is the AM signal with $A_c = aB_c$ and $\mu = 2b/a$. The remaining terms need to be eliminated to yield an amplitude-modulated signal

$$\tilde{x}_{AM}(t) = aB_c\left[1+\frac{2b}{a}\,m(t)\right]\cos\left(2\pi f_c t\right) \tag{7.52}$$

In order to find a way to eliminate the undesired terms, we need to understand what range of frequencies each term occupies in the frequency spectrum of $v_{out}(t)$. Assuming that the message signal $m(t)$ has a bandwidth of W, frequency ranges of individual terms of Eqn. (7.51) are detailed in Table 7.1.

Table 7.1 – Terms in Eqn. (7.51) and the frequencies they occupy.

Index	Term	Frequencies
1	$aB_c\left[1+(2b/a)\,m(t)\right]\cos\left(2\pi f_c t\right)$	$f_c - W \le f \le f_c + W$
2	$a\,m(t)$	$-W \le f \le W$
3	$b\,B_c^2\cos^2\left(2\pi f_c t\right)$	$f = 0$, and $f = 2f_c$
4	$b\,m^2(t)$	$-2W \le f \le 2W$

The frequency spectrum of $v_{out}(t)$ is shown in Fig. 7.19. The terms of the spectrum are numbered based on their indices in Table 7.1.

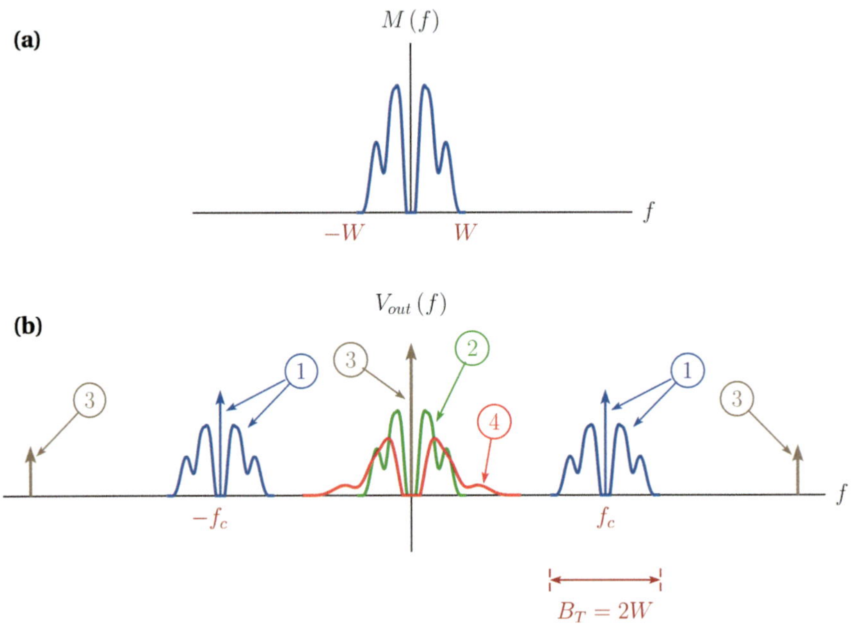

Figure 7.19 – Frequency spectrum of the signal v_{out} of Eqn. (7.51.

An inspection of the spectrum in Fig. 7.19 reveals that the amplitude modulated carrier may be isolated and the undesired terms may be removed by processing the signal $v_{out}(t)$ through a bandpass filter. Thus, the square-law modulator is completed by connecting an appropriate bandpass filter to the output of the nonlinear device. A block diagram for it is shown in Fig. 7.20.

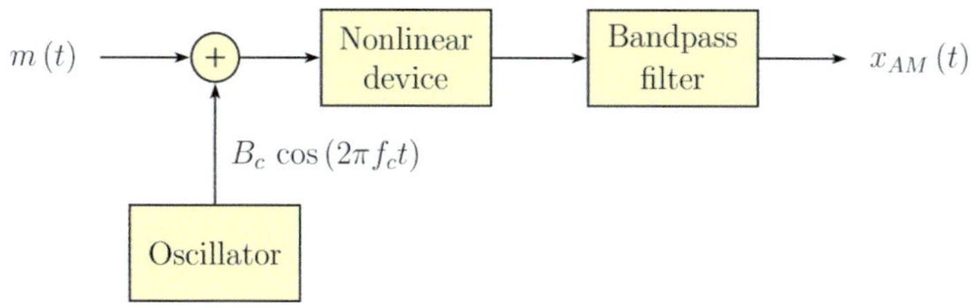

Figure 7.20 – Square-law modulator.

Interactive App: Square-law modulator simulation

The interactive app `ampMod8.m` provides a simulation of the square-law modulator considered in Fig. 7.20, Eqns. (7.47) through (7.51). As in the app `ampMod7.m`, a single-tone modulating signal $m(t) = \cos\left(2\pi f_m t\right)$ is used. Parameters encountered in square-law modulator discussion above, namely f_c, f_m, a and b may be adjusted using the slider controls provided. The signals $v_{in}(t)$ and $v_{out}(t)$ are computed by direct application of Eqns. (7.48) and (7.49), respectively. Afterward, the signal $\tilde{x}_{AM}(t)$ is computed by processing the signal $v_{out}(t)$ through a fifth-order Butterworth bandpass filter with user-specified critical frequencies f_1 and f_2 which are again specified indirectly by means of parameters f_b and f_v as

$$f_1 = f_b - \frac{1}{2} f_v$$

and its higher edge is at

$$f_2 = f_b + \frac{1}{2} f_v$$

The effect of the parameter B_c on the modulator output signal $\hat{x}_{AM}(t)$ is relatively insignificant; therefore this parameter is fixed as $B_c = 1$. The three signals $v_{in}(t)$, $v_{out}(t)$, and $\tilde{x}_{AM}(t)$ are graphed. The modulation index for the synthesized signal is given by $\mu = 2b/a$, and is displayed as the parameters a and b are varied. The ideal envelope

$$x_e(t) = aB_c\left[1 + \mu\, m(t)\right]$$

is displayed superimposed with $\tilde{x}_{AM}(t)$ for comparison. Vary the simulation parameters and observe the following:

a. The relationship between the modulation index μ and the parameters a and b of the nonlinear characteristic,

b. The relationship between the signals $v_{in}(t)$ and $v_{out}(t)$,

c. The time lag between the ideal envelope and the synthesized AM signal that is due to the Butterworth bandpass filter,

d. The effect of the edge frequencies of the Butterworth filter on the quality of the synthesized signal and on the time lag,

e. The relationship between f_c, f_m, and the bandpass filter edge frequencies as dictated by Table (7.1).

Software resource: `ampMod8.m`

Software resource: See MATLAB Exercises 7.5 and 7.6.

7.4.4 Demodulation of AM signals

The term *demodulation* refers to the act of recovering the message signal $m(t)$ as closely to its original form as possible from the modulated waveform $x_{AM}(t)$. It was illustrated in Figs. 7.4 and 7.5 that, as long as the modulation index is less than 100 percent, the envelope of the modulated carrier resembles the message signal. In that case the simplest means of demodulation is by using an envelope detector.

Envelope detector

Consider the simple circuit that consists of a diode, a capacitor, and a resistor as shown in Fig. 7.21.

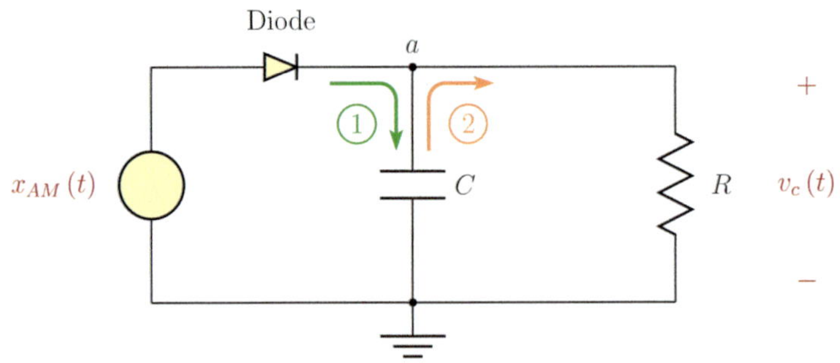

Figure 7.21 – Diode-based envelope detector circuit.

The modulated carrier is applied to this circuit in the form of a voltage signal. The output voltage approximately tracks the envelope of the AM waveform as we will illustrate. Two cases need to be considered:

Case 1: $x_{AM}(t) \ge v_c(t)$

At time instants when the amplitude of the AM waveform is greater than or equal to the capacitor voltage, the diode of the envelope detector circuit is forward biased, and it conducts current. If we assume an ideal diode with no voltage drop between its terminals, the capacitor charges quickly to maintain $v_c(t) = x_{AM}(t)$. This corresponds to current path labeled (1) in Fig. 7.21.

Case 2: $x_{AM}(t) < v_c(t)$

At time instants when the amplitude of the AM waveform is less than the capacitor voltage, the diode becomes reverse biased, and therefore no current flows through it. The capacitor that was charged to the peak value of the carrier must now discharge through the resistor. The discharge current follows the path labeled (2) in Fig. 7.21. Let the peak value of the carrier be $v_c(t_0) = V_0$ just before the capacitor begins its discharge at time $t = t_0$. We know that the current of the capacitor is related to its voltage by $i_c(t) = C\, dv_c(t)/dt$. Writing Kirchhoff's current law at the node labeled "a", we have

$$C\frac{dv_c(t)}{dt} + \frac{v_c(t)}{R} = 0 \tag{7.53}$$

which leads to the differential equation that governs the discharge of the capacitor:

$$\frac{dv_c(t)}{dt} = -\frac{1}{RC}\, v_c(t) \tag{7.54}$$

subject to the initial condition $v_c(t_0) = V_0$. The solution of this differential equation is in the form

$$v_c(t) = V_0 e^{-(t-t_0)/RC} \tag{7.55}$$

where V_0 is the initial value at the beginning of the discharge cycle, that is, the peak value of the carrier at the point it switches from case 1 to case 2. This behavior of the envelope detector is illustrated in Fig. 7.22.

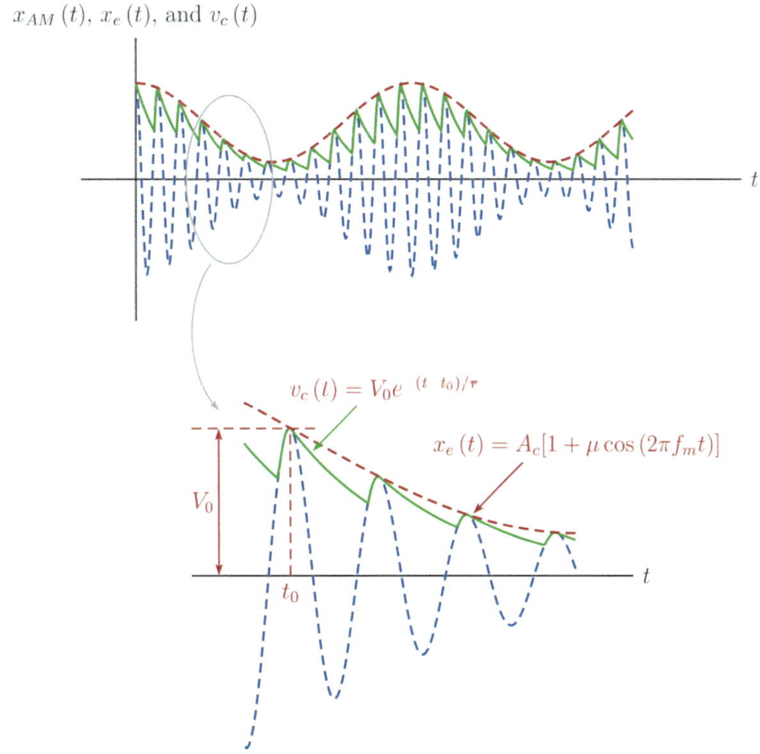

Figure 7.22 – Envelope detector input and output waveforms.

The time constant of the discharge is $\tau = RC$. It is important to note that the rate of decay for the exponential solution in Eqn. (7.55) is critical for the proper operation of the envelope detector. Thus, the time constant must be chosen carefully. If $v_c(t)$ decays too rapidly, the resulting envelope will look quite busy with too much high-frequency detail. If, on the other hand, $v_c(t)$ decays too slowly, then output will have sections where it deviates significantly from the correct envelope behavior as illustrated in Fig. 7.23.

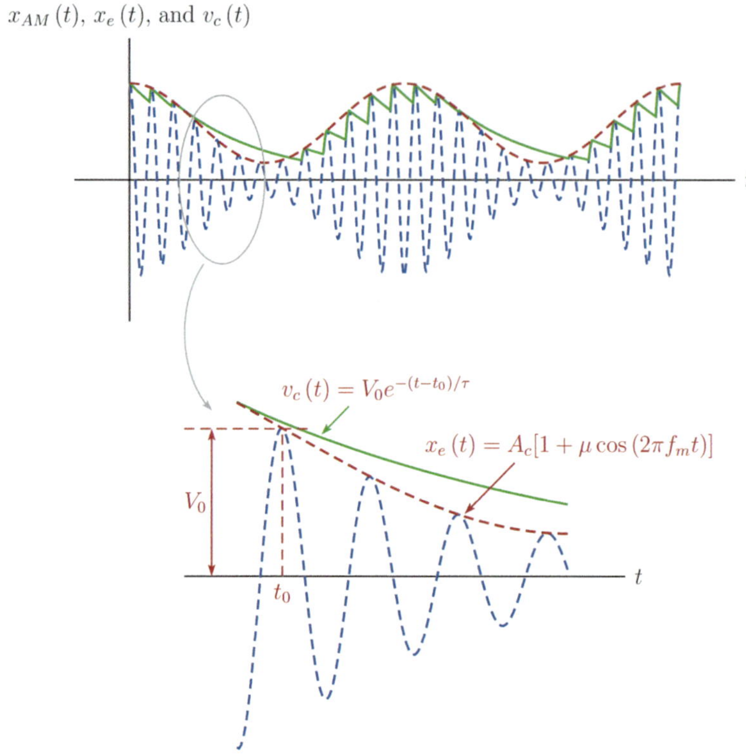

Figure 7.23 – Envelope detector with inappropriate choice of time constant.

We will address the problem of finding a proper value of τ in Example 7.7.

Example 7.7: Envelope detector for tone-modulated AM signal

Consider the envelope detector of Fig. 7.21 with a tone-modulated AM waveform as the input signal. Determine the range of the time constant τ needed to properly recover the single-tone message signal from the modulated waveform.

Solution: The envelope detector fills the gaps between the peaks of the carrier with decaying exponential voltage segments due to the discharging behavior of the RC circuit. In this process, the critical parameter is the time constant τ of the discharge. It was illustrated in Fig. 7.23 that, if the time constant is too large, the envelope detector output voltage fails to track the actual envelope. This happens because the slope of the actual envelope is more negative than the slope of the exponential discharge. Therefore, it would make sense to compare the slopes of the actual envelope and the exponential discharge to come up with the necessary condition that must be imposed on the choice of τ. We will begin with the actual envelope which, for the tone-modulated case, is

$$x_e(t) = A_c\left[1 + \mu\cos\left(2\pi f_m t\right)\right] \tag{7.56}$$

with the assumption that $\mu < 1$. The slope of the envelope is

$$\frac{dx_e(t)}{dt} = -2\pi f_m A_c \mu \sin\left(2\pi f_m t\right) \tag{7.57}$$

Let us pick a time instant $t = t_0$ at one of the peaks of the carrier as shown in Figure 7.22. At that instant, the slope of the envelope is

$$\left.\frac{dx_e(t)}{dt}\right|_{t=t_0} = -2\pi f_m A_c \mu \sin\left(2\pi f_m t_0\right) \tag{7.58}$$

The envelope detector output right after the time instant $t = t_0$ is

$$\begin{aligned} v_c(t) &= x_e(t_0)\, e^{-(t-t_0)/\tau} \\ &= A_c\left[1+\mu\cos\left(2\pi f_m t_0\right)\right] e^{-(t-t_0)/\tau} \end{aligned} \tag{7.59}$$

and the slope of the envelope detector output is

$$\frac{dv_c(t)}{dt} = -\frac{A_c}{\tau}\left[1+\mu\cos\left(2\pi f_m t_0\right)\right] e^{-(t-t_0)/\tau} \tag{7.60}$$

Evaluating this slope at $t = t_0$ where it is maximum, we obtain

$$\left.\frac{dv_c(t)}{dt}\right|_{t=t_0} = -\frac{A_c}{\tau}\left[1+\mu\cos\left(2\pi f_m t_0\right)\right] \tag{7.61}$$

Now we are ready to state the condition for the envelope detector output to correctly track the envelope of the AM signal: We need the slope in Eqn. (7.61) to be more negative than the value in Eqn. (7.58):

$$\left.\frac{dv_c(t)}{dt}\right|_{t=t_0} \le \left.\frac{dx_e(t)}{dt}\right|_{t=t_0} \tag{7.62}$$

Let $\alpha \triangleq 2\pi f_m t_0$ to simplify the notation, and Eqn. (7.62) becomes

$$-\frac{1}{\tau}\left[1+\mu\cos(\alpha)\right] \le -2\pi f_m \mu \sin(\alpha) \tag{7.63}$$

or equivalently

$$\tau \le \frac{1}{2\pi f_m}\left(\frac{1+\mu\cos(\alpha)}{\mu\sin(\alpha)}\right) \tag{7.64}$$

It is important for the inequality in Eqn. (7.64) to be satisfied for all values of α, including when the right side has its smallest possible value. To determine the minimum value of the expression in parentheses, we need to differentiate it with respect to α and set the result equal to 0. This yields $\cos(\alpha) = -\mu$ and, consequently, $\sin(\alpha) = \sqrt{1-\mu^2}$. Substituting these into Eqn. (7.64) we obtain

$$\tau \le \frac{1}{2\pi f_m}\left(\frac{\sqrt{1-\mu^2}}{\mu}\right) \tag{7.65}$$

The time constant $\tau = RC$ should satisfy the inequality in Eqn. (7.65) in order to avoid the loss of tracking for the envelope detector. The result we have derived is for the case when the modulating signal is a single tone. If the modulating signal has a range of frequencies, then Eqn. (7.65) can still be used heuristically, and the inequality should be satisfied for the highest frequency in the input signal, i.e.,

$$\tau \le \frac{1}{2\pi\left[\max\left(f_m\right)\right]}\left(\frac{\sqrt{1-\mu^2}}{\mu}\right) \tag{7.66}$$

Interactive App: Diode-based envelope detector simulation

The interactive app in `ampMod9.m` is based on Example 7.7. It simulates an envelope detector used on an AM signal modulated by a single-tone message signal. Carrier frequency f_c, message frequency f_m, modulation index μ, and envelope detector time constant τ may be adjusted by using the slider controls provided. The program computes and graphs the AM waveform, the correct envelope, and the envelope detector output for the parameter values specified. The minimum time constant needed for proper detection of the envelope is computed using Eqn. (7.66), and is displayed.

a. Observe the dependency of the maximum time constant on the modulation index μ and the message frequency f_m.

b. Explain why the maximum time constant is not dependent on the carrier frequency f_c.

c. Push the time constant beyond its maximum value, and observe the envelope detector output. At what part of the signal does the envelope detector stop tracking the correct envelope?

Software resource: `ampMod9.m`

Coherent demodulation

The simple demodulation technique discussed in the previous section only works if the modulation index is less than 100 percent. For $\mu > 1$ we have seen that the envelope of the modulated carrier is a distorted version of the message signal, and is therefore not useful in the demodulation process. The alternative is to use *coherent demodulation* which requires the availability of a local carrier signal synchronized with the original carrier signal that was used in generating $x_{AM}(t)$. Consider the block diagram shown in Fig. 7.24.

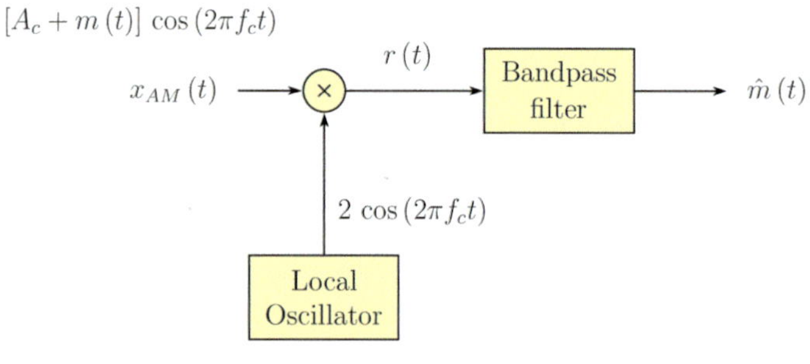

Figure 7.24 – Coherent demodulation of an AM signal.

The product $r(t)$ at the output of the multiplier is

$$\begin{aligned} r(t) =& 2\,x_{AM}(t)\cos(2\pi f_c t) \\ =& 2\left[A_c + m(t)\right]\cos^2(2\pi f_c t) \end{aligned} \tag{7.67}$$

Using the appropriate trigonometric identity Eqn. (7.67) can be written as

$$r(t) = A_c + m(t) + \left[A_c + m(t)\right]\cos(4\pi f_c t) \tag{7.68}$$

Using the sample message spectrum $M(f)$ and the corresponding AM signal spectrum $X_{AM}(f)$ shown in Fig. 7.7, the spectrum of the intermediate signal $r(t)$ is as shown in Fig. 7.25.

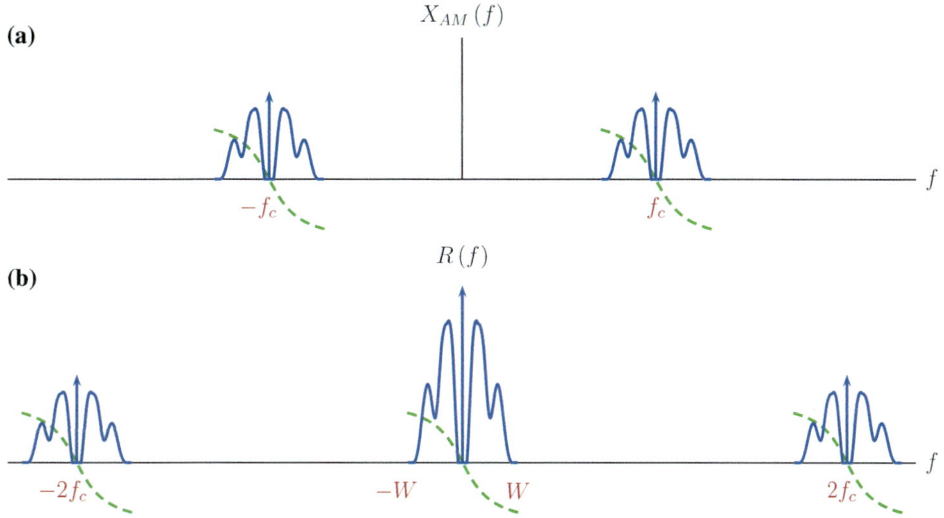

Figure 7.25 – Frequency spectrum $R(f)$ of the intermediate signal in the coherent demodulator shown in Fig. 7.24.

It is clear from Fig. 7.25 that, in order to recover the message spectrum $M(f)$ from $R(f)$ the high frequency component around $2f_c$ and the dc component should be removed (we are assuming that the message signal $m(t)$ does not have a dc component). That is the role of the bandpass filter.

7.5 Double-Sideband Suppressed Carrier Modulation

In Section 7.4.2 we have seen that one of the shortcomings of amplitude modulation is the amount of power wasted on transmitting the carrier along with the two sidebands. Modulation efficiency given by Eqn. (7.34) is theoretically limited to 50 percent, and is much lower in practice. Less than half the transmitted power is used on information bearing signals, and the rest of the power is wasted for transmitting the carrier.

One method of fixing this shortcoming is to simply drop the power-wasting term $A_c \cos(2\pi f_c t)$ from the AM waveform expression in Eqn. (7.3). This results in the signal

$$x_{DSB}(t) = m(t)\cos(2\pi f_c t) \tag{7.69}$$

which is referred to as the *double-sideband suppressed-carrier (DSB-SC)* modulated signal. In the remainder of this text, we will refer to this signal simply as the DSB modulated signal. Fig. 7.26 illustrates the DSB signal in the time-domain.

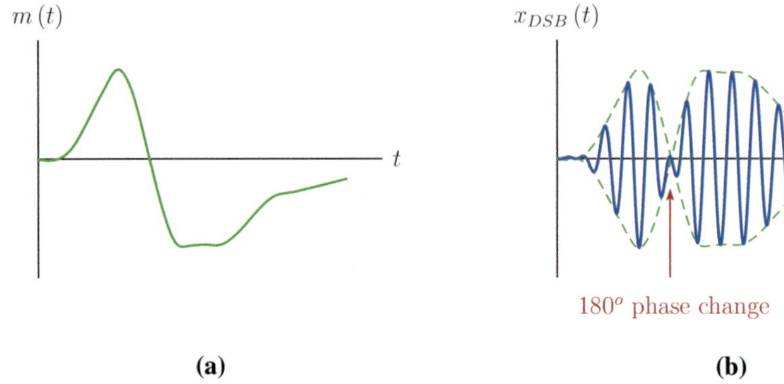

Figure 7.26 – Waveforms involved in DSB modulation: **(a)** The message signal and **(b)** the DSB modulated carrier.

The power efficiency of the DSB modulated signal is 100 percent since no power is being wasted on any non-information-bearing signal components. On the other hand, a comparison of Fig. 7.26 to Fig. 7.4 makes it clear that the envelope is lost in the case of DSB modulation, so we can no longer rely on the simplicity of an envelope detector for recovering the message signal from the modulated signal. A demodulation technique with higher complexity must be employed instead. Specifically, at the receiver end of the communication system, we will need a carrier the frequency and the phase of which are in perfect synchronization with the original carrier used in creating the DSB modulated signal. This is the price paid for improved power efficiency.

Interactive App: DSB-SC modulation

The interactive app in `dspMod1.m` uses the signals in Fig. (7.26). The message signal and the blank carrier are shown on the left side of the graphical user interface. On the right side of the graphical user interface, the DSB modulated signal is shown as computed by Eqn. (7.69). A slider control and an edit control are provided for changing the carrier frequency f_c.

a. Observe the changes in the DSB signal as the carrier frequency f_c is varied.
b. Notice that the envelope no longer resembles the message signal $m(t)$.
c. Pay attention to the 180° phase reversal of the carrier at the point of envelope crossover.

Software resource: `dsbMod1.m`

Example 7.8: Tone modulation with DSB

Determine and graph the DSB-SC signal when the modulating signal is a single tone in the form

$$m(t) = A_m \cos(2\pi f_m t)$$

Solution: Using the specified $m(t)$ the resulting DSB modulated signal is

$$x_{DSB}(t) = A_m \cos(2\pi f_m t)\cos(2\pi f_c t) \tag{7.70}$$

which is shown in Fig. 7.27.

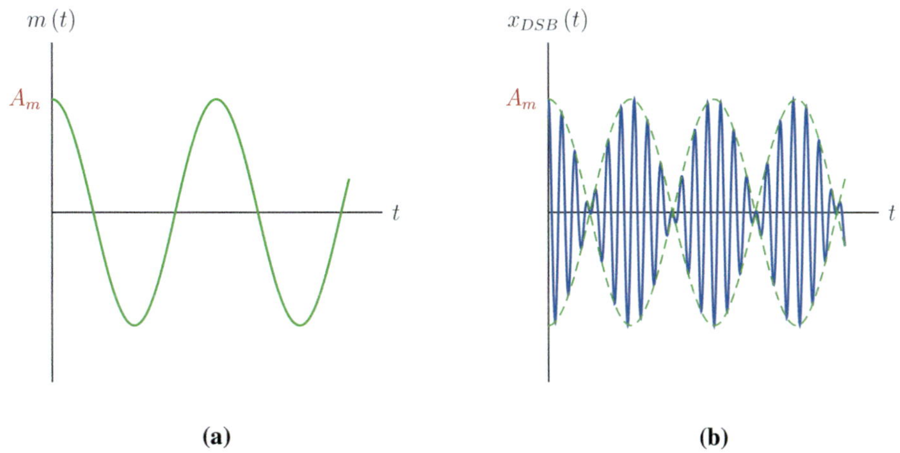

Figure 7.27 – DSB-SC modulation using a single-tone message signal in Example 7.8: **(a)** The message signal, **(b)** DSB-SC signal.

The positive envelope of the DSB signal is

$$x_e(t) = \left|m(t)\right| = \left|A_m \cos(2\pi f_m t)\right| \tag{7.71}$$

As expected, it no longer resembles the message signal $m(t)$.

Software resource: exct_7_8.m

Interactive App: DSB-SC modulation of a carrier with a single tone

The interactive app in `dspMod2.m` is based on Example 7.8 and Fig. 7.27. The DSB modulated signal is graphed with the single-tone modulating signal as given by Eqn. (7.70). Slider controls are provided for changing the values of the message amplitude A_m, the single-tone message frequency f_m, and the carrier frequency f_c. Observe the following:

a. The relationship between the message signal $m(t)$ and the positive envelope $x_e(t)$ of the DSB signal,

b. The effects of changing f_c and f_m, and

c. The 180° phase change that occurs in the carrier when the envelope crosses over the horizontal axis.

Software resource: `dsbMod2.m`

7.5.1 Frequency spectrum of the DSB-SC signal

The mathematical expression for the double-sideband suppressed-carrier signal was given by Eqn. (7.69). Taking the Fourier transform of both sides of Eqn. (7.69) and using the modulation property discussed in Section 3.3.5 of Chapter 3 yields

$$X_{DSB}(f) = \frac{1}{2}M(f + f_c) + \frac{1}{2}M(f - f_c) \tag{7.72}$$

which is illustrated in Fig. 7.28.

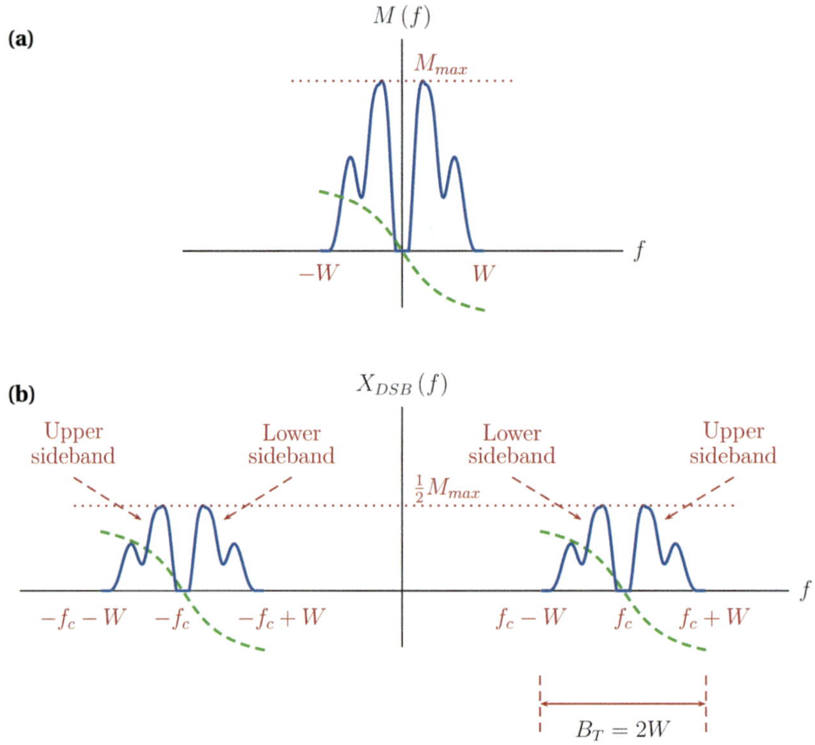

Figure 7.28 – **(a)** Frequency spectrum $M(f)$ of the message signal; magnitude (solid) and phase (dashed) and **(b)** frequency spectrum $X_{DSB}(f)$ of the double-sideband suppressed carrier; magnitude (solid) and phase (dashed).

A quick comparison of Fig. 7.28 and Eqn. (7.72) with their counterparts for amplitude modulation in Section 7.4.1, namely Fig. 7.7 and Eqn. (7.10), reveals that the only difference is the absence of the impulses at $\pm f_c$ from the DSB-SC spectrum. Upper- and lower-sidebands are present as before. The transmission bandwidth, $B_T = 2W$, is also unchanged.

Interactive App: DSB-SC frequency spectrum

The interactive app in `dspMod3.m` follows Fig. 7.28 and Eqn. (7.72). The spectrum of a sample message signal, and the spectrum of the corresponding DSB-SC signal are shown. The sample message spectrum picked for this demo has a bandwidth of 1 kHz and a peak magnitude of $\max\{|M(f)|\} = 1$. A slider control is provided for adjusting the carrier frequency f_c in the range $2\,\text{kHz} \le f_c \le 9\,\text{kHz}$.

a. Observe how the sidebands move inward or outward as the carrier frequency is changed.

b. Compare the spectrum of the DSB signal to that of the AM signal in Fig. 7.7 and the interactive app `ampMod3.m`.

Software resource: `dsbMod3.m`

Example 7.9: Frequency spectrum of the tone-modulated DSB-SC signal

Determine and graph the frequency spectrum of the tone-modulated DSB-SC signal.

Solution: The expression for a tone-modulated DSB signal was given in Eqn. (7.70). Using the appropriate trigonometric identity it can be written as

$$x_{DSB}(t) = \frac{A_m}{2}\cos\left(2\pi\left[f_c + f_m\right]t\right) + \frac{A_m}{2}\cos\left(2\pi\left[f_c - f_m\right]t\right) \tag{7.73}$$

Eqn. (7.73) contains two cosine terms at frequencies $f_c + f_m$ and $f_c - f_m$. For the tone-modulated case these two cosine terms constitute the upper- and the lower-sideband, respectively. Individual Fourier transforms of the two terms in Eqn. (7.73) are

$$\begin{aligned}
\tfrac{A_m}{2}\cos\left(2\pi\left[f_c + f_m\right]t\right) &\overset{\mathcal{F}}{\longleftrightarrow} \tfrac{A_m}{4}\,\delta\left(f + \left[f_c + f_m\right]\right) + \tfrac{A_m}{4}\,\delta\left(f - \left[f_c + f_m\right]\right) \\
\tfrac{A_m}{2}\cos\left(2\pi\left[f_c - f_m\right]t\right) &\overset{\mathcal{F}}{\longleftrightarrow} \tfrac{A_m}{4}\,\delta\left(f + \left[f_c - f_m\right]\right) + \tfrac{A_m}{4}\,\delta\left(f - \left[f_c - f_m\right]\right)
\end{aligned}$$

and the spectrum of the tone-modulated DSB signal is obtained as

$$\begin{aligned}
X_{DSB}(f) = &\frac{A_m}{4}\,\delta\left(f + \left[f_c + f_m\right]\right) + \frac{A_m}{4}\,\delta\left(f - \left[f_c + f_m\right]\right) \\
&+ \frac{A_m}{4}\,\delta\left(f + \left[f_c - f_m\right]\right) + \frac{A_m}{4}\,\delta\left(f - \left[f_c - f_m\right]\right)
\end{aligned} \tag{7.74}$$

Frequency spectra for the single-tone message signal and the amplitude-modulated signal of Eqn. (7.74) are shown in Figure 7.29. The DSB spectrum for a single tone modulating signal is similar to the spectrum of the AM carrier with the same modulating signal which was shown in Fig. 7.8. The main difference is the lack of impulses at frequencies $\pm f_c$.

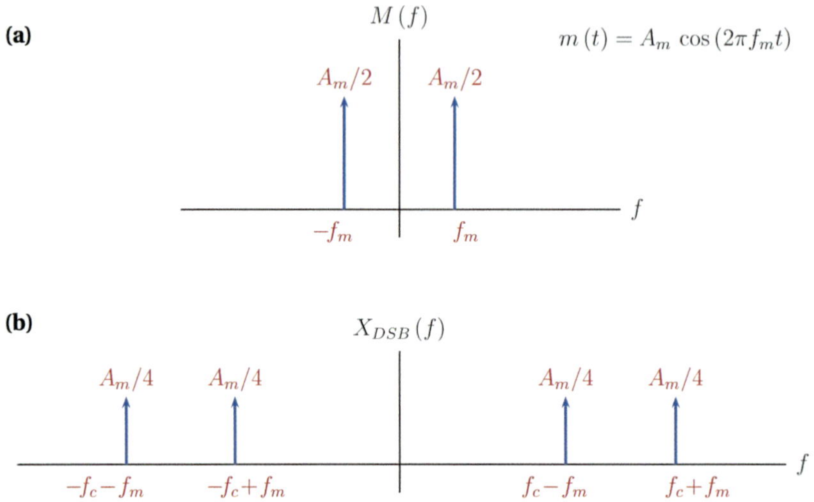

Figure 7.29 – **(a)** Frequency spectrum $M(f)$ of the single-tone message signal and **(b)** frequency spectrum of the tone-modulated DSB-SC signal of Example 7.9.

Interactive App: DSB-SC spectrum for a single-tone message signal

The interactive app in `dspMod4.m` illustrates the concepts explored in Example 7.9 and Fig. 7.29. The two graphs show the spectrum of the single-tone message signal $m(t) = A_m \cos(2\pi f_m t)$ and the spectrum of the corresponding DSB-SC signal. Slider controls allow adjustments to the message amplitude A_m, the message frequency f_m, and the carrier frequency f_c.

a. Pay attention to the frequency shift of the sidebands of the spectrum as the carrier frequency is changed.

b. Observe how the sideband impulses move toward each other or away from each other as the message frequency is changed.

c. Observe the values of the impulses as the message amplitude is changed.

Software resource: `dsbMod4.m`

7.6 Single-Sideband Modulation

One of the shortcomings of amplitude modulation, namely the power efficiency, was addressed in Section 7.5. Dropping the carrier term is the method of improving the power efficiency at the expense of giving up the envelope that resembles the message signal, and the opportunity for simple non-coherent detection afforded by it. Another shortcoming of amplitude modulation is the doubling of the bandwidth. For a message signal with bandwidth W Hz, the transmission bandwidth of both the AM and the DSB-SC waveforms is $B_T = 2W$ as is evident from Figs. 7.7 and 7.28.

Consider a message signal $m(t)$ with the frequency spectrum $M(f)$ shown in Fig. 7.30.

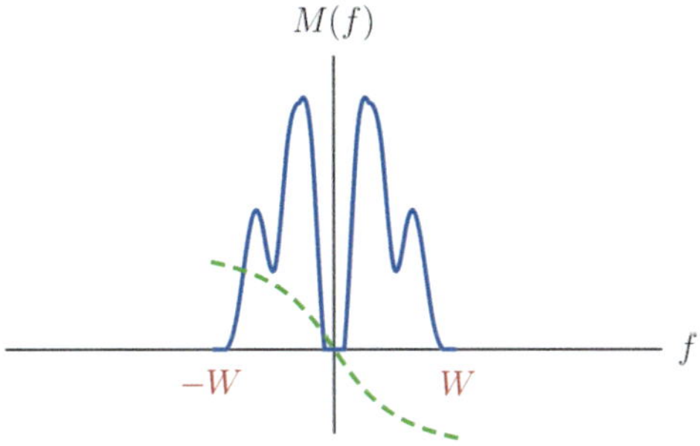

Figure 7.30 – Frequency spectrum $M(f)$ for a sample message signal $m(t)$.

The message signal $m(t)$ is real-valued. Based on the discussion in Section 3.3.5, its frequency spectrum $M(f)$ exhibits conjugate symmetry:

$$M^*(f) = M(-f) \tag{7.75}$$

As a result, the magnitude of the spectrum $M(f)$ is an even function of frequency, and the phase of it is an odd function of frequency.

$$|M(f)| - |M(-f)| \tag{7.76}$$

and

$$\measuredangle M(f) = -\measuredangle M(-f) \tag{7.77}$$

Let us split the message spectrum into two parts defined as

$$M_n(f) = \begin{cases} M(f), & f < 0 \\ 0, & f > 0 \end{cases} \tag{7.78}$$

and

$$M_p(f) = \begin{cases} 0, & f < 0 \\ M(f), & f > 0 \end{cases} \tag{7.79}$$

so that $M(f) = M_n(f) + M_p(f)$. Clearly, $M_n(f)$ represents the part of the message spectrum for negative frequencies, that is, the left side. Conversely, $M_p(f)$ represents the right side. This is illustrated in Fig. 7.31.

Based on the symmetry properties outlined in Eqns. (7.75) through (7.77), the knowledge one of the two components of the spectrum, either $M_n(f)$ or $M_p(f)$, makes the other term redundant. For example, if we know $M_p(f)$, we can obtain $M_n(f)$ as

$$M_n(f) = M_p^*(-f) \tag{7.80}$$

As a result, the transmission of a modulated signal with just one of the sidebands in it is sufficient for the receiver to get the message signal. This is referred to as *single-sideband (SSB)* modulation.

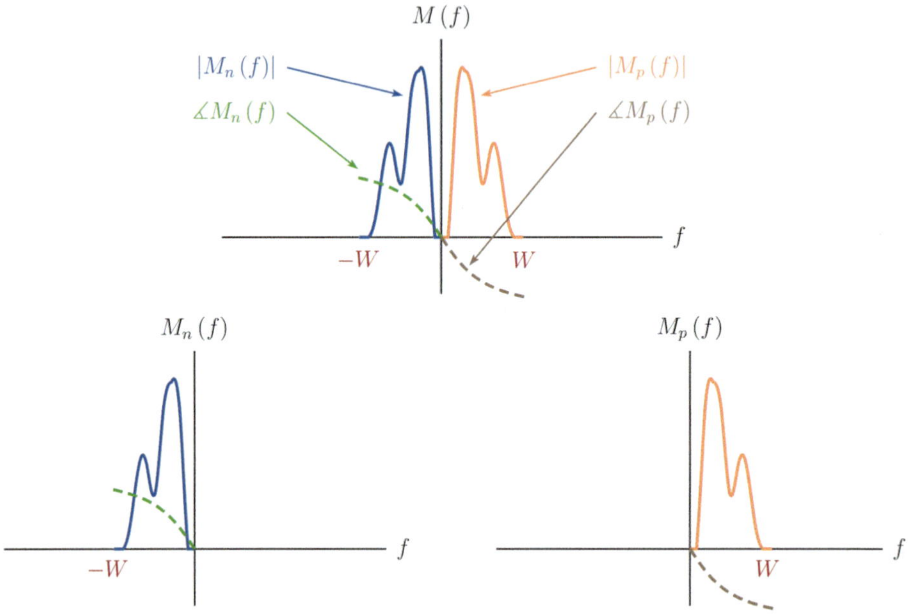

Figure 7.31 – Splitting the message spectrum $M(f)$ into its left and right halves.

Recall that the spectrum of the DSB-SC signal was given by Eqn. (7.72), and includes both lower- and upper-sidebands. The DSB spectrum can be rewritten using the two components of $M(f)$ as

$$\begin{aligned} X_{DSB}(f) &= \frac{1}{2} M(f+f_c) + \frac{1}{2} M(f-f_c) \\ &= \frac{1}{2} M_n(f+f_c) + \frac{1}{2} M_p(f+f_c) + \frac{1}{2} M_n(f-f_c) + \frac{1}{2} M_p(f-f_c) \end{aligned} \tag{7.81}$$

A single-sideband spectrum can be formed by retaining either the upper-sidebands or the lower-sidebands. If the upper-sidebands are retained, the resulting spectrum is

$$X_{SSB,U}(f) = \frac{1}{2} M_n(f+f_c) + \frac{1}{2} M_p(f-f_c) \tag{7.82}$$

We will refer to $X_{SSB,U}(f)$ as the *upper-sideband SSB spectrum.* Alternatively, we can retain the lower-sidebands to obtain

$$X_{SSB,L}(f) = \frac{1}{2} M_p(f+f_c) + \frac{1}{2} M_n(f-f_c) \tag{7.83}$$

which is the *lower-sideband SSB spectrum.* Fig. 7.32 illustrates the results obtained in Eqns. (7.81) through (7.83).

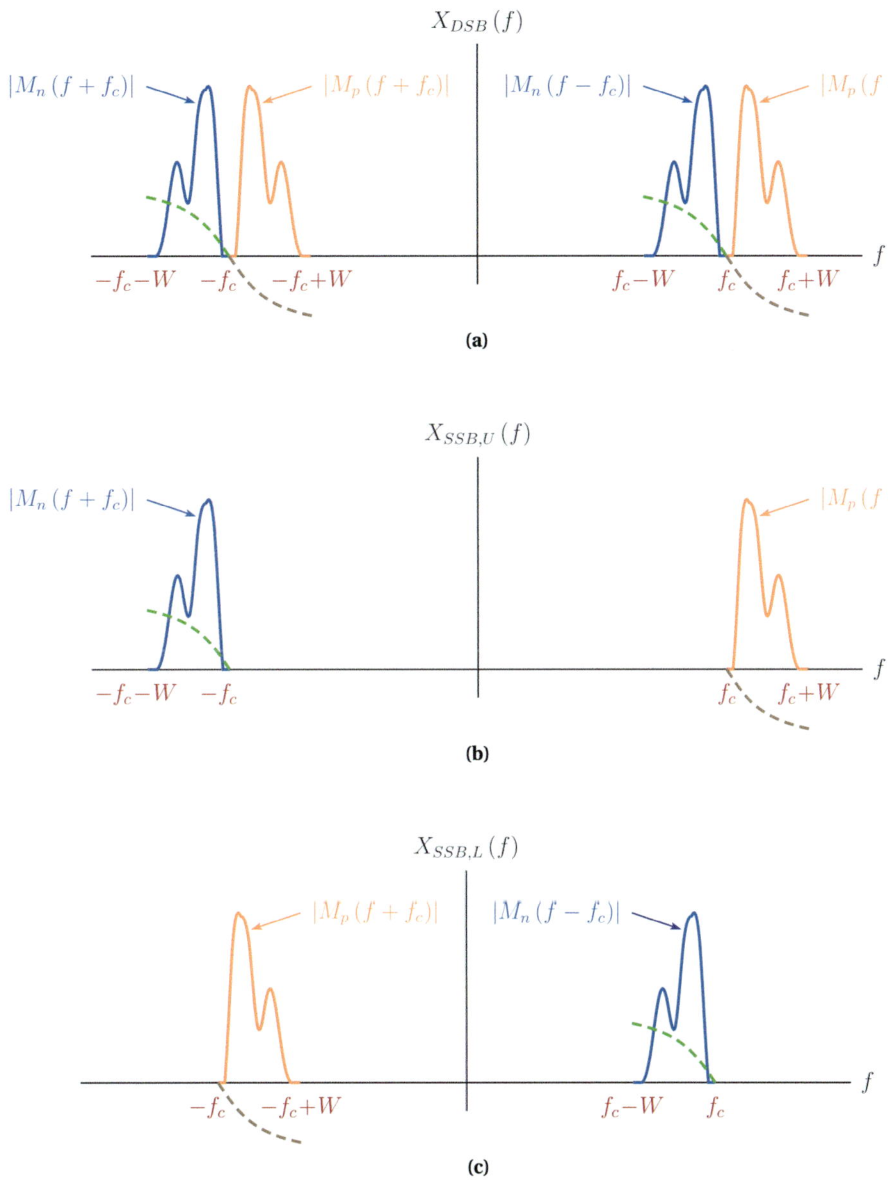

Figure 7.32 – **(a)** The frequency spectrum of a DSB-SC signal expressed using the spectral components $M_n(f)$ and $M_p(f)$, **(b)** the upper-sideband SSB spectrum $X_{SSB,U}(f)$, and **(c)** the lower-sideband SSB spectrum $X_{SSB,L}(f)$.

The time-domain signals that correspond to the two SSB spectra of Eqns. (7.82) and (7.83) would be

$$x_{SSB,U}(t) = \mathscr{F}^{-1}\{X_{SSB,U}(f)\} \tag{7.84}$$

and

$$x_{SSB,L}(t) = \mathscr{F}^{-1}\{X_{SSB,L}(f)\} \tag{7.85}$$

Theoretically, the SSB signals $x_{SSB,U}(t)$ and $x_{SSB,L}(t)$ can be obtained from the DSB signal $x_{DSB}(t)$ through sideband filtering as illustrated in Fig. 7.33.

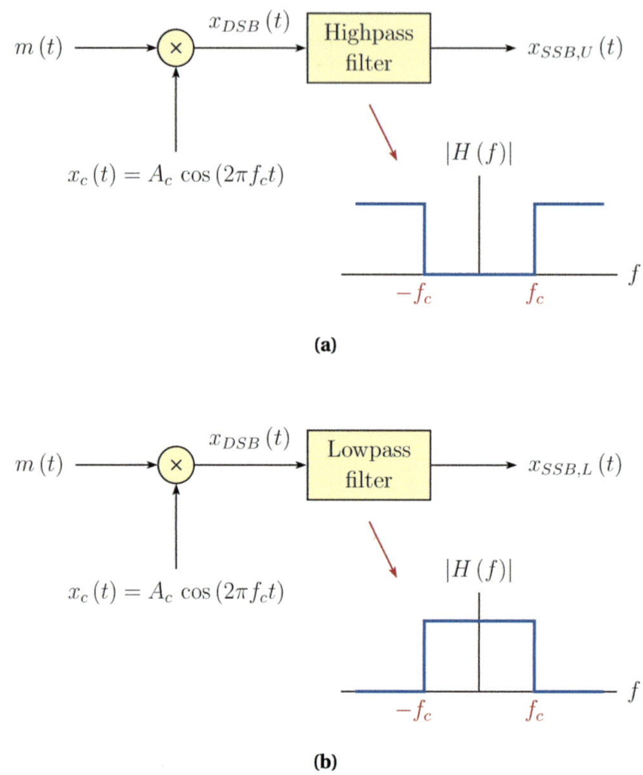

Figure 7.33 – Obtaining SSB signals from the DSB signal through sideband filtering: **(a)** block diagram for obtaining the upper-sideband SSB signal and **(b)** block diagram for obtaining the lower-sideband SSB signal.

The block diagrams in Fig. 7.33 assume that the filters used have ideal characteristics. This approach is not used in practical implementation of SSB modulation since ideal filters are not available and would be near impossible to approximate since the allowed transition band would be extremely narrow relative to the carrier frequency.

One question that we have not yet considered up to this point is: *What does the time-domain SSB signal look like?* We will use Eqns. (7.82) and (7.83) to derive time-domain expressions for the lower-sideband and upper-sideband SSB signals. The message signal $m(t)$ can be written in the form

$$m(t) = m_n(t) + m_p(t) \tag{7.86}$$

where $m_n(t)$ and $m_p(t)$ are the time-domain signals that correspond to the two components of the spectrum, $M_n(f)$ and $M_p(f)$, respectively:

$$m_n(t) \overset{\mathscr{F}}{\longleftrightarrow} M_n(f) \tag{7.87}$$

$$m_p(t) \overset{\mathscr{F}}{\longleftrightarrow} M_p(f) \tag{7.88}$$

Using the frequency-shifting property of the Fourier transform, time-domain equivalents of Eqns. (7.82) and (7.83) can be derived as

$$x_{SSB,U}(t) = \frac{1}{2}\, m_n(t)\, e^{-j2\pi f_c t} + \frac{1}{2}\, m_p(t)\, e^{j2\pi f_c t} \tag{7.89}$$

and

$$x_{SSB,L}(t) = \frac{1}{2}\, m_p(t)\, e^{-j2\pi f_c t} + \frac{1}{2}\, m_n(t)\, e^{j2\pi f_c t} \tag{7.90}$$

Let us first consider $m_p(t)$ and its Fourier transform $M_p(f)$. The transform $M_p(f)$ is non-zero only for positive frequencies, and is not conjugate symmetric. As a result, we expect the time-domain signal $m_p(t)$ to be complex-valued. One way of relating the transform $M_p(f)$ to the message spectrum $M(f)$ is through the use of the signum function:

$$\begin{aligned} M_p(f) =& M(f)\left[\frac{1}{2} + \frac{1}{2}\,\text{sgn}(f)\right] \\ =& \frac{1}{2}\, M(f) + \frac{1}{2}\,\text{sgn}(f)\, M(f) \end{aligned} \tag{7.91}$$

Consider a signal $\hat{m}(t)$ derived from $m(t)$ and $M(f)$ through the Fourier transform relationship

$$\hat{m}(t) = \mathscr{F}^{-1}\{-j\,\text{sgn}(f)\, M(f)\} \tag{7.92}$$

The Fourier transform of $\hat{m}(t)$ is

$$\hat{M}(f) = \mathscr{F}\{\hat{m}(t)\} = -j\,\text{sgn}(f)\, M(f) \tag{7.93}$$

The signal $\hat{m}(t)$ is known as the *Hilbert transform* of the signal $m(t)$. It can be modeled as the output signal of a *Hilbert transform filter* with system function

$$H(f) = -j\,\text{sgn}(f) \tag{7.94}$$

driven by the message signal $m(t)$ as illustrated in Fig. 7.34.

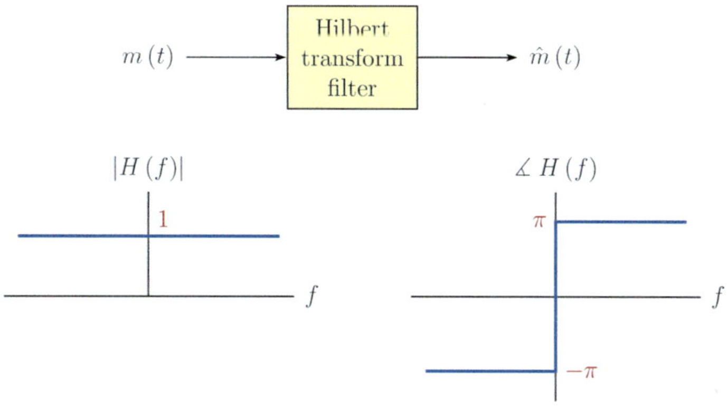

Figure 7.34 – Hilbert transform $\hat{m}(t)$ of the signal $m(t)$ modeled as the output of a Hilbert transform filter.

Substituting Eqn. (7.93) into Eqn. (7.91) we obtain

$$M_p(f) = \frac{1}{2}\, M(f) + j\,\frac{1}{2}\,\hat{M}(f) \tag{7.95}$$

and

$$m_p(t) = \frac{1}{2}\, m(t) + j\,\frac{1}{2}\,\hat{m}(t) \tag{7.96}$$

Mathematical expressions for the left half of the spectrum and the corresponding time-domain signal can also be obtained through similar derivations:

$$M_n(f) = \frac{1}{2} M(f) - j\frac{1}{2} \hat{M}(f) \tag{7.97}$$

and

$$m_n(t) = \frac{1}{2} m(t) - j\frac{1}{2} \hat{m}(t) \tag{7.98}$$

We are now ready to write the time-domain expressions for the upper-sideband SSB and lower-sideband SSB signals in terms of the message signal. Substituting Eqns. (7.96) and (7.98) into Eqn. (7.89) yields the expression for the upper-sideband SSB signal as

$$x_{SSB,U}(t) = \frac{1}{2}\left[\frac{1}{2} m(t) - j\frac{1}{2} \hat{m}(t)\right] e^{-j2\pi f_c t} + \frac{1}{2}\left[\frac{1}{2} m(t) + j\frac{1}{2} \hat{m}(t)\right] e^{j2\pi f_c t} \tag{7.99}$$

which can be simplified to

$$\begin{aligned} x_{SSB,U}(t) =& \frac{1}{4} m(t) \left[e^{j2\pi f_c t} + e^{-j2\pi f_c t}\right] + \frac{1}{4} \hat{m}(t) \left[j\, e^{j2\pi f_c t} - j\, e^{-j2\pi f_c t}\right] \\ =& \frac{1}{2} m(t) \cos(2\pi f_c t) - \frac{1}{2} \hat{m}(t) \sin(2\pi f_c t) \end{aligned} \tag{7.100}$$

Similarly, the lower-sideband SSB signal can be found by substituting Eqns. (7.96) and (7.98) into Eqn. (7.90), and simplifying the result.

$$x_{SSB,L}(t) = \frac{1}{2} m(t) \cos(2\pi f_c t) + \frac{1}{2} \hat{m}(t) \sin(2\pi f_c t) \tag{7.101}$$

Summary of Key Points

☞ Modulation is the process of embedding an information bearing signal, referred to as the *message signal*, in another signal known as the *carrier*. The carrier by itself does not contain any information, yet it may have other features that make it desirable. Some examples of these desirable features are:

1. suitability for transmission through a particular medium,
2. better performance in the presence of noise and interference, and
3. suitability to more efficient utilization.

☞ The type of modulation used in a communication system depends on

1. the type of carrier used, and
2. the particular parameter of the carrier that is utilized for embedding the message signal into it.

☞ Two commonly used types of carrier signals are a namely a sinusoidal signal and a periodic pulse train. A sinusoidal carrier signal in the form

$$x_c(t) = A_c \cos(2\pi f_c t + \phi_c)$$

has three parameters, namely the amplitude A_c, the frequency f_c and the phase ϕ_c. Amplitude modulation is obtained if the amplitude A_c of the carrier is varied as a function of the message signal $m(t)$.

☞ An alternative to the sinusoidal carrier described is a periodic pulse train in the form

$$x_c(t) = \sum_{k=-\infty}^{\infty} A_c \Pi\left(\frac{t-kT_0}{\tau}\right)$$

Pulse-amplitude modulation (PAM) is obtained by varying the amplitude of the pulse train as a function of the message signal.

☞ An amplitude modulated carrier is in the form

Eqn. 7.3: $$x_{AM}(t) = A_c \cos(2\pi f_c t) + m(t)\cos(2\pi f_c t)$$

where $m(t)$ is the message signal.

☞ The *envelope* of the carrier is

Eqn. 7.4: $$x_e(t) = \left|A_c + m(t)\right|$$

The *modulation index* is defined as

Eqn. 7.6: $$\mu = \frac{|m_{\min}|}{A_c}$$

For the envelope to display the correct shape of the message signal, we need $0 \le \mu \le 1$.

☞ The frequency spectrum of the AM signal is

Eqn. 7.10: $$X_{AM}(f) = \frac{1}{2}A_c\,\delta(f+f_c) + \frac{1}{2}A_c\,\delta(f-f_c) + \frac{1}{2}M(f+f_c) + \frac{1}{2}M(f-f_c)$$

where $M(f)$ is the message spectrum. The part of the AM spectrum that occupies frequencies greater than the carrier frequency f_c is referred to as the *upper sideband*. The part that occupies frequencies less than f_c is referred to as the *lower sideband*.

☞ A message signal which occupies a bandwidth of W Hz in its baseband form requires twice that bandwidth when it is embedded into the amplitude of a carrier. Amplitude modulation is wasteful of bandwidth.

☞ Modulation efficiency of the AM signal is defined as the percentage of the useful power within the total power of the AM signal. It is computed as

Eqn. 7.34: $$\eta = \frac{\text{Power in sidebands}}{\text{Total power}} = \frac{P_m}{P_c + P_m} = \frac{\langle m^2(t)\rangle}{A_c^2 + \langle m^2(t)\rangle}$$

If the envelope must be preserved, the maximum theoretical efficiency that can be attained with amplitude modulation is $\eta = 0.5$, or 50 percent. Actual efficiency values obtained in practice are significantly lower. Amplitude modulation is wasteful of power.

☞ The term *demodulation* refers to the act of recovering the message signal $m(t)$ as closely to its original form as possible from the modulated waveform $x_{AM}(t)$. If the envelope has been preserved ($\mu < 1$), the simplest means of demodulation is by using an envelope detector. If the envelope has not been preserved, coherent demodulation techniques can be utilized. These techniques typically require that a local copy of the carrier signal be available at the receiver.

☞ One method of improving the power efficiency of amplitude modulation is to suppress the carrier term. This results in the signal

Eqn. 7.69: $$x_{DSB}(t) = m(t)\,\cos(2\pi f_c t)$$

which is referred to as the *double-sideband suppressed-carrier (DSB-SC)* modulated signal.

☞ The frequency spectrum of the DSB-SC signal is

Eqn. 7.72: $$X_{DSB}(f) = \frac{1}{2}M(f+f_c) + \frac{1}{2}M(f-f_c)$$

Upper- and lower-sidebands are present as before. The transmission bandwidth is unchanged.

☞ Single sideband (SSB) modulation is obtained by removing one of the sidebands of the DSB-SC signal. The bandwidth of the SSB signal is the same as that of the original message signal.

Further Reading

[1] L.W. Couch. *Digital and Analog Communication Systems.* Prentice Hall, 2013.

[2] B.P. Lathi and Z. Ding. *Modern Digital and Analog Communication Systems.* The Oxford Series in Electrical and Computer Engineering. Oxford University Press, 2009.

[3] J.G. Proakis, M. Salehi, and G. Bauch. *Contemporary Communication Systems Using MATLAB, 3rd ed.* Cengage Learning, 2011.

[4] Bernard Sklar. *Digital Communications: Fundamentals and Applications.* Pearson, 2021.

MATLAB Exercises with Solutions

MATLAB Exercise 7.1: Compute and graph the AM signal

Develop a MATLAB script for generating samples of the tone-modulated AM signal given by Eqn. (7.8), using parameter values $f_c = 25$ kHz, $f_m = 2$ kHz, $A_c = 2$, and $\mu = 0.7$. Graph the resulting signal.

Solution: An important issue is the time increment used in evaluating amplitude values of the AM signal. Even though our final graph will mimic the behavior of an analog signal, it is obvious that any signal we generate with MATLAB is discrete by nature. In essence, we will be sampling the analog AM signal of Eqn. (7.8) to produce a discrete-time signal, and then graphing it by connecting the dots so that the result looks like an analog signal. The time increment used in the calculation of AM signal values is the same as the sampling interval $T_s = 1/f_s$. It needs to be chosen sufficiently small compared to the period of the carrier so that the resulting graph is an accurate representation of the AM signal. Nyquist sampling theorem states that the sampling rate f_s needs to at least twice the highest signal frequency. In this example, however, we will go well beyond that requirement so that the resulting analog graph looks smooth. For the parameter values chosen, the highest frequency in the signal $x_{AM}(t)$ is $f_c + f_m = 27$ kHz. We will choose the time increment as $T_s = 10^{-6}$ seconds, corresponding to a sampling rate of $f_s = 1$ MHz. MATLAB code is listed below.

```
% Script: mexct_7_1.m
Ac = 2;
Mu = 0.7;
fc = 25000;
fm = 2000;
t = [0:1e-6:1e-3];          % Vector of time instants.
% Compute the AM signal.
x_am = Ac*(1+Mu*cos(2*pi*fm*t)).*cos(2*pi*fc*t);  % Eqn. (7.8)
% Compute the envelope.
x_env = abs(Ac*(1+Mu*cos(2*pi*fm*t)));            % Eqn. (7.9)
% Graph the AM signal and the envelope.
plot(t,x_am,'-',t,x_env,'m--');
grid;
title('Tone modulated AM signal and its envelope');
xlabel('Time (sec)');
ylabel('Amplitude');
legend('AM signal','Envelope');
```

The graph that is generated by the script `mexct_7_1.m` is shown in Fig. 7.35.

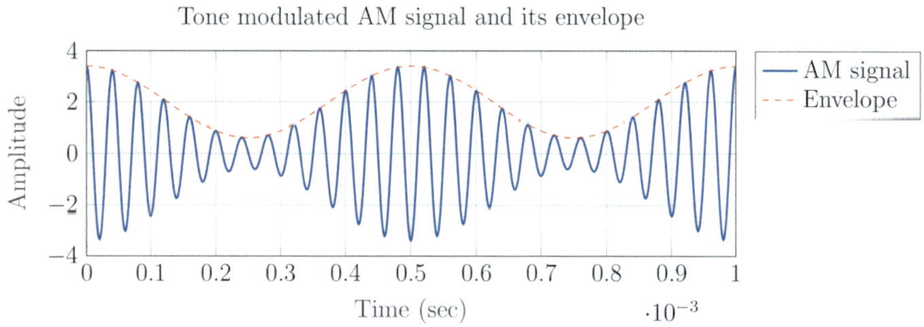

Figure 7.35 – MATLAB graph for single-tone modulated signal.

Software resource: `mexct_7_1.m`

MATLAB Exercise 7.2: EFS spectrum of the tone-modulated AM signal

Refer to the tone-modulated AM signal graphed in MATLAB Exercise 7.1. Determine its fundamental period. Afterward compute and graph its EFS spectrum.

Solution: First we need to determine the periodicity of the signal. With parameter values $f_c = 25$ kHz and $f_m = 2$ kHz, the frequencies contained in the AM signal are $f_c = 25$ kHz, $f_c - f_m = 23$ kHz and $f_c + f_m = 27$ kHz. The fundamental frequency of the signal $x_{AM}(t)$ is $f_0 = 1$ kHz corresponding to a fundamental period of $T_0 = 1$ ms (see the discussion in Section 1.7 of Chapter 1). The script listed below computes samples of exactly one period of the tone-modulated AM signal and then uses the function `ss_efsapprox()`. For details of the function `ss_efsapprox()`, see the companion book *Discrete-Time Signals and Systems: A MATLAB Integrated Approach.*

```
% Script: mexct_7_2.m
Ac = 2;
Mu = 0.7;
fc = 25000;
fm = 2000;
t = [0:999]*1e-6;
x_am = Ac*(1+Mu*cos(2*pi*fm*t)).*cos(2*pi*fc*t);  % Eqn. (7.8)
clf;
k=[-30:30];
ck = ss_efsapprox(x_am,k);
stem(k,ck);
title('EFS coefficients for x_{AM}(t)');
xlabel('Index k');
```

The graph that is generated by the script `mexct_7_2.m` is shown in Fig. 7.36.

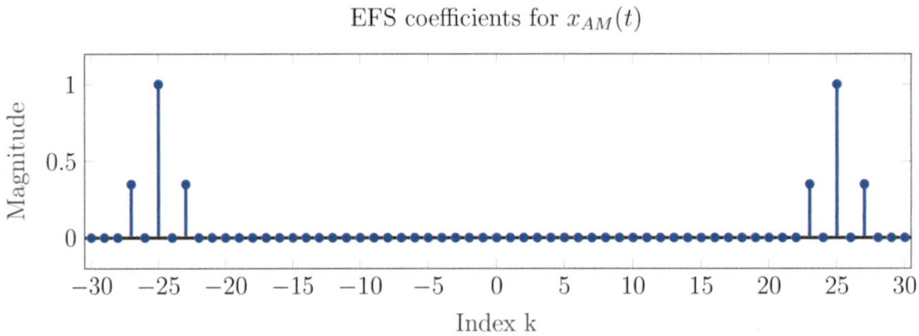

Figure 7.36 – MATLAB graph for line spectrum of tone-modulated AM signal.

Software resources: `mexct_7_2.m` , `ss_switchmod.m`

MATLAB Exercise 7.3: Function to simulate a switching modulator

In this example, we will develop a MATLAB function to simulate the behavior of the switching modulator discussed. In MATLAB, the switching behavior in Eqn. (7.38) can be implemented with the following statements:

```
>> control = (carrier>=0);
>> v_out = v_in.*control;
```

The two statements may also be combined as

```
>> v_out = v_in.*(carrier>=0);
```

The next step is to use a bandpass filter for cleaning up the undesired terms. Function `butter()` will be used for designing a suitable bandpass filter for this purpose (see MATLAB

Exercises 6.1 and 6.9 in Chapter 6).

```
>> [numz,denz] = butter(N,Wn,'bandpass');
```

Once a suitable bandpass filter is designed, function `filter()` is used for processing the switched signal through it. The listing for the function `ss_switchmod()` is given below. The input arguments to the function are

- `msg`: Vector holding samples of the message signal.
- `Bc`: Amplitude of the carrier to be added to the message signal.
- `fc`: Carrier frequency in Hz.
- `Ts`: Time increment in seconds, used for the message signal in vector "`msg`", and also in computing samples of the synthesized AM signal.
- `f1`: Lower edge frequency in Hz for the bandpass filter.
- `f2`: Upper edge frequency in Hz for the bandpass filter.

The function returns a vector holding the samples of the synthesized AM waveform.

```
function x_am = ss_switchmod(msg,Bc,fc,Ts,f1,f2)
  nSamp = length(msg);          % Number of samples in "msg"
  t = [0:nSamp-1]*Ts;           % Vector of time instants
  carrier = Bc*cos(2*pi*fc*t);
  % Compute input to the diode switch.
  v_in = carrier+msg;           % Eqn. (7.37)
  % Simulate the diode switch.
  v_out = v_in.*(carrier>=0);   % Eqn. (7.38)
  % Design the bandpass filter.
  [numz,denz] = butter(5,[2*f1*Ts,2*f2*Ts],'bandpass');
  % Process switch output through bandpass filter.
  x_am = filter(numz,denz,v_out);
```

In this exercise we are designing a discrete-time Butterworth filter so that the action of the switching modulator can be simulated using a discrete step size. The design of discrete-time filters are covered in the companion textbook *Discrete-Time Signals and Systems: A MATLAB Integrated Approach*. For the purpose of this exercise, it is important to understand the relationship between analog frequencies in Hz and the *normalized* frequencies used in discrete-time filter design. Given the frequency interval $f_1 < f < f_2$ Hz that we would like to retain, and a sampling rate of f_s used in the simulation, the normalized edge frequencies of the bandpass filter are

$$F_1 = \frac{f_1}{f_s} \qquad \text{and} \qquad F_2 = \frac{f_2}{f_s}$$

Since MATLAB defines normalized frequencies a bit differently, the values used in the call to function `butter()` are $2F_1 = 2f_1/f_s$ and $2F_2 = 2f_2/f_s$.

Software resource: `ss_switchmod.m`

MATLAB Exercise 7.4: Testing the switching modulator

In this exercise a tone-modulated AM waveform with message frequency $f_m = 2$ kHz, carrier frequency $f_c = 25$ kHz and modulation index $\mu = 0.5$ will be generated in order to test the function `ss_switchmod()` developed in MATLAB Exercise 7.3. Let the sampling interval be $T_s = 1$ ţs corresponding to a sampling rate of $f_s = 1$ MHz.

For the bandpass filter, a passband between $f_1 = 15$ kHz and $f_2 = 35$ kHz is sufficient. MATLAB script to test the switching modulator function is listed below.

```
% Script: mexct_7_4.m
t = [0:999]*1e-6;          % Vector of time instants
msg = cos(2*pi*2000*t);    % Single-tone message signal
Bc = 4/(0.5*pi);           % Achieve a modulation index of 0.5
x_am = ss_switchmod(msg,Bc,25000,1e-6,15000,35000);
% Compute correct envelope for comparison purposes.
x_env = Bc/2*(1+0.5*msg);
% Graph the synthesized AM signal and correct envelope.
plot(t,x_am,t,x_env);  grid;
title('AM signal synthesis using a switching modulator');
xlabel('t (sec)');
ylabel('Amplitude');
legend('Synthesized signal','Correct envelope');
```

The AM signal produced is shown in Fig. 7.37 superimposed with the ideal envelope of the single-tone modulated signal.

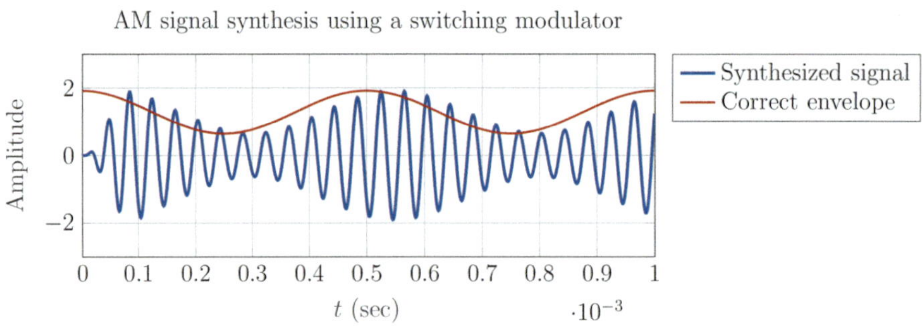

Figure 7.37 – MATLAB graph for switching modulator simulation.

A comparison of the synthesized AM waveform and the correct envelope in Fig. 7.37 indicates a lag in the AM waveform. This is due to the delay characteristics of the fifth-order Butterworth bandpass filter used. This delay could be increased or decreased by changing the filter order or the passband edge frequencies.

Software resource: `mexct_7_4.m`

MATLAB Exercise 7.5: Function to simulate a square-law modulator

The listing for a MATLAB function `ss_sqlawmod()` to implement a square-law modulator is given below. The input arguments to the function are

- `msg`: Vector holding samples of the message signal.
- `Bc`: Amplitude of the carrier to be added to the message signal.
- `fc`: Carrier frequency in Hz.
- `Ts`: Time increment in seconds, used for the message signal in vector "`msg`", and also in computing samples of the synthesized AM signal.
- `a`: The parameter a in the in the input-output characteristic of the nonlinear device in Eqn. (7.47).
- `b`: The parameter b in the in the input-output characteristic of the nonlinear device in Eqn. (7.47).
- `f1`: Lower edge frequency in Hz for the bandpass filter.
- `f2`: Upper edge frequency in Hz for the bandpass filter.

```
function x_am = ss_sqlawmod(msg,Bc,fc,Ts,a,b,f1,f2)
  nSamp = length(msg);            % Number of samples in "msg"
  t = [0:nSamp-1]*Ts;             % Vector of time instants
  carrier = Bc*cos(2*pi*fc*t);
  % Compute input to the nonlinear device.
  v_in = carrier+msg;             % Eqn. (7.48)
  % Simulate the nonlinear device switch.
  v_out = a*v_in+b*v_in.*v_in;  % Eqn. (7.47)
  % Design the bandpass filter.
  [num,den] = butter(5,[2*f1*Ts,2*f2*Ts],'bandpass');
  % Process v_out through bandpass filter.
  x_am = filter(num,den,v_out);
end
```

Software resource: `ss_sqlawmod.m`

MATLAB Exercise 7.6: Testing the square-law modulator

In this exercise we test the function `ss_sqlawmod()` developed in MATLAB Exercise 7.5. A tone-modulated AM waveform with message frequency $f_m = 2$ kHz, carrier frequency $f_c = 25$ kHz, and modulation index $\mu = 0.5$ is used for testing. Based on Eqn. (7.52) the modulation index produced by the square-law modulator using the nonlinear characteristic of Eqn. (7.47) is $\mu = 2b/a$. Choosing $a = 1$ and $b = 0.25$ allows us to obtain $\mu = 0.5$. The sampling interval used in the generation of various signals is $T_s = 1$ micros corresponding to a sampling rate of $f_s = 1$ MHz. MATLAB script `mexct_7_6.m` to test the square-law modulator is listed below.

```
% Script: mexct_7_6.m
t = [0:999]*1e-6;           % Vector of time instants
msg = cos(2*pi*2000*t);  % Single-tone message signal
% Set parameter value for Ac=1 and modulation index of 0.5
Bc = 1;
a = 1;
b = 0.25;
x_am = ss_sqlawmod(msg,Bc,25000,1e-6,1,0.25,15000,35000);
% Compute correct envelope for comparison purposes.
x_env = 1+0.5*msg;
% Graph the synthesized AM signal and correct envelope.
plot(t,x_am,t,x_env); grid;
title('AM signal synthesis using a square-law modulator');
xlabel('t (sec)');
ylabel('Amplitude');
legend('Synthesized signal','Correct envelope');
```

The AM signal produced is shown in Fig. 7.38 superimposed with the ideal envelope of the single-tone modulated signal.

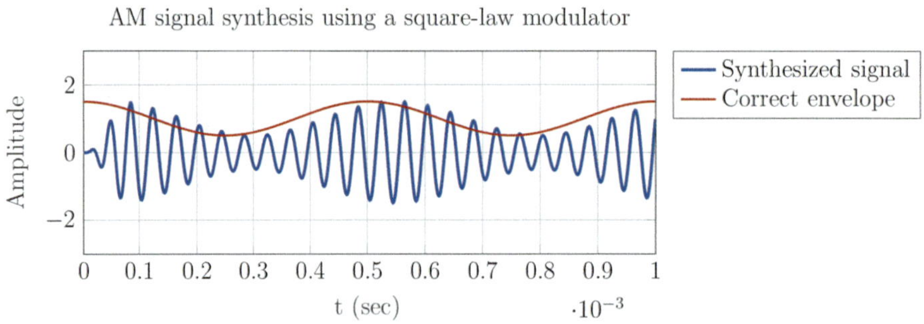

Figure 7.38 – MATLAB graph for square-law modulator simulation.

As in MATLAB Exercise 7.4, comparison of the synthesized AM waveform and the correct envelope in Fig. 7.38 indicates a lag in the AM waveform due to the delay characteristics of the fifth-order Butterworth bandpass filter used.

Software resource: mexct_7_6.m

MATLAB Exercise 7.7: Function to simulate envelope detector

In this exercise we develop a MATLAB function to simulate the actions of the envelope detector described in Section 7.4.4. Refer to the circuit in Fig. 7.21. To mimic the behavior of the envelope detector, two cases need to be considered, namely 1) $x_{AM}(t) \ge v_c(t)$ and 2) $x_{AM}(t) < v_c(t)$. The code for the function ss_envdet() is given below. The input arguments to the function are

x_am:	Vector holding samples of the AM signal.
Ts:	Time increment used in computing the samples of the AM signal.
tau:	Time constant for the RC circuit of the envelope detector.

```
function x_env = ss_envdet(x_am,Ts,tau)
  x_am = x_am.*(x_am>0);    % Rectify the AM signal
  % Create a vector to hold samples of the envelope.
  x_env = zeros(size(x_am));
  nSamp = length(x_am);     % Number of samples in "x_am"
  out = -1;
  % The loop to compute the output.
  for i=1:nSamp,
    inp = x_am(i);
    if (inp>=out)     % Case 1: x_am(t) > vc(t)
      out = inp;
    else              % Case 2: x_am(t) < vc(t),  see Eqn. (7.104)
      out = out*(1-Ts/tau);
    end;
    x_env(i) = out;
  end;
```

Note that, for case 2, we need to solve the differential equation given in Eqn. (7.54) using a numerical approximation technique. Evaluating the differential equation at $t = t_0$ we have

$$\left.\frac{dv_c}{dt}\right|_{t=t_0} = -\frac{1}{\tau}\,v_c(t_0) \tag{7.102}$$

Replacing the derivative on the left side of Eqn. (7.102) with a finite difference approximation at time $t = t_0$ we obtain

$$\frac{v_c(t_0+T_s)-v_c(t_0)}{T_s} \approx -\frac{1}{\tau}\,v_c(t_0) \tag{7.103}$$

Solving Eqn. (7.103) for $v_c(t_0+T_s)$ yields

$$v_c(t_0+T_s) \approx v_c(t_0)\left[1-\frac{T_s}{\tau}\right] \tag{7.104}$$

This result allows us to approximate the output voltage at time $t = t_0+T_s$ from the knowledge of its value at time $t = t_0$, and can be used iteratively to approximate the output at time instants t_0+2T_s, t_0+3T_s, and so on. It is a special case of what is known as Euler's method for numerically solving a first-order differential equation one time step at a time (see MATLAB Exercise 2.4 in Chapter 2).

Software resource: `ss_envdet.m`

MATLAB Exercise 7.8: Testing the envelope detector function

In this exercise the envelope detector function `ss_envdet()` developed in MATLAB Example 7.7 will be tested. For the sake of simplicity, a single-tone modulated AM signal will be used for testing. The single-tone message frequency is $f_m = 1.5$ and the carrier frequency is $f_c =$ 10 kHz. The modulation index is $\mu = 0.6$. Let the time constant of the envelope detector circuit be chosen as $\tau = 0.1$ ms. The MATLAB script that generates an AM waveform with the prescribed parameter values and then simulates an envelope detector to demodulate it is listed below:

```
% Script: mexct_7_8.m
time = [1:999]*1e-6;        % Vector of time instants
fc = 10000;                 % Carrier frequency
fm = 1500;                  % Single-tone message frequency
tau = 0.0001;               % Time constant for envelope detector
mu = 0.6;                   % Modulation index
% Compute the AM waveform.
x_AM = (1+mu*cos(2*pi*fm*time)).*cos(2*pi*fc*time);
% Compute the ideal envelope.
x_env = abs(1+mu*cos(2*pi*fm*time));
% Compute the envelope detector output.
x_det = ss_envdet(x_AM,1e-6,tau);
% Graph the AM waveform, its envelope, and the detector output.
plot(1000*time,x_AM,1000*time,x_env,1000*time,x_det);
title('Envelope detector simulation');
ylabel('Amplitude');
xlabel('t (ms)');
legend('AM waveform','Ideal envelope','Detector output');
```

The MATLAB graph that is produced by the script listed above is shown in Fig. 7.24.

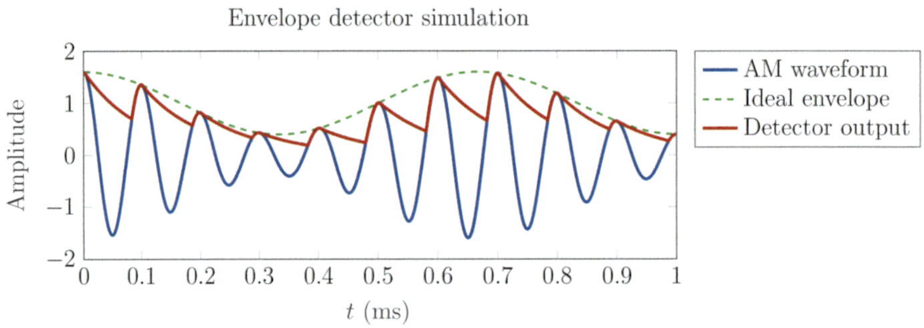

Figure 7.39 – MATLAB graph for envelope detector simulation with $f_c = 10$ kHz, $f_m = 1.5$ kHz, $\mu = 0.6$, and $\tau = 0.1$ ms.

It can be seen that our choice of parameter values above produces a properly detected envelope in Fig. 7.39. (Compare the detector output signal to the correct envelope of the AM waveform.) Let us increase the time constant to $\tau = 0.3$ ms and observe the resulting envelope detector output by simply modifying one line of the code:

```
tau = 0.0003;
```

The new value of the time constant leads to the graph shown in Fig. 7.40.

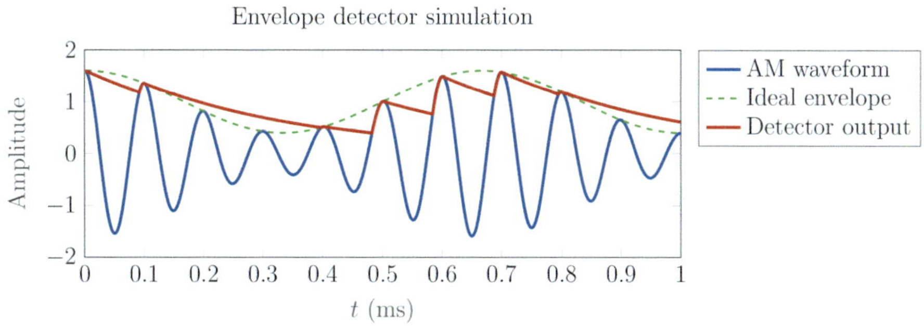

Figure 7.40 – MATLAB graph for envelope detector simulation with $f_c = 10$ kHz, $f_m = 1.5$ kHz, $\mu = 0.6$, and $\tau = 0.3$ ms.

Observe how the envelope detector loses track of the envelope as a result of a time constant that is too large.

Software resource: `mexct_7_8.m`

Problems

7.1. Consider the single-tone modulated AM signal $x_{AM}(t)$ given by Eqn. (7.7). For each set of parameters listed below, determine the missing parameter value.

a. $A_m = 2\,, \quad \mu = 0.7\,, \quad A_c = ?$

b. $A_m = 3\,, \quad \mu = 0.85\,, \quad A_c = ?$

c. $A_c = 2\,, \quad \mu = 0.9\,, \quad A_m = ?$

d. $A_c = 3\,, \quad A_m = 2.5\,, \quad \mu = ?$

7.2. Manually sketch each single-tone modulated AM signal the parameters of which are given below. Indicate critical values on each axis.

a. $A_c = 3\,, \quad \mu = 0.6\,, \quad f_c = 10 \text{ Hz}\,, \quad f_m = 2 \text{ Hz}$

b. $A_c = 3\,, \quad \mu = 1.3\,, \quad f_c = 10 \text{ Hz}\,, \quad f_m = 2 \text{ Hz}$

c. $A_c = 5\,, \quad A_m = 4\,, \quad f_c = 25 \text{ Hz}\,, \quad f_m = 8 \text{ Hz}$

7.3. A sinusoidal carrier is in the form

$$x_c(t) = A_c \cos(20\pi t)$$

For each modulating signal listed below, the modulation index is required to be $\mu = 0.7$. Determine the parameter A_c in each case.

a. $m(t) = \cos(4\pi t) + 2\cos(6\pi t)$

b. $m(t) = \cos(4\pi t) + 3\sin(6\pi t)$

c. $m(t) = 3\sin(4\pi t) + 2\cos(6\pi t - \pi/3)$

7.4. Refer to the tone-modulated AM signals the parameters of which are listed below. For each, draw the phasors $\boldsymbol{X}_{\text{car}}$, $\tilde{\boldsymbol{X}}_{\text{usb}}$, and $\tilde{\boldsymbol{X}}_{\text{lsb}}$ at the time instant indicated.

a. $A_c = 3\,,\quad \mu = 0.6\,,\quad f_c = 10\ \text{Hz}\,,\quad f_m = 2\ \text{Hz}\,,\quad t = 0.35\ \text{s}$

b. $A_c = 3\,,\quad \mu = 1.3\,,\quad f_c = 10\ \text{Hz}\,,\quad f_m = 2\ \text{Hz}\,,\quad t = 0.6\ \text{s}$

c. $A_c = 5\,,\quad A_m = 4\,,\quad f_c = 25\ \text{Hz}\,,\quad f_m = 8\ \text{Hz}\,,\quad t = 150\ \text{ms}$

7.5. Refer to the tone-modulated AM signals the parameters of which are listed in Problem 7.1. Compute and graph the frequency spectrum of each signal.

7.6. Refer to Problem 7.3. For each modulating signal determine and sketch the frequency spectrum of $x_{AM}(t)$. Afterward determine the modulation efficiency η. (Use the value of A_c computed in Problem 7.2 for each case.)

7.7. The frequency spectrum $M(f)$ of a message signal $m(t)$ is shown in Fig. P.7.7.

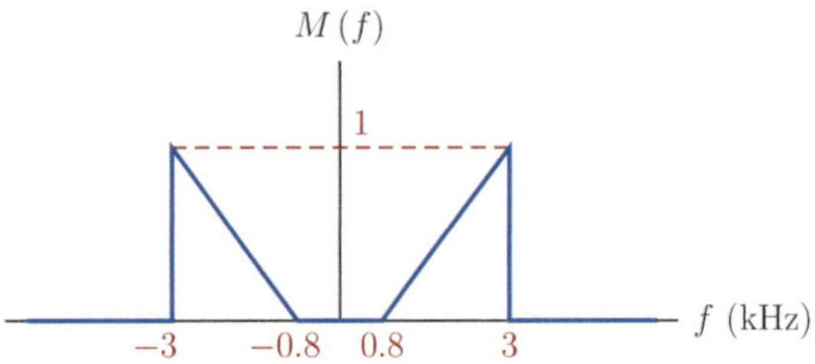

Figure P. 7.7

The signal $m(t)$ is used as input to the AM modulator shown in Fig. 7.14. The carrier amplitude is $A_c = 1.5$ and the carrier frequency is $f_c = 10$ kHz. Sketch the frequency spectrum of the signal $x_{AM}(t)$.

7.8. Refer to the message signal $m(t)$ the spectrum of which is shown in Fig. P.7.7. Assume that this signal is used as input to the AM modulator shown in Fig. 7.20. The carrier amplitude is $B_c = 1.5$ and the carrier frequency is $f_c = 10$ kHz. The nonlinear device has the input-output characteristic

$$v_{out}(t) = 2\,v_{in}(t) + 0.8\,v_{in}^2(t)$$

a. Determine and sketch the frequency spectrum of the output signal at the output of the nonlinear device.

b. Specify the appropriate bandpass filter to produce the AM signal $x_{AM}(t)$.

7.9. A single-tone modulated signal is shown in Fig. P.7.9.

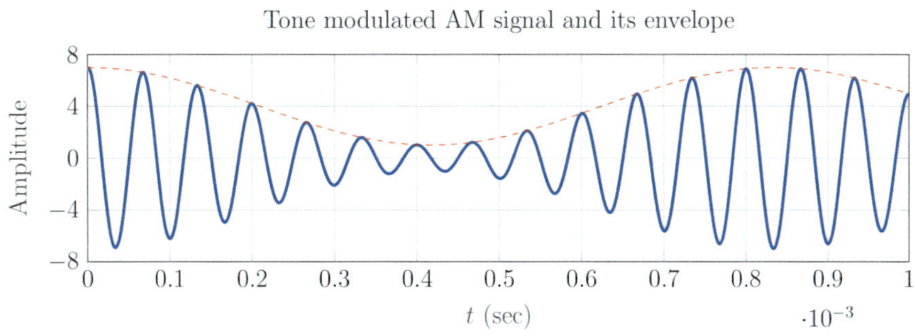

Figure P. 7.9

Estimate the following from the figure:

a. Carrier frequency f_c,
b. single-tone message frequency f_m,
c. the modulation index, and
d. the efficiency of modulation

7.10. Refer to Problem 7.3. For each modulating signal determine the efficiency of modulation using the value of A_c computed in Problem 7.3.

7.11. Consider the switching modulator shown in Fig. 7.15. Let the message signal be the single tone

$$m(t) = 3\cos(200\pi t)$$

The carrier amplitude is $B_c = 5$, and the carrier frequency is $f_c = 800$ Hz.

a. Sketch the signal $v_{in}(t)$.
b. Assuming the diode is ideal, sketch the signal $v_{out}(t)$.
c. Determine the magnitude characteristic of an ideal bandpass filter that would be appropriate for use in this switching modulator.

7.12. Consider the *balanced modulator* block diagram shown in Fig. P.7.12 which can be used for producing DSB with suppressed carrier without the need for filtering.

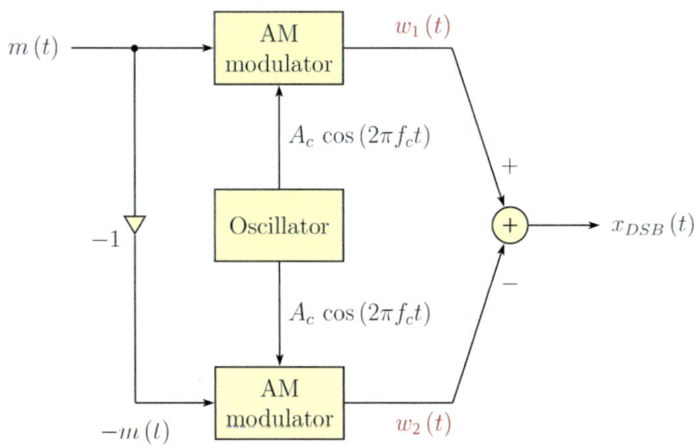

Figure P. 7.12

Determine the intermediate signals $w_1(t)$ and $w_2(t)$. Show that the output signal is the proper DSB modulated signal with the carrier suppressed.

7.13. Refer to the message signal $m(t)$ the frequency spectrum $M(f)$ of which is shown in Fig. P.7.7. A sinusoidal carrier with amplitude is $A_c = 1.5$ and the carrier frequency is $f_c = 10$ kHz is to be used in conjunction with this signal.

- **a.** If the signal $m(t)$ is used for modulating a sinusoidal carrier using DSB, sketch the frequency spectrum of the signal $x_{DSB}(t)$.
- **b.** Repeat part (a) using SSB with the upper sideband retained.
- **c.** Repeat part (a) using SSB with the lower sideband retained.

7.14. A single-tone modulated AM signal given by

$$x_{AM}(t) = [4 + 3.2\cos(200\pi t)]\cos(2000\pi t)$$

is applied to the envelope detector shown in Fig. 7.21. Determine an appropriate time constant for the envelope detector so that the single-tone message signal is recovered correctly. Based on the time-constant found, pick practical values for R and C.

7.15. Refer to the coherent demodulator shown in Fig.P.7.15. Assume that the local oscillator signal is not perfectly synchronized with the original carrier, but rather has a phase error θ as shown.

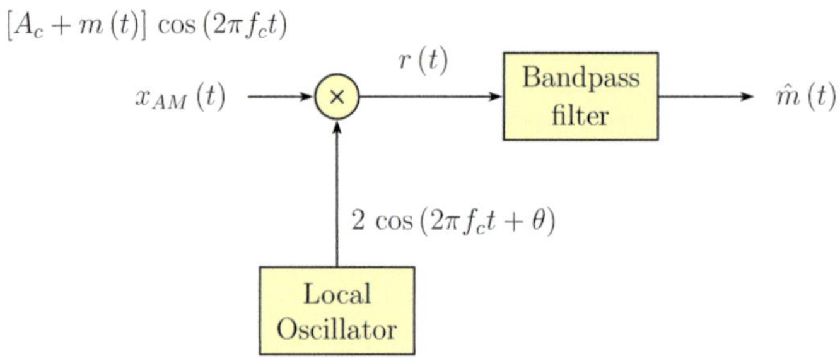

Figure P. 7.15

Determine the signal $r(t)$ and the bandpass filter output $\hat{m}(t)$. Explain the effect of the phase error on these signals.

7.16. Refer to the coherent demodulator shown in Fig. P.7.16. Assume that the local oscillator signal is not perfectly synchronized with the original carrier, but rather has a frequency error so that the local carrier frequency is $f_c + \Delta f$.

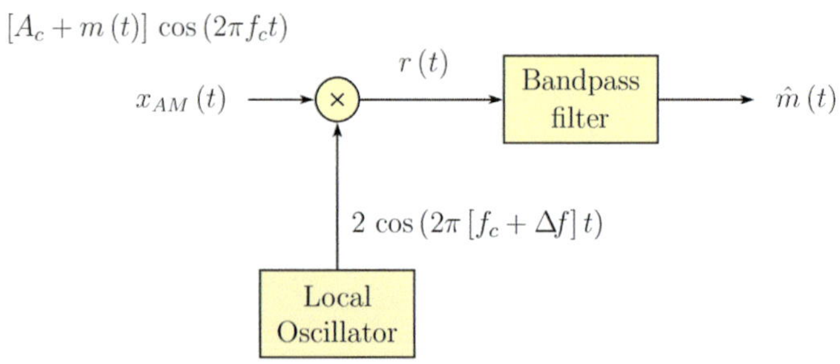

Figure P. 7.16

Determine the signal $r(t)$ and the bandpass filter output $\hat{m}(t)$. Explain the effect of the frequency error on these signals.

MATLAB Problems

7.17. Refer to Problem 7.2. For each set of parameters given, develop a MATLAB script to compute and graph the corresponding AM signal. In each case choose an appropriate sampling interval and produce a graph that spans at least three periods of the modulating tone.

7.18. Refer to Problem 7.3. For each modulating signal write a script to compute and graph $x_{AM}(t)$. Choose the sampling interval appropriately, and graph the corresponding signal in a time interval sufficient to show relevant details.

7.19. Refer to Problem 7.18. Determine the fundamental frequency and the fundamental period for each multitone-modulated AM signal. Write a script for each case to produce 1024 samples of exactly one period. Compute and graph the approximate EFS spectrum using the function `ss_efsapprox()`.

7.20. Refer to Problem 7.11. Write a MATLAB script to

- **a.** Compute and graph the signal $v_{in}(t)$.
- **b.** Compute and sketch the signal $v_{out}(t)$ assuming that the diode is ideal.
- **c.** Use the function `ss_switchmod()` developed in MATLAB Exercise 7.3 to compute the output signal $x_{AM}(t)$. Graph the result.

7.21. Using the `ss_switchmod()` as the starting point, develop a new MATLAB function `ss_switchmod2()` to simulate a switching modulator that uses the more realistic diode model shown in Fig. P.7.21 instead of the ideal diode model.

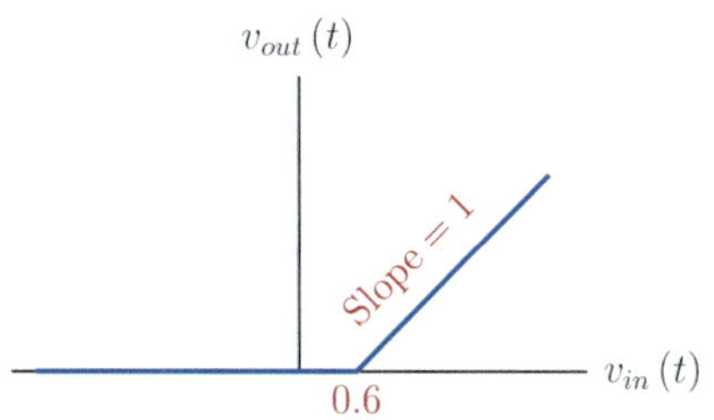

Figure P. 7.21

Afterward rework Problem 7.20 using the function `ss_switchmod2()` instead of `ss_switchmod()`. Compare the results.

7.22. Refer to Problem 7.14. Write a MATLAB script to generate samples of the signal $x_{AM}(t)$ and then to simulate the envelope detector using the function `ss_envdet()` developed in MATLAB Exercise 7.7. First, use the value of τ found in Problem 7.14. Afterward repeat with half the value found and with twice the value found. Graph the envelope detector output signal for each case and compare the results.

MATLAB Projects

7.23.

a. Develop a MATLAB function `ss_coherent()` to simulate the coherent demodulator shown in Fig. 7.24. The syntax of the function should be as follows:

```
>> mhat = ss_coherent(xAM,Ts,fc,theta,deltaf)
```

The input arguments to the function are

`xAM`:	Vector holding samples of the AM signal.
`Ts`:	Sampling interval used for signal in vector "`xAM`".
`fc`:	Carrier frequency in Hz.
`theta`:	Phase error of the local carrier (in radians).
`deltaf`:	Frequency error of the local carrier (in Hz).

The vector `mhat` holds samples of the output signal. Internally, implement the bandpass filter as the cascade of two filters: A fifth-order Butterworth lowpass filter with the appropriate cutoff frequency, and a sixth-order Chebyshev type-I highpass filter to remove frequencies less than 2 percent of the carrier frequency.

b. Test the function `ss_coherent()` with the signal

$$x_{AM}(t) = [4 + 3.2\cos(200\pi t)]\cos(2000\pi t)$$

using a sampling rate of $f_s = 25$ kHz. Set phase and frequency errors equal to zero. Generate samples of the AM signal for $t = 0, \ldots, 100$ ms. Compute and graph the simulated output signal $\hat{m}(t)$.

c. Repeat part (b) using a phase error of $\theta = \pi/10$ radians.

d. Repeat part (b) using a frequency error of $\Delta f = 60$ Hz.

7.24. A SSB modulated signal with lower sideband retained was given by Eqn. (7.101) and is repeated here:

$$\text{Eqn. (7.101):} \qquad x_{SSB,L}(t) = \frac{1}{2}\,m(t)\cos\left(2\pi f_c t\right) + \frac{1}{2}\,\hat{m}(t)\sin\left(2\pi f_c t\right) \tag{7.105}$$

The term $\hat{m}(t)$ represents the Hilbert transform of $m(t)$

a. Using the `fdatool` program in MATLAB, design a length-55 FIR filter to approximate an ideal Hilbert transform filter. Export the impulse response of the designed filter to a data file.

b. Let the message signal be

$$m(t) = 2\cos(200\pi t)$$

and use the carrier frequency $f_c = 1000$ Hz. Write a script to compute the signal $x_{SSB,L}(t)$ as given above. Your script should

- load the length-55 Hilbert transform filter impulse from data file,
- generate samples of $m(t)$ using a sampling rate of $f_c = 25$ kHz,
- process $m(t)$ through the FIR filter designed to obtain an approximation to the Hilbert transform $\hat{m}(t)$, and
- compute an approximation to $x_{SSB,L}(t)$ as specified. Hint: Keep in mind that the length-55 FIR filter has a delay of $(55-1)/2 = 27$ samples, so it produces an approximation to a delayed version of $\hat{m}(t)$.

APPENDIX A

COMPLEX NUMBERS AND EULER'S FORMULA

A.1 Introduction

Complex numbers allow us to solve equations for which no real solution can be found. Consider, for example, the equation

$$x^2 + 9 = 0 \tag{A.1}$$

which cannot be satisfied for any real number. By introducing an imaginary unit[1] $j = \sqrt{-1}$ so that $j^2 = -1$, Eqn. (A.1) can be solved to yield

$$x = \mp j3$$

Similarly, the equation

$$(x+2)^2 + 9 = 0 \tag{A.2}$$

has the solutions

$$x = -2 \mp j3 \tag{A.3}$$

A general complex number is in the form

$$x = a + jb \tag{A.4}$$

where a and b are real numbers. The values a and b are referred to as the *real part* and *imaginary part* of the complex number x, respectively. Following notation is used for real and imaginary

---

[1] The imaginary unit was first introduced by Italian mathematician Gerolamo Cardano as he worked on the solutions of cubic and quartic equations, although its utility was not fully understood and appreciated until the works of Leonhard Euler and Carl Friedrich Gauss almost two centuries later.

DOI: 10.1201/9781032675251-A

parts of a complex number:

$$a = \mathrm{Re}\{x\}$$
$$b = \mathrm{Im}\{x\}$$

For two complex numbers to be equal, both their real parts and imaginary parts must be equal. Given two complex numbers $x_1 = a + jb$ and $x_2 = c + jd$, the equality $x_1 = x_2$ implies that $a = c$ and $b = d$.

It is often convenient to graphically represent a complex number as a point in the *complex plane* constructed using a horizontal axis corresponding to the real part and a vertical axis corresponding to the imaginary part as depicted in Fig. A.1.

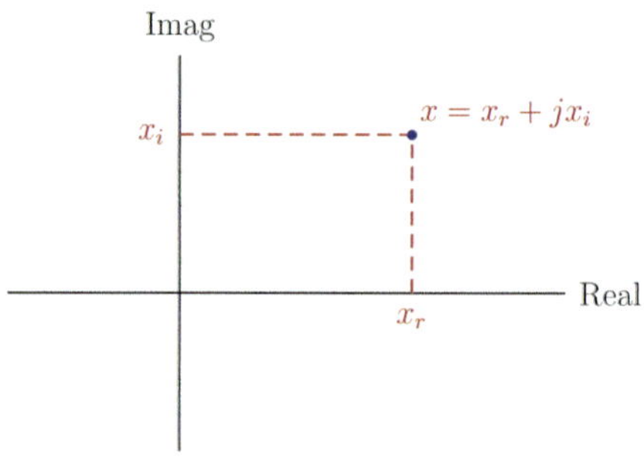

Figure A.1 – Complex number $x = x_r + jx_i$ shown as a point in the complex plane.

If the imaginary part of a complex number is equal to zero, the resulting number is said to be *purely real.* An example is $x = 3 + j0$. Similarly, a complex number the real part of which is equal to zero is said to be *purely imaginary.* An example of a purely imaginary number is $x = 0 + j5$. Purely real and purely imaginary numbers are simply special cases of the more general class of complex numbers.

Another method of graphical representation of a complex number can be found through the use of a vector as shown in Fig. A.2. Vector representation of complex numbers is quite useful since vector operations can be used for manipulating complex numbers.

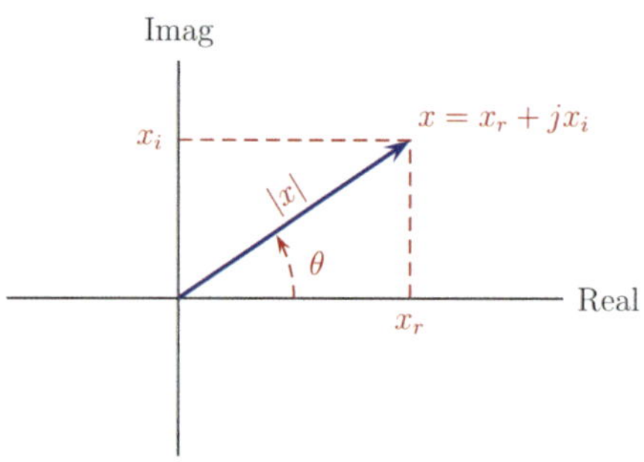

Figure A.2 – Complex number $x = x_r + jx_i$ shown as a vector.

A complex number written as the sum of its real and imaginary parts such as $x = x_r + jx_i$ is said to be in *cartesian form.* An alternative form referred to as a *polar form* is derived from the vector representation of a complex number by using the norm $|x|$ and the angle θ of the vector instead of the real and imaginary parts x_r and x_i. The two forms of the complex number x are:

$$x = x_r + j x_i \,, \quad \text{Cartesian form}$$

$$x = |x| \, e^{j\theta} \,, \quad \text{Polar form}$$

The norm $|x|$ is the distance of the point representing the complex number from the origin. The angle θ is the angle measured counterclockwise starting with the positive real axis.

A complex number can easily be converted from cartesian form to polar form and vice versa. Through simple geometric relationships it can be shown that

$$|x| = \sqrt{x_r^2 + x_i^2} \tag{A.5}$$

and

$$\theta = \tan^{-1}\left(\frac{x_i}{x_r}\right) \tag{A.6}$$

Alternately, real and imaginary parts can be obtained from the norm and the angle through

$$x_r = |x| \cos(\theta) \tag{A.7}$$

and

$$x_i = |x| \sin(\theta) \tag{A.8}$$

A.2 Arithmetic with Complex Numbers

Definitions of arithmetic operators are easily extended to apply to complex numbers. In Section A.2.1 addition and subtraction operators will be discussed. Multiplication and division of complex numbers will be discussed in Section A.2.2.

A.2.1 Addition and subtraction

Consider two complex numbers $x_1 = a + jb$ and $x_2 = c + jd$. Addition of these two complex numbers is carried out as follows:

$$\begin{aligned} y &= x_1 + x_2 \\ &= (a + jb) + (c + jd) \\ &= (a + c) + j(b + d) \end{aligned} \tag{A.9}$$

Real and imaginary parts of the result can be written separately as

$$\begin{aligned} \text{Re}\{y\} &= \text{Re}\{x_1\} + \text{Re}\{x_2\} \\ \text{Im}\{y\} &= \text{Im}\{x_1\} + \text{Im}\{x_2\} \end{aligned}$$

Subtraction of complex numbers is similar to addition. The difference $z = x_1 - x_2$ is computed as follows:

$$\begin{aligned} z &= x_1 - x_2 \\ &= (a + jb) - (c + jd) \\ &= (a - c) + j(b - d) \end{aligned} \tag{A.10}$$

Real and imaginary parts of the result can be written separately as

$$\begin{aligned} \mathrm{Re}\{z\} &= \mathrm{Re}\{x_1\} - \mathrm{Re}\{x_2\} \\ \mathrm{Im}\{z\} &= \mathrm{Im}\{x_1\} - \mathrm{Im}\{x_2\} \end{aligned}$$

Addition and subtraction operators for complex numbers are analogous for addition and subtraction of vectors using the parallelogram rule. This is illustrated in Fig. A.3.

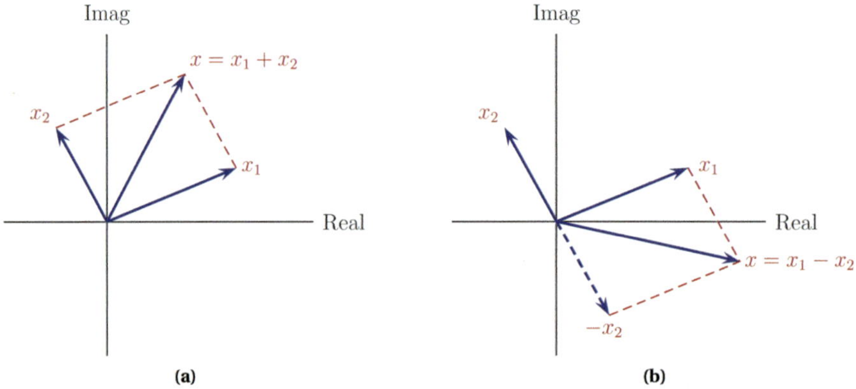

Figure A.3 – Addition and subtraction operators with complex vectors.

If the complex numbers to be added are expressed in polar form, they must first be converted to cartesian form before they can be added. Consider two complex numbers in polar form given as

$$x_1 = |x_1|\, e^{j\theta_1}, \quad \text{and} \quad x_2 = |x_2|\, e^{j\theta_2} \tag{A.11}$$

The sum z of these complex numbers is found as

$$\begin{aligned} z &= x_1 + x_2 \\ &= |x_1| \cos(\theta_1) + j\,|x_1| \sin(\theta_1) + |x_2| \cos(\theta_2) + j\,|x_2| \sin(\theta_2) \\ &= \Big[\, |x_1| \cos(\theta_1) + |x_2| \cos(\theta_2) \Big] + j \Big[\, |x_1| \sin(\theta_1) + |x_2| \sin(\theta_2) \Big] \end{aligned} \tag{A.12}$$

which can be put into polar form as

$$z = |z|\, e^{j\theta_z} \tag{A.13}$$

where

$$|z| = |x_1|^2 + |x_2|^2 + 2\,|x_1|\,|x_2| \cos(\theta_1 - \theta_2) \tag{A.14}$$

and

$$\theta_z = \tan^{-1}\left(\frac{|x_1| \sin(\theta_1) + |x_2| \sin(\theta_2)}{|x_1| \cos(\theta_1) + |x_2| \cos(\theta_2)} \right) \tag{A.15}$$

A.2.2 Multiplication and division

Again consider two complex numbers $x_1 = a + jb$ and $x_2 = c + jd$. The product of these two complex numbers is computed as follows:

$$\begin{aligned} y =& x_1\, x_2 \\ =& (a+jb)\,(c+jd) \\ =& ac + j\,ad + j\,bc + j^2\,bd \\ =& (ac-bd) + j\,(ad+bc) \end{aligned} \tag{A.16}$$

Real and imaginary parts of the result can be written separately as

$$\begin{aligned} \mathrm{Re}\{y\} &= \mathrm{Re}\{x_1\}\,\mathrm{Re}\{x_2\} - \mathrm{Im}\{x_1\}\,\mathrm{Im}\{x_2\} \\ \mathrm{Im}\{y\} &= \mathrm{Re}\{x_1\}\,\mathrm{Im}\{x_2\} + \mathrm{Im}\{x_1\}\,\mathrm{Re}\{x_2\} \end{aligned}$$

Product of a complex number and its own complex conjugate is equal to the squared norm of the complex number:

$$\begin{aligned} x_1\, x_1^* =& (a+jb)\,(a+jb)^* \\ =& (a+jb)\,(a-jb) \\ =& a^2 + b^2 = |x_1|^2 \end{aligned}$$

Division of complex numbers is slightly more involved. Consider the complex number z defined as

$$z = \frac{x_1}{x_2} = \frac{(a+jb)}{(c+jd)} \tag{A.17}$$

The expression in Eqn. (A.17) can be simplified by multiplying both the numerator and the denominator by the complex conjugate of the denominator:

$$z = \frac{x_1}{x_2} = \frac{(a+jb)\,(c-jd)}{(c+jd)\,(c-jd)} \tag{A.18}$$

A.3 Euler's Formula

A complex exponential function can be expressed in the form

$$e^{jx} = \cos(x) + j\,\sin(x) \tag{A.19}$$

This relationship is known as *Euler's formula* and will be used extensively in working with signals, linear systems, and various transforms. If the sign of x is changes in Eqn. (A.20) we get

$$e^{-jx} = \cos(-x) + j\,\sin(-x) = \cos(x) - j\,\sin(x) \tag{A.20}$$

Using Eqns. (A.19) and (A.2) trigonometric functions $\cos(x)$ and $\sin(x)$ may be expressed in terms of complex exponential functions as

$$\cos(x) = \frac{e^{jx} + e^{-jx}}{2} \tag{A.21}$$

and

$$\sin(x) = \frac{e^{jx} - e^{-jx}}{2j} \tag{A.22}$$

Appendix B

Mathematical Relations

B.1 Trigonometric Identities

$$\cos(a \pm b) = \cos(a)\cos(b) \mp \sin(a)\sin(b) \tag{B.1}$$

$$\sin(a \pm b) = \sin(a)\cos(b) \pm \cos(a)\sin(b) \tag{B.2}$$

$$\tan(a \pm b) = \frac{\tan(a) \pm \tan(b)}{1 \mp \tan(a)\tan(b)} \tag{B.3}$$

$$\cos(a)\cos(b) = \tfrac{1}{2}\cos(a+b) + \tfrac{1}{2}\cos(a-b) \tag{B.4}$$

$$\sin(a)\sin(b) = \tfrac{1}{2}\cos(a-b) - \tfrac{1}{2}\cos(a+b) \tag{B.5}$$

$$\sin(a)\cos(b) = \tfrac{1}{2}\sin(a+b) + \tfrac{1}{2}\sin(a-b) \tag{B.6}$$

$$\cos(a) + \cos(b) = 2\cos\left(\frac{a+b}{2}\right)\cos\left(\frac{a-b}{2}\right) \tag{B.7}$$

$$\cos(a) - \cos(b) = -2\sin\left(\frac{a+b}{2}\right)\sin\left(\frac{a-b}{2}\right) \tag{B.8}$$

DOI: 10.1201/9781032675251-B

$$\sin(a) + \sin(b) = 2\sin\left(\frac{a+b}{2}\right)\cos\left(\frac{a-b}{2}\right) \tag{B.9}$$

$$\sin(a) - \sin(b) = 2\sin\left(\frac{a-b}{2}\right)\cos\left(\frac{a+b}{2}\right) \tag{B.10}$$

$$\cos(2a) = \cos^2(a) - \sin^2(a) \tag{B.11}$$

$$\sin(2a) = 2\sin(a)\cos(a) \tag{B.12}$$

$$\tan(2a) = \frac{2\tan(a)}{1-\tan^2(a)} \tag{B.13}$$

$$\cos^2(a) = \tfrac{1}{2} + \tfrac{1}{2}\cos(2a) \tag{B.14}$$

$$\sin^2(a) = \tfrac{1}{2} - \tfrac{1}{2}\cos(2a) \tag{B.15}$$

B.2 Indefinite Integrals

$$\int x e^{ax}\,dx = \frac{ax-1}{a^2}\,e^{ax} \tag{B.16}$$

$$\int \frac{1}{x^2+a^2}\,dx = \frac{1}{a}\tan^{-1}\left(\frac{x}{a}\right) \tag{B.17}$$

$$\int \frac{x}{x^2+a^2}\,dx = \frac{1}{2}\ln\left(x^2+a^2\right) \tag{B.18}$$

$$\int \frac{x^2}{x^2+a^2}\,dx = x - a\tan^{-1}\left(\frac{x}{a}\right) \tag{B.19}$$

$$\int x\cos(ax)\,dx = \frac{1}{a^2}\left[\cos(ax) + ax\sin(ax)\right] \tag{B.20}$$

$$\int x\sin(ax)\,dx = \frac{1}{a^2}\left[\sin(ax) - ax\cos(ax)\right] \tag{B.21}$$

$$\int e^{ax}\cos(bx)\,dx = \frac{e^{ax}}{a^2+b^2}\left[a\cos(bx) + b\sin(bx)\right] \tag{B.22}$$

$$\int e^{ax}\sin(bx)\,dx = \frac{e^{ax}}{a^2+b^2}\left[a\sin(bx) - b\cos(bx)\right] \tag{B.23}$$

$$\int (a+bx)^n\,dx = \frac{(a+bx)^{n+1}}{b(n+1)}, \qquad n>0 \tag{B.24}$$

$$\int \frac{1}{(a+bx)^n}\,dx = \frac{-1}{b(n-1)(a+bx)^{n-1}}, \qquad n>1 \tag{B.25}$$

B.3 Laplace Transform Pairs

Signal	Transform	ROC
$\delta(t)$	1	all s
$u(t)$	$\frac{1}{s}$	$\mathrm{Re}\{s\} > 0$
$u(-t)$	$-\frac{1}{s}$	$\mathrm{Re}\{s\} < 0$
$e^{at}\,u(t)$	$\frac{1}{s-a}$	$\mathrm{Re}\{s\} > a$
$-e^{at}\,u(-t)$	$\frac{1}{s-a}$	$\mathrm{Re}\{s\} < a$
$e^{j\omega_0 t}\,u(t)$	$\frac{1}{s-j\omega_0}$	$\mathrm{Re}\{s\} > 0$
$e^{-\lvert t\rvert}$	$\frac{-2}{s^2-1}$	$-1 < \mathrm{Re}\{s\} < 1$
$\Pi\left(\frac{t-\tau/2}{\tau}\right)$	$\frac{1-e^{-s\tau}}{s}$	$\mathrm{Re}\{s\} > -\infty$
$\cos(\omega_0 t)\,u(t)$	$\frac{s}{s^2+\omega_0^2}$	$\mathrm{Re}\{s\} > 0$
$\sin(\omega_0 t)\,u(t)$	$\frac{\omega_0}{s^2+\omega_0^2}$	$\mathrm{Re}\{s\} > 0$
$e^{at}\cos(\omega_0 t)\,u(t)$	$\frac{s-a}{(s-a)^2+\omega_0^2}$	$\mathrm{Re}\{s\} > a$
$e^{at}\sin(\omega_0 t)\,u(t)$	$\frac{\omega_0}{(s-a)^2+\omega_0^2}$	$\mathrm{Re}\{s\} > a$

B.4 z-Transform Pairs

Signal	Transform	ROC
$\delta[n]$	1	all z
$u[n]$	$\dfrac{z}{z-1}$	$\lvert z\rvert > 1$
$u[-n]$	$\dfrac{-z}{z-1}$	$\lvert z\rvert < 1$
$a^n\, u[n]$	$\dfrac{z}{z-a}$	$\lvert z\rvert > \lvert a\rvert$
$-a^n\, u[-n-1]$	$\dfrac{z}{z-a}$	$\lvert z\rvert < \lvert a\rvert$
$\cos(\Omega_0 n)\, u[n]$	$\dfrac{z\,[z-\cos(\Omega_0)]}{z^2-2\cos(\Omega_0)\,z+1}$	$\lvert z\rvert > 1$
$\sin(\Omega_0 n)\, u[n]$	$\dfrac{\sin(\Omega_0)\,z}{z^2-2\cos(\Omega_0)\,z+1}$	$\lvert z\rvert > 1$
$a^n \cos(\Omega_0 n)\, u[n]$	$\dfrac{z\,[z-a\cos(\Omega_0)]}{z^2-2a\cos(\Omega_0)\,z+a^2}$	$\lvert z\rvert > \lvert a\rvert$
$a^n \sin(\Omega_0 n)\, u[n]$	$\dfrac{a\sin(\Omega_0)\,z}{z^2-2a\cos(\Omega_0)\,z+a^2}$	$\lvert z\rvert > \lvert a\rvert$
$n\,a^n\, u[n]$	$\dfrac{az}{(z-a)^2}$	$\lvert z\rvert > \lvert a\rvert$

APPENDIX C

Closed Forms for Sums of Geometric Series

Summations of geometric series appear often in problems involving various transforms and in the analysis of linear systems. In this appendix we give derivations of closed-form formulas for infinite and finite-length geometric series.

C.1 Infinite-Length Geometric Series

Consider the sum of infinite-length geometric series in the form

$$P = \sum_{n=0}^{\infty} a^n \tag{C.1}$$

For the summation in Eqn. (C.1) to converge we must have $|a| < 1$. Assuming that is the case, let us write Eqn. (C.1) in open form:

$$P = 1 + a + a^2 + a^3 + \ldots \tag{C.2}$$

Subtracting unity from both sides of Eqn. (C.2) we obtain

$$P - 1 = a + a^2 + a^3 + \ldots \tag{C.3}$$

in which the terms on the right side of the equal sign have a common factor a. Factoring it out leads to

$$\begin{aligned} P - 1 &= a\left(1 + a + a^2 + a^3 + \ldots\right) \\ &= aP \end{aligned} \tag{C.4}$$

DOI: 10.1201/9781032675251-C

which can be solved for P to yield

$$P = \frac{1}{1-a}, \qquad |a| < 1 \tag{C.5}$$

C.2 Finite-Length Geometric Series

Consider the finite-length sum of a geometric series in the form

$$Q = \sum_{n=0}^{L} a^n \tag{C.6}$$

Unlike P of the previous section, the convergence of Q does not require $|a| < 1$. The only requirement is that $|a| < \infty$. Let us write Q in open form:

$$Q = 1 + a + \ldots + a^L \tag{C.7}$$

Subtracting unity from both sides of Eqn. (C.7) leads to

$$Q - 1 = a + \ldots + a^L \tag{C.8}$$

As before, the terms on the right side of the equal sign have a common factor a which can be factored out to yield

$$\begin{aligned} Q-1 &= a\left(1+\ldots+a^{L-1}\right) \\ &= a\left(Q-a^L\right) \end{aligned} \tag{C.9}$$

which can be solved for Q:

$$Q = \frac{1-a^{L+1}}{1-a} \tag{C.10}$$

Consistency check: If L is increased, in the limit Q approaches P provided that $|a| < 1$, that is

$$\lim_{L\to\infty} \left[Q\right] = P \quad \text{if} \quad |a| < 1 \tag{C.11}$$

C.3 Finite-Length Geometric Series (Alternative Form)

Consider an alternative form of the finite-length geometric series sum in which the lower limit is not equal to zero.

$$Q = \sum_{n=L_1}^{L_2} a^n \tag{C.12}$$

In order to obtain a closed form formula we will apply the variable change $m = n - L_1$ with which Eqn. (C.12) becomes

$$Q = \sum_{m=0}^{L_2-L_1} a^{m+L_1} \tag{C.13}$$

which can be simplified as

$$Q = a^{L_1} \sum_{m=0}^{L_2-L_1} a^m \tag{C.14}$$

The summation on the right side of Eqn. (C.14) is in the standard form of Eqn. (C.6), therefore

$$Q = a^{L_1}\left(\frac{1-a^{L_2-L_1+1}}{1-a}\right) = \frac{a^{L_1}-a^{L_2+1}}{1-a} \tag{C.15}$$

Consistency check: For $L_1 = 0$ and $L_2 = L$, Eqn. (C.15) reduces to Eqn. (C.10).

APPENDIX D

ORTHOGONALITY OF BASIS FUNCTIONS

Orthogonality properties are used extensively in the development of Fourier series representations for the analysis of continuous-time and discrete-time signals. The use of an orthogonal set of basis functions ensures that the relationship between a periodic signal and its Fourier series representation is one-to-one, that is, each periodic signal has a unique set of Fourier series coefficients, and each set of coefficients corresponds to a unique signal. In this appendix we will summarize several forms of orthogonality properties and carry out their proofs.

D.1 Orthogonality for Trigonometric Fourier Series

Consider the two real-valued basis function sets defined as

$$\phi_k(t) = \cos(k\omega_0 t)\;; \quad k = 1,\ldots,\infty \tag{D.1}$$

and

$$\psi_k(t) = \sin(k\omega_0 t)\;; \quad k = 1,\ldots,\infty \tag{D.2}$$

where ω_0 is the fundamental frequency in rad/s. The period that corresponds to the fundamental frequency ω_0 is

$$T_0 = \frac{2\pi}{\omega_0}$$

It can be shown that each of the two sets in Eqns. (D.1) and (D.2) is orthogonal within itself, that is,

$$\int_0^{T_0} \phi_k(t)\,\phi_m(t)\,dt = \begin{cases} T_0/2\,, & k = m \\ 0\,, & k \neq m \end{cases} \tag{D.3}$$

and

$$\int_0^{T_0} \psi_k(t)\,\psi_m(t)\,dt = \begin{cases} T_0/2\,, & k = m \\ 0\,, & k \neq m \end{cases} \tag{D.4}$$

DOI: 10.1201/9781032675251-D

Furthermore, the two sets are orthogonal to each other:

$$\int_0^{T_0} \phi_k(t)\, \psi_m(t)\, dt = 0\,, \quad \text{for any } k, m \tag{D.5}$$

Writing Eqns. (D.3), (D.4), and (D.5) using the definitions of the basis function sets in Eqns. (D.1) and (D.2), we arrive at the orthogonality properties

$$\int_0^{T_0} \cos(k\omega_0 t)\cos(m\omega_0 t)\, dt = \begin{cases} T_0/2\,, & k = m \\ 0\,, & k \neq m \end{cases} \tag{D.6}$$

$$\int_0^{T_0} \sin(k\omega_0 t)\sin(m\omega_0 t)\, dt = \begin{cases} T_0/2\,, & k = m \\ 0\,, & k \neq m \end{cases} \tag{D.7}$$

$$\int_0^{T_0} \cos(k\omega_0 t)\sin(m\omega_0 t)\, dt = 0\,, \quad \text{for any } k, m \tag{D.8}$$

Proof of Eqn. (D.6):

Using the trigonometric identity in Eqn. (B.4), Eqn. (D.6) becomes

$$\int_0^{T_0} \cos(k\omega_0 t)\cos(m\omega_0 t)\, dt = \frac{1}{2}\int_0^{T_0} \cos\big((k+m)\,\omega_0 t\big)\, dt + \frac{1}{2}\int_0^{T_0} \cos\big((k-m)\,\omega_0 t\big)\, dt \tag{D.9}$$

Let us first assume $k \neq m$. The integrand of the first integral on the right side is periodic with period

$$\frac{2\pi}{(k+m)\,\omega_0} = \frac{T_0}{(k+m)}$$

The integrand $\cos\big((k+m)\,\omega_0 t\big)$ has $(k+m)$ full periods in the interval from 0 to T_0. Therefore the result of the first integral is zero. Similarly the integrand of the second integral on the right side of Eqn. (D.9) is periodic with period

$$\frac{2\pi}{|k-m|\,\omega_0} = \frac{T_0}{|k-m|}$$

The integrand $\cos\big((k-m)\,\omega_0 t\big)$ has $|k-m|$ full periods in the interval from 0 to T_0. Therefore the result of the second integral is also zero. If $k = m$ then Eqn. (D.9) becomes

$$\begin{aligned}\int_0^{T_0} \cos(k\omega_0 t)\cos(m\omega_0 t)\, dt &= \int_0^{T_0} \cos^2(k\omega_0 t)\, dt \\ &= \frac{1}{2}\int_0^{T_0} \cos(2k\omega_0 t)\, dt + \frac{1}{2}\int_0^{T_0} dt \\ &= \frac{T_0}{2}\end{aligned} \tag{D.10}$$

completing the proof of Eqn. (D.6). Eqns. (D.7) and (D.8) can be proven similarly, using the trigonometric identities given by (B.5) and (B.6) and recognizing integrals that span an integer number of periods of sine and cosine functions.

D.2 Orthogonality for Exponential Fourier Series

Consider the set of complex periodic basis functions

$$\phi_k(t) = e^{jk\omega_0 t}\,; \quad k = -\infty, \ldots, \infty \tag{D.11}$$

where the parameter ω_0 is the fundamental frequency in rad/s as in the case of the trigonometric set of basis functions of the previous section, and $T_0 = 2\pi/\omega_0$ is the corresponding period. It can be shown that this basis function set in Eqn. (D.11) is orthogonal in the sense

$$\int_0^{T_0} \phi_k(t)\,\phi_m^*(t),dt = \begin{cases} T_0\,, & k = m \\ 0\,, & k \neq m \end{cases} \tag{D.12}$$

The second term in the integrand is conjugated due to the fact that we are working with a complex set of basis functions. Writing Eqn. (D.12) in open form using the definition of the basis function set in Eqns. (D.11) we arrive at the orthogonality property

$$\int_0^{T_0} e^{jk\omega_0 t}\, e^{-jm\omega_0 t}\, dt = \begin{cases} T_0\,, & k = m \\ 0\,, & k \neq m \end{cases} \tag{D.13}$$

Proof of Eqn. (D.13):

Let us combine the exponential terms in the integral of Eqn. (D.13) and then apply Euler's formula to write it as

$$\begin{aligned} \int_0^{T_0} e^{jk\omega_0 t}\, e^{-jm\omega_0 t}\, dt &= \int_0^{T_0} e^{j(k-m)\omega_0 t}\, dt \\ &= \int_0^{T_0} \cos\big((k-m)\,\omega_0 t\big)\, dt + j \int_0^{T_0} \sin\big((k-m)\,\omega_0 t\big)\, dt \end{aligned} \tag{D.14}$$

If $k \neq m$, both integrands on the right side of Eqn. (D.14) are periodic with period

$$\frac{2\pi}{|k-m|\,\omega_0} = \frac{T_0}{|k-m|}$$

Both integrands $\cos\big((k-m)\,\omega_0 t\big)$ and $\sin\big((k-m)\,\omega_0 t\big)$ have exactly $|k-m|$ full periods in the interval from 0 to T_0. Therefore the result of each integral is zero. If $k = m$, we have

$$\int_0^{T_0} e^{jk\omega_0 t}\, e^{-jm\omega_0 t}\, dt = \int_0^{T_0} e^{j(0)\omega_0 t}\, dt = \int_0^{T_0} dt = T_0 \tag{D.15}$$

which completes the proof for Eqn. (D.13).

D.3 Orthogonality for Discrete-Time Fourier Series

Consider the set of discrete-time complex periodic basis functions

$$w_N^k = e^{-j\frac{2\pi}{N}k}\,; \quad k = 0, \ldots, N-1 \tag{D.16}$$

Parameters N and k are both integers. It can be shown that this basis function set in Eqn. (D.16) is orthogonal in the sense

$$\sum_{n=0}^{N-1} e^{j(2\pi/N)kn}\, e^{-j(2\pi/N)mn} = \begin{cases} 0\,, & k \neq m \\ N\,, & k = m \end{cases} \tag{D.17}$$

Proof of Eqn. (D.17):

The summation in Eqn. (D.17) can be put into a closed form using the finite-length geometric series formula in Eqn. (C.10) to obtain

$$\sum_{n=0}^{N-1} e^{j(2\pi/N)kn}\, e^{-j(2\pi/N)mn} = \sum_{n=0}^{N-1} e^{j(2\pi/N)(k-m)n}$$

$$= \frac{1-e^{j(2\pi/N)(k-m)N}}{1-e^{j(2\pi/N)(k-m)}}$$

which can be simplified as

$$\sum_{n=0}^{N-1} e^{j(2\pi/N)kn}\, e^{-j(2\pi/N)mn} = \frac{1-e^{j2\pi(k-m)}}{1-e^{(j2\pi/N)(k-m)}} \tag{D.18}$$

Let us begin with the case $k \neq m$. The numerator of the fraction on the right side of Eqn. (D.18) is equal to zero. Since both integers k and m are in the interval $k, m = 0, \ldots, N-1$ their absolute difference $|k-m|$ is in the interval $n = 1, \ldots, N-1$. Therefore the denominator of the fraction is nonzero, and the result is zero, proving the first part of Eqn. (D.17).

If $k = m$ then the denominator of the fraction in Eqn. (D.18) also becomes zero, and L'Hospital's rule must be used leading to

$$\sum_{n=0}^{N-1} e^{j(2\pi/N)kn}\, e^{-j(2\pi/N)mn} = N\,, \qquad \text{for } k = m \tag{D.19}$$

APPENDIX **E**

PARTIAL FRACTION EXPANSION

The technique of partial fraction expansion is very useful in the study of signals and systems, particularly in finding the time-domain signals that correspond to inverse Laplace transforms and inverse z-transforms of rational functions, that is, a ratio of two polynomials. It is based on the idea of writing a rational function, say of s or z, as a linear combination of simpler rational functions the inverse transforms of which can be easily determined. Since both the Laplace transform and the z-transform are linear transforms, writing the rational function $X(z)$ or $X(z)$ as a linear combination of simpler functions, and finding the inverse transform of each of those simpler functions allows us to construct the inverse transform of the original function in a straightforward manner. The main problem is in determining which simpler functions should be used in the expansion, and how the weight of each should be.

In this appendix we will present the details of the partial fraction expansion technique first from the perspective of the Laplace transform for continuous-time signals, and then from the perspective of the z-transform for discrete-time signals.

E.1 Partial Fractions for Continuous-Time Signals and Systems

As an example, consider a rational function $X(s)$ with two poles, expressed in the form

$$X(s) = \frac{B(s)}{(s-a)(s-b)} = \frac{k_1}{s-a} + \frac{k_2}{s-b} \tag{E.1}$$

The rational function $X(s)$ is expressed as a weighted sum of partial fractions

$$\frac{1}{s-a}$$

and

$$\frac{1}{s-b}$$

DOI: 10.1201/9781032675251-E

with weights k_1 and k_2 respectively. The weights k_1 and k_2 are called the *residues* of the partial fraction expansion. Before we address the issue of determining the values of the residues, a question that comes to mind is: What limitations, if any, should be placed on the numerator polynomial $B(s)$ for the expansion in Eqn. (E.1) to be valid? In order to find the answer to this question we will combine the two partial fractions under a common denominator to obtain

$$\frac{k_1}{s-a}+\frac{k_2}{s-b}=\frac{(k_1+k_2)\,s-(k_1 b+k_2 a)}{(s-a)\,(s-b)} \tag{E.2}$$

It is obvious from Eqn. (E.2) that the numerator polynomial $B(s)$ may not be higher than first-order since no s^2 term or higher-order terms can be obtained in the process of combining the two partial fractions under a common denominator. Consider a more general case of a rational function with P poles in the form

$$\begin{aligned} X(s) &= \frac{B(s)}{(s-s_1)\,(s-s_2)\ldots(s-s_P)} \\ &= \frac{k_1}{s-s_1}+\frac{k_2}{s-s_2}+\ldots+\frac{k_P}{s-s_P} \end{aligned} \tag{E.3}$$

If the P terms in the partial fraction expansion of Eqn. (E.3) are combined under a common denominator, the highest order numerator term that could be obtained is s^{P-1}.

> As a general rule, for the partial fraction expansion in the form of Eqn. (E.3) to be possible, the numerator order must be less than the denominator order.

If this is not the case, some additional steps need to be taken before partial fraction expansion can be used. Suppose $X(s)$ is a rational function in which the numerator order is greater than or equal to the denominator order. Let

$$X(s)=\frac{b_Q s^Q+b_{Q-1}s^{Q-1}+\ldots+b_1 s+b_0}{a_P s^P+a_{P-1}s^{P-1}+\ldots+a_1 s+a_0} \tag{E.4}$$

with $Q \geq P$. We can use long division on numerator and denominator polynomials to write $X(s)$ in the modified form

$$X(s)=C(s)+\frac{\bar{b}_{P-1}s^{P-1}+\bar{b}_{P-2}s^{P-2}+\ldots+\bar{b}_1 s+\bar{b}_0}{a_P s^P+a_{P-1}s^{P-1}+\ldots+a_1 s+a_0} \tag{E.5}$$

where $C(s)$ is a polynomial of s. Now the new function

$$\bar{X}(s)=X(s)-C(s)=\frac{\bar{b}_{P-1}s^{P-1}+\bar{b}_{P-2}s^{P-2}+\ldots+\bar{b}_1 s+\bar{b}_0}{a_P s^P+a_{P-1}s^{P-1}+\ldots+a_1 s+a_0} \tag{E.6}$$

has a numerator that is of lower order than its denominator, and can therefore be expanded into partial fractions. Examples E.1 and E.2 will illustrate this.

Example E.1:

Consider the function

$$X(s)=\frac{(s+1)\,(s+2)}{(s+5)\,(s+6)}$$

Since both the numerator and the denominator are of second order, the function cannot be expanded into partial fractions directly. In other words, an attempt to express $X(s)$ as

$$X(s)\stackrel{?}{=}\frac{k_1}{s+5}+\frac{k_2}{s+6}$$

would fail since the s^2 term in the numerator cannot be matched by combining the two partial fractions under a common denominator. Instead, we will use long division to extract a polynomial $C(s)$ from $X(s)$ so that the difference $X(s) - C(s)$ can be expanded into partial fractions. Let us first multiply out the factors in $X(s)$ to write it as

$$X(s) = \frac{s^2 + 3s + 2}{s^2 + 11s + 30}$$

Now we can divide the numerator polynomial by the denominator polynomial using long division to obtain

$$\begin{array}{rrrl|rrr} & & & & & 1 & \\ \cline{5-7} s^2 & +11s & +30 & & s^2 & +3s & +2 \\ & & & & s^2 & +11s & +30 \\ \cline{5-7} & & & & & -8s & -28 \end{array} \tag{E.7}$$

Thus, the function $X(s)$ can be written in the equivalent form

$$X(s) = 1 + \frac{-8s - 28}{s^2 + 11s + 30} \tag{E.8}$$

Let

$$\bar{X}(s) = X(s) - 1 = \frac{-8s - 28}{s^2 + 11s + 30} \tag{E.9}$$

We can now expand the function $\bar{X}(s)$ into partial fractions as

$$\bar{X}(s) = \frac{k_1}{s+5} + \frac{k_2}{s+6} \tag{E.10}$$

and write $X(s)$ as

$$X(s) = 1 + \bar{X}(s) = 1 + \frac{k_1}{s+5} + \frac{k_2}{s+6} \tag{E.11}$$

Example E.2:

Consider the function

$$X(s) = \frac{(s+1)(s+2)(s+3)}{(s+5)(s+6)}$$

the factors of which can be multiplied to yield

$$X(s) = \frac{s^3 + 6s^2 + 11s + 6}{s^2 + 11s + 30}$$

Using long division on $X(s)$ we obtain

$$\begin{array}{rrrl|rrrr} & & & & s-5 & & & \\ \cline{5-8} s^2 & +11s & +30 & & s^3 & +6s^2 & +11s & +6 \\ & & & & s^3 & +11s^2 & +30s & \\ \cline{5-8} & & & & & -5s^2 & -19s & +6 \\ & & & & & -5s^2 & -55s & -150 \\ \cline{6-8} & & & & & & 36s & +156 \end{array} \tag{E.12}$$

Thus, the function $X(s)$ can be written in the equivalent form

$$X(s) = s - 5 + \bar{X}(s) = s - 5 + \frac{36s + 156}{s^2 + 11s + 30} \tag{E.13}$$

and can be expressed using partial fractions as

$$X(s) = s - 5 + \frac{k_1}{s+5} + \frac{k_2}{s+6} \tag{E.14}$$

Now that we know how to deal with a function that has a numerator of equal or greater order than its denominator, we will focus our attention on determining the residues. For the purpose of determining the residues, we will assume that the order of the numerator polynomial is less than that of the denominator polynomial. In addition, we will initially consider a denominator polynomial with only simple roots. Thus, the function $X(s)$ is in the form

$$X(s) = \frac{B(s)}{(s - s_1)(s - s_2)\dots(s - s_P)} \tag{E.15}$$

to be expanded into partial fractions in the form

$$X(s) = \frac{k_1}{s - s_1} + \frac{k_2}{s - s_2} + \dots + \frac{k_P}{s - s_P} \tag{E.16}$$

Let us multiply both sides of Eqn. (E.16) by $(s - s_1)$ to obtain

$$(s - s_1)\, X(s) = k_1 + \frac{k_2\,(s - s_1)}{s - s_2} + \dots + \frac{k_P\,(s - s_1)}{s - s_P} \tag{E.17}$$

If we now set $s = s_1$ on both sides of Eqn. (E.17) we get

$$\begin{aligned}(s - s_1)\, X(s)\Big|_{s=s_1} &= k_1\Big|_{s=s_1} + \frac{k_2\,(s - s_1)}{s - s_2}\Bigg|_{s=s_1} + \dots + \frac{k_P\,(s - s_1)}{s - s_P}\Bigg|_{s=s_1} \\ &= k_1 \end{aligned} \tag{E.18}$$

Thus, the residue of the pole at $s = s_1$ can be found by multiplying $X(s)$ with the factor $(s - s_1)$ and then setting $s = s_1$. Using the form of $X(s)$ in Eqn. (E.15) the residue k_1 is found as

$$(s - s_1)\, X(s)\Big|_{s=s_1} = \frac{B(s_1)}{(s_1 - s_2)\dots(s_1 - s_P)} \tag{E.19}$$

which amounts to canceling the $(s - s_1)$ term in the denominator, and evaluating what is left for $s = s_1$. Generalizing this result, the residue of the n-th pole is found as

$$k_n = (s - s_n)\, X(s)\Big|_{s=s_n} \qquad n = 1, \dots, P \tag{E.20}$$

Example E.3:

Find a partial fraction expansion for the rational function

$$X(s) = \frac{(s+1)(s+2)}{(s+3)(s+4)(s+5)}$$

Solution: The partial fraction expansion we are looking for is in the form

$$X(s) = \frac{k_1}{s+3} + \frac{k_2}{s+4} + \frac{k_3}{s+5} \tag{E.21}$$

We will use Eqn. (E.20) to determine the residues. The first residue k_1 is obtained as

$$\begin{aligned} k_1 &= (s+3)\, X(s)\Big|_{s=-3} \\ &= \frac{(s+1)\,(s+2)}{(s+4)\,(s+5)}\Big|_{s=-3} \\ &= \frac{(-3+1)\,(-3+2)}{(-3+4)\,(-3+5)} = 1 \end{aligned}$$

The remaining two residues k_2 and k_3 are obtained similarly:

$$\begin{aligned} k_2 &= (s+4)\, X(s)\Big|_{s=-4} \\ &= \frac{(s+1)\,(s+2)}{(s+3)\,(s+5)}\Big|_{s=-4} \\ &= \frac{(-4+1)\,(-4+2)}{(-4+3)\,(-4+5)} = -6 \end{aligned}$$

$$\begin{aligned} k_3 &= (s+5)\, X(s)\Big|_{s=-5} \\ &= \frac{(s+1)\,(s+2)}{(s+3)\,(s+4)}\Big|_{s=-5} \\ &= \frac{(-5+1)\,(-5+2)}{(-5+3)\,(-5+4)} = 6 \end{aligned}$$

Thus, the partial fraction expansion for $X(z)$ is

$$X(s) = \frac{1}{s+3} - \frac{6}{s+4} + \frac{6}{s+5}$$

Next we will consider a rational function the denominator of which has repeated roots. Let $X(s)$ be a rational function with a pole of order r at $s = s_1$.

$$X(s) = \frac{B(s)}{(s-s_1)^r\,(s-s_2)} \tag{E.22}$$

The partial fraction expansion for $X(s)$ needs to be in the form

$$X(s) = \frac{k_{1,1}}{s-s_1} + \frac{k_{1,2}}{(s-s_1)^2} + \ldots + \frac{k_{1,r}}{(s-s_1)^r} + \frac{k_2}{s-s_2} \tag{E.23}$$

For the multiple pole at $s = s_1$ as many terms as the multiplicity of the pole are needed so that combining the terms on the right side of Eqn. (E.23) can yield a result that matches the function $X(s)$ given by Eqn. (E.22).

The residue k_2 of the single pole at $s = s_2$ can easily be determined as discussed above. We will focus our attention on determining the residues of the r-th order pole at $s = s_1$. Let us begin by multiplying both sides of Eqn. (E.23) by $(s-s_1)^r$ to obtain

$$(s-s_1)^r\, X(s) = k_{1,1}\,(s-s_1)^{r-1} + k_{1,2}\,(s-s_1)^{r-2} + \ldots + k_{1,r} + \frac{k_2\,(s-s_1)^r}{s-s_2} \tag{E.24}$$

If we now set $s = s_1$ on both sides of E.24) we would get

$$\begin{aligned} (s-s_1)^r\, X(s)\Big|_{s=s_1} &= k_{1,1}\,(s-s_1)^{r-1}\Big|_{s=s_1} + k_{1,2}\,(s-s_1)^{r-2}\Big|_{s=s_1} + \ldots + k_{1,r}\Big|_{s=s_1} + \frac{k_2\,(s-s_1)^r}{s-s_2}\Big|_{s=s_1} \\ &= k_{1,r} \end{aligned} \tag{E.25}$$

It can be shown that the other residues for the pole at $s = s_1$ can be found as follows:

$$k_{1,r-1} = \frac{d}{ds}\left[(s-s_1)^r\, X(s)\right]\Bigg|_{s=s_1}$$

$$k_{1,r-2} = \frac{1}{2}\frac{d^2}{ds^2}\left[(s-s_1)^r\, X(s)\right]\Bigg|_{s=s_1}$$

$$\vdots$$

$$k_{1,2} = \frac{1}{(r-2)!}\frac{d^{r-2}}{ds^{r-2}}\left[(s-s_1)^r\, X(s)\right]\Bigg|_{s=s_1}$$

$$k_{1,1} = \frac{1}{(r-1)!}\frac{d^{r-1}}{ds^{r-1}}\left[(s-s_1)^r\, X(s)\right]\Bigg|_{s=s_1}$$

Generalizing these results, the residues of a pole of multiplicity r are calculated using

$$k_{1,n} = \frac{1}{(r-n)!}\frac{d^{r-n}}{ds^{r-n}}\left[(s-s_1)^r\, X(s)\right]\Bigg|_{s=s_1} \qquad n = 1,\ldots,r \tag{E.26}$$

In using the general formula in Eqn. (E.26) for $n = r$ we need to remember that $0! = 1$ and

$$\frac{d^0}{ds^0}\left[(s-s_1)^r\, X(s)\right] = (s-s_1)^r\, X(s) \tag{E.27}$$

Example E.4:

Find a partial fraction expansion for the rational function

$$X(s) = \frac{s}{(s+1)^3\,(s+2)}$$

Solution: The partial fraction expansion for $X(s)$ is in the form

$$X(s) = \frac{k_{1,1}}{s+1} + \frac{k_{1,2}}{(s+1)^2} + \frac{k_{1,3}}{(s+1)^3} + \frac{k_2}{s+2} \tag{E.28}$$

The residue of the single pole at $s = -2$ is easily found using the technique developed earlier:

$$k_2 = (s+2)\, X(s)\Big|_{s=-2} = \frac{s}{(s+1)^3}\Bigg|_{s=-2} = \frac{-2}{(-2+1)^3} = 2$$

We will use Eqn. (E.26) to find the residues of the third-order pole at $s = -1$.

$$k_{1,3} = (s+1)^3\, X(s)\Big|_{s=-1} = \frac{s}{(s+2)}\Bigg|_{s=-1} = \frac{-1}{(-1+2)} = -1$$

$$k_{1,2} = \frac{d}{ds}\left[(s+1)^3\, X(s)\right]\Bigg|_{s=-1} = \frac{d}{ds}\left[\frac{s}{s+2}\right]\Bigg|_{s=-1} = \frac{2}{(s+2)^2}\Bigg|_{s=-1} = 2$$

$$k_{1,1} = \frac{1}{2!}\frac{d^2}{ds^2}\left[(s+1)^3\, X(s)\right]\Bigg|_{s=-1} = \frac{1}{2!}\frac{d^2}{ds^2}\left[\frac{s}{s+2}\right]\Bigg|_{s=-1} = \frac{-2}{(s+2)^3}\Bigg|_{s=-1} = -2$$

Thus, the partial fraction expansion for $X(z)$ is

$$X(s) = -\frac{2}{s+1} + \frac{2}{(s+1)^2} - \frac{1}{(s+1)^3} + \frac{2}{s+2}$$

E.2 Partial Fraction Expansion for Discrete-Time Signals and Systems

The residues of the partial fraction expansion for rational z-transforms are obtained using the same techniques employed in the previous section. One subtle difference is that, when working with the inverse z-transform, we often need the partial fraction expansion in the form

$$\begin{aligned} X(z) &= \frac{B(z)}{(z-z_1)(z-z_2)\ldots(z-z_P)} \\ &= \frac{k_1\,z}{z-z_1}+\frac{k_2\,z}{z-z_2}+\ldots+\frac{k_P\,z}{z-z_P} \end{aligned} \tag{E.29}$$

In order to use the residue formulas established in the previous section and still obtain the expansion in the form of Eqn. (E.29) with a z factor in each term, we simply expand $X(z)/z$ into partial fractions:

$$\begin{aligned} \frac{X(z)}{z} &= \frac{B(z)}{(z-z_1)(z-z_2)\ldots(z-z_P)\,z} \\ &= \frac{k_1}{z-z_1}+\frac{k_2}{z-z_2}+\ldots+\frac{k_P}{z-z_P}+\frac{k_{P+1}}{z} \end{aligned} \tag{E.30}$$

The last term in Eqn. (E.30) may or may not be needed depending on whether the numerator polynomial $B(z)$ has a root at $z=0$ or not. Once the residues of the expansion in Eqn. (E.30) are found, we can revert to the form in Eqn. (E.29) by multiplying both sides with z.

APPENDIX F

REVIEW OF MATRIX ALGEBRA

In this section we present some basic definitions for matrix algebra.

Matrix

A matrix is a collection of real or complex numbers arranged to form a rectangular grid:

$$\mathbf{A} = \begin{bmatrix} a_{11} & a_{12} & \dots & a_{1M} \\ a_{21} & a_{22} & \dots & a_{2M} \\ \vdots & & & \\ a_{N1} & a_{N2} & \dots & a_{NM} \end{bmatrix} \tag{F.1}$$

A total of NM numbers arranged into N rows and M columns as shown in Eqn. (F.1) form a $N \times M$ matrix. The numbers that make up the matrix are referred to as the *elements* of the matrix. Two subscripts are used to specify the placement of an element in the matrix. The element at the in row-i and column-j is denoted by a_{ij}.

Vector

A vector is a special matrix with only one row or only one column. A matrix with only one row (a $1 \times N$ matrix) is called a *row vector* of length N. A matrix with only one column (a $M \times 1$ matrix) is called a *column vector* of length M.

Square matrix

A *square matrix* is one in which the number of rows is equal to the number of columns. An example is

$$\mathbf{A} = \begin{bmatrix} 5 & 2 & 0 & 4 \\ 3 & -1 & 0 & 0 \\ 0 & 1 & 1 & 3 \\ -7 & 4 & 3 & 5 \end{bmatrix} \tag{F.2}$$

DOI: 10.1201/9781032675251-F

Diagonal matrix

A *diagonal matrix* is a square matrix in which all elements that are not on the main diagonal are equal to zero. The main diagonal of a square matrix is the diagonal from the top left corner to the bottom right corner. Elements on the main diagonal have identical row and column indices such as $a_{11}, a_{22}, \ldots, a_{NN}$. An example is

$$\mathbf{A} = \begin{bmatrix} 5 & 0 & 0 & 0 \\ 0 & -1 & 0 & 0 \\ 0 & 0 & 1 & 0 \\ 0 & 0 & 0 & 4 \end{bmatrix} \tag{F.3}$$

Identity matrix

An *identity matrix* is a diagonal matrix in which all diagonal elements are equal to unity. For example, the identity matrix of order 3 is

$$\mathbf{I}_3 = \begin{bmatrix} 1 & 0 & 0 \\ 0 & 1 & 0 \\ 0 & 0 & 1 \end{bmatrix} \tag{F.4}$$

Equality of two matrices

Two matrices **A** and **B** are equal to each other if they have the same dimensions and if all corresponding elements are equal, that is

$$\mathbf{A} = \mathbf{B} \quad \Longrightarrow \quad a_{ij} = b_{ij} \quad \text{for } i = 1, \ldots, M \text{ and } j = 1, \ldots, N \tag{F.5}$$

Trace of a square matrix

The *trace* of a square matrix is defined as the arithmetic sum of all of its elements on the main diagonal.

$$\text{Trace}\,(\mathbf{A}) = \sum_{i=1}^{N} a_{ii} \tag{F.6}$$

Transpose of a matrix

The *transpose* of a matrix **A** is denoted by $\mathbf{A}^T$ and is obtained by interchanging rows and columns so that row-i of **A** becomes column-i of $\mathbf{A}^T$. For example, the transpose of the matrix **A** in Eqn. (F.2) is

$$\mathbf{A}^T = \begin{bmatrix} 5 & 3 & 0 & -7 \\ 2 & -1 & 1 & 4 \\ 0 & 0 & 1 & 3 \\ 4 & 0 & 3 & 5 \end{bmatrix} \tag{F.7}$$

Determinant of a square matrix

Each square matrix has a scalar value associated with it referred to as the *determinant.* Methods for computing determinants can be found in most texts on linear algebra. As an example, the determinant of the matrix

$$\mathbf{A} = \begin{bmatrix} 5 & 2 \\ 3 & -1 \end{bmatrix} \tag{F.8}$$

is computed as

$$|\mathbf{A}| = (5)\,(-1) - (2)\,(3) = -11 \tag{F.9}$$

Minors of a square square matrix

The row-i column-j minor m_{ij} of a square matrix **A** is defined as the determinant of the sub-matrix obtained by deleting row-i and column-j from the matrix **A**. Consider, for example the square matrix

$$\mathbf{A} = \begin{bmatrix} 5 & 2 & 4 \\ 3 & -1 & 2 \\ -1 & 0 & 3 \end{bmatrix} \tag{F.10}$$

The minor m_{23} is found by deleting row-2 and column-3 and computing the determinant of the matrix left behind as

$$m_{23} = \begin{vmatrix} 5 & 2 \\ -1 & 0 \end{vmatrix} = 2 \tag{F.11}$$

Cofactors of a square square matrix

The row-i column-j cofactor Δ_{ij} of a square matrix **A** is defined as

$$\Delta_{ij} = (-1)^{i+j}\, m_{ij} \tag{F.12}$$

Using the matrix in Eqn. (F.10) as an example, the cofactor Δ_{23} is

$$\Delta_{23} = (-1)^{2+3}\, m_{23} = -2 \tag{F.13}$$

Adjoint of a square square matrix

The adjoint of a square matrix **A** is found by first transposing **A** and then replacing each element by its cofactor. For example, the adjoint of the matrix **A** in Eqn. (F.10) is

$$\text{Adj}(\mathbf{A}) = \begin{bmatrix} -3 & -6 & 8 \\ -11 & 19 & 2 \\ -1 & -2 & -11 \end{bmatrix} \tag{F.14}$$

Scaling a matrix

Scaling a matrix by a constant scale factor means multiplying each element of the matrix with that scale factor.

Addition of two matrices

Addition of two matrices is only defined for matrices with identical dimensions. The sum of two matrices **A** and **B** is the matrix **C** in which each element is equal to the sum of corresponding elements of **A** and **B**.

$$\mathbf{C} = \mathbf{A} + \mathbf{B} \quad \Longrightarrow \quad c_{ij} = a_{ij} + b_{ij} \quad \text{for } i = 1, \ldots, M \text{ and } j = 1, \ldots, N \tag{F.15}$$

Scalar product of two vectors

The scalar product of a $1 \times N$ row vector **x** and a $N \times 1$ column vector **y** is defined as

$$\mathbf{x} \cdot \mathbf{y} = \sum_{i=1}^{N} x_i\, y_i \tag{F.16}$$

Multiplication of two matrices

The product of an $M \times K$ matrix **A** and a $K \times N$ matrix **B** is a $M \times N$ matrix **C** the elements of which are computed as

$$c_{ij} = \sum_{k=1}^{K} a_{ik}\, b_{kj} \tag{F.17}$$

In other words, the row-i column-j element of matrix **C** is the scalar product of row-i of of matrix **A** and column-j of matrix **B**.

Inversion of a matrix

The inverse of a square matrix **A** denoted by $\mathbf{A}^{-1}$ is also a square matrix. It that satisfies the equation

$$\mathbf{A}^{-1}\mathbf{A} = \mathbf{A}\mathbf{A}^{-1} = \mathbf{I} \tag{F.18}$$

It is computed by scaling the adjoint matrix by the reciprocal of the determinant, that is,

$$\mathbf{A}^{-1} = \frac{\text{Adj}\,(\mathbf{A})}{|\mathbf{A}|} \tag{F.19}$$

The inverse does not exist if the determinant of **A** is equal to zero. Such a matrix is called a *singular* matrix.

Eigenvalues and eigenvectors

The scalar λ and the $N \times 1$ vector **v** are called an *eigenvalue* and an *eigenvector* of the $N \times N$ matrix **A**, respectively if they satisfy the equation

$$\mathbf{A}\mathbf{v} = \lambda\,\mathbf{v} \tag{F.20}$$

An $N \times N$ matrix **A** has N eigenvalues and associated eigenvectors although they are not necessarily unique.

APPENDIX G

ANSWERS/PARTIAL SOLUTIONS TO SELECTED PROBLEMS

G.1 Chapter 1 Problems

1.2.

a.

$$x_a(t) = \begin{cases} 0\,, & t < -1 \text{ or } t > 3 \\ 2t + 2\,, & -1 < t < 0 \\ -t + 2\,, & 0 < t < 1 \\ 1\,, & 1 < t < 2 \\ -t + 3\,, & 2 < t < 3 \end{cases}$$

1.3.

a.

$$g_1(t) = \begin{cases} 0\,, & t < -1 \text{ or } t > 3 \\ 3.5t + 3.5\,, & -1 < t < 0 \\ -2.5t + 3.5\,, & 0 < t < 1 \\ -1.5t + 2.5\,, & 1 < t < 2 \\ 0.5t - 1.5\,, & 2 < t < 3 \end{cases}$$

DOI: 10.1201/9781032675251-G

1.4.

a.

$$g_1(t) = x(2t-1)$$

Step 1: Time scaling

$$g_{1a}(t) = x(2t)$$

Step 2: Time shifting

$$g_1(t) = g_{1a}(t-0.5) = x(2t-1)$$

b.

$$g_2(t) = x(-t+2)$$

Step 1: Time reversal

$$g_{2a}(t) = x(-t)$$

Step 2: Time shifting

$$g_2(t) = g_{2a}(t-2) = x(-t+2)$$

1.5.

e.

Step 1: Time scaling

$$g_{5a}(t) = x\left(\frac{t}{3}\right)$$

Step 2: Time shifting

$$g_5(t) = g_{5a}(t-1) = x\left(\frac{(t-1)}{3}\right)$$

f.

Step 1: Time scaling

$$g_{6a}(t) = x(4t)$$

Step 2: Time shifting

$$g_6(t) = g_{6a}(t-3/4) = x(4t-3)$$

g.

Step 1: Time scaling

$$g_{7a}(t) = x\left(\frac{t}{3}\right)$$

Step 2: Time reversal

$$g_{7b}(t) = g_{7a}(-t) = x\left(= \frac{t}{3}\right)$$

Step 3: Time shifting

$$g_y(t) = g_{7b}(t-3) = x\left(1 - \frac{t}{3}\right)$$

1.6.

a. $x_3(t) = \frac{1}{2}\cos\left(2\pi\left(f_1 - f_2\right)t\right) - \frac{1}{2}\cos\left(2\pi\left(f_1 + f_2\right)t\right)$

1.18

a. Periodic. $f_0 = \dfrac{1}{\pi}$ Hz.

b. Periodic. $f_0 = \dfrac{\sqrt{5}}{\pi}$ Hz.

c. Not periodic.

1.19.

a. $f_0 = \dfrac{5}{2\pi}$ Hz.

b. $f_0 = 5$ Hz.

1.23.

a. $E_x = \dfrac{1}{2}$

1.24.

a. $E_x = 5$

1.25.

a. $P_x = \dfrac{1}{3}$

1.30

$$x_E(t) = \frac{1}{2}\left(e^{-2t} + e^{2t}\right)e^{j10\pi t}$$

1.31.

a.

$$x_e(t) = \frac{1}{2}\sin(t)\left[e^{-5t}\,u(t) - e^{5t}\,u(-t)\right]$$

$$x_o(t) = \frac{1}{2}\sin(t)\left[e^{-5t}\,u(t) + e^{5t}\,u(-t)\right]$$

b.

$$x_e(t) = e^{-3|t|}\cos(t)$$

$$x_o(t) = 0$$

1.34.

a. $x(t) = 5\cos\left(20\pi t + 14^{\circ}\right)$

1.35.

a.

```
xa = @(t)..
  (2).*((t>=0)&(t<1))+..
  (1).*((t>=1)&(t<2))+..
  (t-1).*((t>=2)&(t<3))+..
  (2).*((t>=3)&(t<4));
t = [-1:0.01:6];
plot(t,xa(t)); grid;
axis([-1,6,-1,3]);
title('The signal x_{a}(t)');
xlabel('t (sec)');
ylabel('Amplitude');
```

1.36.

a.

```
% Define the signal xa(t) using data points in vectors tp and xp
tp = [-2,-1,0,1,2,3,4];
xp = [0,0,2,1,1,0,0];
% Obtain signal xa(t) through interpolation
t = [-2:0.01:4];
xa = interp1(tp,xp,t,'linear');
plot(t,xa,'b-',tp,xp,'ro'); grid;
axis([-2,4,-1,3]);
title('The signal x_{a}(t)');
xlabel('t (sec)');
ylabel('Amplitude');
```

1.38.

a.

```
% Define the signal xa(t) using data points in vectors tp and xp
tp = [-100,-1.5,-0.5,1,3,4,100];
xp = [0,0,1,1,-1,0,0];
x = @(t) interp1(tp,xp,t,'linear');
% Obtain signal xa(t) through interpolation
t = [-10:0.02:10];
plot(t,x(t),'b-',tp,xp,'ro'); grid;
axis([-10,10,-2,2]);
title('The signal x(t)');
xlabel('t (sec)');
ylabel('Amplitude');
```

1.39.

a.

```
x = @(t) (exp(-t)-exp(-2*t)).*(t>=0);
t = [-10:0.02:10];
plot(t,x(t)); grid;
axis([-10,10,-0.2,0.5]);
title('The signal x(t)');
xlabel('t (sec)');
ylabel('Amplitude');
```

1.40.

a.

```
xfw = @(f0,t) abs(sin(2*pi*f0*t));
```

1.43.

a.

```
t = [-2:0.01:8];
x = ss_step(t)+ss_step(t-1)-3*ss_step(t-2)+ss_step(t-3);
plot(t,x); grid;
axis([-2,8,-5,5]);
title('The signal x(t)');
xlabel('t (sec)');
ylabel('Amplitude');
```

b.

```
x = ss_ramp(t)-2*ss_ramp(t-2)+ss_ramp(t-3);
plot(t,x); grid;
axis([-2,8,-5,5]);
title('The signal x(t)');
xlabel('t (sec)');
ylabel('Amplitude');
```

1.44.

a.

```
x = @(t) -ss_ramp(t+1)+3.5*ss_ramp(t)..
        -3*ss_ramp(t-1)-0.5*ss_ramp(t-2)+ss_ramp(t-3);
t = [-5:0.01:5];
plot(t,x(t)); grid;
axis([-5,5,-2,2]);
title('The signal x(t)');
xlabel('t (sec)');
ylabel('Amplitude');
```

G.2 Chapter 2 Problems

2.1.

a. Nonlinear, time-invariant.

b. Linear, not time-invariant.

2.3.

a. Using the first configuration: $y(t) = 3t\,x(t)$
Using the second configuration: $\bar{y}(t) = 3t\,x(t)$

b. Using the first configuration: $y(t) = 3\,x(t) + 5$
Using the second configuration: $\bar{y}(t) = 3\,x(t) + 15$

2.4.

$$\frac{d^2y(t)}{dt^2} + \frac{1}{RC}\frac{dy(t)}{dt} + \frac{1}{LC}\,y(t) = \frac{1}{RC}\frac{dx(t)}{dt}$$

$$y(0) = 2 \quad , \qquad \left.\frac{dy(t)}{dt}\right|_{t=0} = -\frac{2}{RC} - \frac{1}{C} - \frac{1}{RC}\,x(0)$$

2.7.

a. $y(t) = 1 - e^{-4t}\,, \qquad t \ge 0$

b. $y(t) = 1 + 4\,e^{-4t}\,, \qquad t \ge 0$

2.8.

a. $y(t) = \dfrac{1}{4} - \dfrac{5}{4}\,e^{-4t}\,, \qquad t \ge 0$

b. $y(t) = \begin{cases} 1 + e^{-2t}\,, & 0 < t < 5 \\ \left[e^{10} + 1\right]e^{-2t}\,, & t > 5 \end{cases}$

2.9.

a. $y(t) = 6\,e^{-t} - 3\,e^{-2t}\,, \qquad t \ge 0$

b. $y(t) = -\dfrac{5}{2}\,e^{-t} + \dfrac{1}{2}\,e^{-3t}\,, \qquad t \ge 0$

2.10.

a. $y(t) = 2\cos\left(\sqrt{3}\,t\right)\,, \qquad t \ge 0$

b. $y(t) = -2\,e^{-t}\cos(t) - 3\,e^{-t}\sin(t)\,, \qquad t \ge 0$

2.13.

a. $y(t) = -\frac{5}{4}\, e^{-4t}\;, \qquad t \geq 0$

b. $y(t) = \frac{9}{10}\, e^{-2t} + \frac{1}{2}\sin(2t) - \frac{1}{2}\cos(2t) + \frac{8}{5}\cos(t) + \frac{4}{5}\sin(t)\;, \qquad t \geq 0$

2.14.

a. $y(t) = 5\,e^{-t} - \frac{5}{2}\,e^{-2t} + \frac{1}{2}\;, \qquad t \geq 0$

b. $y(t) = 3\,e^{-t} - \frac{8}{9}\,e^{-3t} + \frac{1}{3}\,t - \frac{1}{9}\;, \qquad t \geq 0$

2.16.

a.

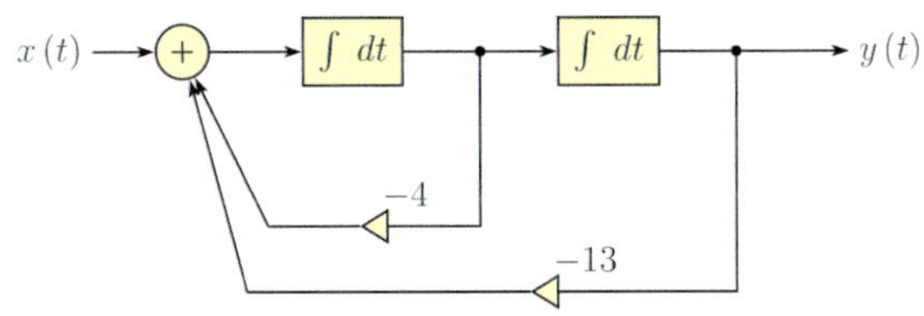

2.20.

a. $h_{eq}(t) = h_1(t) + h_2(t) * h_3(t)$

2.21.

a. $h_{eq}(t) = h_1(t) + h_2(t) * h_3(t) + h_2(t) * h_4(t)$

2.25.

$$y(t) = x(t) + 0.5\,x(t-1) + 0.3\,x(t-2) + 0.2\,x(t-3)$$

2.27.

a. $y(t) = \frac{1}{2}\left(1 - e^{-2t}\right) u(t)$

b. $y(t) = \left[\frac{1}{2} - e^{-t} + \frac{1}{2}\,e^{-2t}\right] u(t)$

2.32.

a. $h(t) = u(t)$ causal, unstable.

2.33.

a.

```
x = @(t) exp(-t).*cos(2*t).*(t>=0);
t = [-1:0.01:5];
```

b.

```
w = @(t) 3*x(t);
plot(t,w(t)); grid;
axis([-1,5,-1,4]);
title('w(t)');
xlabel('t (sec)');
ylabel('Amplitude');
```

2.34.

a.

```
x = @(t) exp(-t).*cos(2*t).*(t>=0);
t = [-1:0.01:5];
```

Compute and graph $w(t)$:

```
w = @(t) 3*x(t);
plot(t,w(t)); grid;
axis([-1,5,-2,4]);
title('w(t)');
xlabel('t (sec)');
ylabel('Amplitude');
```

2.37.

a.

```
t = [0:0.001:2];
% Compute the exact solution
y = 0.25-1.25*exp(-4*t);
% Compute the approximate solution using Euler method
Ts = 1/40;
t2 = [0:Ts:2];
yhat = zeros(size(t2));
yhat(1) = -1;  % Initial value
for k=1:length(yhat)-1,
  g = -4*yhat(k)+1;
  yhat(k+1) = yhat(k)+Ts*g;
end;
% Graph exact and approximate solutions
plot(t,y,'-',t2,yhat,'r.'); grid;
title('Exact and approximate solutions');
xlabel('Time (sec)');
ylabel('Amplitude');
legend('Exact solution','Approximate solution',..
  'Location','SouthEast');
```

2.38.

a.

```
t = [0:0.001:2];
% Compute the exact solution
y = 0.25-1.25*exp(-4*t);
% Compute the approximate solution using function ode45()
Ts = 1/40;
t2 = [0:Ts:2];
ga = @(t,yhat) -4*yhat+1;
[t2,yhat] = ode45(ga,t2,-1);
% Graph exact and approximate solutions
plot(t,y,'-',t2,yhat,'r.'); grid;
title('Exact and approximate solutions');
xlabel('Time (sec)');
ylabel('Amplitude');
legend('Exact solution','Approximate solution',..
  'Location','SouthEast');
```

G.3 Chapter 3 Problems

3.2.

b. $a_0 = 0.3\,A \quad , \qquad a_1 = \dfrac{A\sin(0.6\pi)}{\pi} \quad , \qquad b_1 = \dfrac{A\left[1-\cos(0.6\pi)\right]}{\pi}$

3.5.

b. $a_0 = \dfrac{1}{2} \quad , \qquad a_1 = \dfrac{2}{\pi} \quad , \qquad a_2 = 0$

3.7.

$$a_0 = 1 \quad , \qquad a_k = 0 \;\text{ for } k > 0 \quad , \qquad b_k = \frac{1}{\pi k}\left[2 - \cos\left(\frac{2\pi k}{3}\right) - \cos\left(\frac{4\pi k}{3}\right)\right]$$

3.8.

$$a_0 = 0.4323 \quad , \qquad a_k = \frac{3.4587}{4+4\pi^2 k^2} \;\text{ for } k > 0 \quad , \qquad b_k = \frac{3.4587\pi k}{4+4\pi^2 k^2}$$

3.12.

a. $c_k = -j\,\dfrac{a}{2\pi k} \;\text{ for } k \neq 0 \quad , \qquad c_0 = \dfrac{a}{2}$

3.16.

$$c_k = \begin{cases} \dfrac{2}{\pi\left(1-k^2\right)}\,, & k \text{ even} \\ 0\,, & k \text{ odd} \end{cases}$$

3.21.

$$X(f) = \text{sinc}(f)\left[e^{-j\pi f} - e^{-j3\pi f}\right]$$

3.25.

$$X(\omega) = \text{sinc}^2\left(\frac{\omega}{2\pi}\right)\left[e^{-j0.5\omega} - e^{-j1.5\omega}\right]$$

3.26.

$$\frac{2}{1+4t^2} \stackrel{\mathscr{F}}{\longleftrightarrow} \pi\, e^{-|\omega|/2}$$

3.29.

a. $X(f) = \frac{1}{2}\,\text{sinc}(f+5) + \frac{1}{2}\,\text{sinc}(f-5)$

b. $X(f) = \text{sinc}(2f+10) + \text{sinc}(2f-10)$

3.33.

$$X(f) = \frac{1}{j2\pi f}\left[\text{sinc}(f)\, e^{-j\pi f} - \text{sinc}(2f)\, e^{-j8\pi f}\right]$$

3.35.

a. $X(f) = \frac{1}{2}\left[\text{sinc}(f-0.5) + \text{sinc}(f+0.5)\right] e^{-j\pi f}$

3.39.

c. $M = 1$

d. $M = 3$

e. $M = 21$

3.40.

c. $M = 4$

d. $M = 11$

e. $M = 51$

3.45.

```
% Compute the signal x(t)
xp = @(t) 2*(t<1)+1*((t>=1)&(t<=2));  % One period
t = [-4:0.005:6];
x = xp(mod(t,3));  % Periodic extension
% Compute the coefficients
k = [1:10];
a0 = 1;
a = (sin(2*pi*k/3)+sin(4*pi*k/3))./(pi*k);
b = (2-cos(2*pi*k/3)-cos(4*pi*k/3))./(pi*k);
```

```
% Approximation with M=3 harmonics
xhat = a0*ones(size(t));
for m=1:3,
  xhat = xhat+a(m)*cos(2*pi*m*t/3)+b(m)*sin(2*pi*m*t/3);
end;
plot(t,x,t,xhat); grid;
axis([-3,3,-1,3.5]);
xlabel('t');
legend('Original signal','Approx. with M=3');
% Approximation error for M=3
plot(t,x-xhat); grid;
axis([-3,3,-1.2,1.2]);
xlabel('t');
```

3.50.

```
% Compute the EFS coefficients
d = 0.5;
k = [-25:25];
c = d*sinc(k*d);
offset = -k(1)+1;  % MATLAB index of center coefficient "c0"
% Set f0 = 1 Hz and create a time vector.
f0 = 1;
t = [-2:0.005:2];
% Compute the signal for M=1
m = 1;
xhat = zeros(size(t));
for i=-m:m,
  xhat = xhat+c(i+offset)*exp(j*2*pi*i*f0*t);
end;
plot(t,real(xhat)); grid;
axis([-2,2,-0.2,1.2]);
```

G.4 Chapter 4 Problems

4.1.

a. $X(s) = \dfrac{1}{s+2}\ , \qquad \mathrm{Re}\{s\} > -2$

b. $X(s) = \dfrac{e^{-2}\,e^{-s}}{s+2}\ , \qquad \mathrm{Re}\{s\} > -2$

4.3.

a. ROC: $-1 < \mathrm{Re}\{s\} < 1$

4.4.

a. ROC: $\mathrm{Re}\{s\} > 1$

4.9.

a. $X(s)=\dfrac{-2\,(s-2)}{(s+1)\,(s+3)}\,, \qquad \text{Re}\{s\}>-1$

b. $X(s)=\dfrac{s-11}{(s+1)\,(s-3)}\,, \qquad -1<\text{Re}\{s\}<3$

4.10.

a. $X(s)=\dfrac{e^{-s}}{s+2}\,, \qquad \text{Re}\{s\}>-2$

b. $X(s)=\dfrac{e^{-(s+2)}}{s+2}\,, \qquad \text{Re}\{s\}>-2$

4.13.

a. $X(s)=\dfrac{1}{s}\left(1-2e^{-s}+e^{-2s}\right), \qquad \text{Re}\{s\}>-\infty$

4.17.

a. $x(t)=e^{-t}\,u(t)-e^{-2t}\,u(t)$

b. $x(t)=-e^{-t}\,u(-t)+e^{-2t}\,u(-t)$

4.18.

b. $x(t)=-2\,e^{-2t}\,u(-t)+6\,e^{-3t}\,u(-t)$

4.20.

a.

$$\begin{aligned}
\text{Re}\{s\}<-3 &: \quad x(t)=e^{-2t}\,u(-t)-2\,e^{-3t}\,u(-t)\\
-3<\text{Re}\{s\}<-2 &: \quad x(t)=e^{-2t}\,u(-t)+2\,e^{-3t}\,u(t)\\
\text{Re}\{s\}>-2 &: \quad x(t)=-\,e^{-2t}\,u(t)+2\,e^{-3t}\,u(t)
\end{aligned}$$

4.22.

a. $H(s)=\dfrac{7}{3s+2}$

4.23.

c. $\dfrac{d^2y(t)}{dt^2}+5\,\dfrac{dy(t)}{dt}+6\,y(t)=\dfrac{dx(t)}{dt}+x(t)$

4.25.

a. $\dfrac{dy(t)}{dt}+y(t)=\dfrac{dx(t)}{dt}$

4.28.

a. $y(t)=0.3714\cos(2t+1.9513)$

4.34.

a. $H(s) = \dfrac{2\left(s^2+1\right)}{s^2+3\,s+2}$

4.38.

a. $H(s) = \dfrac{1}{s^2+2\,s+a}$

4.42.

a. The inverse system is both causal and stable. Its differential equation is

$$\frac{dy(t)}{dt} + 5\,y(t) = 2\,\frac{dx(t)}{dt} + 3\,x(t)$$

4.44.

c.

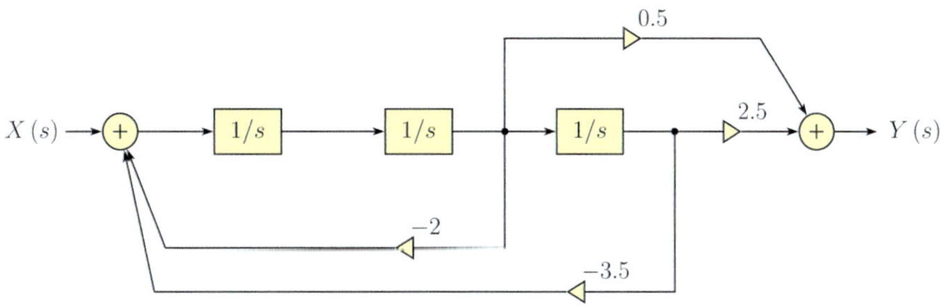

4.45.

c.

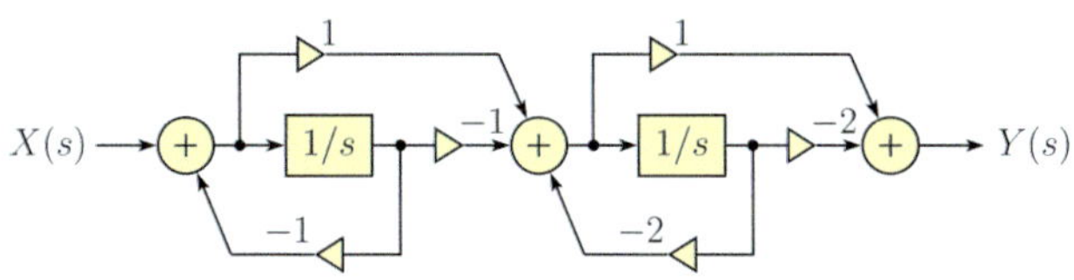

4.47.

a. If $x(t) = 0$ for $-1 < t < 0$, then

$$G_u(s) = e^{-s}\,X_u(s)$$

b. If $x(t) = 0$ for $0 < t < 2$, then

$$G_u(s) = e^{2s}\,X_u(s)$$

4.48.

a. $$\frac{d^2y(t)}{dt^2}+2\frac{dy(t)}{dt}+8y(t)=8x(t)$$

4.49.

a. $$\frac{d^2y(t)}{dt^2}+3\frac{dy(t)}{dt}+y(t)=\frac{dx(t)}{dt}$$

4.50.

b.

```
[sr,si] = meshgrid([-5:0.1:5],[-5:0.1:5]);
s = sr+j*si;
Xs = @(s) s.*(s+2)./((s+1).^2+9);
XsMag = abs(Xs(s));
XsMag = XsMag.*(XsMag<=2)+2.*(XsMag>2);
```

4.53.

a.

```
syms s t
Xs = 1/(s^2+3*s+2);
xt = ilaplace(Xs)
```

4.54.

a.

```
sys = tf([1,1,0],[1,4,13])
```

4.56.

a.

```
num1 = [1,1];
den1 = [1,5,6];
sys1 = tf(num1,den1);
bode(sys1);
grid;
```

G.5 Chapter 5 Problems

5.3.

$$\dot{\mathbf{x}}(t)=\begin{bmatrix}-4 & 2\\ 3 & -3\end{bmatrix}\mathbf{x}(t)+\begin{bmatrix}2\\ 0\end{bmatrix}r(t)$$

$$y(t)=\begin{bmatrix}0 & 1\end{bmatrix}\mathbf{x}(t)$$

5.4.

$$\dot{\mathbf{x}}(t) = \begin{bmatrix} 0 & 1.2 \\ -5 & -5 \end{bmatrix} \mathbf{x}(t) + \begin{bmatrix} 0 \\ 5 \end{bmatrix} r(t)$$

$$y(t) = \begin{bmatrix} 0 & 1 \end{bmatrix} \mathbf{x}(t)$$

5.5.

a.

$$\dot{\mathbf{x}}(t) = \begin{bmatrix} 0 & 1 \\ -2 & -3 \end{bmatrix} \mathbf{x}(t) + \begin{bmatrix} 0 \\ 2 \end{bmatrix} r(t)$$

$$y(t) = \begin{bmatrix} 1 & 0 \end{bmatrix} \mathbf{x}(t)$$

b.

$$\dot{\mathbf{x}}(t) = \begin{bmatrix} 0 & 1 \\ -3 & -4 \end{bmatrix} \mathbf{x}(t) + \begin{bmatrix} 0 \\ 3 \end{bmatrix} r(t)$$

$$y(t) = \begin{bmatrix} 1 & 0 \end{bmatrix} \mathbf{x}(t)$$

5.6.

a.

$$\dot{\mathbf{x}} = \begin{bmatrix} -1 & 0 \\ 0 & -2 \end{bmatrix} \mathbf{x}(t) + \begin{bmatrix} 2 \\ -2 \end{bmatrix} r(t)$$

$$y(t) = \begin{bmatrix} 1 & 1 \end{bmatrix} \mathbf{x}(t)$$

5.11.

c. $$\frac{d^2 y(t)}{dt^2} + 7\frac{dy(t)}{dt} + 12\,y(t) = 5\,r(t)$$

5.15.

$$y(t) = \left(\frac{8}{3} + 4\,e^{-t} - \frac{14}{3}\,e^{-3t}\right) u(t)$$

5.17.

a. $$\boldsymbol{\phi}(t) = \begin{bmatrix} e^{-3t}\,u(t) & \left(e^{-t} - e^{-3t}\right) u(t) \\ 0 & e^{-t}\,u(t) \end{bmatrix}$$

5.22.

$$G(s) = \frac{s^2 + 26}{s^2 + 2\,s + 26}$$

5.25.

a.

```
A = [49,25,-9;-96,-49,17;48,24,-9]
B = [-1;2;-1]
C = [5,3,1]
D = 0
[num,den] = ss2tf(A,B,C,D,1)
```

G.6 Chapter 6 Problems

6.1.

a. $|\omega| \le 142.5$ rad/s.

b. -0.67 percent

6.2.

a. $\omega_0 = 0.3775$ rad/s

b. 4.31 percent

6.3.

a. $H_{LP}(\omega) = \Pi\left(\dfrac{\omega}{400\pi}\right) e^{-j0.1\omega}$

6.4.

a. $H_{BP}(\omega) = \left[\Pi\left(\dfrac{\omega-450\pi}{500\pi}\right) + \Pi\left(\dfrac{\omega+450\pi}{500\pi}\right)\right] e^{-j0.3\omega}$

6.8.

$$H(s) = \frac{6.6667}{s^4 + 2.6044\,s^3 + 7.3915\,s^2 + 9.3217\,s + 6.9602}$$

6.9.

$$H(s) = \frac{0.2873\,s^4 + 9.1951\,s^2 + 36.7805}{s^4 + 5.3571\,s^3 + 16.9916\,s^2 + 23.9481\,s + 36.7805}$$

6.13.

a. $\omega_0 = 10$ rad/s

b. $p_1 = -1.6667$ rad/s

6.19.

```
[num,den] = butter(4,2,'s')
```

6.20.

```
omg = [0:0.01:6];
H = freqs(num,den,omg);
```

6.24.

```
% Set up filter specifications
omg1 = 1;
omg2 = 2.5;
Rp = 1;
As = 30;
% Design the lowpass prototype
[N, omgc] = buttord(omg1,omg2,Rp,As,'s')
[numL,denL] = butter(N,omgc,'s')
% Lowpass to highpass transformation
[numH,denH] = lp2hp(num,den,5)
```

G.7 Chapter 7 Problems

7.1.

a. $A_c = 2.86$

b. $A_c = 3.53$

7.3.

a. $A_c = 3.68$

b. $A_c = 5.71$

7.5.

a.

$$\begin{aligned} X_{AM}(f) = {} & 1.4286\,\delta(f - f_c) + 1.4286\,\delta(f + f_c) \\ & + 0.5\,\delta(f - f_c - f_m) + 0.5\,\delta(f - f_c + f_m) \\ & + 0.5\,\delta(f + f_c + f_m) + 0.5\,\delta(f + f_c - f_m) \end{aligned}$$

7.9.

a. $f_c = 15$ kHz.

b. $f_m = 1.2$ kHz.

7.10.

a. $\eta = 0.1558$

b. $\eta = 0.1330$

7.14. $\tau \le 1.2$ ms.

7.17.

a.

```
xAM = @(t,Ac,mu,fc,fm) Ac*(1+mu*cos(2*pi*fm*t)).*cos(2*pi*fc*t);
t = [0:2e-6:2e-3];
plot(1000*t,xAM(t,3,0.6,10000,2000));
axis([0,2,-8,8]);
xlabel('t (ms)');
```

7.19.

a. $f_0 = 1$ Hz.

7.20.

```
t = [0:0.05:30]/1000;
Bc = 5;
fc = 800;
carrier = Bc*cos(2*pi*fc*t);
message = 3*cos(200*pi*t);
vin = carrier+message;
plot(1000*t,vin); grid;
axis([0,30,-10,10]);
xlabel('t (ms)');
title('v_{in}(t)');
```

INDEX